Tropical Forest Remnants

Tropical Forest Remnants

Ecology, Management, and Conservation of Fragmented Communities

Edited by

William F. Laurance and

Richard O. Bierregaard, Jr.

The University of Chicago Press *Chicago & London*

William F. Laurance is a research scientist for the Biological Dynamics of Forest Fragments Project in Manaus, Brazil. He has served as a senior research fellow at the CSIRO Tropical Forest Research Centre in Atherton, Queensland (Australia), and as director of the Centre for Rainforest Studies at the School for Field Studies in Yungaburra, Queensland.

Richard O. Bierregaard, Jr., is adjunct assistant professor in the Department of Biology, University of North Carolina, Charlotte. He has served as a director of the Biological Dynamics of Forest Fragments Projects, as a member of the editorial board for the Brazilian Ornithological Society, and as a member of the board of directors of the Carolina Raptor Center.

The University of Chicago Press, Chicago 60637
The University of Chicago Press, Ltd., London

Printed in the United States of America

06 05 04 03 02 01 00 99 98 97 5 4 3 2 1

ISBN (cloth): 0-226-46898-4
ISBN (paper): 0-226-46899-2

Library of Congress Cataloging-in-Publication Data

Tropical forest remnants : ecology, management, and conservation of fragmented communities / edited by William F. Laurance and Richard O. Bierregaard, Jr.
p. cm.
Includes bibliographical references (p.) and index.
ISBN 0-226-46898-4.—ISBN 0-226-46899-2 (pbk.)
1. Fragmented landscapes—Tropics. 2. Conservation biology—Tropics. 3. Rain forest ecology. 4. Rain forest—Management. I. Laurance, William F. II. Bierregaard, Richard O.
QH541.15.F73T76 1997
577.34—dc21 96-38038
CIP

This book is printed on acid-free paper.

Contents

Foreword ix
Thomas E. Lovejoy

Preface: A Crisis in the Making xi
William F. Laurance and Richard O. Bierregaard, Jr.

SECTION I: The Scale and Economics of Tropical Deforestation

Introduction 1
William F. Laurance

1 Tropical Forest Disturbance, Disappearance, and Species Loss 3
T. C. Whitmore

2 The Role of Economic Factors in Tropical Deforestation 13
James R. Kahn and Judith A. McDonald

SECTION II: Physical Processes and Edge Effects

Introduction 29
William F. Laurance

3 Edge-Related Changes in Environment and Plant Responses Due to Forest Fragmentation in Central Amazonia 33
Valerie Kapos, Elisa Wandelli, José Luis Camargo, and Gislene Ganade

4 Edge and Aspect Effects on the Microclimate of a Small Tropical Forest Remnant on the Atherton Tableland, Northeastern Australia 45
Stephen M. Turton and Heidi Jo Freiburger

5 The Influence of Edge Effects and Forest Fragmentation on Leaf Litter Invertebrates in Central Amazonia 55
Raphael K. Didham

6 Hyper-Disturbed Parks: Edge Effects and the Ecology of Isolated Rainforest Reserves in Tropical Australia 71
William F. Laurance

SECTION III: Tropical Forest Faunas

Introduction 85
William F. Laurance

7 Disturbance, Fragmentation, and the Dynamics of Diversity in Amazonian Forest Butterflies 91
Keith S. Brown, Jr., and Roger W. Hutchings

8 Phenetic Variation in Insular Populations of a Rainforest Centipede 111
John F. Weishampel, Herman H. Shugart, and Walter E. Westman

9 Fragmentation Effects on a Central Amazonian Frog Community: A Ten-Year Study 124
Mandy D. Tocher, Claude Gascon, and Barbara L. Zimmerman

10 Understory Birds and Dynamic Habitat Mosaics in Amazonian Rainforests 138
Richard O. Bierregaard, Jr., and Philip C. Stouffer

11 Avian Extinction and Persistance Mechanisms in Lowland Panama 156
Kathryn E. Sieving and James R. Karr

12 Frugivorous Birds in Fragmented Neotropical Montane Forests: Landscape Pattern and Body Mass Distribution 171
Carla Restrepo, Luis Miguel Renjifo, and Paul Marples

13 Structure and Conservation of Forest Avifauna in Isolated Rainforest Remnants in Tropical Australia 190
Neil H. Warburton

14 Biomass and Diversity of Small Mammals in Amazonian Forest Fragments 207
Jay R. Malcolm

15 Rapid Decline of Small Mammal Diversity in Monsoon Evergreen Forest Fragments in Thailand 222
Antony J. Lynam

16 Internal Fragmentation: The Effects of Roads, Highways, and Powerline Clearings on Movements and Mortality of Rainforest Vertebrates 241
Miriam Goosem

17 Transitory States in Relaxing Ecosystems of Land Bridge Islands 256
John Terborgh, Lawrence Lopez, José Tello, Douglas Yu, and Ana Rita Bruni

SECTION IV: Plants and Plant–Animal Interactions

Introduction 275
William F. Laurance

18 Tropical Forest Disruption and Stochastic Biodiversity Losses 281
Mark Andersen, Alan Thornhill, and Harold Koopowitz

19 Regeneration of Large-Seeded Trees in Australian Rainforest Fragments: A Study of Higher-Order Interactions 292
Graham N. Harrington, Anthony K. Irvine, Francis H. J. Crome, and Les A. Moore

20 Dispersal and the Dynamics of Genetic Structure in Fragmented Tropical Tree Populations 304
John D. Nason, Preston R. Aldrich, and J. L. Hamrick

21 Plant Dispersal in Fragmented Landscapes: A Field Study of Woody Colonization in Rainforest Remnants of the Mascarene Archipelago 321
Christophe Thébaud and Dominique Strasberg

22 Long-Term Survival in Tropical Forest Remnants in Singapore and Hong Kong 333
Richard T. Corlett and I. M. Turner

SECTION V: Restoration and Management of Fragmented Landscapes

Introduction 347
William F. Laurance

23 Dynamics and Restoration of Forest Fragments in the Brazilian Atlantic Moist Forest 351
Virgílio M. Viana, André A. J. Tabanez, and João Luis F. Batista

24 Rejoining Habitat Remnants: Restoring Degraded Rainforest Lands 366
David Lamb, John Parrotta, Rod Keenan, and Nigel Tucker

25 Measuring Landscape Changes in Remnant Tropical Dry Forests 386
Elizabeth A. Kramer

26 Quantifying Habitat Fragmentation Due to Land Use Change in Amazonia 400
Virginia H. Dale and Scott M. Pearson

SECTION VI: Site Selection and Design of Tropical Nature Reserves

Introduction 411
William F. Laurance

27 Deforestation, Fragmentation, and Reserve Design in Western Madagascar 415
Andrew P. Smith

28 Molecular Perspectives on Historical Fragmentation of Australian Tropical and Subtropical Rainforests: Implications for Conservation 442
Craig Moritz, Leo Joseph, Michael Cunningham, and Chris Schneider

29 Diversity, Differentiation, and the Historical Biogeography on Nonvolant Small Mammals of the Neotropical Forests 455
James L. Patton, Maria Nazareth F. da Silva, Márcia C. Lara, and Meika A. Mustrangi

30 Species Richness and Endemism in South American Birds: Implications for the Design of Networks of Nature Reserves 466
Jon Fjeldså and Carsten Rahbek

SECTION VII: SUMMARY AND NEW PERSPECTIVES

Introduction 483
William F. Laurance

31 Researching Tropical Forest Fragmentation: Shall We Keep On Doing What We're Doing? 485
Francis H. J. Crome

32 Tropical Forest Fragmentation: Synthesis of a Diverse and Dynamic Discipline 502
William F. Laurance, Richard O. Bierregaard, Jr., Claude Gascon, Raphael K. Didham, Andrew P. Smith, Antony J. Lynam, Virgílio M. Viana, Thomas E. Lovejoy, Kathryn E. Sieving, Jack W. Sites, Jr., Mark Andersen, Mandy D. Tocher, Elizabeth A. Kramer, Carla Restrepo, and Craig Moritz

33 Key Priorities for the Study of Fragmented Tropical Ecosystems 515
Richard O. Bierregaard, Jr., William F. Laurance, Jack W. Sites, Jr., Antony J. Lynam, Raphael K. Didham, Mark Andersen, Claude Gascon, Mandy D. Tocher, Andrew P. Smith, Virgílio M. Viana, Thomas E. Lovejoy, Kathryn E. Sieving, Elizabeth A. Kramer, Carla Restrepo, and Craig Moritz

Contributors 527
References 533
Name Index 589
Taxa Index 601
Subject Index 609

Color plates follow page 432

Foreword

Thomas E. Lovejoy

It was a mutual love for the Tropics as well as for exploration and the out-of-doors that first led John Terborgh and me to meet in 1970, when I was a graduate student at Yale and John was on the faculty at the University of Maryland. But our fascination with things tropical was merely the overture for increasing interest in what has now become known as conservation biology as well as conservation per se.

A couple of years later, as habitat fragmentation emerged as a global concern, John analyzed oceanic, continental, and habitat islands and I worried about the sizes of protected areas that World Wildlife Fund-U.S. was helping to establish. A heated debate arose in the literature about the value of island biogeography to conservation, most dramatically revolving around the "single large or several small" (SLOSS) reserves dispute.

The debate was fueled more by lack of data than anything else, and this prompted me and many others to study the problem in a variety of ways. Some of that research has borne fruit in the six chapters in this book that report on our experimental Biological Dynamics of Forest Fragments project in Manaus, Brazil. It is interesting to recall how simple the subject seemed then, relative to our current perspective, and how correct we all were to turn our attentions to what is clearly a major way in which we are affecting the biology of this planet.

The field of habitat fragmentation has a variety of intellectual roots. Obvious among them are Charles Darwin's and Alfred Russel Wallace's pioneering studies of islands and Robert MacArthur and E. O. Wilson's elegant models of island biogeography. To these roots has been grafted the intellectual tradition of wildlife biology, which has to a large extent evolved into conservation biology. In particular, wildlife biologists' interest in habitat edges has enriched a field once preoccupied exclusively by habitat area. It is interesting that Aldo Leopold's "edge effect"—once narrowly relating to game and wildlife abundance at forest edges—has become an enormously complex aspect of fragmentation in itself.

These multiple intellectual roots have ignited a veritable explosion of papers—ranging from population genetics to remote sensing and broad landscape ecology—all focused on fragmented ecosystems. Clearly, the time has come to assess what we have learned, and we all owe a debt of gratitude to Bill Laurance and Rob Bierregaard for organizing this volume on tropical forest fragments, as well as its associated symposium at the 1995 annual meeting of the Ecological Society of America in Utah's magnificent Uinta mountains.

The individual chapters herein speak for themselves, but clearly there is a marvelous richness of understanding relative to what John Terborgh was able to show me about his first island analyses in 1972. The New World Tropics and Australia are particularly well represented. I hope that Africa (represented here by the California-sized island of Madagascar) and Southeast Asia (represented by studies in Thailand, Singapore, and Hong Kong) will loom larger whenever it next makes sense to publish a book of this sort.

This volume would be worthwhile simply as a compendium of so many high-quality chapters, which otherwise would be widely scattered in the literature. Yet the editors and authors took the exercise an important step further by convening a workshop immediately after the Utah symposium. Sequestered for two days, we tried to draw some generalities about what has been learned so far and what research priorities have emerged. Both exercises went considerably beyond what anyone had dared hope and provided the basis for two invaluable concluding chapters. Figure 32.1 in the penultimate chapter, encapsulating edge penetration by a diverse collection of variables, is destined to become a classic and shows at a glance how rich and complex the biology of habitat fragmentation has grown as a subject.

Over the intervening years since those first conversations with John Terborgh, the roster of environmental challenges has expanded, with the notable addition of climatic change. The intersection of climatic change and habitat fragmentation is particularly alarming as we rapidly convert once natural landscapes into obstacle courses for dispersal. These barriers to dispersal constitute an important problem even without anthropogenic influences on climate, because as surely as night follows day, there will be future natural climatic change. This threat calls for two things. One is a great expansion of research on corridors and landscape connectivity. The other is a rapid advance toward an ecosystem management approach in both research and applied conservation. We must not only recognize that, but also behave as if, we live *within* ecosystems, rather than perceiving nature as something confined to a few protected areas isolated within a degraded, human-dominated landscape.

Looking ahead, it is clear that tropical forest fragment science will be enriched by research on connectivity and ecosystem management. It is also likely that there will be greater linkage across the vast spatial scales of different studies, so that assessments like those of David Skole and Compton Tucker of Amazon deforestation and fragmentation can be linked in far more detail to on-the-ground research. The next time an overview of tropical forest fragmentation takes place, it is likely to be twice as interesting as this one. Yet we need to remind ourselves that this is not an isolated academic exercise. As we met and discussed fragmented ecosystems high in the Uinta Mountains, where the gigantic mammals named for the mountains once roamed, habitats around the world continued to be rapidly destroyed and fragmented—driving an unknown number of species to join the Uintatheres in extinction.

Preface

A Crisis in the Making

William F. Laurance and Richard O. Bierregaard, Jr.

THE biologist Norman Myers once called tropical forests "the greatest celebration of life on Earth." This is, of course, the most apt of descriptions, for no other land community rivals the species diversity and ecological complexity of tropical forests.

Yet despite their astounding richness and biotic importance, tropical forests are being cleared, burned, fragmented, logged, and overhunted at rates that lack historical precedent. With deforestation driven by a burgeoning human populace and a rush toward industrialization in developing nations, it is now apparent that much of the world's tropical forest will disappear within our lifetimes.

Sober projections highlight the scale of the crisis. In his incisive analysis of estimates of forest conversion made by the Food and Agriculture Organization of the United Nations, T. C. Whitmore (chapter 1) shows that the overall rate of tropical deforestation throughout the 1980s exceeded 150,000 km^2 per year. If impacts such as selective logging are included, the rate of forest conversion exceeds 200,000 km^2 per year, or about 1.2% of the total extent of tropical forests globally. Making the conservative assumption that this rate will not increase, an additional one-third of the remaining tropical forests will be cleared or modified in the next thirty years.

The rapid pace of deforestation is creating a plethora of deleterious effects. On a regional scale, the effects of forest loss can include massive soil erosion, siltation of streams, destabilization of watersheds, a loss of sustainable forest uses, and threats to indigenous peoples (Poore 1979; Smith 1981). Other effects include the large-scale release of atmospheric carbon dioxide and other gases from the burning or microbial breakdown of forest fuels, exacerbating the greenhouse effect (Houghton 1991).

To biologists and natural resource managers, the most alarming aspect of the tropical forest crisis is the unparalleled threat to biological diversity. Tropical forests are the most ancient, the most diverse, and the most ecologically complex of land communities (Lewin 1986). Though occupying only 7% of the earth's land surface, these forests probably sustain more than half of the planet's life forms (Myers 1984). And in virtually every biological discipline, tropical forests have been grossly understudied (Janzen 1986b).

How fast are species disappearing? Estimating the rate of extinctions requires some very creative guesswork. The most common approach to this problem involves using species-area relationships to estimate the percentage of species lost following a given reduction in habitat area, then multiplying this percentage by the number of species thought to occur in a given ecosystem. The inaccuracy in such predictions arises in part because we are still very uncertain about how many species there are on the planet, and especially in tropical forests (Wilson 1988). Given these uncertainties, current rates of species loss appear to be hundreds or even thousands of times higher than natural levels (Ehrlich and Ehrlich 1981; Lugo 1988a; Wilson 1988). Contemporary yearly extinction rates apparently rival those during mass extinction episodes in the geological record (Simberloff 1986). Perhaps 600,000 species will have disappeared by the year 2000, largely as a result of tropical deforestation (Lovejoy 1980). The evidence that deforestation is effecting a massive extinction wave is compelling (Myers 1988).

THE AFTERMATH OF FOREST CONVERSION

An almost inevitable consequence of large-scale forest loss is habitat fragmentation. Hence, one cannot discuss forest fragmentation in any depth without considering also the broader effects of forest loss and conversion. Collectively, the correlated processes of habitat loss and fragmentation probably pose the greatest single threat to the planet's biological diversity (Wilcox and Murphy 1985; Simberloff 1986).

The fragmented landscape is becoming one of the most ubiquitous features of the tropical world—and indeed, of the entire planet. Tropical nations tend to fall into three rough categories. The first includes those, such as Madagascar and the Philippines, that have already experienced a drastic loss of forest cover. The second includes nations such as Nigeria, Costa Rica, and the Malaysian states of Sarawak and Sabah, which have moderate amounts of forest cover remaining but are experiencing rapid forest conversion. The final category includes nations such as Brazil and Papua New Guinea, which are also being rapidly modified but still retain large tracts of intact forest. Australia, a developed nation that recently protected most of its remaining tropical rainforests (which are very limited in extent), stands out as an obvious anomaly.

THE EXPLOSIVE GROWTH OF FRAGMENT RESEARCH

Regardless of the dominant land uses, as a region is colonized and developed, remnant patches or tracts of forest almost always persist (by happenstance or design) as "islands" surrounded by a "sea" of modified habitats. It is now apparent that the long-term survival of much of the tropical biota will depend on the ability of species to persist in highly modified habitats and on our capacity to manage and conserve such degraded landscapes.

The significance of the deforestation crisis has not been lost upon biologists, resource managers, economists, and others involved in tropical research. Investigations of fragmented ecosystems commenced in earnest nearly three decades ago after the publication of MacArthur and Wilson's seminal monograph, *The Theory of Island Biogeography* (MacArthur and Wilson 1967). Since then the field has grown rapidly—explosively in recent years. One of the most telling indicators of this growth is the many dozens of young

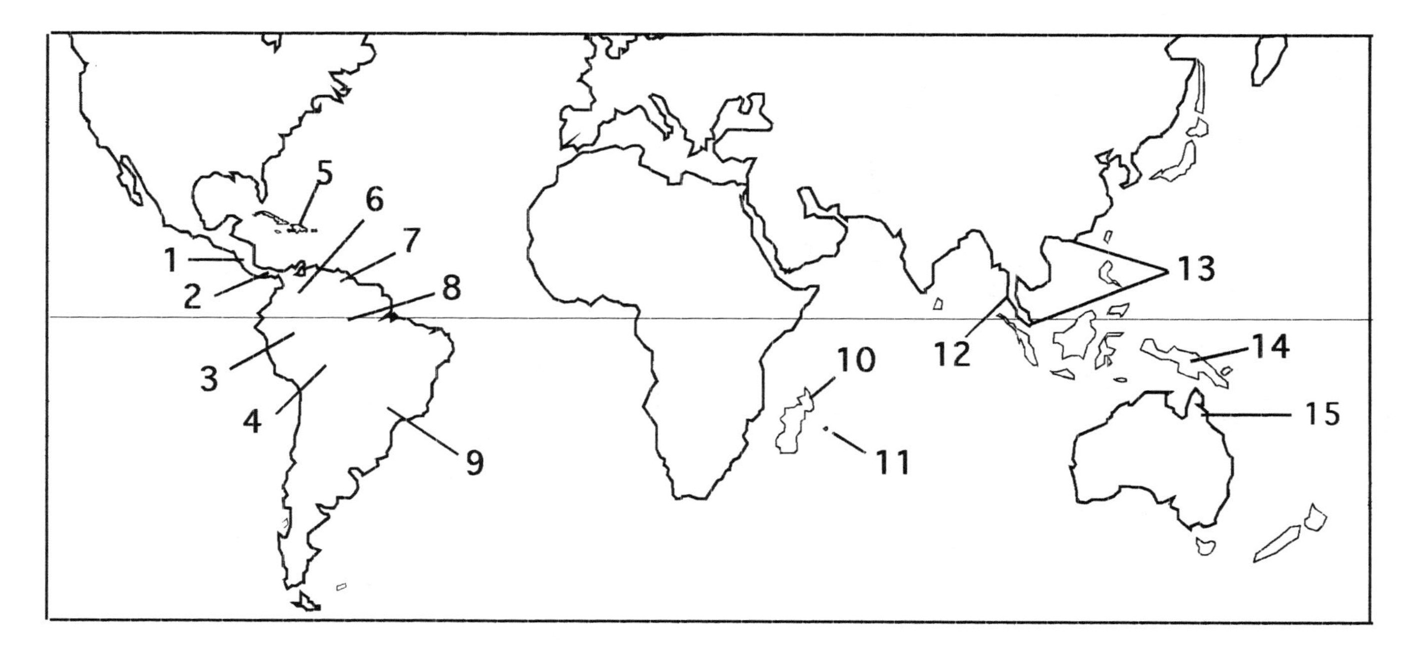

The geographic locations of studies in this volume: 1 = Guanacaste National Park, Costa Rica (chap. 25); 2 = Barro Colorado Island, Panama (chaps. 11 and 20); 3 = forested slopes of the Andes (chap. 30); 4 = Rondônia, Brazil (chap. 26); 5 = Puerto Rico (chap. 24); 6 = the mountains of Colombia (chap. 12); 7 = the Guri Hydroelectric Impoundment, Venezuela (chap. 17); 8 = Manaus, Brazil, site of the Biological Dynamics of Forest Fragments Project (chaps. 3, 5, 7, 9, 10, and 14), and more broadly, the Amazonian lowland forests (chaps. 29 and 30); 9 = the Atlantic coastal forests, São Paulo State, Brazil (chap. 23); 10 = northwestern Madagascar (chap. 27); 11 = the Mascarene Archipelago (chap. 21); 12 = Chiew Larn reservior, Thailand (chap. 15); 13 = Hong Kong and Singapore (chap. 22); 14 = New Guinea (chap. 28); 15 = the Atherton Tableland, Queensland, Australia (chaps. 4, 6, 13, 16, 19, 24, and 28), and more broadly, the coastal rainforests of northeastern Queensland, Australia (chap. 28).

researchers who have recently completed Ph.D.s or other advanced degrees based on studies of fragmented tropical systems. As evidenced by this volume, the labors of these many researchers are beginning to bear considerable fruit.

As the field has grown, it has become increasingly varied and eclectic. The research and applied management projects described herein, for example, vary enormously in terms of their principal goals and hypotheses, as well as their methods, target organisms, forest types, spatial scale of forest remnants, and the nature of the matrix habitats surrounding fragments. These studies encompass much of the tropical world: projects in the Neotropics and Australian Tropics are especially well represented, but studies in other regions—Southeast Asia, Madagascar, the Caribbean, and Oceania—are also presented. Africa is not represented because there has been remarkably little work on fragmentation of tropical forests on that continent (but see Newmark 1991; Medley 1993).

This diversity was both an advantage and a challenge in compiling and editing this book—an advantage because it afforded us a very wide range of perspectives, and a challenge because such diversity can complicate efforts to reach consensus and to develop crucial generalizations about the nature of fragmented ecosystems. We suspect that such trends are typical of fields of scientific endeavor that are experiencing explosive growth.

ORGANIZATION OF THE VOLUME

This book adopts a very broad view of fragmented tropical systems. The chapters focus on the ecology, management, rehabilitation, socioeconomics, and conservation of fragmented systems and the design of regional networks of nature reserves. The structure of the volume reflects our belief that, in a large and diverse compilation such as this, two goals are paramount: the volume must be effectively organized and integrated, and key information must be easily accessible.

The thirty-three chapters in the volume are organized by broad topic into seven sections. Our introduction to each section briefly describes all the chapters therein, highlights their parallels and differences, and places each into an overall context. In addition, a pair of concluding summary chapters provides an integrated perspective on two key questions: What do we really know about fragmented tropical systems? and, What are the priorities for future research? These synthetic chapters were drafted by fifteen of the volume's contributing authors at a stimulating and highly productive meeting that immediately followed our symposium on tropical forest fragmentation at the 1995 annual meeting of the Ecological Society of America in Snowbird, Utah. Finally, to enhance accessibility of information, we required that each chapter provide a final "General Implications" section, which highlights key conclusions of broad general interest and relevance.

ACKNOWLEDGMENTS

Many individuals contributed to the production of this book. We are grateful to the dozens of contributing authors, especially those who participated in the Utah symposium. That meeting would have been impossible without the organizational skills of Jill Baron, who coordinated the conference symposia.

We also thank Thomas Lovejoy for stealing time from his hectic schedule to contribute to the symposium, and for writing the volume's Foreword.

Craig Moritz was initially involved as a co-editor of the volume, but had to withdraw because of pressing time commitments. We appreciate his initial suggestions and enthusiasm for the project.

We are particularly grateful to the dozens of scientists and resource managers who kindly acted as referees (often anonymously) for the thirty-three chapters. Every chapter was improved by their insights, suggestions, and critical thinking. Lucinda McDade offered useful suggestions in response to the original publication proposal. Under the direction of our publisher, four especially brave souls (John Blake, Robert Holt, Doug Stotz, and an anonymous reviewer) read and criticized the entire book manuscript, and we feel indebted to them for many useful editorial suggestions.

Marie Peever and Benjamin Blewitt typed countless corrections on manuscripts during the final editing process. This was an onerous and demanding task, and their help is much appreciated.

Lastly, we wish to thank the staff at the University of Chicago Press. In particular, Christie Henry, the biological sciences editor, took a sincere interest in the volume from its inception through its final production. Her help and advice were invaluable, and we thank her for guiding two novices through the editorial phases of the publication process.

Section I

The Scale and Economics of Tropical Deforestation

INTRODUCTION

THE processes of habitat loss and fragmentation are inextricably linked. It is therefore essential in a volume of this nature to assess both the rate and patterning of deforestation, as well as the economic and social factors that drive forest exploitation in the Tropics.

The chapters in Section I introduce these topics. Both chapters have been written with a general audience in mind, so that, for example, one need have only a rudimentary knowledge of economics to comprehend the arguments presented in the second chapter.

THE SCALE OF DEFORESTATION

In chapter 1, T. C. Whitmore describes the rate of forest loss and conversion on both pantropical and regional scales. He also contrasts rates of clearing and logging in different forest types, such as lowland rainforest, hill and montane forest, and seasonal forest. His database is the United Nations Food and Agriculture Organization's (FAO) assessments of forest cover, which represent the most exhaustive estimates available for the Tropics. By comparing FAO estimates of forest cover in 1980 and 1990, Whitmore has been able to calculate annual rates of loss and conversion throughout the decade of the 1980s.

The figures are alarming. Whitmore estimates that an average of 15.4 million hectares of tropical forests were destroyed each year, with another 5.6 million hectares being selectively logged (most of which was essentially virgin forest). The annual total of forest conversion (21 million ha) means that about 1.2% of the world's tropical forests are being destroyed or degraded *each year.*

On a regional basis, Asia emerges as the greatest concern. Tropical forest cover in Asia (ca. 316 million ha) is more limited than in the Americas (ca. 913 million ha) or Africa (ca. 527 million ha), and it has the highest relative rates of both forest clearing (1.1% annually) and logging (0.7% annually). Much of Asia has now been logged out, and it is alarming to witness the explosive expansion of Malaysian and other Asian logging firms into areas such as Papua New Guinea, the Solomon Islands, and parts of South America.

In the Americas and Africa, the total rate of forest conversion is about 1% annually, but this figure disguises enormous variation among regions and nations. In the Neotropics, for example, forest loss is occurring rapidly throughout Central America and in areas like southern and eastern Amazonia, but in relative terms these losses are still buffered by the vastness of the Amazon Basin.

Whitmore concludes his chapter by highlighting the gross uncertainties involved in predicted rates of species extinctions in the Tropics. He quite rightly points out that many species can survive in timber production forests, especially if efforts are made to mitigate the impacts of harvest methods, and that in the future such forests will undoubtedly play a key role in nature conservation for many tropical nations.

ECONOMICS AND FOREST LOSS

In chapter 2, James Kahn and Judith McDonald explore the economic factors that contribute to deforestation in tropical nations. They begin with an overview of various microeconomic factors that have traditionally been viewed as the main drivers of excessive deforestation rates. Examples include improper government incentives (e.g., poorly designed resettlement schemes and tax incentives for cattle ranching in Amazonia) and poorly constructed timber leasing agreements (e.g., short-term concessions that provide no incentive for timber companies to limit forest damage).

Kahn and McDonald then go on to examine the effects of macroeconomic factors on deforestation, especially the role of foreign debt. Their thesis is that by creating a short-term need for foreign exchange, external debt may play a major role in promoting excessive deforestation. Under continual pressure to service burgeoning debts, many nations are forced to exploit their natural resources at excessive (unsustainable) levels, and invest too little in environmental protection.

Another driving force of deforestation is, of course, rapid human population growth. As Kahn and McDonald point out, population growth often exacerbates both the micro- and macroeconomic factors that lead to rapid deforestation. Unfortunately, the alarming combination of rapid population growth, heavy external debt, and weak incentives for forest conservation is all too common among tropical nations.

Kahn and McDonald conclude by discussing some steps that might be taken to reduce external debt in developing nations and curb deforestation rates. The "debt-for-nature swap" is one such strategy, and they consider both its merits and its weaknesses. They also highlight the need for international cooperation in forest conservation, and for the development of mechanisms by which industrial nations could compensate tropical countries for retaining intact forests. Finally, they urge the development of economic policies that promote sustainable use of forest resources, such as carefully regulated logging methods and extractive reserves.

1

Tropical Forest Disturbance, Disappearance, and Species Loss

T. C. Whitmore

CONTEMPORARY HUMAN IMPACT ON TROPICAL FORESTS

In recent decades there have been several attempts to assess tropical deforestation. At a global scale, a thorough assessment is a gigantic task that requires huge resources. It is necessary to have robust, clear definitions of what is meant by deforestation and of the different kinds of forest, as well as the capacity to glean for one moment in time the extent of forest cover throughout the Tropics.

The Food and Agriculture Organization of the United Nations (FAO), with its network of national contacts, has made two assessments. The first of these was for 1980 (e.g., Lanly 1982) and the second for 1990 (FAO 1993b, 1995). The second assessment is used here as the best overview we have of the amount of tropical forest recently remaining, and of the rate of forest loss during the decade from 1981 to 1990. Notes on my interpretation of FAO's data are given in appendix 1.1.

Forest Area

FAO (1993b) estimated that in 1990 there were 1,756 million ha of natural tropical forests, with the largest forest area (52%) located in the Americas and lesser amounts in Asia (18%) and Africa (30%; table 1.1). By far the largest component (86%) of natural tropical forests are moist forests (rainforest plus seasonal forests: Sommer 1976). In 1990 moist forests covered 1,510 million ha, again with the largest extent in the Americas (58%) and lesser areas in Asia (18%) and Africa (25%; see appendix 1.1).

Lowland evergreen rainforest constituted nearly half (715 million ha) of the total moist forest. Much of this forest type is located in the Americas (63%), especially in the Amazon Basin, with lesser amounts in Asia (25%) and Africa (12%). Seasonal forests in the sense of FAO (i.e., semi-evergreen rainforest plus all monsoon and seasonal forest formations; see appendix 1.1) are least extensive in Asia (7%), compared with 50% and 42% in the Americas and Africa, respectively. Like the other forest types, hill and montane forests are most extensive in the Americas (60%), with smaller areas in Asia (23%) and Africa (17%).

In summary, the Americas have the largest extent of all tropical forest categories. Africa has a relatively large proportion of drier-climate forests, while Asia has a relatively small proportion of drier-climate forests.

Table 1.1 Estimated area of tropical forests in 1990.

	All natural tropical forests		Tropical moist forests: Total		Lowland: Rainforest		Lowland: Seasonal		Hill and montane	
Africa	**527**	**(30)**	**372**	**(25)**	**86.6**	**(12)**	**251**	**(42)**	**35.2**	**(17)**
Continent	512	(29)	360	(24)	82.1	(11)	247	(42)	30.7	(15)
Madagascar	15.8	(1)	12.9	(0.9)	4.5	(0.6)	3.8	(0.6)	4.6	(2)
Asia	**311**	**(18)**	**266**	**(18)**	**178**	**(25)**	**42**	**(7)**	**46.5**	**(23)**
Continent	139	(8)	95.7	(6)	33.5	(5)	36.3	(6)	25.8	(13)
Malay Archipelago	171	(10)	171	(11)	144	(20)	5.7	(1)	20.7	(10)
America	**918**	**(52)**	**870**	**(58)**	**451**	**(63)**	**298**	**(50)**	**122**	**(60)**
Caribbean and Central	115	(6)	111	(7)	44.7	(6)	25.7	(4)	40.9	(20)
Tropical South	803	(46)	759	(50)	406	(57)	272	(46)	80.7	(40)
Global total	**1,756**		**1,510**		**715**		**591**		**203**	

Source: After FAO 1993b, French and Spanish editions.

Note: Area is given in millions of hectares. Percentages of forest in each category (in parentheses) do not always sum to 100 because of rounding errors. Forest types are defined in the Appendix.

Deforestation

The total loss of all natural tropical forests during the decade from 1981 to 1990 was 154 million ha, representing an annual loss of 0.81% of the 1980 total (table 1.2). Nearly half of this deforestation occurred in the Americas (48%), with the remainder divided almost equally between Asia (25%) and Africa (26%).

Tropical moist forests constituted most (85%) of the forest area lost during the decade. Again, the Americas lost the largest amount of forest (67 million ha), followed by Asia (34 million ha) and Africa (30 million ha). However, in terms of the annual percentage loss of moist forests, Asia had the highest rate (1.1%), followed by Africa (0.75%) and the Americas (0.72%).

For lowland rainforests, a total of 19, 22, and 4.7 million ha were lost during the decade from the Americas, Asia, and Africa, respectively. As percentages of lowland rainforest lost annually, these figures constitute 0.4% for the Americas, 1.2% for Asia, and 0.5% for Africa. These figures demonstrate the very high rate of loss (in both area and percentage terms) of lowland rainforests in Asia, and the low area but moderately high annual percentage loss from Africa, where this forest formation is least extensive.

For lowland seasonal forests, the Americas again lost the most during the decade (32 million ha, or 0.96%/yr). Asia, where these formations are least extensive, lost the smallest area (7 million ha), but sustained a high percentage rate of loss (1.4%/yr). Africa, with very extensive seasonal forests, lost 23 million ha, at an average rate of 0.8%/year.

The FAO 1980 assessment had underpredicted the annual pantropical loss of natural forests for 1981–1985 at 11.3 million ha (Lanly 1982). This was updated to 15.4 million ha/year with the hindsight of the 1990 assessment (FAO 1993b). FAO (1995) reported that the annual attrition of tropical forests has been steady at approximately 15 million ha/year since the 1960s and has not accelerated, as had been widely believed. However,

Table 1.2 Estimated annual tropical deforestation (in millions of hectares) and average annual percentage deforestation rates from 1981 to 1990.

		Tropical moist forests			
			Lowland		
	All natural tropical forests	Total	Rainforest	Seasonal	Hill and montane
Africa	**41 (0.72)**	**30 (0.75)**	**4.7 (0.51)**	**22.5 (0.82)**	**2.9 (0.75)**
Continent	40 (0.72)	29 (0.74)	4.4 (0.51)	22.1 (0.82)	2.5 (0.75)
Madagascar	1.3 (0.79)	1.1 (0.77)	0.3 (0.6)	0.4 (0.98)	0.4 (0.78)
Asia	**39 (1.1)**	**34 (1.1)**	**22.3 (1.2)**	**6.8 (1.4)**	**4.9 (0.95)**
Continent	19 (1.2)	13.7 (1.3)	5.1 (1.3)	5.7 (1.3)	3 (1.0)
Malay Archipelago	20 (1.1)	20.4 (1.1)	17.3 (1.1)	1.1 (1.7)	1.9 (0.84)
America	**74 (0.75)**	**67.4 (0.72)**	**19 (0.4)**	**31.8 (0.96)**	**16.6 (1.2)**
Caribbean and Central	12 (0.97)	11.5 (0.94)	2.6 (0.55)	3.8 (.99)	6.1 (1.3)
Tropical South	62 (0.71)	55.9 (0.69)	16.4 (0.29)	28.9 (0.96)	10.5 (1.2)
Global total	**154 (0.81)**	**131 (0.9)**	**46 (0.64)**	**61 (0.94)**	**24 (0.93)**

Source: After FAO 1993b, French and Spanish editions.
Note: Forest types are defined in the Appendix.

during the latter third of the twentieth century, certain countries have been successively depleted (West Africa, Thailand, the Philippines, and now Sabah), so this global figure, although interesting (e.g., to campaigning groups), is not very informative.

Logging

FAO gathered logging statistics for all natural tropical forests aggregated (table 1.3). As noted above, over four-fifths of natural tropical forests are tropical moist forests. About 5.6 million ha were logged each year from 1981 to 1990, or 0.3% of the total forest area. Of this, 2.6, 2.1, and 0.9 million ha were logged in the Americas, Asia, and Africa, respectively, constituting 0.3%, 0.7%, and 0.2% of the respective forest areas. These figures highlight the leading position of Asia in rainforest logging. Within Asia, the greatest impact (1.8 million ha/year) was on the forests of the Malay Archipelago, not on the continent (0.4 million ha/year).

Globally, 84% of the forest was logged for the first time, but this proportion was lower in areas such as Madagascar (30%), continental Asia (66%), and the Caribbean plus Central America (67%). A lower proportion of the logging occurred in primary forests on the African continent (75%) than in the Malay Archipelago (85%) and South America (90%). Repeated logging at short intervals is likely to cause greater damage, and perhaps higher species loss, than a single episode, although much depends on the interval between logging episodes.

In the near future much more second- and third-time logging is likely to take place. For example, considering commercially valuable dipterocarp rainforests, Sabah State in Malaysia has almost depleted the virgin forest outside conservation areas, and Peninsular Malaysia approaches this condition. In contrast, Kalimantan, Indonesia, is likely at cur-

Table 1.3 Annual logging from 1981 to 1990 in all natural tropical forests.

	Millions of hectares		As percentage of all natural forests		Percentage logged for first time	
Africa	**0.91**		**0.2**		**74**	
Continent		0.89		0.17		75
Madagascar		0.02		0.10		30
Asia	**2.15**		**0.7**		**82**	
Continent		0.37		0.26		66
Malay Archipelago		1.78		1.04		85
America	**2.58**		**0.3**		**89**	
Caribbean and Central		0.13		0.1		67
Tropical South		2.45		0.3		90
Global total	**5.64**		**0.3**		**84**	

Source: FAO 1993b, table 5.

rent cutting rates to have virgin dipterocarp forest until perhaps 2020, though re-logging of accessible forest may begin well before then.

In some places logging precedes total deforestation for agriculture, with settlers using the logging roads to gain access to forests. This practice is illegal in some countries, such as the Philippines. Alternatively, logging may be initiated by the landowner with timber extraction being just the first step toward pasture creation, as in Paragominas in Pará state, in the eastern Brazilian Amazon (Veríssimo et al. 1992).

Altered Forest

The area of forest altered is the sum of forest lost (table 1.2) and forest logged (table 1.3). Globally, natural tropical forests were altered at a rate of 15.4 + 5.6 = 21 million ha/year during the decade from 1981 to 1990, or 1.2% annually. The figures for the Americas, Asia, and Africa are 10, 6, and 5 million ha/year, respectively.

National Differences

At the regional and continental levels, the loss of tropical forest during 1981–1990 mostly occurred at a rate of about 1% per year or less, and nowhere at over 2%. However, these averages obscure large differences between nations. On each continent there are examples of countries with high and low rates of forest conversion (table 1.4). For example, were the rate of loss up to 1990 to continue, the Philippines would have no forest left in 25 years' time; by 1990, forest covered only 26% of the land area, in a country once almost completely clothed in forest. At the other extreme, French Guiana lost only 300 ha/yr during the decade and remained 91% covered in forest. Human population pressure lies behind these statistics; the per capita forest area is 87 ha/person in French Guiana, but only 0.1 ha/person in the Philippines (table 1.4).

Forest Fragmentation

The national figures give a clearer picture than the regional and continental ones, but even they obscure the full probable impact of humans on the forest, because no allowance

Table 1.4 Loss of natural forest from 1981 to 1990 in selected tropical countries (in thousands of hectares).

	Area in 1990	Average annual loss	Annual percentage loss	Forest area (ha) per capita
Low rates of deforestation				
Congo	19,865	32.3	0.16	10
Brunei	458	1.8	0.37	1.7
French Guiana	7,997	0.3	0.0	87
High rates of deforestation				
Ghana	9,555	138.4	1.18	0.6
Philippines	7,831	316.0	2.50	0.1
Costa Rica	1,428	49.6	2.27	0.5

Source: FAO 1993b, French and Spanish editions: tables 3 and 4.

is made for the effects of forest fragmentation. This phenomenon has not yet been assessed at the global level. FAO (1993b) introduced two fragmentation statistics, but they have yet to be applied widely. These are the perimeter-area ratio and the edge-core ratio, where edge is defined as the zone within 10 km of the forest boundary.

As FAO (1993b) pointed out, regions often vary in their pattern of fragmentation. For example, the Brazilian Amazon has a pattern of colonization that mainly entails encroachment from the Basin's margins, particularly along the Belém-Brasilia highway in the east and in the states of Acre and Rondônia in the southwest. In contrast, Ghana, the Philippines, Costa Rica (table 1.4), and Peninsular Malaysia, among others, now have forests increasingly persisting as islands in a "sea" of agricultural land. Countries of this second group are at various stages of developing landscapes similar to the more forested areas of Europe, such as Bavaria and Austria.

The heavy impact of damage from fragmentation and edge effects on top of that resulting from deforestation has been computed for Amazonia by Skole and Tucker (1993). They estimated that by 1988, only 6% of closed-canopy Amazonian forest had been cleared, but that 15% was actually affected. Their computation was made using Landsat TM images. They found that during 1978–1988, deforestation averaged 15,000 km^2/year, and the total area deforested increased from 78,000 to 230,000 km^2. However, an additional 30,000 km^2/year was affected by fragmentation or edge effects, and during the decade, the total area classified as being severely affected increased from 208,000 to 508,000 km^2.

SPECIES RICHNESS IN SMALL AREAS

The way in which tree species are packed into lowland tropical rainforest remains largely unexplored, although for Peninsular Malaysia we are beginning to have some insights. This country has over 3,000 tree species of over 30 cm diameter (Whitmore 1972), a figure that can be contrasted with the 50 species indigenous to continental Europe north of the Alps and west of the Urals. Part of this richness is, as expected, due to local endemism. For example, 66 species have never been found outside the northwestern state of

Perak (Ng and Low 1982), a region converted to plantation crops early in the twentieth century. Less easily anticipated is the extreme richness of a single small area—a 50 ha plot at Pasoh in south central Negeri Sembilan, on which all trees larger than 1 cm in diameter have been mapped and identified (Kochummen, LaFrankie, and Manokaran 1992). Among the 340,000 stems enumerated, a total of 802 species were encountered, comprising 25% of the country's total known woody flora and 50% of the flora of this part of the Malay Peninsula. The representation varied greatly between families; for example, all 12 *Mangifera* species of Peninsular Malaysia were present, while 21 of 35 species of Burseraceae and only 30 of 156 species of Dipterocarpaceae were present. Overall, half of all the country's tree genera were represented.

Similar very high representation of the region's flora had earlier been found by Ashton (1969) in a small area at Bukit Raya, Sarawak. On plots totaling 6.6 ha spread across an area of 5 × 2.5 km, there were 711 tree species of at least 10 cm in diameter, representing about half the tree species of this size known from the whole country.

The extent to which small forest areas can sample a regional flora is of course highly relevant to studies of forest fragmentation (see also Andersen et al., chap. 18). It is unknown, however, whether populations in such small areas would be viable in the long term. For example, despite the high numbers of species present at both Pasoh and Bukit Raya, population sizes of trees in these areas are inevitably small.

SPECIES PERSISTENCE IN FOREST FRAGMENTS

There have so far been far fewer recorded extinctions of species in tropical rainforests than predicted from species-area curves (e.g., Simberloff 1986; Heywood et al. 1994). Examples of unexpectedly high survival are accumulating.

The Atlantic Coast rainforest of Brazil has been reduced to only about 8% of its original area (Viana, Tabanez, and Batista, chap. 23), but Brown and Brown (1992) found no documented extinctions among recorded vascular plants, mammals, birds, and butterflies. They believe that this mountainous forest always existed as a mosaic of different habitats, each with its own biota, and of which an aliquot still survives.

Corlett and Turner (chap. 22) and I. M. Turner et al. (1994) have found relatively little extinction among the plants of Singapore. Seventy-one percent of the vascular plant species of inland forest found since records began still exist, despite the loss of 99.8% of inland primary forest. The forest interior avifauna has, however, become depauperate; at least 50% of the species have been extirpated (Corlett and Turner, chap. 22).

Similarly, in Hong Kong, Corlett and Turner (chap. 22) found that, after 400 years of progressive fragmentation, 314 species of native trees still survive, although some have very few individuals. In contrast, in Singapore, nearly half the bird species have been lost, and those that remain today are habitat generalists, not forest-dependent species, and have become abundant (R. T. Corlett, pers. comm.).

On Cebu Island in the Philippines, 7 of 15 endemic bird species survived in 1992 in a total area of 15 km^2, comprising 0.3% of the original forest. This is twice the number of surviving species predicted by the species-area curve (Magsalay et al. 1995). In Central Amazonia, Bierregaard and Stouffer (chap. 10) describe how a number of primary rainforest birds forage in adjacent secondary forest, and have used secondary forest to recolonize small primary forest fragments from nearby large forest tracts. Some species have proved more mobile and flexible in their habitat use than others. The importance of linking forest remnants has been demonstrated in Queensland, Australia, where grass strips along powerline clearings proved impassable to small forest mammals, while clearings spanned by secondary forest were readily crossed (Goosem, chap. 16).

In Sabah, East Malaysia, relict strips and patches of primary forest, albeit logged and then accidentally burned, have retained a rich rainforest mammal and bird fauna that roams out to feed in the surrounding timber tree plantations (Duff, Hall, and Marsh 1984). Finally, a 75 ha lowland rainforest fragment, left intact within an oil palm plantation in Peninsular Malaysia and subsequently protected from hunting, was found after more than sixty years to have an extremely rich, dense primary forest bird and mammal fauna (Bennet and Caldecott 1981).

In Indonesia, the government has decreed that where industrial tree and other plantations are established to replace rainforest, belts of forest must be retained along streams and around water bodies, the intention being to conserve part of the original forest. However, this decree is not yet fully enforced, and no detailed research has yet been undertaken to devise good forest layouts to achieve conservation, or even to assess the extent to which the rainforest biota can persist in such landscapes.

These examples demonstrate differing degrees of survival in forest fragments under different sets of circumstances. One major influence is the proximity of the fragment to large reservoirs of primary forest. Other relevant factors are the time since fragment isolation and fragment size.

Species Committed to Extinction

The survival of more species in forest remnants than is expected from current theory may be a temporary phenomenon. Populations of many species may be too small to be viable in the long term, and numerous species still present in fragments may be committed to extinction (Heywood et al. 1994; Magsalay et al. 1995). In particular, many tree species live for centuries, even in rainforests, and what we are observing today may simply be a lag effect before extinction (see Harrington et al., chap 19). As mentioned above, the sharp decline of birds, but not plants, in Singapore (Corlett and Turner, chap. 22) is an example of this phenomenon.

Moreover, the survival of species per se conceals the largely invisible erosion of genetic diversity as population numbers decline. As Heywood and Stuart (1992) point out, this should be the main focus of conservationists' concern.

SPECIES PERSISTENCE IN LOGGED FORESTS

One of the major impacts on tropical rainforests today is selective timber extraction, which alters but does not destroy forests. Tropical rainforests are dynamic ecosystems. Their canopies experience continual natural disturbance, and the creation of canopy gaps by the felling and removal of trees resembles in many respects natural forest dynamics. The plants and animals that constitute a particular forest ecosystem are those that succeed within the particular disturbance regime the forest experiences. The basic dictum for maintenance of that particular species composition is that human disturbance by timber removal should be similar to natural disturbance regimes (Whitmore 1990). Usually, modern mechanical logging creates greater than natural disturbance. Its overall effect is to favor more disturbance-adapted plants and animals, thus causing a quantitative shift toward those species.

Detailed research on the effects of logging on rainforest plants is scanty. In Queensland, Nicholson, Henry, and Rudder (1988) found that after selective removal of a fraction of the big trees, total tree species richness increased. This was because pioneer species, absent from the original forest, had colonized roadsides and log yards.

Rainforest arboreal vertebrates have been studied in Malaysia, Queensland, and South America (Johns 1986; Crome 1991b; Thiollay 1992; Johns and Johns 1995; Laurance and Laurance 1996). In brief, these animals move to unlogged patches during the actual lumbering operation and then eventually move back to reoccupy the whole forest afterward. If the whole area, including unlogged patches, is censused, the species list remains virtually unaltered by logging. However, the proportions of species change, with edge- and gap-favoring species increasing and disturbance-sensitive species declining.

Ten years after logging, Johns (1986) found that 97% (188/193) and 95% (210/220) of the birds were present in Peninsular Malaysia and Sabah, respectively. However, a few specialist guilds, such as ground-feeding insectivores, had become seriously depleted. There is little detailed evidence yet of the effects of rainforest logging on invertebrates, but Holloway, Kirk-Spriggs, and Chey (1992) found the expected increase in edge species.

These results are for forests logged only once, and in which no particular care was taken to maintain animal habitat (for example, by retaining hollow trees or keystone food species such as fig trees [*Ficus* spp.]). Repeat logging could have greater consequences, but equally, the effects could be reduced if care were taken to conserve animal habitat.

GENERAL IMPLICATIONS

1. We live at a time in which many of the world's tropical forests are being rapidly altered. They are disappearing and becoming fragmented in much the same way that other forest biomes have been changed at the hand of Man in the past.
2. Because of their extreme biotic richness compared with other forest biomes, a much greater number of extinctions is likely to result eventually from the severe loss and fragmentation of tropical rainforests. However, the lag phase for extinctions may be long, and, as this chapter has endeavored to show, the amount of extinction is not easily predicted.
3. The future landscape of most humid tropical nations is likely to contain blocks of conservation forest. Indeed, over the past two decades, rainforest nations have on average designated 5% of their area as national parks or similar reserves (Groombridge 1992, table 29.4). So far, however, most of these reserves receive no or minimal management.
4. Much larger areas are likely to be retained as timber production forests. Today, many of these are being progressively logged for the first time. In some countries little virgin

production forest remains. Second and later logging cycles will exploit much smaller trees and areas than the first cycle.

5. As deforestation proceeds, tropical nations will increasingly contain fragments of forest retained either as islands or as variously linked patches in a predominantly agricultural landscape. Today, the Philippines, Peninsular Malaysia, Ghana, and Costa Rica already approach this state. It has been reached without conscious planning to conserve biodiversity.
6. In some countries land use planning is just starting to demand retention of rainforest within agricultural landscapes. In the future we can expect to see belts and blocks of rainforest deliberately conserved in this manner.
7. Forest loss will result in the extinction of species, and some of these could be of yet unknown or unutilized economic value. The main loci for species survival must be large tracts of rainforest, where the largest populations will remain. The greatest proportion of surviving rainforest is likely to be timber production forest.
8. Research is badly needed to refine timber extraction systems and logging rules so that minimal alteration is made to species composition. We also remain largely ignorant about the ability of plants and animals to maintain viable populations in isolated or linked forest fragments.
9. The agenda for ecologists must be to explore the consequences of disturbance by logging and fragmentation and to design management and land use regimes that maintain biodiversity.
10. The agenda for campaigning organizations should now shift in focus to advocating sustainable logging plus the retention of forest remnants. They should also be advocates for the research that is urgently needed to design and manage fragmented landscapes with biodiversity conservation as the key objective.

ACKNOWLEDGMENTS

I am most grateful to K. Janz of FAO for assistance with obtaining data, and to J. Kikkawa, R. Corlett, W. Laurance, and R. O. Bierregaard, Jr., for comments on earlier drafts of this chapter.

APPENDIX 1.1: Notes on the 1990 FAO assessment of tropical forests

The definitive publication of the 1990 assessment is FAO (1993b), in which the figures differ slightly from various previews (e.g., FAO 1992, 1993a,c). Unfortunately, the tables in the English language edition are misleading and slightly incomplete. Corrected versions, due to be published in the Spanish and French language editions, are used in this chapter.

FAO (1993b) divides forest between ecoclimatic zones, which are not defined. Examination of figure 6 in FAO (1993b) and comparison of the captions of the tables between the English and other editions show that the following meanings can be given. This interpretation is supported by FAO (1989) in a little-known and earlier FAO publication that is not referred to in FAO (1993b).

Among natural forest formations, the *wet zone* of FAO is tropical lowland evergreen rainforest in the sense of Whitmore (1975). This is absent from West Africa and the lower Amazon, which are *moist zone* in figure 6 in FAO (1993b) which also reveals that the moist zone includes semi-evergreen rainforest plus all other seasonal/monsoon forest formations.

Thus, FAO's wet and moist zones together encompass all lowland tropical moist forests (i.e., rainforest plus seasonal/monsoon forests) in the sense of Sommer (1976). To these must be added FAO's *hill and montane zone* (whose lower elevation is not defined) to derive all lowland plus montane moist tropical forest. In tables 1.1 – 1.4 in this chapter, FAO's wet zone is called rainforest and FAO's moist zone is called seasonal forest.

FAO's term *natural forest* includes all vegetation with at least 10% tree cover in the wet and moist zones as well as in the dry, very dry, and desert zones. Primary, disturbed, and old secondary forest are all included in this category, as these cannot be confidently distinguished from remotely sensed images. However, young secondary forest in mosaic with agriculture (i.e., the bush fallow of swidden agriculture) is excluded. For some analyses in this chapter, data for natural forest had to be used because separate figures were not given for the different components of this forest type.

FAO (1993b) omits a few minor occurrences of tropical moist forests, notably in Australia and the Pacific islands east of Papua New Guinea. These are, however, included in the synthesis of global forest resources (FAO 1995) of which FAO (1993b) is just the tropical component. *Deforestation* is defined as total loss of forest cover.

2

The Role of Economic Factors in Tropical Deforestation

James R. Kahn and Judith A. McDonald

THE purpose of this chapter is to discuss the economic factors associated with tropical deforestation, a process that also leads inevitably to the creation of forest remnants. Although forest remnants have several unique characteristics that affect their interaction with economic processes, our discussion will focus on factors that lead to deforestation, while noting differences between forest remnants and forests in general.

At the outset, it is important to note two distinctions between tropical forest remnants and tropical forests per se. First, some set of factors must have halted deforestation activities, forming the forest remnant. These factors could be physical (such as the steepness of the terrain), institutional (such as the establishment of a forest preserve), or economic (the surrounding forest was cut, but the remnant was left standing because its conversion to other uses was not profitable). Second, the forest remnant is generally surrounded by nonindigenous populations and nonindigenous economic practices, so that any changes in the above-mentioned factors could lead to immediate and rapid deforestation activity. For example, all else being equal, if the price of beef increases, some forest remnants will be quickly converted to cattle ranches.

Although deforestation is viewed negatively, some deforestation of a completely forested region may be desirable from a human perspective. The economist is interested in determining the optimal level of deforestation, which is defined as the level that maximizes social welfare. Of course, the optimal level of deforestation will be different from local, regional, national, and global perspectives.

Although it may seem incongruous to many observers, deforestation does generate social benefits. These benefits are equal to the difference between the benefits generated by the deforesting activities (timbering, slash-and-burn agriculture, cattle ranching, mining, etc.) and the cost of those activities. Similarly, there are social benefits associated with the standing forest, which include the net benefits of economic activities such as hunting and gathering and sustainable agroforestry. In addition, there are the benefits associated with the ecological services of the forest. Note that although this is an anthropocentric perspective, it does not ignore the importance of these natural services. Because biodiversity, watershed protection, natural habitat, carbon fixation, and nutrient cycling all contribute to social welfare, these diffuse benefits must be considered in addition to those more directly related to economic processes.

The optimal level of deforestation occurs when the marginal benefits of deforesting another unit of forest are equal to the marginal benefits associated with preserving that unit of forest. An important empirical question is how to define a unit of forest. If one were concerned only with economic production activities, it would be reasonable to use some standard physical unit such as an acre or hectare. However, because ecological services are an important benefit of forest preservation, and because they depend on ecological systems, the definition of the marginal unit of forest becomes more difficult. When looking at forest remnants, the smallest appropriate unit for examination may be the entire remnant.

From a social perspective, some deforestation is desirable because land is needed for other uses, and the remaining forest is capable of providing the important ecological services associated with the standing forest. However, as deforestation continues, ecological services are lost, and these losses will eventually outweigh the economic benefits of forest conversion. Of course, many observers of tropical forests believe that the optimal level of deforestation has already been exceeded.

If we accept the premise that the current level of deforestation is excessive, that it reduces social welfare and the quality of life below what it otherwise would be, then the question that must be answered is "Why?" The question "Why is the level of deforestation excessive?" is fundamentally different from the question "What causes deforestation?" and much more important from a policy perspective.

Deforestation rates may be excessive due to both microeconomic and macroeconomic factors. Microeconomic factors include poorly constructed timber leasing agreements, poorly defined and enforced property rights to forested land, imperfect information about sustainable agricultural practices, lack of access to capital markets, and a variety of incentives for deforestation embedded in public policies. Macroeconomic incentives for deforestation can exist if conditions in the economy create a demand for current income (which is generated by deforestation activities) at the expense of future income (which is generated by keeping the forest intact).

It should be emphasized that economic factors are not the only reason for excessive deforestation. It is the complex interaction of social, economic, political, and ecological processes that shapes the landscape. Macroeconomic conditions may create pressure for more current income, but the implications of this for deforestation depend on the state of the forests, as well as on social, political, and economic conditions in the forested region and the country as a whole. For example, if timber leasing agreements are constructed so that timbering companies have little incentive to harvest in a sustainable fashion, then the economic pressures will cause more deforestation than if the agreements force the timbering companies to protect the sustainability of the forest. Similarly, if forest dwellers are politically disenfranchised or underrepresented in the political process, macroeconomic pressures may lead to more deforestation as outside groups move into the forest in the pursuit of current income.

These factors interact in a complex fashion (fig. 2.1). For example, a change in macro-

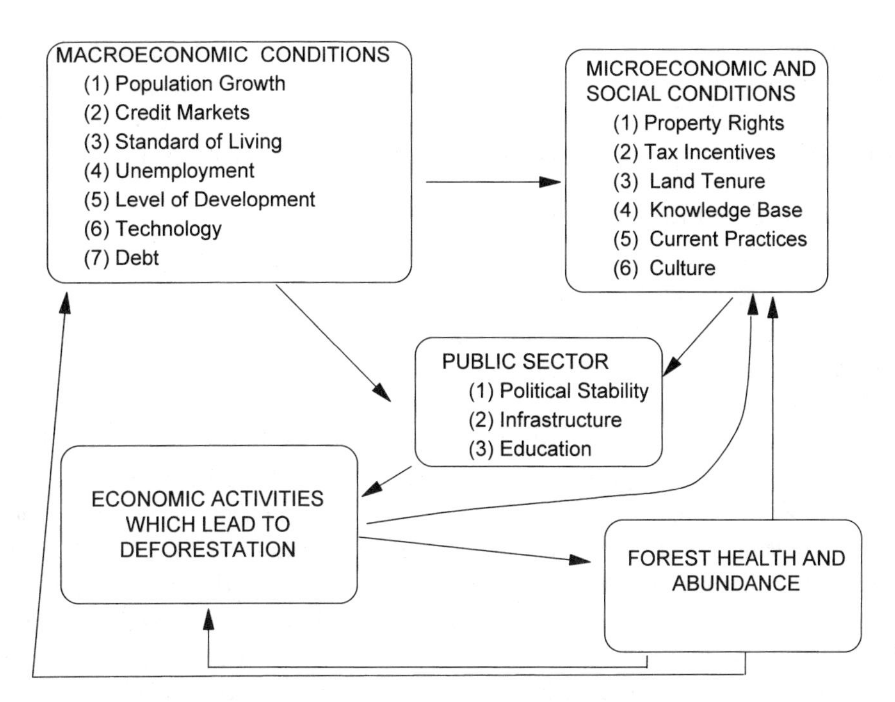

Figure 2.1. Interactions among economic, social, and ecological processes that affect deforestation.

economic conditions (e.g., external debt levels) might directly change economic activities that affect deforestation, but the way in which this change takes place would be influenced by microeconomic and social conditions, the nature of forest resources, the nature of the public sector, and other factors. In addition, there exist many feedback loops (not all of which are depicted in fig. 2.1). Changes in macroeconomic conditions lead to changes in microeconomic conditions, which in turn affect deforestation and the state of the forest, and which have feedback effects on the state of the forest, microeconomic conditions, and macroeconomic conditions.

MICROECONOMIC FACTORS IN DEFORESTATION

Poorly Constructed Timber Leasing Agreements

The leasing of timbering rights is a major contributor to excessive deforestation. The often short-term nature of leasing agreements creates perverse incentives for the harvesting firm. If the firm has no long-term economic interest in the leased forest, it has no incentive to leave the forest in a condition in which it will be productive in the future. Hence, it will not undertake current costs (such as protecting young trees or the soil) when it harvests trees or otherwise manage the forest to promote its long-term viability.

The obvious solution to this problem is to increase the length of the leasing arrangements, making them perpetual and transferable (so that if the forests are degraded, their market value falls). In many circumstances, however, this is not politically feasible, because the harvesting firms are often foreign-owned and granting of perpetual rights is viewed as relinquishing sovereignty. Nevertheless, other strategies are available to deal with this problem by making it costly for the harvesting firm to harvest in a degrading fashion. For example, harvesting firms could be required to post a bond that would be forfeited if they created excessive environmental damage while harvesting.

Poorly Defined and Enforced Property Rights

Perhaps the most important microeconomic factor affecting deforestation rates has to do with poorly defined and poorly enforced property rights. If land tenure is insecure, then farmers do not have an incentive to invest in their land (by leaving the tree cover intact). In fact, depreciating the land (by clearing it) may reduce the probability of someone else seeking to obtain it.

Another important issue is that the way in which property rights are defined does not allow the landowner to capture the benefits from the ecological services of the intact forest. Ecological services are public goods, which are distinguishable from private goods in that there is nonrivalry and nonexcludability in consumption of public goods. Nonrivalry means that one person's consumption of the public good does not diminish the amount available for other people to consume. Nonexcludability means that if the good is available for one person to consume, others cannot be excluded from consuming it. Biodiversity, for example, has both of these properties. One person's enjoyment of the benefits of biodiversity does not reduce the amount of biodiversity available for others to enjoy. Similarly, people cannot be excluded from enjoying the benefits of biodiversity. Because these two properties make it impossible for landowners to require payment from those who benefit from the ecological services of their land, landowners clear forests without considering these social benefits. Consequently, deforestation is excessive.

In order to rectify this problem, policies must be developed that allow the benefits of preservation to flow to the landowners. The problem is made more complex, however, because the benefits of preservation accrue to the global community, and transboundary compensation mechanisms are largely undeveloped. Methods for citizens in developed countries to compensate citizens in developing countries for ecological services must be developed to deal with this issue, but with the exception of the "debt-for-nature swap," these remain largely undeveloped and untested.

Imperfect Information

Many observers of agricultural practices in tropical forests find that sustainable agroforestry, in the long run, is more productive than traditional slash-and-burn practices. For example, Dale and Pedlowski (1992) examined agriculture in Rondônia, Brazil, and found that sustainable agroforestry generates an income approximately twice that of slash-and-burn agriculture. However, many farmers may not engage in the more productive agroforestry because they lack sufficient knowledge of the appropriate techniques. Consequently, they practice what they know, and continue to engage in traditional slash-and-burn practices, which exhaust the productivity of the land in an approximately ten-year cycle and require the clearing of additional forest tracts. Dissemination

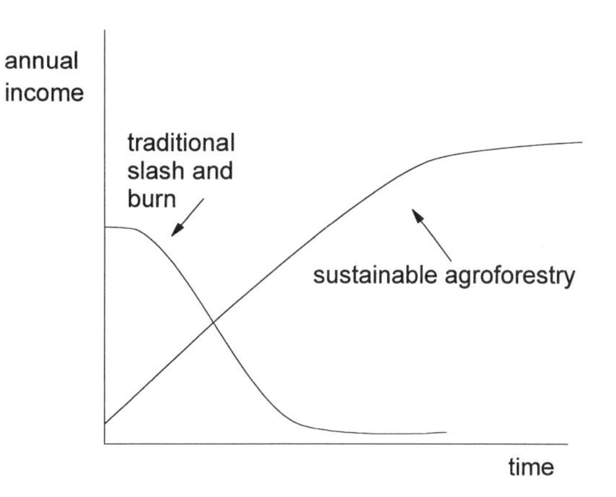

Figure 2.2. Time paths of alternative agricultural practices.

of information about agricultural techniques is an essential strategy for the promotion of sustainable agroforestry.

Lack of Access to Capital Markets

Imperfect information is not the only reason that more profitable, sustainable agroforestry practices remain largely unimplemented. An additional problem is that agroforestry represents an investment, in which initial income is reduced in order to gain large increases in future income. The time paths of income from slash-and-burn agriculture and agroforestry follow opposite trajectories (fig. 2.2). As income from slash-and-burn agriculture declines, new land is cleared, and the cycle repeats. However, farmers who wish to convert to sustainable agroforestry may be unable to do so, as they cannot support their families during the initial low income period needed to establish sustainable practices. In market economies, analogous investments are supported by capital markets when the investing firm or individual borrows against future income. However, the absence of capital markets, imperfections in existing capital markets, and difficulties in storing value over time in subsistence economies may preclude subsistence farmers from engaging in this type of investment. Policies to rectify this problem would include the development of agricultural credit markets, education, and training programs to encourage farmers to adopt sustainable agroforestry, as well as income maintenance programs for farmers who elect to invest in sustainable agroforestry.

Improper Government Incentives

In attempting to deal with certain social problems, governments of forested countries have often developed policies that have exacerbated deforestation. Examples of such policies include those that promote export crops, tax incentives for clearing land, road building, and resettlement schemes. Specific discussion of these incentives can be found in Browder (1988: resettlement schemes, subsidization of cattle ranching, and other government incentives in the Brazilian Amazon), Gillis (1988a: tax, forest leasing, and transmigration policies in Indonesia), and Gillis (1988b: tax and export policies in West

Africa). Justification for adopting these policies may be found in macroeconomic pressures, which are discussed below.

MACROECONOMIC FACTORS IN DEFORESTATION

Macroeconomic conditions affect deforestation primarily by constraining the options that are available to a society. Countries with tropical forests typically are characterized by low per capita income and current consumption needs that virtually exhaust national income. One of the effects of this combination is that there is little savings with which to make investments. Investment in human capital (i.e., education) and human-made capital is an important mechanism for increasing the productivity of labor (which increases personal income). In the 1960s and 1970s, this low level of investment was thought to be the primary cause of poverty in developing nations and was termed "the vicious cycle of poverty." According to this view, low income and high current consumption needs implied low savings, which implied low investment, which led to low productivity of labor (fig. 2.3). This low productivity of labor ensured that per capita income remained low, reinforcing the cycle of poverty.

In the last decade, however, more recognition has been given to the environmental component of the vicious cycle of poverty. Low per capita income and high current consumption needs lead to the consumption of environmental capital to meet current consumption needs (fig. 2.3). This consumption of environmental capital (such as losses

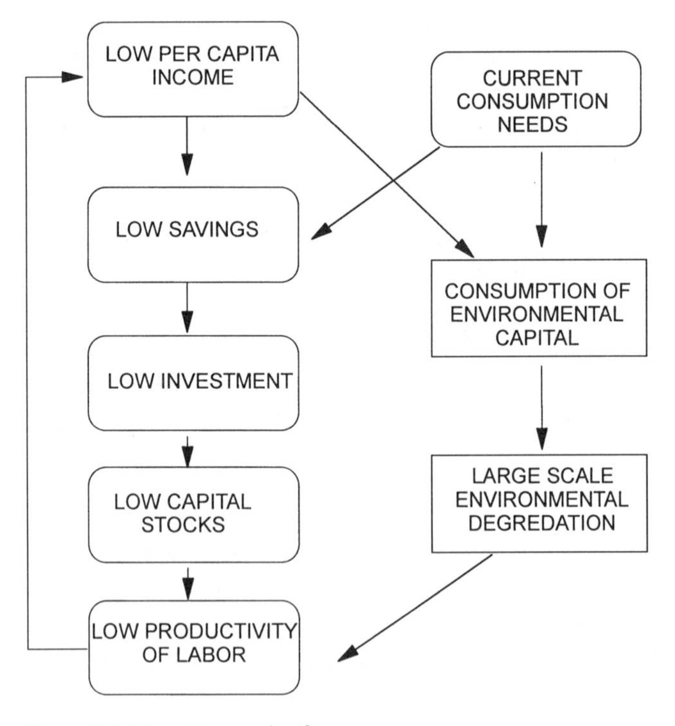

Figure 2.3. The vicious cycle of poverty.

in soil fertility, water pollution, deforestation, desertification, and air pollution) leads to lower productivity of labor, which leads to lower income, perpetuating and reinforcing the cycle of poverty.

There are two reasons why insufficiency of current income to meet current consumption needs causes environmental degradation. First, there is insufficient income to support environmental protection. Second, and most importantly, renewable resources are treated as nonrenewable resources and used to produce current income quickly in a way that degrades their ability to produce future income. Tropical deforestation is a good example of this: slash-and-burn cultivation practices are used to produce a current crop, but the land is quickly degraded and becomes economically unproductive. The important point is that environmental degradation is not only a symptom of poverty, but also a contributor to poverty, because the degradation of natural capital also diminishes the productivity of labor (fig. 2.3).

Macroeconomic factors can contribute to this cycle of environmental degradation in two ways. First, adverse macroeconomic conditions can reduce current income, putting more pressure on environmental resources. Second, the state of the macroeconomy can increase current needs to produce income. This is particularly true when developing countries have an increased need for foreign exchange.

Population growth affects deforestation by exacerbating both microeconomic and macroeconomic factors. Population growth operates through microeconomic forces by making market failures more severe. For example, if one of the sources of deforestation is poorly designed and enforced property rights, population growth may make property rights even more insecure, thereby increasing deforestation. The macroeconomic effects of population growth probably have an even more profound influence on deforestation. As population growth occurs, labor inputs tend to rise, whereas other inputs stay constant. (Labor increases relative to other factors of production because natural capital tends to decline due to overexploitation, and the ability to invest in human and human-made capital is limited by a lack of savings with which to make investments.) As a consequence, the increase in output brought about by an additional unit of labor input, as well as per capita income, will be reduced. Of course, this problem is exacerbated if population growth also changes the age profile of the population, increasing the proportion of children relative to potential workers. In addition, population growth increases current consumption needs, particularly if the age distribution of the population becomes younger. This increase in current consumption needs leads to further exploitation of renewable resource systems and their eventual degradation.

Economists have designed many policies for addressing the microeconomic causes of deforestation, so if microeconomic factors are primarily responsible, then the development of strategies to deal with deforestation may be relatively straightforward. Dealing with macroeconomic factors is much more difficult because small countries' macroeconomies are extremely dependent on the global economy. If macroeconomic factors prove to be important causes of deforestation, then the development of policy will be much more difficult and will require more of a cooperative international effort.

LITERATURE SURVEY

To date, investigations of the economic causes of excessive deforestation rates fall into three categories: (1) case studies of specific countries (which may examine political-economic or human-ecological issues); (2) studies of microeconomic causes; and (3) studies of macroeconomic causes. Our discussion below focuses primarily on studies of the macroeconomic causes of deforestation.

As an example of the case study approach, Repetto and Gillis (1988) assessed how government policies ("forestry" and "non-forestry") in various Third World countries affect the forest sector. They concluded that a variety of inefficient government policies—investment incentives, credit concessions, tax provisions, agricultural pricing policies, trade policies, the nature of lease or sale of forest exploitation rights, and land tenure policies—contribute to "uneconomic and ecologically damaging exploitation" in all ten countries they examined (Repetto and Gillis 1988, xi).

Brown and Pearce (1994) reviewed many different studies that used economic and statistical analyses to examine the factors that may lead to nonoptimal levels of tropical deforestation. Some of these are summarized in the remainder of this section.

Capistrano (1994) divided the period 1967-1985 into four subperiods and identified the principal micro- and macroeconomic variables that were associated with deforestation in each subperiod. During 1967-1971, the value of tropical wood in the export market was the strongest determinant, while in the early 1970s, countries' per capita income and their self-sufficiency in cereals became the most important factors associated with deforestation. In the late 1970s, deforestation was linked most strongly with a country's rate of devaluation of its domestic currency, while from 1981 through 1985, deforestation was associated with the availability of arable land for agriculture.

In addition to Capistrano, many other researchers have investigated the role of per capita income in a country's deforestation rate. Some studies have found, as did Capistrano, that national per capita income (as measured by GNP and GDP) and deforestation rates are positively correlated (e.g., Burgess 1991; Rudel 1994). Others, however, have found a negative (Panayotou and Sungsuwan 1994) or nonsignificant (Shafik 1994) association between these variables. Thus, the relationship between per capita income and national deforestation rates remains debatable.

Palo (1994) examined the relationship between population density and deforestation. He found that, for tropical countries, there is a negative relationship between forest cover and population density. Many other researchers have confirmed this finding (e.g., Palo, Mery, and Salmi 1987; Panayotou and Sungsuwan 1994). Other researchers have used population growth rates rather than population density as a measure of the pressure exerted by a country's population on its resources. For example, Bilsborrow and Geores (1994) discovered a weak positive relationship across countries between population growth rates and the rate of increase in agricultural land and deforestation. Myers (1984) also focused on growing populations of peasant or subsistence cultivators as the chief cause of deforestation. Rudel (1994) similarly concluded that, in some countries, population growth contributes to excessive deforestation.

Southgate (1994), among others, discussed what he sees as the indirect nature of the relationship between population growth and frontier expansion. Population growth leads to increased domestic demand for agricultural products, which increases deforestation. Also, because many tropical countries have a comparative advantage in agricultural goods, as external demand rises, so too may deforestation (assuming that the country's exchange rate and other policies do not counteract this effect). Regression analysis revealed that agricultural export growth and population growth were both positively related to an increase in the amount of land used to produce crops and livestock, and hence positively related to deforestation (Southgate 1994).

Rudel (1994) hypothesized that deforestation occurs automatically as part of the process of development. The reasons for this connection are twofold. First, as stocks of public and private capital increase, so does deforestation. Second, development leads to changes in trade patterns: peripheral nations export agricultural products and raw materials to the core nations to which they are tied by multinational companies. Rudel found a positive relationship between deforestation rates and capital-intensive projects in countries with large forests. In countries with small forests, however, growing peasant populations were strongly associated with deforestation. The empirical relationship between exports and deforestation was also somewhat inconclusive: increasing exports of wood accelerated deforestation in Southeast Asia, while most African and Latin American countries experienced rapid deforestation without significant wood exports. Similarly, while Central America's agricultural exports may have contributed to excessive deforestation, the same effect could not be discerned for most African or Amazon Basin countries. Unlike the previous authors, Shafik (1994) contended that while macroeconomic conditions may exacerbate the incentives to deforest, the root causes of such deforestation are more likely to lie in microeconomic conditions.

HYPOTHESES AND METHODS

In the early 1980s, deforestation rates and external debt both increased dramatically in many developing countries, leading to the speculation that they were causally linked. For example, von Moltke and DeLong (1990, 4) noted that "the fourteen largest debtor countries are also the same countries in which an unprecedented rate of deforestation is occurring." Miller (1991) also claimed that there is a causal relationship between environmental degradation and external debt in many developing countries. Kahn and McDonald (1995) give details on the extent to which debt and debt service dominate the economies of many tropical forest countries, how this debt accumulation interferes with the development process, and the different options available to a country to deal with its debt problem. Kahn and McDonald (1995) also present information on average annual deforestation and public debt levels for their sample of sixty-eight developing countries over the period 1981-1985. Whitmore (chap. 1) details deforestation rates from 1980 to 1989.

It might be hypothesized that the level of deforestation is independent of the level of debt. The argument underlying this position is that whatever level of deforestation maximizes income does so regardless of the debt position of the forested country. An alternative hypothesis, which we investigated, is that debt generates economic pressures that may lead to increased deforestation. A heavily indebted country's ability to invest in its future—either through undertaking conventional investment or through leaving forest and other resources intact—is severely limited by the need to both service its debt and meet current consumption needs.

Von Moltke (1990) discussed the importance of macroeconomic conditions as a constraint on a country's ability to allocate resources between current and future generations. He was also, to our knowledge, the first researcher to articulate the importance of external debt as a potential cause of deforestation. According to him, short-term claims (such as debt servicing) may force countries to adopt short-term policies that run counter to the interests of conservation, which are intrinsically long-term in nature.

Our contribution has been to explicitly model the way in which debt drives a wedge between strategies that must be pursued in the short term and those that are optimal in the long run. Stocks of external debt and the required debt service may lead forested tropical countries to deforest at levels that may not be in their long-term interests. Our research emphasis on the relationship between debt and deforestation should not be construed to mean that we feel debt is the only or even the most important economic factor causing tropical deforestation. Rather, we examined the degree to which external debt may contribute to a country's level of deforestation. As such, our study should be regarded as a complement to both the country case studies and the conceptual microeconomic studies, as all three approaches further our understanding of tropical deforestation. Also, it should be borne in mind that there are many interactions between the various micro- and macroeconomic causes of deforestation.

We used a macroeconomic model that, in addition to the traditional inputs such as capital and labor, includes deforestation as an input to the production process that generates GNP or national output. We assumed that countries maximize the sum of the present value of each period's consumption subject to a constraint that consumption in any period must remain above some minimum acceptable level. In other words, countries cannot reduce consumption to very low levels in some periods in exchange for higher levels of consumption in future periods. This consumption constraint is a critical component of our behavioral model. A fuller description of our model is found in appendix 2.1.

The minimum consumption constraint also provides a mechanism by which debt (and other factors) can affect deforestation. For example, if past debt causes interest payments to be very high, then a country may make a large sacrifice of future consumption so that present consumption can remain at the required minimum level. If capital markets are blocked (because of high debt), then the only way to increase consumption is to increase GNP through higher deforestation rates (even though these actions reduce GNP in the long run). Thus, this minimum consumption constraint, together with the assumption that deforestation generates income (or GNP), implies that *deforestation will increase whenever debt must be serviced* to ensure that consumption will not fall. The hypothesized positive relationship between debt service and deforestation occurs only when the consumption constraint is binding (i.e., consumption is equal to its minimum required level).

A country's ability to consume current goods and services is primarily determined by its current productive capacity (output or GNP). Therefore, if factors of production (or inputs) that generate GNP—such as capital, labor, or forested or nonforested land—increase, then countries can more easily reach the required consumption level without resorting to deforestation (which is also an input into an economy's total output or GNP). However, there are additional demands on a given level of output or GNP aside from meeting a required consumption level. These "competing uses" of GNP include investment spending, government spending, and debt service. If any of these competing uses increases, then there is more pressure on a country to deforest to meet its consumption needs.

Borrowing is a way for a country to increase its competing uses of GNP while still meeting the consumption constraint. If there is a shortfall in GNP, borrowing may allow a country to meet its consumption constraint without increasing deforestation in the short term. However, the implied higher debt service burden in future periods may lead to higher future deforestation because future debt servicing combined with the minimum consumption constraint may increase deforestation pressures. Also, certain countries—such as those that have "overborrowed" in the past, missed interest payments, or defaulted—may be unable to take advantage of this borrowing opportunity. Thus, countries deemed to be uncreditworthy will be more likely to deforest if a GNP shortfall means that the minimum consumption level could not otherwise be attained.

MODELING RESULTS

We examined various scaled measures of deforestation as functions of scaled measures of debt service and other explanatory variables, including inputs and competing uses of GNP. All the regressions are scaled to account for size, as an unscaled regression could establish a relationship between debt and deforestation that is completely driven by country size and has nothing to do with economic relationships. We found a strong positive statistical relationship between debt and deforestation, which was robust across scaling procedures and measures of debt burden. We tried many different specifications to ensure that this result was robust, and have reported typical results in table 2.1. It should, however, be emphasized that, because our results were based on cross-sectional data for 1981-1985, they may not necessarily be applicable in the 1990s. In addition, the quality of the deforestation data we used (from the Food and Agriculture Organization of the United Nations) is often disputed (e.g., Bilsborrow and Geores 1994; Shafik 1994). However, if countries with large deforestation rates underreported deforestation, this would tend to create a conservative bias against rejecting our null hypothesis of no relationship between debt and deforestation.

Several other researchers have also attempted, with somewhat mixed results, to measure empirically the importance of external indebtedness variables in the deforestation process. Capistrano (1994) found that the debt service ratio was negatively related to deforestation in her regressions, which ran counter to her hypothesis that greater debt burdens increased the extent of forest depletion (because of the increased pressure to liquidate forest resources). Shafik (1994) and Gullison and Losos (1991) concluded that

Table 2.1 Two-stage least-squares regressions explaining deforestation (1981–1985) using total debt service relative to exports.

	Direct effects (Investment and change in debt variables included)		Direct and indirect effects (Investment and change in debt variables excluded)	
Independent variables				
Intercept	−0.013	−0.013	−0.022	−0.025
(South America)	(−0.51)	(−0.49)	(−1.28)	(−1.50)
Africa dummy	−0.0022	−0.0054	0.0024	−0.0014
	(−0.14)	(−0.37)	(0.18)	(−0.11)
Central America and Caribbean dummy	−0.013	−0.016	−0.011	−0.015
	(−0.92)	(−1.26)	(−0.82)	(−1.16)
Asia dummy	−0.017	−0.016	−0.013	−0.016
	(−1.04)	(−1.06)	(−0.96)	(−1.14)
Forested land area	0.0011	0.0011	0.0012	0.0011
(10^3 ha/10^6 of US$ real GNP)	(3.01)	(2.91)	(3.10)	(2.98)
Labor force	0.000013	0.000012	0.000012	0.000012
(per 10^6 of US$ real GNP)	(2.80)	(2.74)	(2.88)	(2.91)
Government spending	$-0.23 \cdot 10^{-7}$[a]	$0.59 \cdot 10^{-7}$[b]	$-0.34 \cdot 10^{-7}$[a]	$0.33 \cdot 10^{-7}$[b]
(real US$/$10^6$ of US$ real GNP)	(−0.21)	(0.70)	(−0.37)	(0.46)
Investment	$0.16 \cdot 10^{-7}$	$-0.79 \cdot 10^{-8}$	—	—
(real US$/$10^6$ of US$ real GNP)	(0.23)	(−0.12)		
Relative debt service	0.23	0.22	0.26	0.24
(total debt service/exports)	(2.44)	(2.38)	(3.03)	(2.94)
Annual % change in public external debt	−0.062	−0.074	—	—
	(−0.52)	(−0.64)		
R^2	0.305	0.341	0.304	0.332
Number of observations	55	55	55	55

Source: Kahn and McDonald 1974.

Note: t-statistics in parentheses. Dependent variable: deforestation (1000 ha) per million of real U.S. dollar GNP.

a. Government spending is treated as an endogenous variable.

b. Government spending is treated as an exogenous variable.

debt does not influence deforestation. Although our work and that of Burgess (1991) suggests a positive relationship between debt variables and deforestation, more empirical work is needed to examine the relationship between deforestation and macroeconomic conditions, such as external debt.

DISCUSSION

Our work and that of others suggests that macroeconomic factors are important determinants of deforestation. If this is true, then the development of appropriate policies becomes much more difficult than if deforestation were caused simply by microeconomic factors such as market failure. There are three reasons for this. First, the economies of developing countries are linked to the global economy, so national economic policies could be rendered ineffective by feedback from the global economy. Second, developing countries may not have the internal ability to change macroeconomic conditions, as the "vicious cycle of poverty" is hard to break. Third, macroeconomic conditions are extremely dependent on population levels and population growth, and population policy has proved singularly difficult to implement.

Our research suggests that reducing debt reduces deforestation. This finding may be an argument to offer deforesting, developing countries some form of debt relief, as many deforesting debtor countries have unmanageable debt burdens and are caught in a debt trap. Debt-for-nature swaps appear to be a particularly attractive way to reduce such countries' debt burdens, as they may have a dual effect on deforestation. First, the contractual agreement is designed to preserve forests as part of the swap. Second, the reduction in debt may itself reduce the pressure to deforest, although this indirect effect is small (Kahn and McDonald 1995). The remainder of this section is devoted to the prospects and problems of debt-for-nature swaps.

A debt-for-nature swap is an agreement negotiated between a "donor" (usually, but not necessarily, an international nongovernmental organization) and the government of an indebted forested country. The donor buys the forested country's debt, either from the original creditor or on the secondary debt market, paying the market price, which is usually only a fraction of the face value of the debt (because these debts have often been written down due to a low probability of repayment). The donor then converts this debt into the forested country's local currency, and deposits it at the debtor's central bank. The donor then uses these funds to undertake the agreed-upon conservation project, such as preserving a forest tract. Such swap deals are one of the few market manifestations of a willingness to pay for global environmental benefits (World Bank 1994).

Unfortunately, debt instruments, particularly debt-for-nature swaps, have proven less viable than initially hoped; although many such swaps have occurred (thirty-one had been completed as of December 1993: World Bank 1994), they have been small in scale and occasionally controversial (Pearce and Warford 1993, cited in Center for Strategic and International Studies 1995). As an example of their limited use, as of November 1991, Costa Rica and Ecuador accounted for 90% of the total debt retired through debt-for-nature swaps. According to the World Bank (1994), of the total $46 billion of miscellaneous debt conversions between 1984 and 1992 (equivalent to about 2.5% of total developing country debt stocks), only 0.4% were debt-for-nature swaps.

Perhaps debt-for-nature swaps have not been as numerous as was hoped because of the many potential problems associated with them. First, such agreements are not enforceable, so a country might renege on its agreement to preserve the forest tract. The costs of reneging are high, however, and include damage to that country's international reputation and exclusion from future debt reduction schemes. Second, there is a possibility that swap funds may simply displace existing conservation funding, or that forested countries that negotiate a debt-for-nature swap may do so for a preservation action that would have been undertaken anyway, implying no net preservation gain. Third, countries may use their reduced indebtedness to qualify for more loans, meaning that the net debt reduction of the debt-for-nature swap is small.

The World Bank is pessimistic on the future of debt-for-nature swaps in Latin American and African countries, as above-market gains (brought about by attractive leverage on swaps) are quickly being competed away; also, as with other swaps, returns will decline as their numbers increase. As debtor countries have become less constrained by external debt, they have also become less interested in swapping their debt for 100% in local currency. However, a form of debt-for-nature swap might be used in Eastern Europe; proceeds would probably be devoted to industrial cleanup rather than protected areas (World Bank 1994). All in all, debt-for-nature swaps cannot be considered the sole means by which deforesting developing countries are offered debt relief, nor do they appear very promising as a vehicle for preserving forest tracts.

In addition to debt-for-nature swaps, other innovative policies can help to reduce the economic pressure to deforest. These can be policies to reduce general economic pressures, or to find ways to meet these pressures in a more sustainable fashion. There has been no shortage of proposed financial and trade mechanisms to reduce economic pressures and to allow capital for environmental protection to flow from developed to developing countries: generalized debt relief, debt-for-nature swaps, carbon-offset projects, increased grant aid, concessional financing, trade liberalization, increased foreign investment, and new redistributive international economic instruments (such as internationally allocated pollution permits or taxes:) have all been suggested. (See Center for Strategic and International Studies 1995 for more details on countries' experiences with these various financial and trade mechanisms.)

In addition to these general economic policies, it is important to target the development of sustainable economic activities. Of critical importance in this regard are policies that seek to promote sustainable uses of forest resources. Such strategies include extractive reserves, the development and dissemination of sustainable agroforestry techniques, the creation of markets for sustainable nonwood forest products (fruits, nuts, dyes, rubber, etc.), the implementation of sustainable timbering techniques, aquaculture (in wet forest regions), and the development of more sustainable animal husbandry (raising water buffalo, iguana, turtles, and tropical animals in place of conventional cattle ranching).

GENERAL IMPLICATIONS

1. It is insufficient to attribute deforestation to solely microeconomic factors such as poorly defined leasing agreements, failure to define property rights, and inappropriate government policies. Macroeconomic conditions may be important determinants of deforestation, independent of microeconomic conditions.
2. More empirical work needs to be done in examining the relationship between macroeconomic conditions and deforestation (and environmental degradation in general) if we are to understand better the importance of this relationship and provide appropriate evidence and advice to policy makers.
3. Because our findings and those of others provide preliminary evidence that macroeconomic factors may be important contributors to deforestation, work should begin on developing innovative policies that address both microeconomic and macroeconomic factors leading to deforestation and which provide a mechanism for wealthy nations to compensate developing nations for preserving environmental resources. Debt-for-nature swaps are a good beginning to this process, but more comprehensive policies are direly needed.

ACKNOWLEDGMENTS

The research upon which part of this chapter was based was primarily supported by the Climate Change Division, Office of Policy Analysis, United States Environmental Protection Agency, which is not responsible for the positions taken. The authors are grateful to W. Branson, S. Ferry, M. J. Kealy, W. Laurance, W. O'Neill, and D. Pearce for their comments and suggestions. Several anonymous reviewers also made important contributions to this chapter.

APPENDIX 2.1: Description of the Economic Model

This model is based on the assumption that countries maximize the discounted stream of consumption over their planning horizon. Equation (2.1) represents an identity that shows that consumption *(C)* is equal to GNP less investment *(I)*, government expenditures *(G)*, and interest payments on the stock of debt *(iD)*. The change in debt (δ) is added to this equation since changes in consumption (or competing uses of GNP) can be funded through borrowing.

(2.1) $C(t) = \text{GNP}(t) - I(t) - G(t) + \delta(t) - iD(t-1)$

Equation (2.2) contains the production function for GNP, with time notation eliminated for simplicity. The production function is a multiplicative formulation, in which the greater the amount of any input—level of deforestation *(R)*, forested land *(F)*, nonforested land *(T − F)*, capital *(K)*, or labor *(L)*—the greater the resulting output. Although the forest sector is not modeled explicitly in this production function, it is included in both the *R* and *F* terms. Forestry activities that do not result in the reduction of forested area (such as selective harvesting of certain exotic hardwoods, collecting rubber from wild trees, nondestructive subsistence agriculture, and harvesting of fruit, meat, fiber, nuts, and honey) would be included in *F*. Forestry activities that result in deforestation (clear-cutting and subsequent conversion of the land to nonforested scrub or agricultural land) would be represented by the contribution of deforestation *(R)* to GNP.

(2.2) $\text{GNP} = \alpha_0 R^{\alpha_1} F^{\alpha_2} (T - F)^{\alpha_3} K^{\alpha_4} L^{\alpha_5}, \quad \alpha_0 > 0$

Equation (2.3) contains the maximization problem and the constraints on the maximization process. The control variables in this model are the level of deforestation *(R)*, the change in debt (δ), and the level of investment *(I)*. The constraints on the maximization problem are relatively straightforward, except for the constraint on consumption, $C(t) > C^*$. This constraint stipulates that consumption in each period must remain above some minimum level (C^*). In other words, countries cannot reduce consumption to very low levels in some periods in exchange for higher levels of consumption in future periods, even though such a path may represent an unconstrained maximum. This constraint gives an incentive for short-sighted behavior beyond that specified by Von Moltke (1990, 6), who blames much of the short-sighted behavior on the discounting process.

(2.3) $$Max \int_0^T C(t) e^{-rt} dt$$

$$\text{Subject to } \frac{dF}{dt} = -R \qquad \frac{dD}{dT} = \delta \qquad \frac{dK}{dt} = I$$

$$C(t) \geq C^*$$
$$C(t) \geq 0$$
$$F(0) = F_0$$
$$D(0) = D_0$$
$$K(0) = K_0$$

Table A2.1 Comparative statics: Sign of direct effect on optimal level of deforestation due to changes in inputs and competing uses of GNP.

	Inputs $(F, K, L, T, -F)$	Competing uses of GNP $(I, G, iD, C\star, -\delta)$
Unconstrained solution		
Production function:		
(i) log-linear (see eq. (2.4))	$-$ if $\alpha_1 > 1$	zero
	$+$ if $\alpha_1 < 1$	zero
(ii) quadratic (see eq. (2.5))	$-$	zero
(iii) hybrid (see eq. (2.6))	$-$ if $\beta_2 > 1$	zero
	$+$ if $\beta_2 < 1$	zero
Constrained solution		
Production function:		
(i) log-linear (see eq. (2.4))	zero	$+$
(ii) quadratic (see eq. (2.5))	$+$ or $-$	$+$ or $-$
(iii) hybrid (see eq. (2.6))	$-$	$+$

The Hamiltonian equation associated with this maximization problem is represented by equation (2.4), with time subscripts eliminated for simplicity. Note that consumption, C, has been eliminated by substituting the right-hand side of equation (2.2) into the right-hand side of equation (2.1) and substituting the right-hand side of the resulting equation into equation (2.3) for $C(t)$. The first, second, and third constraints (λ_1 through λ_3) in equation (2.4) represent the state equations for the stock of forests, capital stock, and stock of debt, respectively. The fourth constraint (λ_4) embodies the minimum consumption constraint.

$$(2.4)\quad H = [\alpha_0 R^{\alpha_1} F^{\alpha_2}(T - F)^{\alpha_3} K^{\alpha_4} L^{\alpha_5} - G - I - iD + \delta]e^{-rt} + \lambda_1(-R) + \lambda_2(I) + \lambda_3(\delta) + \lambda_4[\alpha_0 R^{\alpha_1} F^{\alpha_2}(T - F)^{\alpha_3} K^{\alpha_4} L^{\alpha_5} - G - I - iD + \delta - C^*]$$

The first-order conditions for a maximum were generated and solved for R, the rate of deforestation, as a function of the other variables. Solutions were developed for cases in which the minimum consumption constraint was binding and in which it was nonbinding. A similar exercise was performed with production functions of a quadratic form (equation [2.5]) and with a hybrid production function (equation [2.6]).

$$(2.5)\quad \text{GNP} = \gamma_0 + \gamma_1 R + \gamma_2 F + \lambda_3(T - F) + \gamma_4 K + \gamma_5 L + \gamma_6 R^2 + \gamma_7 F^2 + \gamma_8(T - F)^2 + \gamma_9 K^2 + \gamma_{10} L^2 + \gamma_{11} RF + \gamma_{12} R(T - F) + \gamma_{13} RK + \gamma_{14} RL$$

$$(2.6)\quad \text{GNP} = B_0 F^{\beta_0} K^{\beta_4} L^{\beta_5} + B_1(T - F)^{\beta_1} K^{\beta_6} L^{\beta_7} + B_2 R^{\beta_2} K^{\beta_8} L^{\beta_9}$$

Table A2.1 contains the comparative statistics based on these solutions; in other words, it shows the effect of changing different variables on the optimal rate of deforestation. This table contains significant information that was used to structure a regression equation designed to explain the current levels of deforestation. Input variables $(F, K, L, T - F)$ have an effect on deforestation in all three unconstrained solutions and in two of the three constrained solutions. Competing uses of GNP (I, G, iD, C^*, and $-\delta$) are important only when a country's optimization strategies are constrained by the existence of a minimum acceptable consumption level.

Since input variables are relevant in all but one of the six cases, it is not difficult to make a case for including these variables in an estimating equation designed to explain deforestation. However, the variables that constitute competing uses of GNP (I, G, iD, C^*, and $-\delta$) affect deforestation only when a country's optimization possibilities are constrained by a minimum acceptable consumption level. More details on the model and its use in justifying the empirical analysis are contained in Kahn and McDonald (1995).

Section II
Physical Processes and Edge Effects

INTRODUCTION
Edge Effects in Fragmented Forests

One of the most obvious features of fragmented landscapes is a drastic increase in forest edge. In a continuous forest, habitat edges are rare, typically limited to small internal clearings created by landslides, river meanders, or other natural disturbances. But in a heavily fragmented landscape, forest edges become a major—even dominant—feature. The margins of forest fragments are usually abrupt, delineating a sudden transition from forest to pastures, crops, or other modified habitats. Increasingly, researchers are becoming convinced that edge effects—the physical and biotic changes associated with remnant margins—often have a major impact on the ecology of fragmented tropical forests. Species disappearances that were once commonly ascribed to "area effects" (as in classic island biogeography theory) are now being reevaluated, and in many instances it is becoming apparent that edge effects are the principal cause of population declines.

Edge effects can be loosely classified into physical and biotic phenomena. In rainforests, physical edge effects can include elevated wind turbulence and temperature variability, lateral light penetration, and reduced humidity, all of which result from the close proximity of a harsh external climate in the surrounding matrix. Biotic effects can be extraordinarily diverse, and include the proliferation of secondary vegetation along forest margins, invasions of weedy or generalist plants and animals, alteration of ecological processes such as nutrient cycling and energy flows, and myriad other ecological changes.

The four chapters in this section all focus on edge effects. The first two are concerned with forest microclimate, especially the interaction of climatic and biotic changes in remnants, while the third focuses on leaf litter invertebrates, and the final chapter deals with wind damage in fragments. (Several chapters in latter sections also tangentially consider edge effects.)

Forest Microclimate

In chapter 3, Valerie Kapos and her colleagues measure microclimatic changes associated with the margins of a forest fragment in central Amazonia. Their study assessed how relative humidity and soil moisture varied along transects running from the edge to the forest interior. Significantly, they repeated their measurements at a five-year interval, taking them initially soon after the edge was created and later when the edge was five years old. This approach offers an explicit comparison of the effects of edge age on microclimatic parameters.

Kapos and her co-workers found that for the newly cut edge, air humidity and soil moisture increased along simple gradients with distance into the forest, but in the older edge, the patterns were far more complex. One positive aspect of this change is that older forest edges may tend to be "sealed" by a wall of proliferating second growth, reducing the influence of edges on forest microclimate. The fact that two species of specialized understory plants exhibited neither significant responses to forest edges nor any evidence of water stress supports the notion that older edges may help to buffer the forest interior from harsh external conditions. There was, however, an increased frequency of treefall gaps within 70 m of forest margins, demonstrating that wind damage is unlikely to be ameliorated by older edges.

In chapter 4, Stephen Turton and Heidi-Jo Freiburger assess microclimate and dicot seedling densities in a small rainforest remnant in tropical Queensland. The fragment they studied was quite different from those in central Amazonia, occurring at a higher elevation and latitude and being considerably older (more than fifty years). Turton and Freiburger found that several parameters they measured (e.g., soil and ambient temperatures, vapor densities) were influenced as much by edge aspect (the direction in which edges face) as by edge proximity. Seedling densities, however, were depressed near edges, possibly as a result of shading from dense second growth along edges.

Invertebrates

Chapter 5, by Raphael Didham, describes a study with important implications for the conservation of tropical invertebrate communities. Didham examined the abundances of ant, beetle, and other invertebrate taxa along edge-to-interior transects in central Amazonia. He found many intriguing patterns, including complex bimodal patterns of abundance in several taxa as a function of edge distance. Beetle assemblages, however, appeared more strongly influenced by forest isolation than by edge effects, and many taxa that are typical of undisturbed forest were missing even from large (100 ha) remnants. If Didham's conclusions are valid, then many forest remnants will support invertebrate communities that differ drastically from those in intact forests.

Wind Disturbance

In chapter 6, William Laurance assesses levels of structural damage in fragmented and continuous forests in tropical Queensland. Fragments in this region are exposed to strong prevailing winds and windstorms, which can batter smaller remnants. Laurance found that fragments often had reduced canopy cover and elevated abundances of some disturbance-adapted plants relative to continuous forest. The amount of damage appeared to be influenced by the proximity of study plots to the forest edge as well as by local topographic factors.

An important finding of Laurance's study is that structural damage from winds may be elevated as far as several hundred meters from forest edges. This finding suggests that even large (> 1,000 ha) fragments may eventually become chronically disturbed. Overall, these findings indicate that increased disturbance could be an important structuring force for some fragment communities, and may exacerbate the impacts of fragmentation on forest interior species.

SYNTHESIS

Collectively, these four studies demonstrate that edge effects can operate over a wide range of spatial scales. Microclimatic changes seem mainly limited to a zone within 15–60 m of edges (chaps. 3 and 4), but some biotic changes can penetrate much farther than this. For example, functional assemblages of invertebrates may be altered up to 200 m from edges (chaps. 5 and 7), while in some regions elevated wind damage and related changes in forest composition appear detectable within several hundred meters of edges (chap. 6).

It also is becoming apparent that edge effects exhibit much spatial and temporal variation. Such variation may be pronounced in tropical forests, which are notoriously heterogeneous, with patchy species distributions. Invertebrate communities, for example, may be highly dynamic and influenced by fluctuating environmental conditions near edges (chap. 5). A key implication of this idea is that buffer zones surrounding nature reserves should be wide enough to accommodate the maximum penetration of edge effects, not just some average value. This value is likely to vary from community to community, but a buffer zone width of 500–1,000 m or more does not seem overly conservative given present uncertainties regarding the penetration of edge effects.

While much has been learned in recent years, the study of edge effects is still in its infancy, especially in the Tropics. One important phenomenon that has not been explicitly studied in the context of edge effects is human hunting pressure. The synergistic effects of hunting and fragmentation probably have devastating effects on some wildlife species, and as forests are reduced in area, there will be increasing pressures on fauna in isolated reserves and other remnants. For this reason it would be very useful to quantify the distances to which hunters penetrate into forest interiors, and to assess the magnitude of their effects on faunal populations. Studies along such lines could yield many practical principles for the design and management of nature reserves.

3
Edge-Related Changes in Environment and Plant Responses Due to Forest Fragmentation in Central Amazonia

Valerie Kapos, Elisa Wandelli, José Luis Camargo, and Gislene Ganade

One of the major changes brought about by habitat fragmentation is an increase in the proportion of edge exposed to other habitats. The importance of this change depends to some degree on the contrast between the fragmented habitat and the new matrix in which the fragments occur. In the case of tropical forests, which are generally fragmented by clearance for cultivation or pasture, this contrast is very strong. Concern about tropical forest fragmentation has given rise to a body of research on the effects of edges. Work on edge effects in forest fragments was recently reviewed by Murcia (1995), who pointed out that there is as yet no unified understanding of the factors governing variation in their magnitude or extent, nor even of their general patterns. For example, Kapos (1989) reported edge effects, including elevated air temperature and increased vapor pressure deficit, extending at least 60 m into a 100 ha forest fragment, but Williams-Linera (1990b) failed to find measurable edge effects on microclimate more than 15 m away from forest edges in Panama. One possible explanation for this discrepancy is the difference between the two studies in time since edge creation (Camargo and Kapos 1995). Williams-Linera studied four edges more than five years old and a single ten-month-old edge, while Kapos's study focused on edges that were mostly about six months old. Edge age is one of the principal factors that can influence the magnitude and extent of edge effects. In this chapter we report changes in edge effects with increasing age for a single Amazonian forest edge.

Edge effects may be important both for the ecological characteristics of forest fragments themselves and for the local and regional environment. For example, the importance of the Amazon forests in hydrological cycles through their roles in evapotranspiration and soil protection is well known; as much as 50% of the region's rainfall may be water recycled by forest evapotranspiration (Salati, Marques, and Molion 1978; Salati et al. 1979; Salati and Vose 1984; Victória et al. 1991). Deforestation affects these processes principally through reduction of forest area, but understanding changes in the hydrological function of forest remnants may also be important for predicting long-term regional changes. In our studies, we have asked whether factors affecting evapotranspiration balance differ between forest fragments and continuous forest. Depending on the extent to which the influence of desiccating conditions in the clearings penetrates into the forest and on how plants respond to it, forest fragments might be expected to evapotranspire more than equivalent areas of continuous forest.

Therefore, we have focused our research on assessing the edge-related gradients of environmental factors that affect evapotranspiration in forest fragments (temperature, vapor pressure deficit [VPD], and soil moisture) and on plant responses to them. We returned to the edge examined in Kapos's (1989) study to explore changes in edge effects with time and the likely long-term character of the edges of forest fragments in the central Amazon.

STUDY AREA AND DESIGN

These investigations were carried out in the reserves of the INPA/WWF/Smithsonian-sponsored Biological Dynamics of Forest Fragments Project, approximately 80 km north of Manaus (Lovejoy et al. 1986; Lovejoy and Bierregaard 1990; Bierregaard et al. 1992). The reserve used for most of the study (number 2303) was on the Dimona ranch and was a semi-isolated 100 ha forest fragment of which two sides had been isolated by clear-cutting for pasture in 1984 (fig. 3.1; see also fig. 10.1). The study compared measurements made at different distances along transects from the western isolated edge (the same edge studied by Kapos [1989] in 1984) toward the center of the reserve with measurements made in control areas more than 500 m from the forest edge (fig. 3.1). The study was conducted in five overlapping phases from 1988 through early 1990: understory-plant/water relations were monitored in the dry seasons of 1988 and 1989; soil moisture measurements were made over the last ten months of 1988; understory plant distributions were measured in 1988–1989; microclimatic and leaf expansion studies were done in 1989; ^{13}C and vegetation structure studies occurred in late 1989–early 1990.

ENVIRONMENT

The Aerial Environment

In Kapos's (1989) earlier study, manual whirling hygrometers were used at intervals along transects from the edge toward the center of a forest fragment, and the resulting measured VPDs were expressed relative to the VPD measured at the reserve edge. In our subsequent work (Camargo 1993; Camargo and Kapos 1995), we used automatic ventilated psychrometers suspended from pulleys at different distances from the edge and in control areas to measure wet and dry bulb temperatures at five different heights (1.5, 3, 5, 7.5, and 10 m) above the forest floor. The results were expressed relative to simultaneous measurements in the pasture outside the forest, giving a standardized data set comparable to that collected at the same site five years earlier.

The difference between the two sets of results is striking (fig. 3.2). The newly exposed edge clearly increased vapor pressure deficit (VPD) at least 60 m into the forest, but five years later VPDs varied little along an edge-to-interior gradient. As expected, VPD increased with height above the forest floor, and did so more quickly nearer to the forest edge (Camargo 1993).

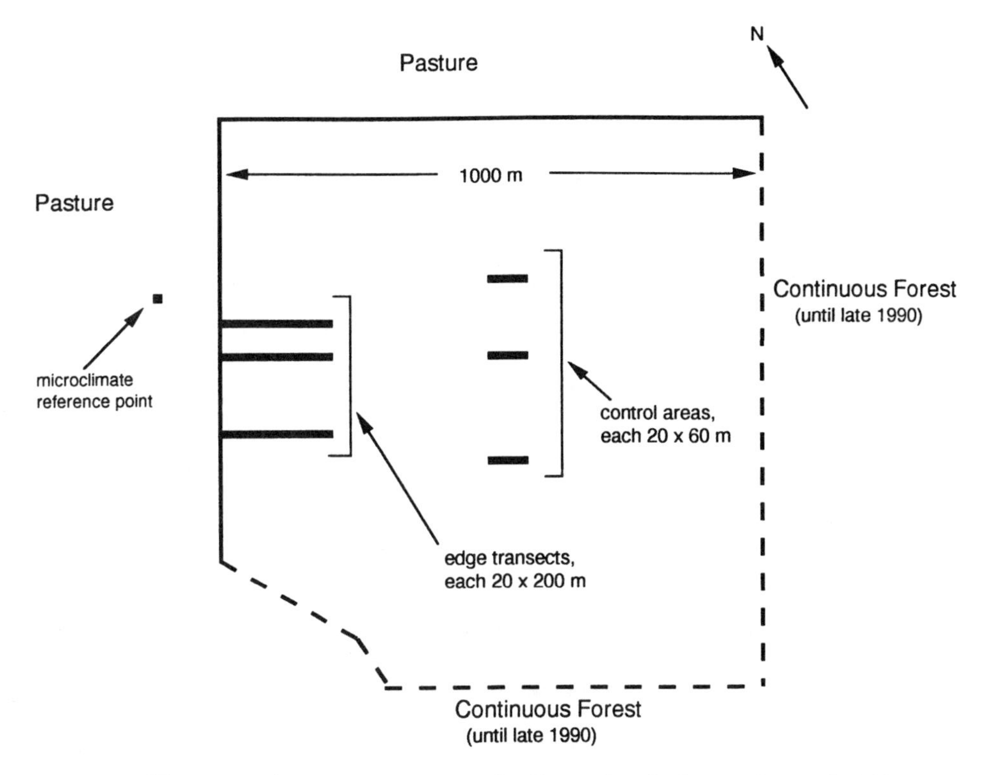

Figure 3.1. The main study area, reserve 2303 on the Dimona Ranch, about 80 km north of Manaus, Brazil. This 100 ha reserve was partially isolated by clear-cutting and burning to create pasture in 1984 and was studied by Kapos (1989) late in that year. In subsequent work, measurements in randomly located transects from the edge toward the center of the reserve were compared with similar measurements from control transects beginning at 500 m from the cut edge and extending farther into continuous forest.

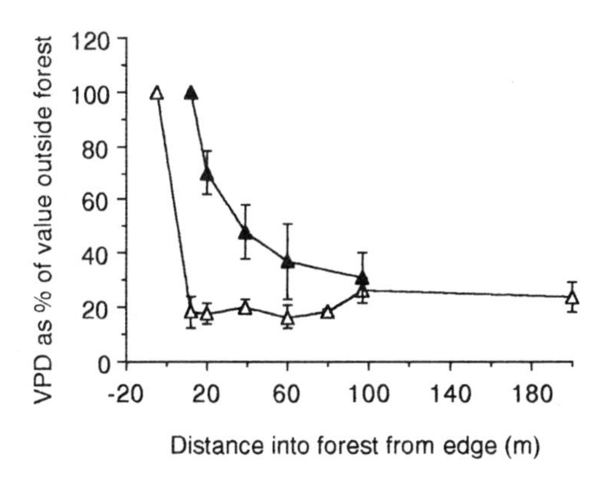

Figure 3.2. The relationship between air vapor pressure deficit (VPD) and distance from the forest edge in Dimona reserve 2303 in 1984, four to six months after edge creation (closed symbols; data from Kapos 1989), and in 1989 (open symbols; data from Camargo and Kapos 1995). Error bars are ±1 standard error of the mean. The data are expressed as a percentage of the VPD measured simultaneously in the open pasture, so that data from different study days can be combined.

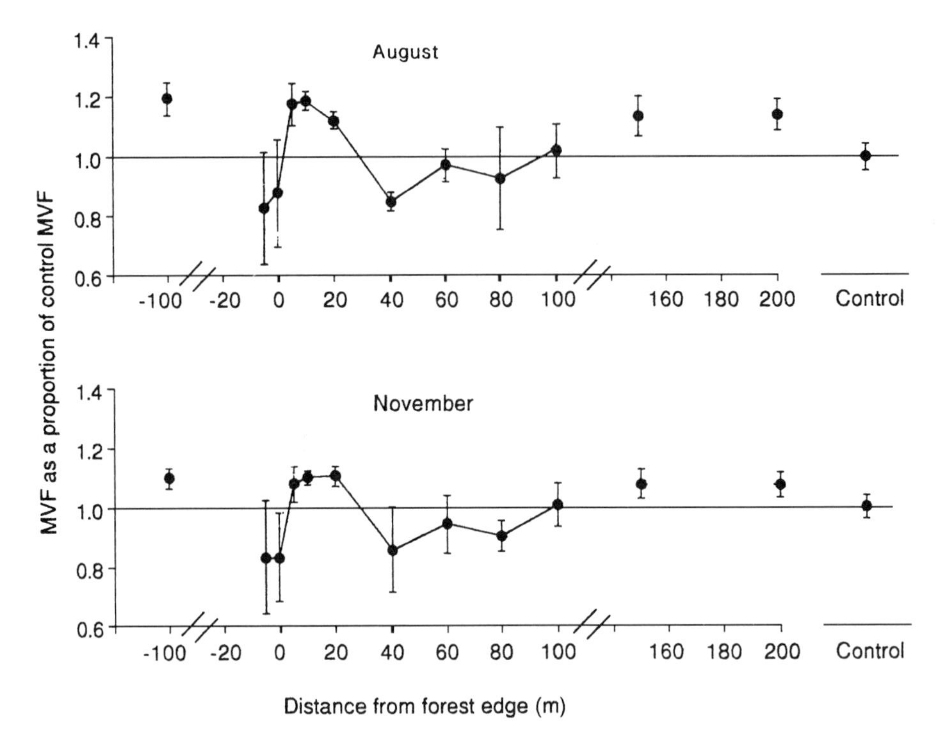

Figure 3.3. The relationship between volumetric soil moisture content (MVF) at 25 cm depth, expressed as a proportion of MVF in control areas, and distance from a five-year-old forest edge during the dry (August) and wet (November) seasons. There were zones of apparent depletion of soil water relative to control areas at the edge itself and from 40 to 80 m from the edge. Soils 10–30 m inside the edge had higher MVFs than control areas. (From Camargo and Kapos 1995.)

Soil Moisture

Kapos (1989), using destructive gravimetric soil moisture samples, showed a marked reduction in soil moisture content within 20 m of the forest edge. Four years later we used a neutron probe to examine spatial and temporal variation in soil water content in the same 100 ha reserve (Camargo and Kapos 1995).

We found in the later study that soil moisture was still affected by proximity to the forest edge, but not in the simple pattern that might have been anticipated from Kapos's (1989) results (fig. 3.3). In both the wet and dry seasons, soil moisture in the edge transects was similar to that in the control areas except at the edge itself and in the region between 40 and 80 m from the edge, where it was highly variable and somewhat depleted relative to control values. Especially during the dry season, soil moisture within 30 m of the reserve edge was slightly higher than in control areas (fig. 3.3; Camargo and Kapos 1995).

Based on moisture characteristic curves and the neutron probe measurements, we found that at 25 cm depth, soil water potentials below wilting point (−1.5 MPa) occurred at some locations in the forest during the dry season (Camargo 1993; Camargo and Kapos 1995), but the driest points were not necessarily near the edge, and we found no evidence of prolonged soil drought. Thus, any drought effects of the edge on plants must be due to the combined effects of reduced soil moisture and higher atmospheric evaporative demand exceeding the supplying power of the vascular systems, rather than to ecosystem-level water shortage.

PLANT RESPONSES

Kapos (1989) found that plants near newly created edges sometimes had lower leaf relative water contents (a measure of leaf water status in terms of water content relative to that at saturation) than those in the interior of a forest reserve, but found no evidence of appreciable water shortage in the relatively wet period when the study was done. To assess the possible longer-term effects of edge proximity and to explore the possible effects of the older edge on plant water status, we studied the distribution and water status of two common understory species, *Astrocaryum sociale* Barb. Rodr. (Palmae) and *Duguetia* aff. *flagellaris* Huber (Annonaceae).

Plant Distribution

Any significant negative influences of the edge on understory plants, whether through water status or other mechanisms, may with time come to be reflected in the distribution of individuals with respect to the edge. Therefore, distributions of the two species were assessed by counting the individuals in edge and control transects. *A. sociale* was also censused along randomly located edge-to-interior line transects: six in reserve 2303 and three perpendicular to each of two eight-year-old forest edges on Fazenda Esteio, approximately 20 km away (see fig. 10.1). *A. sociale* was consistently less abundant near the edge of reserve 2303 and near the other edges studied than in control areas of continuous forest (fig. 3.4; Wandelli 1990). In contrast, *D.* aff. *flagellaris* occurred near the forest edge in abundances similar to those in control areas (fig. 3.4). Though it is possible that gradients in plant distribution existed prior to edge formation (both species are apparently quite long-lived), the consistency in the gradient of *A. sociale* occurrence with respect to the entire 1 km length of the Dimona edge and at two quite different, older, and distant edges suggests that its distribution was indeed affected by edge creation.

Plant Ecophysiological Responses

We assessed the water status of the two understory species by studying diurnal patterns of stomatal conductance and water potential, looking for evidence of restriction of water loss, or significant plant water deficits, or both. Both species had very low conductances (g) at all locations (g_{max} = 80–120 mmol m^{-2} s^{-1}, except in plants in full sun, where g_{max} = 150–200 mmol m^{-2} s^{-1}), as is typical of rainforest understory plants (Roberts, Cabral, and Aguiar 1990). Maximum conductances were usually recorded in late morning, and conductances typically declined continuously thereafter (fig. 3.5). This pattern was similar in plants near the edge and in control areas (Wandelli 1990). Such a response

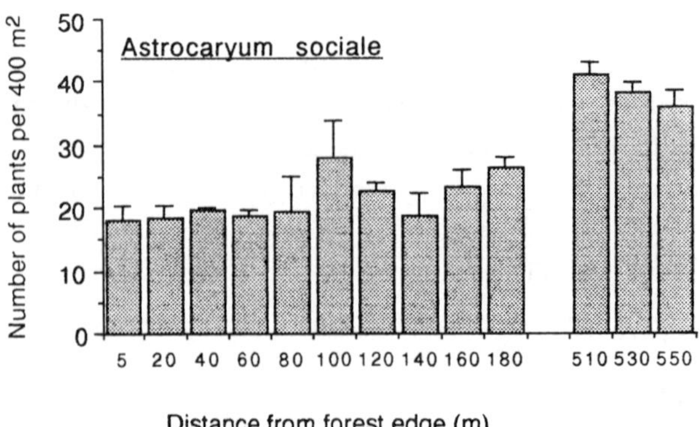

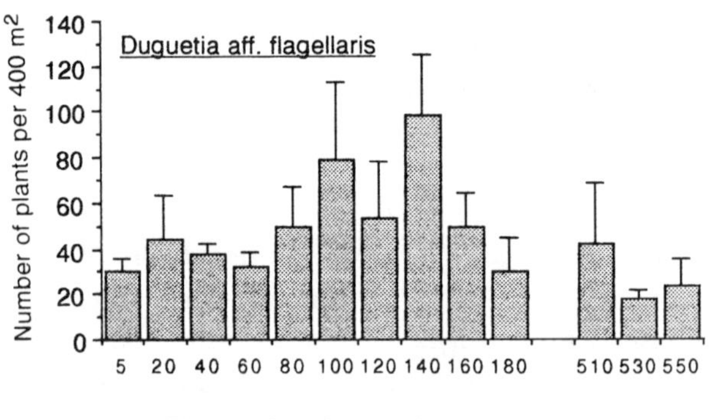

Figure 3.4. Distribution of two understory plants, *Astrocaryum sociale* and *Duguetia* aff. *flagellaris,* relative to distance from the edge of reserve 2303. *A. sociale* occurred at consistently higher densities in control areas than within 200 m of the edge, while density of *D.* aff. *flagellaris* varied more and with little apparent relationship to distance from the forest edge.

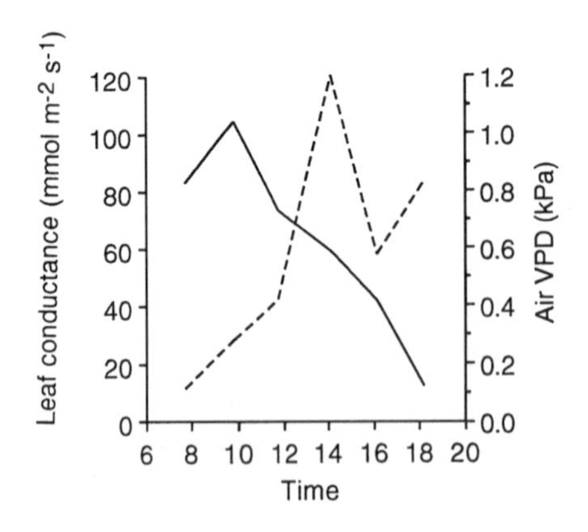

Figure 3.5. Pattern of stomatal conductance (solid line) declining consistently after mid-morning (in this case, for a single individual of *Duguetia flagellaris*), which was typical for all individuals of both *Duguetia flagellaris* and *Astrocaryum sociale* studied (Wandelli 1990). Increasing air VPD (dashed line) may partly explain this pattern, but the values of VPD in the understory were low.

is typical of plants sensitive to increases in VPD and/or with internal water deficits. It may represent a significant restriction of water loss for the individual plants, but understory evapotranspiration is a minor fraction of the total in lowland rainforest (Roberts, Cabral, and Aguiar 1990), and therefore it is unlikely to affect forest water balance.

We never recorded predawn plant water potentials below −0.5 MPa, so it was clear that the plants were not suffering sustained water deficits during the dry season. The midday or minimum leaf water potentials we recorded never approached the turgor loss points determined from pressure-volume relationships, and there was little difference between plants near the edge and those in the forest interior. However, both porometry and pressure chamber measurements are sufficiently cumbersome to reduce the degree of replication possible in the field, so it was difficult to make adequate comparisons between plants in edge and control areas. Therefore, we used two additional approaches to assess plant water status in these different zones.

Leaf Expansion in *Duguetia*

Leaf expansion in dicots is strongly related to plant water status (Dale 1988), so its rate could be a good basis for comparing water status among individuals of the same species growing in different environments. *Duguetia* aff. *flagellaris,* which produced its leaves in a synchronous flush throughout the population in the 100 ha reserve, was an ideal candidate for this assessment. We studied the rates of expansion of several hundred leaves produced by *Duguetia* individuals in control areas and within 40 m of the forest edge by repeatedly measuring leaf dimensions to 1 mm. The expansion of leaf area over a 24-hour period ranged from 0% to 100%, with smaller leaves showing greater relative expansion. For a given leaf size class, however, there was no difference in the rate of leaf expansion between plants in edge zones and those in control areas.

Carbon Isotope Discrimination

Analysis of the isotopic composition of the carbon in leaves is another integrated measure of the water status of plants; it allows better replication and sampling design because leaves can be collected and dried for later analysis. Foliar $\partial^{13}C$ (the difference in the ratio of ^{13}C to ^{12}C in a leaf sample from that in a mineral standard) is generally regarded as an indication of plant water use efficiency because it reflects the internal CO_2 concentration in the leaf and the isotopic composition of the source CO_2 used in photosynthesis. Plants that restrict their water loss through stomatal closure but continue to photosynthesize at the resulting lower internal CO_2 concentrations reduce their discrimination against naturally occurring $^{13}CO_2$ and thus have larger (less negative) foliar $\partial^{13}C$ values (Farquhar, O'Leary, and Berry 1982).

We studied $\partial^{13}C$ of both canopy and understory leaves to determine whether plants near the forest edge respond to that environment by closing their stomata to reduce water loss. For *Licania heteromorpha* Benth. (Chrysobalanaceae) and *Scleronema micranthum* (Ducke) Ducke (Bombacaceae), we compared canopy leaves of individuals growing within 100 m of the edge with leaves from control areas, and found no difference in foliar $\partial^{13}C$ between these two locations (fig. 3.6; Kapos et al. 1993).

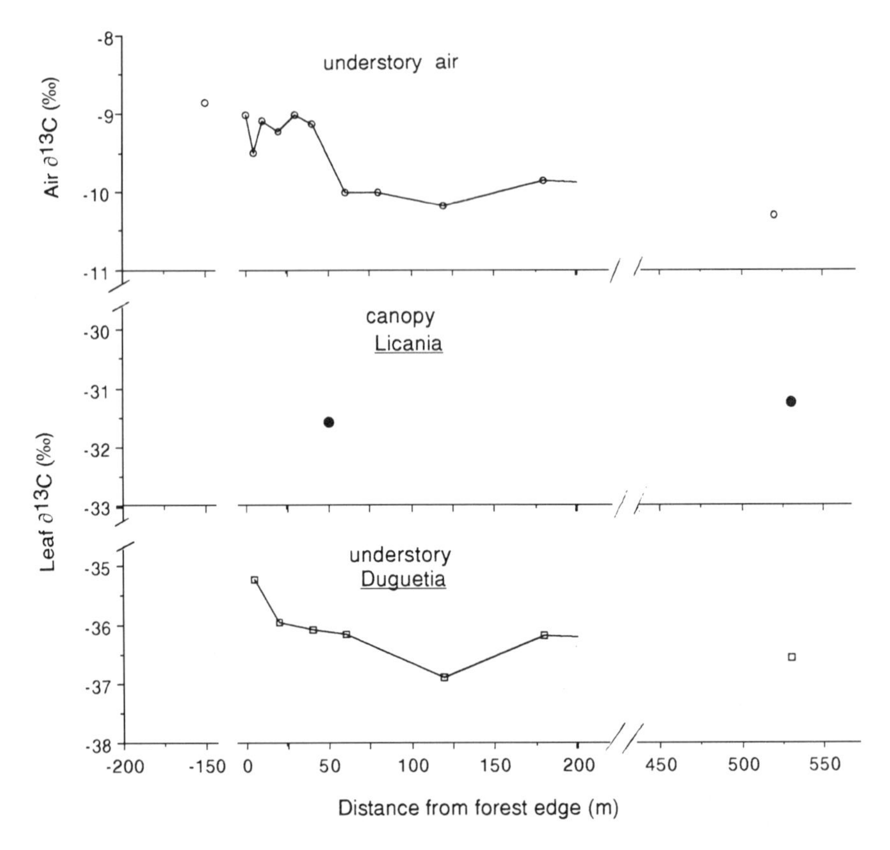

Figure 3.6. Variation in $\partial^{13}C$ of canopy (solid circles) and understory (open squares) leaves and understory air (open circles) in relation to distance from the forest edge. Canopy leaf data are for *Licania heteromorpha* sampled within 100 m of the edge and in control areas. Data shown for understory leaves are for *Duguetia flagellaris.* (Adapted from Kapos et al. 1993.)

In the understory, foliar $\partial^{13}C$ was significantly less negative in *Duguetia* aff. *flagellaris* growing within 40 m of the forest edge than in control areas (fig. 3.6; *t*-test, $P < .05$; Kapos et al. 1993). *Astrocaryum sociale* showed spatial variation in $\partial^{13}C$ similar to that of *D.* aff. *flagellaris,* but with greater variability, so the differences between edge and control areas were not significant.

These results might have suggested that the understory species did indeed have greater water use efficiencies near the forest edge, but we found that the spatial variation in source air $\partial^{13}C$ in the understory was sufficient to account for the patterns of foliar $\partial^{13}C$ we observed. Air $\partial^{13}C$ was significantly higher up to 60 m from the edge than in the control areas (fig. 3.6), indicating that respired CO_2, which has a low $\partial^{13}C$, is a smaller proportion of the CO_2 pool in the zone of edge influence than in undisturbed understory (Kapos et al. 1993). This pattern could be explained by either greater mixing of air from outside the forest with understory air or lower decomposition rates, or both, near the edge.

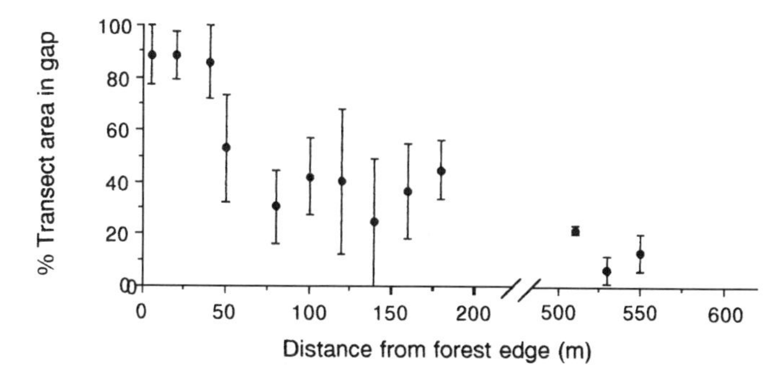

Figure 3.7. The proportion of the area of study transects occupied by gaps in relation to distance from the forest edge. Gaps were identified as areas with no canopy above 3 m, and their areas were measured as the area enclosed by the coordinates of the trunks of the standing trees. Error bars are ± 1SE ($n = 3$).

Gaps and Vegetation Structure

In the four to five years following Kapos's (1989) study, clear edge-related gradients in environmental factors gave way to much more complex patterns, which still suggested some influence of the edge. Although there was little evidence that the edge has any effect on plant water status, the distribution of at least one understory species suggested that proximity to the edge was disadvantageous.

We attribute the environmental change over time to the modification of vegetation structure near the edge. In this zone treefalls are frequent; gaps occupied a substantially greater proportion of the area near the edge than in the control areas (75% of the area in the outer 70 m vs. 14% in the interior; fig. 3.7; Kapos et al. 1993; see also Laurance, chap. 6). Abundant regrowth within these gaps and production of new leaves by remaining standing trees combine to "seal" the vegetation profile near the edge. Camargo (1993) assessed vegetation structure in the areas where the VPD measurements were taken and found a strong ($P < .01$) negative relationship between measured VPD and the total foliage density in the profile above the measurement location (Camargo and Kapos 1995). That is, where large amounts of foliage blocked penetration of solar radiation, VPDs were relatively low. This would be characteristic both of primary forest and of the lower parts of the profile in regenerating treefall gaps. Camargo (1993) also found greater foliage density at lower levels of the profile near the edge than in control areas.

We hypothesize that the pattern of soil moisture variation also reflects the dynamic nature of vegetation near the edge. The forest profile at the edge itself has filled in with many more leaves, which provide a greater total transpirational surface exposed to direct solar radiation. The resulting increased total evapotranspiration could explain the depletion of soil water at the edge in both wet and dry seasons. In the outer 20 m of the reserve is a zone of older gaps, where reduced canopy leaf area and a deeper boundary layer would be expected to cause reduced evapotranspiration and elevated soil moisture (Vitousek and

Denslow 1986). This is seen in the dry season results, but not in the wet season, when it is likely that the soils were near field capacity most of the time, leaving little possibility for increased water contents (Camargo and Kapos 1995). The wet season depletion of soil water between 40 and 80 m from the edge, a zone that is a mosaic of new gaps and relatively intact forest, may reflect a greater total leaf surface area subjected to direct solar radiation because of the oblique penetration of light to the subcanopy of intact forest adjacent to the gaps.

DISCUSSION

Although our study suffers from some of the design faults mentioned by Murcia (1995), such as lack of real replication among edges, our sampling was carefully designed within its restricted range, so that we have confidence in the patterns we describe for the particular edge to which they apply. We also have confidence in the age-related changes discussed here, because our study has followed the evolution of one particular edge.

Alteration of Edge Effects with Time

Although newly created edges do alter environmental conditions within the remaining forest (Kapos 1989; Matlack 1993), our study confirms, for Amazonia, Matlack's (1993) finding that these modifications disappear or become more complex as the edge ages. Changing vegetation structure caused by increased canopy gap formation near the edge (see fig. 3.7) and subsequent regrowth and infilling modifies the direct influence of the edge on microenvironment. The more complex patterns of environmental variation may help to explain patterns of occurrence of some species; for example, the distribution of litter invertebrates reported by Didham (chap. 5) bears a striking similarity to the patterns of soil moisture we found in relation to the older edge (see fig. 3.3).

The greater incidence of gaps near edges is consistent with the edge-related increases in canopy tree mortality reported by Lovejoy et al. (1984) and Rankin-de Mérona, Hutchings, and Lovejoy (1990). The resultant treefalls may be part of the explanation for the reduced occurrence of *Astrocaryum sociale* near forest edges. This species appears to occur less frequently in gaps, either because it is intolerant of gap conditions or because its growth form makes it especially subject to terminal injury by falling trunks and debris (Wandelli 1990). The increased frequency of gaps near edges would also tend to contribute to atmospheric mixing and turbulence (see Laurance, chap. 6), which help to explain the pattern of variation in isotopic composition of understory air.

Laurance (1991b) found evidence of increased disturbance as much as 500 m into an Australian rainforest, with the most striking effects of the edge extending up to 200 m into forest fragments. Although the extent of the edge effects he found may have been exacerbated by a recent cyclone, high winds may also occur in Amazonia (Nelson et al. 1994). Both subcanopy wind penetration and canopy-level turbulence are increased near forest edges; edge effects on wind profiles may extend as far into the forest as six times the forest height (McNaughton 1989). Thus, in central Amazonian forest fragments, winds may be altered up to 180–200 m into the forest, and altered disturbance regimes might be expected to extend nearly that far. Our measurements of gap frequency (see fig. 3.7) do indeed suggest altered canopy structure relative to control areas in the outer 200 m of a forest fragment, a striking similarity to Laurance's (1991b) findings.

The changes in edge effects on environment over time and the importance of changing vegetation structure highlight the necessity of understanding the mechanisms behind edge effects on particular organisms before making irrevocable management decisions. As shown by the two understory species in our study, different species may have quite distinct responses to the same environmental gradients. It is also important to understand the stage in edge evolution at which an evaluation of edge effects is made. For example, an edge function (cf. Laurance and Yensen 1991) based on the distribution of *Astrocaryum sociale* could be interpreted either as a physiological stress-related response to environmental changes at the edge or as an effect of gap formation. Simple edge-related environmental gradients disappear with time, but the dynamic nature of vegetation in the edge zone and related changes in vegetation pattern may persist for decades (Matlack 1994b). If management of this species was an objective, different decisions would be made based on these two different views. Longer-term data on edge effects are essential to understanding the true mechanisms underlying them.

Water Balance and Plant Water Status

Our study provides no basis for predicting a difference in water balance between Amazonian forest fragments and continuous forest. Although the pattern of soil moisture suggests that evapotranspiration may have increased at the forest edge itself and at 40–80 m from the edge, the same data indicate that evapotranspiration may have been reduced in other parts of the study area due to changes in forest structure and composition.

Soil moisture was not severely depleted for any significant length of time anywhere in the study transects, and there was no evidence that plants suffered severe water deficits. However, should regional deforestation lead to a reduction in rainfall, based on evidence from this study it is difficult to say whether such an overall reduction in available water would cause plants to experience water deficits and/or significantly alter their total evapotranspiration. If the increased formation of gaps near edges persists over time (see Laurance, chap. 6), then the additional penetration of solar energy, the filling in of the profile with extra foliage, and changes in species composition that result may be important in the future water balance of forest patches remaining in a deforested landscape.

GENERAL IMPLICATIONS

1. Edge effects on environmental variables, which took the form of simple gradients in new Amazonian forest edges, become more complex as the edge ages.
2. It is likely that these complex patterns and plant responses to them are strongly influenced by the high frequency of gaps near edges, and that changing vegetation structure will continue to alter the nature and extent of edge effects.
3. There is little evidence that exposed edges alter the evapotranspiration balance of forest fragments or the water status of their component plants.
4. Management decisions based on edge effects on particular species should incorporate an understanding of the mechanisms behind those effects. Longer-term studies are necessary to determine such mechanisms and assess accurately their likely changes over time.

ACKNOWLEDGMENTS

This work was supported by International Atomic Energy Agency project BRA/0/010 (Amazonia I) and IAEA fellowships awarded to the latter three authors. The work was also dependent on financial and logistic support from the Biological Dynamics of Forest Fragments (BDFF) Project. This chapter is contribution number 152 in the BDFF Project technical series. We are grateful to the following people for help with field work: Vanildo, Palheta, Sebastião, Cícero. Helpful consultations and comments were provided at various times by C. R. S. D. da Fonseca, B. Forsberg, W. Franken, E. Matsui, M. T. F. Piedade, E. V. J. Tanner, and R. L. Victória. P. Vose provided invaluable support and advice throughout. R. O. Bierregaard, S. M. Turton, and W. F. Laurance commented on an earlier draft of the chapter.

4

Edge and Aspect Effects on the Microclimate of a Small Tropical Forest Remnant on the Atherton Tableland, Northeastern Australia

Stephen M. Turton and Heidi Jo Freiburger

MICROCLIMATE AND EDGE EFFECTS IN FORESTS

EDGE effects result from the interaction of adjacent ecosystems separated by an abrupt transition (Murcia 1995). In many forested areas, fragmentation has produced landscapes in which forest edges are a dominant feature and edge influences are extensive in the remaining tracts of forest (Chen, Franklin, and Spies 1993a,b). According to some theoretical models of forest clear-cutting, interior habitat that is unaffected by edges may constitute less than 50% of the original standing forest (Franklin and Forman 1987), but this depends to a great extent on the pattern of clear-cutting and the size of forest remnants.

In a recent review of edge effects in forest remnants, Murcia (1995) distinguished three types of edge effects: abiotic, direct biological, and indirect biological. Abiotic effects involve changes in physical environmental conditions that result from proximity to a structurally dissimilar matrix. Direct biological effects involve changes in the distribution and abundance of species caused by altered physical conditions near edges, while indirect biological effects result from changes in species interactions (higher-order effects) at or near edges (Murcia 1995).

Rainforest remnants surrounded by cleared agricultural lands are exposed to markedly different solar radiation, wind, water, and nutrient regimes than continuously forested regions (Saunders, Hobbs, and Arnold 1993; Hobbs 1993). Solar radiation, temperature, and moisture drive many biological processes, such as photosynthesis, vegetation development, decomposition, and nutrient cycling (Chen 1993b). Consequently, the spatial and temporal dynamics of microclimate at forest edges are of utmost importance for understanding the biological responses and distribution and abundance of organisms in forest remnants.

Although some significant trends in the microclimate of forest remnant edges have been documented, results have often been inconsistent between studies and environments. After examining the recent literature on edge effects, Murcia (1995) concluded that the lack of generalizations about edge effect patterns can be attributed to the poor design of some studies, the lack of consistency in methodology, and oversimplification of the perception of habitat edges and edge effects. The lack of adequate replication is an important limitation of many studies (Murcia 1995).

The distance to which edge effects penetrate into forest remnants is a common measure used to indicate the intensity of habitat modification (Murcia 1995). A number of investigators have measured air and soil temperatures, vapor pressure deficit (VPD), and photosynthetically active radiation (PAR) to detect edge effects (Matlack 1993; Kapos et al. 1993; Young and Mitchell 1994; Camargo and Kapos 1995; Kapos et al., chap. 3). Some research has led to the general conclusion that edge effects penetrate only about 50 m into the forest, but in some instances no microclimatic gradients in relation to forest edges have been detected (Murcia 1995). It may be unwise to attempt to correlate all effects with distance from the edge because, in all likelihood, different edge effects interact and need not be monotonic (see Didham, chap. 5).

The intensity of edge effects is strongly influenced by edge orientation and forest physiognomy. The amount of solar radiation received determines the range of many physical edge effects, and is essentially dependent upon edge orientation. Variation in solar radiation levels among edge aspects is likely to be greatest in mid- to high latitude forests (Murcia 1995), but some effects are likely in the Tropics, particularly in forests located away from the equator (Turton 1991). Forest physiognomy is also important because it affects the amount of incident solar radiation reaching the forest understory. Matlack (1993) found that for several environmental variables, the greatest edge effects were exhibited by edges with the least lateral protection.

Environmental variables, such as PAR, temperature, VPD, wind speed, and soil heat flux, measured immediately above and within forest canopies depend largely on the vertical distribution of vegetation layers and canopy openness (Fritschen 1985). Malcolm (1994) found that understory foliage thickness increased and overstory foliage thickness decreased toward edges in three primary forest habitats in Brazil. Such changes in understory and overstory foliage density will in turn affect the microclimate at or near ground level.

The size, shape, and age of rainforest remnants will often interact to determine the importance of edge effects (Kapos 1989; Matlack 1994b; Kapos et al., chap. 3). Obviously, external climatic influences are likely to have a greater effect in small or irregularly shaped remnants. Larger and more circular remnants are likely to have a greater core area that is unaffected by edge effects (Laurance and Yensen 1991; Saunders, Hobbs, and Margules 1991). In addition, the character of vegetation at forest edges varies with the successional status of the site (Matlack 1994b). Old, mature edges tend to be closed off or "sealed" by vegetation, whereas recently exposed edges may experience more extreme edge effects that penetrate farther into the forest.

THE NORTHEASTERN AUSTRALIAN CONTEXT

Much of Australia's coastal lowland rainforest has been cleared for agriculture or settlement. Clearing has also been extensive in some upland forests such as those on the Atherton Tableland (see Laurance, chap. 6). Clearing of Tableland forests began about 1909 and proceeded rapidly for the next three decades (Eacham Historical Society 1979). Consequently, most rainforest remnants in this area have developed old, mature edges. By 1983 over 76,000 ha of forest had been removed, leaving about 100 forest remnants ranging from 1 to 600 ha in area, scattered over an area of approximately 900 km^2 (Winter et al. 1987; Laurance 1990). These remnants are important refugia for a wide variety of rare and endemic animals and plants in the region.

Northern Queensland is markedly different from many other tropical forest regions because of its strongly seasonal climate regime (Turton 1991). Due to their latitudinal position (about 17° S), its forests experience significant seasonal variation in solar radiation, rainfall, and temperature, which are major external influences on the internal rainforest microclimate. The rainforests of northern Queensland are thus representative of rainforests on the southern fringe of the "maritime continent" (Ramage 1968) and, as such, differ significantly from equatorial continental rainforests. Consequently, the results of studies based in equatorial rainforest regions such as Amazonia may not be applicable to this region.

The present study focused on horizontal microclimatic gradients in a 20 ha rainforest remnant on the Atherton Tableland. The study had three main aims: (1) to quantify microclimatic gradients along a series of replicated edge-to-interior transects under dry and wet conditions in summer; (2) to compare the distances to which edge effects penetrate into the remnant among edges with differing aspects; and (3) to determine whether seedling densities vary with distance from the edge and edge aspect.

METHODS

The Study Area

The 20 ha rainforest remnant was located between Malanda and Tarzali on the Atherton Tableland (17°22′ S, 145°35′ E), at 760 m elevation. The vegetation type is described by Tracey (1982) as the upland version of complex mesophyll vine forest. Although a detailed history of the remnant is unavailable, selective logging had not occurred within the fragment for 60–70 years. The mean annual rainfall is about 2,000 mm, with most rain falling during the wet season from December through April (Tracey 1982). This remnant was chosen because it was relatively undisturbed, had only a slightly undulating topography, and was large enough to allow us to explore microclimatic gradients and seedling densities in relation to forest edge and edge aspect.

Instrumentation and Sampling

Manual microclimate readings were collected along transects oriented along north-south and east-west gradients. The following variables were measured: soil temperature at three depths (0, 5, and 10 cm); ambient temperature at two levels (20 and 150 cm); and vapor pressure deficit at two levels (20 and 150 cm), which was calculated using wet and dry bulb temperatures. We used an aspirated hygrometer to record wet and dry bulb temperatures and digital temperature probes to measure soil temperature.

The edge-to-interior line transects were established perpendicular to edges on the

Table 4.1 Rainfall (mm) recorded at Malanda Post Office, the closest meteorological station to the study site.

	December 1994	January 1995 (dry conditions)	February 1995 (wet conditions)
Rainfall over the 30 days prior to the study period		40.4	125.8
Rainfall during the eight-day study period		12.4	127.8
Monthly total (mm)	51.2	122.6	554.8

north, east, south, and west sides of the remnant. Spatial variability was assessed by using four parallel transect replicates on each aspect. On the north, east, and west sides the transects were separated by 40 m. However, the transects on the south side were separated by only 20 m due to the shorter length of the side and the presence of an old logging track on the southwest corner of the fragment.

On each transect, flagged stakes marked ten positions for measurements: 5 m outside the remnant, the edge, and 5, 10, 20, 30, 40, 50, 60, and 70 m into the forest. The edge was defined as the location where both rainforest trees of greater than 10 cm diameter at breast height and forest understory species were present. The absence of grass and *Lantana camara* (an exotic weed) also was a factor in determining the exact location of the edge. In most cases the edge was clearly defined by an abrupt change in vegetation.

Due to limitations of time, personnel, and equipment, simultaneous sampling along each aspect was not feasible. Consequently, measurements were taken twice daily using the traverse method and then averaged to minimize temporal variability. The non-simultaneous sampling regime consisted of alternating between north-south and east-west transects, as well as between morning and afternoon, for eight consecutive days of data collection.

Fieldwork was conducted during two eight-day periods, 11–18 January and 7–14 February, 1995. These periods were chosen to coincide with the end of the dry season and the beginning of the wet season, as verified by rainfall data provided by the Bureau of Meteorology (table 4.1).

Hemispherical (160°) photographs of the canopy were used to estimate canopy openness. The 8 × 12 cm photographs were copied to xerox transparencies for analysis using a Delta-T Devices (U.K. Ltd.) leaf-area scanner. It must be noted that canopy openness figures derived by this method cannot be taken as absolute values. However, the resulting percentages are relative and are more accurate than visual estimates.

Seedling counts of dicotyledons less than 150 cm tall were conducted using circular plots with a radius of 2 m around each of the ten position markers per transect. The middle two transects for each aspect were examined, giving a total of eight transects.

RESULTS

Our measures of canopy openness (fig. 4.1) indicated that canopy cover was high throughout the forest remnant, with a clearly defined edge. Canopy openness was negatively

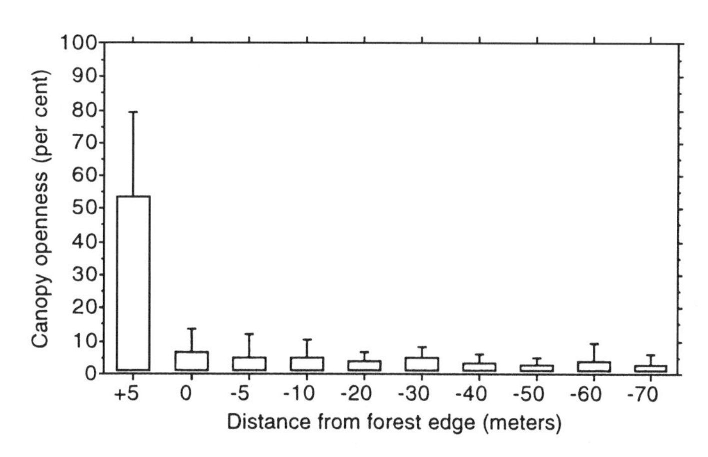

Figure 4.1. Canopy cover (estimated using hemispherical photographs) was high throughout the forest remnant, with a sharply defined edge. Canopy openness was negatively correlated with distance from the edge ($r_s = -.44$, $P < .001$). Error bars indicate one standard deviation from the mean ($n = 32$ for each edge distance value).

correlated with distance from the forest edge ($r_s = -.44$, $P < .001$, Spearman rank correlation).

The effects of distance from the forest edge and edge aspect on various microclimate measurements were tested using two-way analyses of variance (ANOVA; table 4.2). Under dry conditions, none of the seven microclimate parameters varied significantly with distance from the forest edge ($P > .05$), and only ambient temperatures at 20 and 150 cm height were affected by edge aspect ($P < .01$). However, under wet conditions, soil temperatures at 0, 5, and 10 cm depth varied significantly with distance from the edge ($P < .05$), and soil temperatures also were affected by edge aspect at 0 and 10 cm depth ($P < .05$). Under wet conditions, both ambient temperatures and vapor densities at 20 and 150 cm height were unaffected by edge distance ($P > .05$), but varied significantly ($P < .05$) with edge aspect (table 4.2).

Surface soil temperatures were higher near the edge, but tended to stabilize within 30 m of the edge (fig. 4.2). Although the patterns of soil surface and ambient temperatures were similar during the dry and wet periods, the range was not as great under wet conditions (fig. 4.2). The surface temperature was warmer under wet than under dry conditions, but ambient temperatures at 20 cm were similar under dry and wet conditions. In comparison, vapor pressure deficit (VPD) at 20 cm was significantly higher under dry conditions, but it increased toward the edge under both dry and wet conditions (fig. 4.2).

The number of dicotyledonous seedlings increased along the edge-to-interior transect (fig. 4.3). Seedling densities, when grouped into three edge distance categories (0–10 m,

Table 4.2 Two-way analyses of variance showing the effects of distance from the forest edge (factor 1), edge aspect (factor 2), and the interaction of factors 1 and 2 on seven microclimate parameters under dry and wet conditions.

Microclimate parameter	Factor 1: Edge distance	Factor 2: Edge aspect	Interaction: Distance × aspect
Dry conditions			
Soil temp. 0 cm	0.49	2.12	0.59
Soil temp. 5 cm	0.51	0.73	0.44
Soil temp. 10 cm	1.62	1.63	1.02
Ambient temp. 20 cm	0.59	4.02**	0.63
Ambient temp. 150 cm	0.51	4.51**	0.46
Vapor density 20 cm	0.67	0.90	0.19
Vapor density 150 cm	0.01	1.36	0.15
Wet conditions			
Soil temp. 0 cm	4.31**	2.67*	0.34
Soil temp. 5 cm	4.90**	2.26	0.87
Soil temp. 10 cm	11.66**	4.62**	2.89*
Ambient temp. 20 cm	2.20	3.28*	0.62
Ambient temp. 150 cm	1.97	5.15**	0.78
Vapor density 20 cm	0.06	5.51**	0.22
Vapor density 150 cm	0.28	6.97**	0.37

Note: Three categories were used for edge distance (0–10 m, 20–40 m, and 50–70 m). Values depicted are F statistics.

* $P < .05$; ** $P < .01$.

20–40 m, and 50–70 m), were significantly influenced by distance to the forest edge ($P < .01$, one-way ANOVA). The numbers of dicot seedlings increased steadily from the edge to about 30 m into the remnant, where they appeared to level off (fig. 4.3). Numbers of dicot seedlings 0–70 m from the edge were also found to vary significantly with edge aspect (one-way ANOVA, $P < .05$; fig. 4.4).

DISCUSSION

Microclimate and Edge Effects

Canopy openness was remarkably even throughout the forest remnant we studied (see fig. 4.1). Kapos et al. (chap. 3) and Laurance (1991b, chap. 6) discuss the fact that treefalls and canopy gap formation may increase near forest edges. Our study failed to detect such a trend, although it must be emphasized that we examined only a single fragment. Williams-Linera (1990b) noted that canopy openness decreased and became more uniform as the age of the forest edge increased. The canopy openness values found in this study are similar to those reported by Williams-Linera (1990b) for forest edges that were at least twelve years old.

This study has shown that microclimatic edge effects penetrate about 30 m into a small (20 ha) forest remnant under both dry and wet conditions in summer. We also demonstrated that microclimatic edge effects vary not only with distance from the edge, but in some cases with edge aspect as well (see table 4.2, fig. 4.2). In particular, soil temperatures under wet conditions appeared to be strongly affected by distance from the

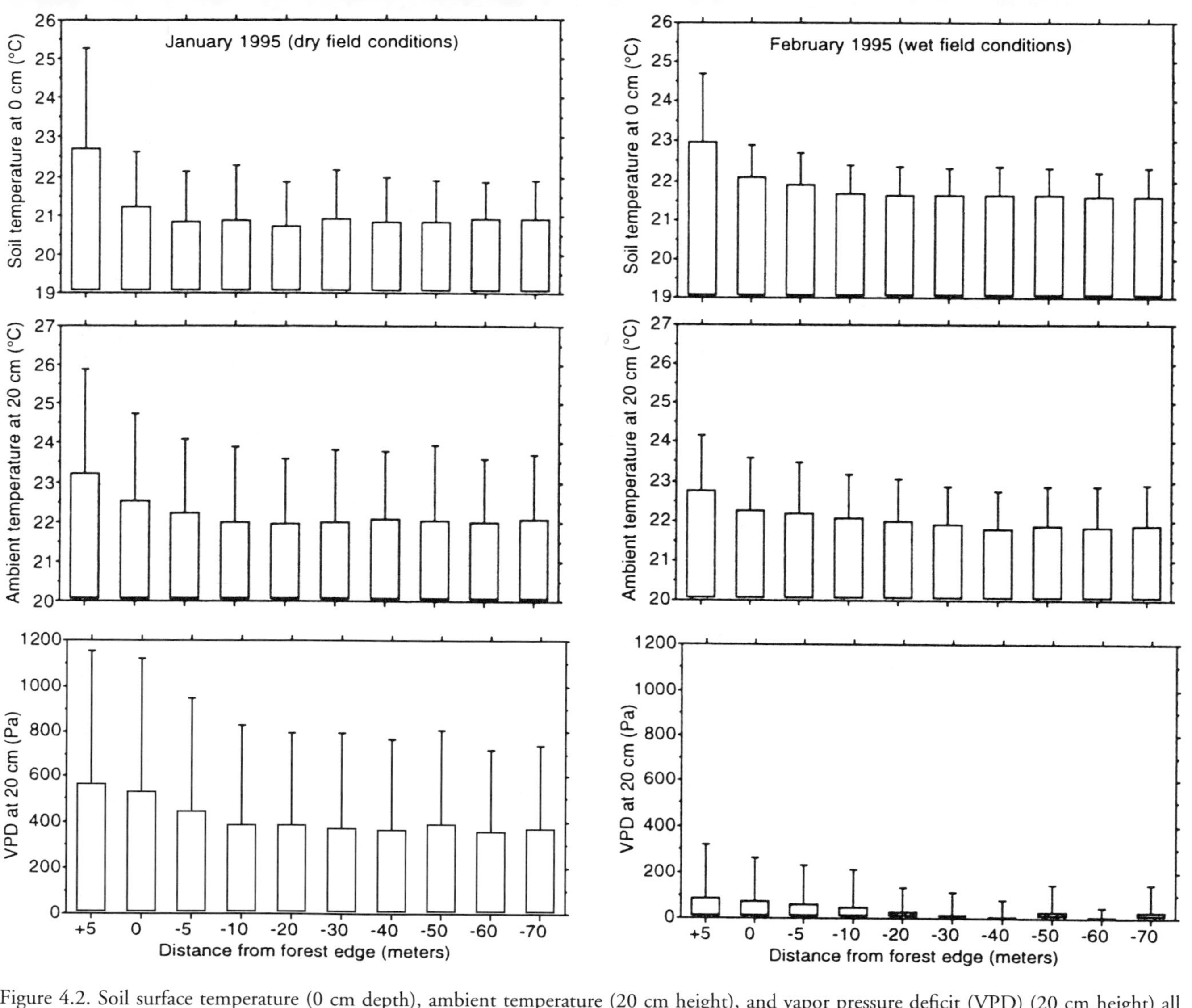

Figure 4.2. Soil surface temperature (0 cm depth), ambient temperature (20 cm height), and vapor pressure deficit (VPD) (20 cm height) all varied significantly with distance from the forest edge during both sampling periods, with higher values recorded toward the edge ($n = 32$ for each edge distance value). Error bars indicate one standard deviation from the mean.

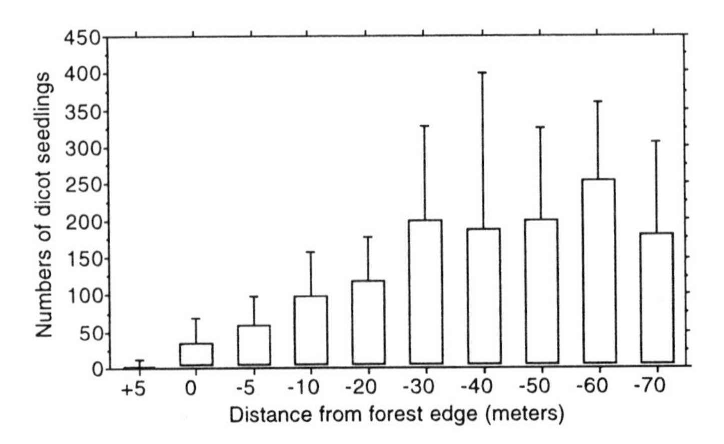

Figure 4.3. The number of dicotyledon seedlings (<150 cm tall) in relation to distance from the forest edge ($n = 8$ for each edge distance value). Error bars indicate one standard deviation from the mean.

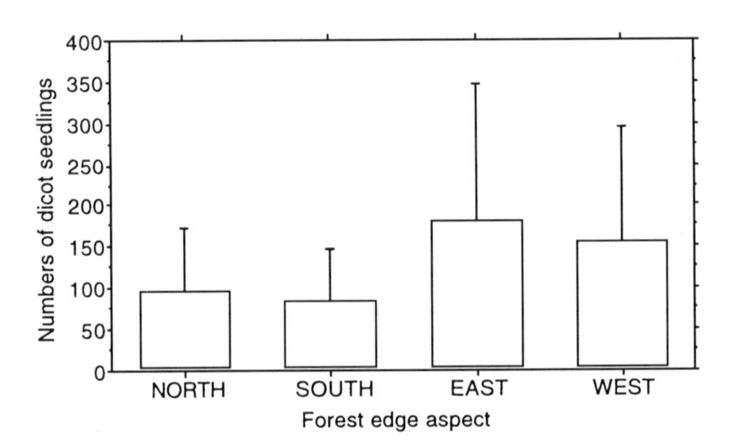

Figure 4.4. The number of dicotyledon seedlings in relation to edge aspect. Error bars indicate one standard deviation from the mean.

edge, and also by aspect. Such trends would have major implications for microbiotic activity in the soil and possibly for seed germination and establishment. In comparison, ambient temperatures at seedling (20 cm) and sapling (150 cm) heights appeared to vary significantly with edge aspect under wet and dry conditions, but were not affected by edge distance. Likewise, vapor density was strongly affected by edge aspect and unaffected by edge distance under wet conditions, while under dry conditions neither aspect nor edge distance significantly affected vapor density.

It is difficult to place the results of this study in the context of previous work because only a few studies have been conducted in this field, particularly in the seasonally wet Tropics. The microclimatic gradients reported in this study are similar to those given by Matlack (1993) for old, closed edges in broadleaf forest remnants in the eastern United

States. The edges examined in this study also were mature, essentially insulating the forest interior from outside conditions.

This study demonstrated that very few species of dicotyledonous rainforest seedlings were able to germinate beyond the protective cover of the rainforest canopy (see fig. 4.3). This pattern may be due to heavy shading produced by the wall of vegetation at the forest edge, together with higher soil surface temperatures at and near the edge. Differences in dicotyledonous seedling densities between edge aspects may be attributable to the penetration of solar radiation into the forest (see fig. 4.4). The highest densities were observed on the east- and west-facing edges, presumably as a result of penetration of the morning and afternoon sun into the remnant. Lower seedling densities on the north- and south-facing edges may be attributable to reduced solar access along these edges.

Although the length of this field study was limited, the sampling period was sufficient to detect trends in microclimate parameters and to examine dry versus wet conditions, together with the effects of aspect. A major criticism of many studies of edge effects has been the lack of adequate replication (Murcia 1995). A strength of this study is its sampling design, which used sixteen replicate transects, with repeated measures during both wet and dry seasons.

Suggestions for Further Research

It should be noted that every forest remnant experiences unique microclimatic conditions based on factors such as remnant size and shape, forest structure, edge orientation, and the successional stage or age of the edge (Matlack 1993; Kapos et al., chap. 3). We recommend, therefore, that a single methodology be replicated in a variety of different-sized remnants at different times of the year. It would also be valuable to measure microclimates in remnant core areas and to sample sites at least 25 m beyond the forest edge along each transect line.

This study did not consider microclimatic edge effects in winter, and it would be very useful to do so. There may be significant differences in mean and extreme minimum temperatures between upland rainforest remnants, such as the one examined in this study, and lowland equatorial forest remnants such as those described by Williams-Linera (1990b) and Kapos et al. (chap. 3). In particular, the effects of near-zero or subzero temperatures, particularly near edges, may be an important disturbance factor in upland remnants.

Likewise, because of the relatively high latitude of northeastern Australian rainforests (14–19° S), there are marked changes in solar angles between the winter and summer solstices (Turton 1991). Such changes are less marked in equatorial regions, where most studies of edge effects and tropical forest remnant microclimates have been performed to date (Kapos 1989; Williams-Linera 1990b; Kapos et al., chap. 3). It is likely that edge aspect and topographic factors will have greater effects on the microclimates of forest remnants in northeastern Australia than in equatorial regions, and these factors certainly warrant further study across a range of forest remnant sizes, shapes, and topographies. Such research seems particularly important in relation to conservation and management of remnants in topographically complex regions, such as the Atherton Tableland.

GENERAL IMPLICATIONS

1. This study suggests that, for some microclimate parameters, edge aspect may be more important than distance from the forest edge, at least in higher-latitude tropical forests. Both edge distance and aspect had significant effects on dicotyledonous seedling densities.
2. Under both wet and dry conditions in summer, microclimatic edge effects penetrated only about 30 m into the remnant we examined. This pattern may be attributable to the protective layer of dense vegetation that had developed near the forest edge in this relatively old fragment. This finding would suggest, in the case of old remnants, that only small forest remnants experience strong microclimatic edge effects on a significant portion of their total area.
3. Microclimate studies have broad implications for designing and implementing effective reforestation and management strategies for rainforests. Investigations of the vegetation dynamics of forest edges, coupled with microclimate studies, will enhance our understanding of rainforest ecology. The effects of microclimatic variation on native fauna in fragments should also be assessed.
4. Quantification of internal forest microclimatic conditions throughout the year is critical if researchers are to accurately interpret the responses of organisms to tropical forest fragmentation. Such studies may be more critical in higher-latitude tropical forest remnants than in equatorial remnants because of pronounced seasonal changes in the regional climate of higher-latitude areas. In particular, strong seasonal variation in solar radiation, rainfall, and temperature in such regions is likely to exert significant effects on microclimate regimes near forest edges, and along edges with differing aspects and topographies.

ACKNOWLEDGMENTS

This project was made possible with the financial support of the Fulbright Commission of the Australian-American Education Foundation to H. J. Freiburger and the Australian Research Council Small Grant Scheme to S. M. Turton. We thank Drs. W. F. Laurance and V. Kapos and one anonymous referee for their constructive comments on the manuscript. We also thank Jen Badstuebner and Ingo Heinrich for assisting with fieldwork.

5

The Influence of Edge Effects and Forest Fragmentation on Leaf Litter Invertebrates in Central Amazonia

Raphael K. Didham

Forest fragmentation occurs when a large area of native forest is transformed into a series of smaller remnant patches isolated by an intervening matrix hostile to forest organisms (Burgess and Sharpe 1981; Harris 1984; Wilcox and Murphy 1985; Saunders, Hobbs, and Margules 1991; Harris and Silva-Lopez 1992; Zipperer 1993). The process of fragmentation has a number of important aspects; namely, loss of original habitat, reduction in remnant patch size, increasing isolation of remnant patches (Andrén 1994), and exposure of forest fragments to edge effects as a result of the abrupt transition between forest and matrix habitats (Murcia 1995). Although the concepts seem straightforward, much confusion exists over the confounding influences of these factors. Many fragmentation studies have simply used area as the surrogate explanation for any change in community structure in small fragments. In particular, edge effects and area effects have been confounded because the amount of edge in a fragment increases in proportion to decreasing patch area. Furthermore, there is often no clear distinction between the proximate and ultimate causes of population declines within isolated fragments (Schonewald-Cox and Bayless 1986; Laurance and Yensen 1991).

Some studies have demonstrated dramatic changes in biotic and abiotic factors at the forest edge (Kapos 1989; Williams-Linera 1990a,b; Laurance 1991b, chap. 6; Kapos et al. 1993, chap. 3; Williams-Linera 1993; Malcolm 1994; Turton and Freiburger, chap. 4), leaving little doubt as to the actual and potential importance of edge effects (Murcia 1995). Early attempts to quantify the importance of edges in forest fragments assessed perimeter length-to-area ratios (p/a ratios: Forman and Godron 1986; Schonewald-Cox and Bayless 1986; Buechner 1987; Stamps, Buechner, and Krishnan 1987). The p/a ratio has been superseded by the core-area model of Laurance and Yensen (1991), which relies on quantifying the edge penetration distance *(d)* in order to calculate the unaffected core area in a fragment of any given size or shape. Malcolm (1994) presents a more realistic model of the additive nature of edge effects affecting a single point within the edge zone, *d*.

Murcia (1995) criticized studies seeking "simplistic and static" patterns of edge effects and advocated exploring new mechanistic approaches to the factors that modulate edge effects. Her most significant contention is a hypothetical bimodal pattern of edge effects that may result from the interaction of two or more different biotic or abiotic factors.

Such patterns have typically been attributed to intrinsic variation in the measured variable (Kroodsma 1982; Kapos 1989; Palik and Murphy 1990; Hester and Hobbs 1992; Williams-Linera 1993). However, diverse evidence suggests that "competition-induced waves of biomass" (i.e., polymodal response curves with alternate peaks and troughs in biomass resulting from competition for a limiting resource) may be common at spatial boundaries in nature (Reichman, Benedix, and Seastedt 1993).

One of the greatest stumbling blocks in the elucidation of fragmentation processes has been the confounding of area and edge effects in small fragments (Schonewald-Cox and Bayless 1986; Laurance and Yensen 1991). As much as anything, this problem stems from the method employed in most fragmentation studies; namely, comparing point samples from interiors of forest fragments with point samples from continuous forest, without controlling for distance from the forest edge. While the results obtained do indicate an effect of fragmentation, they bring us no closer to understanding the mechanisms at work in fragmented systems. Indeed, the postulated effects of fragment isolation on populations and communities (Wilcox and Murphy 1985; Wright 1985; Gilpin and Soulé 1986; Soulé 1987; Bolger, Alberts, and Soulé 1991) have rarely been conclusively demonstrated (Leigh et al. 1993; Adler 1994).

After more than fifteen years of intensive study there are still relatively few published works dealing specifically with the effects of forest fragmentation on invertebrates. It has become clear that invertebrate responses to fragmentation are varied and often species-specific. As a general rule, invertebrate abundance increases toward the forest edge (Webb, Clarke, and Nicholas 1984; Helle and Muona 1985; Webb 1989; Duelli et al. 1990; Shure and Phillips 1991; Buse and Good 1993; Báldi and Kisbenedek 1994; but see Klein 1989; Powell and Powell 1987; Gunnarson 1988). This increase is probably due to the replacement of the natural forest community with invasive (generalist) edge species (Webb, Clarke, and Nicholas 1984; Shmida and Wilson 1985; Webb 1989). At edges and within isolated fragments, species richness and composition are often different from those in continuous forest (Webb, Clarke, and Nicholas 1984; Klein 1989; Souza and Brown 1994). Moreover, the representation of functional groupings is altered (Souza and Brown 1994), and body size patterns of some taxa are modified (Gunnarson 1988; Klein 1989). Klein (1989) noted the adverse effect of a change in invertebrate abundance and species composition on an important ecosystem process, the rate of dung decomposition, in fragmented forest (see also Didham et al. 1996).

Fragmentation studies of invertebrates have so far concentrated on individual insect orders or selected indicator species. Many of the taxa studied are atypical of the majority of invertebrates because they are large (e.g., butterflies: Lovejoy et al. 1986; Brown and Hutchings, chap. 7; dung beetles: Klein 1989), functionally unique (leaf-cutter ants: Vasconcelos 1988), or highly specialized (euglossine bees: Powell and Powell 1987; Becker, Moure, and Peralta 1991). Focusing on particular indicator taxa is undoubtedly valuable in some respects (Sutton and Collins 1991; Holloway, Kirk-Spriggs, and Chey 1992; Simberloff et al. 1992), but the observed patterns may have little general relevance to whole invertebrate communities. In this study, I take the multispecies assemblage approach, using a taxonomically and trophically diverse group of insects (beetles) and a numerically dominant group (ants), so that the results may have greater applicability to other forest invertebrate communities. This chapter focuses primarily on abundance patterns of beetles and ants as a function of distance from forest edge and fragment area. It also examines changes in invertebrate assemblage structure in fragmented forest through a preliminary analysis of species composition within three major beetle families.

METHODS

Study Site

This study was part of the Biological Dynamics of Forest Fragments Project (BDFFP), located 80 km north of Manaus, Amazonas, Brazil (Bierregaard et al. 1992). The forest type was terra firme (upland) forest on moderately rugged terrain. The dominant soil type was a nutrient-poor, yellow alic latosol of high clay content (Chauvel 1983). Forest fragments of different sizes were isolated by cutting and burning for cattle pasture between 1980 and 1984, as described by Lovejoy et al. (1986). Annual rainfall is approximately 2,200 mm, with a pronounced dry season from June to September. A more detailed description of the study area can be found in Bierregaard and Stouffer (chap. 10).

Sampling Protocol

The sampling design was based on a comparison of two independent transects at each of three locations: (1) deep within undisturbed continuous forest (> 10 km from the nearest edge); (2) from the edge to the interior of continuous forest; and (3) from the edge to the interior of two 100 ha isolated forest fragments (fig. 5.1). The two 100 ha fragments

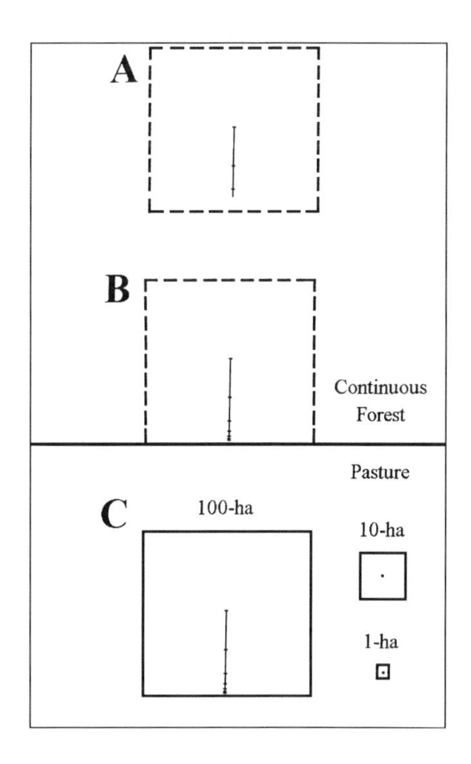

Figure 5.1. Sampling protocol. Transects were sampled at seven distances (0, 13, 26, 52, 105, 210, and 420 m) in *(A)* continuous forest interior, *(B)* edge of continuous forest, and *(C)* 100 ha fragments. Additional samples were taken 105 m into 10 ha fragments and 52 m into 1 ha fragments. The entire design was replicated once.

(BDFFP reserve numbers 2303 and 3304; hereafter designated 100 ha 1 and 100 ha 2, respectively) were located about 20 km apart (details of reserve isolation can be found in Bierregaard and Stouffer, chap. 10, table 10.1). The two continuous forest edges, Edge 1 and Edge 2, were separated by a distance of 2 km (both were on the western edge of continuous forest reserve 1401), and the deep forest control plots, Interior 1 and Interior 2, were 2 km apart (both were in control site reserve 1501). All forest edges abutted well-maintained pasture and were westward-facing, with the exception of 100 ha 2, where the only edge abutting well-maintained pasture was northward-facing.

Along each of the six transects, invertebrates were collected and environmental measurements recorded at seven distances from the forest edge: 0, 13, 26, 52, 105, 210, and 420 m (fig. 5.1). This sampling protocol reflected the a priori expectation that changes in invertebrate community structure would be greatest near the forest edge. In addition, to assess invertebrate populations in small forest fragments, two 10 ha fragments (reserves 3209 and 1202, designated 10 ha 1 and 2, respectively) were sampled at 105 m from the edge, and two 1 ha fragments (reserves 2107 and 2108, designated 1 ha 1 and 2, respectively) were sampled at 52 m from the edge.

Forest edges in Central Amazonia exhibit two basic structural forms: (1) "closed edges" formed by clear-cutting and burning, without fire encroachment into the forest, which regenerate rapidly with a mixed plant assemblage dominated by *Cecropia* spp. (effectively sealing off the edge from deleterious microclimatic effects); and (2) "open edges" subject to fire encroachment (even to a minor degree), which regenerate slowly and are dominated by *Vismia* spp. (maintaining an open edge structure and allowing the movement of hot, dry air into the forest). Edge 1 and 100 ha 1 are examples of closed edges with a dense secondary growth buffer, while 100 ha 2 and Edge 2 are typical fire-encroached edges with an open structure. Unfortunately, sampling constraints and lack of suitable edges prevented me from treating edge type as a properly replicated variable. However, treatment effects were not confounded by edge type.

Invertebrate Sampling

At each of the forty-six sites, twenty random, 1 m^2 leaf litter samples were collected during the January–May 1994 rainy season. Daily sampling was randomly allocated between different transects and sites to prevent bias arising from daily and seasonal variation in activity patterns of invertebrates. All friable leaf litter was scraped rapidly from the quadrat and placed in a large bag sieve to minimize invertebrate escape. The material was immediately sieved over a 9 mm mesh by vigorously shaking the bag sieve for about five minutes. The fine, sieved litter containing invertebrates was then transported to the laboratory in individual cotton bags. Invertebrates were extracted using the Winkler method, whereby sieved leaf litter was carefully placed in coarse mesh bags, which were then suspended inside large sealed cloth bags for three days. As the leaf litter dried out, invertebrates sensitive to desiccation moved downward through the mesh bag and fell into a jar of alcohol below. The author and one field technician operated forty Winkler bags continuously for five months.

The Winkler method proved to be sensitive to climate and collection methods, and hence required strictly standardized application. I collected samples only from plateau forest areas (i.e., transects were not located in gullies or seasonally flooded areas) and only on dry mornings when there had been no rain the previous late afternoon or night. Leaf litter sampling was discontinued if it rained. All samples were hung for three days, and no extra hand sorting of litter was performed. Even with these restrictions, the Winkler method is still inherently a "relative" trapping method and, like most invertebrate sampling methods, does not sample all taxa with equal efficiency. It is a poor method for sampling microinvertebrates such as mites and Collembola, so I restricted my analysis to macroinvertebrates. It is a particularly good method for the rapid and efficient extraction of beetles (Coleoptera) and ants (Hymenoptera: Formicidae) from large numbers of samples (Besuchet, Burckhardt, and Löbl 1987; Nadkarni and Longino 1990; Belshaw and Bolton 1994). These are the taxa considered here.

Because of the strict sampling regime, the majority of invertebrate species sampled in this study were forest leaf litter specialists, although no detailed information is available on their habitat preferences. Thus, there was almost no overlap in species composition with the adjacent pasture invertebrate fauna. The degree of overlap in invertebrate species composition with second-growth forest is unknown.

Analysis

Macroinvertebrates were sorted to higher taxonomic levels (invertebrate class and insect order) and counted. Three beetle families (Staphylinidae, Carabidae, and Scarabaeidae) were further sorted to "recognizable taxonomic units" (also termed "morphospecies"), hereafter referred to as species. Sorting was done by the author and checked by taxonomic experts at The Natural History Museum, London.

Two-way indicator species analysis (TWINSPAN: Hill 1979b) and detrended correspondence analysis (DCA: Hill 1979a; Hill and Gauch 1980) were performed using the PC-ORD package (McCune 1991). TWINSPAN is a polythetic divisive method of classification used to cluster sites of similar species composition and identify indicator species characteristic of different sites or groups of sites. Here, TWINSPAN parameters were set at six levels of subdivision and three pseudospecies cut levels (species abundances of 0, 5, and 25). DCA parameters for axis rescaling and detrending were set to the default levels in the PC-ORD package. Reciprocal averaging (RA) was inappropriate for the heterogeneous data set analyzed here because of a pronounced "arching effect" of site scores (Gauch 1982), such that the second ordination axis was a spurious quadratic distortion of axis 1. The real secondary gradient, shown by RA axis 3, was qualitatively similar to that shown by DCA ordination, although some axis compression was also evident in RA. These problems were overcome by DCA. Multivariate analyses excluded species represented by a single individual ("singletons"), principally because of a 200-species limitation in the PC-ORD package.

The relative positions of sites in DCA ordination space reflect differences in community structure. These differences were quantified by measuring the distance separating

Table 5.1 Summary results of two-level, mixed-model nested ANOVAs for $\log_{10}(X+1)$-transformed ant abundance and beetle abundance per m^2 in two replicate forest fragments each of 1, 10, and 100 ha in central Amazonia ($n = 20$ m^2 per site; see fig. 5.2).

	df effect	MS effect	df error	MS error	*F*	*P*
Ant abundance						
Fragment size	2	2.644	3	1.520	1.740	.315
Replicate (nested subgroup)	3	1.520	114	0.126	12.086	< .001
Beetle abundance						
Fragment size	2	1.3	3	0.171	7.584	.067
Replicate (nested subgroup)	3	0.171	114	0.081	2.114	< .001

Note: P values are statistically significant at $P < .05$.

edge and fragment sites from deep forest sites in DCA ordination space (distances were measured from site coordinates along DCA axes 1 and 2). The background level of dispersion between deep forest control sites was measured as the mean (±1 SE) of all pairwise distances between the fourteen deep forest sites ($n = 91$). Mean (±1 SE) distance from each edge or fragment site to the fourteen deep forest sites was then plotted relative to this background level of dispersion to determine whether beetle community structure in disturbed habitats was significantly different from that in continuous forest. The two 100 ha transects were combined, and the two continuous forest edge transects were combined for clarity (hence, each plotted value represents a mean of twenty-eight pairwise distances).

RESULTS

Abundance Patterns

A total of 79,233 macroinvertebrates were collected and sorted from 920 m^2 of leaf litter. Ants (68.0%) and beetles (10.7%) represented the bulk of total invertebrate abundance. Other orders constituted less than 3% of total invertebrate abundance; the most abundant of these were millipedes (Diplopoda), beetle larvae, Diptera adults, and Diptera larvae.

There was no statistically significant trend in ant abundance with forest fragment size (table 5.1, fig. 5.2A) due to a high within-subgroup variance (i.e., variation in mean-transformed abundance between fragments of the same size). For beetles, the relationship between fragment size and abundance was only marginally nonsignificant (table 5.1, fig. 5.2B). Pooling of the nonsignificant within-subgroup variance with the error term to increase degrees of freedom was not permissible ($F_{0.75[3,114]} = 0.405 < F_{\text{Subgr}}$: Sokal and Rohlf 1995), so the overall treatment effect was formally nonsignificant. The biological reality of the observed trend needs to be more rigorously tested with a greater number of replicate fragments.

Given the varying edge-to-interior distances of different-sized forest fragments, a parsimonious explanation for an increase in beetle abundance with decreasing fragment size would be a simple monotonic decline in abundance with distance from edge (fig. 5.3A). However, in addition to a strong edge peak in abundance along edge-to-interior tran-

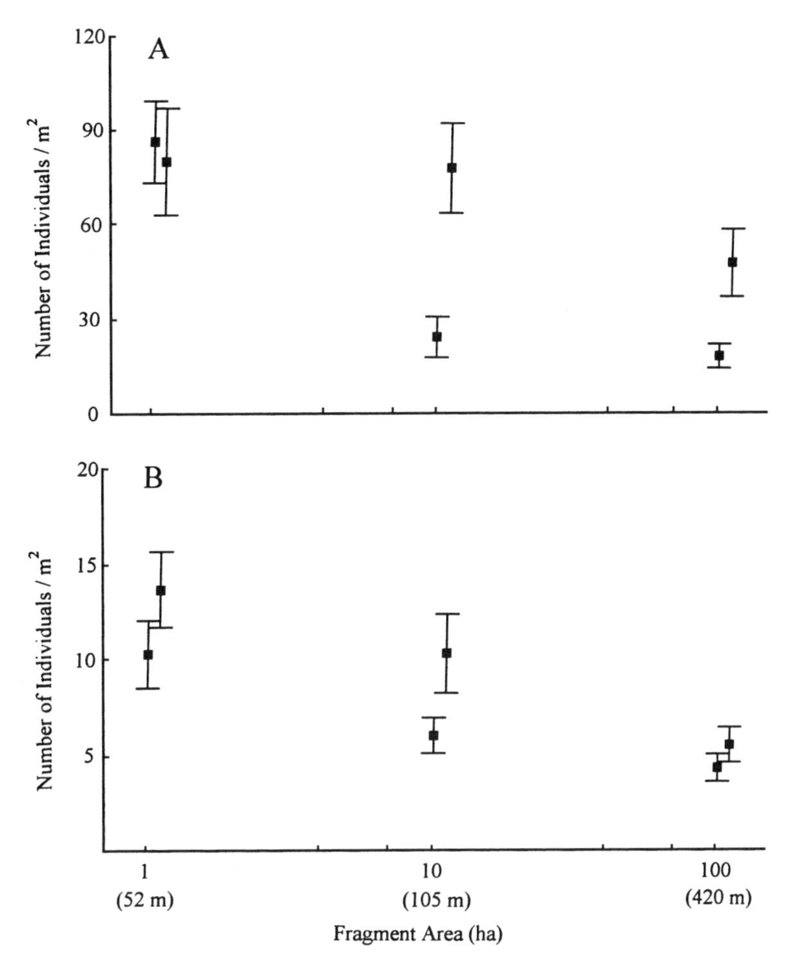

Figure 5.2. Mean (± 1 SE) numbers of individuals of *(A)* ants and *(B)* beetles at the centers of forest fragments (n = 20 m^2 per site). Distances from edge are given in parentheses.

sects, there was also a second, mid-distance peak in abundance at 26–105 m (fig. 5.4), exactly as hypothesized by Murcia (1995) (fig. 5.3B).

One-way ANOVA results for individual transects (using a Bonferroni-corrected P value of $P = .008$) showed that beetle abundance at the edge was significantly higher than deep forest abundance at 0 m and 52 m in 100 ha 1 ($F_{[6,133]} = 5.565$, $P < .0001$, fig. 5.4A) and at 0 m and 26 m in Edge 1 ($F_{[6,133]} = 3.369$, $P < .004$, fig. 5.4B) (Tukey's HSD, all $P < .025$). Ant abundance was significantly higher at 0–26 m in two transects (Edge 2: $F_{[6,133]} = 5.274$, $P < .0001$, Tukey's HSD $P < .0001$; 100 ha 2: $F_{[6,133]} = 3.937$, $P < .0012$, Tukey's HSD $P < .0001$, fig. 5.4C), but was not significantly bimodal (but see fig. 5.4C, D). Significantly, a mid-distance abundance peak was evident along more than one transect and across different taxa. The peak was statistically significant in some cases, but not in others, due to high sample variability.

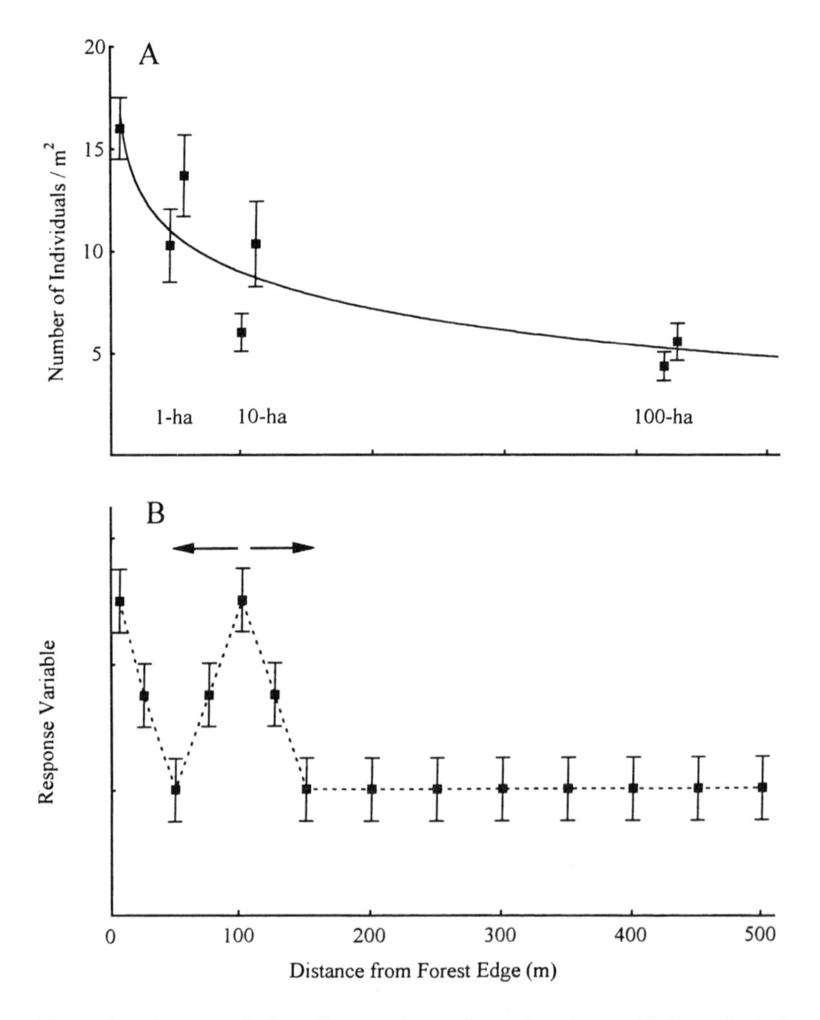

Figure 5.3. Patterns of edge effects on invertebrate abundance. *(A)* Hypothetical monotonic decline in abundance as a function of distance from the forest edge, fitted to empirical data for beetle density in forest fragments. *(B)* Hypothetical, generalized bimodal edge pattern (adapted from Murcia 1995). The mid-distance peak may shift depending on the extent and severity of edge effects.

Explanations for the Mid-Distance Abundance Peak

The mid-distance abundance peak was not a simple function of any measured environmental variable. There was no correlation between mean ant or beetle abundance and mean leaf litter moisture content (%), standing litter biomass (dry weight, grams/m^2), or relative evaporative drying rates (evaporative water loss, ml/h, from a standard experimental apparatus) (ants: $r = .26$, $r = .16$, and $r = -.15$, respectively, all $P > .05$; beetles: $r = -.14$, $r = .25$, and $r = .13$, respectively, all $P > .05$). Much of the between-transect variation in environmental variables was due to the difference between hot, dry,

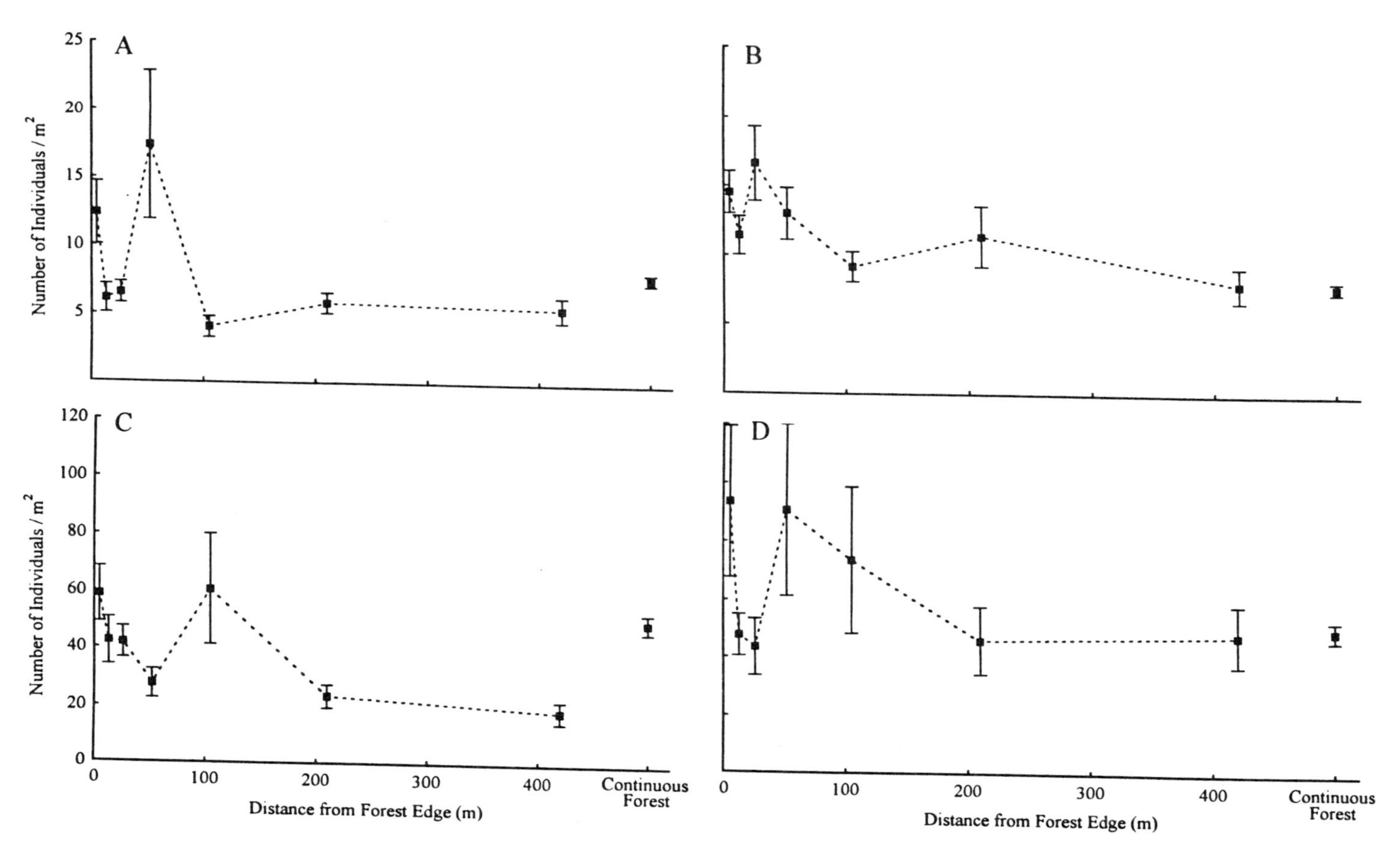

Figure 5.4. Edge abundance patterns (mean ± 1 SE) for *(A)* beetles at 100 ha 1, *(B)* beetles at Edge 1, *(C)* ants at 100 ha 2, and *(D)* ants at 100 ha 1.

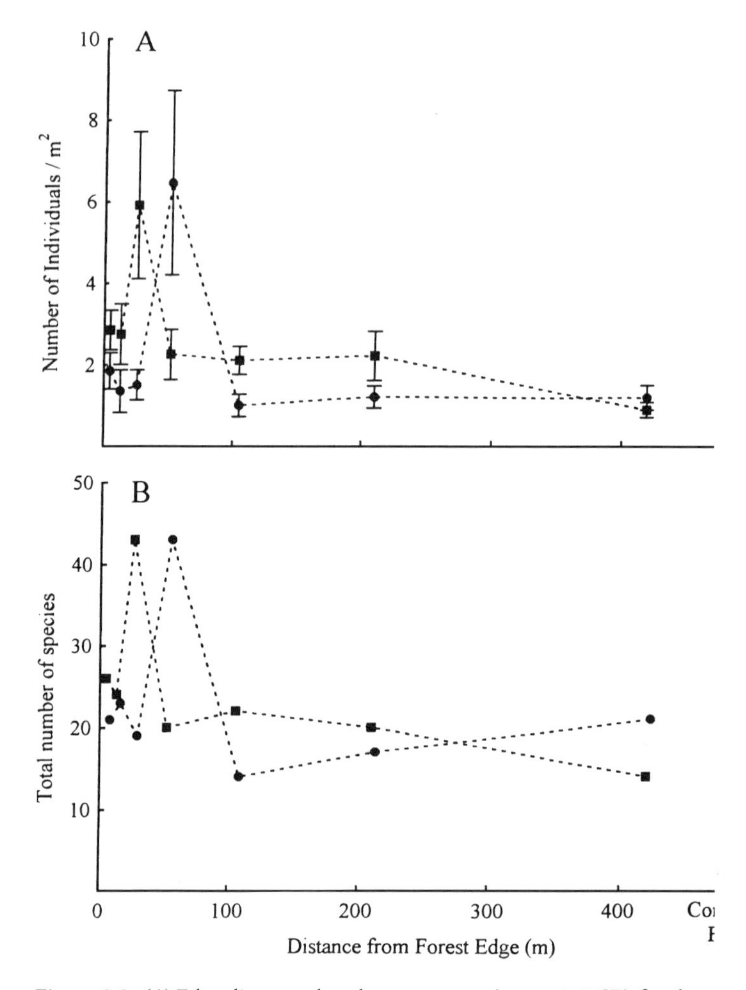

Figure 5.5. *(A)* Edge distance-abundance patterns (mean ± 1 SE) for three beetle families combined (Staphylinidae, Carabidae, and Scarabaeidae) and *(B)* corresponding species richness patterns (circles = 100 ha 1; squares = Edge 1). The species richness value for continuous forest is the mean (± 1 SE) number of species at fourteen forest interior sites. Dashed lines are for illustrative purposes only.

structurally open edges subject to fire encroachment (100 ha 2 and Edge 2) and edges with a cooler, closed structure. Only standing litter biomass at 100 ha 2 showed evidence of a mid-distance peak, but this did not correspond to the observed peak in ant abundance along the same transect (fig. 5.4C).

Species Composition and Community Structure

For the three beetle families sorted to species (Staphylinidae, Carabidae, and Scarabaeidae; total $S = 295$, $N = 2{,}111$), a mid-distance peak in abundance of individuals was evident only at Edge 1 and 100 ha 1 (fig. 5.5A). Species richness patterns at these two sites showed an identical response (fig. 5.5B). The absence of an edge peak in abundance for these three families indicates that other beetle families (Scolytidae, Pselaphidae, and Scydmaenidae) were dominant at the forest edge, and denotes a dramatic shift in family- and species-level representation of beetles along edge-to-interior gradients. These results suggest, then, that the bimodal edge function described above for the total beetle fauna may be a multiple response curve comprising separate (and different) monotonic abundance patterns of different taxa. The mid-distance abundance peak itself comprises an exceptionally high species richness (fig. 5.5B), which is principally due to a large number of "singleton" species.

In an analysis of beetle assemblage structure based on the preliminary data set, TWINSPAN produced a clear division between nonisolated forest sites (interior and edge sites) on the one hand and isolated fragment sites on the other (fig. 5.6). Strong subdivisions were also evident between interior and edge sites; also, among fragments, there was an indistinct clustering of sites based on distance from edge. The dichotomy between nonisolated and isolated sites was based on an almost total absence of many abundant and characteristic primary forest species from isolated fragments (fig. 5.6).

An ordination of study sites (fig. 5.7A) further highlighted the effects of area and edge on beetle species composition in isolated fragments. Species composition appeared to be structured heavily by edge effects, but was more different among isolated fragments than among continuous forest edges, presumably due to the added influence of area effects. In ordination space, the mean separation of edge and 100 ha sites from undisturbed interior sites was greater than the within-continuous forest dispersion, except for continuous forest sites greater than 105–210 m from the edge (fig. 5.7B). In 100 ha fragments, beetle species composition was markedly different from that of interior forest sites, even 420 m into the fragments.

DISCUSSION

Edge Effects and Invertebrates

The standard approach in studies of fragmentation-induced changes in ecological processes has been to compare point samples from forest fragments with those from intact control areas. While instructive, this approach discriminates poorly between the concomitant effects of edge and area. By sampling invertebrates at known distances along edge-to-interior transects, I was able to separate the effects of edge and isolation and demonstrate that an increase in invertebrate abundance in small fragments may be most parsimoniously explained by edge-driven processes. Strong fragmentation effects on population densities of, for example, *Drosophila* spp. (Jaenike 1978; Martins 1989), butterflies (Brown and Hutchings, chap. 7), various insect pollinators (Powell and Powell 1987; Becker, Moure, and Peralta 1991; Aizen and Feinsinger 1994b), and some other invertebrates (e.g., amphipods: Margules, Milkovits, and Smith 1994) may also be attributable to edge effects, rather than area effects as reported. While the increased proportion of edge in a fragment is a function of the fragment's size and shape, the proximate cause

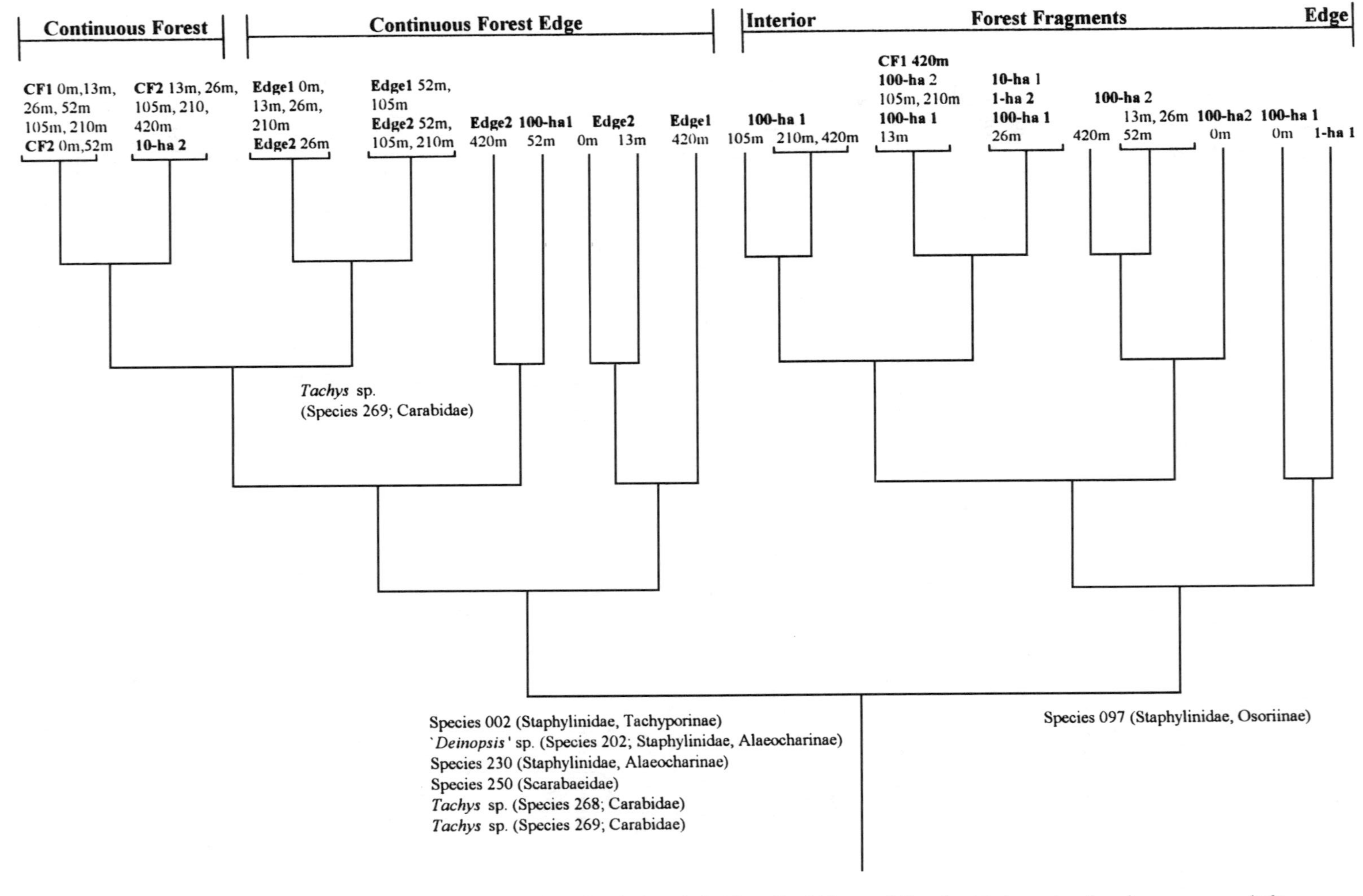

Figure 5.6. TWINSPAN dendrogram of site associations based on beetle (Staphylinidae, Carabidae, and Scarabaeidae) species abundance patterns (161 species, N = 1,977, excluding singleton species). The principal dichotomies between groups of sites are bracketed at the top of the diagram, and indicator species for the major subdivisions are listed. CF = continuous forest (interior) sites.

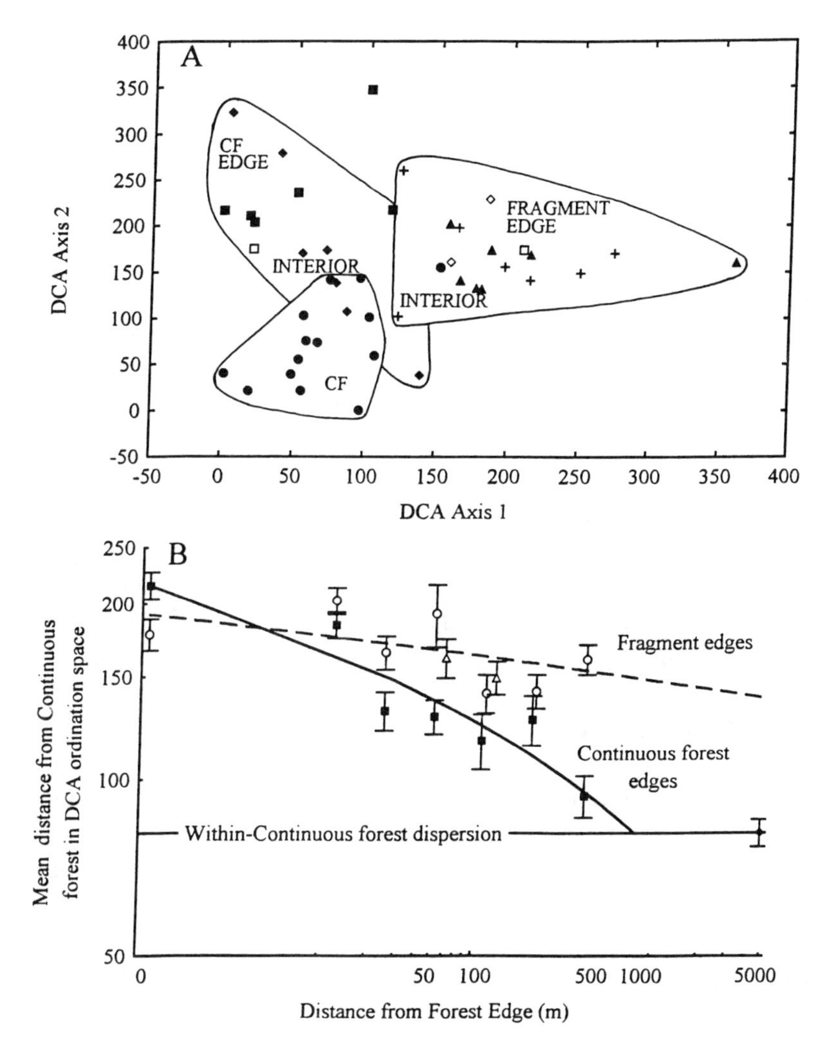

Figure 5.7. *(A)* Decorana ordination of sites for Staphylinidae, Carabidae, and Scarabaeidae beetles (solid circles = interior sites; solid squares = Edge 1; solid diamonds = Edge 2; crosses = 100 ha 2; solid triangles = 100 ha 1; open squares = 10 ha fragments; open diamonds = 1 ha fragments; CF = continuous forest). *(B)* Measurement of the relative dispersion of sites in DCA ordination space (arbitrary units). The within-continuous forest spread of the fourteen interior sites forms the background dispersion expected of sites that are similar to primary forest in species composition (open symbols = fragment edges; closed symbols = continuous forest edges).

of elevated invertebrate populations is almost certainly the altered microclimatic and habitat conditions at the edge, rather than reduced habitat area.

Almost invariably, edge functions have been described by a monotonic increase or decrease in the value of a measured variable with increasing distance from the forest edge (Laurance 1991b; Laurance and Yensen 1991; Malcolm 1994). In an insightful review of edge effects in fragmented forests, Murcia (1995, 60) considered it "unrealistic to expect all edge effects to vary monotonically with distance from the edge." She hypothesized a bimodal pattern of edge effects (see fig. 5.3B) based on a compilation of data measuring variation in abiotic factors (Kapos 1989; Hester and Hobbs 1992), vegetation (Palik and Murphy 1990; Hester and Hobbs 1992; Williams-Linera 1993), and bird density (Kroodsma 1982).

I found that, rather than varying monotonically with distance from edge, terrestrial invertebrate abundance in some cases exhibited a bimodal pattern similar to that hypothesized by Murcia (1995). Although high invertebrate population levels at habitat edges have been well documented (see above), the mid-distance abundance peak has not been shown before for invertebrates. In this study, mid-distance peaks were observed at 26–105 m from the edge across different transects, different edge types, and different invertebrate taxa.

Of key importance is whether the peak abundance zones contain species typical of undisturbed forest (but with greater than usual abundance), or a unique species assemblage. Staphylinidae, Carabidae, and Scarabaeidae showed a mid-distance abundance peak, but no edge peak. Instead, there was an edge peak composed of large numbers of Pselaphidae, Scydmaenidae, and Scolytidae. Thus, the bimodal edge pattern for the total beetle fauna was actually a multiple response curve comprising separate unimodal abundance patterns of different taxa.

The most likely explanation for the mid-distance peak is that 26–105 m into the forest represents a zone in which abiotic and habitat structural features approach interior conditions (Kapos 1989; Kapos et al. 1993; Malcolm 1994; Young and Mitchell 1994; Murcia 1995), creating an edge-interior ecotone. As at more conventional forest-second growth ecotones, there is an overlap of species found on either side of the boundary, producing elevated species richness and abundance (Yahner 1988). The exact position of the mid-distance peak appears to depend to some extent on edge history (see also Kapos et al., chap. 3, Turton and Freiburger, chap. 4). Bearing in mind the limited number of sampling points and difficulty in resolving the peak statistically, in this study the mid-distance peak generally occurred at 52 m from the edge, but up to 105 m at more open-edged sites that had been subject to fire encroachment.

Dynamic Nature of Forest Ecotones

Core-area models of edge effects used by Temple (1986), Temple and Cary (1988), Laurance (1991b), and Laurance and Yensen (1991) describe the edge effect as a discrete function, *d*, that abruptly abuts the undisturbed core area (but see Malcolm 1994). The edge-interior ecotone concept, however, describes dynamic interactions that take place as edge and interior communities grade into one another. The ecotone will, by definition, be both wide and plastic, changing with fluctuating environmental conditions at the edge (see fig. 5.3B). This concept is crucial to our understanding of edge-interior ecotone dynamics, as the undisturbed area within a fragment will be limited by the maximum temporal fluctuations in edge penetration distance. A single sampling event will not characterize this temporal variation and will result in an inadequate description of the penetration and impact of edge effects in a particular fragment.

Minimum Area Requirements of Forest Invertebrates

Given the magnitude of changes in invertebrate abundance, species richness, and community composition with distance from the forest edge, edge effects would be expected to be the primary force structuring invertebrate assemblages in fragmented forest. However, site classification and ordination of a subset of the beetle fauna showed that species composition of this group within isolated fragments was determined more by area than by edge effects. All fragment sites, regardless of distance from edge, were more different from deep forest sites than were continuous-forest edge sites. Thus, there did not appear to be a characteristic "edge community" of beetles. Among the taxa examined, almost all of the dominant species found in primary forest (six species, representing 30% of total individuals) were rare in or absent from the forest fragments, even in the interiors of relatively large (100 ha) patches. The probability that all of these common primary forest species would be absent from all six forest fragments by chance alone is very small. I contend, therefore, that most of these species have declined in abundance, or become extinct, in fragments since isolation. The exact mechanisms driving population declines in these species are unknown. This finding suggests that forest fragments of 100 ha or less fail to preserve an intact terrestrial invertebrate fauna in the Central Amazon.

It is interesting to note that, although invertebrate communities are clearly sensitive to edge effects, primary forest species that disappear from isolated fragments can be moderately abundant near the edge of continuous forest. These peripheral "populations" may be maintained by a large source population within continuous forest. Because fragments do not have a large source population to draw on, a combination of area and edge effects may rapidly push fragment populations to extinction.

GENERAL IMPLICATIONS

1. The mid-distance abundance peak has great theoretical importance. It may provide a useful yardstick by which to measure the extent of edge penetration into forest fragments, and it reveals some of the dynamic processes that occur at the edge-to-interior interface. I have termed this interface the edge-interior ecotone.
2. There is no need to provide conservation managers with an exact distance (to the nearest meter) for the edge-interior ecotone, because by definition it is a fuzzy zone of crossover between edge and interior. The edge-interior ecotone boundary will vary both spatially and temporally, the important factor being the upper limit of variation in edge penetration, which will be the determining factor for core area preservation in the long term. For terrestrial invertebrates in Central Amazonia, a conservative figure for edge penetration distance is 100 m. This is greater than the measured edge penetration distances of most microclimatic factors, which implies that organisms, particularly invertebrates, may be more sensitive indicators of habitat modification than abiotic factors and almost certainly show more complex edge responses.
3. During fragmentation, forest isolation may work in concert with edge effects to drive some populations to extinction. Based on the species abundance patterns observed in this study, I view the edge as a population sink for forest interior species living at the periphery of the core area. These peripheral populations may be maintained indefinitely given a large source population within continuous forest. In isolated fragments, however, the source

population is very limited, and species succumb to the proximate effects of adverse edge conditions and reduced habitat area.

4. The most surprising finding is that an area as large as 100 ha (1 km^2) apparently cannot maintain an intact terrestrial invertebrate assemblage in Central Amazonia. This finding challenges many earlier concepts of minimum habitat area requirements for invertebrates. In the absence of adequate data, I conservatively suggest that the forest area required to preserve the entire local terrestrial invertebrate fauna will be 500–1,000 ha.

ACKNOWLEDGMENTS

I thank J. H. Lawton, N. E. Stork, W. F. Laurance, R. O. Bierregaard, Jr., and five anonymous reviewers for useful comments on an earlier draft of the manuscript. M. D. Tocher, S. J. Hine, and various field technicians at the BDFFP provided assistance in the field. P. M. Hammond gave invaluable taxonomic advice and checked the species sorting. Funding was provided by the Commonwealth Scholarship Commission and The British Council, U.K., The Natural History Museum, London, The Smithsonian Institution, Washington, D.C., the Instituto Nacional de Pesquisas da Amazônia, Manaus, Brazil, the NERC Centre for Population Biology, Silwood Park, U.K., and a University of Canterbury Doctoral Scholarship award, New Zealand. This chapter is publication number 151 in the BDFFP Technical Series.

6

Hyper-Disturbed Parks: Edge Effects and the Ecology of Isolated Rainforest Reserves in Tropical Australia

William F. Laurance

WIND disturbance is an important ecological process in the Tropics, especially in the cyclonic, monsoonal, and hurricane zones from about 7–20° latitude (Whitmore 1975; Lugo et al. 1983; Sousa 1984). Windstorms also affect some equatorial forests, and intense wind blasts from convectional storms can cause severe forest damage in localized areas (Whitmore 1975; Nelson et al. 1994).

When forests are cleared and fragmented, the edges of habitat remnants are typically exposed to increased wind speed, turbulence, and vorticity (Miller, Lin, and Lu 1991; Saunders, Hobbs, and Margules 1991), which may lead to elevated rates of windthrow (Chen, Franklin, and Spies 1992; Esseen 1994) and greater forest structural damage (Laurance 1991b). Winds striking an abrupt forest edge cause an increase in downwind turbulence, resulting in pronounced wind eddies for at least 2–10 times the height of the forest edge (Somerville 1980; Savill 1983; Reville, Tranter, and Yorkston 1990). Greater wind speeds increase the persistence and frequency of wind eddies (Bull and Reynolds 1968), creating an "enhanced gust zone" that often heavily buffets the upper 40% of the forest (Brett 1989).

Increased windthrow and structural damage can negatively affect the biotas and ecology of forest remnants (e.g., Reville, Tranter, and Yorkston 1990; Murcia 1995). In an earlier study using physiognomic and floristic data, I found that many rainforest remnants in tropical Queensland were more disturbed than continuous forest, and suggested that such changes could exacerbate the effects of fragmentation upon forest interior species (Laurance 1991b). Heavy damage to the forest canopy can cause a variety of microclimatic changes in rainforests, including increased temperature variation (Kapos 1989; Williams-Linera 1990b) and shifts in the amount and spectral quality of light reaching the forest floor (Turton 1992). Very small fragments can be devastated by frequent wind disturbance, becoming virtually all edge habitat (Laurance 1991b; Esseen 1994; Malcolm 1994).

In this chapter I compare levels of structural damage in fragmented and nearby continuous rainforest in tropical Queensland. The study described here focused on two isolated forest reserves, Lake Eacham and Lake Barrine National Parks, both about 500 ha in area. These fragments are nearly a century old, and they may yield insights into the fate of similar-sized remnants being created by rapid tropical deforestation elsewhere in the world.

METHODS

Study Area

Lake Eacham and Lake Barrine National Parks (fig. 6.1) are each 490–500 ha in area (Matthews 1993) and are located near the northeastern margin of the Atherton Tableland, a hilly mid-elevation (600–900 m) plateau in northeastern Queensland. These parks are similar in elevation and dominant forest type to nearby Gadgarra State Forest, a large (> 50,000 ha) forest tract contiguous with the extensive forests of the Atherton Uplands. The parks are separated from Gadgarra by a clearing of 1.5–2.5 km in width, containing cattle pastures, second-growth forest, and rural properties. The forests at all three sites are mainly complex mesophyll vine forest on basaltic soils, although simple notophyll vine forests occur in areas with nutrient-poor metamorphic or granitic soils (Tracey 1982). Much of Gadgarra was selectively logged in the 1950s and 1960s, according to Queensland Forest Service records.

Both parks are centered on lakes formed by flooded volcanic craters, which occupy 103 ha at Barrine and 52 ha at Eacham. The parks were initially protected as scenic reserves in 1888, then gazetted as national parks in 1934 (Matthews 1993). Based on Frawley's (1983b) detailed description of land settlement patterns on the Atherton Tableland, both parks are likely to have been isolated from the surrounding forest for 80–90 years.

The parks have experienced some anthropogenic disturbance since isolation. In the early 1900s, both parks were lightly logged for the prized red "cedar" *(Toona australis).* About 15 ha of forest was cleared along the lake margin at Eacham during World War II, then allowed to regenerate subsequently (Matthews 1993). Currently, both parks support recreational facilities for tourists and are bisected by paved roads ranging from 6 to 12 m in width (see fig. 6.1).

Northern Queensland is periodically subjected to high-intensity tropical storms, termed cyclones in Australia. Severe cyclones have struck the Atherton Tableland three times in the twentieth century. The most recent, Cyclone Winifred, on 1 February 1986, achieved maximum wind gust speeds of about 176 km/hour (Unwin et al. 1986). In addition, forests in the region are subjected to strong prevailing winds from the southeast, especially in the late dry season (Partridge 1994).

Habitat Variables and Analysis

From January to September 1994, I and several field assistants recorded fourteen forest physiognomic, edaphic, and landscape features in 10 × 20 m plots. The plots were arrayed either along 200 m "edge transects" situated perpendicular to forest-pasture edges (running from the edge toward the forest interior, comprising twenty plots each); or along 100 m "interior transects" located 200–1,100 m from the nearest forest-pasture edge (comprising ten plots each). Interior transects were accessed by old logging roads,

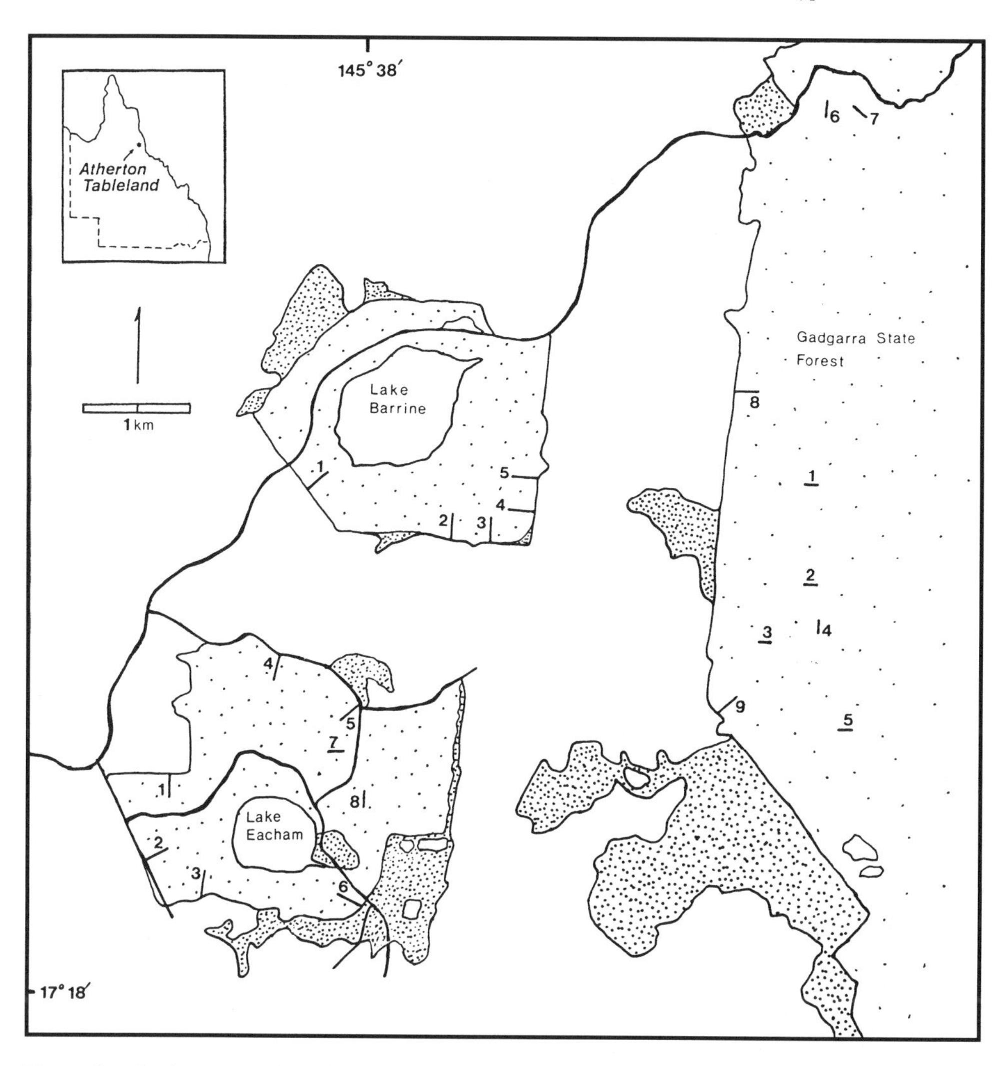

Figure 6.1. Study area on the Atherton Tableland in tropical Queensland, Australia. Study transects are numbered 1–22. Lightly stippled areas are rainforest, darkly stippled areas are 30–50-year-old second-growth forest, and unstippled areas are mostly cattle pastures. Thick wavy lines bisecting the fragments are paved roads.

but were always more than 50 m from the nearest road. Overall, 350 plots were established along twelve edge and ten interior transects (see fig. 6.1). Lake Barrine was sampled with five edge transects, Lake Eacham with five edge and three interior transects, and the control (Gadgarra) with two edge and seven interior transects. Most (20/22) of the transects occurred mainly on basaltic soils.

Eight physiognomic variables were recorded in each plot. Five of these were estimated using a 0–6 ordinal scale: canopy cover (> 15 m height); subcanopy cover (3–15 m height); amount of woody debris on the ground; and the abundances of climbing rattans

(*Calamus* spp.) and lianas (several genera), both of which respond positively to forest disturbance (Laurance 1991b). Canopy cover values were based on the following estimates: 0 = < 30% cover, 1 = 31–50%, 2 = 51–70%, 3 = 71–80%, 4 = 81–90%, 5 = 91–95%, 6 = > 95%. Subcanopy cover was scaled approximately as follows: 0 = < 10% cover, 1 = 11–25%, 2 = 26–40%, 3 = 41–55%, 4 = 56–70%, 5 = 71–85%, 6 = > 85%. Indices for rattans, lianas, and woody debris were on a relative scale designed to encompass the full range of variation encountered in the study area.

The number and size (basal diameter) of treefalls and snapped boles more than 10 cm in diameter and the number of large (> 5 cm) broken limbs also were recorded. The total basal area of treefalls and snapped boles was calculated for each plot by assuming that trees or boles were circular in cross section, then converting diameter estimates to basal areas.

In addition, six edaphic and landscape variables were recorded: distance of the plot from the nearest forest edge; slope (measured with clinometer); aspect; elevation (measured with calibrated altimeter); topographic position (0 = gully, 1 = lower slope, 2 = upper slope, 3 = ridgetop); and an estimate of past logging damage (0–6 ordinal scale), based on the incidence of cut stumps and old logging tracks on or adjoining the plot. Aspect values (compass directions) were converted to a 0–3 ordinal scale that reflected the plot's degree of exposure to prevailing southeasterly winds. All variables were recorded separately in two adjoining 10 × 10 m subplots, then combined to yield a single value for each plot.

Data Analysis

Initially, I used a robust ordination method, Nonmetric Multidimensional Scaling (NMDS: Minchin 1987), to describe major suites of variation in the physiognomic variables. For the analysis I pooled data from individual plots to yield average values for variables on each transect (n = 22 sites). All variables were equally weighted with the "standardization by maximum" method prior to analysis. NMDS was run on the PC-ORD package with all default options, as recommended (McCune 1991).

I then contrasted levels of forest damage in the four main "treatments" (continuous forest interiors, continuous forest edges, fragment interiors, fragment edges), using data from individual plots to increase sample size and hence statistical power. To ensure approximate independence of samples, I selected a random subset of all plots, stratified by transect (n = 105 plots). For interior transects, which were 100 m long, I selected three plots randomly. For edge transects, which were 200 m long, I divided the transect into two 100 m segments, then randomly selected three plots from each segment. Because data were usually non-normal, I used Kruskal-Wallis tests to contrast treatments, followed by Mann-Whitney U-tests to compare sample means.

Finally, I assessed the effects of the six landscape variables on the eight forest structural variables described above, using stepwise multiple regression analyses. For these tests I used both the random subset of plots (n = 105), which described forest variation at a relatively fine spatial scale, and the ordination axes based on transect data ($n = 22$), which described major ecological gradients across the study area.

RESULTS

Ordination of Sites

The ordination of transect data revealed three ecological gradients (table 6.1). Axis 1 discriminated between relatively undisturbed sites with continuous canopy cover and disturbed sites with reduced canopy cover, dense subcanopy cover, and many rattans and broken limbs. Axis 2 described a gradient between disturbed sites with many treefalls, snapped boles, and a high basal area of damaged trees and less disturbed sites with low values for these variables. Axis 3 distinguished sites with continuous canopy cover from those with reduced canopy cover and many rattans, lianas, and dense woody debris.

When sample scores for the transects were plotted, fragment edge ($n = 10$), fragment interior ($n = 3$), and forest edge ($n = 2$) sites were all distinct from interior sites in continuous forest ($n = 7$), especially on axes 1 and 3 (fig. 6.2). *T*-tests revealed that, relative to forest interiors, fragment and edge sites had significantly lower scores on axis 3 ($t = 6.41$, $P < .001$) and moderately higher scores on axis 1 ($t = -1.93$, $P = .091$). Scores on axis 2 did not differ significantly between these sites ($t = -0.27$, $P = .796$). The obvious separation of fragment and edge sites from continuous forest interiors illustrates the general trend toward greater disturbance in fragments and edges.

Comparisons of Fragments and Controls

Kruskal-Wallis tests revealed that three variables, canopy cover, climbing rattans, and lianas, differed significantly between the four main treatments (table 6.2, fig. 6.3). Pairwise comparisons between treatment means generally followed expected trends (table 6.3). Canopy cover was significantly reduced in the edges and interiors of forest fragments relative to continuous forest interiors. Climbing rattans were significantly more abundant in fragment edges and interiors than in continuous forest interiors, and also more abundant in fragment edges than in continuous forest edges. Finally, lianas were more

Table 6.1 Product-moment correlations of rainforest physiognomic features with ordination axes produced by nonmetric multidimensional scaling ($n = 22$ transects).

Variable	Axis 1	Axis 2	Axis 3
Canopy cover	−0.708***	0.329	0.676***
Subcanopy cover	0.718***	−0.460*	−0.253
Climbing rattans	0.595**	−0.005	−0.847***
Lianas	0.035	−0.043	−0.760***
Woody debris	0.169	−0.413	−0.620**
Treefalls/snapped boles	−0.108	−0.907***	−0.018
Basal area of trees/boles	−0.123	−0.887***	−0.126
Broken limbs	0.557*	0.070	0.393

*$P < .05$; **$P < .01$; ***$P < .001$.

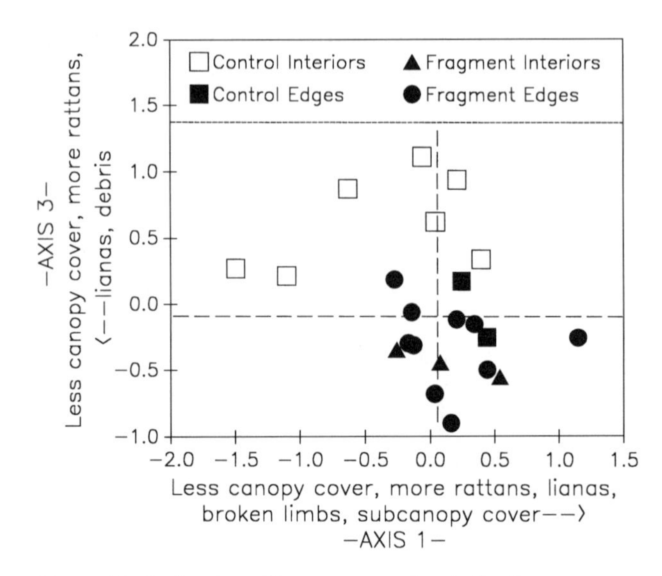

Figure 6.2. Ordination of forest structural variables from twenty-two transects in tropical Queensland. Dashed lines indicate the median value for each axis.

Table 6.2 Comparisons of physiognomic variables between the four treatments using Kruskal-Wallis tests.

Variable	K-W	*P*
Canopy cover	12.13	.033
Subcanopy cover	1.45	.993
Climbing rattans	21.47	< .001
Lianas	22.80	< .001
Woody debris	8.82	.066
Treefalls/snapped boles	7.97	.093
Basal area of trees/boles	7.83	.098
Broken limbs	5.20	.198

Note: The four treatments were continuous forest interiors (n = 21), continuous forest edges (n = 12), fragment interiors (n = 12), and fragment edges (n = 60).

Table 6.3 Pairwise comparisons between the four treatments.

	Comparison[a]					
Variable	CI-CE	CI-FI	CI-FE	CE-FI	CE-FE	FI-FE
Canopy cover	.150	.017	.043	.707	.963	.994
Climbing rattans	.334	< .001	.002	.021	.099	.873
Lianas	< .001	< .001	.003	.143	.330	.838

a. CI = continuous forest interiors, CE = continuous forest edges, FI = fragment interiors, FE = fragment edges. Numbers indicate probability values from Mann-Whitney U-tests.

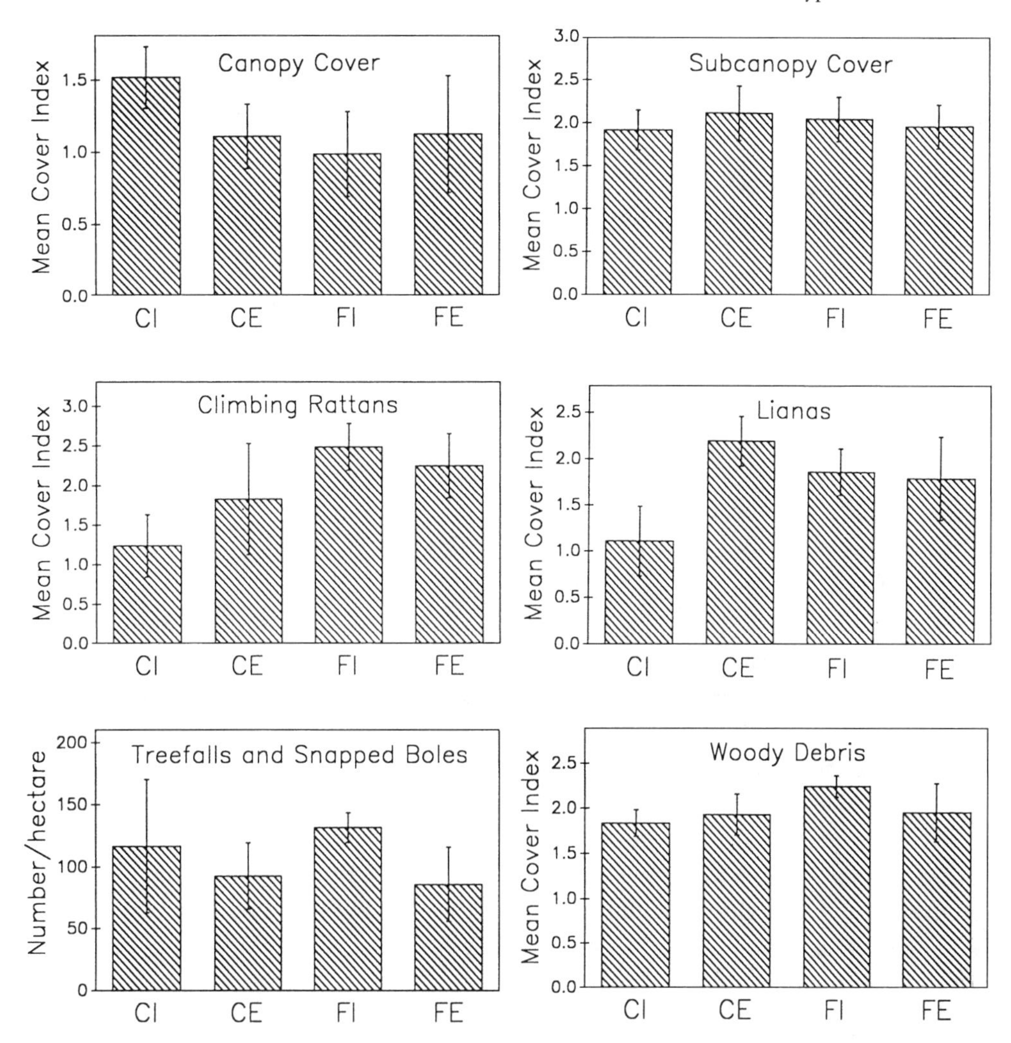

Figure 6.3. Mean estimates ($\bar{X} \pm$ SD) for six key physiognomic variables recorded in fragmented and continuous forest in tropical Queensland. CI = continuous forest interiors; CE = continuous forest edges; FI = fragment interiors; FE = fragment edges.

abundant in fragment edges and interiors than in continuous forest interiors, and more abundant in the edges than in the interiors of continuous forest.

Predictors of Forest Disturbance

Stepwise multiple regressions, using the six edaphic and landscape variables, were employed to search for predictors of forest disturbance (table 6.4). Results were comparable whether data were based on random plots ($n = 105$) or ordination axes ($n = 22$). Correlations between selected predictors were usually nonsignificant ($P > .05$), suggesting that the regression models exhibited few collinearity effects (Sokal and Rohlf 1981).

When random plots were analyzed, canopy cover was found to be influenced by edge distance and elevation, with sites closer to edges and at higher elevations being increas-

Table 6.4 Predictors of forest damage in randomly selected plots (n = 105), using stepwise multiple regression analysis; predictors are also shown for three ordination axes based on transect samples (n = 22).

	Predictors[a]							
Feature	Edge distance	Eleva-tion	Slope aspect	Topog-raphy	Logging	F	R^2 (%)	P
Canopy	+	–		+		2.83	7.8	0.042
Subcanopy		+		–		3.07	5.7	0.051
Rattans	–				–	7.27	12.4	0.001
Lianas	–					17.32	14.4	< 0.001
Debris		+			–	3.87	7.1	0.024
Treefalls			+	+		5.48	9.7	0.006
Basal area			+	+		3.85	7.0	0.024
Limbs						—	—	—
Axis 1	–			–		5.59	37.1	0.012
Axis 2		–				4.01	16.7	0.059
Axis 3	+	–				10.62	52.8	0.001

a. Slopes of axes (±).

ingly prone to disturbance. Subcanopy cover increased at higher elevations, but declined at topographically prominent sites. Climbing rattans and lianas increased near edges, while the number and basal area of treefalls and snapped boles increased on steep slopes and at prominent sites.

Edge distance was the first predictor selected for ordination axes 1 and 3, suggesting that edge effects were substantially responsible for structural differences between fragmented and continuous forest (see fig. 6.2). Elevation was the first and only predictor selected for axis 2 and the second predictor selected for axis 3; in both cases high-elevation sites were more prone to damage. Indices of aspect and past logging were generally weak predictors of forest damage (table 6.4).

DISCUSSION

Wind Disturbance in Fragmented Rainforest

In a previous study of upland rainforest remnants in northern Queensland, I concluded that most of the fragments I surveyed exhibited moderate to heavy structural damage and elevated abundances of disturbance-adapted plants relative to nearby continuous forests (Laurance 1991b). Patterns at Lake Eacham and Lake Barrine National Parks were quite similar. Overall, the fragments had reduced canopy cover and sharply increased abundances of lianas and rattans relative to the interiors of continuous forest. In northern Queensland, the most likely cause of elevated disturbance in forest fragments is wind damage and, in some cases, selective logging (Laurance 1991b; Jenkins 1993). In other regions, anthropogenic disturbances, such as logging, burning, hunting, livestock, and fuelwood gathering, can also be intense in forest remnants (see Viana, Tabanez, and Batista, chap. 23; Smith, chap. 27).

A surprising result was that fragments and continuous forest interiors did not differ in terms of the number (fig. 6.3) and basal area of treefalls and snapped boles, despite indications of elevated disturbance in fragments. However, several factors could have complicated this comparison. First, the continuous forest interiors were significantly steeper ($P < .001$) on average than the fragments (Mann-Whitney *U*-test), and the number and basal area of treefalls and snapped boles increased significantly on steep slopes (see table 6.4). Second, the continuous forests were more heavily and recently logged than the fragments, although the logging index did not appear to be an important predictor of forest damage. Third, intrinsic differences in tree density between sites could complicate comparisons of treefall density. Finally, the proliferation of dense undergrowth, thorny vines, and noxious stinging trees (*Dendrocnide* spp.) in fragments often made access to these sites extremely difficult, and could have obscured a few smaller or older treefalls in fragments.

Multiple regression models (see table 6.4) suggest that proximity of forest edges and topographic factors both influenced indices of forest disturbance. Disturbance was greater at higher elevations and in steep areas, while topographic position also seemed to influence some disturbance indices. Previous studies have demonstrated that rugged topography influences the strength of windstorms and can complicate the effects of winds on forests (e.g., Unwin et al. 1986; Olsen and Lamb 1988; Bellingham et al. 1994; Boose, Foster, and Fluet 1994).

It is noteworthy that the samples in the interior of Eacham were heavily disturbed, despite the location of the study plots 200–680 m from forest edges (see fig. 6.1). The small lakes in the centers of the remnants may have made them more prone to wind disturbance, especially near the lake margins. All but one of the transects, however, were closer to forest-pasture edges than to lake margins, and most (> 80%) interior plots at Eacham were located on lower slopes or gullies, which should not have been prone to wind disturbance. Thus, the heavy disturbance of Eacham interiors seems consistent with my earlier suggestion that wind shear forces may cause elevated structural damage within 200–500 m of fragment edges (Laurance 1991b). A relevant implication of this finding is that even large isolated tracts (2,000–4,000 ha) may be susceptible to increased disturbance, while smaller (< 600 ha) fragments could become exceptionally disturbed (Laurance 1991b).

The interior samples at Eacham were more heavily disturbed than those in continuous forest interiors (see fig. 6.2), despite being situated at comparable distances from forest edge (no interior samples were taken at Barrine). This finding accords with the "additive edge model" of Malcolm (1994), which proposes that the magnitude of edge effects at any point is equivalent to the sum of edge effects originating from all nearby edges, rather than simply the proximity of the nearest edge (Laurance and Yensen 1991). Logically, this model predicts that isolated fragments should be considerably more prone to edge effects than forest tracts having only a single exposed edge (Malcolm 1994).

Disturbance-Adapted Plants

A striking pattern in this study was the marked proliferation of climbing rattans, lianas (fig. 6.4), and other vines and creepers in the two remnants, trends that typify chronically disturbed rainforest (Putz 1984a,b; Tracey 1982; Laurance 1991b; Jenkins 1993). In

Figure 6.4. Examples of (A) undisturbed and (B) heavily disturbed rainforest in northern Queensland. Note the proliferation of rattans and lianas in the disturbed forest.

addition to forest disturbance, however, light-loving plants probably respond positively to other ecological changes in fragments, including (1) the lateral penetration of light near edges; (2) microclimatic changes in fragments, such as increased temperature variability (Kapos 1989; Williams-Linera 1990b), that promote germination in some disturbance-adapted taxa; and (3) an elevated seed rain from disturbance-adapted taxa that proliferate near edges and in surrounding modified habitats (Willson and Crome 1989; Hopkins, Tracey, and Graham 1990; Laurance 1991b). These patterns may not be uncommon; forest remnants in the Atlantic coastal forests of Brazil (Viana, Tabanez, and Batista, chap. 23) and in subtropical areas of eastern Australia (Stockard, Nicholson, and Williams 1985) also exhibit striking proliferations of vines and lianas.

Exotic weeds are far more likely to proliferate in disturbed than intact forest (Humphries and Stanton 1992). Although few floristic data were collected in this study, exotic weeds such as *Lantana camara, Solanum seaforthianum,* and *Rubus alcefolius* were observed in many treefall gaps in the two fragments. Jenkins (1993) identified twenty-seven species of exotic herbs, grasses, vines, and trees growing in gaps in rainforest fragments on the Atherton Tableland. Many of these were ephemeral gap colonists that are quickly overgrown by native species regenerating in gaps, and thus are unlikely to pose a substantial threat to rainforests (see also Olsen and Lamb 1988). Some exotic vines and creepers, however (e.g., *Turbina corymbosa, Solanum seaforthianum, Protoasparagus plumosus*), can blanket and suppress native vegetation and thus may pose a more serious threat (Jenkins 1993). In general, however, rainforests are less susceptible to weed invasions than deciduous communities such as monsoonal forests or tropical dry forests (Hopkins, Tracey, and Graham 1990; Laurance 1991b). In deciduous forests, weeds can often germinate, grow, and reproduce before seasonal canopy closure occurs, progressively altering the floristic composition of the forest (Janzen 1983; Dunphy 1988; Fensham, Fairfax, and Cannell 1994).

Does Disturbance Beget Disturbance?

In chronically disturbed forests, the distribution of treefall gaps may become increasingly contagious in space and time. This can occur for four reasons:

1. When a gap is created, lateral growth of tree crowns at gap margins leaves trees lopsided and prone to falling inward into the gap (Young and Hubbell 1991; Young and Perkocha 1994).
2. Winds frequently form complex vortices in large gaps, resulting in additional windthrow and forest damage (Reville, Tranter, and Yorkston 1990; Thiollay 1992).
3. The proliferation of heavy lianas in disturbed forest can predispose trees to wind damage. In Malaysian and Panamanian forests, liana-infested trees are increasingly susceptible to mechanical damage from strong winds and are more likely to damage adjacent trees if they are felled (Appanah and Putz 1984; Putz 1984a,b).
4. Areas that are regularly disturbed favor short-lived successional trees, which experience relatively rapid mortality and turnover, creating additional gaps (Lieberman et al. 1985).

Consequently, in regions that experience strong winds or other chronic disturbances (see Viana, Tabanez, and Batista, chap. 23), forest remnants may become both heavily disturbed and increasingly prone to subsequent disturbances.

Disturbance and Fragment Biotas

Altered disturbance regimes in fragments are likely to have diverse effects on faunal and floral communities. Floristic changes in isolated stands are expected, with increases in the prevalence of short-lived successional tree species as well as lianas, vines, creepers, and exotic weeds. Long-lived canopy trees, epiphytes, and other taxa adapted to mature forest may progressively decline (e.g., King and Chapman 1983).

Physiognomic and microclimatic changes are inevitable in heavily disturbed remnants. Vegetative cover in the ground and subcanopy layers will increase sharply as gap colonists, vines, and saplings respond to increased light availability in treefall gaps. Resources such as flowers, fruits (Lovejoy et al. 1986), rotting logs, and woody debris (Malcolm, chap. 14) will often increase. Increased temperature variability, reduced humidity (Kapos 1989), and marked shifts in the amount and spectral quality of light reaching the forest floor (Turton 1992) are expected, especially in large gaps. Large-scale deforestation could even alter a region's evapotranspiration budget (Ng 1983; Salati 1985; Saunders, Hobbs, and Margules 1991), possibly resulting in atypically dry winds that could desiccate fragments and cause increased tree mortality (Lovejoy et al. 1986; Kapos 1989).

Faunal communities are likely to respond to the myriad ecological changes in fragments in complex ways (e.g., Bierregaard et al. 1992). Species that favor edges and disturbed forest often increase in fragments (Laurance 1990, 1994). For forest interior species, however, elevated disturbance may act synergistically with habitat loss and isolation, exacerbating the effects of fragmentation (Laurance 1991b). In Queensland, among microchiropteran bats, for example, species adapted for hovering or fluttering in the understory decline in disturbed forest, while fast-flying open-forest species increase (Crome and Richards 1988). The lemuroid ringtail possum *(Hemibelideus lemuroides),* which feeds almost exclusively on the leaves of mature rainforest trees, rapidly declines in fragmented (Laurance 1990, 1991a) and recently logged (Laurance and Laurance 1996) forest, and appears highly sensitive to habitat disturbance. Microhabitat changes associated with disturbance strongly influence small mammal communities, causing species such as *Rattus leucopus, Melomys cervinipes,* and *Antechinus flavipes* to increase near edges and in disturbed habitats, while *R. fuscipes* and possibly others decline (Laurance 1994; see also Malcolm, chap. 14). Many invertebrates are also sensitive to microclimatic changes associated with forest edges (Didham, chap. 5; Brown and Hutchings, chap. 7; Weishampel, Shugart, and Westman, chap. 8).

Such diverse ecological changes could easily cause higher-order changes or "ripple effects" in fragment communities. Omnivorous rodents, for example, which are important predators on small vertebrates (Laurance, Garesche, and Payne 1993; Laurance and Grant 1994), insects, and large seeds (Harrington et al., chap. 19), increase in some fragments (Laurance 1994). Competitive interactions between ecologically similar species may intensify in the small resource base of a forest fragment, with tolerance of disturbance strongly influencing which species persist or decline (Laurance 1994). Given the diversity of ecological interactions in tropical ecosystems, the direction and magnitude of higher-order changes in fragmented forests may be exceedingly difficult to predict (see Terborgh et al., chap. 17; Harrington et al., chap. 19).

Management Options

Some small remnants have high conservation value, often because they protect rare vegetation types or harbor locally endemic species (Shafer 1995). Under such circumstances intensive management to reduce wind disturbance may be justified (e.g., Sheperd 1994). Wind tunnel experiments suggest that the most effective way to reduce gust intensity is to increase the "roughness" of clearings surrounding fragments. This can be achieved by planting belts of fast-growing trees at right angles to prevailing winds along the tops of ridges (Brett 1989). Permeable windbreaks are generally preferable because they cause less downwind turbulence than dense, impermeable windbreaks (Wakefield 1989).

General Implications

1. In tropical and subtropical regions subjected to strong winds or windstorms, fragmented forests may be unusually prone to chronic disturbance. Anthropogenic disturbances, such as logging, hunting, burning, livestock, and fuelwood gathering, can also be intense in forest remnants.
2. Wind turbulence apparently can damage vegetation several hundred meters downwind of forest edges. Even large (2,000–4,000 ha) remnants may be vulnerable to increased wind disturbance, while smaller (< 600 ha) fragments may become hyper-disturbed.
3. Chronic wind disturbance may result in ecological changes in fragments that render them increasingly prone to future disturbance. In particular, the distribution of treefall gaps may become increasingly contagious in space and time.
4. Elevated levels of disturbance may cause diverse changes in the structure, floristic composition, and microclimate of forest remnants.
5. Faunal communities respond to elevated levels of disturbance in complex ways. For forest interior species, increased disturbance is likely to exacerbate the effects of habitat loss and fragmentation.
6. Windbreaks can be used to reduce the effects of wind disturbance in small reserves of high conservation value.

ACKNOWLEDGMENTS

J. H. Connell, P. Green, D. Wilson, J. R. Malcolm, and R. O. Bierregaard, Jr., commented on earlier drafts of the manuscript. E. Sellers, D. Head, E. Scambler, J. Ledger, J. Hall, R. Van Raders, C. Walsh, B. Walsh, and S. Comport kindly assisted with fieldwork. The Queensland Department of Environment and Heritage and Queensland Forest Service allowed access to study sites. This study was supported by a senior research fellowship awarded by the Wet Tropics Management Authority, Cairns, Australia.

Section III
Tropical Forest Faunas

INTRODUCTION
Fragmentation and Forest Fauna

FAUNAS—especially vertebrates—have always been popular subjects among students of habitat fragmentation, for at least two reasons. First, many animal species are both vulnerable to fragmentation and respond quite rapidly to habitat change. Endothermic vertebrates such as birds and mammals—especially species at higher trophic levels—have substantial area and energy needs, which often render them vulnerable to habitat insularization. Many smaller species are also sensitive to fine-scale habitat structure and forest microclimate, and hence may respond strongly to edge effects and other ecological changes in fragments.

Second, faunas—again, usually vertebrates—have been invaluable figureheads for habitat conservation initiatives. It is undeniable that the lay public responds positively to "charismatic megavertebrates"—be they Neotropical monkeys, lemurs in Madagascar, whales, or the Australian cassowary. As the World Wide Fund for Nature discovered in the mid-1980s, the public was far more strongly moved by calls of "Save the primates!" than "Save the rainforests!" That these two objectives are largely synonymous seemed to matter little, so while WWF's field programs addressed habitat or ecosystem preservation, their public appeals focused on charismatic animals.

Despite their unifying focus on fauna, the eleven chapters in this section are highly eclectic, spanning a diversity of taxa, biogeographic regions, and research questions. It is noteworthy that four of the chapters describe studies of birds, which, because of their conspicuous nature and diversity, have always featured strongly in ecological and biogeographic research. There are also four chapters from the Biological Dynamics of Forest Fragments Project (BDFFP) in Brazil, currently the only integrated, long-term study of forest fragmentation in the Tropics.

Invertebrates

In chapter 7, Keith Brown and Roger Hutchings provide an overview of the effects of fragmentation, forest clearing, and other disturbances on Neotropical butterflies. Their summary is based on more than fifteen years of fieldwork in the BDFFP reserves and at other sites in South America. Among their many interesting conclusions is that light-loving butterflies from outside modified habitats can penetrate at least 250 meters into forest remnants—a subtle but nonetheless telling edge effect, as butterflies are excellent ecological indicators of forest disturbance. In general, their results reveal that fragmentation and forest clearing can dramatically alter invertebrate communities, with some species declining or disappearing and others becoming hyperabundant. Because of the strong effects of local microhabitat heterogeneity on butterfly assemblages, Brown and Hutchings advocate a reserve design strategy that involves stratifying multiple reserves—both large and small—across major environmental gradients to maximize the conservation of tropical invertebrate communities.

Chapter 8, by John Weishampel, Herman Shugart, and the late Walter Westman, focuses on a single invertebrate species, a centipede *(Rhysida nuda)* endemic to the rainforests of northern Queensland. Their study is based on the assumption that environmental stresses can cause an animal to exhibit developmental instability during growth, leading to "fluctuating asymmetry"—deviations in the symmetry of bilateral characters, such as a difference in the shape of the left and right antennae. They found that measures of fluctuating asymmetry did in fact increase in fragments relative to large forest tracts, especially in smaller and more irregularly shaped fragments, which are susceptible to such edge effects as increased desiccation. Their findings suggest that microclimatic changes in smaller fragments could indeed lead to increased developmental instability in invertebrate populations, possibly lowering their fitness and increasing the likelihood of local extinction.

Frogs

In chapter 9, Mandy Tocher, Claude Gascon, and Barbara Zimmerman describe another long-term study from the BDFFP, this time on frog communities. As in many BDFFP projects, they compared the composition and diversity of species at the same forest sites before and after fragmentation. Their results, however, were quite unexpected. Instead of finding a reduction in frog diversity in fragments, they found exactly the opposite—frog diversity increased consistently, with fragments gaining an average of ten species each. This increase seemed to result from two factors: many rainforest frogs are unexpectedly resilient to fragmentation, at least over the short term, and fragments are often invaded by generalist frog species that flourish in the surrounding habitat matrix. Taken in context with the other BDFFP studies, their findings demonstrate just how varied the responses of different faunal groups can be.

Birds

The next four chapters all focus on bird communities. In chapter 10, Richard Bierregaard and Philip Stouffer provide a general overview of the BDFFP project, then describe a massive, long-term mark-recapture program to study the effects of fragmentation on understory birds. Their work reveals the highly dynamic nature of bird assemblages throughout the process of forest fragmentation and regeneration of second-growth forest around some once-isolated reserves. In general, many understory insectivores and some frugivores are highly vulnerable to fragmentation, but a number of species will recolonize reserves if they are surrounded by mature second growth. These results highlight the key role of the modified matrix: reserves encircled by denuded pastures or croplands are likely to suffer far more severe changes than those surrounded by mosaics of habitat types that include corridors of mature second growth.

In chapter 11, Kathryn Sieving and James Karr explore factors associated with the vulnerability of forest birds on Barro Colorado Island, Panama (BCI). BCI is a land bridge island, isolated from the surrounding forest over eighty years ago by the inundation of Gatun Lake, part of the Panama Canal. Since isolation, many bird species formerly present on the island have disappeared. Sieving and Karr use ecological, natural history, and genetic data to search for mechanisms driving the highly varied responses of nine species of understory birds to the isolation of BCI. Their tightly reasoned analysis suggests that fecundity and adult survival rates strongly influence bird persistence, and provides a model for those who would aspire to conduct rigorous, integrative research on species' extinction mechanisms.

Chapter 12, by Carla Restrepo, Luis Renjifo, and Paul Marples, is one of the most unusual chapters in this volume. Working in the northern Andes of South America, the authors use a novel statistical technique to identify the structure of "lumps"—aggregations of species of similar body mass—among frugivorous birds inhabiting a wide variety of natural and modified landscapes. They appear to have discovered some interesting patterns. In heavily modified landscapes, for example, there is often a tendency for frugivore assemblages to lose those lumps representing the largest and smallest species. In addition, dramatic changes in lump structure among sites were always associated with dramatic changes in landscape pattern. Although these methods are still exploratory in nature, they merit further consideration by community ecologists, because the lump structure of assemblages appears to capture information not provided by simple measures of species richness or diversity. We await the development of rigorous statistical methods with which to compare the lump structures of different communities.

In chapter 13, Neil Warburton describes the structure of forest avifaunas in over thirty rainforest fragments and additional unfragmented sites in northern Queensland. He uses a nested-subsets analysis to demonstrate that bird assemblages in small or low-diversity reserves often constitute a subset of the species found in larger reserves. Based on these analyses, he considers the implications of this pattern for bird conservation. He also categorizes the responses of sixty bird species to fragmentation and analyzes fifteen ecological and life history traits to identify correlates of local extinction proneness. His analysis suggests that two traits—the presence of a species in certain matrix habitats (riparian strips and planted windbreaks) and its natural abundance—are the most effective predictors of resilience. As did Bierregaard and Stouffer (chap. 10), Warburton suggests that species assemblages in fragments could be strongly influenced by the nature of the surrounding matrix.

Mammals

The next three chapters are wholly or mainly concerned with small mammal communities. In chapter 14, Jay Malcolm assesses the richness, abundance, and biomass of small mammals (rodents and marsupials) in a series of BDFFP fragments and control sites. Malcolm's project is unique because his exhaustive trapping study included not only the forest floor, but a nearly equal effort in the forest canopy. His results suggest that fragmentation alters the terrestrial small mammal fauna, which was increasingly abundant and diverse in smaller fragments, but does not affect the canopy fauna. Notably, few species disappeared from fragments. Like that of certain birds and frogs, the robustness of small mammals to fragmentation seems to result from their ability to exploit second-growth forest surrounding fragments.

In chapter 15, Antony Lynam examines dynamics and local extinction in small mammal communities on recently created land bridge islands in Thailand. The islands he studied had been isolated by a vast hydroelectric reservoir only five years earlier. In the subsequent three years Lynam documented many ecological changes on the islands, including the disappearance of one-fifth of the forest species and a dramatic increase in the exotic house rat *(Rattus rattus).* The islands also became increasingly disturbed by fires set by local fishermen and by logging, resulting in expansion of bamboo thickets and dry, semi-deciduous forest at the expense of evergreen forest. The rapid ecological changes documented by Lynam seem ominous, given ongoing deforestation and the highly fragmented state of many remaining forests in Thailand.

In chapter 16, Miriam Goosem reviews her own and other related research on the effects of linear clearings on rainforest fauna. Linear clearings are created whenever roads, highways, powerlines, or pipelines bisect forests, and they are surely one of the most ubiquitous features of the modern landscape. Working in northern Queensland, Goosem used mark-recapture methods to assess the effects of roads and powerline clearings on the movements of small mammals. A key conclusion was that while smaller (< 15 m wide) clearings appear to have only moderate effects on mammal movements, the effects of larger (> 50 m wide) clearings are far more drastic. Goosem also assessed road mortality of vertebrates along a busy rainforest highway and found that many hundreds of amphibians, reptiles, birds, and mammals are killed per kilometer each year. In her summary, Goosem considers a number of management strategies that could help to mitigate the effects of linear clearings in rainforests.

MECHANISMS OF FAUNAL COLLAPSE

Chapter 17, by John Terborgh and his colleagues, explores the dynamics of faunal extinction on land bridge islands created in 1986 by a large hydroelectric impoundment in Venezuela. As did Lynam (chap. 15), they discovered that smaller islands quickly became impoverished after isolation, and were missing many birds, mammals, reptiles, amphibians, and invertebrates present in mainland areas. Some species, however, that persisted on islands became dramatically hyperabundant. In an intriguing mechanistic hypothesis, Terborgh and his co-workers suggest that the rapid loss of some species and the hyperabundance of others following isolation may create "gross ecological distortions." These ecological imbalances, they suggest, are likely to drive the loss of additional species in habitat isolates. Such changes may occur rapidly after fragmentation, so that the fragment passes through a series of unstable "transient states" until, eventually, it reaches a biologically simplified equilibrium. If their hypothesis is correct, ecological distortions and higher-order effects may play a far more important role in the collapse of insular faunal communities than has previously been suspected.

SYNTHESIS

Despite the varied nature of the studies, there are some recurring themes and patterns in these eleven chapters. The first is that faunal populations often respond rapidly to habitat fragmentation. Invertebrate communities, in particular, can change very quickly (chap. 7), both because of their short generation times, which lead to rapid population processes, and because of their small size and ectothermic nature, which probably render them sensitive to environmental changes in fragments (chap. 8). Vertebrate populations may also change with surprising rapidity, as illustrated by the dynamics of Amazonian bird communities in recently fragmented forests (chap. 10) and the sudden disappearances of vertebrates from newly isolated islands in Thailand (chap. 15) and Venezuela (chap. 17).

A second theme is that the most vulnerable species are often those that avoid the matrix of modified habitats surrounding fragments (chaps. 9, 10, 13, and 14), or respond negatively to edge effects or other ecological changes in fragments (chaps. 7, 8, 10, and 15). Indeed, primary forest specialists commonly exhibit both of these traits. Populations of matrix-avoiding species are often completely isolated within fragments, while those that tolerate or exploit the matrix can be bolstered by the genetic and demographic contributions of immigrants, providing a buffer against local extinction. Matrix-tolerant species can also recolonize fragments should their populations disappear from those areas.

A third trend is that populations on recently created land bridge islands (chaps. 11, 15, and 17) appear to behave differently from those in fragments. Such differences need to be recognized explicitly, but are sometimes glossed over by investigators, possibly because of the historical importance of the theory of island biogeography (MacArthur and Wilson 1967) in studies of habitat fragmentation. Islands are surrounded by an inhospitable matrix (water), which for many terrestrial species is a nearly complete barrier to dispersal. Forest fragments, however, are usually embedded in a mosaic of degraded and semi-natural habitats that can be tolerated or exploited by some forest species, as well as by generalists and exotics.

Differences between fragments and islands can lead to very different effects on faunal populations. For example, small mammal populations on recent land bridge islands in Thailand (chap. 15) seem far more prone to local extinction than those in recent Amazonian fragments (chap. 14). This difference is probably explained by the differences in the remnants' surroundings: many of the Amazonian fragments were encircled by regrowth forest, which supported abundant small mammal populations, while the islands were much more strongly isolated and lacked the productive matrix. Similarly, conclusions regarding the ecological traits associated with extinction proneness in birds differed markedly between Barro Colorado Island (chap. 11) and forest fragments (chaps. 10 and 13).

Researchers studying faunal communities in fragments often emphasize the importance of matrix tolerance and edge effects in determining vulnerability, while those working on islands may emphasize traits such as body size, temporal population stability, or vulnerability to predation. These varied perspectives are at least partly attributable to profound differences in the ecology of islands and fragments.

7

Disturbance, Fragmentation, and the Dynamics of Diversity in Amazonian Forest Butterflies

Keith S. Brown, Jr., and Roger W. Hutchings

To an orbiting satellite or motorized tourist, the structure, function, and points of interest of a tropical forest may seem relatively homogeneous over large areas. A flying butterfly, however, views the same forest as a series of discontinuous patches of variable suitability; the size of these patches is proportional to the butterfly's degree of specialization, acuity of perception, and speed of flight. This patchiness is based on the irregular distribution of essential resources such as light, heat, chemicals, food, mates, and shelter.

A given site of 3–10 km^2 in Amazonian forest is likely to contain 600 to 1,600 species of diurnal Lepidoptera (Brown 1972, 1984, 1991; Robbins et al., in press), each with its own special requirements for physical, chemical, and biological resources. Many biologists have struggled with the question of how to recognize coherent patterns in such excessively complex and diverse communities. Some broad patterns in Amazonian butterfly ecology have been discussed by Bates (1862: mimicry, 1863: spatial variation), Brown and Benson (1974, 1977), and Brown (1979, 1984, 1987a,b: biogeography, endemism, and diversity), while microstructural details of butterfly communities have been described by L. E. Gilbert (1980), Callaghan (1983), and Brown (1991). Edge, area, and other effects of disturbance, fragmentation, and deforestation on tropical butterfly assemblages have been detailed in Lovejoy et al. (1984, 1986), Brown (1991, 1992, in press), Brown and Brown (1992), and (in Mexico) Raguso and Llorente-Bosquets (1992). Most studies agree with the observations of Owen (1971, 1975) in Africa, who suggests that mild disturbance or fragmentation increases the abundance and diversity of tropical butterflies due to edge effects (see Didham, chap. 5).

This chapter reports some trends observed in a rather small (> 500 species) community of true butterflies (excluding skippers) followed during more than 2,000 hours of observations and experiments in low-productivity, botanically diverse forests in the central Amazon. Preliminary data on butterfly assemblages in undisturbed, isolated, and fragmented forest patches in this region (Brown 1984, 1987a,b, 1991; Lovejoy et al. 1984, 1986; Otero and Brown 1986; Hutchings 1991) are here updated and expanded. Special emphasis is placed on patterns observed during and after anthropogenic disturbance and on their implications for forest management.

DATA COLLECTION

True butterflies (Papilionoidea) were censused from 1980 to 1995 in twenty-five reserves (ten of 1 ha, eight of 10 ha, four of 100 ha, and three of 1,000 ha) in the Biological Dynamics of Forest Fragments Project (BDFFP) about 70 km north of Manaus, Amazonas, Brazil (see Bierregaard and Stouffer, chap. 10, for further details on study areas and the experimental design of the BDFFP). Isolation status and exceptional events during the study, such as the creation of large treefall gaps, destructive storms, unusual abundances of spider or bird predators, and long dry seasons, were recorded for each reserve. Hesperioidea (skippers) were very rarely seen in the forest understory or even on forest edges, probably due to the rarity of flowers in these habitats, and thus were not included in the censuses.

The analysis below was based on a total of 161 daily censuses of butterfly species richness conducted before and after forest fragmentation. A census was occasionally spread out over two or three successive days to attain at least five hours of effective observation. Over half of the census lists were recorded during the process of partial (one to three sides exposed) or complete isolation of twelve forest reserves (six of 1 ha, four of 10 ha, and two of 100 ha). The isolation process varied with each reserve, but usually progressed from continuous forest through peripheral removal of undergrowth, tree felling and burning, and then to pasture formation around the reserve. Poor burns usually led to pasture failure and dense regrowth of *Cecropia* spp., *Vismia* spp., *Bellucia* spp., *Solanum* spp., and other early successional bushes and trees.

Each daily census list was compiled during a walk of about 5–20 km along internal trails and edges of a single reserve. A small reserve was covered repeatedly during a day's work, while a large reserve was censused extensively, with few points being visited more than twice (thus covering a more varied array of habitats). Censuses normally ran from dawn to dusk in order to include crepuscular species and represent a similar effort in each daily sample (Brown 1972). Pauses were made at points of greater butterfly activity (such as clearings, ridgetops, rivulets, sunlit edges, flowers, baits, and army ant swarms) until no new species were recorded for five to ten minutes.

All butterflies detected were identified with the aid of binoculars and a field key if possible; difficult-to-identify lycaenids and satyrines were captured and examined in the hand, and occasionally retained for later identification (specimens were deposited at INPA in Manaus, with duplicates in other major Brazilian collections). During the previous evening or in the early morning, fermented banana baits were placed in well-illuminated understory or eye-level positions. Pyrrolizidine alkaloid baits for Ithomiinae (such as drying *Heliotropium indicum* plants) were hung in shaded but open understory locations.

Each daily census list included the numbers and sexes of each butterfly species observed; the weather conditions (because butterflies are sensitive to inclement weather, sunny periods were counted as fully effective hours of observation, cloudy periods added one effective hour per two hours of observation, and rainy periods were not included in the total observation effort); the isolation status of the reserve; and notes on any clearings, special resources, preferred habitats, and activity times and levels of adults or larvae. Additional data on butterfly communities were extracted from nonstandardized species lists and samples made in many of the BDFFP reserves from 1981 to 1984 by Francisco A. da Silva and from intensive biweekly inventories of three 1 ha areas made by R. W. H. in 1986–1988.

Genetic variation was analyzed in irrupting populations through a number of marker genes studied previously (Brown and Benson 1974; Brown 1979, 1980; Sheppard et al. 1985). Population size, individual movements, and survival of butterflies were studied by writing numbers on the wings of individual butterflies with various colors of indelible ink.

RESULTS

Butterfly Distribution in Reserves

The 161 daily lists yielded about 110,000 butterfly records in 1,239 census hours. A total of 455 butterfly species were recorded altogether (table 7.1, fig. 7.1). Many species were rarely encountered: 88 (19%) were recorded in only one of the twenty-five reserves (78 as single individuals), 54 in only two reserves, and almost half of the species (214) in five or fewer reserves (fig. 7.2). In addition, over a hundred common, widespread Neotropical forest butterfly species were never recorded in the reserves, despite their usual conspicuousness in high-productivity or open forests, including sites less than 50 km from the study area. As the sampling effort was spread out over twenty-five reserves, with no more than 90 hours spent in any one (see table 7.2), the individual lists are still far from

Table 7.1 Categories of butterfly species recorded in the Biological Dynamics of Forest Fragments Project reserves.

	Number of species (% of total species in group)					
			Ecological groups			
Taxonomic groups	Total	General forest fauna	Understory shade	Understory sun	Canopy/ clearing	Edge
Nymphalidae						
Danainae	2	1(50)	—	1(50)	—	1(50)
Ithomiinae	19	13(68)	15(79)	4(21)	—	—
Morphinae	9	7(78)	2(22)	3(33)	4(44)	—
Brassolinae	12	5(42)	11(92)	1(8)	—	—
Satyrinae	44	21(48)	28(64)	7(16)	—	9(20)
Charaxinae	24	5(21)	—	8(33)	13(54)	3(13)
Nymphalinae	47	12(26)	2(4)	25(53)	4(9)	16(34)
Heliconiinae	24	9(38)	—	7(29)	10(42)	7(29)
Pieridae	8	4(50)	1(13)	—	4(50)	3(38)
Papilionidae	11	—	1(9)	3(27)	2(18)	5(45)
Lycaenidae						
Theclinae	73	13(18)	25(34)	21(29)	8(11)	19(26)
Riodininae	182	43(24)	23(13)	125(69)	22(12)	12(6.6)
Totals	455	133(29)	108(24)	205(45)	67(15)	75(16)

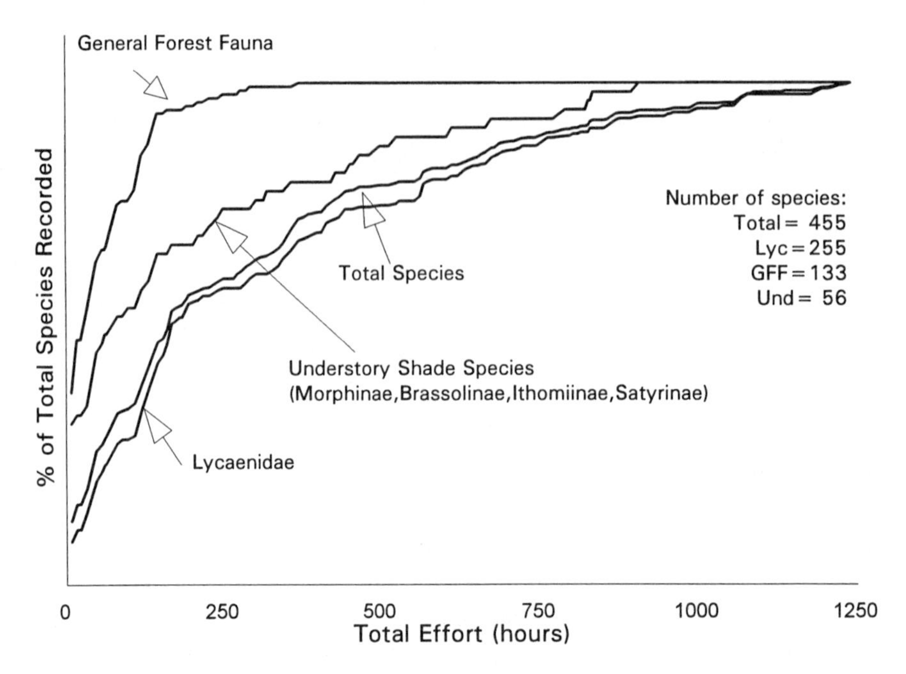

Figure 7.1. Species-sampling effort curves for all butterfly species and for three species groups in twenty-five reserves of the BDFF Project, 1980–1995. To permit species accumulation patterns to be more easily compared, all four curves have been standardized to the same maximum value. The GFF curve includes the first records for all species found in thirteen or more reserves by the end of the sampling; if only the time of recording in the thirteenth reserve is plotted, the curve takes on a near-linear shape.

saturation. Thus, stochastic factors are especially troublesome in direct comparisons and conclusions derived from samples at the daily or reserve level.

A total of 133 species were widespread in the forest fragments (found in thirteen or more reserves) and are termed the "general forest fauna." An important component of this group is the conspicuous, shade-loving understory fauna (48 species), which included many species that were little known from previous inventories in the region. Both the understory shade and general forest faunas exhibited species accumulation curves that were much steeper than those for the total species list or for the highly diversified small Lycaenidae (figs. 7.1, 7.2).

One-sixth of all species recorded (75) were restricted to the edges of isolated reserves, and probably depend on these regionally rare habitats for survival and reproduction.

Area and Isolation Effects

Cumulative species lists for individual reserves contained only a fraction of the total species list; only one reserve (100 ha 3304) included over half the species (after a total of 89 census hours: table 7.2). Even the three 1,000 ha reserves combined (86% of the total area sampled) had only 304 species after 220 hours of sampling; this was fewer species than were recorded in the four 100 ha reserves (329 species in 260 hours), the eight 10 ha reserves (348 in 360 hours), or even the ten 1 ha reserves (327 species in 399

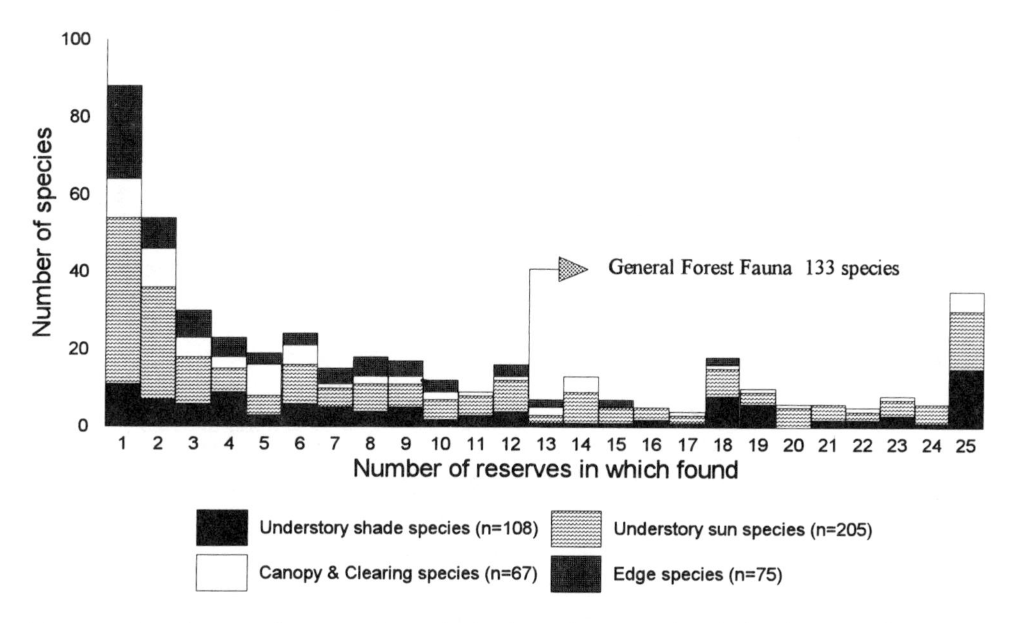

Figure 7.2. Distributions of species richness for 455 butterfly species identified in the twenty-five BDFFP reserves.

hours), which constituted only 3% of the total area. These totals are mostly due to a smaller total sampling effort in the larger reserves (compare species-time curve in fig. 7.1), with only a minor contribution of area.

When daily reserve lists were pooled by degree of forest continuity and observation effort (totaling 30–44 hours), four distinct reserve classes could be distinguished: (1) continuous forest without internal clearings at time of sampling; (2) continuous forest with internal clearings at time of sampling; (3) completely isolated reserves; and (4) and reserves surrounded by high second growth (table 7.2). As shown in figure 7.3, areas inside continuous forest that lacked large internal clearings (solid circles) had lower total species lists, even with greater sampling effort, than those that contained large internal clearings (open circles). Completely isolated reserves (solid triangles) often had lists as small as those for unbroken forest. However, if they had been at some time surrounded by high regrowth forest, the lists were much larger and more like those of reserves with large clearings (fig. 7.3, open triangles).

Area effects (i.e., increasing species richness with greater reserve area) were modest to negligible for many butterfly families (fig. 7.3). The Lycaenidae had the most notable area effect; however, each tenfold increase in reserve size added only about 10–15 species to the base of 40–65 lycaenids.

As sampling effort was increased, large reserves showed a greater increase in species richness than small reserves (see pairs of partial and total lists in same reserve, table 7.2). However, the richness of some groups, notably forest understory Satyrinae and Ithomiinae and light-loving species (fig. 7.3), was affected little by reserve area and sampling effort (most species were detected in the first 30 census hours).

Table 7.2 Reserve lists: Numbers of butterfly species recorded in each Biological Dynamics of Forest Fragments Project reserve, by taxonomic and ecological groups.

Reserve[a]					Number of species																		
				No.	Taxonomic groups[c]													Ecological groups					Sng
Size	No.	Hr	Cond.[b]	lists	Mor	Bra	Sat	Dan	Ith	Hel	Nym	Cha	Pie	Pap	The	Rio	Σ	Shade	Sun	Clear	Edge	GFF	Rec[d]
1 ha	1110	34	F	6	4	3	15	1	10	7	10	6	3	0	14	41	114	40	57	15	2	90	1
	2107p	31	F	3	3	4	20	0	10	7	16	3	4	1	13	33	114	41	50	15	8	86	
	1105	29	Fc	7	6	4	17	1	10	8	14	4	3	1	23	46	137	45	67	20	5	99	2
	1109	28	Fc	6	9	4	20	0	11	8	14	5	5	1	10	39	126	47	56	19	4	99	0
	1113	39	Fc	7	5	5	18	0	13	9	14	9	3	1	13	47	137	45	67	22	3	101	1
	2107	73	F/SI	9	4	5	27	0	10	10	20	5	4	1	20	45	151	46	67	18	20	106	5
	2108	58	F/I	9	3	3	24	1	11	11	8	2	4	1	17	35	120	37	48	14	21	82	5
	1104	42	SI	8	3	3	23	0	9	14	8	4	4	3	18	57	146	37	64	21	24	92	2
	1111	29	SI	5	6	1	18	0	13	8	11	11	5	5	14	45	137	42	64	19	12	96	2
	1112	30	SI	5	2	1	21	1	9	11	11	7	5	2	18	39	127	36	55	12	24	86	4
	3114	37	SI	5	1	1	23	0	10	12	18	6	5	1	24	39	140	41	58	17	24	90	5
	2107p	42	I	6	3	3	18	0	8	10	12	4	4	1	16	27	106	30	45	14	17	81	
	2108p	37	I	6	3	1	20	1	7	8	6	2	4	0	12	24	88	26	34	12	16	62	
Subtotals[e]		399		67	9	8	38	1	15	19	37	19	7	8	57	119	337	83	149	48	57	133	27
10 ha	1205	30	F	6	6	2	14	1	9	8	14	4	4	1	12	45	120	36	65	16	3	90	1
	1210	52	F	6	5	2	27	0	11	5	15	6	3	2	17	51	144	52	66	19	7	102	1
	1210p	35	F	4	5	1	24	0	9	5	11	5	3	2	17	44	126	44	60	15	7	91	
	1204	33	Fc	5	6	6	15	0	13	11	15	10	3	0	18	58	155	48	75	25	7	111	2
	1208	44	Fc	5	4	6	24	1	12	6	18	7	3	1	19	63	164	57	80	21	6	110	4
	3209	45	F/SI	5	3	5	23	1	13	17	19	6	6	2	30	62	181	56	83	18	30	110	7
	2206	52	F/I	8	4	2	23	1	11	8	14	5	3	1	13	50	135	43	66	11	15	92	2
	1202	59	SI	8	3	4	29	1	15	14	20	5	5	4	30	67	197	54	86	22	35	109	7
	1202p	40	SI	5	3	3	29	1	12	12	17	5	4	4	23	58	171	43	75	20	33	101	
	1207	45	SI	6	6	8	23	0	12	13	12	10	4	2	22	66	178	55	79	26	18	113	2
	1207p	35	SI	5	4	2	21	0	11	13	12	5	4	2	20	65	159	44	75	22	18	103	
	3209p	34	SI	3	3	3	22	1	11	17	18	6	6	2	26	55	170	50	73	18	29	101	
	2206p	36	I	6	4	2	23	1	11	8	14	5	3	1	13	44	129	41	62	11	15	90	

Subtotals		360		49	7	8	39	1	18	20	37	15	8	5	52	140	350	94	158	49	49	133	26
100 ha	1301	53	F	8	6	6	22	0	13	10	10	9	3	3	16	60	158	50	76	23	9	113	3
	1301p	32	F	5	6	5	19	0	13	9	6	8	3	3	11	49	132	45	62	20	5	102	
	2303p	34	F	3	4	2	18	1	10	10	11	6	3	4	11	52	132	38	63	20	11	88	
	1302	46	Fc	4	9	6	21	1	11	9	18	9	4	0	22	74	184	62	92	26	4	117	0
	1302p	32	Fc	3	8	4	18	1	11	9	18	8	4	0	21	66	168	57	82	25	4	111	
	3304p	35	F/SI	3	6	3	24	1	12	7	5	3	4	0	18	63	146	54	73	10	9	102	
	2303	72	SI	6	5	7	29	1	11	13	21	8	3	4	15	77	194	54	93	29	18	113	4
	2303p	38	SI	3	4	7	25	1	10	12	17	6	3	3	11	63	162	46	78	24	14	103	
	3304	89	SI	8	6	5	32	2	14	15	22	11	7	5	30	93	242	68	112	25	37	125	14
	3304p	33	SI	3	5	3	25	1	13	13	22	6	6	5	23	65	187	55	82	23	27	114	
Subtotals		260		26	9	9	40	2	16	21	34	19	8	7	41	123	329	87	147	49	46	133	21
1,000 ha	1401	71	F	7	7	6	27	1	12	12	19	7	4	2	27	80	204	60	98	27	19	118	3
	1401p	33	F	4	5	6	22	1	10	7	13	2	3	1	16	58	144	49	73	14	8	103	
	1501	84	Fc	6	7	8	31	1	13	12	18	13	4	7	27	83	224	70	101	34	19	127	6
	1501p	31	Fc	2	5	5	18	1	12	8	13	7	2	2	21	65	159	61	78	20	10	102	
	3402	65	Fc	6	8	7	27	0	12	10	17	13	4	3	27	80	208	65	95	38	10	122	5
	3402p	34	Fc	3	8	5	18	0	10	10	16	10	4	1	23	66	171	48	86	30	7	113	
Subtotals		220		19	8	9	34	1	15	15	31	17	4	7	44	119	304	87	136	49	32	133	14
Totals[f]		1239		161	9	12	44	2	19	24	47	24	8	11	93	182	455	108	205	67	75	133	88

a. Reserves are ordered by size and condition class. p = partial list. Partial lists are subsets of the data from reserves that were sampled under more than one condition and are used for comparative purposes in species-area analysis (see fig. 7.3).

b. F = continuous forest; Fc = forest with large clearings; I = isolated by pasture or low vegetation; SI = semi-isolated by high regrowth.

c. For full names of taxonomic groups see table 7.1. d. Number of species found only in this reserve. e. Totals do not include partial lists.

f. Includes data from 3,490 ha in 25 reserves.

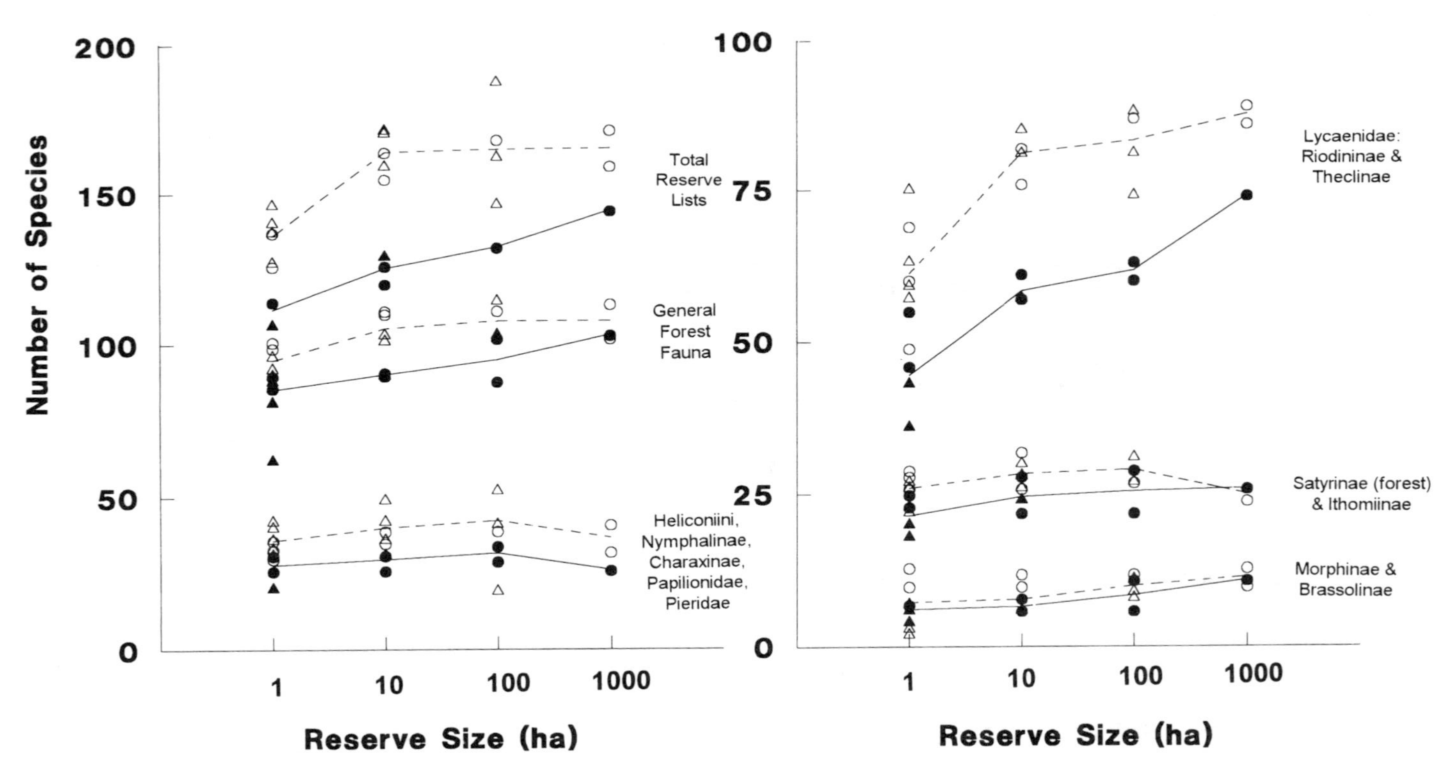

Figure 7.3. Number of butterfly species recorded in reserves of different area and isolation status. Solid circles indicate homogeneous, continuous forest; solid triangles, completely isolated reserves; open circles, reserves with large internal clearings; and open triangles, semi-isolated reserves linked to continuous forest by high regrowth on one or more edges. Lines represent a LOWESS smoothing function for reserves indicated by filled symbols (solid lines) and open symbols (dashed lines). Reserve lists of 30–44 hours are used (see table 7.2).

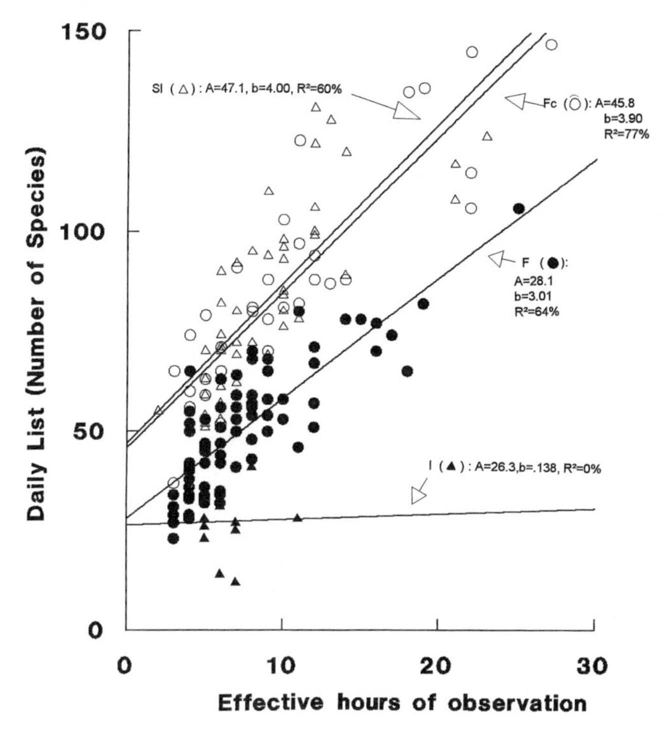

Figure 7.4. Relationship between sampling effort and daily species lists for twenty-five reserves, divided into four categories of disturbance and isolation. Symbols are as in figure 7.3.

Edge Effects

The four categories of reserves described above yielded separate clusters of points, independent of reserve size, when daily census lists were plotted against census effort (fig. 7.4). "Typical" forest interior lists, with only about 60 species recorded in 10 hours of sampling, were obtained in continuous forest without large clearings (solid circles). Even more depressed lists were found in small (1 and 10 ha) reserves after the surrounding forest had been cleared and burned, and in reserves surrounded only by pasture (solid triangles). In contrast, semi-isolated reserves and those having large internal clearings showed much larger daily lists (fig. 7.4, open symbols), reaching 150 species with 22–27 hours of census effort. The penetration of light into the understory of these reserves promotes the growth of new plant tissue and increases microhabitat diversity and flower abundance. This produces a positive "edge effect" that adds over 50% to the number of butterfly species recorded in daily lists (fig. 7.4).

Community Structure and Dynamics

The relative abundances of the 455 species revealed a most unusual pattern (fig. 7.5). A single species, *Hypothyris euclea* (Ithomiinae), constituted almost 30% of all individuals. Although normally common in the region, this species became excessively abundant

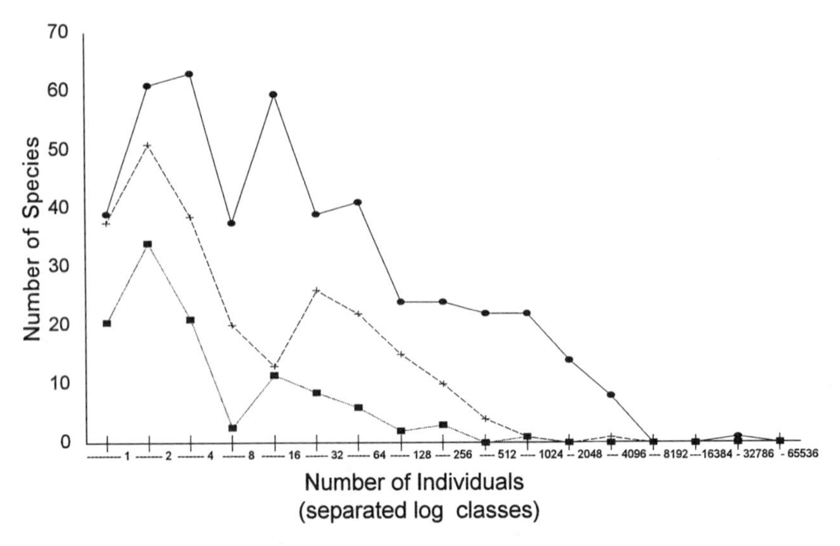

Figure 7.5. Distribution of 455 butterfly species among abundance categories ($\log_2$-transformed data). The upper curve (solid circles) is based on all individuals recorded over a fifteen-year period; the middle curve (crosses) is based on all individuals recorded in a single 100 ha reserve (number 3304) over the same period; and the lower curve (solid squares) represents a single daily census in a 10 ha reserve.

during the Project's early years (1980–1986) following forest isolation and disturbance. The gregarious larvae fed on the common second-growth plant *Solanum asperum.* By the 1990s, however, it was nearly absent, probably due to a great multiplication of its parasitoids.

Next in abundance were eight other species, all small forest-dwelling Lycaenidae or Satyrinae, that each constituted about 2–3% of all individuals; a further nineteen species (sixteen of the forest understory, of which six prefer shade) each constituted about 1% of all records (900–1,850 individuals). The largest single class, however, was the "singletons"—only one individual recorded—which comprised seventy-eight species. Between these two extremes were two to four abundance peaks at various scales (fig. 7.5) that were not due to sampling problems (possibly still present for rare species even after the recording of 110,000 total individuals). These peaks could possibly represent distinctive lifestyles well adapted for competing on host plants and achieving successful matings (butterflies show notoriously variable strategies, solutions, and syndromes in response to both of these problems, greatly increasing the heterogeneity of the community).

Species Turnover

With each new census, each reserve revealed a number of previously unrecorded species, while some of the species recorded in earlier censuses were not encountered. The rate of turnover in species can be estimated as $T_{pq} = (N_p + N_q)/2j$, where j = the number of species shared by any two successive lists p and q taken in the same reserve. Lists taken less than a week apart in the same reserve showed $T = 1.63 \pm 0.15$ (base value), while those separated by 2–24 months showed $T = 1.97 \pm 0.30$ ($N = 65$). After 4–5 years without sampling (1986–1991, 1991–1995), the turnover values were significantly higher ($T = 2.11 \pm 0.31$, $N = 23$; $t = 2.60$, $P > .01$), although still less than those exhibited by reserves that had changed isolation status between samples ($T = 2.40 \pm 0.49$, $N = 27$). With no change in status, isolated and semi-isolated reserves showed a significantly lower turnover (after 2–24 months: $T = 1.76 \pm 0.18$, $N = 23$) than those in continuous forest, with or without clearings ($T = 2.01 \pm 0.32$, $N = 42$; $t = 3.54$, $P > .001$).

Mark-recapture experiments also yielded insights into species turnover processes. Some individuals were recaptured up to 1 km away from their initial capture site, in some cases even moving between different isolated reserves in the same vicinity, suggesting considerable integration of metapopulations in this fragmented landscape. Most individuals were never seen again after their initial capture and marking, and many presumably emigrated to other places. However, a few butterflies persisted for long periods within the same home range. For example, one marked *Heliconius numata* was still in a 1 ha reserve (number 2107) two months after its initial capture, having incredibly survived the whole process of felling and burning of the surrounding forest. Other Heliconiini, Satyrinae *(Pierella),* Nymphalinae *(Catonephele acontius),* and a riodinine *(Stalachtis)* were recaptured at the same site after 54–105 days.

Genetic Liberation in Peaking Populations

Some local butterfly populations in the Manaus area appeared to include color pattern genes more typical of populations in distant regions (see Brown 1979). These unusual "markers" were especially prevalent in irruptions of some gregarious Heliconiini and Ithomiinae, unpalatable insects with few predators, whose larvae feed on abundant, weedy *Passiflora* vines and *Solanum* bushes along newly formed edges (see table 7.4). In 1981–1982, after the initial opening up of clearings in the region, some of these populations included almost all the known color pattern variation of their species. Traits expressed included those typical of races in the southern or upper Amazon, trans-Andean Venezuela and Colombia, and Atlantic Brazil. Because these distinctive color patterns are mostly adaptive in nature—traits related to mimicry that would be subject to strong stabilizing selection—the alternative alleles may be maintained at very low frequencies in the Manaus populations through occasional gene flow from distant populations, and only become apparent in samples during irruptions.

DISCUSSION

Natural Community Dynamics

Both taxonomic and ecological (light-based) groupings of butterfly species were used to classify and compare the daily and reserve census lists (see table 7.2) and to recognize baseline patterns and "signals" in this community. Most species are rare; over 87% of species had less than a 10% chance of being recorded in any two randomly chosen daily census lists, even in large reserves. In tropical forests, it seems that many rare species with rough ecological equivalence (guilds) can replace one another in time or space without implying significant reorganization of forest food webs or the physical environment (see table 7.1, fig. 7.2). Therefore, we analyzed groups of taxonomically and ecologically related species, rather than individual species, to compare sites, seasons, and stages in the fragmentation process.

Ecological and taxonomic groupings were used to discern patterns in butterfly assemblages (see figs. 7.2 and 7.3 and tables 7.1 and 7.2) and to develop a number of taxon- or habitat-related generalizations (see tables 7.3 and 7.4). However, two important processes, unpredictable disturbances and natural species turnover, reduced the effectiveness of this approach somewhat. For example, in 1986, shortly after biweekly samples commenced in two isolated reserves (2107 and 2108), both were severely damaged by a freak windstorm that threw down a third of the trees in each. This major disturbance made later observations in the reserves essentially incomparable with pre-disturbance data. At other times between 1980 and 1995, extensive flooding in parts of reserves led to great plant and insect mortality, independent of reserve isolation status. In addition, population irruptions of spiders, birds, and other predators occasionally wiped out vigorous butterfly populations, and these events appeared unpredictable and not directly linked to fragment isolation. These chance catastrophes made the interpretation and comparison of census lists difficult.

In principle, turnover of resident species in a given area should result in the replacement of those that leave or decline with ecologically equivalent ones. In practice, this was difficult to determine, for at least three reasons:

1. The daily list is usually only a limited sample (25–50%) of the species present in a given area at one time; this causes the "turnover" to have at least three components: stochastic sampling, local population cycles, and vagrant species.
2. Population-level turnover (e.g., replacement of a local deme by a different one of the same species) may be occurring, but is not easy to detect.
3. The highly stochastic nature of populations in this complex, diversified system overwhelms any simple ecological determinism based on resources or available niches. Especially in more desaturated, species-poor butterfly communities such as intact forest interiors, chaos and opportunity act to produce extremely low predictability of community structure and species composition from place to place and even day to day (note the average 40% turnover between lists in the same place and week).

The pronounced turnover between daily lists at a single site is probably attributable both to large sampling variation and to continual ecological changes (e.g., treefalls, succession in clearings, closing in of small light gaps, disappearance or appearance of host plants, flowering and fruiting phenology). Ecological changes in reserves were especially evident after the five-year gap in observations (1986–1991). The pronounced and largely unpredictable turnover in butterfly species must simply be accepted. It results in an inclined rather than a flat asymptote for species-sampling effort curves—a phenomenon well documented in long-term studies of tropical butterfly communities (R. K. Robbins, pers. comm.).

During population peaks, most butterfly populations probably spread out widely over the landscape, occupying any habitat where they find adequate resources and establishing new populations wherever supplies of larval foodplants are adequate. Such movements and migrations are important in maintaining gene flow in scattered metapopulations and permitting survival in unstable or fragmented landscapes. The mechanisms of this process are an important part of the success of butterflies worldwide (see Baker 1984; Singer and Ehrlich 1992).

Effects of Area and Environmental Heterogeneity on Diversity

The very widespread distributions of many species, and the highly erratic occurrence of an even larger number, are typical of Neotropical butterfly faunas. In this system, deep-forest species are relatively abundant while canopy/clearing and edge species are rare, as shown by the distribution of these groups in the reserves (see fig. 7.2). Hence, it is unlikely that any single reserve, even a very large one, could sample as much as 80% of the total regional species assemblage. This scattering of populations supports the notion that a similar scattering of many reserves across the landscape (see tables 7.2–7.3) is a necessary strategy, at least for preserving insect diversity (Simberloff and Abele 1982).

Area effects on butterfly communities in the BDFFP reserves were invariably less than those predicted by island biogeography theory (MacArthur 1969), which would suggest roughly a doubling in species number with each order of magnitude increase in reserve area. Even the strongest area effects observed in this study (in the Lycaenidae: see fig. 7.3) resulted in only a 25–50% increase in species richness for each tenfold increase in area.

Indeed, our observations suggest that local environmental heterogeneity, rather than fragment size per se (although the two are often correlated), is the most important factor in determining the diversity of the herbivorous insects. This pattern probably arises through a combination of diversity of the physical environment and host plants and diversity of ant mutualists of the lycaenid caterpillars. The butterfly community in the BDFFP reserves is generally very sensitive to forest disturbance (see figs. 7.3–7.4), especially to increasing light at ground level, which promotes new growth of understory plants, although some taxonomic groups appear more sensitive to such changes than others (see table 7.4). Essentially, all of the key influences on butterflies (see table 7.3) occur via an increase or decrease of microenvironmental heterogeneity. Important environmental factors include light, temperature, humidity, vegetation structure and composition, and other resources such as shelter, substrate, and stimuli. Increased light and environmental heterogeneity usually lead to a rapid accumulation of species, an effect so well known to tropical lepidopterists that they almost always choose to work in clearings or on edges.

To sustain a large number of species, an isolated reserve should be large enough (100–1,000 ha) both to include a number of internal treefall clearings of reasonable size (0.02–0.1 ha) and varying ages, and to resist negative edge effects from destructive land uses in the surrounding area (see also Didham, chap. 5). Our results indicate that the only answer to the "SLOSS" conundrum (*S*ingle *L*arge *O*r *S*everal *S*mall reserves) is "Yes" (both are necessary), something long suspected by tropical conservationists. The size, number, and scattering of reserves all strongly influcnce species richness via their effects on resources and microhabitats.

Table 7.3 Factors that potentially affect the daily and reserve lists of butterfly species.

Factor	Level and type of effect[a]	Affected groups	Examples (lists or reserves)	Observations
Size of Reserve	D± R±	Understory sun, canopy/clearing, and large species favored, not shade species	See figure 7.3 & table 7.2	10× size increase adds only about 15–30% to species list
Environment of reserve				
Varied topography	D+ R+	Rare species or species that require special, scattered resources	2107 > 2108, 1113 > 1110, 1104, 1112	Provides more niches; more obvious during isolation
Deep ravines	D− R−	Some species added, but in general a poorer fauna	1302 > 1301, 1208 > 1210, 1204 > 1205, 2206	Broad swampy areas are better
More fertile soil	D+ R+	Ithomiinae, Satyrinae, Nymphalinae	Porto Alegre (3xxx) > Dimona (2xxx)	Higher primary productivity
Heterogeneity of microhabitats	D+ R+	All groups	1105 > 1109, 3209 > 2206, 1204 > 1205	Partly a function of variable topography
Light penetration	D+ R+	Understory sun and canopy/clearing—Nymphalinae, Charaxinae, Theclinae, Riodininae	2107 > 2108, 1208 > 1210, 1302 > 1301	Creates new resources in clearings and understory
Large internal clearings	D+ R+	Understory sun and canopy/clearing—dependent on plants that need light	1105 > 1109, 1204 > 1205, 1302 > 1301	Internal edge effects
Permanent water or seasonal flooding	D− R−	All groups, but certain open vegetation Satyrinae and Riodininae favored	1204 > 1205, 3209 > 2206	Inhibits tree growth
Special resources				
Larval foodplants	D+ R+	Any group that can use each particular resource	See table 7.4	New plants are quickly found by mobile butterflies
Sampling methods (daily)				
Effective observation time	D+	All groups, especially Nymphalinae, Riodininae, Theclinae	See figure 7.4	Continual addition of sparse species
Sampling period	D+ D−	Understory sun and canopy/clearing species more affected	Low lists in 1113, 2206, 3402	Best between 10:00 and 15:00
Extensive coverage	D+	All groups, especially rare or local species	See figure 7.3	Related to the environment of the reserve

Intensive coverage	D+	All groups, especially understory shade species	High lists in 1109, 1112, 1113	1 ha reserves
Identifying capacity of observer	D+	Especially Satyrinae, Riodininae, Theclinae, Nymphalinae	Multiple lists in many reserves	Effect of up to 30% on list
Meteorological conditions (daily)				
Cold, rain, clouds	D−	Especially understory sun and canopy/clearing species	Low lists in 1105, 2108, 1110, 2206, 3209, 1301	Especially 11:00–14:00
Strong sunlight between rain	D+	Especially understory sun and canopy/clearing species	Lists in many reserves	Especially 11:00–14:00
Season of year	D+ D−	Lycaenidae (end of dry season), Nymphalinae (rainy season); February, August–September best	Especially see in 1984–1988 (Hutchings 1991)	Can vary between years
Experimental Design				
Trail cutting	D+ R+	Understory sun and canopy/clearing species, Ithomiinae, Satyrinae	All lists	More light in understory
Removal of understory in surrounding forest	D+ R+	All groups, especially understory shade species	Lists on 8/31–9/1/84 (2107, 2108, 2206)	Concentration effect
Cutting of primary forest around reserve	D± R−	All groups, especially understory sun and canopy/clearing and large species	1984 lists in 1112 and 1207; 1985–1986 lists in 2107 and 2108	Depends on the environment of the reserve
Burning	D− R−	All groups	1980 lists in 1104 and 1202; 1983 list in 3304; 1984 lists in 2107, 2108 and 2206	More pronounced effects on 1 ha reserve
Regeneration of low second growth, invading plants	D± R±	All groups, especially understory sun and canopy/clearing species	Porto Alegre (3xxx), Esteio (1xxx)	New resources, light
Presence of high second growth, trees	D+ R+	All groups, especially understory sun and canopy/clearing species	1104, 1112, 1202, 3209, 2303	New microhabitats
Planting of pasture	D+ R±	Understory sun, canopy/clearing, and edge species	1104, 1202, 2303	Sterilizes edge
Population fluctuations of predators	D− R−	Especially more easily caught species	Seen in 1202	Spiders and birds

a. D = daily list; R = reserve list. + indicates factor increases number of species on list; − indicates factor decreases number of species on list.

Table 7.4 Butterfly indicator species or groups and their use in monitoring.

Presence of these butterflies	Indicates existence of these resources or environmental factors	Use in monitoring[a]
Splendeuptychia	Bambuseae stands	P
Morpho menelaus, Euptychia picea, Hypothyris thea, Hyposcada egra, Napeogenes inachia, Euselasia zena, E. erythraea, E. gelon, E. uzita, Cremna actoris, Zelotaea phasma	Humid forested valleys with steep sides	P
Morpho hecuba, M. metellus in canopy	Abundant Menispermaceae vines	A, P
Brassolinae, *Chloreuptychia, Bia, Pierella, Antirrhaea, Haetera, Cithaerias*	Broad valleys with stemless palms	P
Euptychia sensu strictu (four species)	Abundant *Selaginella* and other mosses	P
Magneuptychia libye, Chloreuptychia arnaea, Cissia penelope, Eueides (six species), *Heliconius wallacei, H. Burneyi, H. ethilla, H. sara, Adelpha, Anaea, Historis, Dynamine, Evenus, Arawacus, Mesene, Calospila, Synargis, Thisbe, Menander, Juditha, Nymphidium*	Strong disturbance in vicinity (most likely edges and second growth)	D, E
Yphthimoides, Paryphthimoides, Hermeuptychia, Agraulis vanillae, Eurema, Phoebis sennae, Battus polydamus, Stalachtis phlegia	Grassy fields close by	D
Pareuptychia, Cissia, Lycorea, Methona, Laparus doris, Catonephele, Nessaea, Colobura, Tigridia	Clearings or large penetration of sunlight under the canopy	C
Ithomiinae in general	Abundant *Solanum,* and *Eupatorium* or Boraginaceae flowers, sources of dehydropyrrolizidine alkaloids	P, E
Thyridia, Melinaea, Hyposcada, Methona, Napeogenes	*Cyphomandra, Markea, Juanulloa, Brunfelsia* or *Lycianthes* (Solanaceae)	P, C, E
Heliconiinae in general	Abundant Passifloraceae; flowers with abundant pollen, especially *Gurania*	A, P, C, E
Neruda aoede, Heliconius hermathena, H. demeter	Sandy soils ("campina"), *Dilkea* or *Passiflora* (*Astrophea*) species	A, P, C, E
Hamadryas, Ectima, Eunica, Catonephele, Nessaea	*Dalechampia, Tragia, Alchornea,* and other herbaceous/vine Euphorbiaceae	C, D, E
Colobura, Historis, Tigridia; Marpesia; Callicore, Pyrrhogyra, Temenis	*Cecropia* (regrowth); *Ficus, Brosimum* (Moraceae); *Serjania, Paullinia,* and other Sapindaceae	C, D, E
Parides, Battus	Abundant *Aristolochia*	P, C, E
Eumaeus	*Zamia* (Cycadaceae)	P
Abundant Lycaenidae	Abundance of *Camponotus* and other ants	P, C

a. A = absence of butterflies usually means disappearance of rare resource;

P = presence of butterflies guarantees continuity of rare resource;

C, D, E = presence of butterflies indicates large clearings (C), high levels of disturbance (D), or rich edges (E) nearby.

Environmental Inventory and Monitoring with Butterflies

Singer and Gilbert (1978) have claimed that butterfly communities provide the best rapid indication of habitat quality. As explained above, the butterfly community in the Manaus reserves can be divided either taxonomically or ecologically. The resulting categories are distributed unevenly among the reserve environments (see table 7.2), thereby providing opportunities for inventory and monitoring of human activities in the region using various indicator species and groups (table 7.4).

The understory shade group and general forest fauna can be rapidly sampled in any region (see fig. 7.1) and can be used to assess many aspects of forest ecosystems. Forest-floor satyroids (*Antirrhaea,* Brassolinae, Haeterini) and Ithomiinae, for example, are especially good indicators of habitat diversity in the Neotropics (see Beccaloni and Gaston 1995). These butterflies have been used extensively for environmental inventory and monitoring for over thirty years (see Brown 1972, 1991). General fermented-bait-trapped species other than those mentioned above, although widely recommended for tropical butterfly censuses (Kremen 1992; Sparrow et al. 1994; Daily and Ehrlich 1995), were usually uninformative in our daily censuses and comparisons.

Significantly, butterfly indicator taxa reveal that major community restructuring occurs 100 m into the forest from a new edge, with changes sometimes detectable up to 250 m from edges (Brown 1991). In the Manaus area, east-facing edges appear especially prone to light penetration, plant and insect invasions, the effects of wind, desiccation, and new treefalls. Some effects are evident even along narrow trails in forest interiors. Many sun-loving butterfly species can be observed along these trails, and even understory species fly along trails in search of fallen fruits.

Although still rare in the region, many typical edge species whose larvae feed on grasses or other ruderal plants are now invading the BDFFP forest fragments. The establishment of these butterflies and their host plants up to several hundred meters into reserve interiors is among the most faithful harbingers of major system restructuring occurring after forest fragmentation (see tables 7.3–7.4). A variety of sun-loving species whose larvae feed on second-growth vines, herbs, and shrubs are now common even in clearings far from newly created edges. In addition, a number of shade-loving species typical of the general forest fauna have virtually disappeared from smaller isolated reserves, reappearing only when the reserves are linked to nearby forests by tall regrowth. Also disappearing from small isolated reserves are large, mobile species of *Morpho,* Brassolinae, and Charaxinae, which obtain their resources over a large area (see fig. 7.3).

An Integrated Model of Butterfly Responses to Fragmentation

When all census data are combined from the various stages of isolation and secondary regeneration in the Manaus reserves, a general model can be derived (fig. 7.6) describing the effects of fragmentation on butterfly diversity (see Brown 1991). According to the model, an initial natural or anthropogenic disturbance increases the number of species recorded per day by as much as 50%, whereas very homogeneous vegetation or strong disturbances can reduce this number to less than half that expected, especially in smaller

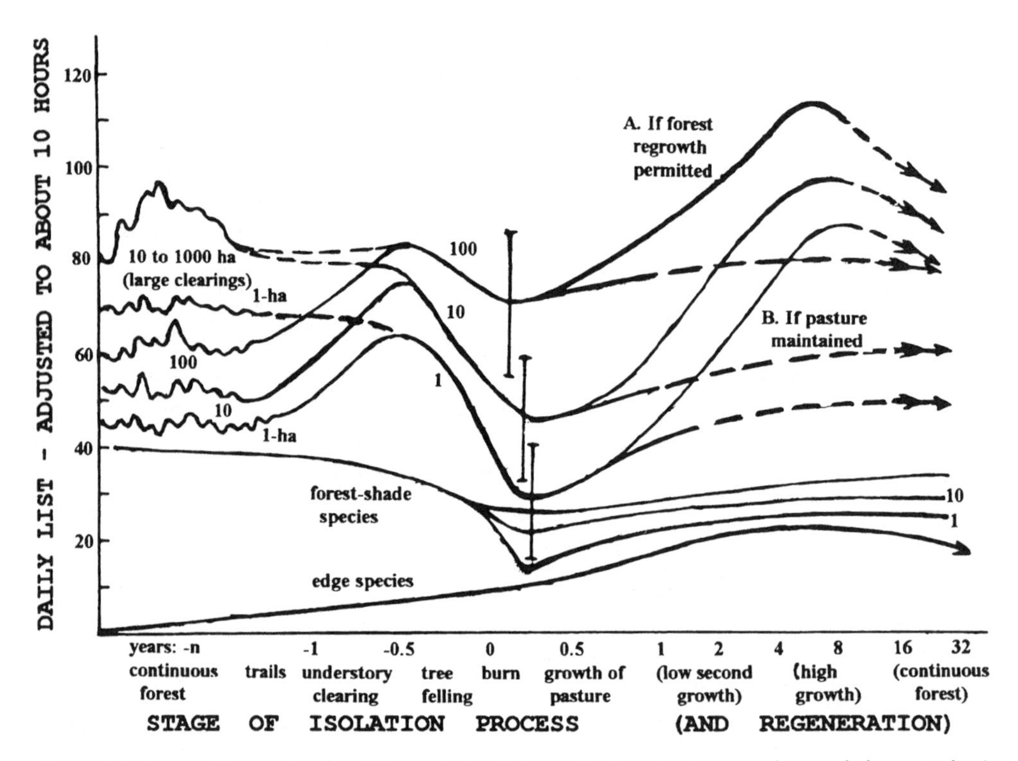

Figure 7.6. An empirical model describing variation in butterfly diversity (estimated using daily species lists) during the process of forest isolation, followed by regrowth of pasture or secondary vegetation along fragment edges. Species richness values are based on about 10 effective hours of census effort. With much more time (probably centuries), the reinvading edge species in the larger reserve total might again be lost, returning the community to an approximation of the original composition in ecological classes, if not in details of species composition.

reserves. Larger, more specialized, or more shade-loving butterflies decline in fragmented reserves, but sun-loving and edge species colonize the area and more than compensate for these declines. If an isolated reserve becomes surrounded by tall regrowth—allowing forest species to recolonize but not creating a system as closed as the original forest—butterfly diversity will climb steeply over a number of years. The increased diversity and abundance of butterflies may provide a greater resource base for native predators such as frogs, birds, and small mammals (see Tocher, Gascon, and Zimmerman, chap. 9: Bierregaard and Stouffer, chap. 10; Malcolm, chap. 14). After several decades some relaxation of the butterfly fauna should occur, yielding a community more diverse than the original forest and with a different species composition (fig. 7.6).

This model emphasizes the importance of small-scale habitat mosaics in the maintenance of butterfly diversity. In large tracts of continuous forest, treefall clearings are recognizable and may be an important factor in maintaining the diversity in this system. In these relatively insect-poor forests, such natural mosaics should be carefully encouraged and maintained in any conservation unit or diversity management program.

Large-scale forest clearing and fragmentation are very different matters. Profound ecological changes revealed by the disappearance after disturbance of common understory species (almost certainly accompanied by many rarer forest-obligate plants and animals) are probably irreversible over ecological time. Such changes can occur because many forest interior species are replaced for long periods by vigorous secondary forest congeners or ecological analogues (see Laurance, chap. 6), which can dominate forest edges, clearings, or successional seres for many decades and perhaps even centuries.

The overall picture of the effects of disturbance on diversity in Amazonian forests, supported by long-term studies in Belém, southern Pará, Rondônia, and Acre, is presented as a model in Brown and Brown (1992) and Brown (in press). This model suggests that when anthropogenic disturbance exceeds natural disturbance regimes in intensity or is very different in quality, it may lead to irreversible regional destructuring of metapopulations, communities, and ecosystem processes. Such changes could effectively preclude possibilities for sustainable resource use in the foreseeable future.

GENERAL IMPLICATIONS

1. For butterflies in central Amazonia, forest fragmentation seems to be simply a special case of general forest disturbance, with edge effects increasing the richness and altering the composition of butterfly communities.
2. Any new penetration of light into the understory leads to an increase in flowers and vegetative growth and to other qualitative changes in the forest system. Microheterogeneity in the physical environment and vegetation constitutes the basis for fine-scale packing of insect species.
3. Fragment size and isolation per se appear to have relatively minor effects on butterfly communities, although area can influence other ecological parameters (such as edge effects) that usually have a major influence on butterflies.
4. Although these systems are relatively robust under low levels of natural or anthropogenic disturbance, rapid and often irreversible changes may occur under more intense disturbance regimes. Thus, attentive monitoring is required to maintain ecological processes and species diversity.
5. A rapid turnover of species in local communities introduces much noise into the system and interferes with the recognition of cause-and-effect relationships. There is no doubt that winged insects move about through the tropical forest landscape much more extensively than has been suspected in the past.
6. Butterfly metapopulations are widely scattered across resource patches in the landscape. Different patches are often occupied by ecologically equivalent but taxonomically diverse species that may occur again only many patches away, without this necessarily implying an interruption of gene flow or a change in the physical environment.
7. At any given place and time, the butterfly community is divided mostly between extremely common and widespread species and extremely rare and localized ones, although several intermediate syndromes are detectable. Under natural disturbance regimes, dozens of common Neotropical forest butterflies are exceedingly rare in low-productivity dense forest systems. The tremendous predominance and importance of small butterfly species (Lycaenidae) is important in characterizing the system and its changes.

8. While this project was originally designed to examine the "minimum critical size of ecosystems," it has now become concerned with several additional minima: *(a)* the minimum heterogeneity in the environment needed to maintain plant and animal diversity; *(b)* the minimum amount of disturbance needed to maintain this heterogeneity; *(c)* the minimum connectedness required to sustain species turnover and population renewal rates across the landscape; and *(d)* the minimum management needed to permit natural processes of diversification to function normally within the complex and unpredictable species assemblages present in the forest mosaic. Concentration on and respect for these "minima" may be more useful and productive for biological conservation in tropical forests than a focus on merely the size of the conservation unit.

ACKNOWLEDGMENTS

We are grateful to Francisco A. Silva, Osmar F. Santos, Manoel J. Pinheiro, Sebastião M. Bezerra, and José A. Souza for field assistance; to R. O. Bierregaard, Thomas E. Lovejoy, Luis Carlos Joels, Ary J. C. Ferreira, and Claude Gascon for logistic support through the Biological Dynamics of Forest Fragments Project, and to the CNPq for fellowships. Identification of butterflies was greatly aided by Lee D. Miller (Allyn Museum of Entomology, Sarasota, Fla.: Satyrinae), Curtis J. Callaghan (Museu Nacional, Rio de Janeiro: Riodininae), and Robert K. Robbins (U.S. National Museum, Washington, D.C.), S. S. Nicolay, and J. Bolling Sullivan III (Theclinae). This study was supported in part by the Instituto Nacional de Pesquisas da Amazônia, The World Wildlife Fund-U.S., and the Smithsonian Institution, and represents publication number 148 in the Biological Dynamics of Forest Fragments Project Technical Series. William F. Laurance, Richard O. Bierregaard, and several anonymous reviewers commented on earlier drafts of the manuscript.

8

Phenetic Variation in Insular Populations of a Rainforest Centipede

John F. Weishampel, Herman H. Shugart, and Walter E. Westman

As a result of rapid clear-cutting, tropical rainforests in many parts of the world resemble archipelagoes of what was previously contiguous habitat. The reduction in habitat area and increase in isolation and edge effects resulting from such fragmentation may alter community structure and affect the demographic and genetic structure of denizen populations (Franklin 1980; Gilpin and Soulé 1986; Gilpin 1987b). Arthropods, in general, have high levels of genetic diversity (Nevo 1978), short generation times, and are sensitive to fine-scale spatiotemporal environmental patterns. Because tropical arthropods tend to have low population densities (Elton 1975), local population stability may be highly susceptible to evolutionarily recent habitat alterations.

Despite the vast diversity of arthropods in tropical rainforests and their potential to serve as bioindicators of environmental quality, only a few studies, mainly from the Biological Dynamics of Forest Fragments Project in Amazonia, have addressed the effects of fragmentation on arthropods. These studies have focused on the effects of recent (< 15 years) fragmentation on species diversity and community composition (e.g., Powell and Powell 1987; Vasconcelos 1988; Klein 1989; Becker, Moure, and Peralta 1991; Souza and Brown 1994; Didham, chap. 5; Brown and Hutchings, chap. 7). Here, we consider the effects of longer-term (70–110 years) fragmentation on phenotypic variation in populations of a scolopendromorph centipede from rainforest patches in northeastern Queensland, Australia. Certain types of changes in phenotypic variation may result from genetic and environmental stresses and are thought to precede decreases in fitness (Clarke 1993), which could increase the probability of local extinction in fragmented habitats.

Phenotypic variation may be examined at four organizational levels: within an individual, among individuals in a population, among different traits, and among populations (Soulé 1982). This study is primarily concerned with phenotypic variation at within-population and within-individual levels. Generalizations, often from anecdotal evidence, that relate patterns of within-population phenotypic variation to biogeographic or landscape-level habitat attributes can be grouped into three overlapping hypotheses:

1. *The niche width variation hypothesis* (Van Valen 1965) posits that, within a population, different phenotypes systematically select different resources and breed preferentially within their resource utilization groups. Thus, a "broader" niche results in more phenotypic variation within the population.

2. *The gene flow variation hypothesis* (Soulé 1971) assumes that between-habitat migration elevates phenotypic variation, and thus more isolated populations should contain less variation. This is consistent with the concept that populations located along the periphery of a species' range tend to have less phenotypic variation than those that are more centrally located (Mayr 1963; Yablokov 1986).
3. *The genetic diversity-phenetic variation hypothesis* is supported by two opposing factions. One group maintains that phenotypic variation decreases with factors such as insularity, decreasing population size, and loss of genetic diversity (Soulé 1973; Bryant 1984; Hanski 1986). The opposing camp believes that phenotypic variation increases following population bottlenecks and a loss of heterozygosity (Lerner 1954; Eanes 1978; Bryant, McCommas, and Combs 1986).

Within-individual phenotypic variation can be measured by fluctuating asymmetry (FA), which we define as minor departures from perfect bilateral symmetry. It has been associated with the erosion of evolutionarily viable genetic variation and the presence of genetic and environmental stress (Leary and Allendorf 1989; Parsons 1992; Markow 1995). FA is a consequence of developmental instability and has been found to correspond to a reduction in fitness in some species (J. S. Jones 1987; Liggett, Harvey, and Manning 1993; Møller 1994; Ueno 1994; Clarke 1995). Therefore, high levels of within-individual variation in a population may indicate a propensity toward extinction. Lerner's (1954) genetic homeostasis hypothesis, which correlates FA with decreased levels of within-population genetic variation (i.e., increased inbreeding and homozygosity), has received support in some (Soulé 1979; Wayne, Modi, and O'Brien 1986) but not most, cases (Fowler and Whitlock 1994; see review by Markow 1995). However, FA has been shown to be a sensitive indicator of environmental parameters such as crowding, temperature stress, and pollution stress (Clarke and McKenzie 1992; Parsons 1992; Clarke 1993). In contrast to within-population phenetic variation, high levels of fluctuating asymmetry have been positively correlated with habitat fragmentation and insularity (Isotomin 1994; Møller 1995).

METHODS

Species and Study Area

Scolopendromorph centipedes are the most widely occurring chilopod family in Australia (Main 1981). The pantropical genus *Rhysida* consists of eighteen species. *Rhysida nuda* (Scolopendridae: Otostigminae), one of three Australian *Rhysida* species, is distributed along the humid eastern and northern coasts of Australia, primarily within Queensland (Koch 1985). Although there is no definitive study of *R. nuda*'s habitat preferences, because of high transpiration rates (as are found in an African *R. nuda* subspecies: Cloudsley-Thompson 1959), this species is probably restricted to moist areas such as rainforest, especially during the dry season.

Twenty to twenty-five *R. nuda* individuals were collected from underneath rocks and woody debris in each of fourteen rainforest remnants (ca. 2.5–430 ha) and two continuous rainforest sites (> 40,000 ha) on the northern end of the Atherton Tableland (fig. 8.1A) during the dry season between June and August 1988 (see Laurance, chap. 6,

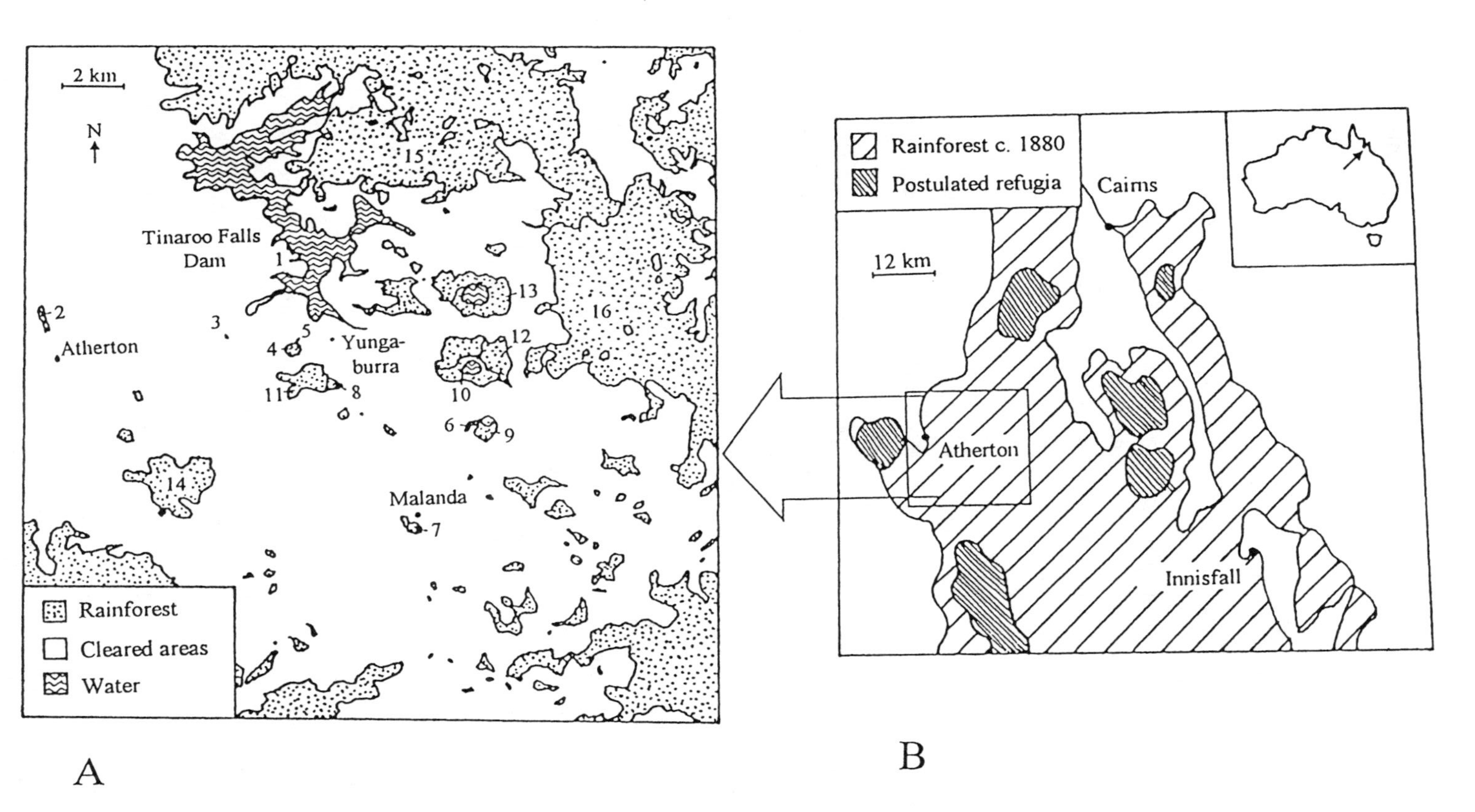

Figure 8.1. *(A)* Map of the study area derived from the February 1988 Landsat MSS image. Numbers indicate the sixteen study sites (numbered in order of ascending patch size). *(B)* Distribution of rainforest prior to clearing (Frawley 1983a), with Pleistocene refugia (from Webb and Tracey 1981). The rectangle surrounding Atherton (17°05′ S, 145° 40′ E) indicates the study area.

and Warburton, chap. 13, for additional research conducted in some of these same fragments). Rainforest remnants in this region are patchily distributed and typically restricted to lands of little agricultural utility, such as highlands and rocky outcrops.

The majority of forest clearing on the northern Atherton Tableland occurred from 1880 to 1920 (Frawley 1983a). Scolopendrids typically reach sexual maturity after 1.5–2 years and have a lifespan of up to 6 years (Lewis 1981). Thus, the time of isolation for each remnant population has been roughly 35–75 generations. However, as a result of climatic fluctuations during the Pleistocene, the rainforests of northern Queensland underwent periods of contraction into refugial pockets as recently as ca. 12,000 years B.P. (fig. 8.1B), followed by expansion during interglacial periods (see Moritz et al., chap. 28). These dramatic changes in rainforest distribution could have influenced the present geographic distribution of phenotypes.

Morphometric Analysis of Centipedes

Twenty-four exoskeletal measurements from the ventral side of the head segment (fig. 8.2) were taken from each centipede using a truss network technique (Bookstein et al. 1985). This involved photographing the head segment of each centipede under 10× –20× magnification and digitizing the appropriate landmarks of each image. Using this procedure, measurements were highly repeatable, and thus measurement errors were relatively small. Metric measurements were used because scolopendrids are subject to individual variation of some meristic characters (i.e., discrete features such as the number of antenna segments) due to changes in age, wear, and inaccurate regeneration (Lewis 1978). Additionally, the use of continuous measures avoids threshold problems associated with discrete characters when used to estimate developmental stability (Swain 1987). Chitinous mouthparts were measured because they are durable and could readily be compared with existing museum collections.

Scolopendrids grow isometrically and are relatively sexually monomorphic; hence, individuals were not separated by sex or age. However, because the measured characters increase in absolute size, they are all positively correlated. To reduce this growth effect, we followed Dodson (1978), taking the logarithm of the ratio of each measurement to the measurement having the highest average raw correlation to the other measurements (i.e., the maxilliped base-plate base, distance 22 in fig. 8.2). This transformation reduced the highest average correlation of each measurement to the other measurements from .907 to .336. The common use of ratios in this manner to compensate for size differences associated with growth has received criticism (Atchley and Anderson 1978).

To examine the level of similarity of morphological variation across populations, we compared profiles of the coefficient of variation (CV), defined as the standard deviation/mean × 100, for each transformed measurement using Kendall's coefficient of concordance (*W*: Sokal and Rohlf 1981). This graphical method of comparing CVs for a suite of characters was further employed to assess the overall level of evolutionary divergence by comparing the average within-population CV profile and the average among-population CV profile (Sokal and Braumann 1980). The phenetic structure among rem-

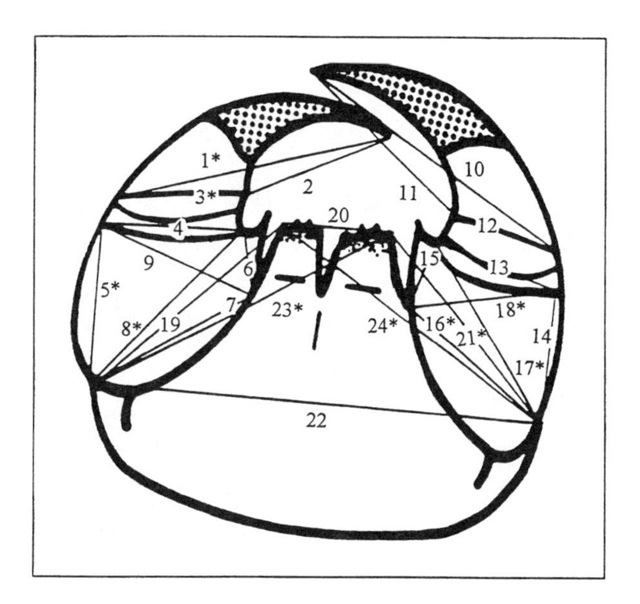

Figure 8.2. Ventral side of centipede head configured with truss network. Asterisks represent measures that exhibited heterogeneous variances among populations ($P < .05$). (Adapted from Koch 1983.)

nant populations was determined with an unweighted pair-group average (UPGMA) cluster method. This method was chosen because it clusters more faithfully to phenetic distances than other commonly used methods (Farris 1969). A Mantel test (Smouse, Long, and Sokal 1986) was used to compare phenetic similarity of populations based on the geographic proximity of fragments (measured as the linear distance from each remnant's edge to the nearest edges of all the other remnants).

An estimator of FA for each centipede was derived by averaging the absolute differences (Palmer and Strobeck 1986) between the eleven bilateral measurements (1 and 10, 2 and 11, 3 and 12, etc., from fig. 8.2) previously adjusted for size effects. Following Wayne, Modi, and O'Brien (1986), Duncan's multiple range test was used in conjunction with an ANOVA to determine the significance of interpopulation differences in the mean values of CV for the heterogeneous measurements ($\overline{CV}$) based on Bartlett's test of homogeneity of variance (fig. 8.2), and of FA ($\overline{FA}$) and mass for each individual.

Landscape Analysis of Phenetic Variation

An exploratory stepwise regression procedure was conducted to quantify the relationships of certain habitat and landscape features to within-population and within-individual phenotypic variance of the *R. nuda* populations, in a fashion similar to that used for island populations of lizards (Soulé 1972). This heuristic technique attempts to find the most parsimonious combination of predictor variables that explains the highest variance of the dependent variable (Sokal and Rohlf 1981). Thus, highly correlated predictors generally do not appear together in the same model because the addition of an intercorrelated variable would not significantly increase the R^2 value from that of the previous model and hence would not be permitted to enter the model. For each population, the depen-

Table 8.1 Selected landscape indices derived from satellite imagery and used to characterize forest remnants in northeastern Queensland.

Area = forest remnant area (ha)
Log Area = $\log_e$(Area)
ΔArea = percentage change in Area from 1981 to 1988
Distance = linear distance of remnant from the nearest relatively continuous forest tract (Danbulla or Gadgarra) (km)
Distance^2 = the square of Distance
Refugium Distance = distance of remnant from the nearest hypothesized refugium (km)
Refugium Distance^2 = the square of Refugium Distance
Shape = shape of the remnant or fractal index, $\log_e$[edge(m)]/Area (Krummel et al. 1987)
ΔShape = percentage change in Shape from 1981 to 1988
Interaction = the patch interaction index (Forman and Godron 1986):

$$\text{Interaction}_i = \sum_{j=1}^{n} A_j / d_j^2$$

where n is the number of neighboring patches within 5 km from the edge of patch i and d_j is the linear distance from the edge of patch i to the edge of patch j (m) and A_j is the area of patch j (ha)
Log Interaction = $\log_e$(Interaction)
Habitat Diversity = heterogeneity of the remnant based on the Shannon Index of the reflectance values of the pixels that constitute the remnant:

$$\text{Habitat Diversity}_i = \sum_{j=1}^{n} P_j \log_e(P_j)$$

where P_j is the proportion of the jth pixel type in remnant i.

dent variables were $\overline{CV}$ and $\overline{FA}$, representing levels of within-population and within-individual variation, respectively.

Landscape variables were derived using a Landsat MSS image from February 1988. The scene was grouped into rainforest and non-rainforest classes using a supervised method in which cover classes known a priori are used as training sites for classification of each pixel based on its spectral signature. Changes in cover class from a 1981 image were used to provide estimates of recent, pre-collection clearing around each remnant. In addition to non-rainforest vegetation, roads and water bodies also acted as remnant boundaries. Remnants separated solely by an elevated, paved road were given the same site name with a different site designation number (e.g., Curtain Fig 1 and Curtain Fig 2). Twelve variables (table 8.1) quantifying relevant landscape attributes, (Forman and Godron 1986; Gilpin 1987b) such as remnant size, degree of remnant isolation, boundary discreteness, habitat diversity, distance from refugium population, and disturbance level, were chosen for study. Because phenotypic variation may not be linearly related to landscape measurements (as found with species-area relationships: MacArthur and Wilson 1967), several landscape variables were squared or log-transformed (table 8.1). Three other variables used in preliminary studies (edge length, log edge length, and change in pixel diversity values from 1981 to 1988) were dropped because they did not contribute to any stepwise regression model.

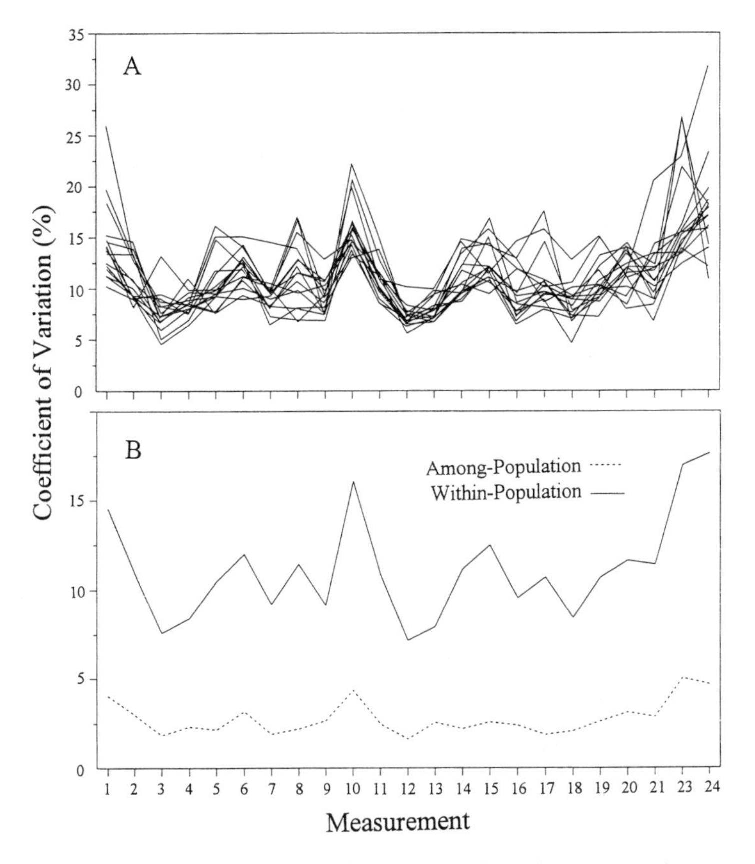

Figure 8.3. *(A)* Superimposed variability profiles from the sixteen study populations. *(B)* Comparison of average within-population and among-population variability profiles.

RESULTS

Despite range differences in ten of the twenty-two measurements (see fig. 8.2), there was significant concordance among the CV profiles for the sixteen populations ($W = .64$, $P < .001$). The superimposed profiles (fig. 8.3A) confirmed that "peak" and "trough" measurements were generally "in phase," suggesting that factors responsible for phenetic variation were acting consistently throughout each phenome (Soulé 1972). The average within-population variability profile was consistently greater than the average among-population profile (fig. 8.3B); thus, there has been little evolutionary divergence among the *R. nuda* populations in remnants. Furthermore, the two lines are significantly parallel ($W = .89$, $P < .05$), supporting Kluge and Kerfoot's (1973) hypothesis that characters that exhibit greater variation in a local population should show greater variation among populations.

The phenogram produced from the UPGMA cluster analysis of the study populations (fig. 8.4) had a cophenetic correlation coefficient of .73, somewhat below the desired

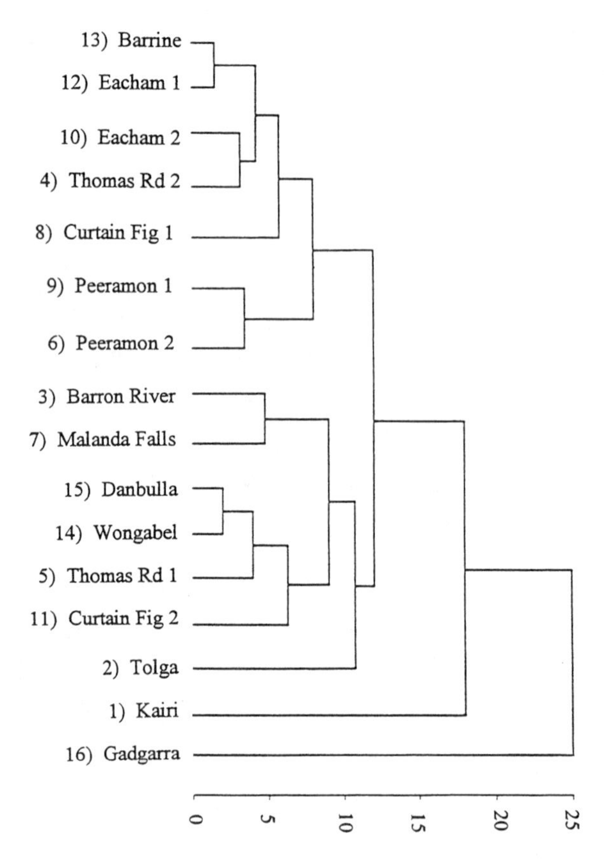

Figure 8.4. Phenogram derived from the UPGMA cluster analysis of the transformed centipede head measurements. The number in front of each site name indicates the site number (see fig. 8.1A). Similarity distances are rescaled.

standard of .85 (Sokal 1986), indicating that there was not a strong agreement between Euclidean distances based on morphological similarities and hierarchical relations derived from clustering. The two smallest remnants, Kairi and Tolga, and the largest tract, Gadgarra, were the most dissimilar to the other remnants. In several cases (i.e., Peeramon and Eacham, but not Thomas Road), adjacent populations separated by an elevated road were phenetically very similar. However, the overall comparison of similarity values based on remnant population phenetics and interpatch geographic distances using the Mantel test proved nonsignificant.

The overall $\overline{\text{CV}}$, $\overline{\text{FA}}$, and average body mass values did not differ significantly (ANOVA). However, the $\overline{\text{CV}}$ value for the population from the largest tract, Gadgarra, was significantly ($P < .05$) greater than those for fourteen of the smaller fragments (Duncan's multiple range test, fig. 8.5). Additionally, the $\overline{\text{FA}}$ and body mass values for the smallest rainforest fragment, Kairi, were greater ($P < .05$) than those for five and three of the larger fragments, respectively. However, individual centipede mass and asymmetry were not significantly correlated over the entire collection.

In the stepwise regression analyses using $\overline{CV}$ as the dependent variable (table 8.2), both Area and Log Area were significant predictors of $\overline{CV}$, although Area accounted for 52% of the variation of $\overline{CV}$, nearly three times as much as Log Area; Area and Log Area were both positively correlated with $\overline{CV}$. After Area, the squared distance from the nearest refugium (Refugium Distance^2) accounted for the largest amount of residual variation and was negatively correlated with $\overline{CV}$. Together, these two variables accounted for 69% of the variation; these R^2 values, however, may be somewhat inflated due to the small number of populations (Rencher and Pun 1980).

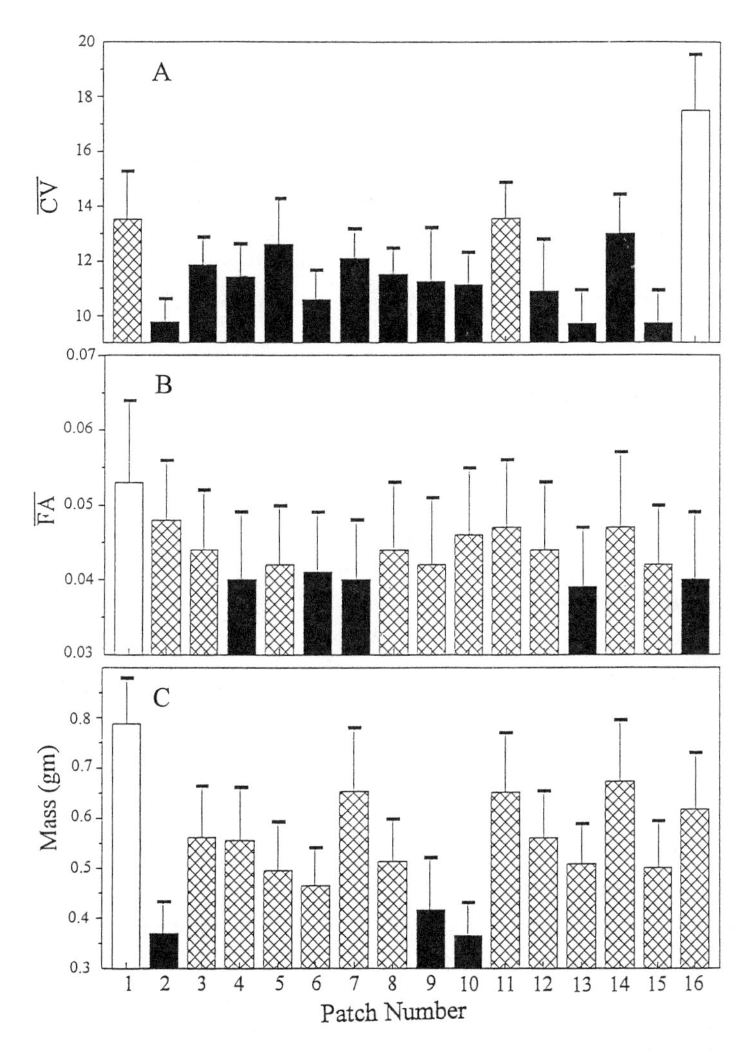

Figure 8.5. *(A)* Average within-population variation, *(B)* average within-individual variation, and *(C)* mean body mass for each remnant population. Extensions above the bars represent standard errors. Means represented by open bars differ significantly ($P < .05$) from those shown by solid bars. Means represented by hatched bars do not differ significantly from either solid or open bars. Remnants are displayed in order of ascending patch size from left to right.

Table 8.2 Stepwise multiple regression models for predicting $\overline{CV}$.

Number of predictors	R^2	P	Predictors	Slope
1	.52	< .01	Area	+
2	.69	< .001	Area	+
			Refugium Distance^2	−
3	.83	< .001	Area	+
			Distance^2	−
			Refugium Distance^2	−
4	.91	< .001	Area	+
			Shape	+
			Distance	−
			Refugium Distance^2	−

Note: Only models in which all predictor variables have significant ($P < .05$) partial regression coefficients are shown.

Table 8.3 Stepwise multiple regression models for predicting $\overline{FA}$.

Number of predictors	R^2	P	Predictors	Slope
1	.366	< .05	Shape	+
2	.605	< .01	ΔShape	−
			Log Interaction	+

Note: Only models in which all predictor variables have significant ($P < .05$) partial regression coefficients are shown.

With $\overline{FA}$ as the dependent variable, the fractal index (Shape) accounted for the most variation (37%; table 8.3). This variable was positively correlated with $\overline{FA}$ for the first two steps in the stepwise regression. The Log Interaction index and squared distance from the nearest "mainland" fragment (Distance^2) were also significant variables that had positive partial regression coefficients. After the fractal index was dropped, percentage change in the fractal index (ΔShape), which had a negative partial regression coefficient, also became significant.

DISCUSSION

Rapid phenotypic divergence of species on small islands or in forest remnants is not an uncommon observation in the evolutionary literature (e.g., Soulé 1972; Kozakiewicz and Konopka 1991). In this study, however, the overall level of evolutionary divergence among remnant centipede populations was found to be low (average among-population levels of phenotypic variation were less than the average within-population levels). Nevertheless, the two phenetic outliers, representing the largest and smallest remnants, exhibited extreme values of within-population and within-individual phenetic variation, which may be indicative of threshold effects associated with fragmentation.

Landscape Correlates of Within-Population Phenetic Variation

The association of high values of within-population variation with remnant area supports both the niche width variation hypothesis of Van Valen (1965) and the gene flow variation hypothesis of Soulé (1971). A larger area or increased niche width may provide a richer array of prey species for the scolopendrids, which are generalized predators. Such an increase in diet item diversity should lead to a relaxation in stabilizing selection (Soulé and Stewart 1970), resulting in more phenotypic variation. Larger remnants typically have higher species richness than smaller remnants (MacArthur and Wilson 1967) and hence a more diverse prey base, and also exhibit greater habitat diversity (Seagle and Shugart 1985). The correlation between Log Area and habitat heterogeneity (estimated by Habitat Diversity; table 8.1) in this study was .51.

Area accounted for more within-population variation than Log Area, in contrast to the relationship found by Soulé (1972), but similar to that found by Soulé and Yang (1973). However, actual centipede populations were not demarcated, and it is possible that the remnant samples may reflect either several isolated Mendelian populations or metapopulations with limited gene flow. The vagility of this centipede is assumed to be low, suggesting that *R. nuda* is more likely to migrate from habitat to habitat within a forest remnant than between remnants. Larger remnants with more habitat types should provide more diverse gene sources, facilitating the mechanisms postulated by the gene flow variation hypothesis.

The rainforests of northern Queensland are believed to have contracted into a series of small upland refugia during the Pleistocene (see fig. 8.1B; Webb and Tracey 1981; Moritz et al., chap. 28). If the *R. nuda* populations found in the study area originated from the putative Herberton refugium, located to the west of Atherton, populations in the western portion of the study area would represent central populations and those to the east would represent peripheral populations. The Gadgarra mainland population may have exhibited the greatest dissimilarity to other populations (see fig. 8.4) because it was the farthest from the Herberton refugium and the closest to other forest refugia. If phenetic distance relates to the evolutionary divergence or age of a population, this could account for the negative correlation between distance from the Herberton refugium and $\overline{\text{CV}}$ (Mayr 1963; Yablokov 1986).

Although the landscape variables incorporating distance do not consider geographic boundaries, such as rivers, that would impede dispersal, the signs of coefficients of other variables in the regression models are generally consistent with the ideas embodied in the gene flow hypothesis. The partial regression coefficients (see table 8.2) suggest that if interpatch migration occurs and is more likely among closer populations, then within-population variation will increase. Analogous studies of insular lizard populations (Soulé 1972; Soulé and Yang 1973) also found habitat area and log area to be positively correlated, and squared distance from a mainland source to be negatively correlated with within-population variation.

Landscape Correlates of Within-Individual Phenetic Variation

Although centipede populations in the smallest remnant exhibited the highest level of asymmetry, chilopods often occur in high densities (e.g., Frith and Frith [1990] found chilopod densities of $4/m^2$ in nearby Australian rainforest, and Lewis [1981] found African scolopendrids in densities of $0.16/m^2$). Thus, even the smallest (2.5 ha) remnant in this study could have contained on the order of $10^3 - 10^5$ *R. nuda* individuals, a population too large to imply inbreeding. Furthermore, without electrophoretic data, it is difficult to evaluate the genetic-phenetic relationship (Soulé 1973). Another possible genomic contributor to fluctuating asymmetry is hybridization, which can impose stress by producing a breakdown of intrinsic genomic coadaptations (Leary and Allendorf 1989). The Log Interaction index was positively correlated with $\overline{FA}$ and may be indicative of hybridization between remnant populations. However, if migration between the patches occurred after fragmentation, then it probably would also have occurred prior to fragmentation. This would not have allowed the populations much opportunity to evolve intrinsic genomic coadaptations.

Within-individual phenetic variation is also related to environmental stress, which may vary with the fractal index (Shape) and percentage changes in area (ΔArea) and shape (ΔShape). The higher the fractal index, the more likely it is that the rainforest patch is influenced by the outside environment, potentially altering the relatively stable, cool, humid microclimate of the rainforest interior (see Turton and Freiburger, chap. 4). Because of desiccation limits associated with small body size, drier habitats tend to select for larger body size in arthropods (Young 1982; Didham, chap. 5). This effect could explain why the Kairi remnant, with the highest fractal index, exhibited the highest centipede mass (column 1, fig. 8.5). Furthermore, centipede populations undergo seasonal fluctuations from the wet to the dry season (Levings and Windsor 1985; Frith and Frith 1990). Populations in remnants with higher edge-to-area ratios may undergo more severe waxing and waning than those in remnants with a smaller fractal index. Such seasonal bottlenecks could also produce genomic stress, which may yield higher levels of fluctuating asymmetry.

Undoubtedly, many factors interact to account for phenotypic variation in patch populations; as such, breeding experiments would be required to untangle the contributions of environmental and genetic variation. Ideally, for a study like this, pre- and post-fragmentation samples should be compared. Most probably, the phenotypic patterns exhibited by *R. nuda* reflect the overlapping of macroevolutionary and anthropogenic microevolutionary processes. Driving forces at the macroevolutionary scale include the contraction and expansion of rainforest and its constituent populations over periods of thousands of years. At the microevolutionary scale, formative processes include the fragmentation of rainforest and its populations into insular patches. With fragmentation, habitat area decreases and the ratio of forest edge to interior increases. Results from this study indicate that, when their habitat is fragmented, some sedentary invertebrates may exhibit a corresponding decrease of within-population variation and an increase of within-individual variation. Such signals have been associated with a decrease in fitness in certain species and hence could point to an increased likelihood of local extinction.

GENERAL IMPLICATIONS

1. Measures of phenetic variation, at both the population and individual levels, may serve as bioindicators of a remnant population's viability and of the quality of its habitat. These

measures may be sensitive indicators of the presence of genomic and environmental stress induced by fragmentation.

2. Within-population levels of phenetic variation in the centipede *Rhysida nuda* were positively associated with habitat area, suggesting that as patch size decreases, evolutionarily viable variation, which permits populations to track environmental changes, also decreases. This variation also declined with distance from the central or "mainland" population.
3. Within-individual levels of phenetic variation (fluctuating asymmetry), as well as total body mass, increased in *R. nuda* as the edge-to-area ratio of remnants increased, perhaps in response to some environmental stress such as increased desiccation. Such changes have resulted in reduced fitness in some organisms.
4. Thus, rainforest fragmentation, by decreasing climate buffering, may contribute to the local extinction of drought-sensitive organisms in small, isolated patches. This may be preceded by decreases in within-population and increases in within-individual levels of phenetic variation in ecologically sensitive species.

ACKNOWLEDGMENTS

This research was a part of a larger study described in Westman, Strong, and Wilcox (1989). We wish to thank the CSIRO Tropical Forest Research Center and the Queensland Museum for their assistance, and G. M. Clarke, W. F. Laurance, M. E. Soulé, D. O. Wallin, J. A. Yeakley, and five anonymous referees for reviewing earlier drafts of this chapter. This research was supported in part by an NSF grant presented under Interagency Agreement No. BSR-8718168 with the Department of Energy, and by the NASA Earth Sciences Division (UPN 677-80-06-05).

9

Fragmentation Effects on a Central Amazonian Frog Community: A Ten-Year Study

Mandy D. Tocher, Claude Gascon, and Barbara L. Zimmerman

Deforestation almost always results in fragmentation of the original forest into isolated patches of habitat embedded in a modified matrix. The most obvious physical change in the landscape is a reduction in total forest area. Species richness and population sizes of forest-dependent animals and plants usually decline as a result of forest loss and fragmentation (Franklin and Forman 1987).

Several factors are important when considering the fate of small populations in a fragmented landscape (Gilpin and Soulé 1986; Simberloff 1986). Although some species can exist at low population numbers, small populations are often highly susceptible to stochastic demographic fluctuations that can greatly increase their probability of extinction and contribute to the loss of genetic variation (Reh and Seitz 1990; Wayne et al. 1992). Losses of heterozygosity due to inbreeding depression in small populations, or founder effects, associated with a sharp drop in population size can result in impaired reproductive success (Ralls and Ballou 1983; Packer et al. 1991). Hence, fragmentation will often result in lower species richness and may reduce the breeding success of some species that persist in fragments.

Negative effects of forest fragmentation have been documented for many animal groups (invertebrates: Klein 1989; Margules, Milkovits, and Smith 1994; reptiles: Dodd 1990; birds: Gates and Gysel 1978; Kitchener et al. 1982; Lovejoy et al. 1984, 1986; Yahner 1988; Powell 1989; Harper 1989; Quintela 1985; Bierregaard and Lovejoy 1988, 1989; Potter 1990; Katten, Restrepo, and Giraldo 1994; Willson et al. 1994; mammals: Malcolm 1988; Rylands and Keuroghlian 1988; Schwarzkopf and Rylands 1989; Laurance 1991a; Lawton and Wooddroffe 1991; Vanapeldoorn et al. 1992; also see reviews by Saunders, Hobbs, and Margules 1991 and Bierregaard et al. 1992). Little is known, however, about the effects of fragmentation on frog communities (but see Reh and Seitz 1990).

The Biological Dynamics of Forest Fragments Project (BDFFP) in Manaus, Brazil, has provided an opportunity to study the effects of tropical forest fragmentation on a wide range of taxa (see Kapos et al., chap. 3; Didham, chap. 5; Brown and Hutchings, chap. 7; Bierregaard and Stouffer, chap. 10; Malcolm, chap. 14; and many of the references cited above). The frog community in primary, continuous forest in this region has been studied for a decade, providing important baseline data on the abundance and distribution of many species (Zimmerman and Bierregaard 1986; Gascon 1990, 1991b; Zimmerman and Simberloff 1996). Forest fragments in the area suffer significant edge-related microclimatic changes, such as elevated temperature and reduced humidity (Kapos 1989; Kapos et al., chap. 3). In addition, breeding habitats for frogs are not homogeneously distributed in primary forest. Hence, isolated fragments, selected without knowledge of the distribution of habitats, may not contain crucial breeding habitats for all frog species (Zimmerman and Bierregaard 1986).

In this chapter we amass data from over a decade of work on the anuran community in the Manaus area to determine how frog assemblages have been affected by forest fragmentation. First, we compare surveys of adult frogs before and after isolation to assess changes in species richness as a function of fragment size. Second, we contrast the breeding success and abundance of selected aquatic and terrestrial breeders in fragments and primary forest. Our basic prediction is that frog richness, abundance, and breeding success will be negatively affected by fragmentation and will be strongly influenced by fragment size.

METHODS

Site Description

All work was carried out in the BDFFP reserves, which are situated about 70 km north of Manaus, Amazonas, Brazil (Lewin 1984; Lovejoy and Bierregaard 1990; Bierregaard and Stouffer, chap. 10). Field surveys commenced in primary forest in 1983 and continued until May 1994. Work in forest fragments was conducted over two consecutive rainy seasons (November–May 1992–1993 and 1993–1994). Ten forest fragments were surveyed: two of 100 ha; four of 10 ha; and four of 1 ha (table 9.1; see Lovejoy et al. 1986; Lovejoy and Bierregaard 1990). The forest is *terra firme* (not subject to river flooding).

Table 9.1 Forest remnants of the Biological Dynamics of Forest Fragments Project surveyed during the study.

Fragment no.	Date of isolation	Fragment chacteristics
100-hectare fragments		
3304	8/83	Surrounded by high second-growth forest on three sides and pasture on the fourth side
2303	10/90	Surrounded by regenerating second growth
10-hectare fragments		
3209	8/83	Surrounded by high second-growth forest
2206	8–9/84	Isolated
1202	7–8/80	Isolated
1207	8/83	Surrounded by mature second-growth forest
1-hectare fragments		
3114	7–8/83	Surrounded by mature second-growth forest
2108	9/84	Isolated
1104	8/80	Isolated
1112	8/83	Surrounded by mature second-growth forest

The topography is characterized by rolling hills and plateaus. Many first- and second-order streams dissect the area, forming gullies with small flooded areas in associated lowlands. Permanent and semipermanent isolated ponds and wallows are patchily distributed throughout the entire area (see Zimmerman and Bierregaard 1986).

Frog Species Composition and Abundance

From 1983 to 1990, frog species composition and relative abundances were studied throughout 2,000 ha of primary forest (Zimmerman 1991; Zimmerman and Bierregaard 1986; Zimmerman and Rodrigues 1990; Gascon 1990, 1991b; Zimmerman and Simberloff 1996). Included in the survey area were four reserves that were subsequently isolated from surrounding primary forest; these were later surveyed as forest fragment isolates to provide a direct pre- and post-isolation comparison.

Four sampling techniques were used: transect counts of calling frogs, visual counts of frogs along transects, surveys of tadpole incidence in pools and streams, and intensive visual and aural searches at two breeding pools (Zimmerman 1994).

Transect counts included aural "spot checks" for calling frogs at selected breeding pools. During the counts, all calling frogs were identified to species and their calls counted. The time of the call and the habitat from which frogs were calling (general forest, stream gully, or pool) were also recorded. One of us (B. L. Z.) performed these baseline surveys with the aid of one or two technicians.

Abundance estimates (calling frogs/km) for each species were obtained by dividing the number of kilometers walked during the peak breeding months by the number of calls recorded. Transects were walked at an average rate of 2.3 km/hour (Zimmerman 1991). Months in which zero calls were recorded were included in the estimate. The aural transect counts were performed in combination with visual counts (Zimmerman and Simberloff 1996). Tadpole incidence in pools and streams was recorded by dip-netting with a fine mesh net (0.5 mm mesh size). All tadpoles were identified to species (Hero 1990; Gascon 1991b).

From 1992 to 1994, survey techniques identical to those described above were used to measure frog species composition and abundance in forest fragments. All work was carried out by the senior author.

Although different observers could possibly bias the comparison of two different survey periods, previous tests of observer bias suggest that this is not the case here (Zimmerman 1991). Moreover, the total sampling effort in the pre-isolation sampling period is one order of magnitude higher than in the post-isolation surveys, which suggests that any observed difference may actually be underestimated.

Breeding Success

Artificial pools (45 cm diameter plastic basins) were randomly placed throughout primary forest. The basins were surveyed bimonthly between 1987 and 1989 by dip-netting. In forest fragments, artificial pools were placed in groups of three along a central trail bisecting each forest fragment. Before the start of the first rainy season (1991–

1992), twelve basins were placed in each 100 ha fragment and nine basins in each 10 ha fragment. At the beginning of the second rainy season (1992–1993), a further nine basins were placed in the 1 ha forest fragments. Basins were surveyed every two weeks during the first rainy season, and monthly during the second rainy season. Tadpoles present were either identified to species in the field or, on occasion, collected for future identification. Breeding success of larval cohorts was measured by dividing the total number of cohorts initiated in the basins by the number of cohorts that developed into late-stage, four-legged tadpoles (Gascon 1990).

The breeding success of at least one species of Amazonian frog is strongly influenced by rainfall (Gascon 1991a). To assess between-year differences in rainfall, rainfall was compared between years in which breeding success was measured in primary forest and those in which it was measured in the fragments. Similarly, to control for possible differences in the location of basins, we measured the number of days taken by each species to colonize each basin in the fragments and compared it with that found in primary forest. We assumed that similar rates of basin colonization indicated an absence of significant influences on basin use.

Measurement of Environmental Variables

Environmental variables used in this study were measured in 1991–1994 by the senior author. In forest fragments, 5 × 5 m quadrats were randomly located along the census transects. In 100 ha fragments and in primary forest, eighty quadrats were sampled; in 10 ha fragments, forty were sampled; and in 1 ha fragments, fifteen quadrats were sampled. Within each quadrat the following variables were measured: canopy cover at 0–2 m, 2–5 m, and > 5 m height (at each level a score from 0 to 3 was assigned for canopy cover, representing 0%, 0–25%, 25–50% and > 50% cover, respectively); canopy height; litter cover; litter depth; number of epiphytes in the vicinity of the quadrat; and ground complexity. Ground complexity was measured by extending a 2 m pole from the middle of the quadrat toward the outside in four equidistant directions. The percentage of the pole near or touching the following items was estimated: decomposing log, decomposing litter, twigs, dry litter, tree roots, saplings, tree trunk, vine, bare ground, and decomposing branch. The percentage scores for each sample (four per quadrat) were converted to an index of complexity by using Simpson's index of diversity (Sokal and Rohlf 1981). Canopy cover and litter depth were measured at each corner of the quadrat, and values were averaged to obtain a mean value per quadrat.

Statistical Analysis

Analysis of covariance (ANCOVA) was employed to compare species-area relationships. *T*-tests were used to compare regression parameters. One-way analysis of variance (ANOVA) was used to test for differences between sites in rates of basin colonization and relative abundances of terrestrial breeders. Following a significant ANOVA, Tukey's HSD tests for unequal sample sizes were performed to determine where significant variation lay (Sokal and Rohlf 1981).

Ordination of the species/site data matrix was carried out using detrended correspondence analysis (DCA) on the DECORANA program (Hill 1979a). Only species presence or absence data were used; thus all species were weighted equally. This technique projects the sites onto a number of ordination axes, so that more similar communities occur closer together. The relative strengths of DECORANA axes were given as eigenvalues, and the relative importance of each axis in explaining variance in the data set was found by dividing the value for that axis by the sum of all three eigenvalues. Groupings of sites were also assessed using a cluster analysis (euclidean distance, group-average method) to test for concordance between analyses.

Both least-squares and all-subsets multiple regression were carried out to investigate the relationship between the measured environmental (explanatory) variables and the relative abundance estimates for the terrestrial-breeding frog species (response variable).

RESULTS

Community Analysis

Species richness decreased with fragment size ($P < .05$, $F_{[1,16]} = 16.94$, ANCOVA). Surprisingly, however, species richness was higher in forest fragments than in primary forest samples of the same size (constants not significantly different: $P = .05$, $F_{[1,16]} = 4.40$, ANCOVA; fig. 9.1). This observed increase in species richness was consistent for fragments of all sizes (slopes not significantly different: $P > .05$, $t_{.05(2)13} = 2.160$, t-test; fig. 9.1). In other words, fragments "gained" an average of 5.51 species over continuous forest (fig. 9.1). Moreover, the fragments surveyed both before and after isolation (numbers 3304, 3209, 1112, and 1104: 100 ha, 10 ha, 1 ha, and 1 ha, respectively) exhibited an average increase of 10 species after isolation (fig. 9.2).

Very few species were lost from the fragments after isolation. A comparison of species lists before and after fragmentation mainly shows an influx of new species (table 9.2). Species that were lost *(Atelopus pulcher, Leptodactylus riveroi, Leptodactylus pentadactylus, Ctenophyrne geayi, Synapturanus miranderibo, Phyrnohyas resinifictrix, Bufo marinus,* and *Colostethus marchesianus)* were rare, uncommon (<1 individual/ha), or stream breeders.

Detrended correspondence analysis (DCA) grouped fragments of the same size into nonoverlapping clusters (fig. 9.3). Clusters and forest tracts surveyed prior to their isolation (the pre-isolation cluster) are ordered along Axis 1 with respect to fragment size. Axes 1 and 2 together accounted for 86% of the variation captured by the first three axes (59% for Axis 1, 23% for Axis 2). Similar groupings were found using cluster analysis, with the exception of fragment 1202 (10 ha), which was grouped with the 1 ha fragments.

When the four fragments surveyed before and after isolation are compared on the DCA plot, all four post-isolation samples have higher values on Axis 2 than the pre-isolation samples (fig. 9.3). For all four fragments, the ordering along Axis 1 remained unchanged after isolation, indicating that there was a similar shift in species composition regardless of fragment size.

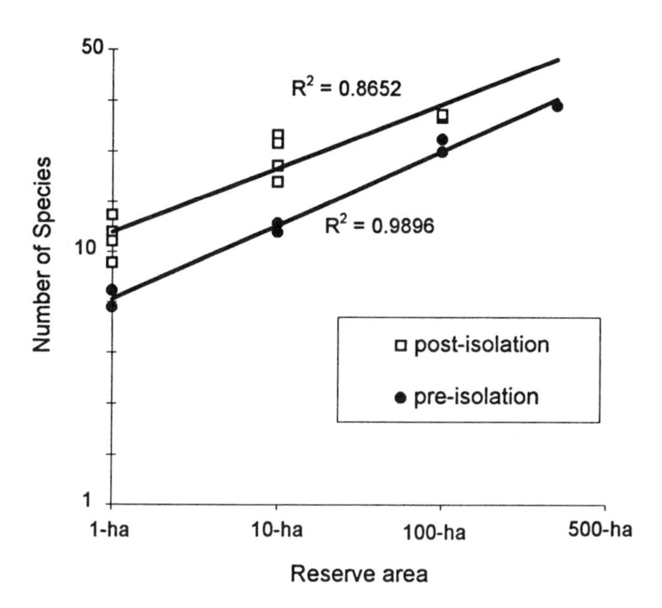

Figure 9.1. Species-area relationships for frog communities in experimental Amazonian reserves surveyed both before and after isolation. (Pre-isolation data from Zimmerman and Bierregaard 1986.) Some isolated fragments were surrounded by second-growth forest (see table 9.1).

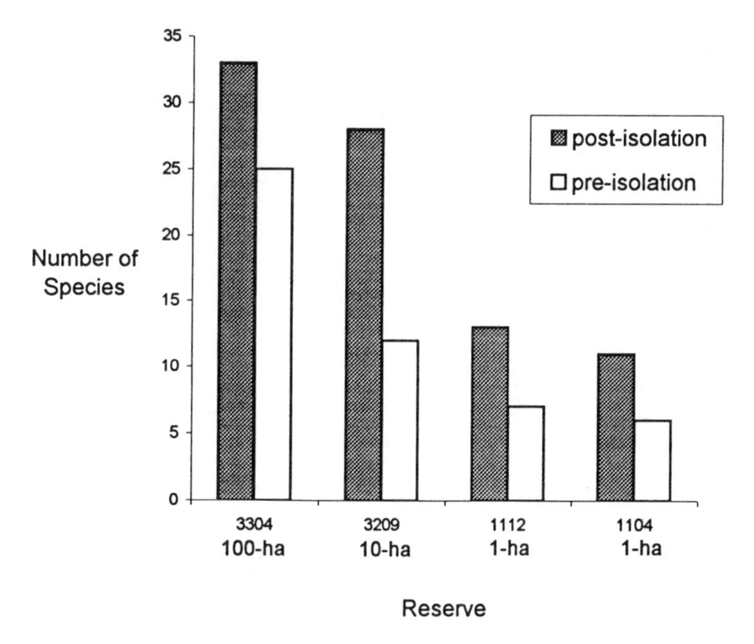

Figure 9.2. Number of frog species detected in fragments before and after isolation from surrounding forest.

Table 9.2 Species lost or gained from four fragment reserves studied before and after isolation.

	Fragment[a]			
	1104	1112	3209	3304
Species	1 ha	1 ha	10 ha	100 ha
Bufonidae				
Atelopus plucher				−
Bufo granulosus			−+	
Bufo marinus	+	−	+	+
Bufo typhonius-like			+	+
Dendrophryniscus minutus		+		+
Dendrobatidae				
Colostethus marchesianus	−	+	+	+
Colostethus stepheni	+	+	+	+
Epipedobates femoralis		−+	−+	−+
Centrolenidae				
Centrolenella oyampiensis			+	
Hylidae				
Hyla gr. *brevifrons*		−+	−+	−+
Hyla geographica				+
Hyla granosa				+
Hyla marmorata			−+	−+
Hyla minuta	−+	−+	−+	−+
Osteocephalus sp.	+	+	+	+
Osteocephalus taurinus			+	+
Phyllomedusa bicolor				+
Phyllomedusa tarsius			−+	−+
Phyllomedusa tomopterna			−+	−+
Phyrnohyas resinifictrix		−	+	+
Scinax cruentomma			+	+
Scinax rubra	−+	−+	−+	
Leptodactylidae				
Andenomera andreae	+	+	+	+
Andenomera hyladactylae				−+
Ceratophrys cornuta		+	+	+
Eleutherodactylus fenestratus	+	+	+	+
Eleutherodactylus zimmermaneae		+	+	+
Leptodactylus knudseni	−+		−+	−+
Leptodactylus leptodactyloides	−+			−+
Leptodactylus mystaceus	+	+	+	+
Leptodactylus pentadactylus	−	−	−	+
Leptodactylus rhodomystax			+	+
Leptodactylus riveroi				−
Leptodactylus stenodema			+	+
Microhylidae				
Chiasmocleis shudikarensis			−+	−+
Chiasmocleis sp. A			−+	−+
Ctenophryne geayi				−
Synapturanus miranderiboi	+			−
Synapturanus salseri			+	+
Microhylidae sp. A			−+	

a. + = species present before and after isolation; − = species lost from reserve after isolation; −+ = species originally absent but gained after isolation.

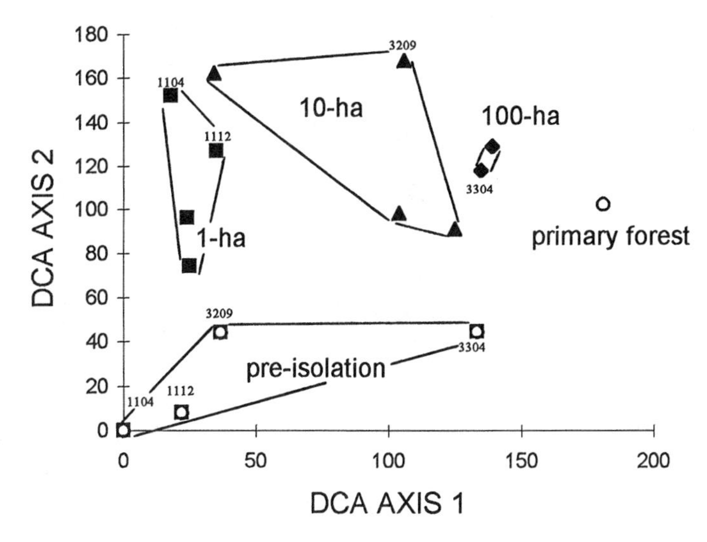

Figure 9.3. DCA ordination of frog communities using species presence/absence data. Note that clusters of sites are arrayed along Axis 1 in order of reserve size. The pre-isolation sites contained fewer frog species.

Breeding Success

Rainfall was high during November and May in 1988–1989, but overall there was no significant difference in rainfall between 1987–1989 (when primary forest breeding success was measured) and 1992–1994 (when fragment breeding success was measured; $P > .05$, $F_{[6,14]} = 3.28$, two-way ANOVA).

Three species, *Colostethus marchesianus, Epipedobates femoralis,* and *Osteocephalus taurinus,* were common colonizers of artificial basins. Basins in two fragments (3304 [100 ha], and 2206 [10 ha]) took significantly longer to be colonized by all three species than those in primary forest and in the other fragments ($P < .05$; $F_{[7,18]} = 20.83$, $F_{[5,39]}$, $F_{[5,27]} = 13.88$ for *C. marchesianus, E. femoralis,* and *O. taurinus,* respectively: Tukey's HSD test; fig. 9.4). Basins in fragment 3209 (10 ha) also took significantly longer to be colonized by *E. femoralis* (fig. 9.4) than did the other sites.

Only *C. marchesianus* colonized basins in 1 ha fragments, and breeding was not successful there (fig. 9.5). However, breeding success of *C. marchesianus* in 10 and 100 ha fragments was generally higher than in primary forest (fig. 9.5). For *E. femoralis,* breeding success was variable in fragments (fig. 9.5), with fragments 2206 and 1207 (10 ha) having higher breeding success than primary forest and 100 ha fragments. For *Osteocephalus taurinus,* breeding success was marginally lower in the fragments, with lower success in the 10 ha than in the 100 ha fragments.

Abundance Estimates of Terrestrial Breeders

Colostethus stepheni had higher abundances in primary forest than in fragments ($P < .05$, $F_{[3,7]} = 7.79$, ANOVA; Tukey's HSD test). *Eleutherodactylus fenestratus* showed an opposite trend, with higher abundances in 100 ha and 1 ha fragments than in primary

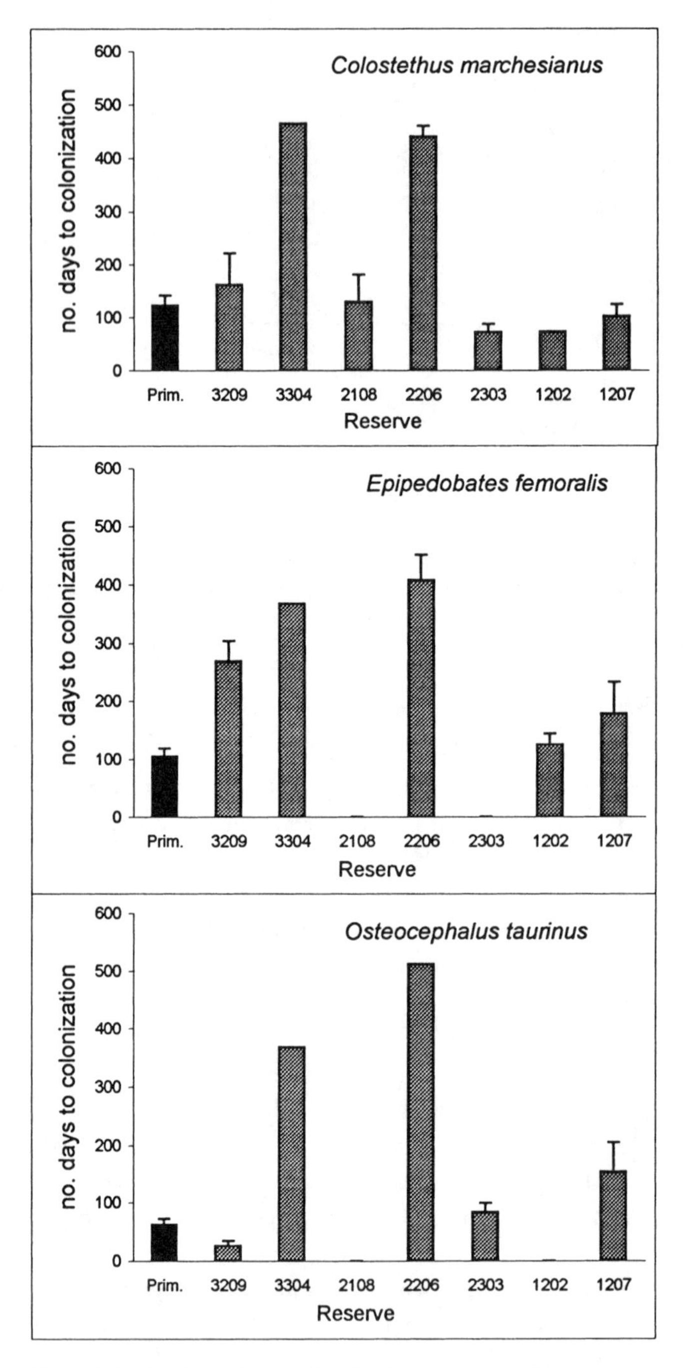

Figure 9.4. Number of days taken by three rainforest frog species to colonize experimental basins in fragments and in primary forest ($\bar{X} \pm$ SE). Sites with no vertical bar were never colonized.

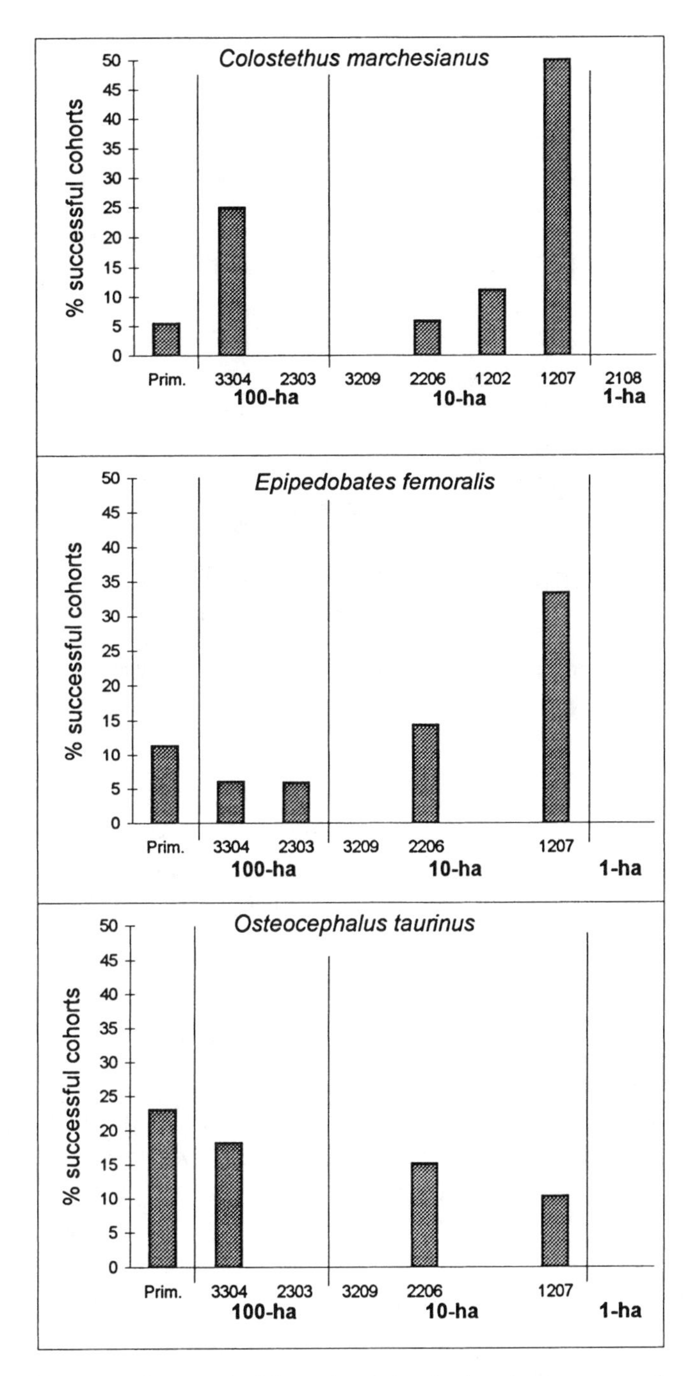

Figure 9.5. Percentage of frog cohorts that developed to late-stage tadpoles, for all sites where cohorts were initiated.

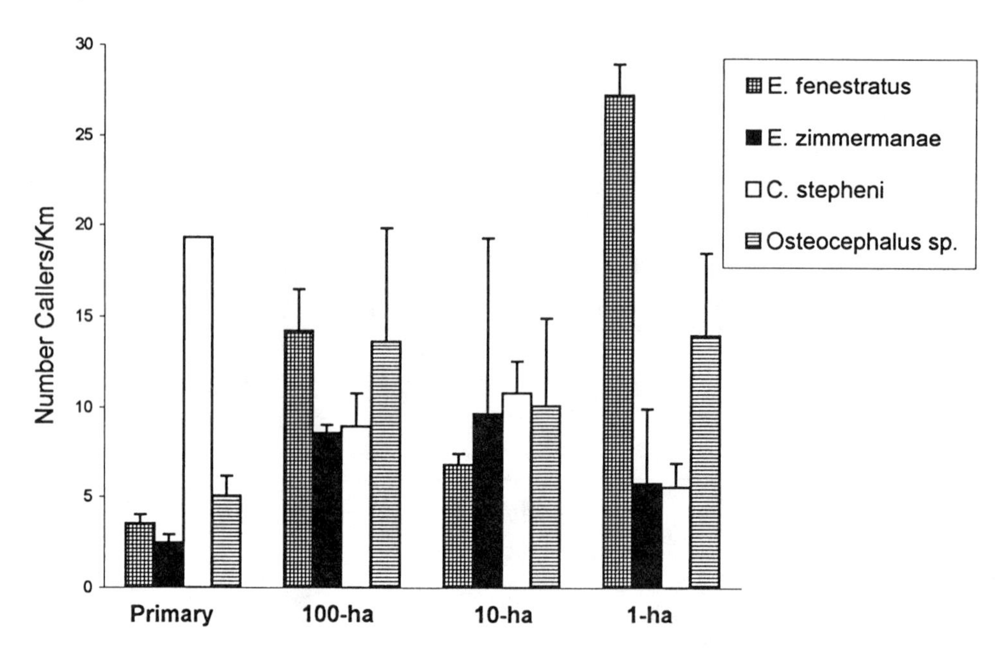

Figure 9.6. Abundance estimates for terrestrial-breeding frog species in fragmented and primary forest ($\bar{X}$ ± SE). For *Colostethus stepheni,* the value shown was calculated from surveys during 1992–1994 only.

forest ($P < .05$, $F_{[3,10]} = 32.35$, ANOVA; Tukey's HSD test), although there was no difference between 10 ha fragments and primary forest (Tukey's HSD test). Neither *E. zimmermaneae* nor *Osteocephalus* sp. differed significantly in abundance between fragments and primary forest ($P > .05$, $F_{[3,10]} = 4.31$ and 1.41 for *E. zimmermaneae* and *Osteocephalus* sp., respectively, ANOVA) (fig. 9.6).

Mid-story canopy cover (2–5 m) and litter depth generally increased in smaller fragments, while litter cover decreased somewhat in smaller fragments, but the trend was variable (table 9.3). Using least-squares regressions, both mid-story canopy cover ($P = .02$) and litter depth ($P = .04$) were significant predictors of *E. fenestratus* abundance. In the all-subsets regression, mid-story canopy cover, litter depth, and ground complexity together explained 72% of the total variation in *E. fenestratus* abundance. There were no significant predictors of abundance for *C. stepheni, Osteocephalus* sp., and *E. zimmermaneae.*

DISCUSSION

Community Analysis

In central Amazonia, species richness of frogs was strongly and positively related to fragment area. Surprisingly, however, compared with pre-isolation data, species richness consistently increased in forest fragments following isolation. This increase was nearly constant in magnitude across the range of fragment sizes (1–100 ha) examined in this study.

This pattern contrasts with those observed in some other taxa studied in the Biological

Table 9.3 Averaged values of environmental variables for fragments of 1, 10, and 100 ha and two primary forest sites.

		Fragments		
	Primary forest	100 ha	10 ha	1 ha
Understory cover (0–2 m)	0.65 ± 0.02	0.96 ± 0.04	0.80 ± 0.06	0.95 ± 0.07
Mid-canopy cover (2–5 m)	1.07 ± 0.02	1.07 ± 0.04	1.10 ± 0.05	1.33 ± 0.19
High canopy cover (>5 m)	1.84 ± 0.07	1.81 ± 0.06	1.78 ± 0.09	1.87 ± 0.13
Canopy height (m)	19.6 ± 3.7	17.8 ± 0.05	21.1 ± 0.9	17.4 ± 0.85
Litter cover (%)	76.8 ± 4.5	67.2 ± 6.6	66.6 ± 1.4	66.4 ± 6.5
Litter depth (cm)	2.0 ± 0.46	2.4 ± 0.12	2.2 ± 0.22	2.5 ± 0.35
Ground complexity	1.9 ± 0.09	2.3 ± 0.0	2.2 ± 0.1	2.0 ± 0.44
Epiphytes (no.)	1.2 ± 0.42	2.1 ± 0.36	1.7 ± 0.30	0.6 ± 0.13

Note: Values shown are the mean ± 1 SE. See Methods section for definition of variables.

Dynamics of Forest Fragments Project (BDFFP), especially groups such as insectivorous and frugivorous birds (Bierregaard and Stouffer, chap. 10) and some primates (Lovejoy et al. 1986). These results are somewhat more similar to trends observed in small mammals (Malcolm, chap. 14) and butterflies (Brown and Hutchings, chap. 7), both of which exhibited an increase in species richness in fragments during the years immediately following isolation. However, few frog and small mammal species were lost from fragments, whereas butterfly assemblages changed substantially, with light-loving species often displacing forest interior species. Much of the observed increase in species richness of frogs in fragments can be explained by the invasion of matrix-associated species not normally found in primary forest.

It should be noted that sampling effects could also have contributed to the elevated richness of frogs after fragment isolation. The initial (pre-isolation) sampling of frogs was carried out by the third author, while the post-isolation sampling was carried out by the senior author, who had the benefit of the third author's prior species lists for the sites. Under these circumstances, it would be natural for a few new species to be added to reserve or fragment lists as the site is studied over progressively longer periods of time and by different observers. However, most of the species added to the reserve lists after isolation were associated with matrix habitats, so their addition is likely to have resulted from the close proximity of matrix habitats after reserve isolation, rather than from a serious sampling bias.

A core group of species was involved in the post-isolation community shift. Species regarded as opportunists of disturbed areas, such as *Hyla minuta* (also considered a forest species), *Hyla* gr. *brevifrons,* and *Scinax rubra* (Duellman 1978; Zimmerman and Simberloff 1996), constitute this core group, and all three of these species migrated into at least three of the four fragments surveyed before and after isolation (see table 9.2). *Epipedobates femoralis,* although not considered an opportunist, also appears capable of long-distance migration (see table 9.2), and was a common colonizer of artificial pools in matrix habitats. A comparable shift in community composition was observed for small mammals in small BDFFP fragments (Malcolm, chap. 14).

In contrast to several other animal groups, few forest frogs were lost from fragments following isolation. Most forest species that were not detected in fragments typically have very low abundances in primary forest (e.g., *Phyrnohyas resinifictrix, Bufo marinus, Colostethus marchesianus, Atelopus pulcher, Leptodactylus riveroi, Ctenophyrne geayi*: Zimmerman and Simberloff 1996). Their absence in fragments may simply be an artifact of the small area sampled, or it may have resulted from an absence of appropriate breeding habitat in fragments.

Because our pre- and post-isolation surveys of fragments were separated by seven years, it is impossible to determine whether some frog species initially declined after fragmentation, then subsequently recovered, possibly after immigration or recolonization from nearby source populations. Many of the forest species in fragments are capable of crossing and breeding in the surrounding pastures (C. Gascon, pers. obs.), which suggests that the fragments were not true isolates with respect to the frog community. Significantly, isolates surrounded by dense second growth exhibited a greater influx of species than those surrounded by cattle pasture (see table 9.1). The nature of matrix habitats has a strong influence on movements of small mammals and birds (Bierregaard and Lovejoy 1989; Gates 1991; Bierregaard and Dale 1996; Malcolm, chap. 14). Malcolm (chap. 14) showed that the community composition of small mammals in small fragments was strongly influenced by the type of matrix (pasture vs. second growth) surrounding the fragment. The nature of the matrix also seemed to influence frog movements strongly, with fragments surrounded by high, shady second growth being more accessible to colonizing and invading species than were sites surrounded by pasture.

The consistent increase in frog species numbers regardless of fragment size means that small fragments had a relatively higher increase in foreign species than large fragments. The consequences of this influx of species and individuals are unknown. In the same study area, Lovejoy et al. (1986) found that displaced forest birds crowded into remaining fragments after the surrounding forest had been cut and displayed greater persistence in fragments than did the former residents. This increase in species and individuals, however, was a short-term phenomenon (Lovejoy et al. 1986). For forest frogs, the long-term consequences of an increased species load in fragments (shifts in species interactions, changes in resource use) are difficult to predict, especially because few frogs in the community are territorial. There is, however, no evidence that resident forest frogs had disappeared from fragments within 7–10 years of isolation.

Breeding Success

Our prediction that fragmentation would have a negative effect on frog breeding success and abundance was only partially supported. For both *Colostethus marchesianus* and *Epipedobates femoralis,* breeding success in artificial pools was higher in 10 and 100 ha fragments than in primary forest (see fig. 9.5), while breeding success of *Osteocephalus taurinus* in these pools declined somewhat in fragmented forest. This may have resulted because *O. taurinus* was less effective than other species in moving through matrix areas. Prior to the experiment, a lack of appropriate breeding habitat in fragments may have inhibited breeding by *C. marchesianus* and *E. femoralis,* and the addition of artificial pools 7–10 years after isolation could have induced an intense breeding effort and the subsequent success of cohorts in fragments.

Abundances of Terrestrial Breeders

There were few consistent trends in terms of frog abundances before and after fragmentation. The abundance of calling *Eleutherodactylus fenestratus* increased in small fragments relative to the other sites, whereas *Colostethus stepheni* showed an opposite response. *E. fenestratus* abundance appeared to be influenced by canopy cover and litter depth, both of which were high in 1 ha fragments (see table 9.3). Deep litter, which maintains high moisture levels, may be important for *E. fenestratus,* whose terrestrial nests are prone to desiccation and predation (Hodl 1990).

Of the four species examined closely, only *C. stepheni* declined in fragments (although the abundance estimate for primary forest was based on only two seasons' surveys, from 1992 to 1994; see fig. 9.6). Although *C. stepheni* appears strongly affected by forest fragmentation, the absence of any relationship with measured habitat variables indicates a need for more research on this species to identify the proximate causes of its population declines.

In summary, species richness of frogs increased after isolation, regardless of fragment size. The four species we studied intensively exhibited highly individualistic patterns of breeding success and abundance in fragments. Only one of these appeared to decline after fragmentation, and no species exhibited reduced breeding success in fragments. These results suggest that the frog community in the Manaus area is not as severely affected by fragmentation as other faunal groups. This may be the case because many frogs can use the matrix surrounding fragments, and also because frogs are heterothermic, and thus have much smaller energetic and spatial requirements than homeothermic vertebrates such as birds and mammals (Duellman and Trueb 1986).

General Implications

1. The frog community in central Amazonia appears to be less affected by localized and recent habitat loss and fragmentation than other vertebrate groups.
2. For the majority of frogs in the Manaus area, forest fragments are probably not truly isolated because many species can utilize and migrate through the surrounding matrix.
3. As in undisturbed continuous forest, the composition of frog assemblages in fragments can largely be predicted by the availability of breeding habitat and, for matrix species, by the presence of forest edge habitat.
4. More pre- and post-isolation comparisons from other locations and long-term monitoring are required before we can conclude that forest fragmentation has only a limited effect on frogs.

ACKNOWLEDGMENTS

This study was supported by a grant from the Smithsonian Institution and logistic support from the Instituto Nacional de Pesquisas da Amazônia and the Zoology Department, University of Canterbury, New Zealand. We thank Ocirio Pereira and R. Didham for help in the field. This chapter is publication no. 150 of the Minimum Critical Size of Ecosystems Project.

10

Understory Birds and Dynamic Habitat Mosaics in Amazonian Rainforests

Richard O. Bierregaard, Jr., and Philip C. Stouffer

Although many tropical forests are subject to frequent disturbance caused by river meanders, treefalls, windstorms, and possibly region-wide climatic cycles, our society has imposed a regime of deforestation upon these natural dynamic patterns that is clearly unprecedented in its rate and extent. By definition, deforestation reduces the amount of primary forest available for native flora and fauna. A nearly inevitable result of this reduction in area is the fragmentation of the remaining forest. In a complex fashion, habitat fragmentation affects the native biota (Saunders, Hobbs, and Margules 1991; Laurance 1994), local and regional hydrological cycles (Salati and Nobre 1991), and social and economic conditions of the local human inhabitants (Schelhas and Greenberg 1996).

Because most human activities supplanting Neotropical forests since European colonization of South America have proven unsustainable, much land that was deforested in Amazonia over the past century has now been abandoned. Silva, Uhl, and Murray (1996) estimated that 30% of the pastures cleared in eastern Amazonia are fallow, and Uhl et al. (1991) stated that millions of hectares of pastures have been abandoned in Amazonia. In the Manaus region, satellite imagery shows that there was little new deforestation from 1985 to 1991, while much of the agricultural land already cleared was left unattended, resulting in an increase in the proportion of secondary forest (Lucas et al. 1993).

These abandoned lands undergo some form of succession, the nature of which depends upon the human activities carried out on them (Uhl, Buschbacher, and Serrão 1988). In turn, succession can be expected to affect any remnant patches of primary forest isolated in the midst of the abandoned pasture. Like the rainforest it replaces, the landscape humans create is never static, but instead highly dynamic.

Much initial research into habitat fragmentation focused on the relationship between fragment size and the number of species persisting in the fragment (Connor and McCoy 1979; Shafer 1990). Subsequently, researchers have focused on other aspects of fragmentation, especially the creation of habitat edges (Laurance 1991b, Laurance and Yensen 1991; Malcolm 1994) and the interaction between the biota of fragments and that of the surrounding matrix of human-altered vegetation, including other nearby fragments (Malcolm 1991b; Fahrig and Merriam 1994).

In this chapter, we review long-term experimental studies of birds in a dynamic system of small forest remnants surrounded by pasture or abandoned pasture undergoing secondary succession, conducted in our study area north of Manaus, Amazonas, Brazil. Data from thirteen years of a mark-recapture program focusing on understory birds reveal changes in species composition and activity levels in a series of 1 and 10 ha fragments and a single 100 ha fragment, originally isolated by clear-cutting.

We have performed separate sets of analyses based on broadly defined ecological guilds: nectar feeders (Stouffer and Bierregaard 1995a), insectivores (Stouffer and Bierregaard 1995b), and frugivores (this chapter). Very few, if any, "frugivorous" species feed exclusively on fruit. Frugivores, like hummingbirds, supplement their diets to some extent with insect prey. Herein, we refer to these species simply as frugivores, while recognizing this dietary niche breadth.

Over the course of the study, most of the pastures surrounding the experimental remnants were abandoned. Depending upon the intensity of ranching activity in the pastures, one of two distinct secondary vegetation types developed. We relate changes in the fragment avifauna to remnant size, time since isolation, and the nature of the surrounding vegetation.

METHODS

Project Background

In 1979, researchers from Brazil's National Institute for Research in Amazonia (INPA) and the World Wildlife Fund-U.S. launched the Minimum Critical Size of Ecosystems Project, a multidisciplinary experimental study of the effects of habitat fragmentation on an Amazonian rainforest ecosystem. The principal goal of the project was to measure the relationship between forest remnant size and the number of species that could persist in the remnant over time. Census data on a broad range of organisms, collected before and after specific tracts of forest were isolated by deforestation, were to be extrapolated in an estimate of the minimum area required to preserve the species and processes of the Amazonian rainforest ecosystem (Lovejoy and Oren 1981). The experimental design permitted a quantified assessment of ecosystem collapse (or relaxation) as species disappeared from a series of different-sized forest remnants. The project, now called the Biological Dynamics of Forest Fragments Project (BDFFP), is currently administered jointly by INPA and the Smithsonian Institution in Washington, D.C.

As originally conceived, the BDFFP was to establish a replicated series of experimental forest reserves of 1, 10, 100, and 1,000 ha. The experimental design took advantage of a Brazilian law that requires landowners in Amazonia to leave the forest standing on half of any land under development. Twenty-four reserves were delineated in primary forest on four cattle ranches in the Distrito Agropecuário, a 500,000 ha agricultural research and development area some 30–90 km north of the city of Manaus, Amazonas (2°30′ S, 60° W; fig. 10.1). The selection of experimental reserves and execution of the project was carried out with the enthusiastic cooperation of the ranchers.

Project researchers collected baseline natural history data on a range of organisms prior

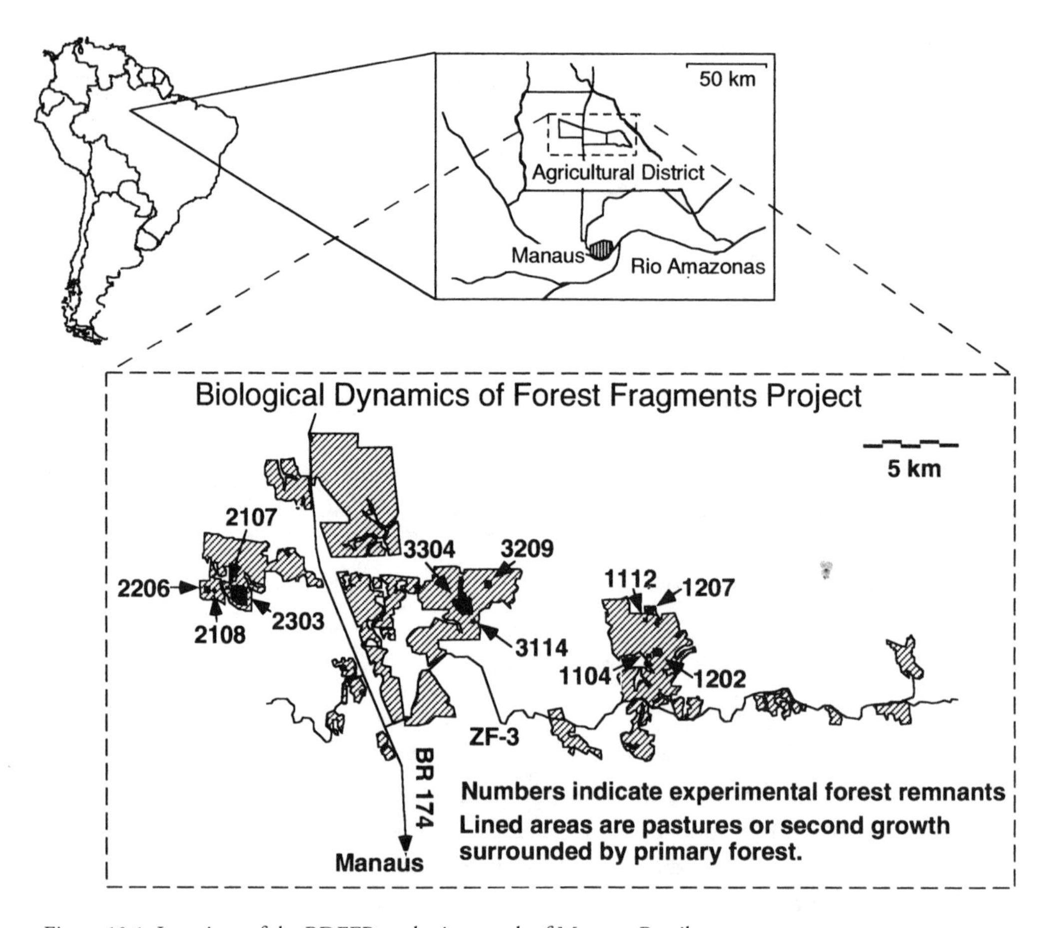

Figure 10.1. Locations of the BDFFP study sites north of Manaus, Brazil.

to isolation of the reserves by clear-cutting by the ranchers. These studies continued during and after the felling of the forest around the experimental reserves. They offer, to the best of our knowledge, the only replicated studies of ecosystem change in fragmented tropical forests that include pre-isolation data from the fragments. Focal groups included, but were not restricted to, canopy trees (Rankin-de Mérona et al. 1992), leaf litter insects (Didham, chap. 5), butterflies (Brown and Hutchings, chap. 7), amphibians (Tocher, Gascon, and Zimmerman, chap. 9), nonflying mammals (Malcolm, chap. 14), and, as discussed here, birds. In addition, after the first reserves were isolated, V. Kapos and her colleagues began studying physical changes in microclimate occasioned by habitat fragmentation (Kapos et al., chap. 3). Reviews of the BDFFP results are found in Lovejoy et al. (1984, 1986) and Bierregaard et al. (1992).

The Primary Forest

The cattle ranches at the BDFFP site each have several thousand hectares of cleared land, interspersed with patches of primary forest (including the Project's experimental reserves). These ranches are themselves surrounded by primary forest, which extends unbroken for hundreds of kilometers to the north, east, and west. To the south, the forest is largely intact for some 60 km, where it meets the Rio Negro and the sprawling city of Manaus. The forest in the study area is classified as tropical moist (Holdridge 1967). Soils are relatively nutrient-poor, sandy or clayey xanthic ferralsols (FAO/UNESCO 1971), called yellow latosols in the Brazilian terminology (Camargo 1979). Annual rainfall in the region averaged 2,186 mm in one thirty-year period (Ministério de Minas e Energia, 1978a), but has been variously reported at 1,900 to 2,500 mm (Stouffer and Bierregaard 1993). There is a pronounced dry season from June through October.

The canopy of the forest near Manaus is about 30–37 m tall, with emergents as high as 55 m. The flora is remarkably diverse in tree species (Prance 1990; Rankin-de Mérona et al. 1992), but the understory is depauperate, both in number of species and density of flowering and fruiting plants, relative to other Amazonian forests (Gentry and Emmons 1987).

Floristically, the Manaus forests differ from those of western Amazonia and Central America. For example, the families Lecythidaceae, Burseraceae, Chrysobalanaceae, Myristicaceae, and Bombacaceae are among the twenty most species-rich families at Manaus, but are far less common at three other well-studied field stations in Amazonia and Central America, each of which has its own particular combination of dominant families (Gentry 1990a). A more detailed description of the BDFFP study areas can be found in Lovejoy and Bierregaard (1990).

Reserve Isolation

By prior arrangement with the cattle ranchers, reserves were demarcated in primary forest so that they would be isolated by at least 200–300 m from adjacent forest after the surrounding forest was cleared. However, last-minute changes in ranch plans left three reserves isolated by only 70–100 m, with the remainder surrounded by at least 150–1,000 m of felled forest (table 10.1).

In the early 1980s, three or four years after the start of the BDFFP, the Brazilian economy was suffering, and cattle ranching in Amazonia was proving to be an ill-fated venture. The government agencies that were subsidizing 75% of the cost of establishing cattle ranches in the Distrito Agropecuário began to divert scarce funds to other development projects in the region. Consequently, the owners of the ranches where the BDFFP reserves are located stopped clearing forest in 1984, after isolating only ten of twenty-four planned reserves (two reserves were isolated in 1980, five in 1983, and three in 1984). An additional reserve was isolated in 1990 (table 10.1).

The felled forest around the five reserves created in 1983 was never burned, and within a year of clear-cutting, supported a robust flush of second-growth vegetation and stump resprouts (fig. 10.2). The surrounding landscape today is dominated by *Cecropia sciadophylla,* which by 1991 had reached 12–15 m in height. In contrast, second growth on fallow pasture is dominated by trees in the genera *Vismia* (Clusiaceae) and *Bellucia* (Melastomataceae), which by 1991 had reached a height of 3–7 m around the 1984 isolates and 5–10 m around the 1980 isolates. The vertical profiles of these two second-growth types are roughly similar, although the *Cecropia* areas have a more continuous canopy

Table 10.1 Characteristics of experimental reserves of the Biological Dynamics of Forest Fragments Project.

Reserve number	Fazenda (ranch)	Size (ha)	Year of isolation	Distance to forest	Burn?	Surrounding vegetation after 1 year	3 years	5 years	7 years
1104	Esteio	1	1980	70	Yes	2nd growth	*Vismia*	Recleared	*Vismia*/pasture
1112	Esteio	1	1983	250	No	2nd growth	*Cecropia*	*Cecropia*	*Cecropia*
3114	Porto Alegre	1	1983	150	No	2nd growth	*Cecropia*	*Cecropia*	*Cecropia*
2107	Dimona	1	1984	150, 300	Yes	Pasture	Pasture	Pasture/*Vismia*	Pasture/*Vismia*
2108	Dimona	1	1984	400	Yes	Pasture	Pasture	Pasture/*Vismia*	Pasture/*Vismia*
1202	Esteio	10	1980	70, 600	Yes	2nd growth	*Vismia*	Recleared	*Vismia*/pasture
1207	Esteio	10	1983	100	No	2nd growth	*Cecropia*	*Cecropia*	*Cecropia*
3209	Porto Alegre	10	1983	650	No	2nd growth	*Cecropia*	*Cecropia*	*Cecropia*
2206	Dimona	10	1984	200	Yes	Pasture	Pasture	Pasture/*Vismia*	Pasture/*Vismia*
3304	Porto Alegre	100	1983	Corridor	No	2nd growth	*Cecropia*	*Cecropia*	*Cecropia*
2303	Dimona	100	1990	150	Yes	2nd growth	*Vismia/Cecropia*	*Vismia/Cecropia*	*Vismia/Cecropia*

Figure 10.2. Eight-year-old second growth dominated by *Cecropia sciadophylla* surrounding a 10 ha experimental forest fragment. This area was clear-cut but never burned because the 1983 rainy season started earlier than usual.

cover than do the *Vismia* areas (Borges 1995). This was due in part to continued grazing by some cattle at the *Vismia*-dominated sites, which kept some areas of pasture open.

Features of the Amazonian Avifauna

The avifauna of South America numbers over 3,000 species, or roughly one-third of the known bird species on the planet. The heart of this species richness lies in rainforests (but see Fjeldså and Rahbek, chap. 30). Over 550 bird species have been reported from at least two individual forest sites in western Amazonia (Karr et al. 1990), while in the BDFFP area, where there is no large lake or river, over 400 species are known to occur (Stotz and Bierregaard 1989; Cohn-Haft, Whittaker, and Stouffer, in press).

Perhaps the most consistent feature of Neotropical forest avifaunas is the preponderance of rarity (Lovejoy 1975). In general, no one species or group of species dominates the community, either in numbers or biomass, and most species are rare. In the BDFFP study, over 90% of the bird species individually represented less than 2% of the total captures, and the most abundant species accounted for only about 10% of all captures (Bierregaard 1990).

Long-term studies of avian communities at four field stations in forests of Central and South America indicate a broad ecological similarity throughout the region. A large number of species are proportioned roughly similarly across broadly defined ecological niches (Karr et al. 1990). Insectivores are usually the most species-rich group, although a small frugivore is often the most abundant species in capture samples at a given site (Karr et al. 1990). Species exhibit substantial vertical stratification in foraging; canopy species rarely descend to the understory, and vice versa (Pearson 1971).

Two highly specialized groups of insectivores are especially conspicuous in the understory. One of these comprises three to five species found in congregations feeding exclusively over swarms of army ants (*Eciton* spp.), preying upon insects flushed by the marauding ants (Willis and Oniki 1978). The other group of about thirteen species forages through the forest in organized mixed-species flocks, with one pair of each species defending conterminous territories from conspecifics in other such flocks (Munn 1985; Powell 1985).

The vast majority of Amazonian forest bird species do not migrate. During the boreal winter, there is a substantial influx of species migrating to the forests of Central America and western Amazonia from their breeding areas in more northern forests (Karr et al. 1990). However, very few migrants, either austral or boreal, are found in the central Amazonian forests (Stotz et al. 1992). The sedentary nature of the species studied here makes it easier to interpret their responses to habitat fragmentation than has been the case for migratory species, whose temperate breeding areas and wintering grounds are both being reduced and fragmented by human activity (Hagan and Johnston 1992).

Census Techniques

Birds were censused with fine black nylon mist nets (12 × 2 m, with a 36 mm mesh, NEBBA type ATX). Nets were strung in 100 m transects (eight nets) in 1 ha reserves, or 200 m transects (sixteen nets) in 10 and 100 ha reserves. With the exception of hummingbirds, all birds were banded with uniquely numbered aluminum leg bands.

Net lines were opened for one day (0600 to 1400 hours) at somewhat regular intervals. Initially, we sampled each reserve about once a month. In the later years of the study, the period between netting sessions averaged six to eight weeks.

Control data were collected in two ways: (1) from mist-net lines in the isolated reserves prior to isolation, and (2) from mist-net lines run in continuous forest located a few hundred meters to a few dozen kilometers from the experimental isolates throughout the course of the study.

Mist netting does not provide an unbiased estimate of abundance, and hence interspecific comparisons of capture rates must be undertaken very cautiously. In this study, most analyses are based on intraspecific comparisons of capture rates in the same reserves at different times.

In our analyses (Bierregaard and Lovejoy 1988, 1989; Stouffer and Bierregaard 1995a,b; this chapter) we do not maintain that our data are necessarily an indication of population size. Changes in capture rates might reflect either changes in population densities or changes in various aspects of social or foraging behavior. However, for the purposes of these analyses, we are interested in finding quantifiable indications of ecological change, whether it be in population levels or in foraging behavior, and consequently find mist-net samples to be useful, while recognizing their limitations. Our goal is to demonstrate measurable changes in avian biology and then to correlate these changes with our experimental perturbations: the size of a reserve or the habitat conditions surrounding it.

Captures from five 1 ha and four 10 ha isolates provide the bulk of the data reviewed in this chapter. Most analyses were based on mist-net capture rates, reported here as captures/1,000 net-hours, with same-day recaptures excluded. During the course of the experiment, the percentage of birds captured on a given sample day that had previously been banded (recapture rates) was also monitored.

Data were grouped into five time periods. All samples prior to the isolation of a given reserve were grouped in period T_0. Period T_1 is the first year after isolation, T_2 is 2–3 years after isolation, T_3 is 4–6 years after isolation, and T_4 is 7–9 years after isolation.

Captures from T_1 were ignored for insectivores and frugivores because these data were confounded by an influx of birds seeking refuge from the felled forest around the isolates (Bierregaard and Lovejoy 1988). Data from hummingbirds, however, suggest that there was no substantial influx of these birds at the time of isolation. Consequently, these data have been included in analyses of their response to isolation (Stouffer and Bierregaard 1995a).

The data presented here on frugivores, not previously reported, were analyzed primarily with the Wilcoxon signed-ranks test, a nonparametric test of paired observations (Sokal and Rohlf 1981). The paired observations were pre- and post-isolation capture rates for species (individually or grouped) from the same fragment in time periods T_0 (pre-isolation), T_2, T_3, and T_4.

We also used repeated-measures analyses of variance (ANOVAs) to isolate and assess the effects of reserve size, surrounding matrix type, and time since isolation on capture rates. Because of small sample sizes for many species, we limited the ANOVAs to six of the seven species for which we recorded at least fifty captures in the pre- and post-isolation samples. Our analyses of the seventh common species, the nomadic ruddy quail-dove *(Geotrygon montana),* have previously been reported (Stouffer and Bierregaard 1993).

RESULTS

Early in the dry seasons of 1980, 1983, and 1984, ranch workers isolated the reserves with machetes, axes, and chainsaws. This drastic alteration of the local habitat initiated a cascade of changes in the understory bird communities. Initially, birds whose forest territories were razed were displaced (with some finding at least temporary refuge in the experimental reserves). Eventually, some of the fragments were reconnected to adjacent virgin forest as some of the cleared land was covered by exuberant second-growth vegetation. We review the effects of habitat loss and edge formation on all species, then present a summary of changes witnessed for understory insectivores, frugivores, and nectarivores.

Physical Changes

As the forest around the reserves was felled, capture rates increased, and the percentage of previously banded birds in mist-net samples decreased. The increase in capture rates was greatest in 1 ha reserves, intermediate in 10 ha isolates (two of four reserves showed no effect), and least, but still clearly discernible, in the 100 ha fragment isolated in 1983. About 200 days after isolation, however, capture rates in the 1 and 10 ha isolates declined precipitously, including those in the 10 ha reserves that showed no initial increase at the time of isolation. As capture rates fell, the percentage of previously banded birds simultaneously increased (Bierregaard and Lovejoy 1988).

Quintela (1985) showed that avian activity levels (as estimated from capture rates) were lower at 100 m from forest edges than in deep-forest control areas. Species that are specialized for treefalls in continuous forest, such as the black-headed antbird *(Percnostola rufifrons),* persisted or increased along such edges. Most continuous forest species, however, shunned the edge, and species typical of second growth very rarely penetrated intact forest. In our regular mist-net samples, capture rates for the black-headed antbird and another gap specialist, the warbling antbird *(Hypocnemis cantator),* increased with time since isolation in 1 ha isolates and remained constant in isolated 10 ha reserves (Stouffer and Bierregaard 1995b).

Insectivores

Abundance and species richness of the thirty-five most frequently captured insectivorous birds declined significantly in post-isolation reserves, including those isolated by only 70 and 100 m from adjacent continuous forest. The pre-isolation capture rate for these species was 156.9 captures/1,000 net-hours. In periods T_2–T_4, the capture rate of insectivores was 65.4—only 41% of the pre-isolation rate (fig. 10.3; Stouffer and Bierregaard 1995b).

The three obligate army ant-following species—the white-plumed antbird *(Pithys albifrons),* rufous-throated antbird *(Gymnopithys rufigula),* and white-chinned woodcreeper *(Dendrocincla merula)*—disappeared completely from 1 and 10 ha isolates within two years after isolation. Mixed-species flocks disintegrated, with all but two species dropping out of the 1 and 10 ha reserves (Bierregaard and Lovejoy 1989).

By five years after isolation, the second growth around the reserves had developed a closed canopy and its own understory vegetation, a structure that proved to be suitable to the movement of some birds of the primary forest understory (Borges 1995). Isolates surrounded by *Vismia* remained depauperate, but those surrounded by *Cecropia* were recolonized by many species. Obligate ant followers, the first species to disappear from newly isolated 1 and 10 ha reserves, returned five years after isolation. By seven to nine years after isolation, mixed-species flocks were reassembling, using both the 10 ha isolates and the surrounding second-growth areas for foraging. The most vulnerable species proved to be some solitary terrestrial feeders, such as leafscrapers (*Sclerurus* spp.), antthrushes (*Formicarius* spp.), and the wing-banded antbird *(Myrmornis torquata),* all of which forage in the leaf litter (Stouffer and Bierregaard 1995b).

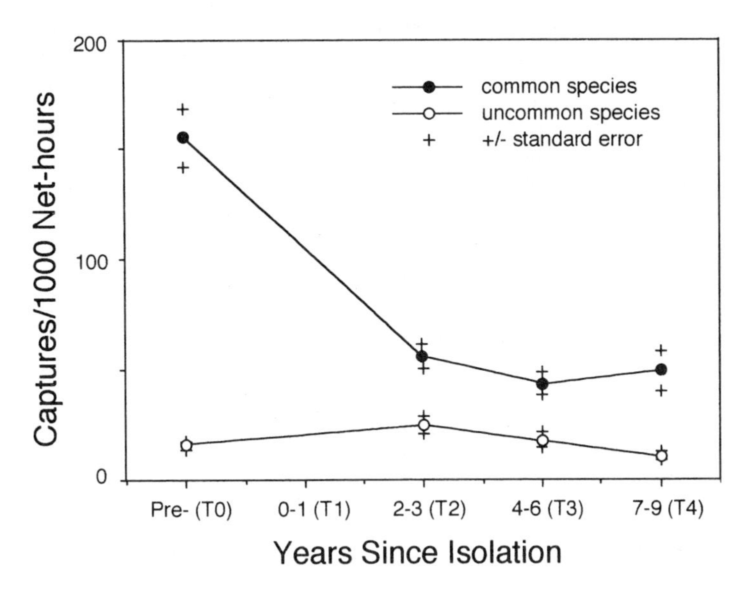

Figure 10.3. Capture rates of thirty-five common and forty-nine uncommon insectivorous birds, including nonforest species, before and after isolation in 1 and 10 ha forest fragments. (Adapted from Stouffer and Bierregaard 1995b.)

An analysis of the understory insectivore community by detrended correspondence analysis, using the DECORANA program (Hill and Gauch 1980), demonstrated that communities in 1 ha fragments diverged from the pre-isolation condition much more strongly than did communities in 10 ha fragments. By seven to nine years after isolation, communities in 10 ha fragments surrounded by *Cecropia* began to resemble those in pre-isolation reserves, while 1 ha reserves surrounded by either *Cecropia* or *Vismia,* as well as 10 ha reserves in *Vismia,* continued to diverge (Stouffer and Bierregaard 1995b).

There was no significant influx of nonforest insectivores (included in fig. 10.3) into the isolates during the course of the study. Only three second-growth specialists—the coraya wren *(Thryothorus coraya),* house wren *(Troglodytes aedon),* and dusky antbird *(Cercomacra tyrannina)*—appeared in the isolates in T_2 and T_3, but they had abandoned the reserves by T_4.

Frugivores

Twenty-nine predominantly frugivorous bird species were captured in the pre- and post-isolation samples. As with insectivores, capture rates for the twelve most common frugivores declined significantly after isolation. Capture rates in years 2–3 (T_2) and 7–9 (T_4) post-isolation were significantly lower than the pre-isolation rates ($P < .001$, Wilcoxon signed-ranks test; table 10.2, fig. 10.4).

Reserve Size Effects

Repeated-measures ANOVAs for only two of the six common frugivores showed significant reserve size effects: McConnell's flycatchers *(Mionectes macconnelli)* and silver-

Table 10.2 Number of captures, capture rates, and habitats (forest strata) for the twenty-nine species of predominantly frugivorous birds captured prior to isolation (T_0) and two to nine years after isolation (T_2–T_4).

				T_0		T_{2-4}	
Species	Common name	Stratum[a]	Total captures	Number of individuals	Capture rate[b]	Number of individuals	Capture rate
Lipaugus vociferans	Screaming piha	Mid	2	1	0.09	1	0.03
Arremon taciturnus	Pectoral sparrow	Mid	2	1	0.09	1	0.03
Coereba flaveola	Bananaquit	Canopy	4	1	0.09	3	0.10
Trogon rufus	Black-throated trogon	Mid	5	1	0.09	4	0.14
Pitylus grossus	Slate-colored grosbeak	Upper	11	1	0.09	10	0.34
Tinamus guttatus	White-throated tinamou	Ground	1	1	0.09	—	—
Catharus fuscescens	Veery	Under	1	1	0.09	—	—
Psophia crepitans	Gray-winged trumpeter	Ground	1	1	0.09	—	—
Catharus minimus	Gray-cheeked thrush	Under	5	2	0.18	3	0.10
Crypturellus variegatus	Variegated tinamou	Ground	6	4	0.36	2	0.07
Pipra erythrocephala	Golden-headed manakin	Mid	11	6	0.54	5	0.17
Cyanocompsa cyanoides	Blue-black grosbeak	Canopy	29	6	0.54	23	0.79
Corapipo gutturalis	White-throated manakin	Mid	20	13	1.16	7	0.24
Tachyphonus surinamus	Fulvous-crested tanager	Mid	28	14	1.25	14	0.48
Geotrygon montana	Ruddy quail-dove	Ground	66	26	2.32	40	1.38
Mionectes macconnelli	McConnell's flycatcher	Under	138	34	3.03	104	3.58
Schiffornis turdinus	Thrush-like manakin	Under	80	57	5.09	23	0.79
Lepidothrix serena	White-fronted manakin	Mid	79	58	5.17	21	0.72
Pipra pipra	White-crowned manakin	Mid	190	61	5.44	129	4.44
Turdus albicollis	White-throated thrush	Under	118	70	6.25	48	1.65
Leptotila verreauxi	White-tipped dove	Second	1	—	—	1	0.03
Pterglossus viridis	Green aracari	Upper	1	—	—	1	0.03
Vireo olivaceus	Red-eyed vireo	Upper	1	—	—	1	0.03

Euphonia cayannensis	Golden-sided euphonia	Canopy	1	—	—	1	0.03
Cyanerpes caeruleus	Purple honeycreeper	Canopy	1	—	—	1	0.03
Tyranneutes virescens	Tiny tyrant-manakin	Mid	4	—	—	4	0.14
Tachyphonus cristatus	Flame-crested tanager	Upper	5	—	—	5	0.17
Oryzoborus angolensis	Lesser seed-finch	Second	6	—	—	6	0.21
Ramphocelus carbo	Silver-beaked tanager	Second	83	—	—	83	2.86
Number of individuals			900	359		5.41	
Capture rate					32.05		18.58
Number of species			29	20		26	

a. Strata of forest most commonly used (heights approximate): Upper = 20 m to canopy; Mid = 5–20 m; Under(story) = 0–5 m; Second = second growth.

b. Captures/1,000 net-hours.

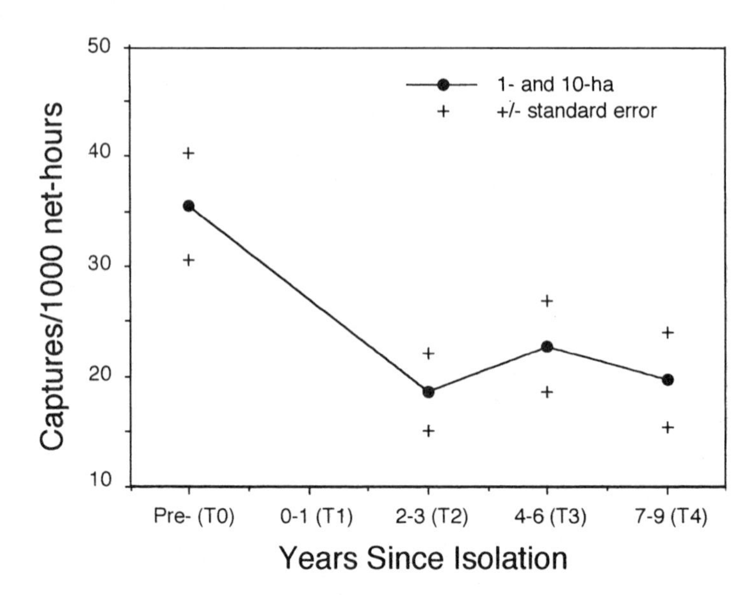

Figure 10.4. Capture rates of the twelve most common frugivorous birds before and after isolation of 1 and 10 ha forest fragments.

beaked tanagers *(Ramphocelus carbo)* were captured more often in 1 ha isolated reserves than in 10 ha reserves (table 10.3). With the exception of a single capture, the silver-beaked tanager, a second-growth specialist, was found only in 1 ha reserves. In contrast, two of three medium-sized (20–100 g) primary forest frugivores—the thrush-like manakin *(Schiffornis turdinus)* and white-throated thrush *(Turdus albicollis)*—avoided almost all of the 1 ha isolated reserves. The thrush-like manakin was never caught in an isolated 1 ha reserve, and the white-throated thrush was captured only very sporadically in the 1 ha isolates. Despite the striking absence of the thrush-like manakin in 1 ha isolates, neither it nor the white-throated thrush showed a significant reserve size effect in the repeated-measures ANOVA.

Upon arriving in the area in years following particularly wet rainy seasons, ruddy quail-doves, which are nomadic, settled into 10 ha isolates, but apparently found 1 ha remnants too small for reasons we cannot define (Stouffer and Bierregaard 1993).

Time Since Isolation

Four species showed significant effects of time since isolation: the white-throated thrush, thrush-like manakin, white-fronted manakin *(Lepidothrix [Pipra] serena),* and silver-beaked tanager (table 10.3). Based on the results of Wilcoxon signed-ranks tests, two common small manakins, the white-crowned *(Pipra pipra)* and white-fronted, showed significant declines in capture rates in T_2. The white-fronted manakin disappeared from all but one isolate in T_2, but returned in T_3 and T_4, primarily to 10 ha reserves. Capture rates for both the white-throated thrush and thrush-like manakin declined over the course of the study. The silver-beaked tanager was present almost exclusively in T_2 and T_3; there were no pre-isolation captures, and as second growth grew taller, it all but disappeared from the remnants by seven years after isolation (T_4).

Table 10.3 Significance levels of the effects of reserve size, type of surrounding vegetation (matrix), and time since isolation on capture rates for six common frugivorous bird species, based on a repeated-measures analysis of variance.

	Species					
Factor	*Turdus albicollis*	*Schiffornis turdinus*	*Mionectes macconnelli*	*Pipra pipra*	*Lepidothrix serena*	*Ramphocelus carbo*
Reserve size	ns	ns	$< .05$	ns	ns	$< .05$
Matrix type	ns	ns	ns	ns	ns	ns
Time	$< .001$	$< .001$	ns	ns	$< .001$	$< .001$
Matrix-size	ns	ns	ns	ns	ns	ns
Matrix-time	ns	$< .05$	ns	ns	ns	ns
Size-time	$< .01$	ns	ns	ns	$< .001$	$< .001$
Matrix-size-time	ns	ns	ns	ns	ns	ns

Matrix Effects

None of the six species studied showed a significant matrix effect in the repeated-measures ANOVAs (see table 10.3), although the white-fronted manakin was found only in those 1 ha reserves surrounded by *Cecropia.*

Replacement by Second-Growth Species

As was the case for insectivores, the decline in activity levels of primary forest species did not trigger any compensation in increased activity among frugivore species specialized for second-growth habitats. Nine frugivore species were captured only in isolated reserves, but six of these are species common in the canopy or upper stages of primary forest, and three are exclusively second-growth or pasture species (*Leptotila verreauxi, Oryzoborus angolensis,* and *Ramphocelus carbo;* see table 10.2). Over 80% of frugivores captured only after isolation were silver-beaked tanagers, a second-growth specialist; moreover, all but one of these captures were in 1 ha isolates, particularly those substantially damaged by windthrows. By T_4, silver-beaked tanagers no longer remained in fragments surrounded by well-developed second growth, so their presence in the community was ephemeral. This species would be expected to persist, however, in areas where the matrix is subjected to continued disturbance.

Nectarivores

Understory hummingbirds proved to be less vulnerable to fragmentation than insectivores or frugivores. Three species—the fork-tailed woodnymph *(Thalurania furcata),* long-tailed hermit *(Phaethornis superciliosus),* and straight-billed hermit *(P. bourcieri)*—were captured in pre-isolation samples in roughly equal proportions. After isolation, there was no detectable decline in activity by the fork-tailed woodnymph or straight-billed hermit, while the long-tailed hermit became twice as abundant in mist-net samples (Stouffer and Bierregaard 1995a). Five other species that were detected in the pre-isolation sample were captured after isolation, and overall capture rates for all eight

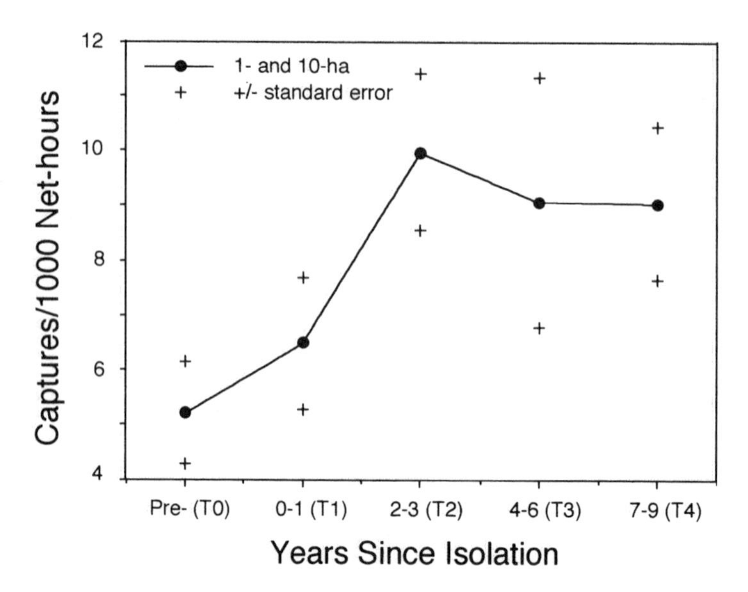

Figure 10.5. Capture rates of eight hummingbird species in 1 and 10 ha forest reserves before and after isolation. (Adapted from Stouffer and Bierregaard 1995a.)

hummingbirds were significantly greater in $T_1 - T_4$ than in pre-isolation samples ($P <$ .025, $G = 5.66$, *G*-test for independence; fig. 10.5). Five other species caught only after isolation do not represent nonforest species moving into the isolates, but rather are species of the primary forest canopy that are occasionally captured in control samples in continuous forest.

In contrast to insectivores and frugivores, reserve size and the nature of the surrounding second growth did not affect the presence or relative abundances of the three forest interior hummingbirds sampled (Stouffer and Bierregaard 1995a).

DISCUSSION

Mist nets are probably essentially free of biases associated with the observer (although this has not been tested). However, the nets themselves introduce biases in that birds of different sizes and with differing behavior patterns have different capture probabilities (Heimerdinger and Leberman 1966; Remsen and Good 1996). Mist-netting data cannot be interpreted without an understanding of the biology of the species in question and supplemental observations—mist nets provide the numbers, biologists supply the interpretations. In the case of some insectivorous species, we are confident that declines in capture rates reflect the disappearance or reduced population densities of the species in question (see fig. 10.3). Obligate army ant followers and terrestrially foraging antbirds must forage near the ground, so their absence from mist-net samples accurately reflects their absence in fragments (Stouffer and Bierregaard 1995b). In contrast, supplemental observations allowed us to interpret an increase in captures of some mid-story insectivores in recently isolated remnants as a change in behavior rather than in densities (Bierregaard and Lovejoy 1989).

A substantial influx of individuals into recently isolated reserves, as reported by Bierregaard and Lovejoy (1988), is likely to be a general phenomenon in habitat islands created by humans. The extent of crowding will depend upon the suitability of the new matrix to species of the original habitat. The resulting elevated population densities are likely to be ephemeral, but may affect the course of "ecosystem decay" and thus the eventual species composition in the remnant (see also Lynam, chap. 15; Terborgh et al., chap. 17). As such, this phenomenon lends further importance to long-term studies of habitat fragments encompassing the entire period of isolation.

Small sample sizes make it impossible to understand how the many rare species in Neotropical forests respond to fragmentation. In fact, the very rarity of these species is a mystery that perplexes students of the Neotropical avifauna (but see Fjeldså and Rahbek, chap. 30). Not knowing why species are rare in the first place indicates a rather profound lack of understanding of their basic natural history and population dynamics. Without understanding these basic parameters, we are unlikely to be able to predict effectively how populations of these species will react to large-scale habitat alteration.

Although many of the first successional plants to appear in pastures are bat-dispersed (e.g., *Solanum* spp.), frugivorous birds are important seed dispersal agents whose presence in second-growth areas promotes succession (Guevara, Purata, and Van der Maarel 1986; McClanahan and Wolfe 1993; Silva, Uhl, and Murray 1996; Lamb et al., chap. 24). The relative scarcity of medium-sized frugivores moving through second growth surrounding the BDFFP reserves suggests that early stages of succession may be skewed toward smaller-fruited plant species. An additional asymmetry exists in the preference of some smaller frugivores for *Cecropia*-dominated second growth, which develops on cleared land that is not subjected to heavy use prior to being left fallow. Fewer species move through the more disturbed *Vismia*-dominated areas, and fewer still travel in or over pastures. Thus, the influx of primary forest seeds that are dispersed by frugivorous birds should be inversely related to the degree of original disturbance (see also Silva, Uhl, and Murray 1996). In addition to this reduction of seed dispersal in areas where pastures are used heavily, forest succession may be severely retarded, and in some cases even arrested, by periodic fires (Nepstad, Uhl, and Serrão 1991).

Hummingbirds are active in second growth and seem to be quite plastic in their habitat requirements. In other Neotropical habitat mosaics, hummingbirds use edges and move between primary and secondary habitats to take advantage of spatial and temporal variation in flower production (Stiles 1975, 1980; Feinsinger 1976). Thus, the plant-pollinator relationships involving these birds may not be substantially altered in a highly modified habitat mosaic.

Very few second-growth species in any of the three guilds studied appear to colonize fragments as assemblages of primary forest species decline in richness and density. The loss of primary forest species in a fragmented habitat mosaic therefore represents a real decrease in species richness and overall abundance of small understory birds.

The appearance of some canopy hummingbirds and frugivores in the post-isolation netting samples probably reflects changes in the physiognomic structure of the reserves themselves. Most of the 1 and 10 ha isolates have suffered moderate to severe wind damage (see also Laurance, chap. 6). In areas where trees have been blown down, the vegetation that fills these gaps has its own canopy, which in early stages of succession may be only a few meters tall. Canopy bird species do not seem to care whether they have three or thirty meters between them and the ground, and hence will move downward to investigate or forage through the "canopy" of a regenerating edge or treefall (see also Stiles 1983). In so doing, they move to the level of our nets.

CONCLUSION

With increasing habitat fragmentation will come the need to manage regional mosaic landscapes. The most important tool we require to actively manipulate fragmented landscapes is biological knowledge. Because birds play an integral role in tropical rainforest ecosystems and are arguably the best-studied group of organisms in these forests, they provide an excellent opportunity to understand faunal responses to habitat fragmentation. Our studies have identified both particularly sensitive and apparently insensitive groups of species, and suggest that some seemingly vulnerable species are more resilient than we had supposed. Sieving and Karr (chap. 11) have begun to unveil the biological underpinnings of avian sensitivity to fragmentation. These are but the first steps in developing the knowledge needed to minimize the effects our species is having on tropical rainforest ecosystems.

GENERAL IMPLICATIONS

1. Habitat fragments and their biotas cannot be understood without knowing what is going on around them.
2. We cannot understand these systems without long-term studies. For example, the insectivorous birds that were the first to disappear after fragment isolation were among the first to return, and have proven themselves much less sensitive to fragmentation than initially thought.
3. A landscape of forest remnants, second growth, and human activity is not devoid of biological activity. Some species of understory birds commonly found in primary forest will at least move through highly disturbed areas (see the similar results for frogs reported by Tocher, Gascon, and Zimmerman, chap. 9). Such a habitat mosaic may serve a number of conservation purposes (see also Schelhas and Greenberg 1996; Laurance et al., chap. 32).
4. Understory hummingbirds and small frugivores appear to move quite readily through tall, complex second growth. This would seem to augur well for natural succession, but the rate and trajectory of forest regeneration will certainly be affected by many other factors as well.

ACKNOWLEDGMENTS

The mist-netting operation has been staffed by over two dozen enthusiastic and dedicated young field interns and a small group of skilled Brazilian woodsmen, all of who will understand if we use our allotted word quota to include a bit more data instead of thanking them individually. This project would not have been possible without their hard work, and we are grateful to them. The chapter benefited from suggestions by L. Barden, J. Blake, F. Crome, R. Holt, W. Laurance, D. Stotz, and two anonymous reviewers. Financial support has come from public and private sources including the World Wildlife Fund-U.S., the Smithsonian Institution, the Instituto Nacional de Pesquisas da Amazônia (INPA), the Instituto Brasileiro de Desenvolvimento Florestal (IBDF), a grant from the National Park Service (cooperative agreement CX-0001-9-0041), the Man and the Biosphere Program of US-AID, the Pew Charitable Trust, and the Barbara Gauntlet Foundation. This chapter is publication number 149 in the Biological Dynamics of Forest Fragments Project technical series.

11

Avian Extinction and Persistence Mechanisms in Lowland Panama

Kathryn E. Sieving and James R. Karr

A PRINCIPAL goal of conservation biology is to understand the dynamics of extinction and to use that knowledge to define societal policies that will prevent anthropogenic extinction. Typically, researchers study insular extinction by first examining patterns of extinction in relation to habitat fragmentation. They then group species by their respective responses (e.g., highly, moderately, or not prone to extinction). Finally, they analyze traits shared by group members in relation to environmental factors, generating mechanistic hypotheses and explanations for the observed patterns. Traits identified for extinction-prone tropical forest birds in fragmented landscapes include large body size (Leck 1979; Willis 1979), restricted mobility (Diamond 1981), specialized patterns of resource use (Willis 1974; Lovejoy et al. 1986), low annual survival rates (Karr 1990a), high population variability (Karr 1982b), and terrestrial foraging and nesting (Terborgh 1974; Willis 1979; Karr 1982a; Lovejoy et al. 1986; Stouffer and Bierregaard 1995b). Environmental characteristics of insular habitats where forest birds go extinct include small area (Faaborg 1979; Lovejoy et al. 1986; Bierregaard and Lovejoy 1989), isolation (Opdam, Van Dorp, and Ter Braak 1984; Haas 1995), and increased mesopredator populations (Terborgh 1974; Terborgh and Winter 1980).

Such findings suggest many proximate causes of forest bird extinctions following habitat fragmentation. Terrestrial insectivores, for example, share traits such as low dispersal capability and near-ground nesting habits; logical mechanisms causing extinctions in these species might include reduced rescue effects (Wright 1985) and higher nest mortality from medium-sized predators (Terborgh 1974; Wilcove 1985). Large species may be more subject to extinction than smaller species because their larger territories dictate smaller, more labile populations in remnants (Shaffer 1981).

This correlative community-level approach, however, reveals a substantial number of inexplicable exceptions. Although terrestrial insectivorous birds, for example, are widely recognized as an extinction-prone guild in the Neotropics (Willis 1979; Karr 1982b), some of these species persist in relatively small forest remnants, and some small-bodied species disappear as readily as larger ones (Leck 1979; Willis 1979; Willis and Eisenmann 1979; Karr 1982b; Sieving 1992). Therefore, generalizations based on univariate correlations cannot fully represent the complexity of life histories or environmental conditions. When a substantial minority of species with extinction-prone traits persists and species without such traits disappear, factors not addressed by community-level correlation must be at work.

To identify the mechanisms accounting for species that persisted or disappeared where correlational analysis indicated that they should not have, we take a guild-centered approach and evaluate autecological data available for nine species of undergrowth forest birds found in central Panama. Our information comes primarily from fieldwork at two sites: Barro Colorado Island, a lowland Neotropical land bridge island in Gatun Lake; and Soberania National Park on the nearby mainland. All nine species are predicted to be highly sensitive to eighty years of isolation on Barro Colorado Island because they are terrestrial insectivores (Karr 1982a), yet four of them still occur there. The five that disappeared from the island (table 11.1) did not do so in a predictable sequence based on body size or population density. All nine occur in nearby mainland forest.

To weigh several interactive hypotheses concerning extinction or persistence mechanisms for each species, we integrate published demographic, natural history, and genetic information about the species with ecological information about the BCI environment. The key mechanism defining insular extinction or persistence among these species appears to be interactions between environmentally induced mortality and species-specific survival and fecundity rates. Although inherent traits of species cannot be altered to reduce the likelihood of extinction, it might be possible to influence extinction probability through habitat management—or, more important, through the management of human influence on habitats—with a better understanding of the interactions between intrinsic species traits and extrinsic environmental factors determining species responses (e.g., Martin 1992).

STUDY AREA

Barro Colorado Island (BCI) is a 1,500 ha land bridge island created when the Chagras River was dammed to form Gatun Lake and the Panama Canal (fig. 11.1). The present

Table 11.1 Selected terrestrial insectivorous birds of central Panama forest.

Species	Code	Family	
BCI-extinct			Last BCI record[a]
Ocellated antbird (*Phaenostictus mcleannani*)	OCAB	Formicariidae	1981
Black-faced antthrush (*Formicarius analis*)	BFAT	Formicariidae	1951
Spectacled antpitta (*Hylopezus perspicillatus*)	SPAP	Formicariidae	1971
White-breasted woodwren (*Henicorhina leucosticta*)	WBWW	Troglodytidae	ca. 1950
Song wren (*Cyphorhinus phaeocephalus*)	SOWR	Troglodytidae	1961
BCI-extant			BCI status
Scaly-throated leaftosser (*Sclerurus guatemalensis*)	STLT	Furnariidae	Uncommon
Bicolored antbird (*Gymnopithys leucaspis*)	BCAB	Formicariidae	Declining
Spotted antbird (*Hylophylax naevioides*)	SPAB	Formicariidae	Common
Chestnut-backed antbird (*Mymeciza exsul*)	CBAB	Formicariidae	Common

a. Willis and Eisenmann 1979.

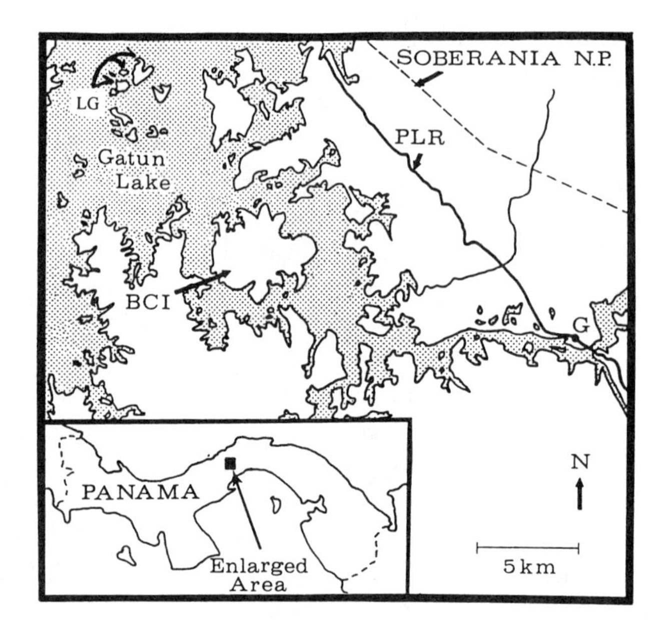

Figure 11.1. Map of Barro Colorado Island and Soberania National Park, Republic of Panama. Abbreviations: BCI = Barro Colorado Island; N.P. = National Park; PLR = Pipeline Road (a maintenance road built to service the transisthmian oil pipeline built during World War II); G = the town of Gamboa; LG = the Los Gatos island group. The dashed line marks the eastern boundary of Soberania National Park.

shore of BCI lies from 200 to 1,000 m from surrounding mainland forest. The average temperature in central Panama forest is 26° C; average rainfall is 2,600 mm/year, with a dry season from late December to early April. Both mainland and island forests in our study areas (9°9′ N, 79°51′ W) are classified as tropical moist forest (Croat 1978; Leigh, Rand, and Windsor 1982). Construction of the primary dam across the Chagras River began in 1903, and water impoundment began in 1911. The isolation of Barro Colorado from (now) mainland forest took three years. No biological studies were done at BCI during this period. (Chapters in Leigh, Rand, and Windsor 1982 and Gentry 1990b provide more complete historical perspectives on the Barro Colorado Island ecosystem.) Today, BCI supports 1,500 hectares of lowland forest, but nearly half the island was under cultivation at the time of isolation (Leigh, Rand, and Windsor 1982). The habitat area available to forest-requiring birds on BCI was thus half as large as it is today. Succession has covered the cleared areas with well-developed second-growth forest, and the remaining area supports old-growth forest at least 500 years old. BCI does not contain all varieties of forest habitat available in the region (e.g., palm swamp: Karr 1982a; Sieving 1990), and most streams on the island are seasonal (Leigh, Rand, and Windsor 1982).

Microhabitat selection does vary among the study species. For example, chestnut-backed antbirds *(Myrmeciza exsul)* prefer treefall tangles, song wrens *(Cyphorhinus phaeocephalus)* avoid steep slopes, and black-faced antthrushes *(Formicarius analis)* prefer open undergrowth (K. E. S. and J. R. K, pers. obs.). Nevertheless, most of the island area, including old and young forest, is now used by the four BCI-extant species and was used by the five BCI-extinct species prior to their extirpation (Willis and Eisenmann 1979).

BCI is ecologically distinct from continental forest remnants in that it is surrounded by water, not by successional habitats that could harbor invasive and detrimental nest predators and parasites (e.g., Robinson et al. 1995) or provide habitat for movement and foraging by forest-requiring vertebrates (Laurance 1990, 1991a; Bierregaard et al. 1992; Bierregaard and Stouffer, chap. 10; Malcolm, chap. 14). And although physical edge effects occur on BCI, most notably in the form of windthrows (see Laurance, chap. 6), their influences on birds have probably been minor (Willis 1974). Genetic evidence (Sieving 1991) and general perceptions of the mobility of the families involved (Diamond 1981) indicate that these species are either incapable of crossing or strongly disinclined to cross the water barrier. Consequently, island biogeographic processes of recolonization (Wilson and Willis 1975), rescue effects, and species turnover (Wright 1985) are probably not principal determinants of species composition within the terrestrial insectivore guild on BCI. The species we examine here have experienced true physical isolation on BCI and have responded primarily to interactions between their autecology and the internal dynamics of the BCI ecosystem.

Mainland Soberania National Park encompasses 22,000 ha of unfragmented lowland forest lying along the eastern shore of Gatun Lake (fig. 11.1). Some areas of virgin forest persist in the park, but much of the forest is regrowth after disturbance (farming, logging, military training, etc.); minimum stand age is sixty to seventy-five years. Wildlife has been hunted with varying intensity since the canal was constructed. Until the early 1980s, most hunting was done by U.S. military personnel; since then, most hunters have been *campesinos* hiking in from settled regions on the eastern border of the park (Karr 1971; J. R. K. and K. E. S., pers. obs.). Even now, the park's avifauna continues to change because of its recent isolation from forests to the east, especially foothill forests, by construction of the transisthmian highway and subsequent human colonization. Nevertheless, the avifauna of Soberania Park serves as a natural experimental control because it encompasses large, noninsular populations of the study species (Karr 1982a,b; Loiselle and Hoppes 1983; Sieving 1992).

AVIFAUNAL HISTORY AND STUDY SPECIES

Fifteen years after BCI's isolation, Frank M. Chapman conducted the first concerted ornithological studies on the site (Chapman 1928, 1929, 1935, 1938). Eugene Eisenmann (1952) compiled the first comprehensive bird species list for BCI and later revised and annotated this list in collaboration with E. O. Willis (Willis and Eisenmann 1979). Willis and Eisenmann made special efforts to document the loss of species known to occur on the island since its isolation. Karr (1990b) provided an overview of the regional avian community.

Nearly a hundred species of birds have disappeared from BCI, apparently for two primary reasons: successional loss of second-growth habitats (thirty-two species) and island effects (fifty to sixty species: Karr 1995). All analyses of the BCI avifauna demonstrate that extinction patterns are not statistically random: species characteristic of earlier successional habitats have disappeared, as have certain insectivorous species of forest undergrowth. Mobile canopy species and frugivores have not disappeared from the island as fast as one might expect if extirpations from BCI were random events (Karr 1982a).

Table 11.2 Biological descriptors of study species.

BCI species[a]	Status	Population		Body mass[d]	Nest type	Relative nest safety[e]	Annual survival[f]	Annual recruitment[g]	No. of broods	No. of renests[h]
		BCI[b]	Mainland[c]							
OCAB	Extinct	50[i]	—	51	Unknown	(Vulnerable)	.73	7.7	2 (3–4)	14
BFAT	Extinct	—	270	57	Cavity	(Safe)	.42	3.5	—	—
SPAP	Extinct	—	705	42	Cup-shrub	Extremely safe	—	—	—	6
SOWR	Extinct	—	1,100	25	Dome-shrub	Safe	.46	9.7	—	—
WBWW	Extinct	—	525	17	Dome-shrub	(Safe)	—	—	—	—
STLT	Present	—	810	34	Bank hole	(Safe)	.49	4.6	—	—
BCAB	Present	90[i]	—	30	Cup-shrub	Vulnerable	.61	14.8	3	10
SPAB	Present	600	1,050	17	Cup-shrub	Extremely vulnerable	.60	18.5	3	12
CBAB	Present	900	645	27	Cup-ground	Vulnerable	—	—	1 (2–3)	—

a. See Table 1 for species codes. b. Estimates from Willis (1974) for all of BCI, expressed as number of birds/1,500 ha.

c. Estimates from mainland study sites (Sieving 1991), expressed as number of birds/1,500 ha.

d. Average for species from mark-recapture studies on mainland (Sieving 1991).

e. Based on results of artificial nest predation experiments conducted on BCI and nearby mainland (Sieving 1992) (not in parentheses), and on reviews by Skutch (1985) and Ricklefs (1969) (in parentheses).

f. Source: Karr et al. 1990. g. Source: Brawn, Karr, and Nichols 1995.

h. Estimates of maximum number of unsuccessful clutches produced in a breeding season, based on observed interclutch intervals (Willis 1974; Sieving 1991).

But even among groups with extinction rates higher than predicted, many species persist. Why?

We focus on a subset of one guild (ground insectivores): nine species in the families Formicariidae (antbirds), Furnariidae (ovenbirds), and Troglodytidae (wrens). All persist in abundance on the mainland; five species have been lost from BCI (table 11.1). The pre-isolation status of all nine species on BCI is known with near certainty, as are their responses to insularization and many aspects of their natural history (Johnson 1954; Skutch 1960, 1966; Willis 1967, 1972, 1973, 1974; Wetmore 1972, 1984; Karr 1971, 1982a,b, 1990a,b; Willis and Oniki 1972, 1978; Morton 1978; Wright 1985; Sieving 1991).

TYPES AND SOURCES OF DATA

Population Status and Density: Mainland and Island

The present status of our study species on BCI is summarized best by Willis (1974), Willis and Eisenmann (1979), and Karr (1982a, 1990a), based on the last reported observations of the species on the island by BCI researchers (table 11.1). Population density estimates (table 11.2) were obtained in all cases using territory mapping of color-banded birds on BCI by Willis (1967, 1972, 1973) and Willis and Oniki (1972), and in Soberania Park by Sieving (1991). All density estimates reported are scaled to 1,500 ha, the present size of available habitat on BCI.

Demography: Annual Survival and Recruitment Rates

Species-typical annual recruitment and survival estimates were obtained from Jolly-Seber mark-recapture models applied to many years of semiannual mist net sampling data from traditional sites in Soberania National Park (Karr 1990a; Karr et al. 1990; Brawn, Karr, and Nichols 1995). Although the potential breeding season for these birds is 8–10 months, the number of successful broods, the number of nesting attempts, and renesting responses to nest predation vary (Willis 1974; Willis and Oniki 1972; Skutch 1950, 1960, 1966, 1969, 1981; Wetmore 1972, 1984; Sieving 1991). Relevant observations are primarily opportunistic and usually for only one or a few pairs of birds. Nevertheless, they suggest upper limits to fecundity.

Nest Predation, Nest Design, and Fecundity

The possibility that the removal of large, top-order predators has led to increased population sizes of medium-sized predators ("mesopredator release") has been discussed extensively in relation to the BCI ecosystem and tropical forests in general (Terborgh 1974). The apparent hyperabundance of coati mundi *(Nasua narica)* along the island trails suggests that this phenomenon may have occurred, but there has been no formal quantification. Recent analyses of mammal census data for BCI do not show markedly high mesopredator densities relative to other Central American sites (Glanz 1990; Wright, Gompper, and DeLeon 1994). However, historical records for BCI suggest that mesopredators have cycled supra-annually, and that mammal populations have been high in several multiyear periods since isolation (Glanz 1982, 1990). An increasing number of studies indicate that hyperabundance of mammal populations is characteristic of newly formed tropical land bridge island systems (e.g., Terborgh et al., chap. 17); therefore, it is very likely to have been an important environmental factor on BCI.

Two experimental nest studies in the 1980s documented high predation rates on artificial nests placed on BCI relative to nearby mainland forest (Loiselle and Hoppes 1983; Sieving 1992), but did not elucidate the specific causes of the elevated predation rates. Therefore, we assume that BCI nest predation pressure in the understory is greater (on average) than on the mainland, regardless of the causative functional or numerical phenomenon.

Assuming that BCI does have higher nest predation rates in the understory, Sieving (1992) tested the hypothesis that BCI mesopredators cause differential mortality based on differences in nest design by conducting replicated artificial nest experiments on five of the nine study species on BCI and in Soberania Park. Nests of the two BCI-extinct species were significantly less vulnerable to predation overall than the BCI-extant nests (table 11.2). And although BCI predation was higher than mainland predation for all five nest types, it was disproportionately so for the BCI-extinct nest types. Thus, perhaps species with very safe nest types, such as the cup-shrub nest of the spectacled antpitta *(Hylopezus perspicillatus),* cannot renest very rapidly, because over evolutionary time predation pressure was too weak to select for high fecundity. Conversely, the experiments indicated that spotted antbirds *(Hylophylax naevioides)* frequently lose nests to predators both on the island and mainland; thus natural selection could have generated high fecundity in this species. Given a high renesting rate, this species may persist on BCI because it can compensate for increases in predation associated with insularization.

Although it has been discussed for temperate birds (Martin 1995), the hypothesis that nest vulnerability determines components of fecundity (e.g., renesting ability) has not been explicitly tested with natural nests. Clutch-removal experiments are called for to determine whether species with relatively safe nests are less capable of producing replacement clutches on demand than species with more vulnerable nests. Nests of the other four study species were not tested in experiments, but were designated safe or vulnerable based upon studies of real nest mortality rates reviewed by Ricklefs (1969) and Skutch (1985), who established generalized schemes of nest safety determined by nest design and site (table 11.2). For example, open cup nests are typically more vulnerable to predation than cavity or covered nests.

Population Genetics: Inbreeding and Island Size

Sieving (1991) used DNA fingerprinting analysis to assess whether populations of three of the study species isolated on islands in Gatun Lake were experiencing population-level inbreeding relative to mainland conspecifics in Soberania Park. Blood samples of bicolored *(Gymnopithys leucaspis),* spotted, and chestnut-backed antbirds were collected on BCI, on three small (< 30 ha) islands in Gatun Lake (collectively identified as Los Gatos), and at two distinct mainland sites in Soberania Park. The Los Gatos islands are close to one other (within 100 m shore to shore) but more than 1 km from BCI or mainland forest (see fig. 11.1).

Comparisons of genetic similarities between island and mainland populations for each species, calculated after Wetton et al. (1987), indicated that bicolored antbirds on BCI are experiencing statistically significant, but not dramatic, inbreeding. Spotted and chestnut-backed antbirds on BCI showed no inbreeding relative to mainland conspecifics, but dramatic inbreeding on the Los Gatos islands. Additionally, comparisons of genetic similarity between individual chestnut-backed antbirds living on different islands in the Los Gatos group revealed no interbreeding of individuals from different islands; average genetic similarities of these pairs were similar to those calculated for BCI and mainland populations of this species. Thus, inbreeding was detected in small (< 100 individuals) island populations, but not in large ones. And although there were fewer than fifty individual chestnut-backed antbirds on all three Los Gatos islands (Sieving 1991), low between-island genetic similarity indicates that even short-distance dispersal over water is undetectably rare in this, and presumably other, terrestrial insectivorous species.

DISCUSSION

BCI-Extinct Species

Black-faced Antthrush

The black-faced antthrush is the largest of our study species and occurs at relatively low densities on the mainland. Although these traits alone suggest insular intolerance, the most telling evidence we have is that annual recruitment and survival rates for this antthrush are the lowest among the six species for which we have data (table 11.2). Low values for recruitment and adult survival could be related to this species' large size, solitary habits, and consequent vulnerability to forest raptors—perhaps adult and fledgling predation rates are higher than for the other species considered. In contrast, the next largest species we studied, the ocellated antbird *(Phaenostictus mcleannani),* forages exclusively in vigilant flocks of army ant followers (Willis 1973) and has high values for recruitment and survival. Because black-faced antthrushes are cavity nesters, presumably suffering lower predation pressures than the open cup-nesting antbirds, we predict that their renesting rates are also limited. Additionally, suitable nest sites (rotted snags: Skutch 1960; Sieving 1991) were probably rare when the forest was young on the cleared portion of BCI. Thus, this species has a problematic combination of traits: low survival and recruitment rates plus probably limited fecundity. Any increase in nest or adult mortality on BCI, or any inhibition of reproduction due to nest site limitation, could explain this species' disappearance.

Spectacled Antpitta

Insufficient capture data were obtained to allow estimation of survival and recruitment rates for the spectacled antpitta. Our best insights into its fate come from Sieving's (1992) test of nest susceptibility on BCI versus mainland forest. Of the five species tested (see above), the antpitta nest type was the safest overall, though with the highest ratio of island to mainland predation. Thus, selection for rapid renesting is probably weak for this species because it normally does not need to replace many clutches due to predation. Yet to compensate for estimated increases in nest loss on BCI would require a doubling or more of this species' renesting rate. One renesting pair observed by Sieving (1991) took sixteen days to lay a new clutch after egg predation occurred. This interclutch interval would limit this species to approximately six attempts per season (table 11.2), about half the number for BCI-persistent antbirds studied by Willis (1974). Based primarily on indirect experimental information, we predict that if good recapture or observational data could be obtained for this species, fecundity or adult survival (or both) would be relatively low and insufficient to support a viable population in the BCI environment.

Ocellated Antbird

The relatively large-bodied ocellated antbird is an obligate ant follower, ranging widely over swarms of raiding army ants in pursuit of invertebrate prey (Willis 1973). Thus, its space requirements are great (Willis 1974). Its post-isolation population size on BCI (Willis 1973, 1974) was at or below accepted estimates of short-term minimum viable population size (Soulé 1987; Lande 1995). Thus, ocellated antbirds probably experienced marked inbreeding on BCI, given that bicolored antbirds are mildly inbreeding at approximately twice the ocellated antbird's average population density prior to its disappearance (see below). Following isolation, ocellated antbirds would have been highly susceptible to environmental and demographic stochasticity and loss of genetic variability; most elements of Gilpin and Soulé's (1986) extinction vortex model probably applied. However, this antbird survived on BCI longer than any of the other forest species that are now extinct (until 1981: Karr 1982a) despite its small post-isolation population size (table 11.2).

This species' sustained tolerance of BCI was probably due to its extraordinarily high annual survival rate (Karr 1990a; table 11.2) and, possibly, to a very high rate of renesting after nest failure (Willis 1974). Karr et al. (1990) reported that survival rates of ocellated antbirds were the highest of ten temperate and twenty-five tropical species they considered. Additionally, Willis (1974) estimated that a single pair of ocellated antbirds could renest up to fourteen times in a season following repeated nest loss on BCI, more than he estimated was possible for either bicolored or spotted antbirds on BCI. Unfortunately, ocellated antbird nests have not been described, and therefore have not been tested experimentally for susceptibility to predation.

Song Wren

The song wren occurs at very high densities in some areas of mainland forest (table 11.2) and, before its first disappearance in 1961, was fairly common on BCI (Willis and Eisenmann 1979). Because song wrens occur in extended family flocks of four to seven individuals, census population size is distinctly larger than genetic effective population size. Despite this, starting densities on the forested half of BCI (if comparable to recent mainland estimates, table 11.2) should have been at least three hundred breeding birds—sufficient to prevent short-term inbreeding.

The song wren nest type was safer than the others tested (Sieving 1992), except for the spectacled antpitta nest. Thus, selection for replacement nesting may have been weak in this species. Additionally, the wrens may have been nest site-limited on BCI. At least 50% of song wren nests encountered at mainland sites where the species was common were in bullhorn acacia trees (*Acacia melanocephala:* Sieving 1991), which are protected by stinging *Pseudomyrmex* ant species (Janzen 1969). In an experiment, artificial nests placed in acacia trees were safer, over several weeks, than nests placed in non-acacia trees (Sieving 1991). This finding suggests that natural selection may favor acacia nest sites for song wrens in central Panama. For unknown reasons, bullhorn acacias are uncommon on BCI, and lack of favorable nest sites could have reduced reproductive success.

Perhaps most important, however, is the finding that artificial song wren nests were preyed on disproportionately heavily on BCI (Sieving 1992). If natural nests experienced the same phenomenon, there might have been unique consequences for song wrens (and white-breasted woodwrens, *Henicorhina leucosticta;* see species account below), because adults roost in nest structures every night of their lives (Skutch 1969). Elevated nest predation on BCI, therefore, could severely increase mortality of roosting juveniles and adults in addition to eggs and nestlings. Given that annual survival rates of song wrens are relatively low and that annual recruitment is only average for the group (table 11.2), combined increases in adult and juvenile mortality on BCI could have crippled the population's reproductive potential. Limitation of preferred nest sites on BCI would have compounded these problems. Interestingly, both species of wrens were reintroduced onto BCI in the 1970s. Fewer than ten individuals of each species survived and nested for approximately two years, then the species disappeared again (Morton 1978; Sieving 1991).

White-breasted Woodwren

The factors determining the extirpation of white-breasted woodwrens from BCI are probably similar to those for the song wren. Their nests are bulky and domed like song wren nests, though closer to the ground and smaller. We predict that these nests are relatively safe, but did not test them in experiments. Insufficient recapture data prevented estimation of recruitment and survival rates. However, nocturnal roosting in nests would have encouraged elevated BCI mortality at all life stages. Unlike the song wren, this species was classified as rare on BCI (Willis and Eisenmann 1979) and is uncommon and patchily distributed on mainland study sites (Sieving 1991). Thus, it may have been affected by stochastic demographic and environmental factors to a greater degree than the song wren. This argument is consistent with its more rapid island disappearance (see table 11.1).

BCI-Persistent Species

Scaly-throated Leaftosser

The scaly-throated leaftosser *(Sclerurus guatemalensis)* nests in burrows deep in dirt banks. Such a nest would keep most mesopredators (*Didelphis* spp., *Cebus* spp., *Nasua narica*, etc.) from feeding on leaftosser eggs and nestlings. Snakes should be more important as predators on this species, but there is no information about snake populations in the region, except that pit vipers have been exterminated from BCI. Given that leaftossers exhibit a survival rate (Karr et al. 1990) nearly as low as that of song wrens and the second lowest recruitment rate of the species studied (table 11.2), their survival on BCI seems incongruous, unless the species is unaffected by elevated predation rates by virtue of its bank nest.

Though mainland densities of the leaftosser can be high (table 11.2; Sieving 1991), the species has never been common on BCI (Willis and Eisenmann 1979). Low island density could be tied to nest site limitation, as this species has an affinity for streamside nest sites (Skutch 1960), which would be rare on BCI but abundant on the mainland (Sieving 1991). Examination of population genetic structure and nest susceptibility on the island and mainland might clarify the nature of this species' persistence on BCI.

Bicolored Antbird

The bicolored antbird, a medium-sized obligate ant follower, has been declining on BCI (Willis 1967, 1974). Less than a hundred individuals occur on the island now, and the number appears to fluctuate from year to year (Willis 1974; Karr 1982b). The BCI population is experiencing genetic inbreeding, though it is not extreme (Sieving 1991). Whether inbreeding depression (i.e., lowered offspring viability) is causing the decline can be determined only by comprehensive analyses of reproductive success in both island and mainland populations. The most obvious explanation for this species' persistence on BCI is its extraordinary recruitment rate (table 11.2; Brawn, Karr, and Nichols 1995). At almost double that of the ocellated antbird, the high recruitment rate of bicolored antbird young is a strong indication of high fecundity. Two other findings are consistent with this hypothesis. First, the bicolored antbird nest type was the second most vulnerable to predation in Sieving's (1992) experiment, after the extremely vulnerable spotted antbird nest type. Second, Willis (1974) estimated that bicolored antbirds can renest up to ten times per season under pressure from repeated nest loss on BCI (table 11.2), almost as many times as spotted antbirds (see below), which are very successful on BCI.

Coupled with the second highest survival rate of the species studied (table 11.2), bicolored antbird fecundity is apparently sufficient to compensate for predation losses and any potential (though unlikely) juvenile mortality associated with genetic inviability (Lovejoy 1978). The bicolored antbird could easily succumb to stochastic environmental effects and increased demographic stochasticity during low population cycles (Karr 1982b) and go the way of the ocellated antbird. At present, however, it represents a unique opportunity to study a population on the verge of local extinction.

Spotted Antbird

The spotted antbird and the chestnut-backed antbird (considered next) are the masters of insular survival in our system. The spotted antbird is small-bodied (table 11.2), and though it habitually follows army ants, it is not dependent on the ants for food. Consequently, it has smaller area requirements than obligate ant-following species. The BCI population is abundant and stable, and along with the chestnut-backed antbird, this species also persists on many of the smaller, more isolated forested islands in Gatun Lake

(Wright 1985). The spotted antbird exhibits an extraordinary recruitment rate and a relatively high adult survival rate (table 11.2). Its nest is extremely vulnerable to predation (Sieving 1992), and its renesting abilities are, predictably, prodigious—up to twelve replacement clutches per season (Willis 1972, 1974). Loss rates of natural nests on BCI and artificial nests frequently averaged more than 90%.

Replacement nesting capability must be the key to continued spotted antbird survival on other islands in Gatun Lake, where census population sizes are less than fifteen birds per island and genetic similarity (i.e., inbreeding) is very high (Sieving 1991). The probability of overcoming inbreeding depression increases with the number of opportunities to draw new combinations of gametes. Thus, long life and rapid, repeated renesting ability are excellent survival traits in insular environments where juvenile mortality is high. Spotted antbirds clearly have a suite of traits enabling them to survive in insular forest patches.

Chestnut-backed Antbird

Fewer demographic data are available for chestnut-backed antbirds than for spotted antbirds because mist net capture rates were too low to allow estimation of survival and recruitment. However, chestnut-backed antbird experimental nest types were safer overall than the other two BCI-extant species' nest types tested, but still more vulnerable than the two BCI-extinct nest types (Sieving 1992). Therefore, we predict that if sufficient data were obtained, chestnut-backed antbird recruitment and clutch replacement capabilities would prove to be at least average. Willis and Oniki (1972) concluded that chestnut-backed antbirds can produce up to three successful broods per year, but could not speculate about their renesting capability. Population density estimates for the species (table 11.2) suggest that BCI may provide better habitat than the mainland. Alternatively, the extinction of a competitor may have triggered density compensation and niche expansion (MacArthur, Diamond, and Karr 1972). In fact, the disappearance of the larger-bodied black-faced antthrush from the island seems to have allowed the chestnut-backed antbird to become more evenly distributed, to occur at higher densities in island undergrowth (table 11.2), and not to be restricted to brushy tangles as it is on the mainland. In any case, the population size on BCI is large, and genetic inbreeding was not detected (Sieving 1991). Overall, the species appears to be thriving there.

DNA fingerprinting analyses have revealed that chestnut-backed antbirds on the much smaller Los Gatos islands are experiencing extreme levels of inbreeding. This is consistent with census population sizes of only two to fifteen individuals (Sieving 1991). Yet despite these suggestions that inbreeding depression must have been severe at some point in these tiny island populations (Templeton and Read 1984), they persist. If chestnut-backed antbirds can renest rapidly, and no data suggest they cannot, then genetic deaths (inviable eggs and nestlings) may have been replaced with viable offspring before population demise. Because the Los Gatos islands are too small to support major nest predators, such as coati mundi and monkeys (Sieving 1991), perhaps freedom from severe predation mortality has enabled these birds to compensate for genetic mortality. Thus, once the period of inbreeding depression passed for Los Gatos birds, nest mortality may have dropped below rates elsewhere, allowing persistence despite continued extreme inbreeding and loss of fitness.

CONCLUSIONS

The information we used in these analyses was obtained over more than twenty years of fieldwork by various researchers applying disparate techniques. Even if techniques could be standardized and efforts streamlined through coordination, this type of species-by-species analysis would be impractical to apply broadly to species-rich tropical avifaunas. The extinction and persistence mechanisms we identified were largely consistent among the species we selected, and were broadly supportive of other findings concerning insular extinction pattern and process for tropical vertebrates (Terborgh 1974; Laurance 1990, 1991a; Bierregaard and Stouffer, chap. 10; Warburton, chap. 13). Thus, we offer some general conclusions that will, we hope, encourage more efficiency in future studies of insular extinction mechanisms and in the design of management strategies for avian diversity in tropical forest isolates.

Low survival rates and indications of low fecundity emerged as consistent predictors of island extinction, with the exception of ocellated antbirds, which disappeared despite high values for both. The disappearance of this species was probably due to both environmental and demographic stochasticity acting on a tiny population size. Insufficient information exists to pinpoint a more defined cause. Conversely, with the exception of scaly-throated leaftossers, persistence on BCI is consistently explained by high annual survival and (suggested) rapid renesting rates. Many of our inferences about species-specific fecundity came from indirect assessment via the relative susceptibility of artificial nests. Although a direct link between nest predation intensity and renesting capability remains to be established, our basic finding that high fecundity reduces extinction probability is wholly consistent with current understanding of extinction resistance (Lovejoy 1978; Terborgh and Winter 1980; Karr 1990a; Laurance 1991a). Thus, identification of extinction-prone species can be directly addressed using any techniques that quantify fecundity and annual survival rates. Survival rates would best be obtained using mist net and mark-recapture techniques (Karr et al. 1990). Fecundity could be addressed in (at least) two ways. Nest searching and monitoring (Ralph et al. 1993) is direct, but labor- and time-intensive, particularly in tropical forests, where intraspecific nest densities are low and nesting is asynchronous. Artificial nest predation experiments could be used to assess relative nest susceptibility, which, indirectly, may indicate relative renesting rate, as it appeared to do for five of our study species.

Terrestrial insectivorous birds on Gatun Lake islands represent an excellent opportunity to assess the effects of inbreeding in small isolated populations, specifically because there is little or no exchange with source populations. What we learned is instructive regarding the importance of inbreeding as a cause of extinction. Given that genetic similarity based on DNA fingerprinting does reflect levels of inbreeding (Kuhnlein et al. 1990; Lynch 1990), we know that inbreeding has occurred in the bird populations of the Gatun Lake islands (Sieving 1991). Given that outbred vertebrate species should experience inbreeding depression for one to several generations following isolation at extremely low population sizes (Templeton and Read 1984), Los Gatos birds probably did, but continue to persist in tiny breeding populations. Other bird populations are known to exist at extremely low genetic effective population sizes (Walter 1990), and although this condition almost certainly depresses long-term population fitness (Jiménez et al. 1994; Lande 1995), it apparently does not guarantee extinction.

Theoretically, high fecundity is an important safeguard against insular extinction in small populations experiencing inbreeding. The mechanism involved is rapid, repeated replacement of inviable offspring in sexually reproducing species (i.e., with genetic segregation and recombination), which affords abundant opportunities to draw genotypes free of deleterious recessive alleles in the homozygous state (Lovejoy 1978; Lynch and Gabriel 1990) that can produce developmentally and reproductively competent individuals (Templeton and Read 1984). This may be the mechanism whereby the spotted and chestnut-backed antbirds persisted on Los Gatos during the first few generations of isolation when they would have experienced inbreeding depression.

Given the persistence of two of our species on Los Gatos, we conclude that inbreeding depression must have been relatively unimportant in initiating population declines on BCI, and that its involvement in the process of final extirpation is very unclear. Thus, short-term ($<$ 100 years) causes of extinction among our species on BCI are more likely to be ecological factors involving interactions between the BCI environment and demography (Lande 1988). Although much remains to be learned about the genetics of small populations in the wild (Lacy 1992), we recommend that inbreeding not be regarded as a dominant cause of insular avian extinction over ecological time scales.

In closing, we note again that the system we examine here is not representative of most tropical forest remnants or avifaunas. BCI is a true island, and our study species represent the least mobile of avifaunal guilds. Thus, our findings will be most applicable to other land bridge island ecosystems and to continental remnants embedded in habitat matrices that are hostile to understory birds, including dispersing transients. As Bierregaard and Stouffer (chap. 10) show, the ability to move among remnant forest patches via the landscape matrix redefines avian population dynamics within patches. Most fragmented forest landscapes contain varying degrees of connectivity with variable matrix composition. We examined one extreme of the complex continuum from patches that are truly insular to those bearing little contrast with the surrounding habitat matrix.

GENERAL IMPLICATIONS

1. Insular extinction and persistence mechanisms are defined by interactions between species' biology and characteristics of the insular environment.
2. High fecundity, survival, and recruitment rates are adaptive in insular ecosystems.
3. Environmental factors increasing mortality rates at any life stage should be a focus for management in insular reserves.
4. Nest predation plays an important role in determining insular extinction probability for many bird species.
5. Over evolutionary time, bird species that adapt to high rates of nest loss will be less

extinction-prone in high-mortality, fragmented ecosystems than species using safe nest types.
6. Changes in nest predation pressure over ecological time scales can differentially determine the extinction probability of different species.
7. For tropical forest birds with small, fixed (two-egg) clutch sizes, the ability to renest rapidly and repeatedly is a key component of fecundity conferring insular tolerance.
8. Inbreeding does not appear to be an important cause of insular avian extinction over ecological (< 100 year) time scales.
9. Ecological (extrinsic) factors that elevate mortality and decrease reproductive success can initiate avian population declines leading to insular extinction.

ACKNOWLEDGMENTS

We thank the Smithsonian Tropical Research Institute for its efforts to provide a logistic and intellectual support base for researchers interested in tropical biology. Support for these studies came from the National Science Foundation (DEB 82-06672 to J. R. K.; pre-doctoral and dissertation improvement grants to K. E. S.), National Geographic Society (J. R. K.), American Philosophical Society (J. R. K.), Smithsonian Institution (K. E. S. and J. R. K.), Exxon Corporation (K. E. S.), Alexander Wetmore Memorial Fund of the American Ornithologist's Union (K. E. S.), the Center for Field Research/ Earthwatch (J. R. K.), and the University of Illinois (K. E. S. and J. R. K.).

12

Frugivorous Birds in Fragmented Neotropical Montane Forests: Landscape Pattern and Body Mass Distribution

Carla Restrepo, Luis Miguel Renjifo, and Paul Marples

TRANSFORMATION of tropical landscapes by humans has influenced plant and animal assemblages in many ways. Most studies have examined how species abundance and richness change with increasing forest fragmentation (Quintela 1985; Bierregaard and Lovejoy 1989; Klein 1989; Newmark 1991; Estrada et al. 1993; Kattan, Alvarez-López, and Giraldo 1994; Laurance 1994; Malcolm 1994; Didham, chap. 5; Warburton, chap. 13; Lynam, chap. 15), or with the transformation of native forests into second-growth or managed ecosystems (Holloway, Kirk-Spriggs, and Chey 1992; Johns 1992b; Lambert 1992; Thiollay 1992; Andrade and Rubio 1994; Escobar 1994). The results of these studies, however, have varied considerably, probably reflecting the complex relationships between habitat modification and biodiversity loss, the inherent differences among study sites, and possibly a mismatch between the scale of the problems being addressed and the methods used. As a consequence, it has often been difficult to establish patterns reflecting the effects of habitat modification on biodiversity and, moreover, how these factors interact to influence ecosystem processes (Vitousek 1990; Kruess and Tschnarntke 1994; Pimm and Sugden 1994; Tilman and Downing 1994; Turner, Gardner, and O'Neill 1995).

Holling et al. (in press) presented a novel conceptual framework that may help to integrate the above issues. First, long-term research has shown that ecosystems are structured by a few processes that generate discontinuities at various spatial and temporal scales (e.g., Clark, Jones, and Holling 1979; Harris 1980; Gunderson 1992). Second, the discontinuous nature of ecosystems influences the behavior and morphology of organisms. In particular, Holling (1992) found that the body masses of boreal birds and mammals are discontinuously distributed, such that species of similar mass tend to aggregate or lump together, in a manner suggesting that they exploit a common suite of resources (but see Manly 1996). Third, differences in the lump structure of animal body masses reflect differences in ecosystem structure. Holling (1992) showed that the distribution of body mass lumps of boreal forest and boreal prairie birds and mammals has a characteristic signature, reflecting the structure of those ecosystems. Hence, a focus on body mass lump structure offers a new way to investigate the complex relationships between landscape pattern, the structure of animal assemblages, and ecosystem processes at broad scales.

In this chapter we ask how changes in landscape pattern affect assemblages of frugivorous birds in Neotropical mountains. We concentrate on frugivores because seed dispersal by birds is especially important in Neotropical mountains as compared with the lowlands (Terborgh 1977; Gentry 1983; Stiles 1985; Renjifo et al., in press), offering an opportunity to relate the process of seed dispersal to landscape pattern. We focus on mountains because natural and human disturbances have generated complex landscapes over small areas (Haslett 1994), offering an ideal opportunity to assess how changes in landscape pattern influence animal assemblages.

We present four comparisons representing two different scales of inquiry to assess how lump structure in body masses of frugivorous birds changes from areas covered mostly by forest to areas covered by open vegetation. The first scale is defined by elevational zones within the mountains of Colombia, and the second by sites within these elevational zones that have been differently affected by human activities. We use body mass as an attribute that reflects information on life history traits, such as fecundity and dispersal ability, that can influence both a species' response to habitat fragmentation (Laurance 1991a; Lawton et al. 1994; Brown 1995; Gaston and Blackburn 1995; Sieving and Karr, chap. 11) and aspects of its foraging behavior, such as the size of seeds it can ingest and disperse (Moermond and Denslow 1985).

METHODS

Study Area

Our study focused on the Andes of Colombia, South America. The Andes consist of three mountain ranges of different geological origin, each running in a north-south direction (Irving 1975). We defined montane habitats as those above the 800 m topographic contour line (fig. 12.1). In the absence of human disturbance this region would be covered by forest, except for the páramo (above 3,400 m) and small areas affected by rain shadows (Cuatrecasas 1958). Forest composition, structure, and physiognomy change with elevation, from the complex lowland tropical forest to the simpler páramos (Cuatrecasas 1958; Espinal et al. 1977).

The area encompassed by this study represents less than 35% of the total area of Colombia (1,380,000 km^2), yet harbors one of the richest biotas in all of the Neotropics (Duellman 1979; Henderson, Churchill, and Luteyn 1991; Gentry 1992a,b; Renjifo et al., in press). It has been postulated that the high levels of diversity and endemism in this area are the result of an intense disturbance regime (Gentry 1992a; J. Luteyn, pers. comm.). A complex topography and geology, combined with high precipitation, generates landslides, mudflows, avalanches, and volcanic eruptions, which continuously transform these mountains (Mejía et al. 1994; Velásquez et al. 1994).

Superimposed on this natural disturbance regime are the effects of human activities. At least 50% of the total population of Colombia (37 million people) has settled in montane areas (Banguero 1993). Presently, less than 30% of this area is covered by forest, most of which is found either at elevations above 2,500 m or on the wetter slopes of the cordilleras (Cavelier and Etter 1995). The remaining area has been transformed into pastures, cultivated fields, coffee and tree plantations, and urban areas. A recent surge in

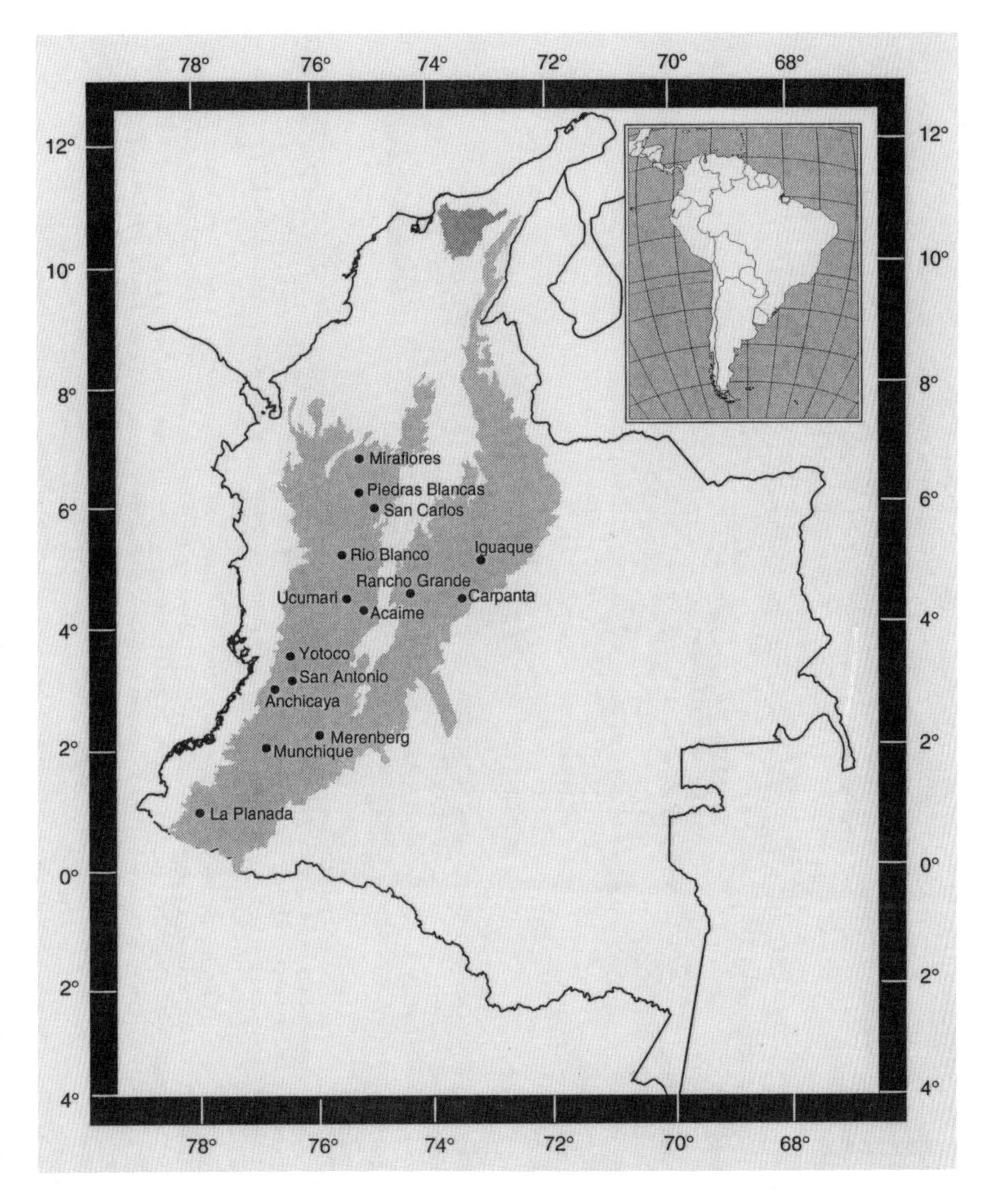

Figure 12.1. Montane habitats of Colombia (shaded areas indicate land above 800 m), showing the locations of sites included in this study.

Table 12.1 Sites included in this study.

Site	Coordinates	Elevation (m)	Life zone[a] (*sensu* Holdrige)	Land use[b]	References[c]
Upper lowland (600/800 to 1,400/1,600 m)					
Anchicaya-Alto Yunda, PNN, Farallones de Cali (AN)	3°32′ N 76°48′ W	1,050	TP-rf	F/F, SG	7, 28
Reserva Forestal Yotoco (YO)	3°52′ N 76°33′ W	1,500	TP-df/mf	F/P, CT	14, 30
Represa San Carlos (SC)	6°13′ N 74°51′ W	750	TP-mf/wf	SG[d]/F, P	22
Finca La Esmeralda (LE)		1,250	TP-wf	CT/CT, P	4
Finca El Ocaso (EO)		1,000	TP-wf	P/F[e]	4
Lower montane (1,400–1,600 to 2,300/2,600 m)					
Reserva Natural La Planada (LP)	1°10′ N 78°00′ W	1,800	TP-rf	F/SG	6, 15, 18, 26
Parque Regional Ucumari, Ucumarí Bajo (UB)	4°47′ N 75°32′ W	1,850	TLM-wf	F/SG, TPE[f,h]	6, 13, 16, 25, 27
Bosque Protector, San Antonio (SA)	3°29′ N 76°38′ W	2,000	TLM-wf	F/SG, U	6, 9, 10, 25, 26, 29
Empresas Publicas de Manizales, Río Blanco (RB)	5°28′ N 75°32′ W	2,400		SG, TPN[g], P	21, 24, 27
Finca Merenberg (ME)	2°14′ N 76°08′ W	2,300	TLM-wf	P/F	6, 19, 31
Represa Miraflores (MI)	6°45′ N 75°20′ W	2,130	TLM-wf	SG/P	5
Finca Rancho Grande (RG)	4°36′ N 74°20′ W	1,700		P/CT, SG, U	12
Finca Mirador, Munchique (MU)	2°30′ N 76°59′ W	2,300	TLM-wf	TPE[h]/SG, P	11
Piedras Blancas (PB)	6°18′ N 75°30′ W	2,350	TLM-wf	TPE[h]	8
Upper montane (2,300/2,600 to 3,100/3,400 m)					
Reserva Natural Carpanta-Estacion Sietecuerales (CA)	4°34′ N 73°41′ W	2,700		F/SG	2
Reserva Natural Alto Quindio, Acaime (AC)	4°37′ N 75°20′ W	2,800	TM-wf	SG/TPN[g], F, P	3, 17
Parque Regional Ucumari, Ucumarí Alto (UA)	4°47′ N 75°32′ W	2,500		SG/TPN[g], F, P	6, 13, 16, 20, 25, 27
Santuario de Flora y Fauna Iguaque-Cañon, Mamarramos (IG)	5°40′ N 73°30′ W	2,600	TLM/TM-mf	SG/P, F[d]	1, 20, 23, 24
Páramo (3,100/3,400 to 4,800 m)[i]					

a. Life zones: TP-df = tropical premontane dry forest; TP-mf = tropical premontane moist forest; TP-wf = tropical premontane wet forest; TP-rf = tropical premontane rain forest; TLM-wf = tropical lower montane wet forest; TLM-mf = tropical lower montane moist forest; TM-wf = tropical montane wet forest.

b. Land uses; F = native forest; SG = second growth; P = pasture; CT = shaded coffee plantation; TPE = tree plantations with exotic species; TPN = tree plantations with native species; U = weekend cottages.

c. References: (1) Acevedo 1987; (2) Andrade 1993; (3) Arango 1994; (4) Corredor 1989; (5) Cuadros 1988; (6) Gentry 1992a; (7) Hilty 1980; (8) Johnels and Cuadros 1986; (9) Kattan, Restrepo, and Giraldo 1984; (10) Kattan, Alvarez-Lopez, and Giraldo 1994; (11) Mondragón 1989; (12) Munves 1975; (13) Naranjo 1994; (14) Orejuela, Raitt, and Alvarez 1979; (15) Orejuela and Cantillo 1990; (16) Rangel 1994; (17) Renjifo 1988; (18) Restrepo 1990; (19) Ridgely and Gaulin 1980; (20) Rosas 1986; (21) Uribe 1986; (22) Velásquez 1992; (23) Velez 1987; (24) N. Arango, pers. comm.; (25) G. Kattan, pers. comm.; (26) C. Restrepo, pers. obs.; (27) L. M. Renjifo, pers. obs.; (28) S. Hilty, unpublished list; (29) G. Kattan, H. Alvarez, and M. Giraldo, unpublished list; (30) H. Alvarez, unpublished list; (31) G. Kattan, H. Alvarez, and E. Buttkus, unpublished list.

d. Selectively logged native forest. e. Native forest dominated by the giant bamboo *Bambusa guadua*.

f. Plantations with *Fraxinus sinensis*. g. Plantations with *Alnus acuminata*. h. Plantations with *Pinus patula* and *Cupressus lusitanicus*.

i. Because no site-specific bird species inventories were available for páramo, we did not compare sites within this elevational zone.

demand for opium derivatives has prompted forest clear-cutting at higher altitudes to grow poppies (*Papaver somniferum*: Cavelier and Etter 1995).

Frugivorous Birds

We included in our analysis all bird species (both residents and migrants) that were reported to consume fruits or seeds to any degree (Fitzpatrick 1980; Hilty and Brown 1986; Isler and Isler 1987; Renjifo 1988; Ridgely and Tudor 1989; Stiles and Skutch 1989; Fjeldså and Krabbe 1990; Velásquez 1992; Arango 1993, 1994; Ridgely and Tudor 1994; L. M. Renjifo and C. Restrepo, pers. obs.) and were found in the Andes at or above an elevation of 800 m. Thus, our data set (476 species) combines both seed dispersers and seed consumers.

Body mass data were obtained from published records (Goodwin 1976; Isler and Isler 1987; Stiles and Skutch 1989; Hoyo, Elliott, and Sargatal, 1992; Dunning 1993; Arango 1993), museum specimens (Colección de Ornitología, Universidad del Valle, Cali, Colombia), and our own field observations. For bird species for which we could not obtain mass measurements, we averaged the recorded masses of congeners of the same length. We could not estimate body mass for a small fraction (2%) of the species, and these were not included in the analyses.

To explore the relationship between body mass and landscape pattern along the elevational gradient, we classified birds into four groups (table 12.1) based on their elevational ranges: upper lowland (ca. 800–1,500 m), lower montane (ca. 1,500–2,400 m), upper montane (ca. 2,400–3,400 m), and páramo (ca. 3,400–4,800 m). Some species fell into more than one zone and were entered into the analyses two or more times. Elevational zones follow Chapman (1917); elevational ranges were taken from Hilty and Brown (1986).

To explore the relationship between body mass and landscape pattern along a gradient of land use, we compared subsets of the birds found in each elevational zone (eighteen subsets or sites: table 12.1). Inventories at each site were conducted by experienced ornithologists over periods of at least one year and included visual, auditory, and mist-netting observations. The authors of these lists reported that few species were being added to their lists in the latter periods of their studies, suggesting that the lists are fairly complete inventories. The eighteen sites were grouped according to elevational zone and type of land use. Within each elevational zone we arranged sites from those covered mostly by forest to those covered by open vegetation (table 12.1).

Our approach relies on two assumptions; first, that there is a common pool of species for elevational zones and sites, but historical, geographic, and climatic events, and more recently human activities, have determined the set of species found today at any one site; and second, that within elevational zones, the less disturbed sites represent the conditions that existed at the other sites prior to human intervention.

Data Analysis

We used the "lump analysis-gap rarity index" (LA_{GRI}) technique to identify aggregates or lumps in the distribution of body masses of frugivorous birds (Holling et al., in press).

The LA_{GRI} index, which is being developed by one of the authors (P. M.) in collaboration with C. S. Holling, relies on the generation of a null distribution from input data ($\log_{10}$[body mass] in this case) and the calculation of the gap rarity index statistic (GRI). This statistic tests whether any observed discontinuity or gap in the distribution of rank-ordered data (fig. 12.2A) occurs by chance alone in data sets sampled from a continuous, unimodal null distribution fit to the observed data. Significant gaps between individual data points separate lumps.

The null distribution is obtained by constructing a normal kernel density estimate that uses the smallest window width (h) that smooths an observed frequency distribution into a continuous, unimodal distribution (Silverman 1986). GRI values are generated by sampling the null distribution 10,000 times and calculating for each data point the absolute discontinuity value (d_i) as $d_i = S_{i+1} - S_i$, where S_i is the $\log_{10}$ of body mass of the ith species in a rank-size ordered data set. The GRI represents the proportion of simulated absolute discontinuities that are smaller than those observed.

The significance of GRI values is established by comparing them against a critical value (fig. 12.2B) for various alpha values. These critical values are associated with probabilities under the null hypothesis that gaps in an observed frequency distribution are artifacts of random sampling.

Sample size and the trade-off between Type I and Type II errors influence lump structure. In data sets with a large number of data points, small values of α will reveal only the most pronounced gaps (large GRI values), whereas large values of α will reveal many more gaps (fig. 12.2B). The detection of many more gaps, however, might increase the probability of detecting gaps that are sampling artifacts (increase of Type I error). In data sets with a small number of data points, small values of α will reveal no gaps, whereas large values of α will reveal gaps. In this situation, increasing α reduces the probability of not detecting gaps that do exist (reduction of Type II error).

This interplay between sample size and the two types of statistical error should be taken into account when comparing multiple data sets. (See Lipsey 1990 for an excellent discussion of the importance of using different values of α when detection of pattern is important). For simplicity, we kept the α level constant within each comparison. We set α to 0.05 when the average number of species within the comparisons was 81 or more (range 30–395), and set α to 0.1 when the average number of species was less than 81 (range 30–141).

LA_{GRI} generates two types of outputs. The first is the distribution of GRI values against body mass (fig. 12.2B). The second, derived from the first, is the distribution of lumps and the proportion of species falling within each lump for a given data set (e.g., fig. 12.3). The lump structure for a given data set can be described by the number and size of the lumps and the proportion of species falling within each lump. The "size" of a lump represents the range of body masses between two significant gaps and varies with the chosen α level (fig. 12.2B). The proportion of species falling within a lump largely varies independently of lump size; in some data sets, large lumps contain relatively few species, whereas in other data sets large lumps contain

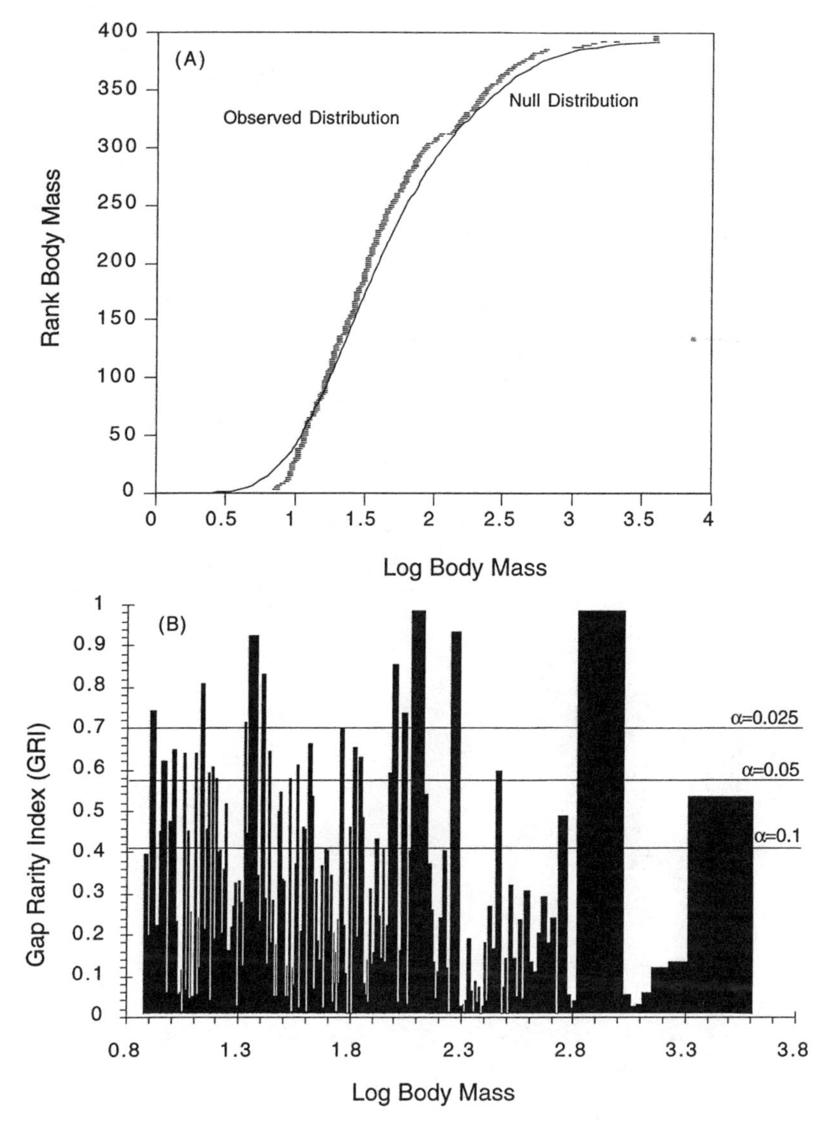

Figure 12.2. *(A)* Observed and smoothed rank-ordered body mass distributions and *(B)* rank-ordered body mass versus gap rarity index (GRI) values for frugivorous birds in upper lowland forest in Colombia. In *(B)*, potential gaps between lumps (shaded bars) are represented by GRI values that exceed the criterion lines (α values). The lump structure generated from this distribution for $\alpha = 0.05$ is the upper lowland species group shown in figure 12.3.

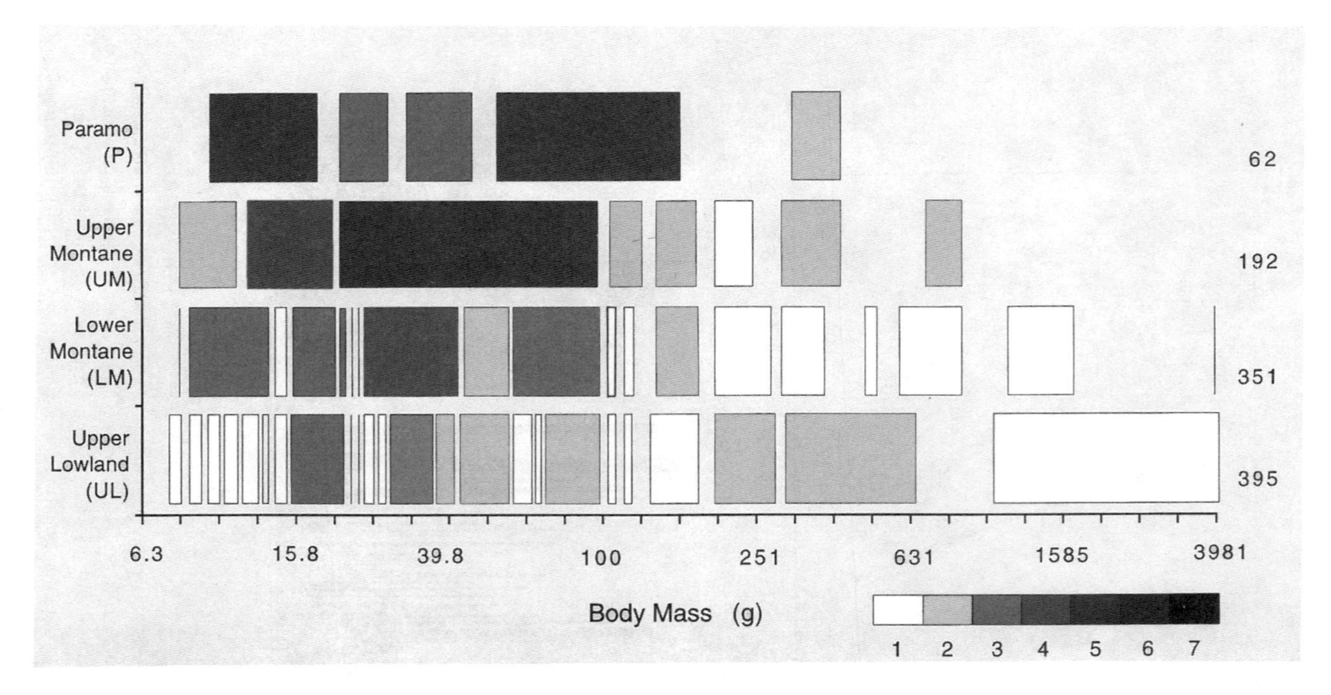

Figure 12.3. Lump structures of Colombian montane frugivorous birds according to elevational zone from forest (bottom) to páramo (top). UL = upper lowland; LM = lower montane; UM = upper montane; P = páramo. Each box represents a lump, and the spaces between boxes represent gaps in the distribution of body mass. The different shades indicate the proportion of species falling within lumps: (1) 0–5%, (2) 5–10%, (3) 10–20%, (4) 20–30%, (5) 30–45%, (6) 45–60%, and (7) 60–100%. Vertical lines represent less than 5% of species. Numbers on the right indicate the number of species for the corresponding data set.

many species. Graphically, lumps are depicted as boxes along the axis of body mass, with the proportion of species within each lump depicted by different shades of gray (e.g., fig. 12.3).

While the identification of lumps and gaps in individual data sets marks the extent of our statistical analysis, a much more revealing analysis can be accomplished by comparing lump structure among multiple data sets in an attempt to find a consistent pattern. The statistical comparison of lump structure among multiple data sets, however, is still in development, so in this chapter we compare lump structure by eye. In so doing we look at the distributions of lumps over the entire body mass range, the locations of gaps within the data sets, and the proportions of species falling within lumps.

For our comparisons we derived four continuous unimodal distributions. The four elevational zones (upper lowland, lower montane, upper montane, and páramo) were compared using a null distribution generated from the database on all frugivorous montane birds, while sites within three of the elevational zones were compared using a null distribution generated from the database on frugivorous birds for the corresponding zone.

RESULTS

Elevational Zones and Body Mass Distributions

We examined the lump structure of the birds of the four elevational zones ($\alpha = 0.05$) and found that the number of lumps decreased sharply from the upper lowland (twenty-four lumps) to the páramo zone (five lumps: fig. 12.3). Across this gradient, many lumps were lost from the upper range (> 316 g) and few from the lower range (< 10 g) of body mass. The lumps at both extremes of the body mass range contained the smallest proportions of species for the upper lowland, lower montane, and upper montane zones, but not for the páramo.

The positions of gaps and lumps in the body mass range from 12 to 575 g showed some similarities in the upper lowland and lower montane zones. However, the proportions of species in these lumps differed between the two zones. The lump structure of the upper montane zone resembled that of the upper lowland and lower montane zones in the body mass range of 83–316 g, but not above or below that range. The páramo zone shared only one lump (> 301 g) with the other three zones.

Sites and Body Mass Distribution

Upper Lowland Zone

The number of lumps ($\alpha = 0.05$) decreased from the site covered extensively by forest, Anchicaya (AN: eight lumps), to the site dominated by pastures, El Ocaso (EO: three lumps; fig. 12.4). Lumps were lost only at the upper end of the body mass range (> 316 g).

The positions of gaps and lumps were most similar between two pairs of sites—Anchicaya (AN) and Yotoco (YO), and San Carlos (SC) and La Esmeralda (LE)—even though AN and SC have twice as many species as YO and LE, respectively (fig. 12.4). The similarities spanned almost the entire range of body masses. In contrast, the lump

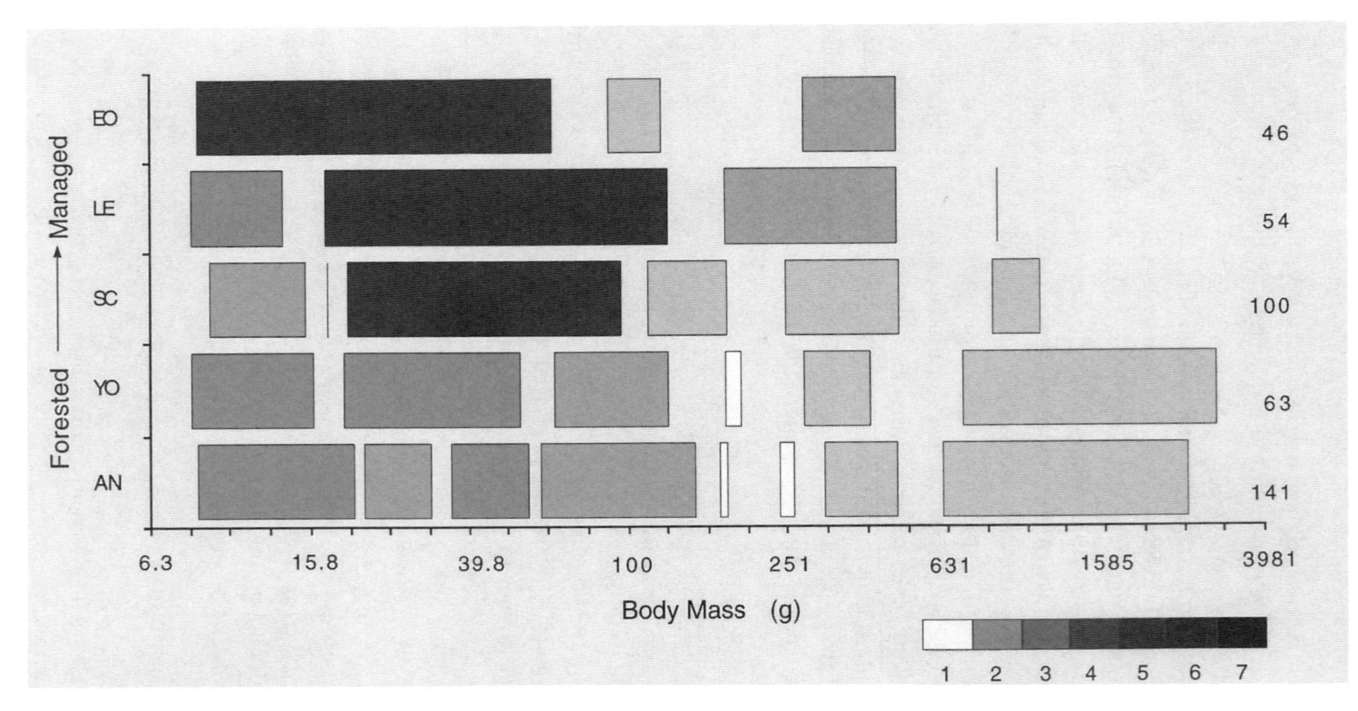

Figure 12.4. Lump structures of Colombian frugivorous birds within the upper lowland zone. Sites are arranged from those covered mostly by forest (bottom) to those highly transformed by human activities (top). AN = Anchicaya; YO = Yotoco; SC = San Carlos; LE = La Esmeralda; EO = El Ocaso. Each box represents a lump, and the spaces between boxes indicate gaps in the distribution of body mass. The different shades indicate the proportion of species falling within lumps: (1) 0–5%, (2) 5–10%, (3) 10–20%, (4) 20–30%, (5) 30–45%, (6) 45–60%, and (7) 60–100%. Vertical lines represent less than 5% of species. Numbers on the right indicate the number of species for the corresponding data set.

structures of LE and EO showed important differences, even though these two sites have a similar number of species and are separated by only 7 kilometers.

Lower Montane Zone

The number of lumps ($\alpha = 0.1$) decreased from sites covered extensively by forest (La Planada, LP, Ucumarí Bajo, UB, and San Antonio, SA), with an average of eight lumps, to sites in which the original forest has been replaced by orchards (Rancho Grande, RG) or forestry plantations of exotic species (Munchique, MU, and Piedras Blancas, PB), with an average of only four lumps (fig. 12.5).

We grouped the lower montane sites according to major land use types. Three of the sites are covered extensively by forests in which second growth (LP), second growth and forestry plantations (UB), or second growth, pastures, and weekend cottages (SA) cover less than 50% of the land (see table 12.1). The lump structure of LP and UB was similar over almost the entire range of body masses, despite the fact that these sites had different numbers of species. The lump structure of SA was similar to that of UB in the 63–313 g range, but differed below 63 g, despite the fact that these sites had a similar number of species.

Three sites are covered half by native forests and half by pastures (Merenberg, ME), second growth and selectively logged forests (Miraflores, MI), or tree plantations established for watershed restoration (Río Blanco, RB). The lump structures of RB and ME showed similarities below 39 g and above 301 g, and they differed greatly from MI over most of the 15–251 g range of body masses.

In the last three sites, native forest has been replaced almost entirely by coffee plantations and orchards (Rancho Grande, RG), or by tree plantations of exotic species for wood production (Munchique, MU, and Piedras Blancas, PB). RG and MU showed different lump structures, even though they had the same number of species. The lump structures of MU and PB were very similar, but the lump representing birds over 398 g was much smaller in PB than in MU.

Upper Montane Zone

We did not find marked variation in the number of lumps among the four upper montane sites ($\alpha = 0.1$). Iguaque (IG), the most disturbed site, had four lumps, and the remaining three sites had five each (fig. 12.6). Lumps at the upper end of the body mass range for two sites (Carpanta, CA, and Ucumarí Alto, UA) showed a high proportion of species, but the opposite was true of the other two sites (Acaime, AC, and Iguaque, IG). More interesting, however, was the finding that lumps at the lower end of the body mass range (< 12 g) tended to decrease in size.

Carpanta (CA), the least disturbed site, is covered by native forest and second growth, and its lump structure differed from those of AC and UA, particularly for species above 25 g. It is noteworthy that the highest proportion of species in CA was in the lump representing species over 63 g; for little-disturbed sites in the other elevational zones, lumps representing the largest birds always contained the smallest proportion of species. The lump structures of AC and UA were similar; both of these sites have been planted

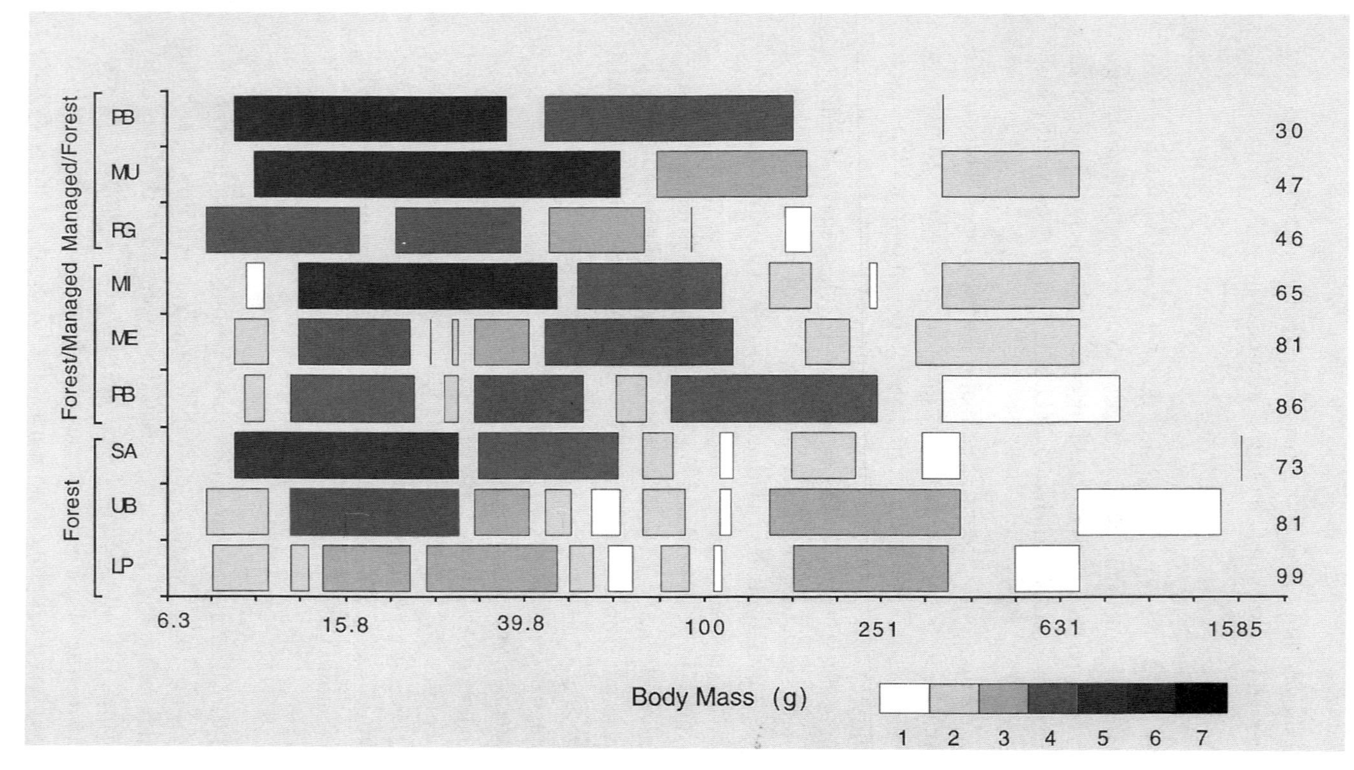

Figure 12.5. Lump structures of Colombian frugivorous birds within the lower montane zone. Sites are arranged from those covered mostly by forest (bottom) to those highly transformed by human activities (top). LP = La Planada; UB = Ucumarí Bajo; SA = San Antonio; MI = Miraflores; ME = Merenberg; RB = Río Blanco; RG = Rancho Grande; MU = Munchique; PB = Piedras Blancas. Each box represents a lump, and the spaces between boxes indicate gaps in the distribution of body mass. The different shades indicate the proportion of species falling within lumps: (1) 0–5%, (2) 5–10%, (3) 10–20%, (4) 20–30%, (5) 30– 45%, (6) 45–60%, and (7) 60–100%. Vertical lines represent less than 5% of species. Numbers on the right indicate the number of species for the corresponding data set.

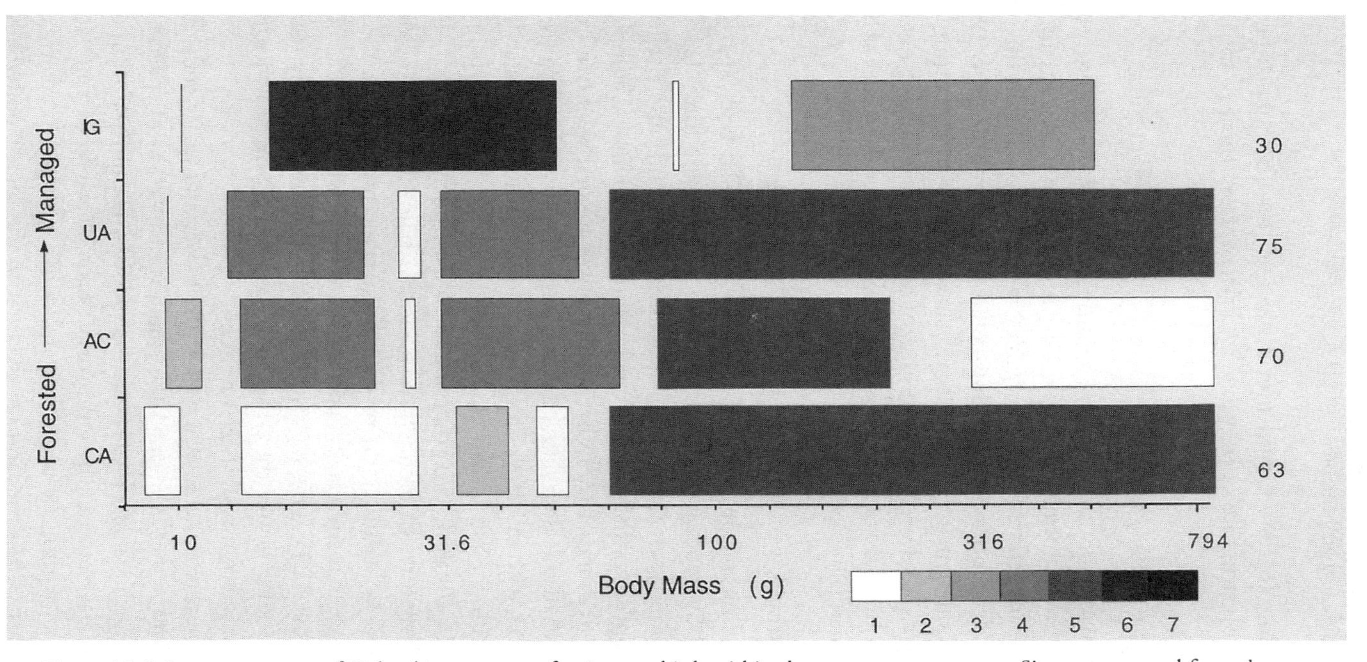

Figure 12.6. Lump structures of Colombian montane frugivorous birds within the upper montane zone. Sites are arranged from those covered mostly by forest (bottom) to those highly transformed by human activities (top). CA = Carpanta; AC = Acaime; UA = Ucumarí Alto; IG = Iguaque. Each box represents a lump, and the spaces between boxes indicate gaps in the distribution of body mass. The different shades indicate the proportion of species falling within lumps: (1) 0–5%, (2) 5–10%, (3) 10–20%, (4) 20–30%, (5) 30–45%, (6) 45–60%, and (7) 60–100%. Vertical lines represent less than 5% of species. Numbers on the right indicate the number of species for the corresponding data set.

with native trees in an effort to restore land previously used for cattle ranching (see table 12.1). Finally, the lump structure of IG was very different from those of the other three sites.

DISCUSSION

Because of the exploratory nature of this work, and because techniques for testing ecosystem "lumpiness" are still in the formative stages, our interpretations are intended as working hypotheses rather than conclusions. Our analyses are exploratory in the sense that we made use of available information and thus could not control for many factors that might confound the results, including size of area surveyed, effects of hunting, and differences in vegetation types. However, the repetition of some patterns among our four analyses suggests that lump structure of body mass in Andean frugivores changes as landscapes are modified by human activities.

Patterns in Lump Structure

In general, the number of lumps (aggregates of species with a similar body mass) decreased from areas covered by continuous native forest to those where forest has been replaced by simpler vegetation types. It might be argued that this trend simply reflects a decrease in the number of species, which in turn results from reduced habitat complexity (e.g., Karr and Roth 1971; Terborgh 1977); that is, with fewer species, we might expect to find fewer lumps. Several of our data sets, however, suggest other relationships. For example, three pairs of sites with differing frugivore richness exhibited similar lump structures (Anchicaya vs. Yotoco, San Carlos vs. La Esmeralda: fig. 12.4; Carpanta vs. Ucumarí Alto: fig. 12.6). In contrast, two pairs of sites with similar frugivore richness exhibited different lump structures (Merenberg vs. Ucumarí Bajo, Rancho Grande vs. Munchique: fig. 12.5). A close examination of the information available for the sites shows slight differences between members of a pair for the former, whereas the contrary is true for the latter (see table 12.1). This suggests that lump structure of body mass in frugivorous birds reflects to some degree the structure of landscapes and provides information not contained in simple counts of species richness.

Major changes in lump structure were observed at the upper and lower ends of the body mass range, in the lumps that contained the smallest proportion of species. For example, lumps representing the largest bird species were lost when moving from areas covered mostly by native forest to those where the forest was replaced by páramo (across elevational zones) or pasture (within the upper lowland zone). Lumps representing the smallest bird species were lost across the land use gradient in the upper montane zone. The lower montane zone exhibited an intermediate pattern (see figs. 12.3–12.6). These results suggest that human activities in Neotropical montane landscapes affect the smallest and largest frugivorous species, often to the point of eliminating entire aggregates of species.

Other changes in lump structure were represented by lumps that fused or broke down among sites representing different landscapes. This pattern indicates that gaps in the distribution of body mass are being closed and opened, probably as a result of the replacement of species across different elevational zones (Terborgh 1971; Hooghiemstra 1984; Gentry 1992a), as well as local extinctions and invasions by species that were formerly rare or absent at a site (Pacheco et al. 1994; Lynam, chap. 15).

Similarities in lump structure across elevational zones and across sites within elevational zones resulted from the persistence of some lumps. The structure of these lumps remained relatively unchanged, probably due to the presence of the same species across the elevational or land use gradient and/or the replacement of species by congeners of similar size. For example, the twenty-first upper lowland (UL), thirteenth lower montane (LM), and fifth upper montane (UM) lumps contain a total of seventeen, thirteen, and ten species respectively (see fig. 12.3). Five species are shared by UL, LM, and UM, three by UL and LM, five by LM and UM, and nine are exclusive to UL, suggesting that some species—those shared by UL, LM, and UM—maintain the structure of these lumps, while other species—those exclusive to one or two zones—are not crucial to the maintenance of lump structure. In other words, these two groups of species may represent the drivers (keystone species) and passengers (redundant species), respectively, mentioned by Walker (1992) when discussing the role of biodiversity in ecosystem processes.

The presence of lumps in distributions of animal body masses, however, is a matter of debate, mainly because of the lack of well-known statistical techniques to identify them. For example, Manly (1996) reanalyzed Holling's (1992) data and found lumps in the body mass distributions of boreal forest and prairie birds, but not as many as Holling reported. Manly advocated the use of Silverman's (1981) "bump-hunting" technique, but recognized that it is difficult to identify lumps even if they are in fact present in a distribution of body mass. We believe that the "bump-hunting" technique is extremely prone to Type II errors; the development of LA_{GRI} is an attempt to overcome this limitation while avoiding some of the problems associated with Holling's (1992) and Manly's (1996) approaches.

Causes of Lump Structure

The comparisons that we set up among different elevational zones, and among different sites within the same elevational zone, represent two different scales of inquiry, yet produced similar results. That is, at both scales we found lumps in the body mass distributions of frugivores and a decrease in the number of lumps from complex to simple ecosystems. Elevational zones, and sites within elevational zones, were arranged from those covered mostly by native forest to those in which the native forest has been replaced by open vegetation (páramo) or by managed ecosystems (pastures or tree plantations) in which scattered forest fragments remain. Thus, changes in lump structure seem to reflect a common causality best explained by landscape complexity, which is probably a function of the vertical and horizontal structure of vegetation. Our proposition is supported by work in other regions (August 1983; Thiollay 1992; Lescourret and Genard 1994).

The vertical structure of vegetation refers to the height of vegetation and the diversity of growth forms, and it becomes simpler as one moves from the lowlands to the páramo, or from forest to pasture. Along the elevational gradient, fog, air temperature, and radiation climate are proximate factors that explain changes in the vertical structure of the vegetation (Leigh 1975; Grubb 1977a). Along the gradient of land use, soil and rainfall distribution account for changes in patterns of land exploitation and hence the structure of the vegetation (e.g., Holdridge et al. 1971). In addition, within forest fragments, wind shear forces, atmospheric humidity, and soil moisture can affect the vertical structure of forest vegetation (Esseen 1994; Laurance 1991b, chap. 6; Kapos et al., chap. 3).

The horizontal structure of vegetation refers to the spatial distribution of vegetation types resulting from changes in abiotic conditions or disturbance (Wiens, Crawford, and Gosz 1985). In Colombia, horizontal structure becomes simpler as one moves from the lowlands to the páramo, or from more natural to highly modified landscapes. At lower and middle elevations, landslides result locally in the development of stands of second growth (Gardwood, Janos, and Brokaw 1979; Mejía et al. 1994; Velásquez et al. 1994), while at higher altitudes, climate and soil determine the presence of elfin forest and páramo (Cuatrecasas 1958; Espinal et al. 1977). Along the gradient of land use, abiotic and socioeconomic factors determine not only rates of deforestation but also the type of matrix in which forest fragments are embedded.

Two sites in the lower montane zone, Rancho Grande (RG) and Munchique (MU), show how vertical and horizontal vegetation structure may relate to lump structure. Rancho Grande and Munchique represent highly modified landscapes where fragments of degraded native forest are interspersed with orchards and pastures (RG) or even-aged pine plantations (MU: Munves 1975; Mondragón 1989). The spatial distribution of vegetation types at the two sites is similar, but the vertical structure of the vegetation types differs. The presence of pastures and orchards with shade trees suggests that RG has a more complex vertical structure than MU, where high densities of pine trees inhibit the development of other plant species. The two sites have similar numbers of species, but RG has more lumps, and species are more evenly distributed among lumps, than at MU (see fig. 12.5).

Even though there are reasons to believe that lump structure results from the interplay of the vertical and horizontal structure of the vegetation, we cannot discard other possibilities. Lumps in body mass distributions of frugivores may result from the response of frugivores to other features of landscapes (Trophic-Through Hypothesis: Holling 1992) or from phylogenetic constraints (Urtier-Historical Hypothesis: Holling 1992). Features that are correlated with landscape structure and that can generate lumps in body mass distributions include seed and fruit patch size. For example, seed size influences the behavior of frugivores (Moermond and Denslow 1985), but is also correlated with both the dispersal and regeneration modes of plants (Salisbury 1974; Martin 1985; Hughes et al. 1994; Osunkoya et al. 1994). Phylogeny may constrain the range of body masses within a family and may explain why closely related species are distributed more often among lumps within a given range of body mass. However, it does not explain the presence of discontinuities in a frequency distribution of body mass, since gaps do not separate lumps composed exclusively of species belonging to the same family.

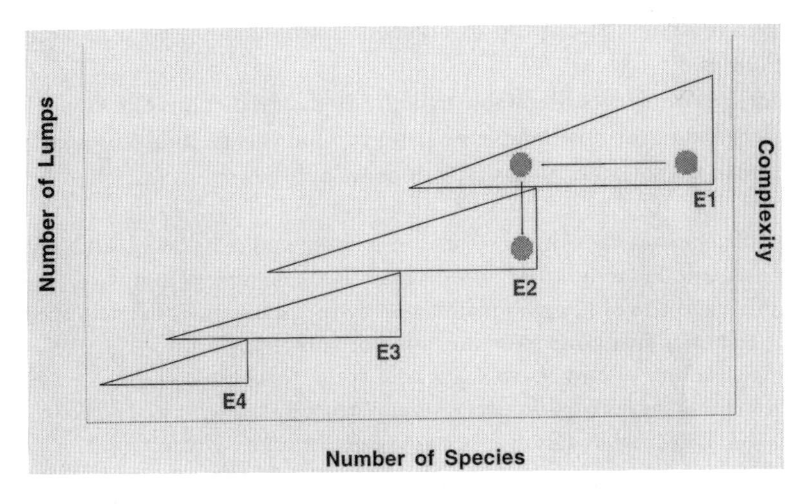

Figure 12.7. Relationship between species richness and lump structure in landscapes of varying complexity. The triangles represent ecosystems from the most complex (E1) to the simplest (E4). Circles within a triangle indicate changes in species richness while lump number remains relatively constant; circles in two different triangles indicate important changes in lump number while species richness remains relatively constant.

A Model Linking Lumps and Species Diversity in Landscapes

Our results reveal relationships between lump structure and number of species along a gradient of landscape structural complexity. In addition, they hint at a relationship between lump structure and ecosystem resilience. Resilience, as defined by Holling (1973), is a measure of the amount of disturbance or change that an ecosystem can absorb before turning into a different one.

We summarize our findings in a graphical model (fig. 12.7) that links numbers of species and lumps. In the model, the triangles represent different ecosystems (E1–E4), arranged from the most complex (top) to the simplest (bottom). The arrangement of these triangles along a diagonal reflects a general trend in which the number of lumps tends to decline as species richness declines. Ecosystems are depicted as triangles to reflect the same positive association between species richness and lump number, with historical, climatic, and edaphic factors and human activities contributing to the variability in species richness and lump number within each ecosystem.

In a given ecosystem, lump structure (the vertical dimension of each triangle, simplified here to just the number of lumps) can remain relatively unchanged despite marked differences in the number of species between sites (the horizontal dimension of each triangle). In our data sets, changes in species richness, but not lump number, were associated with changes in species numbers within a given type of land use. However, as one approaches the apex of each triangle, the probability that important shifts in lump structure will occur increases, even though species richness can remain nearly constant. In our data sets, such shifts were associated with changes in land use. Each jump into a successively simpler ecosystem (from top to bottom) is accompanied by a reduction in the number of lumps.

The model generates a set of testable hypotheses and predictions that could contribute to our understanding of how landscape pattern, biodiversity, and ecosystem processes interact at broad spatial scales. The predictions are as follows:

1. Among different sites, there is a threshold in species number below which lump structure changes dramatically, as indicated by a decrease in the number of lumps.
2. The number and size of lumps is maintained by the persistence of certain species. Their disappearance may alter lump structure in important ways.
3. The lump structure of bird assemblages reflects the resilience of a given ecosystem. Removal of species may have less impact on lump structure in highly diverse ecosystems (top triangles, fig. 12.7) than in less diverse ecosystems (bottom triangles, fig. 12.7).

Natural and anthropogenic disturbances, either alone or in concert, can affect landscapes over spatial and temporal scales ranging from hundreds of meters to hundreds of kilometers and years to decades, respectively. The inherent complexity of landscapes on this spatiotemporal scale justifies a search for new approaches and methods for understanding the relationships among biodiversity, landscape patterns, and ecosystem processes. Rather than focusing on individual parts, these new approaches should concentrate on aggregates of parts and the key processes that structure ecosystems (Hay 1994; Turner, Gardner, and O'Neill 1995; Holling et al., in press). The "lump" approach discussed here represents one of these new lines of inquiry.

GENERAL IMPLICATIONS

1. The lump structure (distribution of body mass aggregates, or lumps) of animal assemblages provides a measure of diversity that captures information not contained in simple counts of species richness.
2. In assemblages of frugivorous birds in Colombia, there were distinct relationships between lump structure and landscape complexity, and these were quite consistent at two different scales of inquiry (between elevational zones, and at sites within elevational zones). This finding suggests that similar processes are responsible for generating lump structure in different regions and at different spatial scales, and that management should focus on these processes.
3. There is some indication that assemblages of Neotropical montane frugivorous birds are robust to human disturbance, provided that landscapes are not severely modified.
4. In Neotropical montane ecosystems, the disappearance of particular lumps (species of common body mass) of frugivorous birds may reflect important changes in seed dispersal, and thus regeneration trajectories of vegetation after disturbance.
5. The fragmentation and transformation of Neotropical montane ecosystems does not seem to generate the same patterns in assemblages of frugivorous birds at low, middle, and high altitudes. These differences may have important consequences for the conservation and management of ecosystems along the elevational gradient.

ACKNOWLEDGMENTS

We are indebted to M. Giraldo, N. Gomez, S. Arango, R. Samudio, and G. Kattan, who provided us with useful references and unpublished observations on Colombian montane birds. We appreciate very much the help provided by R. Forrest (data uploading) and J. Meisel and J. Sendizimir (solving computer-related problems). Many ideas in this chapter were developed as a result of exciting discussions with N. Arango, G. Peterson, and C. S. Holling. This chapter benefited from comments by R. Edwards, G. Kattan, J. Kikkawa, and three anonymous reviewers. In particular, we want to thank R. Bierregaard, J. Ewel, R. Green, W. Laurance, and D. Levey for their valuable criticisms.

13

Structure and Conservation of Forest Avifauna in Isolated Rainforest Remnants in Tropical Australia

Neil H. Warburton

Clearing of natural vegetation for agriculture, urban development, and other purposes creates fragmented landscapes containing remnant vegetation patches surrounded by matrices of altered vegetation and human land use. The ability of such landscapes to conserve a region's biota is of concern to all those interested in biological conservation. The obvious similarity between such landscapes and archipelagoes of true islands prompted researchers to apply island biogeography theory (MacArthur and Wilson 1967) in an effort to explain the distribution of species in insular habitats. Studies of island biogeography have usually focused on determinants of species richness rather than on the composition of the biota or the detailed processes affecting species composition. The results of such studies have been applied to the conservation of insular biotas in fragmented landscapes and the design of nature reserves (Diamond 1975a; Wilson and Willis 1975).

Some researchers, however, have raised questions about this approach. Zimmerman and Bierregaard (1986) questioned the value of island biogeography theory in the absence of autecological studies, arguing that it provides no special insights relevant to conservation. Saunders, Hobbs, and Margules (1991) critically reviewed the use of island biogeography theory by managers of conservation reserves and suggested that it has little value because it provides no information about the types of species fragmented habitats will support. Further, they pointed out that most managers are faced with a fait accompli, rather than being involved in the initial planning of landscapes before fragmentation. They called for research and discussion focused on practical issues relating to the effects of fragmentation on ecosystems and the management of remnant vegetation for conservation (see also Kramer, chap. 25).

More recent investigations have endeavored to elucidate patterns of species composition (e.g., Patterson and Atmar 1986) and to determine the processes that give rise to such patterns (e.g., Karr 1990a; Laurance 1991a; Sieving and Karr, chap. 11). For example, a nested subset pattern often occurs in which the species that inhabit depauperate islands are a subset of those on richer islands (Patterson and Atmar 1986). Such patterns have been documented for mammal (Patterson and Atmar 1986) and bird (Patterson 1987) faunas on archipelagoes and in a number of other insular situations (Patterson

1990). The occurrence of a nested subset pattern has important implications for conservation: in particular, it implies that a series of smaller reserves or islands will sustain only a fraction of the total species pool contained in large fragments or islands, and that a highly nonrandom component of the biota will be prone to extinction in small remnants (Patterson 1987).

Species vary in their responses to habitat fragmentation. Populations of some increase in fragments, some are unaffected, and yet others decline or disappear. A variety of ecological and life history traits have been associated with local extinction proneness in insular populations of tropical forest birds, including natural rarity (small population size), large body size (Karr 1990a), low adult survival rates (Karr 1982a,b), low fecundity (Sieving and Karr, chap. 11), and low tolerance of the modified habitat matrix surrounding forest remnants (Diamond, Bishop, and Balen 1987; Bierregaard and Stouffer, chap. 10; Tocher, Gascon, and Zimmerman, chap. 9).

In this chapter I use nested subsets and other analyses to assess the structure and composition of the avifauna in remnant rainforest patches on the Atherton Tableland in northeastern Australia. I also search for ecological characteristics of species that influence their response to fragmentation. Finally, I analyze environmental features of fragments to determine which are the best predictors of bird species richness and assess the relevance of these findings to the so-called SLOSS debate (single large vs. several small reserves: Simberloff and Abele 1982; Simberloff 1988).

METHODS

Site Selection Criteria

The study area was on the Atherton Tableland, in the humid tropical region of northeastern Queensland, Australia. The Atherton Tableland and adjacent Evelyn Plateau resulted from uplifting by granite intrusions. They were covered by broad lava plains of tertiary basalts that have weathered to produce fertile deep red or chocolate brown soils. The Tableland, ranging from 600 to 900 m in elevation, has a cool upland climate (mean average temperature is 20° C). Rainfall is seasonal, with the wettest months being from January to April. The Tableland exhibits a gradation in rainfall from southeast (annual mean = 2,625 mm) to northwest (mean = 1,275 mm: Australian Bureau of Meteorology records 1914–1987). Most arable land on the Tableland has been cleared, leaving only fragments of the original vegetation (fig. 13.1). The process of clearing began in 1909 and proceeded until recent times (Eacham Historical Society 1979). The only substantial areas of rainforest are on the surrounding hills and escarpment to the east (see Laurance, chap. 6, for further details).

Thirty patches of rainforest were identified that were at least 500 m from continuous rainforest and from other fragments. Two control sites were established in continuous rainforest. The vegetation at these sites is classified as either complex mesophyll or complex notophyll vine forest (Tracey 1982). The forests are typified by multiple tree layers, a high diversity of tree species, and abundant vines and epiphytes.

The sizes of the forest fragments studied ranged from 0.5 ha to 620 ha. Fragments were selected, where possible, so that their sizes were equally distributed among six size

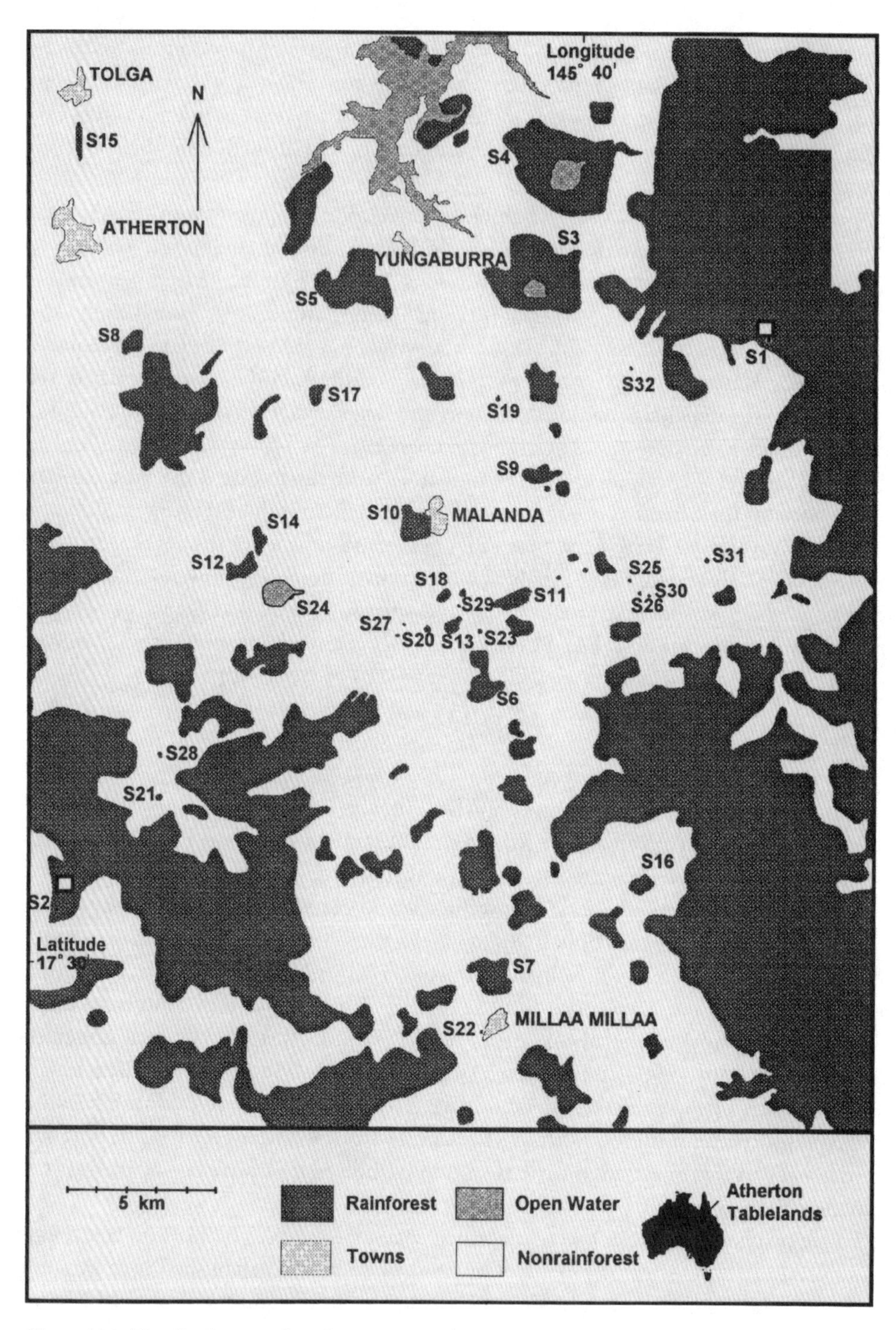

Figure 13.1. The distribution of rainforest remnants and continuous forest tracts on the Atherton Tableland, northern Queensland. Cleared areas are mainly cattle pastures with some agriculture and residential development, interspersed with narrow corridors of regrowth forest along streams. Study sites are labeled S1–S32.

classes (see fig. 13.2). The forest containing the control sites was at least 100,000 ha in area. Study sites were private properties, fauna reserves, national parks, or state forests. Many fragments had been subjected to past or present disturbances by logging, human visitation, and livestock.

Census Methods

Sixty bird species that are restricted to, or reach their greatest abundance in, rainforests of the Atherton Tableland were selected for this study (Warburton 1987) on the basis of a literature review (MacDonald 1973; Readers' Digest 1976; Pizzey 1980; Blakers, Davies, and Reilly 1984). Widespread species that use a variety of habitats throughout their range were selected if they occurred frequently in rainforest on the Atherton Tableland. No attempt was made to differentiate between the little *(Chrysococcyx malayanus)* and rufous-breasted *(Chrysococcyx russatus)* bronze-cuckoos because their ecology and behavior is indistinguishable (Ford 1981); both were lumped as the former species.

Following the recommendations of Dawson (1981), sites were sampled in approximate proportion to their area during each census: sites smaller than 2 ha for 0.5 hour; 2–5 ha sites for 1.0 hour; 5–20 ha sites for 1.5 hours; and sites larger than 20 ha for 2 hours. Each site was repeatedly censused until no more than one new species was recorded during three consecutive censuses. The number of censuses was strongly correlated with site area ($r_s = 0.82$; Spearman rank correlation). All censuses were conducted by the author. Censuses were made in each site at all times of the year and at all times of the day from April 1982 to September 1986.

Several of the species studied mimic the calls of other birds (e.g., spangled drongo, *Dicrurus hottentottus,* and golden bowerbird, *Prionodura newtoniana*), making identification of birds by call alone subject to error. Consequently, every effort was made to obtain sight records of birds, although records of calls were used in determining the number of censuses required for each site.

Nested Subsets Analysis

Nested subsets analyses (Patterson 1987) were performed on the species' presence/absence data. The procedure is a five-stage process:

1. Sites are ranked in order of species richness, and species are ranked according to the number of sites at which they occur.
2. For each species in turn (species i), the least species-rich site where that species occurs is identified.
3. All sites with greater species richness are examined for the presence of species i.
4. The number of sites of greater species richness where species i is absent is tallied to give N_i.
5. This process is repeated for all species in the study, and N_is are then summed to give an "index of nestedness," N, for the set of sites.

This index is a measure of the departure of the archipelago of sites from perfect nestedness. I calculated three different indices, N_{all}, N_{sed}, and N_{mov}, using all species, sedentary species, and species that undertake regular seasonal movements, respectively. Because N_is are derived in the context of other species, the sum of N_is calculated using only sedentary species and those calculated using nonsedentary species does not equal those calculated using all species.

To test the significance of the *N* index, simulated archipelagoes, created using the RANDOM programs described by Patterson and Atmar (1986), were compared with the observed pattern. The latest versions of these programs (April 1995) were implemented for GW BASIC on IBM and compatible machines by Jin Jou Hwang Scientific Support Services, Field Museum of National History, Chicago, USA. $RANDOM_0$ simulations constrained species richness of sites to actual values; for site faunas, species were drawn from a uniform probability distribution as described in Patterson and Atmar (1986). $RANDOM_1$ uses a different randomization routine to select species in simulated faunas, but yields comparable results. This simulation constrained species richness of sites to actual values and the frequency of species occurrence in sites to actual frequencies. Student's *t*-tests were used to determine the probability that the observed value could have been drawn from the symmetric, approximately normal distributions of simulated values.

Responses to Fragmentation

Bird species were ranked in terms of their apparent response to fragmentation. To accomplish this, incidence functions were plotted for each species (fig. 13.2), using the method described by Diamond (1975b). These incidence functions were arranged into groups of species having similar-shaped curves, yielding five categories. Category 1 species are those that show little response to fragmentation; they either occur in all sites or show no consistent response to fragment area. Category 2 species are those that occur in most sites, but occur less frequently in very small fragments than in large or medium-sized fragments. Category 3 species occur in all or almost all very large sites, are missing from all or almost all very small sites, and occur in only some medium-sized sites. Category 4 species occur in very large sites, but are missing from some large sites and occur in fewer sites as fragment size decreases. The white-browed scrubwren *(Sericornis frontalis)* is included in this group because it shows a similar pattern, but in the reverse direction. Category 5 species occur only in some large and very large fragments. The category to which a species belonged was used as the response to fragmentation variable. The study species and their categories are listed in table 13.1.

Ecological Traits of Birds

Data on fifteen ecological traits of each species were recorded from the literature or field observations (see table 13.3). Four features of nests were recorded from the literature: minimum nest height, maximum nest height, location of nest (ranked by perceived vulnerability of nests to predators: 1 = on or near the ground; 2 = in leaves or vines; 3 = suspended; 4 = in tree trunks or hollows), and nest shape (ranked by nest complexity: 1 = platform; 2 = saucer; 3 = bowl; 4 = cup; 5 = goblet; 6 = domed; 7 = hollow; 8 = mound). Nest traits for parasitic breeders were taken to be those of the host. Three

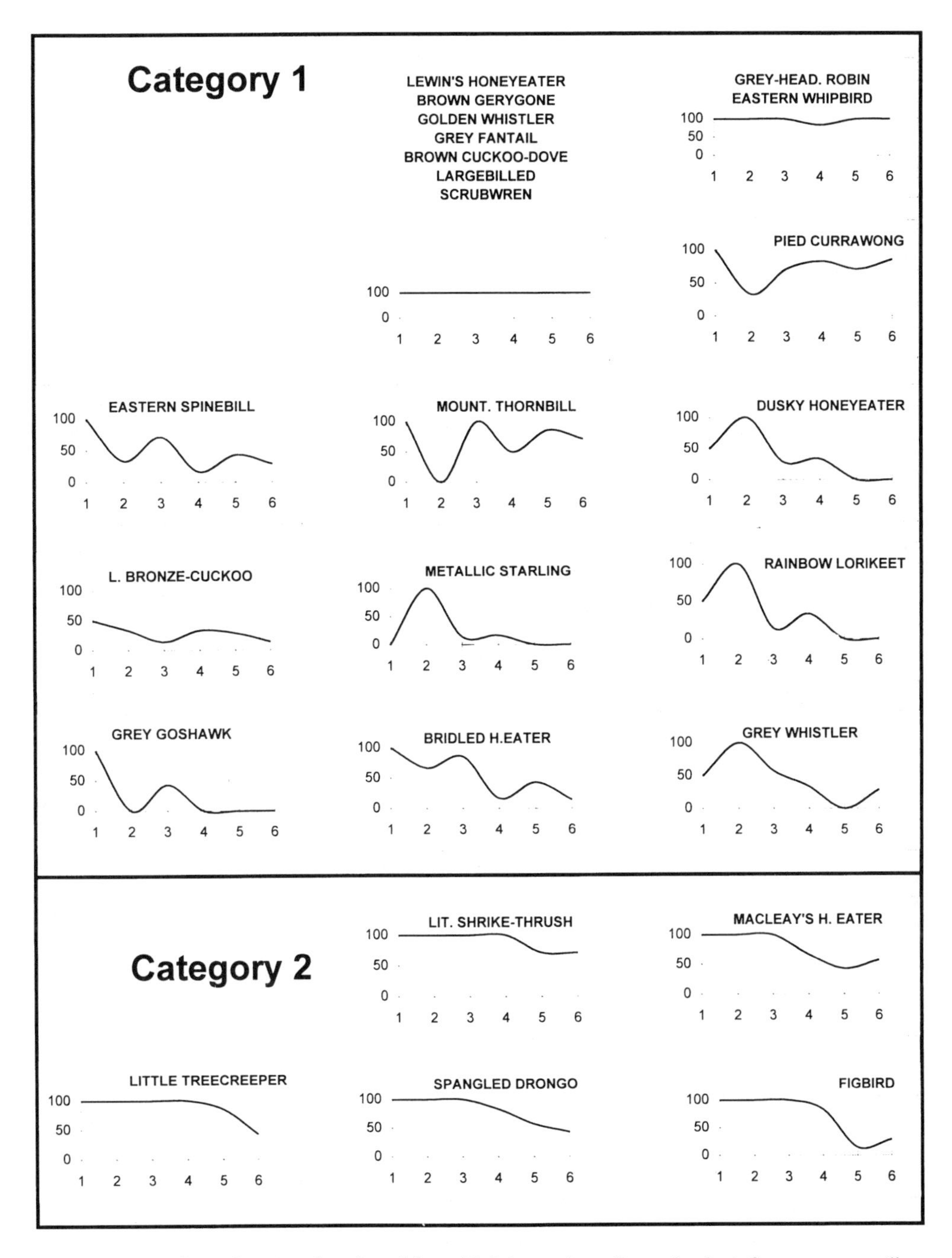

Figure 13.2. Incidence functions for selected forest birds in northern Queensland rainforest remnants, illustrating five types of response to forest fragmentation. Horizontal axes indicate fragment size categories (1 = > 665 ha; 2 = 150–665 ha; 3 = 33–149 ha; 4 = 4.5–32 ha; 5 = 1–4.4 ha. Vertical axes indicate the percentage of sites in each size category in which that species was detected.

(continued)

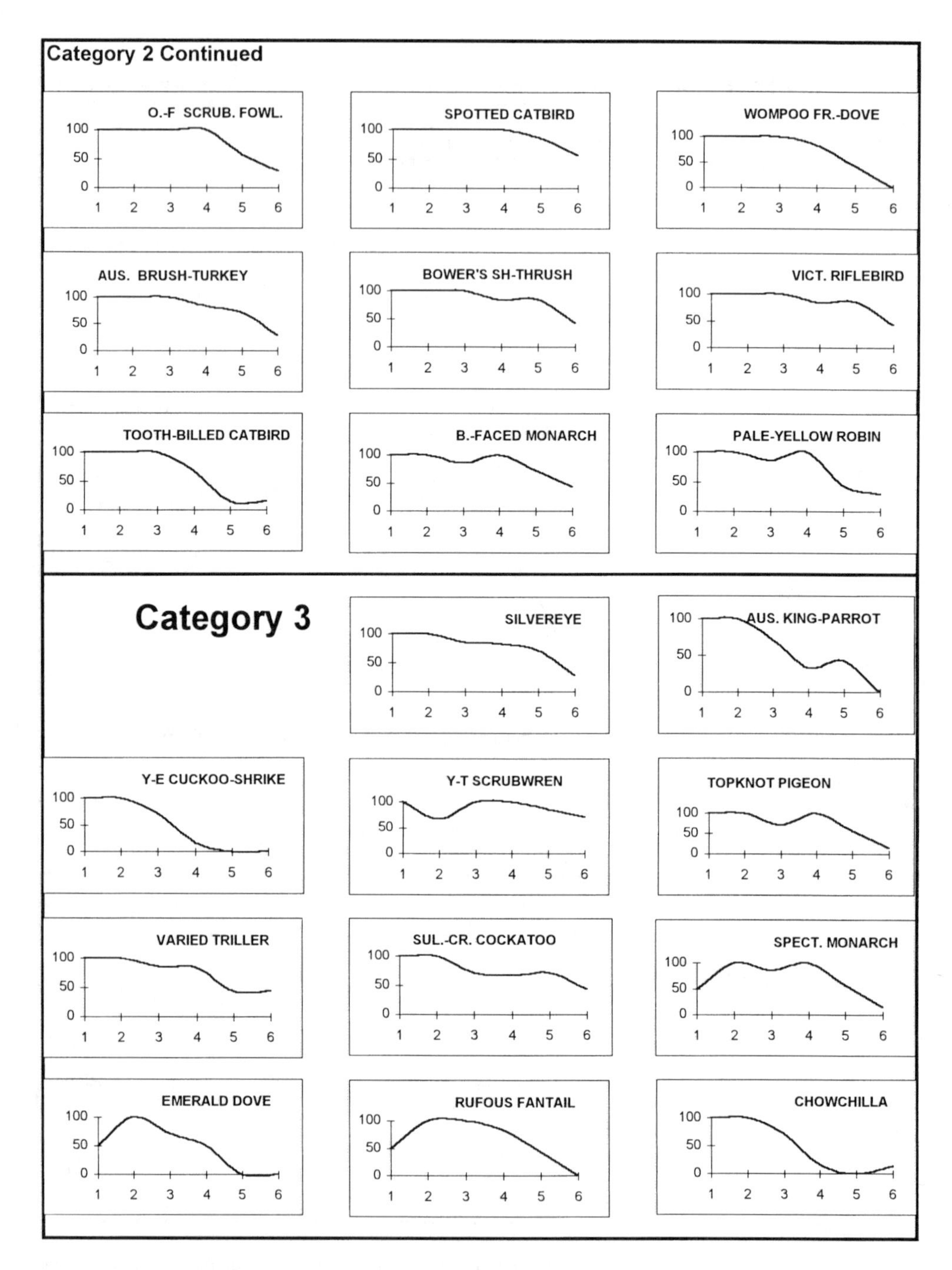

Figure 13.2. (*Continued*)

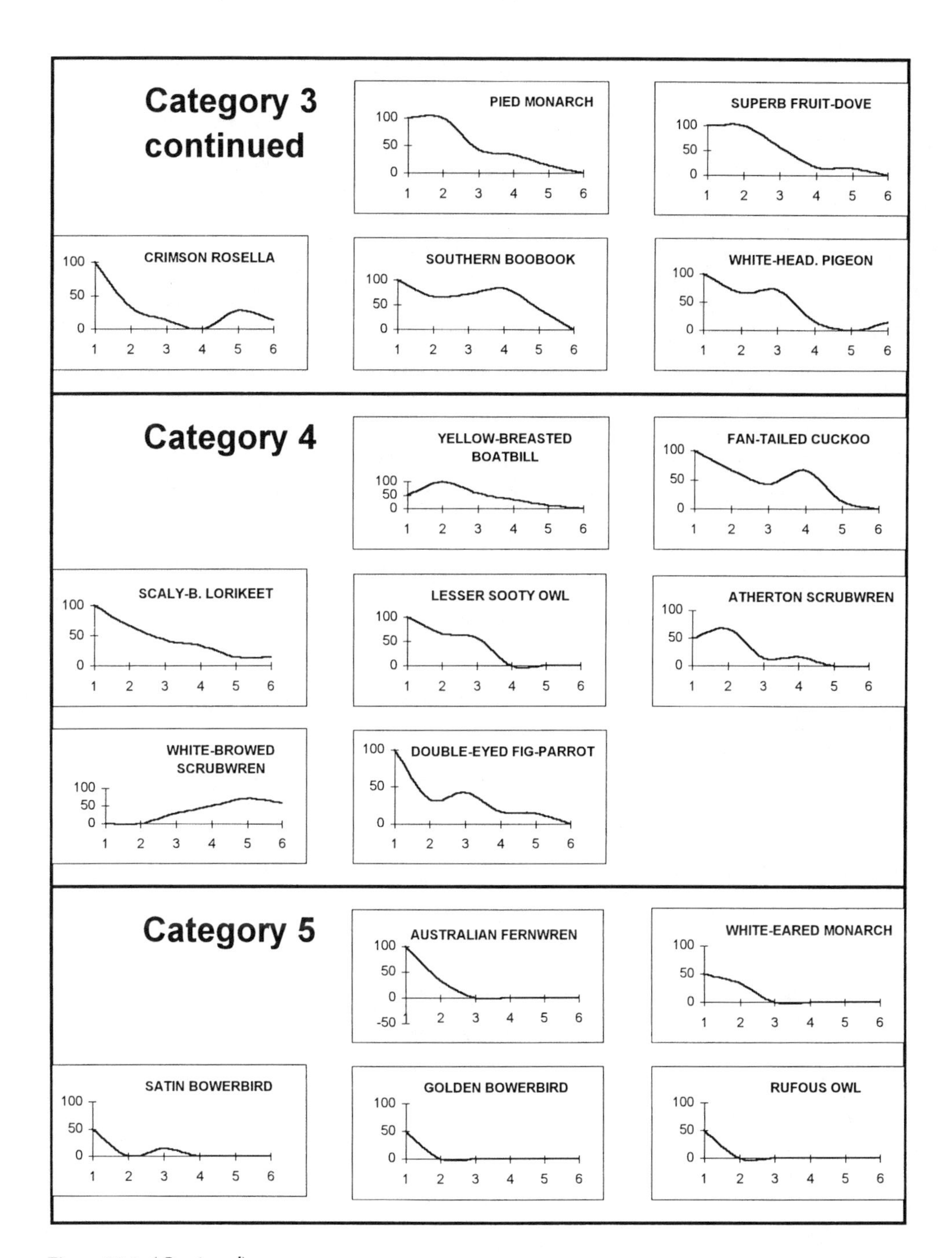

Figure 13.2. (*Continued*)

Table 13.1 Species used in this study and their responses to forest fragmentation.

Common name	Scientific name	Response to forest fragmentation
Grey goshawk	*Accipiter novaehollandiae*	1
Orange-footed scrubfowl	*Megapodius reinwardt*	2
Australian brush-turkey	*Alectura lathami*	2
Wompoo fruit-dove	*Ptilinopus magnificus*	2
Superb fruit-dove	*Ptilinopus superbus*	3
Topknot pigeon	*Lopholaimus antarcticus*	3
White-headed pigeon	*Columba leucomela*	3
Brown cuckoo-dove	*Macropygia amboinensis*	1
Emerald dove	*Chalcophaps indica*	3
Sulphur-crested cockatoo	*Cacatua galerita*	3
Rainbow lorikeet	*Trichoglossus haematodus*	1
Scaly-breasted lorikeet	*Trichoglossus chlorolepidotus*	4
Double-eyed fig-parrot	*Psittaculirostris diophthalma*	4
Australian king-parrot	*Alisterus scapularis*	3
Crimson rosella	*Platycercus elegans*	3
Fan-tailed cuckoo	*Cuculus pyrrhophanus*	4
Little bronze cuckoo	*Chrysococcyx malayanus*	1
Southern boobook	*Ninox novaeseelandiae*	3
Rufous owl	*Ninox rufa*	5
Lesser sooty owl	*Tyto tenebricosa*	4
Yellow-eyed cuckoo-shrike	*Coracina lineaata*	3
Varied triller	*Lalage leucomela*	3
Pale-yellow robin	*Tregellasia capito*	2
Grey-headed robin	*Poecilodryas albispecularis*	1
Golden whistler	*Pachycephala pectoralis*	1
Grey whistler	*Pachycephala simplex*	1
Little shrike-thrush	*Colluricincla megarhyncha*	2
Bower's shrike-thrush	*Colluricincla boweri*	2
Yellow-breasted boatbill	*Machaerirhynchus flaviventer*	4
Black-faced monarch	*Monarcha melanoposis*	2
Spectacled monarch	*Monarcha trivirgatus*	3
White-eared monarch	*Monarcha leucotis*	5
Pied monarch	*Arses kaupi*	3
Rufous fantail	*Rhipidura rufifrons*	3
Grey fantail	*Rhipidura fuliginosa*	1
Chowchilla	*Orthonyx spaldingii*	3
Eastern whipbird	*Psophodes olivaceus*	1
Australian fernwren	*Crateroscelis gutturalis*	5
Atherton scrubwren	*Sericornis keri*	4
Largebilled scrubwren	*Sericornis magnirostris*	1
Yellow-throated scrubwren	*Sericornis citreogularis*	3
White-browed scrubwren	*Sericornis frontalis*	4
Brown gerygone	*Gerygone mouki*	1
Mountain thornbill	*Acanthiza katherina*	1
Little treecreeper	*Climacteris minor*	2
Macleay's honeyeater	*Xanthotis macleayana*	2
Lewin's honeyeater	*Meliphaga lewinii*	1
Bridled honeyeater	*Lichenostomus frenatus*	1
Eastern spinebill	*Acanthorhynchus tenuirostris*	1

Table 13.1 (*Continued*)

Common name	Scientific name	Response to forest fragmentation
Dusky honeyeater	*Myzomela obscura*	1
Silvereye	*Zosterops lateralis*	3
Metallic starling	*Aplonis metallica*	1
Figbird	*Sphecotheres viridis*	2
Spangled drongo	*Dicrurus hottentottus*	2
Golden bowerbird	*Prionodura newtoniana*	5
Satin bowerbird	*Ptilonorhynchus violaceus*	5
Tooth-billed catbird	*Ailuroedus dentirostris*	2
Spotted catbird	*Ailuroedus melanotis*	2
Victoria's riflebird	*Ptiloris victoriae*	2
Pied currawong	*Strepera graculina*	1

traits quantified bird fecundity: average number of eggs per nest; number of breeding cycles per year; and annual fecundity (the product of the former two traits). If a range was given in the literature, the average was recorded. No data were available on clutch sizes of cuckoo species; hence, a value of five was used based on the clutch size of the pheasant coucal *(Centropus phasianinus),* a nonparasitic member of the cuckoo family (Slater, Slater, and Slater 1992).

Three measurements of habitat specialization were used: a rainforest specialization score (0–10) (from Crome 1990), ranging from 0 for species that are independent of rainforest to 10 for those that are highly dependent upon rainforest; preferred habitat type (1–3) (from Crome, Isaacs, and Moore 1995: 1 = totally dependent upon rainforest; 2 = rainforest species, but also tolerate other habitats; 3 = mixed-habitat species); and presence/absence of each species in corridors and windbreaks (0–2) (from Crome, Isaacs, and Moore 1995: 2 = occurs in linear forest remnants along streams and planted windbreaks; 1 = occurs in linear forest remnants, but not planted windbreaks; 0 = occurs in neither linear forest remnants nor planted windbreaks).

Five general traits were recorded: average body size (distance in cm from bill tip to tail tip); geographic range size (1 = 10,000–30,000 km^2; 2 = 30,000–100,000 km^2; 3 = 100,000–300,000 km^2; 4 = 300,000–1,000,000 km^2; 5 = >1,000,000 km^2: Lindsey 1992); trophic level (1 = herbivore; 2 = omnivore; 3 = invertebrates; 4 = some vertebrates in diet); natural abundance (estimated as the percentage of times a species was detected during repeated censuses of sites above 500 ha in area); and movement pattern (1–6). Species' movement patterns are confused in the literature because the movement patterns of a single species can vary in different parts of its range, and because some species have both sedentary and migratory individuals. Consequently, a relatively complex index was used to rank species' movement patterns (1 = sedentary; 2 = sedentary and locally nomadic; 3 = locally nomadic; 4 = locally nomadic and migratory; 5 = sedentary, locally nomadic, and migratory; 6 = migratory).

Spearman rank tests and multiple regression analyses were used to search for associations between ecological traits of species and their responses to fragmentation and to identify key predictors of vulnerability.

Environmental Variables

An exhaustive set (148) of environmental and landscape variables was measured for each site, based partly on Webb, Tracey, and Williams (1976). These included measures of site location, topography, soil type, presence of water, indicator plants, and many forest structural features. Nineteen of the variables were used to describe the degree of isolation, spatial configuration, area, and shape of each forest fragment (see Warburton 1987 for further details). A subset of variables that showed sufficient diversity across sites and seemed likely to affect bird distributions was selected for analysis. When pairs of variables were strongly intercorrelated, the variable deemed least likely to influence bird distributions was removed from further analysis. A stepwise multiple regression analysis was used to determine which combination of environmental variables best predicted bird species richness in fragments. A simple regression with double-log data was used to determine the slope *(z)* of the species-area relationship.

RESULTS

Nested Subsets

Overall, the avifauna in forest remnants on the Atherton Tableland exhibited a highly nested pattern of species distributions (the data are available from the author upon request). The *N* index for these sites (total deviation from perfect nestedness) was 420. The average *N* value and standard deviation for 1,000 simulated archipelagoes generated using $RANDOM_0$ was 704.8 ± 10.9; hence the observed *N* value was 26.2 SD from the mean of the simulated *N* and had a vanishingly small chance of belonging to that distribution of values ($P \ll .0001$: table 13.2). The *N* index for the $RANDOM_1$ simulations was lower (631.4 ± 22.3) because the simulations were to some degree nested (because species richness of the simulated archipelagoes was made equal to the observed values). However, the observed *N* value was still 9.5 SD from the simulated mean, which again is highly significant (P < .0001; table 13.2).

Table 13.2 Nested subsets analysis, showing parameters of simulated archipelagoes (Mean and SD of *N*, the index of nestedness) and comparisons with observed nestedness values.

Simulation	Mean *N* index	SD	Observed value	SD from simulation mean	*P*
All species					
$RANDOM_0$	704.8	10.9	420	26.2	< .00001
$RANDOM_1$	631.4	22.3	420	9.5	< .00001
Sedentary species					
$RANDOM_0$	292.3	6.7	125	25.0	< .00001
$RANDOM_1$	249.2	17.5	125	7.1	< .00001
Mobile species					
$RANDOM_0$	385.9	9.8	285	10.3	< .00001
$RANDOM_1$	354.4	15.7	285	4.4	< .00001

Note: Probability values indicate the likelihood that the observed *N* differs significantly from the simulated random distribution of *N* values. See text for the explanation of $RANDOM_0$ and $RANDOM_1$ models and definitions of sedentary and mobile species.

Interestingly, species recorded in the literature as being sedentary (twenty-eight species) tended to have lower N_i values (and hence showed greater nestedness) than locally nomadic or migratory birds (thirty-two species). These differences were highly significant ($P = .0075$; Mann-Whitney *U*-test). When the nested subsets analysis was repeated for the two groups of birds separately, both groups showed a nested pattern, but the pattern was stronger among sedentary species than for those that undergo seasonal movements (table 13.2).

Ecological Correlates of Vulnerability

Four of the fifteen traits tested were significantly correlated with bird responses to fragmentation (table 13.3). The two most strongly correlated were the presence of species in windbreaks or linear forest corridors along streams ($r_s = -.416$) and bird abundance ($r_s = -.403$); hence, species that used secondary matrix habitats and were naturally abundant appeared least vulnerable to fragmentation. The number of breeding cycles per year ($r_s = -.287$) and fecundity ($r_s = -.313$) also were significant, suggesting that more fecund species are less vulnerable (see Sieving and Karr, chap. 11).

Similarly, two of the fifteen variables—presence of species in corridors and windbreaks, and bird abundance—were selected in a stepwise multiple regression analysis, yielding a highly significant multiple regression ($F = 11.07$, df $= 57$, $r^2 = 0.280$, $P = .0001$). As expected, coefficients of both variables were negative, again indicating that species that use corridors or windbreaks and are naturally abundant are least vulnerable to fragmentation. Conversely, uncommon species that are strongly rainforest dependent appear most vulnerable to fragmentation.

Table 13.3 Spearman rank correlations (r_s) between ecological and life history attributes of sixty rainforest bird species and their responses to fragmentation.

Bird attribute	r_s
Minimum nest height	−.015
Maximum nest height	.091
Location of nest	.061
Nest shape	.112
Number of eggs/nest	−.216
Breeding cycles/year	−.287*
Fecundity	−.313*
Habitat specialization	.103
Preferred habitat	−.237
Presence in corridors and windbreaks	−.416**
Body size	.028
Movement index	−.003
Geographic range size	−.040
Trophic level	.004
Natural abundance	−.400**

Note: The response to fragmentation variable is each species' incidence function category, as illustrated in figure 13.2.

*$P < .05$ **$P < .01$

Table 13.4 Spearman rank correlations (r_s) between selected environmental and landscape variables and species richness of forest birds in thirty rainforest remnants (0.5–620 ha) in North Queensland.

Variable	r_s
Landscape variables	
Total area of fragments with edges within 1 km	.024
Fragment altitude	−.118
Fragment isolation index	.332
Distance to nearest large rainforest tract	.471*
Fragment area	.940***
Habitat variables	
Density of ground cover	.210
Forest disturbance	.358
Presence of permanent water	.631**
Presence of rocks on soil surface	.371*
Epiphytes on tree trunks	.390*

*$P < .05$ **$P < .01$ ***$P < .001$

Environmental Correlates of Bird Diversity

Most (114) of the 148 habitat and landscape variables recorded showed little variation among study sites. Of the remaining 34 variables, 10 were selected for further analysis; these showed reduced intercorrelations and also represented the parameters judged to be most meaningful ecologically.

Five of these ten parameters were significantly correlated with avian species richness (table 13.4). By far the strongest correlation was between species richness and fragment area ($r_s = .94$). When the effects of fragment area (natural log-transformed) were removed with partial correlation analyses, none of the other variables was significantly correlated with species richness.

A stepwise regression analysis confirmed the strength of the species-area relationship. Only log area was selected as a significant predictor, explaining 88% of the total variation in species richness ($F = 195.91$, df = 28, $P < .0001$). A simple regression indicated that the slope *(z)* value of the log species-log area relationship was 0.149.

DISCUSSION

Nested Subsets Analysis

The avifauna of remnant patches of rainforest on the Atherton Tableland exhibits a highly nonrandom distribution pattern. Species assemblages in small, low-diversity isolates generally constitute a subset of those in larger, more species-rich isolates (fig. 13.3).

As a group, sedentary species exhibited greater nestedness than those that undergo regular seasonal movements. A similar pattern was observed among birds in woodlots near São Paulo, Brazil (Patterson 1990). Presumably this pattern arises because migratory or nomadic species move more readily between forest remnants than do sedentary species, and more frequently occupy depauperate sites. Hence, they are less strongly affected by forest fragmentation.

In North America, Blake (1991) found that two groups—birds that require forest habitat for breeding and Neotropical migrants—showed highly nested distributions in fragments, while short-distance migrants and species that breed in forest edge habitat showed more variable patterns. This too is broadly concordant with my findings, at least

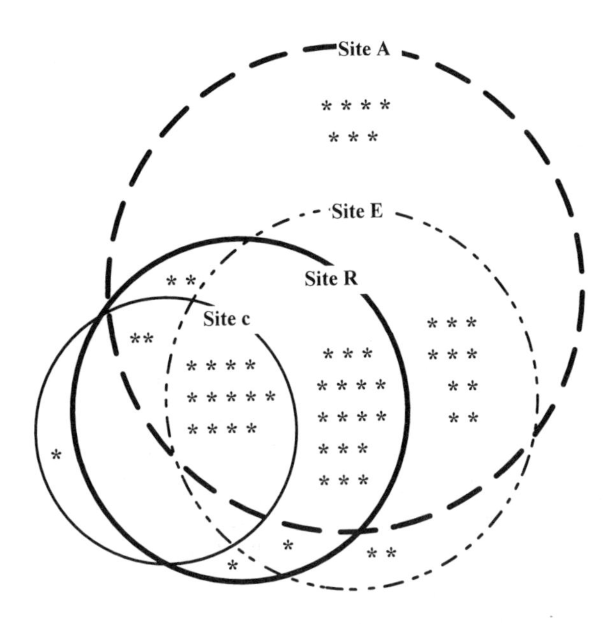

Figure 13.3. Distribution of fifty-five bird species among four rainforest remnants on the Atherton Tableland. The sites were arbitrarily chosen from among thirty remnants censused. The number of bird species recorded for the sites are: Site A > Site E > Site R > Site C.

in terms of the forest-breeding species and short-distance migrants (there was only a single long-distance migrant in my study, the metallic starling, *Aplonis metallica*).

The standard explanation for the nested subset pattern in islandlike habitats is differential extinction: extinction-prone species are progressively lost from fragments over time, with smaller fragments usually exhibiting the greatest losses. If the sequence of species loss is roughly comparable between isolates, the result should be a nested subset pattern (Patterson 1987).

The deviation from perfect nestedness in this study (21.9%) is comparable to that observed among birds of the New Zealand archipelago (20.8%: Patterson 1987). Deviations from perfect nestedness could have arisen for several reasons; attributes of my study sites and the bird species may both have had an effect. Although my isolates were relatively uniform in habitat, they did span altitudinal and rainfall gradients, which probably influenced the bird species found within them: for example, several species occurring locally (e.g., golden bowerbird; satin bowerbird, *Ptilonorhynchus violaceus;* grey whistler, *Pachycephala simplex;* yellow-breasted boatbill, *Machaerirhynchus flaviventer;* white-eared monarch, *Monarcha leucotis*) are strongly influenced by altitude.

The nested subset analysis is highly relevant to the SLOSS debate—whether a conservation strategy should favor one or a few large reserves or multiple, smaller reserves of equal area (Simberloff and Abele 1982; Simberloff 1988). Despite departures from perfect nestedness, there is a strong tendency for small remnants (< 20 ha) to support only a subset of the total avifauna, consistently lacking some sensitive species found only in larger forest tracts. On the Atherton Tableland, for example, small reserves often fail to conserve golden bowerbirds, rufous owls *(Ninox rufa),* lesser sooty owls *(Tyto tenebricosa),* Australian fernwrens *(Crateroscelis gutturalis),* chowchillas *(Orthonyx spaldingii),* and southern cassowaries (*Casuarius casuarius*) (Warburton 1987). Small remnants tend to converge in composition, supporting locally common species that often survive well in modified habitats.

Ecological Correlates of Vulnerability

In this study, a species' degree of tolerance of certain matrix habitats (linear strips of secondary forest along streams and planted windbreaks) and its natural rarity (in large patches and continuous forest) were the strongest correlates of vulnerability. To a lesser extent, more fecund species also appeared less vulnerable. These findings accord well with Laurance's (1990, 1991a, 1994) conclusions regarding ecological correlates of extinction proneness in nonvolant mammals on the Atherton Tableland. As in this study, Laurance (1991a) found that matrix intolerance was a key predictor of local vulnerability, and he suggested that species detected in the matrix are most effective at dispersing between fragments or between mainland forests and fragments. Populations of such species in fragments can be bolstered by the demographic and genetic contributions of immigrants (Brown and Kodric-Brown 1977). Matrix-tolerant species are also likely to tolerate ecological changes in fragmented rainforests, such as edge effects and elevated forest disturbance (Laurance 1991b). Hence, despite large differences in mobility, there appear to be similarities between mammals and birds in terms of the traits that influence their responses to forest fragmentation. Birds and small mammals in Amazonian forest fragments showed similar patterns of vulnerability (Bierregaard and Stouffer, chap. 10; Malcolm, chap. 14).

Rare or uncommon species also appear more vulnerable to fragmentation. In western North America, Bolger, Alberts, and Soulé (1991) found a strong correlation between the ability of bird species to persist in fragments of chaparral and their density in undisturbed habitat. Soulé et al. (1988), working in the same area, found low density and small body size to be good predictors of extinction proneness. Other researchers (e.g., Terborgh 1974; Willis 1974; Wilcox 1980) have suggested that trophic level and body size are good predictors of vulnerability, but there was no clear indication of such trends in this study.

Ecological traits that may affect the susceptibility of bird nests to predation, such as minimum and maximum nest height and nest location, were not significantly correlated with species' vulnerability. Laurance, Garesche, and Payne (1993) and Laurance and Grant (1994) showed significant predation by omnivorous rodents on artificial avian ground nests. This pattern, along with the fact that forest fragments often support high rodent densities (Laurance 1994), suggests that some bird species could experience elevated nest predation in fragmented rainforest. Although little is known at present about rates of predation on real bird nests in forest remnants, Sieving and Karr (chap. 11) argue that birds with nests that are vulnerable to nest predation have evolved high fecundity and consequently are less vulnerable to habitat fragmentation.

Environmental Predictors of Species Richness

For bird assemblages in remnant rainforest patches on the Atherton Tableland, the major determinant of species richness clearly is fragment area (fig. 13.4). Although several other

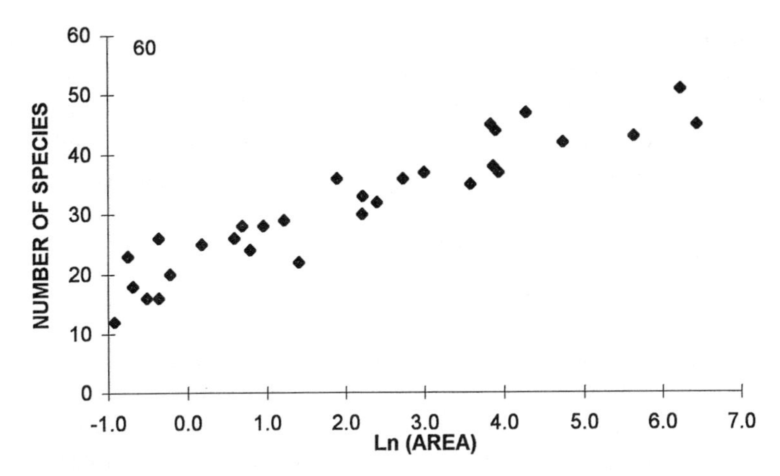

Figure 13.4. Species-areas relationship for the thirty remnants censused.

variables were significantly correlated with bird species richness (see table 13.4), none of these correlations was nearly as strong, and no other predictor was significant when the effects of fragment area were removed with partial correlation analyses.

Such strong correlations between species richness and remnant area have often been seen to provide support for island biogeography theory (F. S. Gilbert 1980; Kent 1987). The species-area relationship, however, is at least partly epiphenomenal (Boecklen and Gotelli 1984) because area is confounded with many other variables, such as vegetation diversity, that clearly influence bird species diversity. Indeed, Boecklen and Gotelli (1984) argue that remnant area alone is inadequate for modeling species number, and that species-area relationships demonstrate only that species number generally increases with area. However, in a study of thirty forest islands in central New Jersey, USA, Galli, Leck, and Forman (1976) found that bird richness was likely to result from remnant area itself, not from internal environmental heterogeneity. In general, the utility of remnant area as a predictor of species diversity probably depends at least partly on the degree of correspondence between area and measures of habitat diversity.

The slope of the log species-log area relationship *(z)* reflects the rate at which species richness declines with isolate area (MacArthur and Wilson 1967). Preston (1962) predicted values of 0.20–0.40 for true isolates and lower values for less isolated systems, while MacArthur and Wilson (1967) suggested that nonisolated samples within continents would range from 0.12–0.19. The value of z obtained in this study (0.149) is within the predicted range for nonisolated sites, and suggests that the species richness of smaller fragments is being inflated by transient or generalist species. The Australian rainforest avifauna has a low number of specialist species compared with other tropical avifaunas (Crome 1990). Because many species are relatively generalized in their habitat requirements, it is likely that, on the whole, this avifauna will be more tolerant of forest fragmentation than avifaunas in other tropical rainforests.

Even the largest forest remnant I examined (620 ha) failed to conserve populations of some forest-dependent species (e.g., southern cassowary; rufous owl: Warburton 1987). Additional species may disappear from remnants over time; most fragments surveyed in this study have been isolated for less than sixty years and have reached their present size and degree of isolation even more recently. In addition, ecological changes in forest fragments may be considerable, with many remnants exhibiting elevated levels of disturbance (Laurance 1991b, chap. 6), and this is likely to influence the structure of avian communities in fragmented landscapes.

GENERAL IMPLICATIONS

1. The species-area relationship can be used to determine (and predict) the number of species a given fragment is likely to support, while incidence functions can be used to determine the probability that a given species will be present in fragments of a certain size.
2. Nested subsets analysis reveals that, on the Atherton Tableland, the distribution of bird species among fragments is highly nonrandom. The avifaunas of smaller fragments tend to converge in composition and become dominated by generalist (non-rainforest specialist) species.
3. The species found in small rainforest remnants were those that most readily tolerated the matrix of modified habitats (such as linear forest remnants along creeks and planted windbreaks) surrounding the remnants.
4. Although fragment area was by far the best predictor of bird species richness, other features, such as fragment habitat diversity and degree of isolation, may also influence bird assemblages in fragments.
5. I conclude that while even small remnants (< 20 ha) of rainforest can support a significant fraction of the local avifauna, substantial areas (well over 600 ha) are required to conserve the total rainforest avifauna on the Atherton Tableland. The relatively high species diversity in small patches is probably due in part to immigration from nearby forested areas. A relative paucity of ecological specialists may also make the Australian rainforest avifauna more tolerant of habitat modification than are avian assemblages in other tropical regions.

ACKNOWLEDGMENTS

Chief thanks are owed to Bill Laurance for his interest, stimulation, criticisms, and suggestions, without which this chapter would not have been written. The chapter was greatly improved by editing and suggestions by Bill Laurance, David Westcott, and Richard Bierregaard. Peter Valentine, my thesis supervisor, is warmly thanked for his patience, advice, and assistance.

14

Biomass and Diversity of Small Mammals in Amazonian Forest Fragments

Jay R. Malcolm

INCREASINGLY, large blocks of continuous rainforest are being reduced to remnant patches in a sea of pasture and other secondary habitats. A direct result of this rapid loss of primary forest habitat is the extinction of tropical species (Anderson, Thornhill, and Koopowitz, chap. 18). Estimates of species loss by the end of the twentieth century are highly variable (e.g., Lugo 1988a; Whitmore, chap. 1). To refine these estimates, it is necessary to obtain information on species richness and distribution patterns in intact rainforest (Wilson 1988) and to examine responses of tropical species and ecosystems to landscape modification (Lugo 1988a). An understanding of how fragmented communities and ecosystems are structured can potentially lead to more efficient designs for reserves and reserve clusters and to strategies for maintaining biological diversity and natural ecosystem integrity in human-dominated ecosystems. Studies of the effects of fragmentation are thus vital because they provide us with empirical information on species loss during fragmentation and give us insights into methods of alleviating species loss.

At the Biological Dynamics of Forest Fragments Project (BDFFP) site in the central Amazon, fragmentation occurred in the late 1970s and 1980s after a series of cattle ranches were established about 80 km north of Manaus, Brazil. Several square primary forest reserves of 1, 10, and 100 ha were delineated prior to deforestation and were subsequently isolated when the surrounding forest was converted to pasture via clear-cutting (Lovejoy et al. 1984, 1986). Simultaneously, many sites were established in continuous, undisturbed forest that could be used as controls to distinguish the effects of fragmentation.

Research on nonvolant small mammals in the area began in February of 1982 (Emmons 1984; Lovejoy et al. 1984) and continued for six months. During a second study of seven months' duration in 1983–1984, I compared terrestrial small mammal communities between 10 ha fragments and continuous forest sites (Malcolm 1988). One 10 ha fragment isolated for about three and a half years exhibited a small mammal community very different from that of any other site, but communities in other fragments (in all cases isolated from continuous forest for less than a year) were indistinguishable from those in continuous forest. Thus, any effect of fragmentation in the "oldest" fragment stood without replication.

In this chapter, I present results from repeated surveys from 1985 to 1989 of a larger number of fragments (four fragments each in 1 ha and 10 ha size classes) isolated for one to eight years. In addition, I describe an intensive effort to census the little-known small mammal fauna of the forest canopy. My purpose here is to discover whether forest fragmentation significantly affected the small mammal community, and whether any effect varied systematically with fragment area. Area-related variation is predicted both by island biogeography theory (MacArthur and Wilson 1967) and by models of edge effects (Levenson 1981; Malcolm 1994; Kapos et al., chap. 3; Laurance, chap. 6). I used three measures to quantify small mammal community structure: total abundance, biomass, and species richness. Abundance and biomass provide simple measures of the ecological importance of a group of organisms from an energy flow standpoint: a shift in biomass indicates changes in the resource base (Eisenberg 1980) and may signal altered patterns of interactions. Species richness in concert with abundance data provides information on ecological dominance—the relative contributions of various species to overall diversity.

MATERIALS AND METHODS

Study Site

The study, part of the Biological Dynamics of Forest Fragments Project (Lovejoy et al. 1984, 1986), took place on three cattle ranches 80 km north of Manaus, Brazil. Primary forest in the area is upland (or *terra firme*) on moderately rugged terrain, and is dissected by small creeks in the headwaters of three small rivers (the Cuieiras, the Preto da Eva, and the Urubú) that drain into the Rios Negro and Amazon. Most soils are nutrient-poor, yellow, alic latosols of high clay content (Chauvel 1983, cited in Klein 1989). Annual rainfall near Manaus averaged about 2,200 mm during a 70-year period, with a dry season of less than 100 mm/month from July to September (Ministério de Minas e Energia 1978b, cited in Klein 1989). Additional details about the study site are provided by Lovejoy and Bierregaard (1990) and Bierregaard and Stouffer (chap. 10).

I censused small mammal populations in four 1 ha fragments (reserve numbers 1104, 1112, 2107, 3114) and four 10 ha fragments (1202, 1207, 2206, 3209; see Bierregaard and Stouffer, chap. 10, for a general description of each fragment). The fragments were created when surrounding forest was clear-cut to create cattle pastures, and were isolated from continuous forest by 100–800 m of secondary habitat. Isolation of the fragments took place in the dry seasons of 1980 (1104 and 1202), 1983 (1112, 3114, 1207, and 3209), and 1984 (2107 and 2206). Live-trapping data were collected during a forty-seven-month period from May 1985 to March 1989; thus, fragments were censused at one to eight years post-isolation.

The secondary habitats surrounding the fragments varied from site to site. Forest cut in 1980 was burned, and some areas close to the fragments were maintained as pasture. Others were abandoned to secondary forest, which was eventually cut in 1987. Clear-cuts created in 1983 were never burned or recut, and by the end of the study (1989), secondary forest in those areas was about 12 m high. The clear-cut created in 1984 was burned and maintained as pasture.

Trapping Methods

Census methods differed between two sampling periods, as described below.

Phase 1

From May 1985 to July 1987, I repeatedly censused three 1 ha fragments, three 10 ha fragments, and eleven sites in continuous forest (usually > 1 km from clear-cut areas). Most sites were censused more than once, at intervals of six to eight months, and more effort was devoted to terrestrial than to canopy trapping (table 14.1). Trap stations were spaced at 20 m intervals, but the exact trap configuration varied from site to site according to whether a BDFFP 1, 10, 100, or 1,000 ha trail system delineated the site (fig. 14.1).

Phase 2

From September 1987 to March 1989, in each of four different spatial "blocks," I repeatedly censused a 1 ha fragment, a 10 ha fragment, and a site in continuous forest (a randomized block design). Traps were configured in 1 ha subsampling units, one per 1 ha fragment and four per 10 ha fragment or continuous forest site. Thus, each block had nine 1 ha units, for a grand total of 36 1 ha units. Within each 1 ha unit, traps were configured to maximize sampling coverage while minimizing the number of traps, and effort was equally devoted to terrestrial and canopy trapping (fig. 14.1). Each block was censused once during each of three periods: (1) September 1987 to February 1988, (2) March 1988 to September 1988, and (3) October 1988 to March 1989.

Terrestrial stations consisted of a Tomahawk wire mesh trap (14 × 14 × 40 cm) and a Sherman folding aluminum trap (8 × 8 × 23), placed 2–4 m apart on the forest floor. Canopy stations consisted of a Sherman on top of a Tomahawk at about 14 m height, and were raised and lowered using a pulley system (see Malcolm 1991a for details). Traps were baited with bananas and peanut butter; in addition, canopy traps had a small cloth sack filled with raisins and peanut butter. Terrestrial traps were rebaited daily, whereas canopy traps were rebaited only when they had a capture or were otherwise disarmed. Traps were usually left open for nine consecutive nights in phase 1 and for eight consecutive nights in phase 2. Captured animals were identified, measured, and released (Malcolm 1988, 1992). Voucher specimens are lodged at the Instituto Nacional de Pesquisas da Amazônia in Manaus and the U.S. National Museum of Natural History in Washington, D.C.

In one of the 1 ha fragments (1112) censused in phase 2, each station consisted of two steel snap traps (9 × 15 cm) instead of two live traps. Also, because of extensive damage to the canopy of this fragment during a windstorm in 1987, five of the twelve canopy stations were at only 2 m height; the remaining seven were at close to 14 m.

Data Analysis

Abundance

Terrestrial and canopy captures were analyzed separately. I standardized effort in phase 1 by combining all captures at each site and calculating the mean number of individuals

Table 14.1 Trapping effort from May 1985 to July 1987 (phase 1) in two sizes of forest fragments and in continuous forest.

		Terrestrial trapping			Canopy trapping		
	Site code[a]	Number of censuses	Number of traplines per census	Total effort (station-nights)	Number of censuses	Number of traplines per census	Total effort (station-nights)
1 ha fragments							
	1104[b]	2	1.2	324	2	0.8	216
	2107[b]	2	1.2	324	2	0.8	216
	3114[b]	2	1.2	324	2	0.8	216
				972			648
10 ha fragments							
	1202	2	3	810	2	1	270
	2206	3	3	1,215	3	1	405
	3209	4	3	1,620	2	1	270
				3,645			945
Continuous forest							
	1101[b]	2	1.2	324	2	0.8	216
	1105[b]	2	1.2	324	2	0.8	216
	0234	2	2	540	2	2	480
	1201	2	3	810	2	2	540
	1204	1	3	405	3	1.67	675
	1205	1	3	405			2,127
	1208	1	3	405			
	1301	2	8	2,160			
	1302	2	8	2,160			
	2303	2	8	2,160			
	3402	2	8	2,160			
				11,853			

Note: Each trap station consisted of two traps (a Sherman and a Tomahawk) placed on the ground (terrestrial trapping) or at about 14 m height (canopy trapping).

a. The second digit in the four-digit site code signifies the area of the trail system that delineated the site: 1 = 1 ha; 2 = 10 ha; 3 = 100 ha; 4 = 1,000 ha.

b. Sites delimited by 1 ha trail systems (1 ha fragments and two sites in continuous forest) were censused using a terrestrial grid (1.2 traplines) or canopy lines (0.8 traplines).

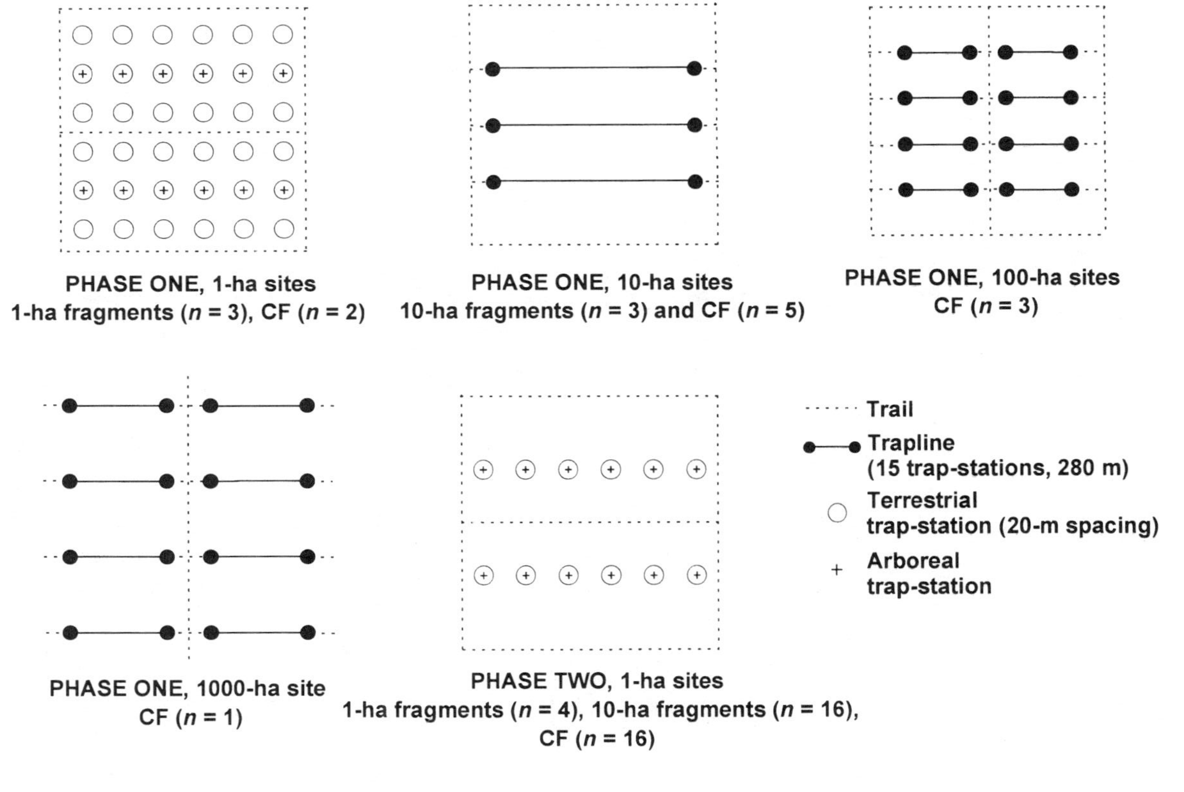

Figure 14.1. Trap configurations used to census small mammals at the BDFFP site during two sampling periods (phases 1 and 2). Each trap station consisted of two traps (a Sherman and a Tomahawk) placed on the ground (terrestrial) or in the canopy (arboreal). During phase 2, a statistical block design was used.

per census per trapline (135 station-nights). At 1 ha sites, I assumed that terrestrial grids sampled an area equivalent to 1.2 traplines and that canopy lines sampled an area equivalent to 0.8 traplines. These conversion constants assume that "trap radii" (see biomass calculations below) are about 25–75 m (Malcolm 1991b). ANOVA and Duncan's test ($\alpha = 0.05$) comparing means among treatments (1 ha fragment, 10 ha fragment, continuous forest) were performed on rank-transformed data.

In phase 2, I combined captures within treatment-by-block combinations and calculated the mean number of individuals captured per census per 1 ha unit (96 station-nights). Means were ranked within blocks, and mean ranks were compared via ANOVA and Duncan's test. This procedure is equivalent to Friedman's two-way analysis for block designs (SAS 1985).

Biomass

To calculate biomass, an estimate of the population density and body weight of each taxon was required. Densities were calculated by dividing the number of individuals captured in a trap session by the area that the traps sampled. In a previous publication (Malcolm 1990), I used distributions of maximum distances between recaptures to estimate the area that a trap sampled. I assumed that an individual used a circular area of radius r during a trapping session in such a way that traps set within the circle would catch the animal, whereas traps set outside it would not (note that r is a measure of the area used by an individual during an eight- or nine-night trapping session, not an estimate of home range size). Thus, r was also the radius of the circular area sampled by a trap. For canopy captures, the circular area was assumed to be in the x–y plane. The model also assumed that if several traps were placed within the circle, the individual would be caught in at least the two traps most distant from each other. Given this second assumption, the distribution of maximum distances between recaptures of individuals can be used to calculate r. Earlier, I solved for the expected distribution of maximum distances given r and a trapline of t traps, and used least squares to estimate r from observed distributions (Malcolm 1990). I used the same method here for the phase 1 data. The geometry of the trap configuration used in phase 2 was more complex, so I calculated the expected distributions numerically. Again, least squares was used to estimate r from the observed distributions of maximum recapture distances. Given an estimate of r, numerical methods were used to calculate the area sampled by a 1 ha unit.

Average maximum recapture distances appeared to vary little between fragments and continuous forest (Malcolm 1991b); hence, to increase sample sizes for calculation of r, I combined all phase 1 and 2 captures regardless of treatment. The final r estimates used in biomass calculations were weighted means of the phase 1 and 2 estimates (weighted by the number of individuals recaptured). Separate estimates were obtained for terrestrial and arboreal captures of the same species. For species that had fewer than four recaptured individuals at one trap height or the other, I used estimates from other species of similar size that showed a similar degree of arboreality.

Estimated r values, maximum distances between recaptures, and body weights are given in appendix 14.1. Because of large *r* values and the resultant overlap of subsampling units within a site, *Didelphis marsupialis* and *Philander opossum* were not included in the total biomass estimates. Hence, biomass estimates included only those species with average adult weight of < 300 g. Statistical tests were as described previously (see abundance calculations above), except that phase 2 estimates were not rank-transformed and a randomized-block ANOVA was performed.

Species Richness

I used only the phase 2 data because trap effort could be easily standardized. I combined canopy and terrestrial captures and calculated two measures of richness: (1) the average number of species per census per 1 ha unit, and (2) the average number of species per 1 ha unit (censuses combined). Data from the 1 ha units in each treatment-by-block combination were averaged, and ANOVA and Duncan's test were performed on within-block rankings.

RESULTS

Total Number of Individuals

Terrestrial capture rates in phase 1 varied significantly among the three treatments ($P < .01$). According to Duncan's test, average mammal abundance did not differ between 1 and 10 ha fragments, but was significantly greater in both fragment sizes than in continuous forest (table 14.2). Similarly, in phase 2, the ANOVA comparing rank-transformed terrestrial abundance among the three forest types was highly significant ($P < .01$), and according to Duncan's test on the ranks, average abundance did not differ between 1 and 10 ha fragments, but was significantly greater in fragments than in continuous forest. In both phases 1 and 2, means and variances for terrestrial abundance data decreased in the following sequence: 1 ha fragment, 10 ha fragment, continuous forest.

Mean arboreal abundances from phase 1 did not differ significantly among forest types ($P = .16$). However, one of the 1 ha sites in continuous forest (site 1101) had no arboreal captures. When this site was excluded, the arboreal mean for continuous forest increased

Table 14.2 Abundances of small mammals in primary forest fragments (1 and 10 ha) and continuous forest.

			Mean ± SD	
Census period	Treatment	No. of sites	Terrestrial trapping	Canopy trapping
Phase 1	1 ha fragments	3	10.1 ± 6.8	2.7 ± 1.3
	10 ha fragments	3	4.1 ± 1.0	6.3 ± 2.0
	Continuous forest	11	1.9 ± 0.9	5.5 ± 4.1
Phase 2	1 ha fragments	4	4.0 ± 2.7	2.8 ± 1.8
	10 ha fragments	4	2.4 ± 1.5	3.2 ± 2.1
	Continuous forest	4	1.2 ± 0.6	1.9 ± 0.4

Note: Live-trapping was conducted during two time periods (phases 1 and 2). The average number of captures per trapline (phase 1) or 1 ha sampling unit (phase 2) was calculated for each site and the mean calculated across sites.

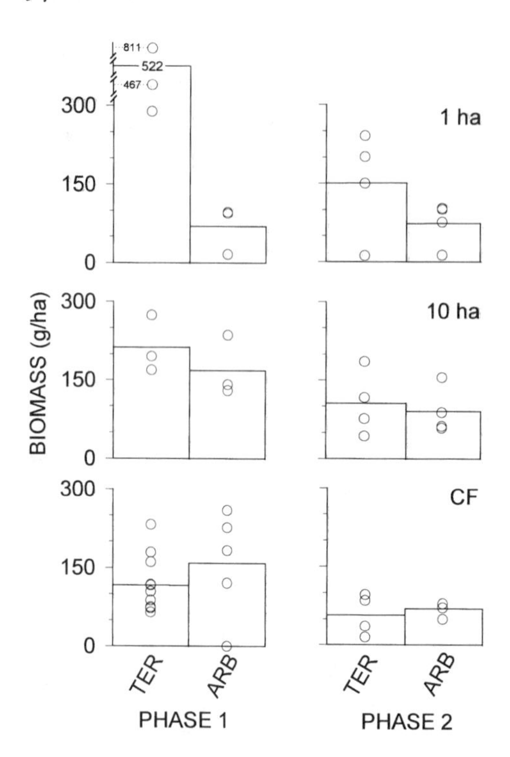

Figure 14.2. Estimated arboreal and terrestrial biomass of small mammals in 1 ha and 10 ha fragments and continuous forest near Manaus, Brazil. Data are from two live-trapping periods (phases 1 and phase 2). Histograms indicate the overall means; circles represent biomass from the individual sites.

from 5.5 ± 4.1 to 6.8 ± 3.1 ($\bar{X} \pm$ SD), and the ANOVA was significant ($P < .03$), with Duncan's test indicating significantly fewer arboreal captures in 1 ha fragments than in 10 ha fragments or continuous forest. In phase 2, however, differences among treatment means were not even close to significant ($P = .75$).

Biomass

Comparisons of biomass among the three treatments gave results qualitatively similar to those obtained for mammal abundances (fig. 14.2). Terrestrial biomass during phase 1 differed among forest types ($P < .01$), and according to Duncan's test, average biomass was significantly greater in fragments than in continuous forest, but did not differ between 1 and 10 ha fragments. Although the ANOVA on phase 2 terrestrial biomass was not quite significant ($P = .06$), the means followed the same sequence as in phase 1. According to Duncan's test, biomass in continuous forest and in 1 ha fragments differed significantly, but neither treatment differed significantly from 10 ha fragments.

Canopy biomass during phase 1 did not differ significantly among treatments ($P = .18$), but when the continuous forest site with no canopy captures was removed, the ANOVA was significant ($P = .03$), and Duncan's test indicated significantly lower canopy biomass in 1 ha fragments than in continuous forest or 10 ha fragments. However, there was little evidence of variation in canopy biomass in phase 2 ($P = .64$).

In both phases 1 and 2, canopy biomass in continuous forest slightly exceeded terrestrial biomass, whereas in 10 ha fragments, the reverse was true (fig. 14.2). In 1 ha fragments, terrestrial biomass greatly exceeded canopy biomass. Interestingly, overall biomass estimates from phase 1 were nearly twice as large as those from phase 2 (except for canopy biomass in 1 ha fragments: fig. 14.2).

Species Richness

During nearly four years of trapping, I captured nineteen species of small mammals (listed in Malcolm 1990). All of these species were captured in the 10 ha fragments, while three of the species *(Philander opossum, Caluromys lanatus,* and *Neacomys guianae)* were never caught in continuous forest, and two *(Mesomys hispidus* and *Isothrix pagurus)* were never caught in 1 ha fragments. During subsequent trapping in May of 1990, an *Oecomys regalis* was captured in a 1 ha fragment, bringing the project total to twenty species. Species accumulation curves suggest that few additional species will be found in upland forest in the area (J. R. Malcolm, pers. obs.). An additional rodent *(Echimys chrysurus)* was captured by hand in continuous forest and a 10 ha fragment, but was never trapped.

ANOVAs on the two measures of richness—per census and combined across the three censuses—from phase 2 were close to significant ($P = .08$). According to Duncan's test, species richness in 10 ha fragments was significantly greater than in continuous forest, whereas that in 1 ha fragments did not differ significantly from that in either of the other two treatments. Means and variances in species richness decreased in the following sequence: 1 ha fragment, 10 ha fragment, continuous forest. Thus, patterns of variation in species richness paralleled those observed for small mammal abundance. Moreover, there was little evidence to suggest that the relationship between number of species and number of individuals differed among the three treatments (fig. 14.3: analysis of covariance, $P = .68$ in part A and $P = .40$ in part B), indicating similar patterns of numerical dominance in the communities. Abundance and species richness varied dramatically among individual 1 and 10 ha fragments (fig. 14.3); for example, one 1 ha fragment yielded two individuals of two species during a trapping session; another yielded thirty-four individuals of eight species.

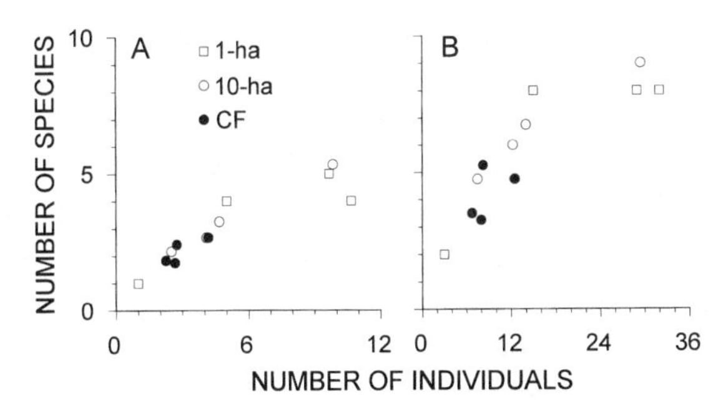

Figure 14.3. Number of species and individuals per 1 ha unit in four 1 ha fragments, four 10 ha fragments, and at four sites in continuous forest. *(A)* Census-specific means (96 station-nights at each of two trap heights). *(B)* Data from the three censuses combined (288 station-nights at each of two trap heights).

DISCUSSION

Abundance and Diversity of Small Mammals

Recently isolated forest fragments in central Amazonia had a more abundant and diverse terrestrial small mammal fauna than did sites in continuous forest (see table 14.2, fig. 14.2). In each of two intensive trapping periods, terrestrial small mammal abundance, biomass, and species richness were significantly greater in 1 and 10 ha fragments ranging from one to eight years old than in continuous forest. This "fragmentation effect" was most extreme in the smallest fragments: 1 ha fragments had consistently greater abundance, biomass, and diversity than 10 ha fragments, although significantly so in only a few cases.

Increased small mammal abundance in the fragments did not result from an increase in just a few taxa; in fact, the relationship between number of individuals and number of species appeared constant across fragments and continuous forest (see fig. 14.3), indicating a general increase across many species. Nor was the increased abundance and diversity in fragments attributable to the presence of species not normally found in continuous forest; all of the abundant species in fragments were commonly caught in continuous forest, and the three species *(Philander opossum, Caluromys lanatus, Neacomys guianae)* caught only in fragments during the study were neither widespread nor abundant.

In contrast to the terrestrial fauna, the abundance, biomass, and richness of the canopy fauna was not convincingly affected by fragmentation. There was some indication that arboreal abundance was lower in 1 ha fragments than in the other two treatments in phase 1, but the same trend was not evident in phase 2. The net effect of these changes in arboreal and terrestrial species was a marked increase in the proportion of the small mammal biomass close to the ground in fragments, especially in 1 ha fragments.

Ecological Changes in Forest Fragments

The elevated abundances of terrestrial small mammals in fragments appeared to result from edge-induced habitat changes, coupled with an overflow of individuals from abundant populations in the secondary habitats surrounding fragments. In fact, 1 ha fragments, which had the greatest proportion of edge-modified habitat, had small mammal communities virtually identical to those in the abutting secondary habitats (Malcolm 1991b). The 1 ha fragments surrounded by pasture had relatively few small mammals, just as the pasture itself did, whereas those surrounded by secondary forest had a superabundant and diverse small mammal fauna, just as the secondary forest did (Malcolm 1991b, 1995). Insularization itself appeared unimportant (Malcolm 1991b, 1995).

In young secondary forests and along edges, understory plants grew rapidly, presumably in response to increased light levels and reduced root competition (Kapos et al., chap. 3; Laurance 1991b; Malcolm 1994), which in turn led to more abundant arthropod populations (Malcolm 1991b). In addition, terrestrial microhabitat diversity was higher in secondary habitats due to large quantities of fallen timber (felling of the original forest resulted in enormous quantities of fallen timber, which remained to a large extent even after burning). The explosion of terrestrial small mammal populations in fragments apparently occurred in response to this rich and diverse resource base. Studies of the mammalian fauna in fragments in northeastern Queensland came to very similar conclusions; the most useful indicator of extinction proneness was often the relative abundance of a species in forest regrowth (Laurance 1990, 1991a, 1994). In the central Amazon, small mammal abundance in the canopy was not influenced by fragmentation, possibly because habitat loss or modification resulting from elevated treefall rates in fragments (Lovejoy et al. 1986; see also Kapos et al., chap. 3; Laurance, chap. 6) was balanced by increases in productivity in the remaining canopy.

Density Compensation and Species Interactions

Small mammal abundances may also be affected by interspecific interactions, as suggested in models of density compensation (MacArthur, Diamond, and Karr 1972) and "excess" density compensation (Case 1975; Case, Gilpin, and Diamond 1979). In these models, a loss of some of the species in a guild in fragments is compensated for by increases in the abundance of other species. In the special case of excess density compensation, removal of the more efficient consumers in the guild will lead to higher resource levels, and hence greater densities within the guild. Alternatively, a loss of one or more dominant competitors that engage in either interspecific or intraspecific interference (or both) will lead to more efficient use of resources by the community, and consequently greater guild density (Case, Gilpin, and Diamond 1979).

If the terrestrial small mammals constitute a discrete guild, then neither form of compensation would seem to be a factor in the present case, because none of the species were lost or declined substantially as a result of fragmentation. However, defining a guild is problematic (Adams 1985); for example, insectivorous birds may be using the same food resources as many of these largely insectivorous small mammals, and a decrease in their density could contribute to the increased density of small mammals. Nevertheless, it seems reasonable to define terrestrial small mammals as a discrete guild; all are apparently insectivore/omnivores (Charles-Dominique et al. 1981; Guillotin 1982; Robinson and Redford 1986), their foraging microhabitats probably differ greatly from those of other groups of animals, and their nocturnal activity may lead to a predominance of night-active insects in their diets. If they do define a guild in its entirety, the reduction in abundance of a single terrestrial species *(Oryzomys macconnelli)* could hardly be expected to result in density increases in a host of others.

Fonseca (1988) suggested that a form of "reverse" density compensation explained the lower diversity of small mammals in small (ca. 80 ha) fragments of Atlantic coastal forest in Brazil. According to his hypothesis, fewer predators in small fragments led to increased densities of dominant generalist species such as the opossum *Didelphis marsupialis,* which, through competition for food and nest sites, decreased the species richness of other small mammals. His hypothesis seems unlikely to apply to the present study, however, because fragmentation did not convincingly influence the density of *D. marsupialis* or of other numerically dominant species such as *Proechimys* spp. For example, the average density of *D. marsupialis* in phase 1 was greater in fragments than in continuous forest, whereas the opposite was true in phase 2 (Malcolm 1991b). Nor did fragmentation convincingly influence dominance patterns, a prediction of his hypothesis. However, a less diverse or less abundant predator community in secondary habitats and small fragments could contribute to increased small mammal densities, a hypothesis that I am unable to test.

Indirect Effects

The ecological role of small mammals in intact rainforest has been poorly studied. However, the altered vertical distribution of small mammal biomass in fragments could have important implications for forest regeneration, especially close to fragment edges. Terrestrial small mammals move between forest fragments and the matrix (Malcolm 1991b), and seeds transported from secondary forest to primary forest could alter the primary forest seed bank and, ultimately, the nature of succession within primary forest. In addition to their possible role as seed dispersers, rodents are important seed (Osunkoya 1994; Harrington et al., chap. 19) and seedling predators (Lwanga 1994). Putz, Leigh, and Wright (1990) suggested that reduced seed predation led to changes in the flora of small islands in Gatun Lake in Panama. They found that the flora of small islands, in contrast to the mainland, was dominated by just a few tree species that tended to have large seeds. They argued that the absence of seed-eating mammals on the islands led to lower overall seed predation rates and set the stage for competitive dominance by the seedlings that came from large seeds. Small mammals can also be expected to have important effects on populations of animal prey species, including birds (Laurance and Grant 1994) and arthropods. Further study of the biology of rodent and marsupial species will help to identify "second-order" effects that may ripple through fragment communities (see Terborgh et al., chap. 17; Harrington et al., chap. 19).

Differing Responses of Faunas to Fragmentation

The elevated diversity and abundance of terrestrial small mammals in fragments is surprising in light of the responses of other faunal groups to fragmentation. In these same fragments, bird, primate, coprophagous beetle, and leaf-litter beetle communities were less abundant and diverse than in continuous forest (Bierregaard and Lovejoy 1989; Klein 1989; Schwarzkopf and Rylands 1989; Bierregaard and Stouffer, chap. 10; Didham, chap. 5). For these groups, species richness decreased with fragment area, as classically predicted by island biogeography theory. The variation in responses is dramatically illustrated by two groups that apparently rely on the same resource base: insectivorous understory birds that specialize on treefall gaps (which declined in fragments) and terrestrial small mammals (which increased in the same fragments).

The responses of vertebrates to fragmentation appear to correspond to their tolerance of anthropogenic secondary forest. For birds, primates, and coprophagous beetles, young secondary forests were an inhospitable habitat in which species diversity and abundance were low. For these taxa, the fragments themselves may become increasingly inhospitable near forest edges (Didham, chap. 5; Bierregaard and Stouffer, chap. 10). In contrast, I found that small mammal abundance and diversity was high in young secondary forests (Malcolm 1995) and along edges (Malcolm 1991b). Other investigators had found that marsupials are typically more abundant in secondary than in primary Neotropical forests (Charles-Dominique 1983; Fonseca 1988; Stallings 1988), but I detected the same trend for rodents as well (Malcolm 1991b). In fact, only two species were clearly less abundant in fragments and secondary forest than in continuous forest (the rodent *Oryzomys macconnelli* and the marsupial *Caluromys philander*).

Why are so many of the small mammal species of primary forest able to successfully exploit secondary forest? One idea is that, relative to other taxa, small mammals are adapted to early successional seres, or perhaps to ecosystems with relatively high disturbance regimes and hence more secondary habitats (Stallings 1988). This hypothesis implies that, compared with other faunal groups, small mammals have for some reason experienced a significantly different suite of habitats during their evolutionary history, a situation that is difficult to envision.

An alternative idea is that, because of their nocturnal habits, small mammals perceive only slight differences between primary and secondary forest. Insects figure prominently in the diets of most small mammal species in the Manaus area (Charles-Dominique et al. 1981; Guillotin 1982; Robinson and Redford 1986), and we might expect that ecological separation among insectivores is achieved by differential exploitation of foraging microhabitats (Dickman 1988). Although empirical tests of this prediction are lacking, a preliminary analysis of data from spool-and-line devices indicated possible microhabitat partitioning with respect to fallen timber and treefall gaps. For example, short-tailed opossums *(Monodelphis brevicaudata)* usually foraged beneath fallen wood and other litter; spiny rats (*Proechimys* spp.) used fallen logs extensively as conduits; and a rice rat *(Oryzomys macconnelli)* appeared to nest in areas devoid of fallen wood (J. R. Malcolm, pers. obs.). Arboreal species also may partition microhabitats, possibly by selecting different diameters and orientations of support branches (Charles-Dominique et al. 1981). Presumably, nocturnal small mammals are able to select these microhabitat features by using cues perceived over short distances. From a small mammal's perspective, primary and secondary forests may differ only in degree: secondary forests may be more or less indistinguishable from large treefall gaps in primary forests. In contrast, gross visual cues perceived over long distances, presumably of more importance to diurnal animals such as birds and primates, identify large tracts of young secondary forest as environments radically different from even the largest gaps in primary forest. According to this hypothesis, the way in which small mammals perceive primary forest habitats preadapts them for intensive use of secondary forests.

Obviously, when fragments are surrounded by pastures and other habitats devoid of arborescent vegetation, communities in fragments may be dominated by extinction and immigration dynamics, as predicted by classic island biogeography theory (MacArthur and Wilson 1967). In these circumstances, understanding the responses of species to secondary habitats may be relatively unimportant for predicting faunal responses to fragmentation. Edge effects will still create secondary habitats within the fragments themselves, however, and thus the area available to primary forest specialists will usually be smaller than the total forested area (Levenson 1981; Laurance and Yensen 1991; Laurance, chap. 6).

Clearly, more information on resource use and partitioning is needed to understand the responses of primary forest faunas to secondary habitats and to the mosaic of different-aged secondary habitats and forest remnants that constitute fragmented landscapes. The suitability of secondary forest for various groups of organisms will depend not only on the forest's physical structure and resource base, but also on the cues and "search images" that organisms use in habitat selection.

GENERAL IMPLICATIONS

1. Forest fragmentation at a central Amazonian site systematically altered the vertical distribution of small mammal abundance, diversity, and biomass. This change is likely to have important second-order effects, including altered patterns of predation on seeds, seedlings, arthropods, and bird nests.
2. The primary mechanism structuring these changes in the small mammal community appeared to be edge effects, including in situ changes in the forest along the edge and ex situ changes in adjoining secondary habitats. Evidently, fragment communities can be managed to some extent by managing edge effects. For example, edge effects can be reduced by decreasing the perimeter-area ratios of remnants—by creating larger, rounder fragments—and can also be modified by managing the habitats adjoining fragments.
3. Different faunal groups in primary forest often respond differently to fragmentation. Understory insectivorous birds, for example, decline sharply in forest fragments, while terrestrial small mammals increase in abundance. I suggest that the responses of faunal groups to fragmentation can be predicted by examining their responses to anthropogenic secondary forests. Some primary forest taxa may be preadapted to secondary forests because of the types of cues they use to select habitats.

ACKNOWLEDGMENTS

R. Bierregaard, W. Laurance, J. Ray, and an anonymous reviewer made helpful comments on an earlier draft of this chapter. Fieldwork benefited greatly from the assistance of R. Cardoso, A. Cardoso, J. Santos, J. Lopes, M. Santos, L. Reis, and O. Souza. I would also like to thank L. Manzatti, C. Martins, D. Oliveira, and J. Voltolini for their help during six-month field internships. R. Bierregaard, T. Lovejoy, and J. Eisenberg made the research possible. Additional assistance was provided at one time or another by M. Carleton, L. Emmons, C. Gascon, L. Joels, J. Rankin, A. Rylands, and B. Zimmerman. Funding came from World Wildlife Fund-U.S., the Instituto Nacional de Pesquisas da Amazônia, the National Geographic Society, the Tinker Foundation, and Sigma Xi. Additional support was provided by a postgraduate scholarship from the Natural Sciences and Engineering Research Council of Canada and graduate assistantships from the Department of Wildlife and Range Sciences and the Katharine Ordway Chair of Ecosystem Conservation of the University of Florida. This chapter is publication number 148 in the Biological Dynamics of Forest Fragments Project Technical Series.

Appendix 14.1 Mean body weight ± SD *(n)*, maximum distance between recaptures (MDR) ± SD (number of individuals recaptured), and radii of trappability *(r)* of small mammals captured.

		Terrestrial trapping					Canopy trapping				
		Phase 1[a]		Phase 2[b]			Phase 1[c]		Phase 2[b]		
Taxon	Body weight (g)[d]	MDR (m)[e]	*r*	MDR (m)	*r*	*r*[e]	MDR (m)	*r*	MDR (m)	*r*	*r*[e]
Marsupials											
Marmosa cinerea	92 ± 31 (376)	38.2 ± 33.7 (56)	39	22.9 ± 10.7 (14)	27	37	54.8 ± 51.5 (42)	47	49.5 ± 28.4 (32)	63	54
M. parvidens	17 ± 6 (93)	30.0 ± 10.7 (8)	30	20.0 ± 0.0 (1)	24	29	0	—	0	—	—
M. murina	41 ± 15 (341)	28.0 ± 17.9 (5)	23	35.4 ± 22.8 (7)	37	31	0	—	0	—	—
Monodelphis brevicaudata	58 ± 25 (198)	44.0 ± 56.7 (15)	26	20.0 ± 20.0 (3)	30	27	0	—	0	—	—
Didelphis marsupialis	793 ± 527 (167)	63.0 ± 66.8 (61)	126	63.6 ± 32.3 (4)	92	124	190.0 ± 99.0 (2)	76	0	—	124
Philander opossum	355 ± 179 (12)	0	—	0.0 ± 0.0 (2)	<10	124	0	—	0	—	—
Metachirus nudicaudatus	245 ± 100 (116)	32.9 ± 32.9 (56)	23	53.8 ± 0.0 (1)	70	26	0	—	0	—	—
Caluromys philander	169 ± 53 (168)	40.0 ± 40.0 (3)	53	0	—	37[f]	68.7 ± 47.9 (32)	65	58.6 ± 29.2 (21)	72	68
C. lanatus	278 ± 71 (11)	0	—	0	—	37[f]	0	—	53.8 ± 0.0 (1)	70	68[f]
Rodents											
Oryzomys capito	48 ± 14 (292)	27.6 ± 28.5 (98)	27	30.0 ± 24.5 (5)	40	28	0	—	0	—	—
O. macconnelli	68 ± 16 (167)	30.6 ± 29.6 (62)	27	40.0 ± 0.0 (1)	33	27	0	—	0	—	—
Oecomys paricola	37 ± 11 (84)	20.0 ± 28.3 (4)	17	33.5 ± 23.2 (4)	33	25	50.0 ± 42.4 (2)	54	56.9 ± 4.3 (2)	67	61
O. bicolor	23 ± 7 (44)	0.0 ± 0.0 (1)	<10	0	—	25[f]	0	—	0	—	61[f]
Neacomys guianae	16 ± 2 (45)	0	—	0	—	25[f]	0	—	0	—	—
Rhipidomys mastacalis	51 ± 19 (166)	40.0 ± 28.3 (2)	43	0.0 ± 0.0 (1)	<10	25[f]	35.0 ± 29.7 (16)	37	37.1 ± 24.9 (38)	37	37
Proechimys spp.	156 ± 68 (532)	26.7 ± 26.8 (114)	25	47.3 ± 26.7 (6)	56	27	0	—	0	—	—
Mesomys hispidus	111 ± 28 (31)	0	—	0	—	25[f]	0.0 ± 0.0 (1)	<10	60.0 ± 0.0 (1)	43	61[f]
Isothrix pagurus	202 ± 65 (8)	0	—	0	—	25[f]	0	—	20.0 ± 0.0 (1)	27	37[g]

a. Within trapline recaptures during May 1985–July 1987 in continuous forest, 10 ha fragments, and a 100 ha fragment.
b. Within hectare recaptures during October 1987–March 1989 in continuous forest, 10 ha fragments, and 1 ha fragments.
c. Within trapline recaptures during January 1984–September 1986 in continuous forest and 10 ha fragments. Radii used in biomass calculations (see text).
d. All individuals captured, regardless of capture date or location.
e. Radii of trappability used in biomass calculations.
f. Estimated from *M. cinerea, O. paricola,* or *C. philander.*
g. Estimated from *R. mastacalis.*

15

Rapid Decline of Small Mammal Diversity in Monsoon Evergreen Forest Fragments in Thailand

Antony J. Lynam

The tropical forests and forest biotas of Southeast Asia are highly threatened because of unsustainable land use practices. Current rates of deforestation in countries of this region are on average higher than anywhere else in the world (Whitmore, chap. 1), and losses of species associated with these forests are likely to be large compared with those in other tropical regions (Reid 1992). At current rates of forest loss, up to one-tenth of Asian tropical species are likely to be committed to extinction within fifty years (Reid 1992; Simberloff 1992).

To understand which species are most at risk from forest loss and to predict the expected magnitude of species loss in a given reserve, it is necessary to comprehend the local-scale processes that lead to species extirpation following forest disturbance. However, there have been few opportunities to observe ongoing changes in local species composition during and after habitat fragmentation (but see Bierregaard et al. 1992; Margules 1992) because usually only the aftermath of fragmentation is observed. Information about pre-fragmentation species distributions, as well as the rate, sequence, and mechanisms of species disappearance, is unavailable in most studies. Ironically, certain development projects in Southeast Asia, such as logging and hydroelectric reservoirs, have provided opportunities to investigate experimentally the responses of tropical species to habitat loss and disturbance (Marshall and Swaine 1992).

In 1986, 165 km^2 of lowland rainforest was inundated following the damming of the Saeng River (Khlong Saeng) to create a hydroelectric reservoir in Surat Thani Province, southern Thailand (Nakasathien 1989). Almost a hundred permanent islands (formerly hilltops in the landscape) were formed within the reservoir. These islands are perfect isolates with no terrestrial connections to the large forest areas on the adjacent mainland. Because the time of isolation and disturbance history of these islands are known, the archipelago is a natural laboratory in which responses of various organisms to fragmentation can be critically assessed.

The Khlong Saeng Biodiversity Project was established in 1990 to coordinate studies of this fragment system (Woodruff 1990). Research initially focused on the distributional patterns of small mammals on island fragments. Separate studies of genetic variability in rodent populations and of distributions of litter and canopy invertebrates and trees will be published elsewhere. In this chapter, I discuss temporal changes in small mammal distributions during the first few years following fragmentation. I address three questions: (1) How rapidly has species loss occurred, and how does this rate compare with patterns observed in other fragmented systems? (2) Do small mammals vary in their temporal responses to fragmentation? and (3) Are there autecological traits that make some species more prone to local extinction than others? Using new data and drawing upon the results of a previous study (Lynam 1995), I show that fragmentation and associated habitat disturbance, especially on small islands, has led to rapid changes in the composition of these small mammal assemblages.

STUDY AREA

Regional Setting

The tropical forests of Southeast Asia have fluctuated considerably in size and composition during the recent geological past. Tropical forests, including rainforests and seasonally wet forests, have been the dominant vegetation type over all of Southeast Asia for at least 20 million years (Heaney 1991). Sea level fluctuations during the geologically recent past (Hutchison 1989) caused major changes in rainfall and temperature, which probably induced expansions and contractions of the region's rainforests (Heaney 1991). During marine transgressions in the Miocene (19–10 million years B.P.) and until as recently as 18,000 years B.P., the Thai and Malay Peninsula was covered with seasonally dry forest and savanna woodland (Heaney 1991). The present-day rainforests of the region developed following inundation of the Sunda Shelf landmass and increasing rainfall after the last glacial period.

Peninsular Thailand is the land area that lies between 6° and 12° N latitude. It is a region of high national conservation significance because it contains all of the Thai Kingdom's last remaining lowland rainforests, the most biologically diverse and threatened of forest types in Southeast Asia (Brockelman and Baimai 1993). Less than 20% of the Thai Peninsula remains under forest cover (Collins, Sayer, and Whitmore 1991; Arbhabhirama et al. 1988)—an area of only 8,400 km^2—probably the smallest amount since the Miocene (Heaney 1991). These lowland forests have been mainly converted for rubber, fruit, and rice cultivation. Surviving forest occurs as isolated fragments (Brockelman and Baimai 1993), mostly located on upland slopes or limestone peaks unprotected by law (Graham and Round 1994). The median size of these remnants is only 102 km^2, which is smaller than those in other areas of Thailand (Brockelman and Baimai 1993).

The study sites of the Khlong Saeng Biodiversity Project are located at the Chiew Larn Hydroelectric Reservoir (9°00′ N, 98°45′ E), which lies just south of the Isthmus of Kra (10°30′ N), the narrowest part of the Thai Peninsula (fig. 15.1). A considerable turnover of vertebrate species occurs over a transition zone near the isthmus, where faunas of Sundaic origin meet Indo-Burmese faunas (Chasen 1940; Medway and Wells 1976). This disjunction, in combination with a major zone of transition between Malayan rainforest and monsoon evergreen forest 400 km to the south, makes this an area of considerable biogeographic importance.

Rainfall averages about 1,700 mm per annum at the study area. The local climate is

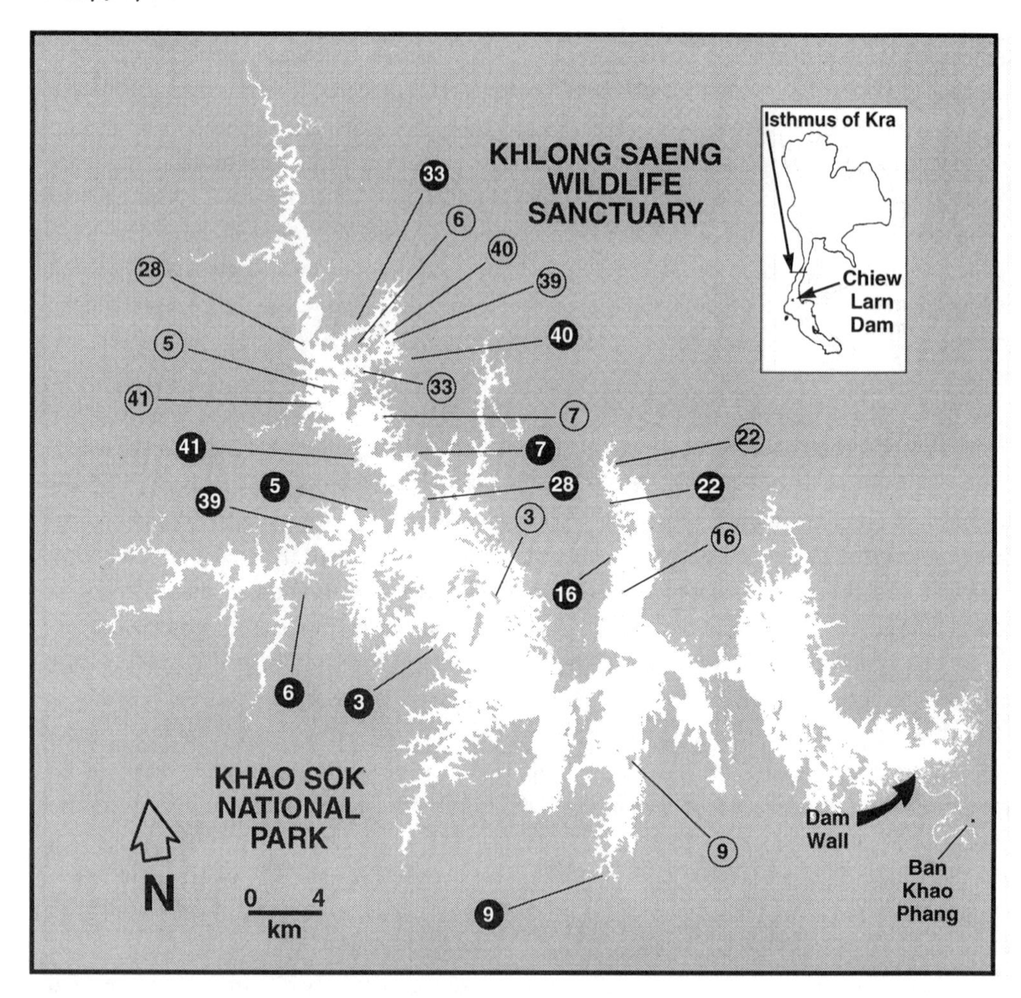

Figure 15.1. Locations of study sites at the Chiew Larn reservoir, Surat Thani Province, Thailand (9°00′ N, 98°45′ E). Open circles indicate island sites in the reservoir; solid circles indicate continuous forest sites on mainland.

dominated by monsoon winds that bring rainfall from April to November, with a short, hot dry season during the remaining four months of the year. Precipitation averages 60 mm/month during the dry season, and increases to 180 mm/month in the wet season (Electricity Generating Authority of Thailand 1980).

Local Landscape History

The Chiew Larn Reservoir lies within the boundaries of two protected areas: Khlong Saeng Wildlife Sanctuary and Khao Sok National Park. Together with four adjoining protected areas (Khlong Yan and Khlong Nakha Wildlife Sanctuaries, and Kaeng Krung and Sri Phangnga National Parks), they constitute the largest contiguous protected area in the region (3,430 km^2).

Prior to 1982, 30% of the natural vegetation in the catchment area had been converted for agriculture. From 1982 to 1986, land up to 100 m above sea level (ASL) was clear-cut or selectively logged (Electricity Generating Authority of Thailand 1980). In 1986 the dam was sealed, and water had filled the reservoir to 90 m ASL by early 1987. Areas above 90 m ASL were left as islands inside the reservoir; these islands range from 0.7–109 ha in area and are isolated from the mainland by distances of 25–1,000 m. Since inundation in 1987, 25–30 km^2 of forest bordering the reservoir has been permanently damaged or destroyed by fire and illegal timber cutting.

A total of 165 km^2 of land was flooded by the reservoir, an area larger than nearly three-quarters of the individual remnant forest tracts in Peninsular Thailand (Brockelman and Baimai 1993). Remnant tracts in Peninsular Thailand are usually not continuous forest, but are composed of many isolated forest patches, which are often comparable in size to the islands examined in this study. Hence, studies at this spatial scale are highly relevant to forest conservation in the region.

METHODS

Hypotheses

Equilibrium and nonequilibrium models of island biogeography (MacArthur and Wilson 1967; Crowell 1986) provide testable predictions about fragmentation. First, both models predict that fragmentation should lead to an overall decline in species richness due to increased extinction rates in habitats of reduced area. Second, they predict that species numbers should stabilize after fragmentation, at a point determined by the rates of extinction and colonization. A third prediction, unique to the equilibrium model and less easily tested, is that species composition should change over time, or exhibit "turnover," at this point of balance. Not specified by either model are two things: the expected time frame during which such changes should take place, and the order and identity of species extinctions.

Because species vary in their responses to fragmentation (Robinson et al. 1992; Margules, Milkovits, and Smith 1994), it is useful to identify the characteristics that predispose some organisms to extinction in small fragments (Pimm, Jones, and Diamond 1988; Laurance 1991a). Hypotheses concerning extinction proneness tend to fall into two categories: those directly related to a species' natural abundance, and those concerning ecological traits other than natural abundance. For example, one popular hypothesis states that species that naturally occur at low densities should have small population sizes in fragments, and hence be more susceptible to local extinction due to stochastic environmental or demographic fluctuations (MacArthur and Wilson 1967; Terborgh and Winter 1980).

Extinction proneness may, however, have less to do with rarity than with a species' tolerance of habitat and microclimatic changes occurring within fragments or along fragment margins (Laurance 1991a; Saunders, Hobbs, and Margules 1991). A corollary of this idea is the prediction that small fragments should lose many species because they have the highest perimeter-area ratios and hence are mostly habitat edge (Wilcove,

Table 15.1 Descriptions of small mammals studied at the Chiew Larn Reservoir, Thailand.

Species	Group	Weight (g)[a]	Habitat[d]	Diet[d,e]
Echinosorex gymnurus	Insectivora: Erinaceidae	950[b]	Forest edges close to water	Insectivorous
Sundamys muelleri	Rodentia: Muridae	397[b]	Forest edges close to water	Omnivorous
Leopoldamys sabanus	Rodentia: Muridae	343[b]	Primary forest specialist	Omnivorous
Menetes berdmorei	Rodentia: Sciuridae	180[c]	Primary/secondary forest	Omnivorous
Maxomys surifer	Rodentia: Muridae	155	Primary forest specialist	Omnivorous
Tupaia glis	Scandentia: Tupaiidae	141[c]	Primary/secondary forest	Omnivorous
Rattus rattus	Rodentia: Muridae	140[b]	Forest edges close to water	Omnivorous
Niviventer bukit	Rodentia: Muridae	75[b]	Primary/secondary forest	Omnivorous
Niviventer cremoriventer	Rodentia: Muridae	67[c]	Primary/secondary forest	Omnivorous
Maxomys whiteheadi	Rodentia: Muridae	44[b]	Primary/secondary forest	Omnivorous
Chiropodomys gliroides	Rodentia: Muridae	26[b]	Primary/secondary forest	Omnivorous
Hylomys suillus	Insectivora: Erinaceidae	15[b]	Primary/secondary forest	Insectivorous

a. Mean adult body weight for species. b. Nowak 1991. c. Lynam, unpublished data.
d. Medway 1978; Lynam 1995. e. Langham 1983.

McClennan, and Dobson 1986), while large fragments should actually gain species because their core areas are large enough to support resident species while new, disturbance-tolerant taxa colonize altered habitats along the perimeter (Margules 1992).

Study Species

Most medium-sized and large mammals have disappeared from or maintain only tenuous populations on the Chiew Larn islands due to their large area requirements (Wanghongsa 1989). Small mammals were chosen for study because they are easily identifiable, relatively abundant (Medway 1978), and readily censused using mark-recapture techniques. Small mammals are also important predators of seeds of tropical trees (Langham 1983; Harrington et al., chap. 19), key dispersers of mycorrhizal fungi (Janos, Sahley, and Emmons 1995), and form the basis of mammalian food chains (Rabinowitz and Walker 1991), all important functions in forest ecosystems.

Seventeen species make up the small mammal community at Chiew Larn. Of these, three arboreal squirrels *(Callosciurus caniceps, C. notatus,* and *Sundasciurus tenuis)* could not be reliably censused at ground level, and two terrestrial shrews *(Crocidura fuliginosa* and *Suncus etruscus),* which were detected with pitfall traps, were extremely rare and are not considered here. The remaining twelve species include eight murid rodents, one sciurid rodent, two insectivores, and one tree shrew (table 15.1). All but one species, the house rat *(Rattus rattus),* are indigenous to the forests of Peninsular Thailand. House rats were introduced to the Peninsula from northern Asia or Europe (Musser and Newcombe 1983) and have invaded forest areas inside the Chiew Larn Reservoir from nearby human settlements.

Trapping Methods

Small mammals were censused from five to seven years after fragmentation at twenty-four forest sites: twelve islands ranging in size from 0.7 to 109 ha, and twelve continuous forest areas of equivalent size on the adjacent mainland. For statistical purposes, it was assumed that the mainland areas were effective controls for the islands and contained small mammal populations similar in abundance and composition to the prefragmentation populations on the islands. Four surveys were conducted and are named to indicate the time since the islands were created: Year 5 Dry Season (January–March, 1992), Year 6 Dry Season (December–March, 1993), Year 6 Wet Season (September–December, 1993), and Year 7 Dry Season (January–March, 1994).

Mammals were surveyed along 135 m traplines, each with ten ground-set cage traps and four Sherman box traps positioned 1–2 m above the ground on lianas, small trees, or bamboo culms. At each site, traplines were placed so as to sample available microtopographies (ridges, slopes, and streambeds) and broad habitat types (primary and secondary forest) for small mammals. From one to twelve traplines were used per site, more at larger sites, making a total of sixty-four traplines. Voucher specimens are lodged at the West Australian Museum in Perth, Australia.

Data Analyses

Standardized measures of diversity and abundance (Magurran 1988) were used to facilitate comparisons of sites between years and seasons. Two measures of diversity were used: species richness and the Shannon index of diversity (H), where $H = -\Sigma p_i \ln(p_i)$, and p_i is the proportion of individuals contributed by the ith species. Species richness at each site was determined when the cumulative species number approached an asymptote with trapping effort (Lynam 1995). Changes in species numbers on islands of different sizes suggested the degree to which relaxation (*sensu* Brown 1971) had occurred across the archipelago. Extinction events were defined when neither a single juvenile, pregnant female, nor pair of adult individuals of a given species was detected on an island after having been present in a previous survey. Colonization events were defined by the reverse situation. Islands were grouped into three categories for analysis: small (1–10 ha) islands (nine replicates), medium-sized (10–50 ha) islands (two replicates), and large (> 50 ha) islands (one replicate).

Relative abundance estimates for each site (mean number of individuals captured per trapline per night of sampling) were square root-transformed to help normalize the data. Analysis of covariance (ANCOVA) was performed on the species abundance data. Factors examined were Site (two levels: island or mainland) and a repeated measure, Year (four levels: Year 5, Year 6 Dry, Year 6 Wet, Year 7). To control for differences between replicates due to size of the sampling area, Area ($\log_{10}$ of island or sampling area) was used as a covariate in the analyses.

Several analyses were used to assess the responses of species to fragmentation. First, temporal changes in abundance for each species were determined using ANCOVA. Second, to assess changes in the composition of small mammal assemblages, the proportional abundance of each species was compared between trapping surveys. Third, the relative susceptibility to extinction of individual species was quantified using an index, (S), calculated as $(M - I)/M$, where M is the number of mainland sites where a species was recorded and I is the number of islands where the species occurred. This index was used to assess the relationship between extinction proneness and species abundance characteristics. Finally, nonparametric tests were used to assess temporal changes in species richness on islands.

RESULTS

Community-Level Changes Following Fragmentation

A total of 4,986 captures of twelve small mammal species were recorded from 31,892 trap-nights of sampling on islands and mainland sites. The significance of changes in species richness on larger (> 10 ha) islands could not be determined due to the low number of replicates, although species numbers increased on the largest (109 ha) island in accordance with Margules' (1992) prediction. Species numbers on medium-sized islands had not stabilized by seven years after isolation (fig. 15.2). On small (< 10 ha) islands, species numbers declined, though not significantly ($P > .05$, $H = 5.01$, df $= 3$, Kruskal-Wallis test). On five of the nine small islands (where at least one native species was present in Year 5), only the exotic *Rattus rattus* was detected in Year 7.

When islands and paired mainland sites were compared using Year 5 data, species richness and diversity (table 15.2, appendix 15.1) did not differ significantly ($P > .05$, Wilcoxon paired-sample test). By the Year 6 Dry season survey, diversity was lower on islands ($P < .05$), but species richness did not differ significantly ($P > .05$). By the Year 6 Wet and Year 7 surveys, both richness and diversity had declined on islands relative to mainland sites.

To assess temporal changes associated with fragmentation, nonparametric tests were performed on the differences in richness and diversity between the twelve pairs of mainland and island sites. The disparity in total species richness and richness of native species

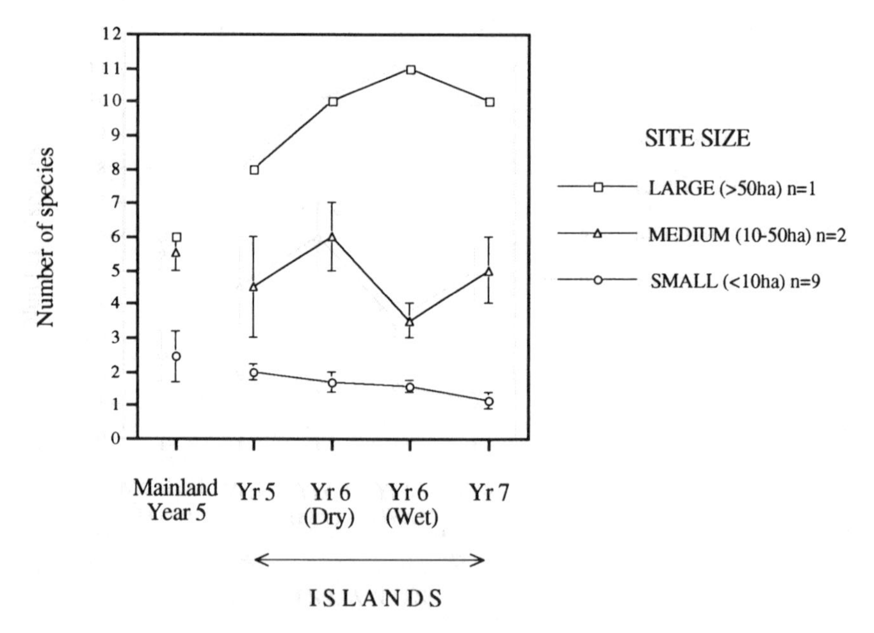

Figure 15.2. Temporal changes in species numbers of small mammals on Chiew Larn islands. Values shown are mean species numbers ± one standard error.

Table 15.2 Significance of temporal changes in richness and diversity of small mammals.

(A) Paired comparisons of island and mainland sites using Wilcoxon paired tests

Index	Sampling period (years post-isolation) 5	6 (Dry)	6 (Wet)	7
Species richness	NS	NS	< .05	< .05
Shannon diversity	NS	< .05	< .01	< .01

(B) Temporal trends in the difference in diversity between paired island and mainland sites

Type of difference[a]	Kruskal-Wallis *(H)*	*P*
N_{tot}(ML) − N_{tot}(IS)	7.485	.06
N_{nat}(ML) − N_{nat}(IS)	6.428	.09
H(ML) − H(IS)	9.156	< .05

a. ML = mainland; IS = island; N_{tot} = total number of species; N_{nat} = number of native species; H = Shannon diversity index.

(i.e., excluding *R. rattus*) did not change significantly between Year 5 and Year 7 ($P >$ .05, Kruskal-Wallis tests; table 15.2). However, island diversity declined during this period relative to mainland controls ($P < .05$, $N = 46$, df = 3), suggesting that mammalian diversity on islands eroded over time. The result was similar when the Year 6 Wet survey data were excluded and the analysis re-performed using dry season data only.

Changes in Species Composition

One species was apparently absent on all islands by Year 5: the insectivore *Hylomys suillus* was never caught on islands, but occurred at low abundance on the mainland (five individuals were captured in 16,000 trap-nights). The murid rodent *Leopoldamys sabanus* was abundant in continuous forest, but was represented by only a single female on the largest island (109 ha). A third species, the insectivore *Echinosorex gymnurus,* was caught only once on the mainland, but was frequently captured on the largest islands in Year 6 and Year 7.

In mainland forest, five common species, *Maxomys surifer, Leopoldamys sabanus, Maxomys whiteheadi, Rattus rattus,* and *Tupaia glis,* each accounted for more than 5% of all individuals captured; *M. surifer* was the most common species. All other species were defined as rare (< 5% of all captures). In Year 5, *Rattus rattus* was the most abundant species on islands, while *Chiropodomys gliroides, T. glis, M. surifer,* and *M. whiteheadi* were also common. *T. glis* became relatively rare on islands following the Year 6 dry season, while *Sundamys muelleri* and *Niviventer bukit* had become relatively common by Year 7.

Temporal Changes in Species Abundances

The total abundance of small mammals differed between surveys ($P < .05$, ANCOVA; table 15.3), but did not differ between mainland and island sites ($P > .05$). However, species varied in their individual patterns of abundance. Densities of *M. surifer* and *M. whiteheadi* were higher at mainland sites than on islands ($P < .05$ and $P = .05$, respectively). *Chiropodomys gliroides, T. glis,* and *M. whiteheadi* all tended to decline on islands between Year 5 and Year 7, though not significantly so. *Niviventer cremoriventer* tended

Table 15.3 Effect of site and temporal differences on species abundances by analysis of covariance.

	C. gliroides	*M. berdmorei*	*M. surifer*	*M. whiteheadi*	*N. bukit*	*N. cremoriventer*	*R. rattus*	*S. muelleri*	*T. glis*	Native species only	All species
Effects between factors											
Site	0.931	1.857	62.004***	4.292	1.542	4.101	27.491***	7.133***	1.608	38.837	1.575
Area	1.367	5.665**	0.774	3.615*	0.029	2.368	0.767	0.378	22.997***	6.441**	0.228
Effects within factors											
Year	0.381	1.669	0.038	0.216	0.768	3.764**	6.286**	2.023	0.442	1.212	3.544**
Year * Site	0.885	0.656	0.143	0.977	2.547*	1.487	0.440	0.723	1.281	3.081**	0.432
Year * Area	0.245	0.838	0.165	0.516	0.208	2.373*	0.831	1.055	0.596	0.313	0.479

*$P < 0.10$, **$P < 0.05$, ***$P < 0.001$.

to be more abundant on the mainland than on islands, but declined both there and on the islands over time ($P < .05$). Finally, *N. bukit* tended to increase in mainland forest ($P = .07$), but not on islands. As a group, native species were less abundant on islands than on the mainland, and declined on islands over time ($P < .05$; fig. 15.3). In contrast, *R. rattus* was more abundant on islands than on the mainland, and increased on both islands and the mainland over time (table 15.3; fig. 15.3). These results were qualitatively similar when the analyses were run using data from the three dry season surveys only.

Natural Abundance and Extinction Proneness

The rarity of small mammals in continuous forest was estimated using rank-order abundances of species at mainland sites in Year 7. *Rattus rattus* was excluded from the analysis because it probably was absent from mainland forest areas prior to dam construction. Although relative extinction proneness *(S)* was significantly correlated with a species' natural rarity ($P < .05$; fig. 15.4), the regression slope was unexpectedly negative.

The rarity of a species, however, may be related to the number of mainland sites *(M)* where it was detected; for example, a species occurring at only one or two sites, even if locally common, might have a low overall abundance. Because *M,* the number of mainland sites where a species was detected, is intrinsic to the equation for extinction proneness (i.e., $S = (M - I)/M$), the regression may have been confounded. To assess this possibility, residuals of the regression were themselves regressed against *M.* This regression was not significant ($P > .05$), and thus the relationship holds: species that were more common in continuous forest were prone to extinction, while rarer species were less likely to be absent from islands.

DISCUSSION

Temporal Changes in Small Mammal Diversity

Unless the pre-fragmentation species composition of fragments is known (e.g., Bierregaard and Stouffer, chap. 10), or can be approximated based on knowledge of spatial distributions of species in continuous habitat, the loss of biodiversity from fragments cannot be adequately estimated. At Chiew Larn, it was possible to infer the pre-fragmentation species richness of small mammals on islands by rigorously examining the numbers and distributions of species in mainland control areas, and thereby to estimate the overall rate of species loss.

Two small mammals that are mostly confined to forest interiors, *Hylomys suillus* and *Leopoldamys sabanus* (Medway 1978; Kemper and Bell 1985), were rarely if ever captured on islands (*H. suillus* was never caught on islands, and only one *L. sabanus* individual was recorded). Both species had random spatial distributions in unfragmented habitat (Lynam 1995) and thus were likely to occur on most or all of the islands at the time of fragmentation. Consequently, their absences on most or all islands suggest regional extinctions in the archipelago, rather than sampling effects. Hence, within a five-year period corresponding to about twenty small mammal reproductive generations (Medway 1978), two of the twelve species in the assemblage (17%) were effectively lost from the islands. Further reductions in species richness and diversity were detected on islands after

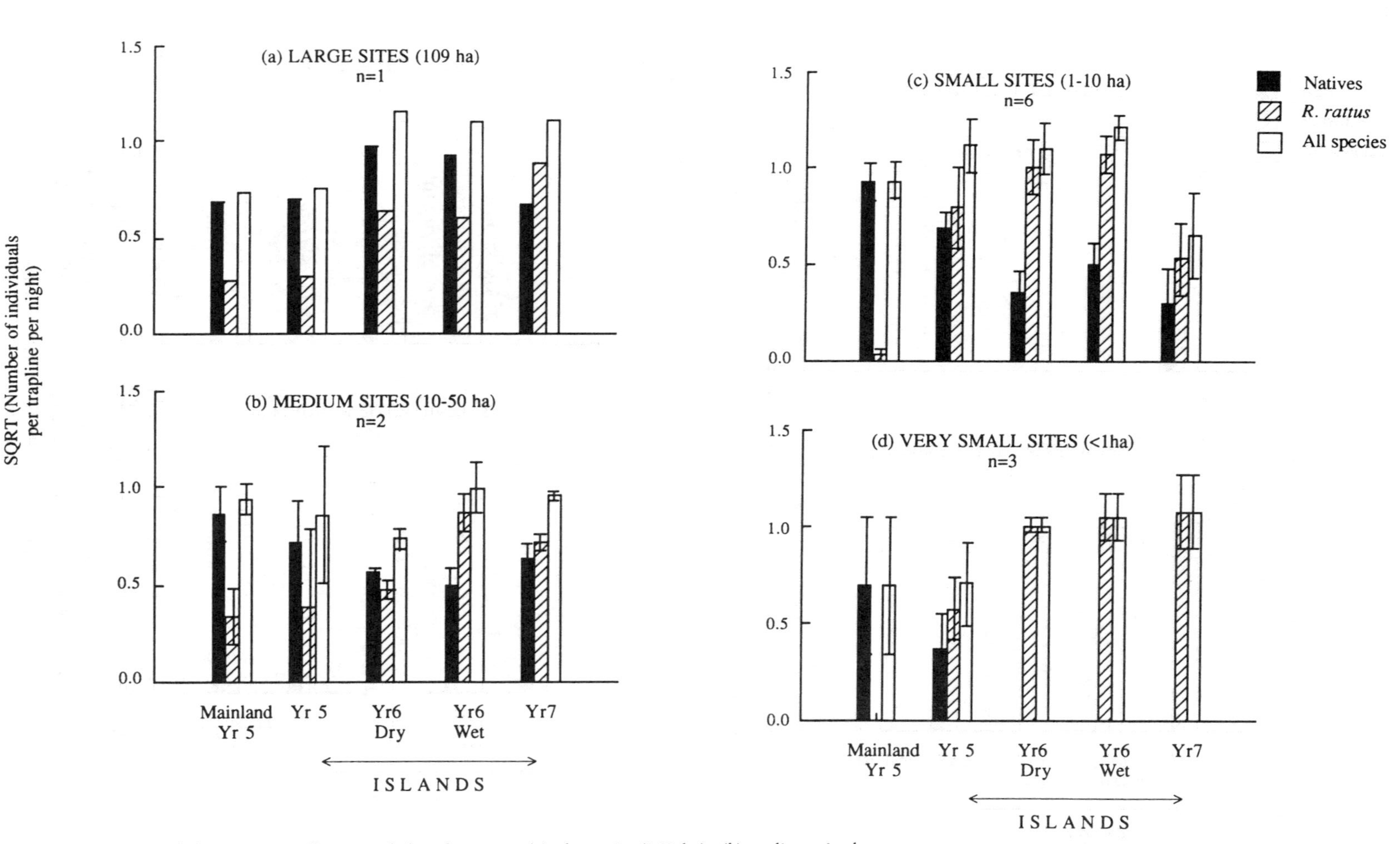

Figure 15.3. Temporal changes in small mammal abundances on *(a)* a large site (109 ha); *(b)* medium-sized sites (10–50 ha); *(c)* small sites (1–10 ha); and *(d)* very small sites (< 1 ha).

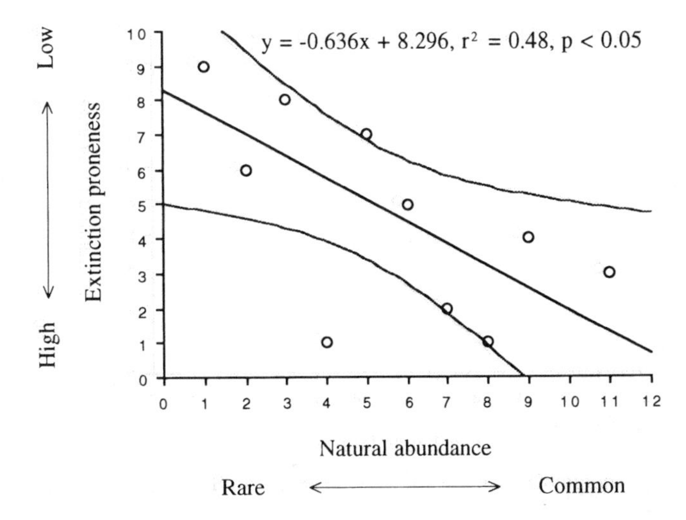

Figure 15.4. Relationship between extinction proneness and natural abundance for Chiew Larn small mammals. The curved lines indicate 95% confidence limits for the regression.

Year 5 (see table 15.2), and species numbers had not stabilized even on medium-sized and large islands by Year 7 (see fig. 15.2), indicating that further extinctions are likely.

Comparative data on extinction rates of small mammals in other fragmented forest landscapes are scarce. Most other studies have examined species distributions in older (> 50 years) fragments. In central Panama, for example, only one of sixteen putative indigenous rodent species remained on small (0.3–17.1 ha) islands 80 years after the inundation of Gatun Lake (Adler and Seamon 1991). In southern California, all of seven native rodent species had disappeared from thirteen of twenty-five urban chaparral fragments of less than 100 ha by 18–79 years after fragmentation (Bolger et al., in press). Although the initial rate of loss of mammals in these landscapes is unknown, the average rate was one species per 3–11 years.

Studies of young Amazonian rainforest remnants (Bierregaard et al. 1992) provide a temporally equivalent comparison to the present study. Fragmentation actually led to increases in species richness of small mammals in 1 ha and 10 ha forest remnants in the first two years after isolation (Malcolm, chap. 14). Average species richness in those remnants was almost twice as high as that in continuous forest, due to invasions of several rodent species from outside the remnants (Malcolm 1988). Similarly, no species were lost from fragments of temperate forest in central Pennsylvania during the first five years after isolation, and the abundances of some species actually increased (Yahner 1991).

Initial losses of mammals on the Chiew Larn islands have been relatively rapid compared with those in recently fragmented landscapes. True islands probably lose species more rapidly than terrestrial fragments because mammalian dispersal is much less frequent across water than via the secondary habitat matrix surrounding fragments (Leung, Dickman, and Moore 1993). Consequently, immigrants are less likely to "rescue" small island populations on the brink of extinction (Brown and Kodric-Brown 1977) or to recolonize islands from which populations have disappeared.

Figure 15.5. The contrast between understories of *(A)* undisturbed forest and *(B)* recently isolated Chiew Larn islands.

Figure 15.5. (*continued*)

Forest Disturbance and Extinction Proneness

In this study, forest islands were subjected to major habitat changes due to selective logging and frequent anthropogenic burning. These disturbances have entirely converted the evergreen forest on most small (< 1 ha) islands to open forest with a dense bamboo understory (fig. 15.5), and have reduced the core areas (*sensu* Temple 1986) of evergreen forest on large islands (fig. 15.6). Thus, the islands are not simply isolates of formerly continuous forest; they contain habitats that differ in structure and composition from mainland forests and hence may not function as true land bridge islands.

These ecological changes clearly influence small mammal assemblages. For example, thirteen local extinction events were detected on islands during this study (Lynam 1995): two on large islands, six on medium-sized islands, and five on small islands. Although these extinction events were balanced by an equivalent number of colonizations, this was not turnover in the equilibrium sense because disturbance-adapted mammals increased on islands at the expense of forest interior species. For three large-bodied (> 150 g) species that utilize a variety of forest habitats, including streams and forest edges, and are good overwater dispersers (*Echinosorex gymnurus, Rattus rattus, Sundamys muelleri:* Harrison 1957, 1958), the number of colonizations on islands outweighed local extinctions (nine versus four, respectively: Lynam 1995). Eight of the remaining species are less likely to cross forest edges and water gaps because they tend to avoid disturbed forest (e.g., *Leopoldamys sabanus* and *Maxomys* spp.: Medway 1978; Kemper and Bell 1985; Lynam 1995). For these species, colonizations of islands occurred less frequently than extinctions (three versus nine events).

Disturbance-sensitive mammals were common in mainland forest, but exhibited lowered abundance and were prone to extinction on islands. Although unlikely to cross water gaps, the arboreal mouse *Chiropodomys gliroides* was tolerant of habitat changes associated with fragmentation because it readily uses bamboo (Rudd 1979). Despite being naturally rare, this and other disturbance-adapted species were favored by fragmentation and associated disturbances, and consequently increased on islands. Therefore, for mammals at Chiew Larn, as in some other fragmented landscapes (Lovejoy et al. 1984; Laurance 1990, 1991a), natural abundance was not a meaningful predictor of extinction proneness. Instead, the fate of mammals in fragments is apparently linked to their relative tolerance of modified habitats (Malcolm, chap. 14).

Changes in mammal abundances on islands could be mediated by changes in food resources because fire and logging may reduce the availability of seeds and fruit on the forest floor (Kartawinata, Riswan, and Soedjito 1980) and alter invertebrate abundance (Burghouts et al. 1992; Holloway, Kirk-Spriggs, and Chey 1992). Preliminary evidence suggests that the diets of island *Maxomys surifer* differ from those of mainland populations (Lynam 1995) and that the composition of litter invertebrates differs between forest edges and interiors on islands (L. Lebel, pers. comm.). Thus, habitat disturbance can alter the abundance and distribution of resources on islands, which may favor mammals that can exploit secondary forest (Malcolm, chap. 14) but contribute to the loss of disturbance-sensitive forms.

Figure 15.6. Two small (ca. 1 ha) Chiew Larn islands. *(A)* An island that has not been logged or burnt. *(B)* An equivalent island after logging and burning. Note the dense bamboo edge and the scattered, tall trees that constitute the core forest.

Hyperabundance or Hypoabundance in Fragments?

Forest fragmentation can lead to increased population densities of some species in fragments (e.g., Gliwicz 1984; Laurance 1994; Terborgh et al., chap. 17). This phenomenon may be a temporary effect resulting from movements of individuals into fragments as the surrounding matrix is cleared or otherwise disturbed (Bierregaard and Lovejoy 1989). Alternatively, it may be a permanent outcome associated with the creation of new microhabitats and resulting changes in food resources (Yahner 1991; Malcolm 1995), reduced densities or losses of predators or competitors (Terborgh and Winter 1980), or the frustrated dispersal of individuals between habitat isolates (Adler and Levins 1994). If habitat changes associated with fragmentation are unfavorable, however, permanent reductions in populations may occur in fragments (Adler and Levins 1994).

Native mammals were less abundant on islands than in mainland areas at Chiew Larn (see table 15.3), although the overall abundance of small mammals on islands was not significantly depressed because of increases in *Rattus rattus* on islands. During the course of this study, native species declined on medium-sized and small islands and went extinct on very small (< 1 ha) islands (see fig. 15.3). These changes appear to be permanent and not a result of temporary crowding effects. In contrast, *R. rattus* was hyperabundant on islands relative to the mainland (Site effect $P < .001$; see table 15.3), and particularly so on smaller (< 10 ha) islands where native species had declined or disappeared (see fig. 15.3). Habitat changes, and possibly the associated loss of native species, have provided favorable conditions on small islands for *R. rattus,* although this exotic species has had less success invading mainland forest (fig. 15.3; Lynam 1995).

An important point is that the pattern of species loss depicted in figure 15.3 may eventually lead to a situation similar to that which has occurred in some older fragment systems. For example, in Gatun Lake, Panama, only a single native rodent *(Proechimys semispinosus)* survived on small islands following fragmentation (Adler and Seamon 1991). Likewise, a single species, *R. rattus,* albeit an exotic, is rapidly taking over small islands at Chiew Larn. If current trends persist, native mammals may eventually become confined to the few large islands (> 10 ha) in the archipelago.

Implications for Reserve Design

Most forest reserves elsewhere in Peninsular Thailand are isolated and internally fragmented (see Goosem, chap. 16), and many contain secondary forest in a matrix of heavily degraded forest or scrub. Primary forest fragments of 100 ha or less within these designated reserves may be too small and disturbed to maintain intact assemblages of small mammals, unless they are connected to larger, relatively undisturbed tracts of forest. The degree to which these reserves can support viable populations of small mammals is likely to be strongly influenced by the extent to which dispersal between fragments is possible across the disturbed habitat matrix (Laurance 1990; Bierregaard et al. 1992). Future research in this region should seek to identify the relative dispersal abilities and matrix tolerances of mammals in fragmented landscapes. Current management efforts in reserves should focus on lowering the frequency of dry season fires and localizing their effects where possible in order to maximize the spatial heterogeneity of different forest types.

GENERAL IMPLICATIONS

1. Disturbance-tolerant species often become atypically abundant in fragmented habitats. In this study, species that favored forest edges and other disturbed habitats were rare in primary forest, but increased markedly on islands.
2. Rarity may be a misleading indicator of local extinction proneness on islands or in fragments. At Chiew Larn, several naturally rare species increased on islands following fragmentation, while some naturally abundant species declined or disappeared on islands.
3. Faunal collapse may be very rapid in some fragmented habitats, especially those, such as islands, that are surrounded by a matrix that is hostile to forest species. Island mammal assemblages at Chiew Larn are rapidly becoming dominated by the disturbance-adapted exotic rodent *Rattus rattus.* Large islands support one or two other disturbance-tolerant native species and are the last refuge for forest interior dwellers. Similar patterns have been documented in fragment archipelagoes where faunal collapse presumably occurred long ago.
4. Not all islands surrounded by water behave as islands in the equilibrium sense. On the Chiew Larn islands, rapid turnover has resulted in the replacement of disturbance-sensitive species with disturbance-adapted forms, and this transition is likely to occur only once.
5. Consideration of habitat changes associated with fragmentation is essential for predicting changes in community composition.

ACKNOWLEDGMENTS

I am indebted to the sixteen local Thais who assisted me in conducting the surveys. I thank the Royal Forestry Department and the National Research Council of Thailand for permitting me to conduct research in Thailand, and the Chiefs of Khao Sok National Park and Khlong Saeng Wildlife Sanctuary and the Electricity Generating Authority of Thailand for allowing the field team to live at the site. The project was made possible by an NSF Conservation Biology grant (NSF-BSR 9000486) to David Woodruff. Bill Laurance, Ted Case, Richard Corlett, James Patton, Greg Witteman, David Woodruff, and four anonymous reviewers made useful comments on earlier versions of this chapter. This chapter is contribution no. 1 to the Khlong Saeng Biodiversity Project.

Appendix 15.1 Numbers of species and Shannon diversity indexes (in parentheses) of small mammals at sites at Chiew Larn, Thailand, from five to seven years after fragmentation.

	Area	Year 5		Year 6 (Dry)		Year 6 (Wet)		Year 7[a]	
Site	(ha)	Island	Mainland	Island	Mainland	Island	Mainland	Island	Mainland
6	109.4	8 (1.81)	6 (1.43)	10 (1.81)	9 (1.54)	11 (1.70)	7 (1.60)	10 (1.42)	8 (1.64)
5	22.6	3 (1.00)	6 (1.52)	7 (1.31)	7 (1.75)	4 (0.73)	7 (1.49)	6 (1.41)	6 (1.41)
9	10.6	6 (1.45)	5 (0.86)	5 (1.66)	3 (1.11)	3 (0.78)	7 (1.33)	4 (1.24)	7 (1.33)
3	6.4	2 (0.50)	8 (1.56)	2 (0.69)	5 (1.46)	2 (0.69)	4 (1.35)	2 (0.69)	7 (1.61)
41	3.4	3 (0.74)	2 (0.69)	3 (0.54)	6 (1.61)	2 (0.45)	5 (1.31)	—	—
7	2.9	3 (0.71)	3 (1.00)	3 (0.29)	2 (0.45)	1 (0.00)	4 (1.48)	—	—
39	2.1	1 (0.00)	1 (0.00)	2 (0.50)	3 (1.27)	2 (0.90)	3 (0.80)	1 (0.00)	3 (0.45)
28	1.9	2 (0.33)	2 (0.50)	1 (0.00)	1 (0.56)	2 (0.52)	5 (1.49)	0 (0.00)	3 (0.97)
22	1.1	2 (0.69)	2 (0.60)	1 (0.00)	2 (0.60)	2 (0.26)	4 (0.54)	2 (0.60)	4 (1.24)
16	0.9	2 (0.63)	2 (0.41)	1 (0.00)	3 (0.74)	1 (0.00)	3 (0.88)	1 (0.00)	1 (0.00)
40	0.8	2 (0.69)	2 (0.69)	1 (0.00)	5 (1.47)	1 (0.00)	4 (1.22)	1 (0.00)	2 (0.69)
33	0.7	1 (0.00)	0 (0.00)	1 (0.00)	2 (0.68)	1 (0.00)	2 (0.53)	1 (0.00)	3 (0.97)

a. Surveys of island and mainland sites 7 and 41 were not completed in Year 7 due to logistical difficulties.

16

Internal Fragmentation: The Effects of Roads, Highways, and Powerline Clearings on Movements and Mortality of Rainforest Vertebrates

Miriam Goosem

"INTERNAL fragmentation" occurs when natural habitat is fragmented and wildlife populations subdivided by linear clearings such as roads and powerlines. Roads, highways, and powerline clearings are an integral feature of the modern landscape and, outside urban areas, are one of the most obvious anthropogenic impacts on the natural environment. The provision of such infrastructure to service growing human populations can result in a network of linear clearings subdividing conservation areas into much smaller habitat blocks (fig. 16.1).

The construction and maintenance of roads and powerline clearings has a variety of effects on native fauna, including: (1) destruction or alteration of habitats, with consequent reductions in population size; (2) disturbances, edge effects, and intrusions of fauna alien to the natural habitat; (3) increased mortality due to vehicle traffic; and (4) fragmentation of habitats and wildlife populations (Andrews 1990; Bennett 1991).

Given the paucity of published information on the effects of linear clearings in tropical rainforests, I briefly review a range of linear barrier studies in other vegetation communities. I then critically examine the effects of roads and powerline clearings in tropical rainforest in relation to fragmentation of wildlife populations. The intuitive expectation that wildlife of the structurally complex tropical rainforest understory should be strongly influenced by linear clearings is confirmed for many of the species examined.

HABITAT LOSS AND ALTERATION

The most direct impact of road and powerline construction is the destruction or alteration of wildlife habitat. The United States Council on Environmental Quality (1974) estimated that each kilometer of interstate highway can require the alienation of up to 13.5 ha of wildlife habitat. Powerline clearings can also result in major losses of habitat; for example, from 1960 to 1970, over 1.6 million ha were acquired in the United States for powerline clearings (Johnson, Schreiber, and Burgess 1979). In northern Queensland, Australia, considerable habitat has been alienated by linear clearings within the 9,000 km^2 Wet Tropics of Queensland World Heritage Area—a total of 1,316 ha in 324 km of powerline clearings and 608 ha in 1,427 km of roads and highways (fig. 16.1). This represents 0.2% of the total habitat of the World Heritage Area (WTMA 1995).

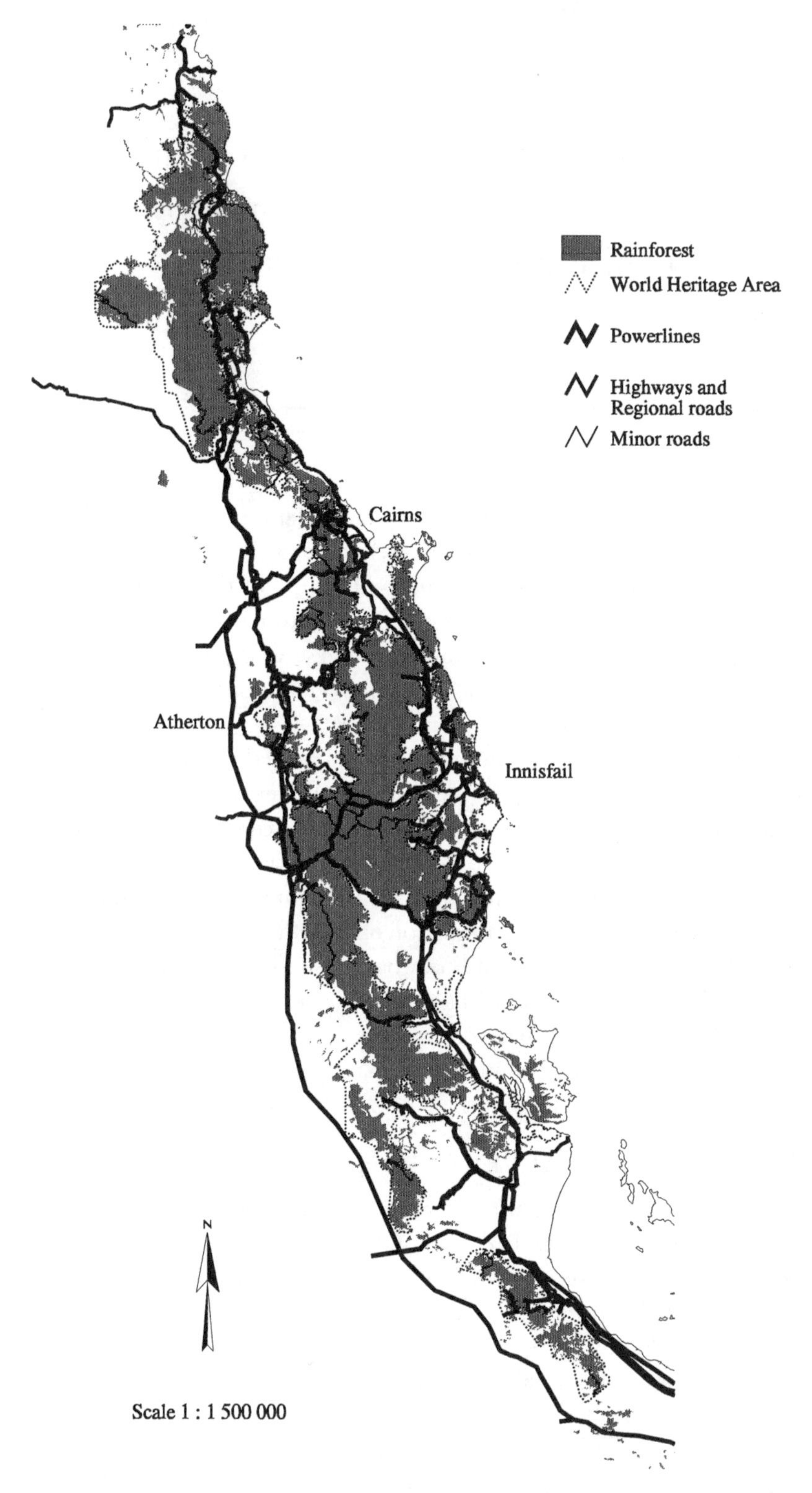

Figure 16.1. Subdivision of rainforest habitat by the network of linear clearings for highways, roads, and powerlines traversing the Wet Tropics of Queensland World Heritage Area in northeastern Australia.

Habitat alteration is not limited to the area cleared. Streams and rivers near roads can be affected because erosion, sedimentation, and altered flow patterns can have substantial negative effects on downstream aquatic life (McElroy et al. 1975; Broadbent and Cranwell 1979; Campbell and Doeg 1989). Drainage of adjacent wetlands (Patrick 1973) and changes in patterns of groundwater distribution may also cause modification of terrestrial habitats (Parizek 1971; Department of Main Roads NSW 1980). Streams may be polluted by chemicals from road surface runoff and clearing maintenance.

EDGE EFFECTS, DISTURBANCE, AND FAUNAL INTRUSIONS

Most road verges and powerline rights-of-way are maintained as grasslands or low, weedy shrublands by cutting regrowth, spraying with herbicides, mowing, burning, or grading. These practices have the dual effects of maintaining a structurally different plant community in the clearing and creating an unstable edge between this community and the native vegetation.

Edge Effects

A swath cut through natural vegetation for a linear clearing has twice its length in edges, allowing edge effects to penetrate into the surrounding natural habitat from both sides of the clearing. Different types of edge effects may penetrate different distances into the forest. For example, Lovejoy et al. (1986) found that many forest interior birds stay at least 50 m away from edges of Amazonian rainforest fragments, while microclimatic variables may be affected for distances of up to 100 m, and light-loving butterflies penetrate at least 300 m from the edge. From his analysis of structural damage in northern Queensland rainforests, Laurance (1989) concluded that edge effects often penetrate 200 m into rainforest and may be detectable up to 500 m from the edge. Further discussion of edge effects in tropical rainforests is presented elsewhere in this volume (see chaps. 3–6).

Because of edge effects, the long forest edges typical of powerline and road clearings can have hidden effects, causing changes in various habitat attributes for varying distances beyond the clearing and substantially increasing the areal extent of habitat alteration. For example, if edge effects extend only 100 m from the edges of clearings, then an additional 6,480 ha of rainforest is affected along the 324 km of powerlines in the Wet Tropics of Queensland World Heritage Area—about five times the area actually cleared for powerlines (1,316 ha). If edge effects extend as far as 500 m (Laurance 1989), then an additional 16,200 ha is affected—a total of 1.8% of the World Heritage Area.

Edge effects include the introduction of edge or generalist species into forest habitats. Species with excellent dispersal abilities, capable of invading and colonizing disturbed habitats, are attracted to edges and may penetrate into the core of natural habitats via the edges of intruding linear clearings (Andrews 1990). Ferris (1979) demonstrated that recognized "edge" species constituted 16% of the bird community within 100 m of a major highway in Maine, but only 2–3.5% of the bird community at 100–400 m from the edge. Although forest interior bird species did not avoid edges along New Jersey powerline clearings of 8–23 m in width, edge species were strongly attracted to these clearings, particularly nest predators and brood parasites that had the potential to reduce breeding success of forest interior species (Rich, Dobkin, and Niles 1994).

Faunal Intrusions

Forest powerline clearings and roadsides maintained as grassland or as low, shrubby swaths allow the penetration of fauna that are alien to the surrounding forest habitat. In the United States, several studies have demonstrated faunal intrusions along powerline clearings and roadsides. In Tennessee, Schreiber and Graves (1977) detected populations of two small mammal species *(Peromyscus leucopus* and *Blarina brevicauda)* in a 91 m wide powerline clearing that were distinct and separate from those found within adjacent forest habitat. At the same site, Johnson, Schreiber, and Burgess (1979) demonstrated that the altered habitat of the clearing allowed the penetration of two species not normally present in the forest *(Reithrodontomys humilis* and *Sigmodon hispidus),* resulting in a more diverse small mammal community than occurred in the adjacent intact forest. The construction of an interstate highway in Illinois with an uninterrupted 5 m wide grassy verge allowed one rodent species *(Microtus pennsylvanicus)* to expand its range by 100 km in six years by using the highway verge as an intrusion route (Getz, Cole, and Gates 1978).

A distinct group of invading bird species inhabited an 80 m wide powerline clearing in Tennessee deciduous forest, but penetrated only a few meters into the edge, while another group of edge species reached their peak density about 60 m from the edge (Kroodsma 1982). When differing widths of powerline clearings were compared in Tennessee deciduous forest, a 12 m wide clearing had a low density and diversity of birds, composed mainly of forest edge species, while corridors of 31, 61, and 91 m width allowed the intrusion of many more grassland species, causing the displacement of forest interior birds (Anderson, Mann, and Shugart 1977).

Where a 60 m wide powerline clearing traverses tropical rainforest in northeastern Queensland, Middleton (1993) demonstrated that the small mammal community of the grassy clearing was composed of grassland species identical to those in nearby pastures and distinct from those in adjacent forest. The grassland species were well established in the powerline and associated access road clearings, but were not captured in traps located 100 m inside the surrounding rainforest. Rainforest species were only infrequently captured in clearings (fig. 16.2A). Goosem and Marsh (in press) confirmed this pattern of faunal invasion into powerline clearings and demonstrated that small grassland mammals rarely penetrated even 10 m into the rainforest. Conversely, rainforest species were very seldom captured within the powerline clearing, and then almost invariably where shrubby weeds occurred along the edge. Where narrow strips of remnant and regrowth rainforest grew in gullies spanning the clearing, only rainforest species were captured (fig. 16.2B). It is possible that one rainforest rodent *(Melomys cervinipes)* avoided grassy clearings because of interspecific competition with a congeneric grassland species *(M. burtoni). M. cervinipes* has been commonly recorded in pastures and grasslands near rainforest at sites where the congener is absent (Laurance 1989; Williams 1990).

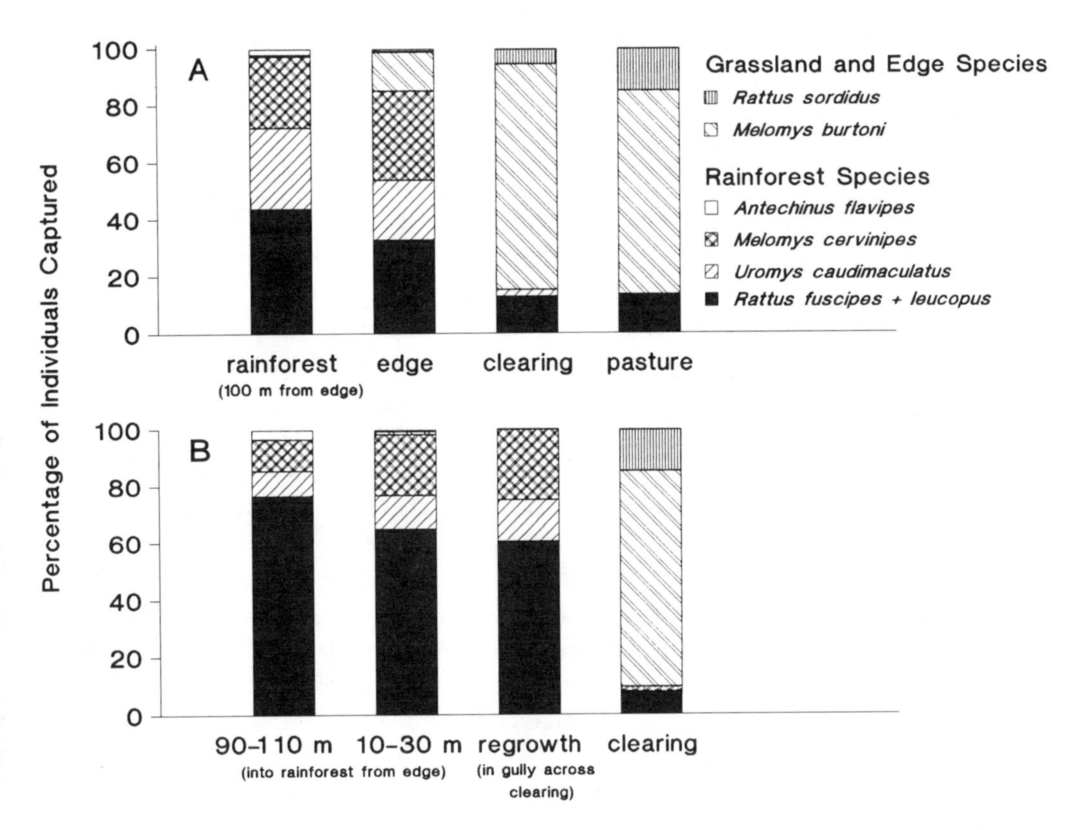

Figure 16.2. Invasion of grassland small mammals along a powerline clearing in tropical rainforest. *(A)* Mammal captures compared between traps located 100 m inside the rainforest, at the forest edge, and within the clearing. Also shown are data from nearby cattle pastures. (Adapted from Middleton 1993.) *(B)* Mammal captures compared between traplines located 90–110 m and 10–30 m from the edge of the rainforest and in the grassland or regrowth rainforest of a powerline clearing. Regrowth rainforest extended across the clearing in gullies, providing continuous canopy cover. (Adapted from Goosem and Marsh, in press.)

Disturbance

Disturbance occurs in the vicinity of linear clearings due to visual, acoustic, and mechanical stimuli, which include noise, movement, dust, headlights, car exhaust fumes, and human presence. Avoidance of areas near roads has been demonstrated by a variety of fauna, including large mammals such as elk (Ward 1973), deer (Rost and Bailey 1979), antelope (Pienaar 1968), wolves (Thurber et al. 1994), and bears (McLellan and Shackleton 1988; Brody and Pelton 1989). A density-depressing effect of roads was demonstrated for two species of meadow birds (the lapwing, *Vanellus vanellus,* and black-tailed godwit, *Limosa limosa*) in the Netherlands. As traffic volume increased, the distance at which the maximum density of the birds was reached increased from 200 m to 2,000 m from the road, reducing populations by 30–65% in the habitat degraded by chronic disturbance (van der Zande, ter Keurs, and van der Weijden 1980). Likewise, Madsen

(1985) found that roads in Denmark with traffic volumes as low as 20–50 vehicles per day seriously inhibited habitat utilization by geese up to 500 m from the road. Vocal communications of amphibians near a Texas highway were masked by traffic noise, altering and restricting their reproductive behavior (Barass 1985).

Using self-triggering cameras in a Sumatran tropical rainforest, Griffiths and van Schaik (1993) found that frequent human foot traffic depressed the intensity of trail use by other mammals and the diversity of mammal species using trails, and caused native pigs to alter their behavior by becoming nocturnal. In Venezuela, forest interior bird species avoided edge vegetation along a 10 m wide gravel road, even though the road edge vegetation was structurally similar to that of the forest interior (Mason 1995).

FRAGMENTATION OF POPULATIONS

A number of studies have examined the effects of linear clearings on populations of vertebrates in desert, grassland, agricultural land, and conifer and hardwood forests, mainly in North America and Europe. Many of these studies have demonstrated a barrier effect in which animals are inhibited from crossing linear clearings (table 16.1). Barrier effects have been demonstrated in large mammals such as reindeer and bears (Klein 1971; McLellan and Shackleton 1988; Brody and Pelton 1989), many species of small mammals (Oxley, Fenton, and Carmody 1974; Wilkins 1982; Garland and Bradley 1984; Mader 1984; Swihart and Slade 1984; McEuen 1995), amphibians (Maynardier and Hunter 1995), and beetles (Mader 1984). When a barrier completely inhibits faunal movements, subdivided populations may become increasingly prone to local extinction and the loss of genetic variability (e.g., Nason, Aldrich, and Hamrick, chap. 20).

In a major trapping study in conifer and mixed forests of Canada, Oxley, Fenton, and Carmody (1974) found that only 3.1–3.3% of individuals of two small mammal species would cross roads with a clearing width of 31 m or greater (fig. 16.3A), despite the fact that individuals often moved similar or greater distances within the adjoining forest. Many more road crossings by small mammals were directly observed during daylight and by using a nightscope, but were almost always restricted to narrower (11–19 m wide) clearings. A road mortality survey supported these findings: small mammals were never killed where clearing widths were very wide (> 90 m), and were seldom killed where clearings were moderately wide (30–90 m). Traffic volume did not appear to be a key inhibiting factor on wide roads, because no small mammals crossed a wide highway (137 m clearing width) with low traffic volume (4 vehicles/hour). Road surface was also not a key factor, as mammals crossed both paved and graveled roads. Overall, these results indicated that road clearing width was the most important factor in the inhibition of movements by small forest mammals (fig. 16.3A; Oxley, Fenton, and Carmody 1974). In Europe, Mader (1984) found that a highway with an 8 m wide clearing and a traffic volume of 60 vehicles per hour completely isolated populations of one forest-dwelling beetle species and two species of forest-dwelling mice (despite the fact that the mice were highly vagile and theoretically able to cross the road within seconds). When mice were translocated to the opposite side of a 6.5 m wide road, only 14.2% of individuals crossed

Table 16.1 Studies of the effects of linear clearings on faunal movements.

Reference	Location	Habitat	Clearing habitat	Clearing width (m)	Traffic volume (cars/day)	Taxa	Effect
Oxley, Fenton, and Carmody 1974	Ontario	Conifer forest	Paved	137	96	Small rodents	Complete movement inhibition
		Temperate forest	Paved	118	14,400	Small rodents	Complete movement inhibition
		Temperate forest	Paved	30/31	4,800	Small rodents	Severe movement inhibition
		Temperate forest	Paved	19/27	288	Small rodents	Severe movement inhibition
		Temperate forest	Paved	11/15	120	Small rodents	Slight movement inhibition
		Temperate forest	Paved	118–137	96–14,400	Medium-sized mammals	Complete movement inhibition
		Temperate forest	Paved	19–31	288–4,800	Medium-sized mammals	Severe movement inhibition
		Temperate forest	Gravel	11–15	120	Medium-sized mammals	Little movement inhibition
Schreiber and Graves 1977	Tennessee	Conifer/hardwood forest	Low shrubs	49		Small rodents	Not a barrier; distinct population in clearing
		Conifer/hardwood forest	Grassland/ low shrubs	104		Small rodents	Not a barrier; distinct population in clearing
Mader 1984	Germany	Temperate forest	Paved	8	5,160	*Apodemus flavicollis*	Complete movement inhibition
		Temperate forest	Paved	6.5	480	*A. flavicollis*	Complete movement inhibition; not a complete barrier (several translocated individuals crossed)
		Temperate forest	Paved	3	48	*A. flavicollis*	Movement inhibition; not a complete barrier
Bakowski and Kozakiewicz 1988	Poland	Temperate forest	Gravel	15	72	*Clethrionomys glareolus*	Significant movement inhibition; not a barrier
		Temperate forest	Gravel	15	72	*A. flavicollis*	Slight movement inhibition
Swihart and Slade 1984	Kansas	Abandoned farmland	Gravel	3.6		*Sigmodon hispidus*	Slight movement inhibition
		Abandoned farmland	Gravel	3.6		*Microtus ochrogaster*	Slight movement inhibition

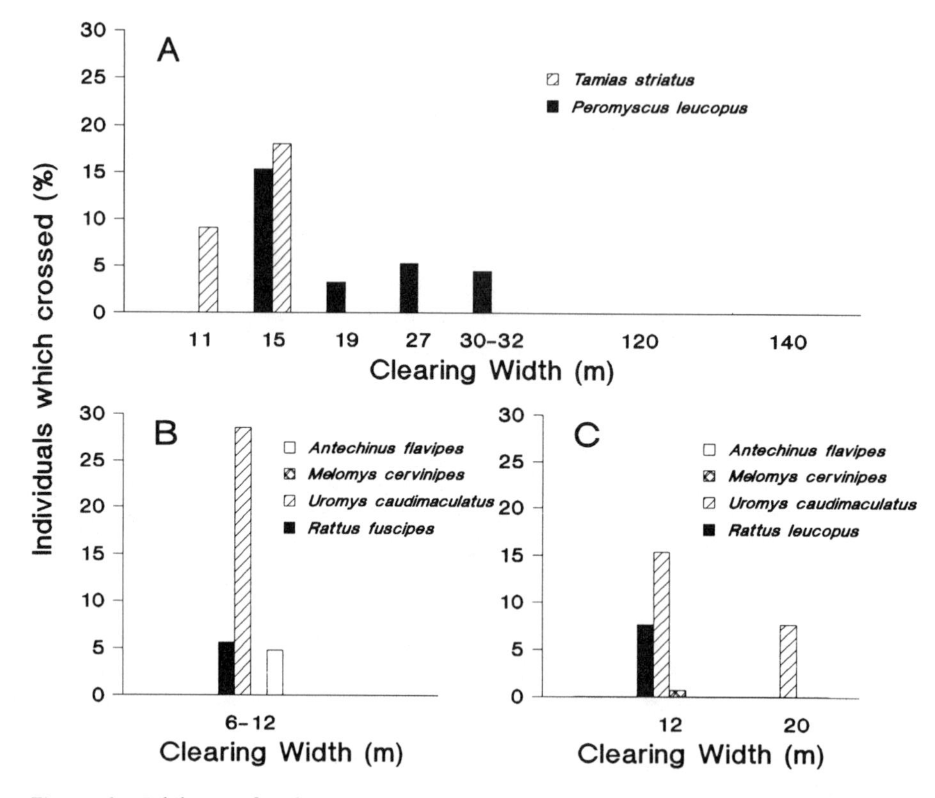

Figure 16.3. Inhibition of road-crossing movements by small mammals increases with clearing width. The linear barrier effect is species-dependent. Data shown are the number of individuals that crossed at least once as a percentage of the total number of individuals known to be alive. *(A)* Adapted from Oxley, Fenton, and Carmody 1974. *(B)* Adapted from Burnett 1992. *(C)* Data from the author.

to return to their normal home range. Even forest roads with minimal traffic (2 vehicles/hour) and a clearing width of only 3 m strongly inhibited mouse crossings (Mader 1984).

FRAGMENTATION IN TROPICAL RAINFORESTS

Intuitively, one would expect that wildlife adapted to habitats of high structural complexity, such as the understory and ground layer of closed-canopy rainforests, would find a linear clearing with a dense grassland or paved surface a more formidable barrier than would species of more open, structurally simpler habitats. The extreme variability of light, temperature, and other microclimatic factors in the clearing in comparison to the more constant microclimate of the rainforest understory (Lovejoy et al. 1986; Turton and Freiburger, chap. 4) would compound the perceived environmental discontinuity. Added to these factors is environmental instability caused by clearing maintenance procedures such as burning, mowing, grading, and spraying of herbicides. Edge effects, especially changes in plant species composition and vegetation structure near clearings (e.g., Kapos et al., chap. 3; Laurance, chap. 6), also affect animal movements and habitat use. The degree of inhibition of movement across clearings is determined by the mobility, habitat specificity, and behavior of each species.

Several recent studies using road mortality statistics and live trapping (Burnett 1989, 1992; Mason 1995; Goosem and Marsh, in press) have begun to elucidate the magnitude of the barrier effect in tropical rainforests.

WILDLIFE MORTALITY ON RAINFOREST ROADS

One method of determining whether roads cause a linear barrier effect is to assess the incidence of wildlife mortality due to vehicle-wildlife collisions. If an animal has attempted to cross the road, even though it is subsequently killed, it can be concluded that the species does not perceive the road as a complete barrier. By comparing the species present in an area with those recorded as roadkill statistics, it is possible to determine which species are most susceptible to road mortality. Species not found as roadkills may be using alternative crossing routes, such as canopy connections or underpasses; may be fast or intelligent traffic avoiders; or may simply avoid roads entirely. It is the last group that could be cause for conservation concern. The extent to which road features such as traffic volume and clearing width influence mortality rates can also be assessed.

In a study on the Kennedy Highway in northeastern Queensland, I examined wildlife mortality on a weekly basis for 38 months from 1989 to 1992 to determine which of a variety of road factors influenced wildlife mortality rates. Where the highway traverses rainforest at 100–400 m altitude, I chose four 0.5 km long transects, incorporating varying clearing widths and culvert designs and dimensions. Traffic volume data (mean = 3,500 vehicles/day) was collected daily by a Queensland Department of Transport traffic counter. Over the 38-month period, more than 4,000 vertebrates of over 100 species were recorded as roadkills on the 2 km of road examined. These included about 500 mammals, almost 90 birds, 450 reptiles, and more than 3,000 amphibians.

Preliminary results demonstrated that increasing traffic volume had little effect on mammal and bird mortality but was significantly and negatively correlated with mortality in reptiles and amphibians. This finding suggests that reptiles and amphibians exhibit stronger road avoidance in response to greater disturbance from noise, vibrations, headlights, and vehicle numbers. Unsustainable road mortality rates could also depress local population sizes of amphibians (Fahrig et al. 1995), although declines of this magnitude did not appear to occur in this study.

Clearing width was a significant factor in roadkill numbers for all major vertebrate groups. As clearings became wider, there were fewer attempts to cross, as demonstrated by a reduction in wildlife mortality for all major groups (fig. 16.4). A large proportion (59%) of the fauna killed were understory and ground-dwelling species, indicating that many of these species will cross linear barriers, particularly where the clearing is narrow. As clearing width increases, however, it seems likely that most rainforest species cease their attempts to cross (fig. 16.4). For example, Bierregaard and Stouffer (chap. 10) demonstrated that many Amazonian rainforest understory birds will not cross 100 m wide clearings.

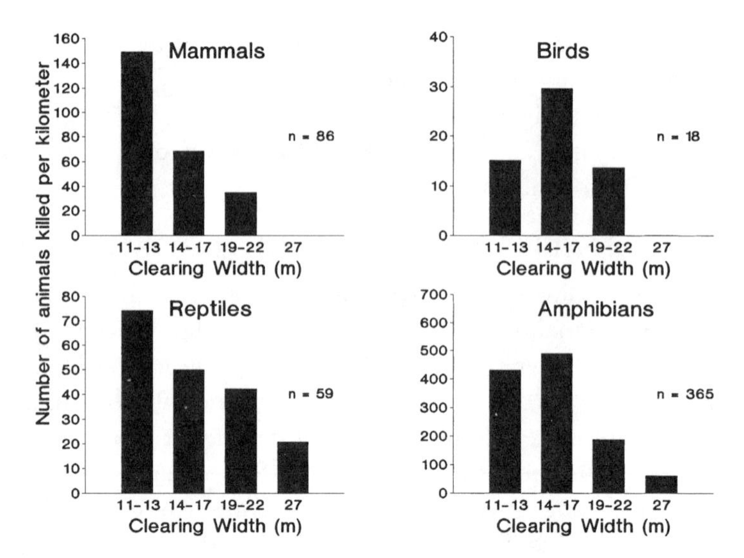

Figure 16.4. Vertebrate road mortality is reduced as clearing width increases, probably because fewer crossings are attempted. Data are from the author's nine-month survey of four 0.5 km long transects along a highway traversing rainforest in northeastern Australia. Mean traffic volume was 3,500 vehicles per day.

A number of species known to occur in the study area were underrepresented in or absent from roadkill statistics (table 16.2). These species may use existing culverts or bridges to cross below the road, or may use arboreal canopy connections to cross above it. They may also be better at crossing roads quickly or at choosing times of the day when traffic is minimal. Alternatively, these species could be avoiding the vicinity of roads and hence could be suffering severe barrier effects.

Many species underrepresented in roadkill statistics are of special conservation or evolutionary interest, such as certain ground-dwelling marsupials and birds, small understory birds, litter-dwelling frogs, and arboreal marsupials. For example, a small marsupial, the musky rat-kangaroo *(Hypsiprymnodon moschatus),* and a ground-dwelling bird, the chowchilla *(Orthonyx spaldingii),* are both common in the study area but have yet to be observed close to the highway. Arboreal marsupials such as possums and tree kangaroos were also underrepresented in road mortality statistics. Arboreal species may use the few remaining canopy connections as crossing points, or avoid the road altogether. Canopy connections are probably crucial for arboreal mammals such as the lemuroid ringtail possum *(Hermibelideus lemuroides),* a species that almost never ventures to the ground (Laurance 1990).

In the case of rare or threatened species, road mortality must be considered a threatening process. For example, the roadkill statistics include one individual of the endangered southern cassowary *(Casuarius casuarius johnsonii),* a species thought to number only about 1,300 adults in the Wet Tropics of Queensland (F. H. J. Crome, pers. comm.). A total of seventeen cassowaries were killed on roads at Mission Beach (one of the species' remaining strongholds) over a three-year period (Bentrupperbaumer 1988), constituting 14% of known individuals in the area at that time.

Road mortality is seen as a threat to amphibians and reptiles in North America (Bernardino and Dalrymple 1992; Rosen and Lowe 1994; Fahrig et al. 1995), particularly to those species that undertake seasonal migrations. There is a similar potential for heavy mortality in many rainforest species. For example, one of the frog species found in large numbers in this study *(Litoria xanthomera)* is attracted to roadside drains for spawning, while others find a film of rainwater on paved road surfaces attractive. The significance of faunal mortality on rainforest highways is reinforced by the fact that an average of 1,365 road-killed vertebrates were detected on the 2 km section studied each year. This is necessarily an underestimate of total mortality because collections were made only weekly, and the longevity of carcasses is less than a week for many of the recorded species.

Table 16.2 Species absent or underrepresented in the roadkill statistics from the Kennedy Highway in northeastern Australia.

Species	Roadkill numbers	Potential cause of absence/ underrepresentation
Mammals		
Hypsiprymnodon moschatus Musky rat-kangaroo	0	Complete avoidance of highway vicinity
Pseudocheirops archeri Green ringtail possum	2	Use of canopy overpasses; avoidance—reluctance to descend to ground
Cercartetus caudatus Long-tailed pygmy possum	1	Use of canopy overpasses; avoidance—reluctance to descend to ground
Dendrolagus lumholtzi Lumholtz's tree kangaroo	1	Avoidance of highway vicinity; use of canopy overpasses
Birds		
Psophodes olivaceus Eastern whipbird	0	Complete avoidance of highway vicinity
Orthonyx spaldingii Chowchilla	0	Complete avoidance of highway vicinity
Reptiles		
Lampropholis basiliscus a litter skink	3	Avoidance of highway surface
Carlia rubrigularis a litter skink	12	Avoidance of highway surface
Lygisaurus laevis a litter skink	2	Avoidance of highway surface
Varanus scalaris Spotted tree-monitor	5	Avoidance of highway surface; possible fast traffic avoidance
Physignathus lesueurii Eastern water dragon	4	Use of highway bridge underpasses
Amphibians		
Cophixalus ornatus a litter microhylid frog	1	Avoidance of highway surface; destruction of carcass by traffic
Sphenophryne pluvialis a litter microhylid frog	4	Avoidance of highway surface, destruction of carcass by traffic

LIVE-TRAPPING STUDIES IN TROPICAL RAINFOREST

Linear barrier studies in tropical rainforest using live trapping have focused on small mammals. The exception is a study by Mason (1995), who examined understory birds in a Venezuelan rainforest and demonstrated an inhibition of their movements at the forest edge. These birds crossed a 10 m wide unpaved road significantly less often than would be predicted from their movements in continuous forest.

In northern Queensland upland rainforest, Burnett (1992) conducted a mark-recapture study to assess the effects of 6 m and 12 m wide roads on the movements of three small mammal species. A grid of traps was used to show that road crossings constituted only 1.8% of the total movements of bush rats *(Rattus fuscipes),* which weighed up to 150 g, while only 5.6% of individuals crossed. Only 5.2% of movements of *Antechinus flavipes,* a small marsupial that weighed up to 85 g, were crossings, and only 4.7% of individuals crossed. A larger rodent, *Uromys caudimaculatus,* which can weigh up to 1 kg, was not so severely affected; 20% of its movements were road crossings, and 28.6% of individuals crossed (fig. 16.3B,C). Because the distance required to cross the road was only one-third of the average distance between successive recaptures of the two smaller species, Burnett concluded that the road had a significant inhibitory effect on the movements of all three species. However, individuals of all three species could easily be induced to cross the road by translocating them to the opposite side, or by placing baited traps on one side of the road only.

The physical barrier effect of a road may be similar to that caused by a wide expanse of water, where crossings of water-shy species, apart from occasional, stochastic movements, are prevented (Burnett 1992). Although the road in Burnett's study was not a physical barrier, as demonstrated by the ease with which individuals could be induced to cross it by baits or translocation, it did strongly inhibit movements in some species. Two explanations were offered for the demonstrated linear barrier effect. Some species might tend to avoid the open space of the road due to the extreme contrast in microhabitat that it represents. Alternatively, small mammals may align their territories or home ranges along physical features of the environment, including roads, and the presence of territorial individuals could inhibit movements throughout the area.

Using similar grid-trapping techniques, I examined the potential linear barrier effect of a wider (12 m and 20 m clearing width) gravel road in northeastern Queensland with a low traffic volume (20–40 vehicles/day). In all cases the distance between rainforest margins was less than or equal to the majority of daily small mammal movements perpendicular to the road. However, crossing movements constituted only 1.8% of recaptures for *Rattus leucopus,* 0.6% for *M. cervinipes,* and 6.8% for *U. caudimaculatus.* Crossing individuals comprised only 7.7%, 0.9%, and 15.4% of the known population for *R. leucopus, M. cervinipes, and U. caudimaculatus* respectively at the 12 m clearing width, whereas the only species to attempt to cross a 20 m clearing was *U. caudimaculatus* (7.7% of individuals) (fig. 16.3C). Thus, crossing movements of rodents were inhibited by a road with 12–20 m clearing widths. Areas of activity of smaller species were restricted to one side of the road, and the relatively few crossings were mostly by dispersing individuals that were not subsequently recaptured. Increased clearing width did not cause an increased inhibition of movements, although baiting traps on only one side of the road induced more crossings on the narrower part of the road (12 m) than on the wider one (20 m). Visual observations suggested that many individuals used culverts under the road as a crossing route. Where culverts with suitable ground cover near their entrances were lacking, comparatively few crossings occurred. This observation suggests that, at clearing widths ranging from 12 to 20 m, the presence or absence of culverts may have a stronger influence on across-road movements than clearing width.

Goosem and Marsh (in press) demonstrated that movements of small rainforest mammals were strongly inhibited by a 60 m wide powerline clearing passing through rainforest in northeastern Queensland. Vegetation in the clearing was either dense grassland with occasional patches of woody shrubs near the rainforest edge, or low regrowth (and remnant) rainforest in gullies, which created a continuous canopy connection that spanned the clearing. Distinct rainforest and grassland small mammal communities were detected, with each being almost completely confined to its preferred habitat. The regrowth rainforest in gullies supported a small mammal community typical of rainforest (fig. 16.2B).

Grid trapping over fourteen months revealed that the grassy swath of the powerline clearing completely inhibited crossings by rainforest species, even though the distance required to cross the clearing was well within their capabilities for movement (Goosem and Marsh, in press). This finding demonstrated a severe fragmentation effect caused by the presence of the linear clearing. In contrast, strips of regrowth rainforest in gullies provided an effective crossing route for ground-dwelling rainforest mammals.

A shade-cloth tunnel (with a light regime similar to that in intact rainforest) was constructed across the powerline clearing to test experimentally whether small mammals could be induced to cross clearings in this manner. The tunnel was unsuccessful. Regrowth vegetation in gullies, however, did help to mitigate barrier effects by allowing small mammal movements across the clearing, thereby reducing the isolation of populations.

HYPOTHESES CONCERNING THE CAUSE OF LINEAR BARRIER EFFECTS

Four hypotheses may be proposed for the causative mechanism(s) of linear barrier effects. First, an alignment of home ranges of small mammals may occur along a linear environmental discontinuity, restricting movements to only one side of the barrier (Burnett 1992). For example, Oxley, Fenton, and Carmody (1974) visually observed a large number of road crossings by eastern chipmunks *(Tamias striatus)* in Canada, although only 3.3% of animals recaptured during trapping were road crossers. This finding suggests that the road-crossing animals may have been quickly repulsed by resident individuals with established territories on the other side of the road (Burnett 1992).

Second, some forest animals avoid open spaces (Burnett 1992). For example, King (1978) demonstrated in laboratory enclosures that individuals of the marsupial *Ante-*

chinus stuartii avoid areas of open space following an initial period of exploration. Several rainforest species (e.g., *Rattus fuscipes, Melomys cervinipes, Uromys caudimaculatus*) have been shown to prefer dense, complex rainforest vegetation over open forest or pine plantations (Barnett, How, and Humphries 1978; Hockings 1981; Smith 1985; Williams 1990). Traffic disturbances and edge effects could increase the avoidance of road-edge habitats by some species.

Third, when an invading habitat such as a grassy swath is present, rainforest species may be excluded by species better adapted for the invading habitat. As discussed previously, some support for this hypothesis comes from the fact that one of the rainforest rodents I studied, *Melomys cervinipes,* occupies pastures and natural grasslands where its congener, *M. burtoni,* is absent, but completely avoided a powerline clearing where *M. burtoni* was present.

Finally, linear barrier effects might be caused by intensive road mortality if such a high proportion of potential road crossers were killed that across-road movements were strongly reduced (Fahrig et al. 1995). Such effects could conceivably occur among amphibians that congregate on wet roads and hence are highly vulnerable to traffic.

In many cases, one or more of the mechanisms proposed above could be acting synergistically to produce the linear barrier effect.

GENERAL IMPLICATIONS

1. Linear environmental discontinuities such as roads and powerlines may fragment rainforest wildlife populations. Arboreal species and those of the rainforest understory and ground layer are most likely to be affected.
2. The mobility and behavioral characteristics of species, particularly those relating to open-space avoidance, largely determine the magnitude of the linear barrier effect. In contrast, species that do not avoid open space may be prone to road mortality or predation along roads.
3. As the width of linear clearings increases, road-crossing attempts by many rainforest species rapidly decline. Eventually, clearings are likely to form a complete or almost complete barrier, apart from rare stochastic crossings.
4. In clearings wide enough to support grassy swaths, the presence of a distinctive grassland faunal community may compound structural and microclimatic differences to result in an almost complete exclusion of rainforest species.
5. To mitigate fragmentation effects of rainforest roads and highways, the widths of clearings should be minimized, particularly near streams and gullies, where many species attempt to cross.
6. Canopy connectivity above a road should be maintained wherever possible to provide potential crossing points for arboreal species. In addition, the presence of some canopy cover produces a microclimate on the road surface that is somewhat similar to that of the adjacent rainforest, and provides some cover for understory species attempting to cross.
7. Underpasses and culverts form crossing routes for a number of rainforest mammals and other vertebrates. A range of sizes and designs of culverts may increase their utility as faunal crossing routes.
8. Allowing rainforest regrowth to develop along powerline clearings clearly helps to mitigate the barrier effects of these clearings for small mammals.

9. The effects of new powerlines can be reduced by clearing only tower footprints, and subsequently maintaining the powerline by helicopter. Alternatively, strips of forest can be retained in gullies along the powerline route.
10. When planning for new roads and powerlines, resource managers should strive to avoid critical habitats for rare or threatened species as well as areas of high biological diversity and abundance.

ACKNOWLEDGMENTS

I would like to acknowledge the constructive comments provided by Scott Burnett, John Winter, Steve Goosem, Bill Laurance, and Helene Marsh during the preparation of this chapter. My own studies reviewed in this chapter were supported by research grants from the Wet Tropics Management Authority, the Queensland Electricity Commission, and James Cook University. Greatly valued field and laboratory assistance was provided by Mike Trenerry, Les Moore, Steve Goosem, Mick Godwin, and Keith McDonald. Data from traffic counters was provided by the Queensland Department of Transport. Finally, I thank my supervisor, Helene Marsh, for her many valuable suggestions during my dissertation research.

17
Transitory States in Relaxing Ecosystems of Land Bridge Islands

John Terborgh, Lawrence Lopez, José Tello, Douglas Yu, and Ana Rita Bruni

---

LAND bridge islands are landscape fragments that have been isolated by rising water levels with a concomitant reduction in habitat area (Diamond 1972). Numerous studies have documented that land bridge islands and forest fragments lose species after isolation (Diamond 1972, 1984; Terborgh 1974, 1975; Willis 1974; Wilcox 1978; Terborgh and Winter 1980; Heaney 1984; Case and Cody 1987). Large-bodied vertebrates, especially those high on the trophic ladder, are particularly vulnerable to habitat contraction and are typically among the first species to disappear (Brown 1971; Heaney 1984; Lomolino, Brown, and Davis 1989). Other species persist for varying periods, depending largely on the population size supported by the fragment or island (Brown 1971; Terborgh and Winter 1980; Pimm, Jones, and Diamond 1988). Overall, species numbers on land bridge islands decline along a roughly exponential trajectory toward a dynamic equilibrium, one characterized by simplified community structure and reduced species diversity (Diamond 1972; Terborgh 1974).

Between the initial "supraequilibrial" condition and final equilibrium, land bridge island ecosystems decay, or "relax," through a series of transitional states (Simberloff 1976). Isolation by water reduces recolonization to low levels, and extinction prevails over colonization, at least initially. Although little studied, these transitional states must include faunal and floral assemblages that would rarely, if ever, occur on the original mainland. For example, mainland ecosystems almost invariably include top predators (felids, canids, eagles), while the faunas of land bridge islands rarely do (Brown 1971; Heaney 1984, 1986). What, then, are the effects on land bridge island ecosystems of the absence of trophic levels or entire ecological functions, and what consequences do such imbalances have for the surviving species? These are largely unanswered questions, but ones that are amenable to investigation in a set of land bridge islands undergoing relaxation.

Here we report on the transient ecological states encountered on a series of twelve land bridge islands created by a hydroelectric impoundment, Lago Guri, located in the state of Bolivar, Venezuela (7°20′ N, 62°48′ W). The associated company town of Guri is located about 100 km south of Puerto Ordaz (Ciudad Guyana on some maps), a major industrial and port city on the Orinoco River. Further details about Lago Guri, its climate, islands, and vegetation are given in Alvarez et al. (1986).

The following is a progress report based on four field sessions in the Lago Guri islands (1990, 1993, 1994, 1995). The islands were four years old when our work began in 1990, and therefore represent an early stage in the relaxation process. Our preliminary objectives were to select a group of islands for study and then to survey the islands, as well as a mainland control site, for selected vertebrate and invertebrate taxa. The results reveal high levels of interisland variability in the presence and absence of taxa and in the abundances of both individual species and some higher taxa (e.g., birds). The faunal assemblages present on many Lago Guri islands are distinct from any likely to be encountered on the mainland, and some fundamental ecological processes (e.g., herbivory) are dramatically overrepresented, while others (e.g., predation) are underrepresented. We speculate that such ecological distortions drive the relaxation process.

THE LAGO GURI ISLANDS

The Raul Leoni Dam that forms Lago Guri was built in two stages. The first was completed in 1968 and raised the water to 217 m above sea level (ASL) from a base of 120 m ASL. Construction of the second stage followed several years later and began, in 1983, to raise the water again. The final level of 270 m ASL was attained in October 1986, by which time the impoundment had inundated an area of 4,300 km^2 (Morales and Gorzula 1986). It should be noted that all but perhaps one (Miedo) of the islands we selected for study are isolated from the mainland by water less than 50 m deep, providing assurance that they were not islands prior to 1986.

Lago Guri and a several-kilometer-wide watershed protection zone surrounding it are administered by EDELCA (Electrificación del Caroni), the Venezuelan energy company that operates the hydroelectric project. Access is strictly controlled, and no hunting is allowed, although some illegal poaching occurs despite regular aerial and lacustrine patrolling by the national police.

The formation of Lago Guri has created several dozen forested islands that range in size over nearly four orders of magnitude, from less than 0.1 ha to about 650 ha (Alvarez et al. 1986). The large number of available islands offers the possibility of selecting replicated small, medium, and large islands located both near and far from larger landmasses. From a large number of candidate islands, we selected twelve for study: four small (ca. 1 ha) "near" islands; four small "far" islands; three medium islands (ca. 10 ha; one "near," one "far," and one at an intermediate distance); and one large island (350 ha) at an intermediate distance. Our operational definitions of "near" and "far" are as follows: "near" is 0.5 km or less from the nearest larger landmass; "far" is 1.0 km or more from the nearest larger landmass.

A nearby mainland control site was located less than 2 km from some of the study islands (table 17.1). The mainland site was located on a peninsula rimmed by shoreline on two sides. We chose this situation to mimic the proximity to edge and exposure to prevailing winds that are typical of islands. The perimeters of all small and medium islands were surveyed with a hand-held compass and mapped to determine the areas occupied by forest. Trail grids were cut and mapped on all twelve islands and at the mainland control site.

Table 17.1 Some characteristics of the censused Lago Guri islands and the mainland control area.

Locality	Area (ha)[a]	Distance to nearest larger landmass (km)	Distance to mainland (km)
Small, far islands			
Colon	0.6	1.8	4.2
Iguana	1.4	1.9	4.5
Miedo	0.7	2.6	3.2
Triángulo	2.3	1.0	4.9
Small, near islands			
Búmeran	0.9	0.1	1.5
Palizada	1.8	0.4	0.4
Perímetro	1.7	0.5	3.4
Rocas	0.6	0.1	0.1
Medium islands			
Corral (near)	12.3	0.2	0.2
Lomo (intermediate)	12.0	0.5	2.2
Panorama (far)	11.1	1.5	1.9
Large island			
Danto Machado	23.1[b]	2.2[c]	2.2
Mainland			
Terra firme	26.0[d]	0.0	0.0

a. Area of island; entire island censused unless otherwise noted.
b. Area censused; area of Danto Machado is ca. 350 ha.
c. A similar-sized island lies 0.4 km to the west. d. Area censused.

The habitat of all study islands and the mainland is semi-deciduous tropical forest. The canopy is low, between 15 and 20 m, and only the largest trees rise above 25 m. The understory is laced with wiry vines, making passage difficult except on trails. The climate is markedly seasonal, with a dry season from November through April and a single wet season from May through October. Annual rainfall at the Raul Leoni Dam is roughly 1,100 mm, but the climate becomes progressively wetter southward (Alvarez et al. 1986). Our sites had gentle topography and well-drained soils, with the exception of two islands, Miedo and Lomo, which were steep and somewhat rocky. The forests of the Lago Guri mainland had all been high-graded for prime timber species before EDELCA took control of the watershed, but the presence of large trees at all of our sites indicated that none of them had been clear-cut. Further information on the structure and composition of the forest of Lago Guri can be found in Huber (1986).

METHODS

Birds

We censused birds on all twelve study islands and on the mainland during the 1993 breeding season, beginning in May and ending in July. We employed the standard spot-mapping technique, recording the locations of singing males in relation to mapped trail grids at each site (Terborgh et al. 1990). Parallel legs of a trail grid were generally 100–150 m apart, in keeping with the observation that the territorial song of most forest-dwelling bird species can be heard to distances of at least 100 m (Emlen and DeJong 1981). Small and medium islands were censused in their entireties. The bird communities of the large island and mainland were sampled in plots of 23 and 26 ha, respectively. Each site was censused four times. All censuses were conducted in the early morning hours, beginning at first light and ending at about 0900 hours, when vocal activity tended to subside.

In tabulating the results, we distinguished five categories of birds: (1) resident forest-dwelling and presumably breeding species; (2) visitors of presumably nonbreeding species; (3) species with large spatial requirements that often visited islands but were not resident on them; (4) edge species; and (5) aquatic species.

Males or pairs were considered to be resident breeders only if they were recorded in the same area on two or more censuses. In a large majority of cases, resident individuals were recorded on three or all four censuses. Individuals or species that were recorded only once were regarded as visitors, with the exception of some night birds and hummingbirds that were not readily detected in the regular censuses and were counted as residents. Species with large spatial requirements, such as raptors, vultures, parrots, pigeons, swifts, and swallows, occurred daily on or over nearly all islands, even the most isolated ones, but in most cases were probably not resident breeders. Such species were not included in the tallies unless there was evidence of resident status (lingering near probable nest site, or flushed from nest or nest hole).

We distinguished edge species as a separate category because of the presence of "ghost forest" around the Lago Guri islands. Around every island and the mainland there is a fringe, extending from a few meters up to more than a kilometer, of dead trees, killed by inundation, the crowns of which remain emergent from shallow areas of the lake. (This phenomenon is typical of hydroelectric impoundments in the Tropics, where forests are rarely cleared prior to flooding.) A number of bird species, including woodpeckers, flycatchers, swallows, house wrens, and a few others, are strongly attracted to the dead trees and use them for both foraging and breeding. Edge species were not tallied with resident breeders because we felt that their presence was attributable to the ghost forest rather than the living forest of the islands. Aquatic species, such as cormorants, anhingas, herons, rails, terns, and kingfishers, were not included within the reported census results.

Other Vertebrates and Invertebrates

All twelve study islands and the mainland were surveyed for mammals, reptiles, amphibians, and certain invertebrates in February and March, 1994 (dry season). All sites were surveyed on three occasions, twice by day (once in early morning and once near midday) and once by night. Small islands were searched for one hour on each occasion. The procedure was to walk slowly along the trails (mean distance ca. 200 m), listening for vocalizations or sounds made by moving animals and looking for animal signs. When the trails had been covered, the observer(s) then covered the areas between trails until the entire island had been searched. On medium and large islands and the mainland, the procedure differed in that all searching was conducted from trails. Trails covering the three medium islands ran around the margin, 25–50 m in from the edge, and across the center, for total lengths of 1,100–1,500 m. On each survey, two teams of observers walked around the trails in opposite directions. On the large island and mainland, the surveys were conducted in the same countercurrent fashion on the trails that had been used to spot-map birds the year before. Observers exchanged routes or directions on successive censuses.

Species were recorded as present on a given island if fresh diagnostic sign was detected. Such sign included: vocalizations (e.g., primates); feces (tapir, *Tapirus terrestris;* brocket deer, *Mazama americana;* ocelot, *Felis pardalis;* capybara, *Hydrochaeris hydrochaeris;* the common iguana, *Iguana iguana*); footprints (felids, tapir); active burrows (armadillo, *Dasypus novemcinctus;* paca, *Agouti paca;* a teiid lizard, *Ameiva ameiva,* tarantulas); runways (leaf-cutter ants, *Atta sexdens*); and quills (porcupine, *Coendou prehensilis*).

Perimeter-Strip Counts

A rough index of capybara and iguana densities was obtained from strip counts in which an observer walked a measured distance and recorded all fecal deposits found within 5 m of the path. For small islands, the distance walked was the perimeter of the island. For medium and large islands and the mainland, the transect length was arbitrarily set at 10,000 m, except for Panorama, where it was 5,000 m (because of difficult terrain on the exposed side of the island).

Artificial Nest Assays

Following standard methods, wicker nests were tied to low shrubs and saplings and stocked with two quail (*Coturnix* sp.) eggs each (Loiselle and Hoppes 1983; Wilcove 1985). The nests were checked after one week. Any nest from which one or more eggs had been removed or broken was scored as having been raided.

RESULTS

Already by mid-1990, less than four years after the water had first reached its maximum level, a large majority of the vertebrate species present on the nearby mainland could not be found on the 1 ha class of islands. The large island, in contrast, still retained a nearly intact fauna, including all of the primates, ungulates, and large frugivorous birds known to occur in the region. Large predators (jaguar, *Panthera onca,* puma, *Puma concolor,* harpy eagle, *Harpia harpyja*) occurred on the mainland, but were not detected on any island in the formal surveys. However, footprints of a large felid were noted on the large island in June 1995, although it is unlikely that an animal so high on the trophic scale could be resident in a mere 350 ha. EDELCA biologists have observed both jaguar and puma swimming in the lake, so near islands are probably visited by these predators occasionally (L. Balbas, pers. comm.).

Birds

Bird densities varied greatly among islands. Small islands had densities that ranged from far below to far above the level registered on the mainland (table 17.2). However, the mean density (no./ha) of forest-dwelling resident pairs found on the eight small islands

Table 17.2 Population densities and species numbers of birds on the Lago Guri islands and mainland.

Locality	No. resident pairs[a]	No. resident pairs/ha	No. edge pairs[b]	No. edge pairs/ha	No. resident species	No. edge species	No. nonresident species[c]	No. nonresident edge species	Total no. species
Small, far islands									
Colon	10.0	16.7	4.5	7.5	8	4	6	3	21
Iguana	14.0	10.0	4.5	3.2	13	4	4	2	23
Miedo	5.0	7.1	3.0	4.3	5	3	2	0	10
Triángulo	14.5	6.3	5.0	2.2	12	4	11	5	32
Small, near islands									
Búmeran	10.0	11.1	4.0	4.4	10	3	13	1	27
Palizada	5.5	3.1	4.0	2.2	6	4	10	1	22
Perímetro	5.5	3.2	5.0	2.9	5	4	10	2	21
Rocas	8.0	13.3	3.0	5.0	8	3	6	0	17
Medium islands									
Corral (near)	14.5	1.2	10.0	0.8	10	5	17	3	34
Lomo (intermediate)	10.0	0.8	0.0	0.0	7	0	6	1	15
Panorama (far)	48.0	4.3	3.0	0.3	8	3	8	3	32
Large island									
D. Machado	191.0	8.3	9.5	.4	58	5	16	3	82
Mainland									
Terra firme	127.5	4.9	11.0	.4	43	7	18	5	73

a. Resident = resident species as defined in text. b. Edge = edge species inhabiting margins and ghost forest, but not forest interior.
c. Nonresident = species detected on one or two censuses, but presumed not resident.

was 8.9, versus 4.9 on the mainland, while the mean number of resident species on the small islands was 8.3. Therefore, nearly every resident species on the small islands was represented by only a single pair.

In contrast, bird densities were strikingly low on two of the three medium islands, with one, Lomo, supporting only 0.8 resident pair/ha, equivalent to one-sixth of the mainland density. Another, Corral, supported only one-fourth of the mainland density. We shall return to these anomalous cases below.

A 23 ha plot on the large island contained 8.3 resident pairs/ha, about twice the density (4.9 pairs/ha) recorded in a 26 ha plot on the mainland. The large island contained more species as well (58 versus 43 on the mainland), but the higher species count on the large island is attributable, at least in part, to the greater total number of birds censused. These numbers, of course, reflect only what was recorded within the respective census plots. Altogether, more than a hundred forest-dwelling species are resident on the large island and mainland, so the regional species pool is substantially larger than the number of species recorded on any of the censuses.

Other Vertebrates and Invertebrates

The faunas of small islands consistently contained a subset of the species that were present elsewhere. These included unidentified small rodents, capybaras, howler monkeys, *Alouatta seniculus,* the lizards *Ameiva ameiva* and *Iguana iguana,* the tortoise *Geochelone carbonaria,* a poison-arrow frog, *Dendrobates leucomelas,* leaf-cutter ants, *Atta* sp., and two species of tarantulas that occupied arboreal cavities and burrows, respectively (table 17.3).

One or more medium islands supported additional mammals, including olive capuchin monkeys, *Cebus olivaceus,* armadillo, porcupine, agouti, *Dasyprocta leporina,* and paca. Fresh felid (ocelot?) scat was found on two of the medium islands, but given that an individual home range exceeds 100 ha on the mainland (Emmons 1988), it is doubtful that the scats represented resident individuals. In one case (Panorama), we found ocelot tracks leading up from the water's edge into the forest, indicating that the animal had swum over from a neighboring island. Similarly, tapir sign was found on a small, far island (Miedo), where the animal could hardly have been resident.

Additional mammal species occupied the 350 ha island: the bearded saki monkey, *Chiropotes satanus,* brocket deer, tapir, peccary, *Tayassu tajacu,* anteater, *Tamandua tetradactyla,* and tayra, *Eyra barbara.* We did not find mammals on the mainland that were not also recorded on the large island.

The herpetological data included only common and conspicuous species that were encountered frequently on the formal surveys.

We recorded only a few invertebrates judged capable of having an effect on other species. Leaf-cutter ants (*Atta sexdens*) were notably abundant on small and medium islands. We counted six large colonies on Búmeran (0.9 ha) and five on Iguana (1.4 ha), in addition to numbers of incipient colonies (table 17.4). All remaining small and medium islands had at least one colony, and most had two or more. In contrast, a systematic survey of 20.6 ha on the large island, conducted by M. Rao in June 1995, revealed the presence of only two *Atta* colonies, a density of one per 10 ha. No *Atta* colonies were detected along a 2.5 km transect on the mainland. *Atta* densities on the small islands are thus elevated by at least an order of magnitude over those on larger landmasses. The large island and mainland supported colonies of both major species of army ants, the swarm raider *Eciton burchelli* and the column raider *E. hamatum.* Neither species was found on any of the small or medium islands. Tarantulas (counted at night when they were active) were abundant on several of the small islands, but were not conspicuously abundant on larger landmasses.

Fecal Counts

Both iguanas and capybaras frequent edges around Lago Guri. Iguanas bask in trees at the periphery of the forest, preferentially overhanging water, and capybaras feed on the grasses and forbs that flourish on exposed banks during dry periods when the lake level is drawn down for power generation. Judging from the numbers of feces encountered in circumferential transects of the islands, iguana numbers are high on small islands, where we found a mean of 4.5 feces/1,000 m^2, and low on the mainland, where the corresponding value was only 0.4 (table 17.5).

During the February–March period in which the measurements were conducted, counts of capybara fecal deposits suggested that most Guri capybaras were using the large island or mainland. The water level in the impoundment is high at this time of year, and capybaras appeared largely to have withdrawn from the small islands. Unfortunately, we do not have a parallel set of counts from June and July, when the drawdown reaches its lowest level and when the exposed banks of the impoundment are flush with grasses and forbs. Capybara sign is then plentiful on most small and medium islands, suggesting that the animals undertake an annual migration to and from the smaller islands in response to the availability of fresh forage.

Artificial Nest Experiments

To test the possibility that high rates of nest predation might be responsible for the low densities of resident birds recorded on some islands, we set out artificial nests stocked with *Coturnix* eggs, as described above. We found that the eggs in most nests survived the one-week experiment on the mainland and on all islands, with the single exception of Lomo, where 100% of the nests were raided (table 17.6).

DISCUSSION

Previous work on land bridge islands has focused principally on assessing the dynamics of species loss through measurements or reconstructions of faunal composition before and after isolation (Diamond 1972, 1984; Terborgh 1974, 1975; Wilcox 1978; Heaney 1984). Generally, the mechanism(s) of species loss have been treated as a black box in quantitative models (Diamond 1972; Terborgh 1974). We suspect that gross ecological distortions that quickly develop on small land bridge islands, and more slowly on larger ones, provide the driving mechanisms in species loss, but these mechanisms have scarcely been touched upon in previous work (but see Davidson, Samson, and Inouye 1985; Brown et al. 1986).

Table 17.3 Presence/absence of some nonflying vertebrates and selected invertebrates on Lago Guri islands and mainland.

Taxon	Landmass[a]												
	Ml[b]	DM	Cor	Lom	Pan	Búm	Pal	Per	Roc	Col	Igu	Mie	Tri
Mammals													
Edentates													
Dasypus novemcinctus	p	+	+	+	+						+		
Tamandua tetradactyl[a]	p	+											
Primates													
Alouatta seniculus	+	+		+	+					+	+	+	
Cebus olivaceus	+	+		+									
Chiropotes satanus	p	+											
Carnivores													
Eyra barbara	p	+											
Felid sp.	+	+	+	+	+		+						
Ungulates													
Mazama americana	+	+	+					+					
Tapirus terrestris	+	+			+							+	
Tayassu tajacu	+	+											
Rodents													
Agouti paca		+		+	+								
Dasyprocta punctata	+	+	+	+	+	+					+		+
Coendou prehensilis	p	+		+					+				
Hydrochaeris hydrochaeris	+	+	+	+	+	+	+	+	+	+	+		+

Small rodent spp.	+	+	+		+	+	+	+	+	+	+		+
Reptiles													
Ameiva ameiva	p	+	+	+	+	+	+	+	+		+	+	+
Caiman crocodilus	+	+	+		+						+		+
Geochelone carbonaria	+	+		+	+	+	+		+	+	+	+	+
Gonatodes (vittatus)	+	+	+	+	+		+	+		+	+		
Iguana iguana	+	+	+	+	+	+	+	+	+	+	+	+	+
Microteiid sp.	+	+	+	+	+	+	+	+	+	+		+	+
Snake sp.	+	+	+	+	+	+	+	+	+			+	
Thecadactylus sp.	+	+				+			+	+	+		+
Amphibians				+									
Bufo marinus	+	+		+	+	+		+			+	+	
Dendrobates leucomelas	+	+	+	+	+	+		+	+	+	+	+	+
Invertebrates													
Atta sp.	+	+	+	+	+	+	+	+	+	+	+	+	+
Eciton burcheli	+												
Eciton hamatum	+	+	+										
Tarantula spp.	+	+	+	+	+	+		+	+	+	+	+	+

a. Landmasses: Ml = mainland; DM = Danto Machado (the large island); Cor = Corral; Lom = Lomo; Pan = Panorama; Búm = Búmeran; Pal = Palizada; Per = Perímetro; Roc = Rocas; Col = Colon: Igu = Iguana; Mie = Miedo; Tri = Triángulo.

b. p = Probably on mainland, though not detected.

Table 17.4 Counts of selected species on Lago Guri islands and mainland; data normalized to 1 ha.

	Island class[a]												
	Small, far				Small, near				Medium			Lg	Ml
Taxon	Col	Igu	Mie	Tri	Búm	Pal	Per	Roc	Cor	Lom	Pan	DM	TF
Mammals													
Dasypus	0	+[b]	0	0	0	0	1	0	2	1	1	1	+
Rodent, small	2	1	0	1	4	6	6	+	2	1	7	1	1
Reptiles													
Amieva ameiva	0	5	24	6	11	3	5	8	1	0	4	2	1
Geochelone carbonaria	0	1	0	1	1	0	1	2	0	2	2	1	0
Amphibians													
Bufo marinus	2	0	1	0	1	0	2	0	0	+	0	1	0
Dendrobates leucomelas	10	3	4	6	11	0	0	7	1	10	3	2	0
Invertebrates													
Atta sp. (colonies)	3	5	1	0	6	1	1	2	2	2	1	0	0
Tarantula spp.	8	3	20	9	2	0	6	7	5	9	10	9	6

a. See table 17.3 for key to landmass abbreviations. b. + = recorded on landmass, but not on formal census.

Table 17.5 Transect counts of capybara and iguara feces along shorelines of Lago Guri islands and mainland.

Locality	Area censused (m ^{2})	No. capybara feces/1,000 m ^{2}	No. iguana feces/1,000 m ^{2}
Small, far islands			
Colon	3,000	0.0	16.7
Iguana	4,500	0.0	5.3
Miedo	2,910	0.0	3.1
Triángulo	3,990	0.8	2.3
Small, near islands			
Búmeran	5,740	1.0	1.6
Palizada	5,180	1.0	0.2
Perímetro	6,880	0.9	1.9
Rocas	2,960	0.0	5.1
Medium islands			
Corral (near)	10,000	0.7	0.9
Lomo (intermediate)	10,000	0.8	0.0
Panorama (far)	5,000	0.0	0.6
Large island			
Danto Machado	10,000	2.6	2.8
Mainland			
Terra firme	10,000	2.2	0.4

Table 17.6 Artificial nest assays, Lago Guri, Venezuela, February–March, 1994.

Locality	No. nests	No. destroyed	Percentage destroyed
Small, far islands			
Colon	7	1	14
Miedo	7	0	0
Small, near islands			
Búmeran	7	0	0
Perímetro	7	0	0
Rocas	7	0	0
Medium islands			
Corral (near)	14	2	14
Lomo (intermediate)	8	8	100
Panorama (far)	14	2	14
Large island			
Danto Machado	14	2	14
Mainland			
Terra firme	14	4	29

The types of ecological distortions that develop in forest fragments embedded in a mainland habitat matrix and on true land bridge islands may differ. In North American forest fragments, for example, high rates of nest predation due to raccoons, opossums, and other "mesopredators" and nest parasitism due to cowbirds have been attributed to elevated populations of nest predators and parasites in the surrounding matrix (Wilcove 1985; Wilcove, McClennan, and Dobson 1986; Soulé et al. 1988; Robinson et al. 1995).

("Mesopredator" is a term coined by Soulé et al. [1988] to describe a guild of medium-sized omnivorous mammals that, when abnormally abundant, can adversely affect populations of birds and other small vertebrates.) Mesopredators and cowbirds are habitat generalists that move freely between matrix and fragments. Such freedom of movement is severely curtailed in the case of land bridge islands.

Nevertheless, land bridge islands also exhibit ecological pathology, as demonstrated by the case of Barro Colorado Island, Panama, where extinction of forest bird populations has been associated with elevated levels of nest predation (Morton 1978; Karr 1982a,b; Loiselle and Hoppes 1983; Sieving and Karr, chap. 11). Recently, investigation of much smaller (ca. 1 ha) land bridge islands in Lake Gatun has revealed the presence of drastically simplified tree communities, in which one or two species dominate (Putz, Leigh, and Wright 1990; Leigh et al. 1993). The extinction of most tree diversity on Lake Gatun islets in eighty years (roughly one tree generation) suggests dramatically altered recruitment dynamics after isolation.

Our faunal surveys of Lago Guri islands indicate that a large fraction of the vertebrate species found on the nearby mainland quickly vanished from the small and medium islands, if indeed they were present at all subsequent to isolation. Nevertheless, virtually all species except top predators remain on the large island, despite its relatively modest size (350 ha). If, as a working hypothesis, one adopts the axiom that what happens quickly on small islands happens more slowly on larger islands, then one can regard the small islands as offering a preview of the future as it may unfold on larger landmasses. We have therefore directed our attention to the small islands, where ecological distortions of various kinds are especially evident.

In the absence of much of the regional fauna, the densities of many of the species surviving on small Lago Guri islets are conspicuously elevated above those observed on the nearby mainland. It has been customary to refer to this phenomenon as "ecological release" (Crowell 1962; MacArthur, Diamond, and Karr 1972; Cody 1983; Faeth 1984). Most commonly discussed in the context of exploitation competition, ecological release, as it occurs in Lago Guri, probably involves a number of distinct phenomena. Although densities certainly could have increased simply in response to a paucity of competitors, population increases might also have resulted from reduced predation or reduced parasitism. Whether competition, predation, or parasitism was the principal factor in a particular instance could be affirmed only through a detailed analysis of the individual case.

Birds

Bird densities varied radically between islands, from 0.8 to nearly 17 resident pairs per hectare, versus 4.9 on the mainland. However, small islands generally supported higher avian densities than the mainland control site, an observation that echoes previous island studies (Crowell 1962; MacArthur, Diamond, and Karr 1972). In contrast, avian densities consistently declined after isolation of experimentally created 1 and 10 ha forest fragments in the Brazilian Amazon, suggesting that the processes regulating avian densities on islands may differ from those operating in habitat fragments (Bierregaard and Lovejoy 1989; Stouffer and Bierregaard 1995b; Bierregaard and Stouffer, chap. 10).

The causes of this high between-site variation in bird densities remain largely undiagnosed, with the possible exception of the extremely low density found on the 11 ha island, Lomo, the only medium Guri island known to support a mesopredator, the olive capuchin monkey. In 1993 there were at least four adult capuchins on Lomo, and quite possibly more. Olive capuchins on the mainland have been found to occupy home ranges in excess of 100 ha (Robinson 1986). On Lomo, where a group is confined to an area one-tenth as large, it is reasonable to expect the species' ecological impact to be proportionately amplified. Artificial nests set out on Lomo were raided 100% of the time, while no more than 14% were destroyed on the other islands, even those having anomalously low bird densities (e.g., Corral). In other island studies, depressed and elevated bird densities have been correlated with abnormally high and low pressure of nest predators, respectively (Møller 1983; George 1987; Savidge 1987). The varying avian densities recorded here, along with the results of artificial nest assays, suggest that the Guri islands may differ in nest predator pressure and possibly in other factors that regulate bird densities.

Generalist Herbivores

Small Lago Guri islands consistently supported four or five generalist herbivore species, frequently at elevated densities. Four of these were vertebrates (the arboreal howler monkey and iguana, and the terrestrial capybara and tortoise) and one was an invertebrate (the leaf-cutter ant).

One or more howler monkeys were still present in 1995 on three of the four small, far islands, but on none of the small, near islands. It is possible that howler monkeys were present on most or all of these islands when the lake level was rising in the mid-1980s. In the case of the near islands, we suspect that they simply swam off to the nearest larger landmass. The only small, far island not to retain howlers is occasionally connected to other islands when the lake is drawn down more than 15 m. Any howlers initially trapped on this island could have walked off. Howlers currently remain only where they could not escape, namely, on far islands that are isolated by deep channels. Where they do remain, their densities are very high, the equivalent of 100–300/km^2. Howler monkeys persisted in an experimentally created 10 ha forest fragment in Brazil at slightly higher densities than in intact forest, perhaps benefiting from the increased net productivity typical of pioneer vegetation, or from the loss of other competing large frugivorous primates (Lovejoy et al. 1984). Barro Colorado Island in Panama, an old land bridge island, supports approximately 100 howlers/km^2 (Milton 1982). Mainland densities are typically lower, often in the range of 15–30/km^2 (Emmons 1984; Crockett and Eisenberg 1986).

Leaf-cutter ants provided the most spectacular case of "ecological release," exhibiting an elevated abundance on small islands, where potential competitors, predators, and parasites may have been absent. All eight small Guri islands supported leaf-cutter ant densities at least an order of magnitude higher than those at control sites. In a parallel case, four new colonies of these ants became established within three years of isolation in an experimentally created 10 ha forest fragment in Brazil (Lovejoy et al. 1984).

Iguanas and tortoises also exhibit hyperabundance on several islands. What the combination of as many as five hyperabundant generalist folivores may be doing to the vegetation of small Guri islands is an intriguing, unanswered question, but clearly a gross ecological imbalance is implied if the mainland is taken as the frame of reference.

Seed Predators

The vertebrate seed predator guild at Lago Guri consists of small rodents and parrots (on all islands), agouti and paca (on the medium and large islands and mainland), peccary, brocket deer, and bearded saki monkey (on the large island and mainland). Squirrels are strangely absent. Agoutis are particularly significant in this guild because they serve both as seed predators and seed dispersers (Smythe 1986).

Certain trees, such as *Hymenaea courbaril* (Leguminosae), appear to be wholly dependent on agoutis for dispersal (Hallwachs 1986). *H. courbaril* seedlings and saplings are extremely abundant on the two medium islands where there are agoutis (Panorama and Lomo), but uncommon and of patchy occurrence on the large island and mainland (J. Terborgh, pers. obs.). Seedlings of another large-seeded legume, *Copaifera pubiflora* (Leguminosae), are also abundant on Panorama and Lomo. Are these findings early indications of an eventual substitution of a small number of large-seeded tree species for a highly diverse mixed forest, as has been found on eighty-year-old 1 ha islets in Lake Gatun, Panama (Putz, Leigh, and Wright 1990; Leigh et al. 1993)? At this point it is too early to say, but these observations are suggestive.

Micropredators

Following Soulé et al. (1988), we shall apply the term "micropredators" to a set of small predatory creatures that have become hyperabundant on several of the small and medium Lago Guri islands. These micropredators include a mammal (armadillo), a lizard *(Ameiva ameiva),* a toad *(Bufo marinus),* a poison-arrow frog *(Dendrobates leucomelas),* and two species of tarantulas. All of them forage for arthropods or small vertebrates in leaf litter. One can only speculate as to why these particular species have survived and prospered on the Guri islands. One potential factor is that army ants (*Eciton* spp.) are absent on the islands where these micropredators have become hyperabundant. *Eciton* spp. were quick to disappear from experimentally created 1 and 10 ha forest fragments in Brazil (Lovejoy et al. 1984). Army ants are major predators of leaf litter arthropods, so in their absence other micropredators may increase in response to augmented food resources.

However, there are probably other factors at work as well. Armadillos, a major prey item of jaguar in Belize, may profit from the absence of big cats (Rabinowitz and Nottingham 1986; Fonseca and Robinson 1990). On large islands and the mainland, *Ameiva ameiva* is restricted to sunny forest edges, but on some small islands where the species is particularly abundant, individuals regularly forage in the shaded forest understory. A paucity of predators or a superabundance of prey could be contributory factors here. The persistence of *Bufo marinus* and *Dendrobates leucomelas* on small islets must, in part, be attributable to peculiarities of their reproductive biology. *Bufo marinus* has poisonous tadpoles and is capable of breeding in lakes with piraña and other fish. In contrast, the eggs and early larvae of *Dendrobates* spp. are carried on the backs of males and later develop in tank bromeliads, so that reproduction is independent of standing water. Other amphibians appear to be excluded by an absence of suitable breeding sites, as pools and streams do not occur on most Guri islands. As for the tarantulas, their success on Guri islets could be due to increased densities of leaf litter arthropods or to the absence of the tarantula hawk (*Pepsis* sp., Pompilidae), a giant wasp that is a specialist predator on adult tarantulas. These possibilities await further testing.

Small Island Faunas: Chance or Determinism?

Some previous authors who have studied the faunas of land bridge islands and forest fragments have documented nested sets, implying highly deterministic patterns of species loss (Diamond 1975a; Case 1983; Bolger, Alberts, and Soulé 1991; Warburton, chap. 13). There is evidence for nested sets in the Guri islands, too, in that the species that occur on small islands are also generally found on medium islands, and those that are found on medium islands also occur on the large island and mainland. What is noteworthy to us about the Guri situation, however, is the great variation that we see between islands, both in the densities of their bird faunas and in patterns of presence and absence of nonavian vertebrates on islands of a given size class.

Complex patterns of presence and absence on islands of a given size class can result from a number of causes, which we shall briefly review.

1. *Accidents of sampling.* Today's islands are the tops of erstwhile hills. Some are steep-sided, while others are gentle, and some are connected to other islands by shallow saddles, while others are isolated by deep channels. Consequently, the degree of area contraction experienced as the water level rose must have varied greatly between islands (Diamond 1972). The initial post-isolation faunas of different islands must therefore have varied in accordance with accidents of sampling and the degree of isolation and contraction in area. Larger land bridge islands, such as those near New Guinea studied by Diamond (1972), can be expected to show less variation in faunal composition because larger areas will initially contain a greater fraction of the total available species pool. Subsequent "relaxation" from more similar starting conditions will therefore be more likely to generate nested sets among the residual faunas.
2. *Residual effects of concentration.* By now, eleven years after isolation, most species will have succeeded or failed on their own merits, erasing some of the sampling variance generated during the contraction phase. The hyperabundant small island populations of some long-lived species (e.g., the howler monkey and the tortoise *Geochelone carbonaria*) might, however, still be a vestige of initial concentration.
3. *Voluntary escape.* Some species that might have been able to persist on certain islands may simply have departed under their own power. Birds, bats, and flying insects are the most obvious candidates, but circumstantial evidence mentioned above suggests that howler monkeys may have swum away from small, near islands. Evidence cited above also suggests that capybaras migrate annually from large, near islands and the mainland to small, far islands in response to the availability of forage. Another observation that suggests volun-

tary escape is that small, far islands appear to harbor depauperate butterfly communities (G. Shahabuddin, pers. comm.). It is possible that freshly emerged adults disperse, or are blown away, from small islands, quickly depleting the island populations.

4. *Involuntary entrapment.* Just as howler monkeys are absent from small, near islands and from small islands that are occasionally interconnected to other islands at low water, howlers have persisted for eleven years on small, far islands isolated by deep channels. These are clearly doomed populations, and it is astonishing that they have survived so long. The capuchins on Lomo are another case in point. A further example was uncovered in 1990 when we visited a roughly 20 ha island that is one of the most remote in the lake, located more than 4 km from any other landmass. It was occupied by a large herd of capybaras. The ground was littered with capybara dung, and large patches of bare earth, free of any vegetation, attested to the intense grazing pressure exerted by the apparently trapped herd.
5. *Recurrent colonization.* Flying vertebrates and invertebrates (and possibly some nonflying mammals) are capable of recolonizing islands after populations have gone extinct (Brown and Kodric-Brown 1977). In our bird censuses, we noted many species that we classified as transients because they appeared only once on a given island. Given that about 90% of the bird populations of small Guri islands consisted of a single pair, it is highly unlikely that these populations are self-perpetuating. Recolonizations by birds, at least, must therefore be commonplace.
6. *Short-term visitation.* As just mentioned, transient birds are frequent, even on the most remote Guri islands, and some nonflying mammals (e.g., tapir, ocelot) have been found to swim between islands at least occasionally. Capybaras appear to move on an annual schedule between large and small landmasses. Short-term visitation thus increases both within- and between-island variance in faunal composition.
7. *Interisland differences in habitat quality.* No two 1 ha samples of a highly diverse tropical forest will contain identical tree communities. The habitat available on small Guri islets thus inevitably varies, even though the islands in the study were selected to minimize habitat heterogeneity. How important, then, is habitat variation? One can point to the similarities and consistencies in the faunas of 1 ha islets and claim that habitat variation is of no consequence, or one can point to the differences and claim that habitat variation is the cause. Given that points 1 through 6 above provide alternative explanations for interisland faunal differences, our feeling is that habitat differences exert only a minor influence.

Patterns of Extinction and Survival

It should now be clear that which species persist on a given island, and which do not, depends on a great many factors (see Lynam, chap. 15), some of which are more stochastic (accidents of sampling) and some of which are more deterministic (insufficient area or inappropriate habitat; voluntary departure). A pattern that clearly emerges above the noise level, however, is that species loss from land bridge islands is nonrandom with respect to trophic role, so that, other things being equal (e.g., body size), predators are lost before herbivores and parasites before hosts. We suspect, in general, that it is the selective loss of predators and parasites that has led to the conspicuous hyperabundance of so many of the species that have persisted on small Guri islets (Terborgh 1988). Their hyperabundance, in itself, may then create further ecological distortions that subsequently result in a cascade of secondary extinctions. At least this will be our working hypothesis as we pursue more detailed studies of the bizarre ecology of small Lago Guri islands.

CONCLUSIONS

The Venezuelan energy company EDELCA has created a billion-dollar laboratory that provides an unrivalled opportunity for conducting measurements and experiments that can offer powerful insights into the workings of ecological systems. The collapsing ecosystems of land bridge islands harbor ephemeral assemblages of plant and animal species that would not otherwise occur in nature. These transitory assemblages are strongly imbalanced from a functional standpoint. They contain too few predators and too many herbivores, not enough seed dispersers and too many seed predators.

After an initial bout of extinctions directly associated with contraction in area, we anticipate that ecological imbalances will lead to a cascade of secondary extinctions, and that most of these extinctions will occur within the first tree generation subsequent to isolation, when altered recruitment dynamics will lead to forests dominated by just one or two tree species, as has already occurred on small islets in Lake Gatun, Panama. The conditions we are observing now in Lago Guri are dynamic and transitional. One distortion—reduction in area—has led to others, and those secondary distortions are bound to generate still more. In the end, diversity will be the loser.

GENERAL IMPLICATIONS

1. Habitat fragmentation of a kind that impedes the normal movements of animals through the landscape results in serious ecological imbalances.
2. Imbalances arise because certain trophic guilds (predators, parasites) are more sensitive to fragmentation than others (generalist herbivores) and selectively disappear, while other species or guilds persist.
3. The reduction or loss of key species and/or ecological functions, such as predation, leads to "ecological release" of the surviving species, resulting in hyperabundance.
4. Hyperabundance of ecologically important species, in turn, can result in further ecological distortions.
5. We deduce that such ecological distortions drive a process of species loss, the end point of which is a greatly simplified ecological system lacking much of the initial diversity.
6. Insidious from the point of view of public perception is that ecological imbalances and their consequences develop slowly, especially on larger landmasses. In the absence of monitoring, the appearance will always be that of normality.
7. Our findings reaffirm that large protected areas are needed to maintain normal ecological balances, and therefore to conserve biodiversity.

ACKNOWLEDGMENTS

We dedicate this chapter to the memory of Warren Kinzey, whose unstinting generosity and encouragement were indispensable to establishing our program in the Lago Guri islands. We are also profoundly grateful to Blgo. Luis Balbas of EDELCA for generous support with myriad logistic and administrative details. We thank EDELCA, CONICIT, and PROFAUNA for permits necessary to conduct the research in the Guri impoundment. Marilyn Norconk has kindly shared many logistic facilities. Clemencia Rodner, Ana Rita Bruni, and Viviana Salas of Audubon de Venezuela are warmly thanked for generous hospitality and vital assistance of many kinds. Finally, we thank the MacArthur Foundation for essential financial support, and Phil Stouffer, Rob Bierregaard, and Bill Laurance for comments on an earlier draft of the chapter.

Section IV
Plants and Plant-Animal Interactions

INTRODUCTION
Fragmentation and Plant Communities

PLANTS define the physical structure of forests, and in the humid Tropics, forests can achieve a structure of unparalleled complexity. In rainforests this complexity results from a rough vertical stratification of vegetation layers and a remarkable profusion of life forms—towering canopy trees, smaller treelets, shrubs, forbs, vines, lianas, epiphytes, stranglers, and even parasitic saprophytes.

Floristically, perhaps the most striking feature of wet tropical forests is their extraordinary diversity of species. It is not uncommon for botanists to encounter more than three hundred species of woody plants within a single hectare of tropical rainforest. A second notable feature is that many species are rare—often occurring at densities of less than one individual per hectare. Further, in many regions a sizable fraction of the flora appears to have highly restricted geographic ranges, and these local endemics may be especially vulnerable to forest loss and fragmentation.

Of course, plants do more than simply define the structure of the forest. In the Tropics more than anywhere else, plants are involved in a rich array of mutualistic and antagonistic interactions with animals, fungi, and microorganisms. The majority of rainforest canopy trees, for example, require animals (usually insects) for cross-pollination, and they generally attract a different suite of animals for seed dispersal. A surprising number of tropical plants are defended by aggressive ants, in return for which the ants receive both food and shelter. Insect herbivores in the Tropics are often specialized upon a very small number of host plant species. These myriad relationships form the basis of the remarkably complex ecological webs that are perhaps the most distinctive feature of wet tropical ecosystems.

Given the central role of plants in forest structure and functioning, it is perhaps surprising that they have been less actively studied than animals in fragmentation research. There are, however, some valid reasons for this. Many species, especially large trees, can have very long generation times and thus may respond slowly to fragmentation. Daniel Janzen coined the term "the living dead" to describe those trees that, though they may persist for centuries, are doomed to eventual extinction because their coevolved pollinators or seed dispersers have vanished. Plant ecologists also are faced with formidable taxonomic constraints: in the Tropics, plant diversity is so high that rarely more than a handful of scientists ever achieve a general knowledge of a region's flora. Finally, for most people, plants lack the charisma of large vertebrates, and hence are less likely to galvanize the public's interest in nature conservation.

The five chapters in this section encompass both plants and plant-animal interactions (in addition, chaps. 3, 6, 23, 24, and 25 are concerned with certain aspects of plant or vegetation ecology). As in earlier sections, the chapters herein are highly eclectic, spanning a range of biogeographic regions, taxa, and research questions.

Predicting Rates of Plant Extinctions

In chapter 18, Mark Andersen, Alan Thornhill, and Harold Koopowitz use a stochastic model based on a "distribution profile" of Neotropical plant species (i.e., a frequency distribution of the number of sites at which species in the region are found) to assess the effects of habitat loss on rare endemic plants. Their model also attempts to predict the extent to which deforestation will turn species with moderate-sized geographic ranges into rare endemics. The predictions were generated using a distribution profile for over 5,000 Neotropical plant species. The authors' innovative approach provides an alternative to species-area curves for predicting the pace of species extinctions; at current rates of deforestation, for example, the model predicts that 71–95 species of Neotropical plants are disappearing each year. This modeling approach has important implications for species conservation and reserve design strategies, especially for local endemics.

Trees and Seed Predators

In chapter 19, Graham Harrington and his colleagues describe a major field study on the interactions of rainforest trees and their seed predators and dispersers in northern Queensland. Because fragmentation disrupts many animal populations, they set out to compare rates of seed loss, dispersal, and seedling recruitment among thirteen large-seeded tree species in both fragmented and continuous forests. The seeds of several of these tree species were preyed upon by a rodent, the giant white-tailed rat *(Uromys caudimaculatus),* which decimated their seed populations, while the other species had far lower rates of seed loss. As expected, the vulnerable species had very low ratios of seedlings to adults, indicating that few seeds successfully germinated. Surprisingly, however, the authors detected no effect of fragmentation on seed predation or seedling to adult ratios, even though the rats had declined markedly in small fragments. They also found no significant differences between fragments and continuous forest in terms of populations of trees species likely to be dispersed by southern cassowaries *(Casuarius casuarius),* despite the fact that these massive ratite birds have disappeared from most fragments. Their results highlight the inherent difficulties of studying higher-order interactions in tropical forests, and suggest that the outcomes of such interactions may be extremely difficult to predict.

Genetics of Tropical Trees

In chapter 20, John Nason, Preston Aldrich, and James Hamrick review the processes that can affect the genetic fate of fragmented tropical tree populations. Their analysis demonstrates that genetic diversity within and among remnant populations is influenced by several factors: the number of individuals within each forest patch, the spatial scale of patches relative to preexisting genetic structure, and pollen and seed dispersal, which in the Tropics are commonly animal-mediated. This reliance on animal vectors means that changes in pollinator and frugivore assemblages in fragments could strongly affect patterns of gene flow and genetic variation within remnant tree populations.

An important conclusion of Nason and his colleagues is that, in continuous forests, the genetic neighborhoods of tropical trees may span areas of tens to hundreds of hectares, which is larger than many contemporary forest fragments. There may, nevertheless, be gene flow among trees in different forest fragments because of between-fragment movements by some pollinators and seed dispersers. This suggests that forest fragments will often be worthy of conservation because they contribute to the genetic connectivity of a larger metapopulation and act as "stepping stones" for pollinators and seed dispersers.

Recolonization of Matrix Habitats

Chapter 21, by Christophe Thébaud and Dominique Strasberg, examines the recolonization of matrix habitats by rainforest plants from small forest fragments. The study was conducted on La Réunion Island in the Indian Ocean, and in this case the matrix was composed of recent lava flows—an ideal experimental design because the matrix was of a uniform age and composition. Thébaud and Strasberg found that most species expanded slowly from fragment margins out into the matrix. Surprisingly, the rate of colonization was lower for fleshy-fruited species (which normally are dispersed by frugivores) than for wind-dispersed species, which is the reverse of the normal pattern. The explanation seems to lie in the dramatic extinction rate of vertebrate frugivores on La Reunion during the past three hundred years, due mainly to habitat destruction and introductions of exotic species. Apparently, fleshy-fruited species have been unable to adjust to the loss of their dispersers, and this could easily affect their prospects for long-term survival.

Long-Term Persistence in Forest Remnants

In chapter 22, Richard Corlett and Ian Turner compare the persistence of higher plants and vertebrates in Singapore and Hong Kong, islands that contain some of the oldest known anthropogenic forest fragments in the Tropics—from 150 to over 350 years old. They describe the history of deforestation on both islands, then demonstrate that while bird and mammal extinctions have been severe, there have been considerably fewer floral extinctions than expected. A rich forest flora has persisted even in small (< 50 ha) fragments, although some extinctions of plants and animals continue at present. These striking differences suggest that plants are either generally less vulnerable to fragmentation than vertebrates, or that they respond more slowly to habitat losses owing to the long-lived nature of some species.

SYNTHESIS

Although these chapters share only a few common themes, they highlight several issues of interest to fragmentation researchers. For example, taken as a whole, vascular plant populations do in fact appear less vulnerable to fragmentation than those of many vertebrates (chap. 22). In general, plants may be more resilient than vertebrates because they have smaller area requirements and because many species are long-lived and have many reproductive episodes. An alternative idea is that, given their long generation times, trees and long-lived plants simply take much longer than animals to disappear from fragments. There is little compelling evidence for this latter proposition at present, although information on the long-term dynamics of fragmented plant populations is scarce (see chap. 22 and references therein).

Despite the general resilience of plants, certain plant taxa or functional groups are probably strongly affected by fragmentation. Epiphytes and orchids, for example, seem unusually vulnerable (chap. 22), possibly because they are sensitive to microclimatic changes in fragments. Fragments are exceptionally prone to certain natural and anthropogenic disturbances, and such changes are likely to favor disturbance-adapted taxa at the expense of mature-phase species (see chaps. 6 and 23). Another possibility is that many rare species will ultimately disappear from fragments because their local populations are maintained by immigration from other source areas (chap. 20). In tropical forests, the loss of rare species could sharply erode the floristic diversity of fragments because many species are naturally rare.

Plants that have obligate mutualisms with vulnerable pollinators or seed dispersers may also be vulnerable to fragmentation (chaps. 19 and 20). There has, however, been relatively little work on plant-animal interactions in tropical forest remnants, despite much emphasis by researchers on the importance of coevolved relationships in the Tropics. In Singapore and Hong Kong, the disappearance of many forest-dependent vertebrates from fragments has not resulted in a correspondingly large loss of plant species, suggesting that few plants there require only a single or a few species of pollinators or seed dispersers (chap. 22). In tropical Queensland, the loss of cassowaries from fragments also seems to have had little effect on tree species known to be dispersed by this bird, at least over the short term (chap. 19).

However, as noted above, there may be considerable time lags between fragmentation and the loss of plant populations from remnants. On La Réunion Island, plants that require now-extinct vertebrate seed dispersers are much less effective at recolonizing matrix habitats than are wind-dispersed species (chap. 21). This mechanistic analysis suggests that a large-scale loss of frugivores could eventually lead to the collapse of dependent plant populations. Clearly, the long-term dynamics of fragmented populations require much further attention.

Another topic of general interest is the notion that regions that have been unstable over geological or ecological time scales are intrinsically more resilient to fragmentation because the most extinction-prone species will already have disappeared. Hong Kong, for example, which experiences frequent typhoons and periodic frosts, may be more resilient to forest disruption than climatically stable regions such as Singapore (chap. 22). Likewise, areas that experienced severe Pleistocene forest contractions may contain fewer vulnerable species than those that have remained largely intact (see chaps. 28, 30, and 33). If valid, this concept has important implications, because it suggests that conservation efforts should be focused on climatically or geologically stable regions, which will contain concentrations of extinction-prone species.

Finally, there is a growing consensus that small forest remnants can have important conservation value, especially in regions that have already been heavily degraded. Although they are very prone to edge effects (see section II), small fragments may have surprisingly high floristic diversity (chap. 22) and may be important sources of propagules and animal seed dispersers for forest regeneration in matrix habitats (chap. 21). In severely fragmented landscapes, small fragments can serve as "stepping stones" for faunal movements and thus help to maintain some degree of ecosystem connectivity (chap. 20). Small fragments can also harbor populations of local endemics (chap. 18), which may not be represented in larger reserves (see chap. 32). This line of thinking suggests that land managers should adopt a holistic approach to forest conservation, and not limit their attention to large forest tracts or the most pristine areas.

18

Tropical Forest Disruption and Stochastic Biodiversity Losses

Mark Andersen, Alan Thornhill, and Harold Koopowitz

DEFORESTATION in the humid Tropics is one of the major threats to global biodiversity (Park 1992; Whitmore and Sayer 1992). Two important phenomena exacerbate this threat. First, many areas with high rates of forest clearance also harbor many endemic plant and animal species. Second, many of the areas with the highest rates of deforestation are also those with the least forest remaining (Park 1992).

The species-area relationship has been widely used both to predict the effects of habitat loss on extinction rates (Reid 1992) and to guide reserve design recommendations (Shafer 1990). This approach has been criticized, however (Simberloff 1992), largely because the theory cannot predict which species are most likely to go extinct and tends to treat individual species as interchangeable units. We have recently proposed a promising alternative way of using biogeographic information to study and predict biodiversity losses (Koopowitz 1992; Koopowitz, Thornhill, and Andersen 1993, 1994; Koopowitz, Andersen et al. 1994).

Our objectives in this chapter are (1) to show how we use our particular biogeographic approach to assess potential biodiversity losses in different regions or taxa; (2) to examine the effects of different taxon-wide or regional distribution patterns on the sensitivity of a flora or taxon to habitat disruption; (3) to propose some possible biological explanations for different taxon-wide or regional distribution patterns; and (4) to formulate some specific conservation and research recommendations based on points (2) and (3) above.

BACKGROUND

The relative frequencies of narrow endemic species (i.e., species with highly restricted geographic ranges) form the core of our approach to the biogeography of tropical plant species. This idea originated with Wilson (1988) and was extended by Koopowitz (1992; see also Koopowitz, Thornhill, and Andersen 1993, 1994; Koopowitz, Andersen et al. 1994). Consider the frequency distribution, for a particular taxon or region, of the number of distinct localities at which each species in that taxon or region has been collected. We call this frequency distribution the *distribution profile* for that taxon or region. To compile a distribution profile for 5,136 species and subspecies of Neotropical plants from thirty-eight families, we recorded the number of collection localities reported for each

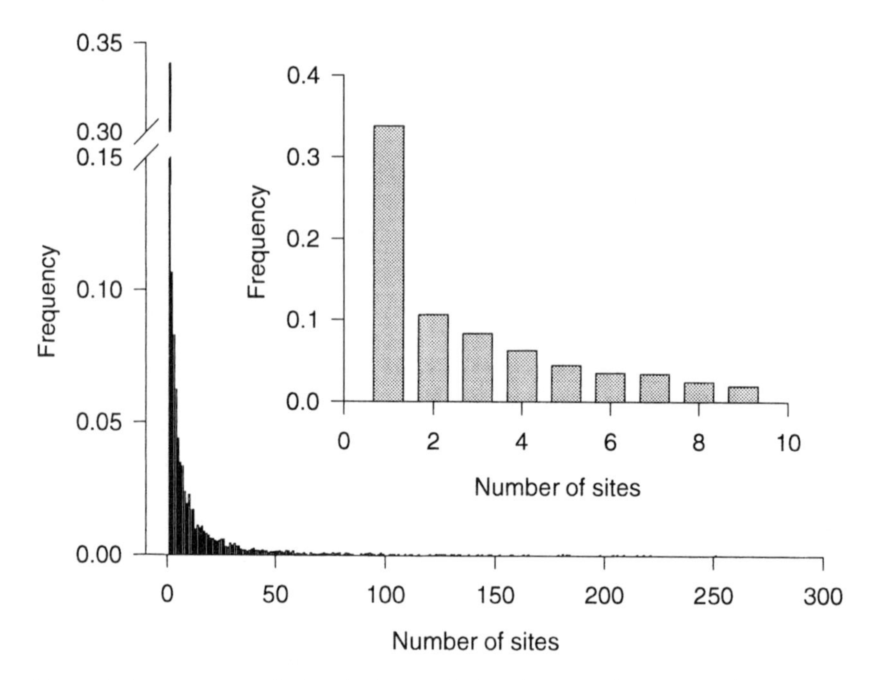

Figure 18.1. An example of a distribution profile (the frequency distribution of the number of sites at which each species is found) for 5,136 Neotropical plants. The inset shows a detailed view of the first nine categories of the distribution profile.

species in forty volumes of the *Flora Neotropica* series (volume citations given in Koopowitz, Thornhill, and Andersen 1994) (fig. 18.1). We analyzed a subset of this data set to obtain the results reported here. Of these species and subspecies, 33.8% are known from only a single locality, and 52.7% from three or fewer localities. Our underlying assumption, of course, is that species known from few localities have highly restricted distributions, while species recorded from a large number of localities have broad distributions.

Distribution profiles can be used to predict losses of biodiversity due to deforestation (or other forms of habitat conversion). To do so requires distribution profiles for the taxa or regions in question, estimates of rates of deforestation, and the total number of species in each taxon or region of interest. For a single episode of deforestation, the number of species going extinct will be

$$E_t = a\sum_{n=1}^{y} b_n c^n \tag{18.1}$$

where E_t is the number of species going extinct (at time t), a is the original total number of species in the taxon or region of interest, b_n is the proportion of species in the taxon or region found at n localities, c is the deforestation rate, and y is the maximum possible number of localities.

With a little additional work, we can estimate the effects of continuing deforestation. Specifically, after each episode of deforestation (assumed to occur annually: Koopowitz,

Thornhill, and Andersen 1994) the values of b_n and a will change. The value of a can be updated by the formula

$$(18.2)\qquad a_{t+1} = a_t - E_t,$$

and the b values will evolve according to the formula

$$(18.3)\qquad b_{n,t+1} = b_{n,t} - \sum_{x=1}^{n} b_{n,t}c^x + \sum_{z=n}^{y} b_{z+1,t}c^{(z-n+1)}.$$

Using this model, Koopowitz, Thornhill, and Andersen (1994) estimated that, at current deforestation rates (FAO 1991), the entire Neotropics loses between 71 and 95 plant species per year. This modeling approach, of course, contains a number of potentially troublesome assumptions (several of these are reviewed in Koopowitz, Thornhill, and Andersen 1994). We discuss several of our most crucial assumptions below.

Perhaps the most crucial assumption is that our estimates of Neotropical distribution profiles correspond to the real geographic ranges of plants in the field. Our database consists of a large number of species representing a wide variety of habitat preferences, dispersal mechanisms, pollination strategies, and growth forms. Nevertheless, our sampling of taxa may be biased simply because the authors of the taxonomic monographs to which we referred focused their attention on highly diverse taxa. However, in other data that we have compiled, distribution profile shapes are remarkably consistent across taxa and across regions. For example, a distribution profile for several East African orchid taxa is remarkably similar in shape to a worldwide orchid distribution profile that did not include the East African taxa (Koopowitz, Andersen et al. 1994). Even more remarkable, a distribution profile for tropical East African epiphytic orchids has a shape very similar to the distribution profile for Neotropical epiphytic bromeliads (Koopowitz, Andersen et al. 1994). Thus, if our distribution profiles are merely artifacts of the work habits of plant taxonomists, they are at least highly structured and consistent artifacts.

EXTINCTION PRONENESS AND THE SHAPE OF THE DISTRIBUTION PROFILE

The modeling and data analysis results presented here are based on a subset of taxa chosen using two criteria (table 18.1). First, the taxon had to include a reasonably large number of species (at least fifteen). Second, the taxon had to be somewhat uniform in its ecology and growth habits; for example, *Rinorea* spp. are explosively dispersed forest shrubs, *Costus* spp. are shade-loving forest interior herbs, and *Swartzia* spp. are canopy trees. In many cases, judgments of ecological uniformity were necessarily subjective.

Many of the distribution profiles we have examined have a highly skewed shape. Therefore, it is much easier to compare distribution profiles if they are plotted on a semilogarithmic scale (i.e., number of species vs. logarithm of the number of sites). If we choose $\log_2$, these plots amount to plotting the distribution profiles in "octaves" (as in Preston 1962). Preliminary studies show that our distribution profiles follow an approximately normal distribution on the octave scale (Dial, 1995).

Table 18.1 Plant taxa used in analyses and modeling.

Taxon	*Flora Neotropica* volume	No. of species
Brunellia	2	126
Pitcairnioideae (Bromeliaceae)	14 (I)	785
Bromelioideae (Bromeliaceae)	14 (III)	646
Swartzia	1	164
Trigonia	19	30
Tillandsioideae (Bromeliaceae)	14 (II)	926
Cavendishia	35	107
Lecythidaceae	21 (II)	66
Costus	8	54
Renealmia	18	57
Rhamnus	20	28
Myrceugenia	29	49
Carlowrightia	34	20
Parkia	43	17
Rinorea	46	51
Calceolaria	47	208
Hedyosmum	48	41
Sapotaceae	52	395
Cecropiaceae	51	139
Dimorphandra	44	89
Connaraceae	36	139
Chrysobalanaceae	9	400
Pilocarpinae (Rutaceae)	33	61
Lauraceae	31	54
Meliaceae	28	132
Bignoniaceae	25	17
Voyria and *Voyriella*	4	20
Krameria	49	17

Side-by-side comparisons of distributions are simpler still if we plot the distributions as boxplots rather than as histograms (Velleman and Hoaglin 1981). A boxplot shows the middle half of the data (that portion between the first and third quartiles) as a box. A line across the box shows the location of the median, and a notch on the box shows an approximate 95% confidence interval for the median. "Whiskers" extend from the box out to 1.5 times the interquartile range on both ends of the box. Points beyond the whiskers may be considered outliers, or at least extreme values.

The data for the twenty-eight taxa chosen (see table 18.1) are plotted in octaves, as described above (fig. 18.2). The taxa are plotted in order of increasing median $\log_2$ (number of sites). While no natural categories of distribution profiles suggest themselves from examination of figure 18.2, it is possible to impose artificial categories on these taxa. We recognize four basic groups of taxa:

Group 1: *Brunellia,* the Pitcairnioideae, and the Bromelioideae. Taxa in this group have a median $\log_2$(number of sites) equal to zero.

Group 2: *Swartzia, Trigonia, Cavendishia,* the Tillandsioideae, the Lecythidaceae, and *Calceolaria.* Taxa in this group have a median $\log_2$(number of sites) between zero and two.

Group 3: *Costus,* the Sapotaceae, the Bignoniaceae, *Renealmia, Hedyosmum, Dimorphandra,* the Pilocarpinae, the Lauraceae, the Cecropiaceae, the Connaraceae, the Meliaceae, *Rhamnus, Carlowrightia, Rinorea, Myrceugenia,* and *Parkia.* Taxa in this group have a median $\log_2$(number of sites) between two and four.

Group 4: *Voyria, Voyriella,* and *Krameria.* Taxa in this group have a median $\log_2$ (number of sites) greater than four.

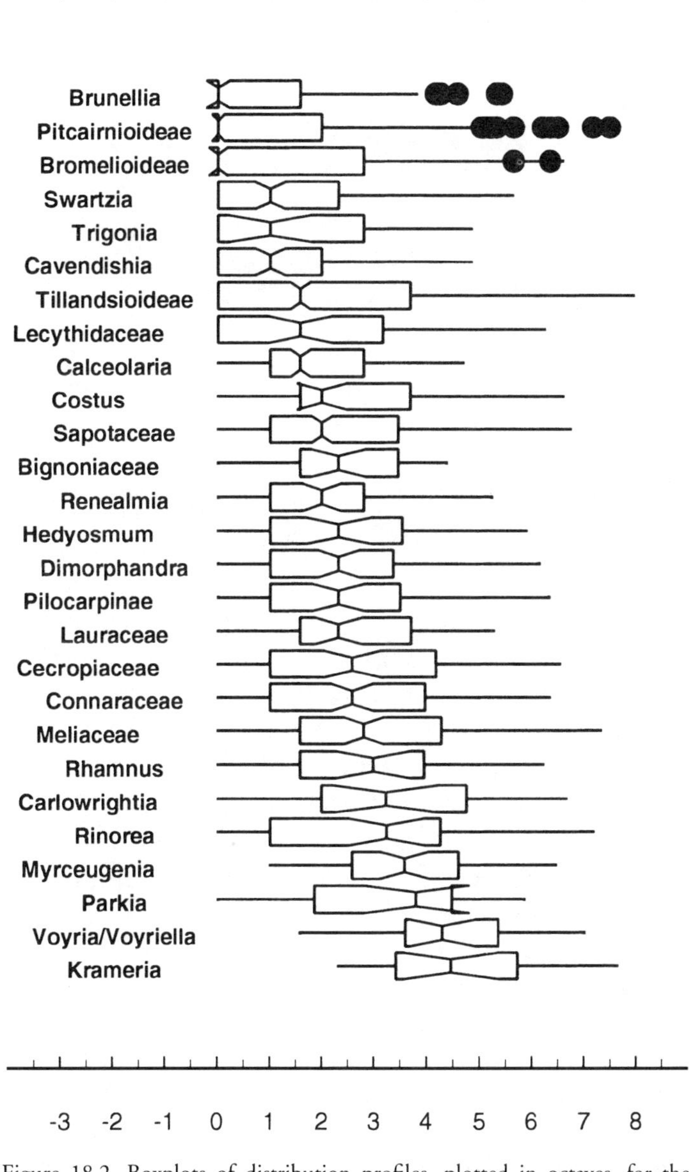

Figure 18.2. Boxplots of distribution profiles, plotted in octaves, for the twenty-eight taxa listed in table 18.1. Filled circles indicate outhers.

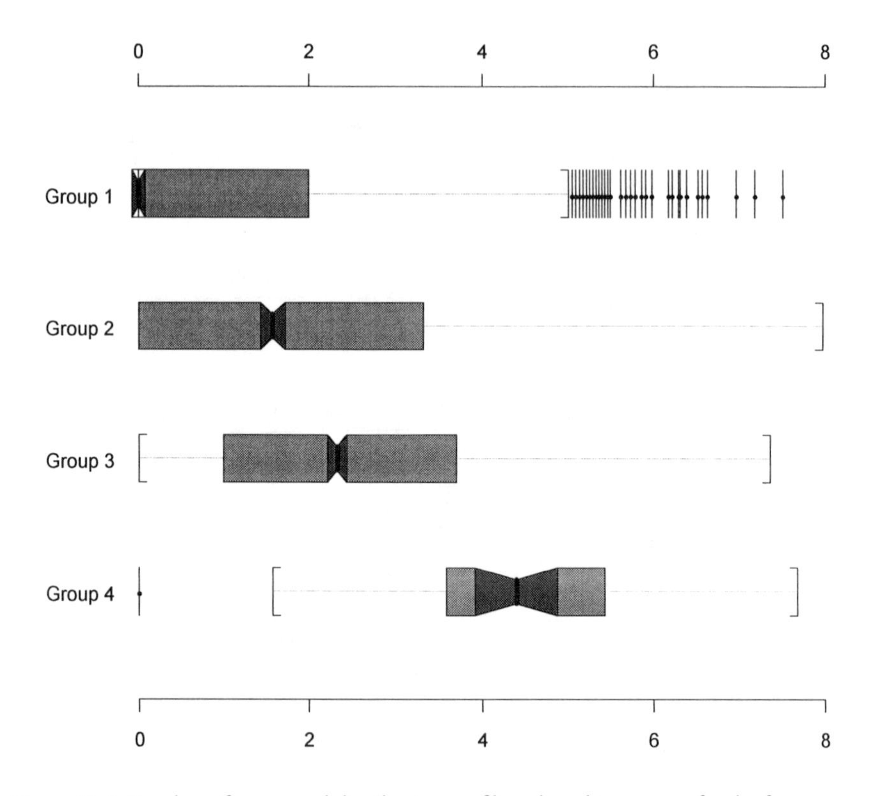

Figure 18.3. Boxplots of aggregated distribution profiles, plotted in octaves, for the four groups of species.

These groups, although admittedly artificial, are at least distinct. Note that the confidence intervals about the medians of the four groups do not overlap (fig. 18.3).

One might reason that taxa in group 1 would be particularly likely to lose species due to deforestation and that those in group 4 would be especially resistant to such losses. One might also expect that the shape of the distribution profile would influence the rate at which species' geographic ranges become restricted as a result of deforestation (i.e., the rate at which the number of sites where species are found decreases as those sites become deforested). Below we present some simulation results that test these hypotheses.

We generated four distribution profiles, using the means and standard deviations of the actual distribution profiles for the four groups shown in figure 18.3. First, we generated a normal distribution with the desired mean and standard deviation, and then transformed it back to the original scale of sites rather than octaves. (Recall that the distribution profiles are approximately normally distributed on a $\log_2$ scale). We refer to these distribution profiles as type 1 through type 4, corresponding to the observed distribution profiles in groups 1 through 4. We then simulated continuing deforestation by applying the formulas of equations (18.2) and (18.3) to these simulated distribution profiles, and at each time step kept track of the total (i.e., cumulative) number of species lost to extinction, as well as the distribution profiles themselves.

One can make two assumptions about the deforestation process in such simulations. Specifically, one can assume either that a constant fraction of the remaining forest is cut each year, or that a constant area is cut each year. If a constant absolute amount of the remaining forest is cut each year, the fraction cut each year increases steadily. We ran our simulations under both assumptions.

In our simulations, we used a value of $c = 0.005$ (corresponding to 0.5% of extant forest cut each year under the "constant fraction" assumption discussed above). This estimate comes from analyses (described in Koopowitz, Thornhill, and Andersen 1994) of the regional and national deforestation data published by the United Nations Food and Agriculture Organization (FAO 1991). Note that this is a conservative estimate of global deforestation; Whitmore (chap. 1) gives a figure of 0.8% per year. Thus, our predictions of biodiversity loss may be conservative.

The cumulative number of extinctions over time for simulations using both assumptions varies among distribution profile types (fig. 18.4). Under both assumptions, type 1 loses species to extinction most rapidly while type 4 loses species most slowly. This finding confirms the first hypothesis presented above—that is, that the sensitivity of taxa to species loss due to deforestation increases from group 1 to group 4.

The fraction of species found at only one site (i.e., $b_{1,t}$) changes over two hundred years of simulated deforestation (fig. 18.5). Although the $b_{1,t}$ category of the type 1 distribution profile seems to grow a bit more rapidly than the others, the difference is small. In general, though, it appears that the proportions of species in a particular distribution profile category changed at approximately equal rates for the four groups. Thus, we conclude that the shape of the distribution profile does not greatly influence the rate at which species' ranges become restricted due to deforestation.

DISCUSSION

We have shown that taxa or regions characterized by many narrow endemic species are much more prone to species losses due to deforestation than are taxa or regions with more flattened distribution profiles. We have also shown that the shape of the distribution profile does not influence the rate at which species' ranges become restricted due to deforestation. This finding is puzzling, since it seems that both the extinction rate and the rate of range restriction should differ for the four groups. We conjecture that the difference in extinction rates is due to disproportionately large numbers of extinctions among narrow endemic species in the highly skewed distribution profiles (groups 1 and 2). In the flatter distribution profiles (groups 3 and 4), however, there are proportionately more species in the tail of the distribution (i.e., found at more sites) initially; as these species undergo range restriction, the initial differences between the shapes of the distribution profiles decrease. Therefore, the difference in rates of extinction is not reflected in a difference in rates of range restriction.

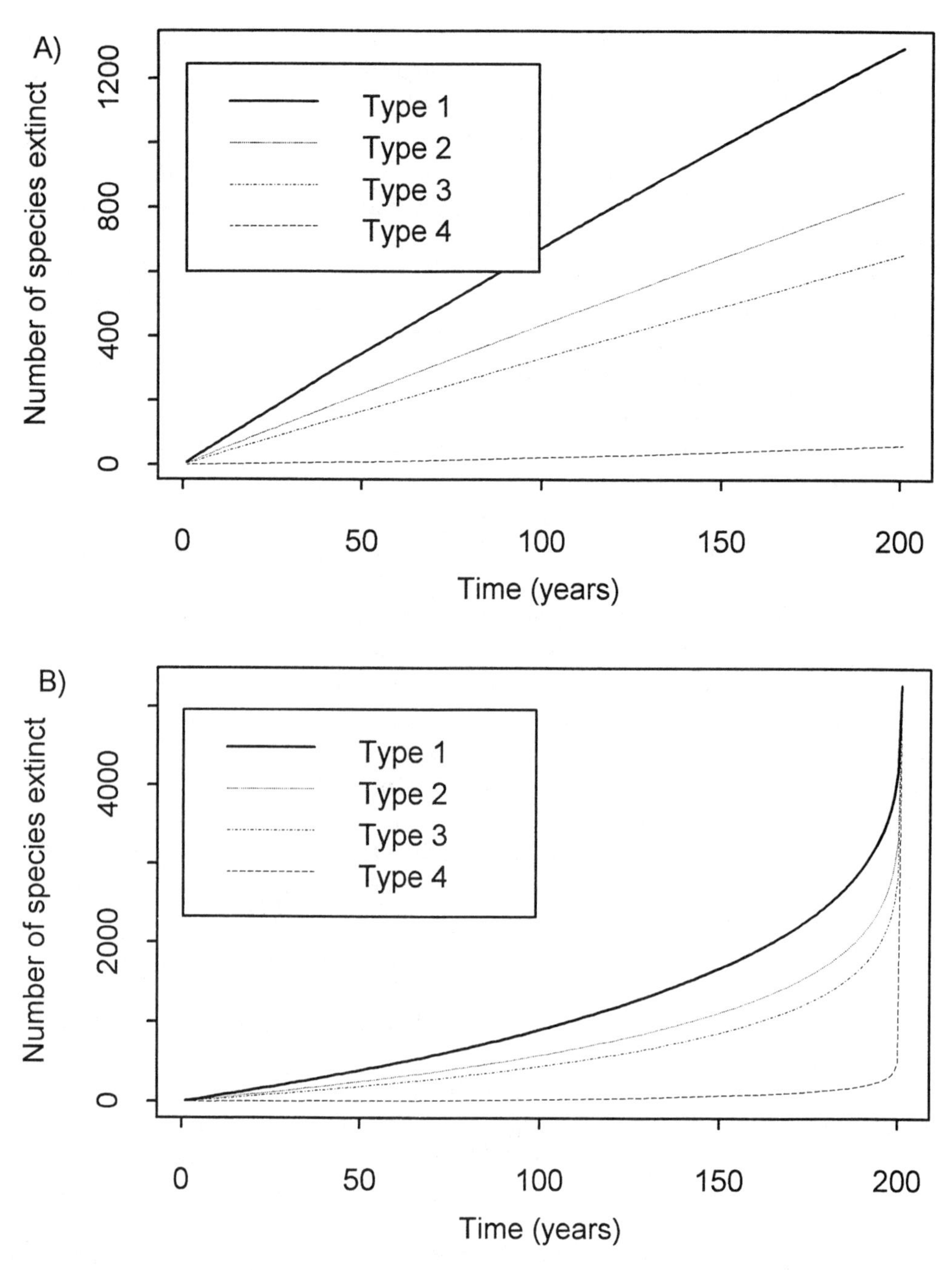

Figure 18.4. Cumulative number of species going extinct as a function of time: *(A)* assuming that a constant fraction of the forest is cut each year; *(B)* assuming that a constant area of forest is cut each year.

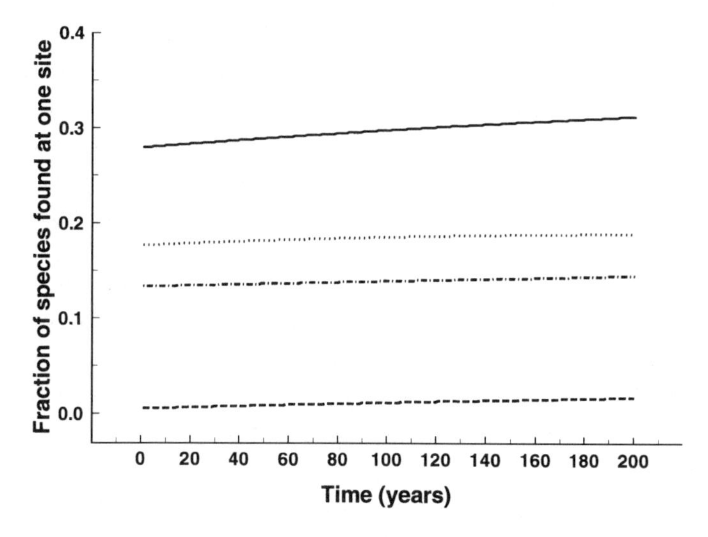

Figure 18.5. Fraction of the total number of species in a taxon or region found at only one site as a function of time. In the simulation shown in this graph, we assumed that a constant fraction of the remaining forest was being cut each year. Results for the "constant area cut" assumption were similar. Distribution types are as in figure 18.4.

Assumptions of the Model

These results, of course, are not without their limitations. We have assumed that (1) species are distributed randomly with respect to forest conversion, (2) the spatial pattern of deforestation is random relative to the distribution of forest types, and (3) the pattern of forest conversion is independent of the pattern of species occurrence. If any of these assumptions are violated, expected numbers of extinctions may be higher than reported here. In addition, if forest types dominated by widespread species are preferentially cut, actual extinction rates may be lower than our estimates. In fact, we believe that the spatial pattern of deforestation is often quite nonrandom (Whitmore, chap. 1; Smith, chap. 27) and is usually intimately related to patterns of land use and ownership as well as landscape patterns (Dale and Pearson, chap. 26).

We also have assumed that (1) deforestation is equivalent to sterilization (but see Corlett and Turner, chap. 22; Nason, Aldrich, and Hamrick, chap. 20; Whitmore, chap. 1); (2) the areas occupied by local populations (the "sites" of the distribution profile) are less than or equal to the size of deforested patches; and (3) new local populations are not established over the time scale of deforestation events. If any of these assumptions are violated (again, as we expect will often be the case), expected numbers of extinctions may be lower than reported here.

We also assume that the consistent patterns seen in aggregate distribution profiles, including many taxa or a broad region, imply that the differences we observe between distribution profiles for single taxa (such as those plotted in figure 18.2) reflect real biological differences. If differences in distribution profile shape are random, not reflecting ecological or evolutionary differences between the taxa represented, then our four groups of distribution profile shapes (see figure 18.3) are purely artificial. However, this does not necessarily negate the usefulness of our classification of distribution profiles into four groups; it just means that the four groups cannot be interpreted as having a biological basis. We are currently studying a large set of distribution profiles to see whether ecological differences among taxa are reflected in differences among their distribution profiles.

Thus far, we have studied only distribution profiles compiled from published taxonomic monographs. If one were to collect field data explicitly for the purpose of compiling distribution profiles, many sampling problems would have to be addressed. These problems are well beyond the scope of this chapter, but would certainly have to include consideration of strategies for stratification with respect to the geographic ranges of individual taxa and environmental heterogeneity.

It is worthwhile to compare our model's predictions with those of the species-area relationship (Preston 1962). If 90% of a region's habitat is removed, our model predicts a loss of 48% of the original species. The species-area relationship predicts a loss of 65% of the species if $Z = 0.45$ (a moderately high value), and 56% of species if $Z = 0.35$ (an intermediate value). The difference between the predictions of the two models can be interpreted as the number of species lost due to long-term processes, or the "relaxation effect" (Reid 1992; Whitmore, chap. 1).

Biology of the Distribution Profile

If the distribution profiles we have compiled capture some aspect of biological reality, then it is logical to ask what sorts of ecological or evolutionary characteristics determine the shapes of these distribution profiles. At this time, we can only speculate on possible reasons for these patterns.

It seems plausible that plants with long-distance pollination or seed dispersal should have flattened distribution profiles. There is some evidence for this view; for example, note that the wind-dispersed Bignoniaceae fall into group 3, and that wind-pollinated plant taxa seem to include few narrow endemics (Balslev 1979). Moreover, plant taxa that show a high degree of endemism, such as orchids (Dressler 1990), often have highly obligate and specific plant-pollinator relationships.

Taxa that exhibit high rates of speciation should have distribution profiles characterized by high frequencies of narrow endemism (as in group 1). Specific mechanisms leading to these patterns could include founder effects and the evolution of cohesive coevolutionary complexes. Phylogenetic constraints on physiology and morphology may cause species within particular taxa to have flattened or skewed distribution profiles.

Highly specialized species may tend to have restricted distributions, and thus to have distribution profiles similar to that of group 1. On the other hand, weedy invasive species are likely to have more flattened profiles. Note that the gap-invading Cecropiaceae fall into group 3.

Growth habit also seems to be important. Epiphytic taxa, at least, seem to include more narrow endemic species than nonepiphytic taxa. A distribution profile for 236

African epiphytic orchid species included significantly more narrow endemic species than a distribution profile for 284 African terrestrial orchid species (Koopowitz, Thornhill, and Andersen 1994). Bromeliads also show unusually high levels of endemism (Koopowitz, Andersen et al. 1994).

GENERAL IMPLICATIONS

1. Researchers and conservation planners should focus their attention on taxa and regions that contain large numbers of narrow endemic species. Regions fitting this description probably correspond to the "biodiversity hotspots" identified by other researchers (e.g., McNeely et al. 1990). In situ conservation efforts (e.g., establishment of forest preserves) should be focused on such regions, while ex situ conservation efforts (conservation outside protected areas, such as seed banks and captive propagation) should focus on taxa with a high proportion of narrow endemics.
2. Because of economic and political constraints on the numbers and sizes of tropical forest reserves that can be purchased, maintained, and protected, in situ conservation will not be able to protect the bulk of tropical plant biodiversity, as witnessed by the huge numbers of narrow endemic species our analysis revealed. Ex situ conservation will therefore be necessary to preserve our planet's biodiversity.
3. We believe that it is important to search for ecological and evolutionary characteristics that influence the shape of a taxon's distribution profile. The speculations discussed above concerning such characteristics need to be subjected to rigorous empirical testing.
4. We submit that a biogeographic approach, particularly one that explicitly recognizes the importance of endemism, can make a major contribution to our understanding of current threats to tropical biodiversity. Future development of this modeling approach should include full consideration of the consequences of violating its simplifying assumptions. Particular emphasis should be focused on the relationship between patterns of deforestation and landscape patterns.

19

Regeneration of Large-Seeded Trees in Australian Rainforest Fragments: A Study of Higher-Order Interactions

Graham N. Harrington, Anthony K. Irvine, Francis H. J. Crome, and Les A. Moore

This chapter summarizes a study of small mammal assemblages and the effects of seed dispersal and predation on populations of large-seeded trees in fragmented and continuous forests in northern Queensland. Our aim was to assess whether isolated forest fragments in northern Queensland had long-term conservation value, or whether they were likely to degrade ecologically over time. We also wished to identify any management actions that might be required to limit the degradation of forest in fragments. Because a comprehensive study of forest function was beyond our resources (and indeed may not feasible under any circumstances; see Crome 1994; Crome, chap. 31), we elected to study an ecological process vital to forest function that we deemed to be potentially vulnerable to fragmentation.

The process we chose was the regeneration of trees with large (> 2 cm diameter) seeds. We reasoned that if native long-lived trees were unable to survive and regenerate in isolated fragments, then there was little hope for maintaining any semblance of normal forest functioning in the long term (see Janzen 1986b). The sizes of seeds and fruits in our target trees make dispersal between fragments by volant animals unlikely. Many of the terrestrial animals large enough to move such seeds do not readily cross the hostile matrix of agricultural fields between forest fragments (Crome and Moore 1990; Laurance 1994). Furthermore, we knew that populations of some animals that disperse and eat these seeds, such as southern cassowaries (*Casuarius casuarius*) and white-tailed rats *(Uromys caudimaculatus),* are affected by forest fragmentation (Crome and Moore 1990; Laurance 1994).

In choosing to focus on this particular forest process, we assumed that seed dispersal was essential to the long-term maintenance of populations of tree species in forest fragments (cf. Grubb 1977b; Howe and Smallwood 1982; Clark and Clark 1984; Mack 1993). Because large seed size has coevolved with particular animal dispersers (Jordano 1995), it may be a severe disadvantage to the plant if the disperser population is disrupted (see Thébaud and Strasberg, chap. 21). If fragmentation affects production, predation, or dispersal of propagules, or the establishment from seed of any species, then in all probability it will eventually affect recruitment into the adult population (Howe and Smallwood 1982). Because the youngest plants in a previously fragmented forest are mainly the result of regeneration from seed after the fragmentation process has occurred, the density and pattern of seedlings may provide clues to the possible effects of fragmentation on tree regeneration.

Consequently, we sought to determine whether forest fragmentation affected (1) the rate of seed predation, (2) the pattern of seed dispersal, and (3) regeneration patterns in space and time of selected large-seeded tree species. We also compared the abundance of small mammals between continuous and fragmented forests, as these animals are thought to be important predators of seeds of several of the target tree species.

METHODS

Study Area

The study was carried out on the Atherton Tableland, a hilly, mid-elevation (600–900 m) plateau in northern Queensland (fig. 19.1) (see Laurance, chap. 6, for a more detailed description of the Atherton Tableland). A cluster of four fragments (3, 8, 20, and 97 ha in area), all of which had been isolated for at least forty years, was selected for study. The fragments were separated from each other by cattle pastures, with patchy regrowth of shrubs and trees along gullies. Study plots in continuous forest were subjectively chosen to approximate the fragments in terms of their soil type, topographic position, distance from forest edge, and presence of thirteen large-seeded target tree species. The fragments were 0.7–1.5 km apart from each other, 12 km from the nearest continuous forest, and 13–18 km from the study plots in continuous forest. All study sites ranged from 720 to 750 m elevation. The rainforest at these sites is classified as complex mesophyll vine forest (Tracey 1982).

Tree Demography

Square plots of 4 ha were established in the three larger (8–97 ha) fragments and at matching continuous forest sites. A square plot of 2 ha was situated in both the 3 ha fragment and a matched site in continuous forest.

Thirteen "target" tree species with seeds larger than 2 cm diameter were selected for study (table 19.1). All plants of these species with a stem diameter of more than 2.5 cm were mapped in each plot and their stem diameters recorded. Smaller (< 2.5 cm diameter) plants were counted on twenty subplots of 400 m^2 that were randomly located within each plot.

Small Mammal Populations

To obtain an indication of the abundance and flux of potential seed predators, small mammals at each site were live-trapped, tagged, and released over a three-day period every third month from March 1990 to March 1992. Eight rectangular trapping grids (twenty traps in two parallel lines 5 m apart) were arranged equidistantly over each of the 4 ha plots (four grids were used in the 2 ha plots). Each trapline consisted of eight Elliot box traps (30 × 10 × 10 cm) and two cage traps (50 × 20 × 20 cm) at 2 m intervals. A test of the efficiency of the trapping method indicated that there was no significant increase in the number of animals caught when trap density was increased by 350%. We

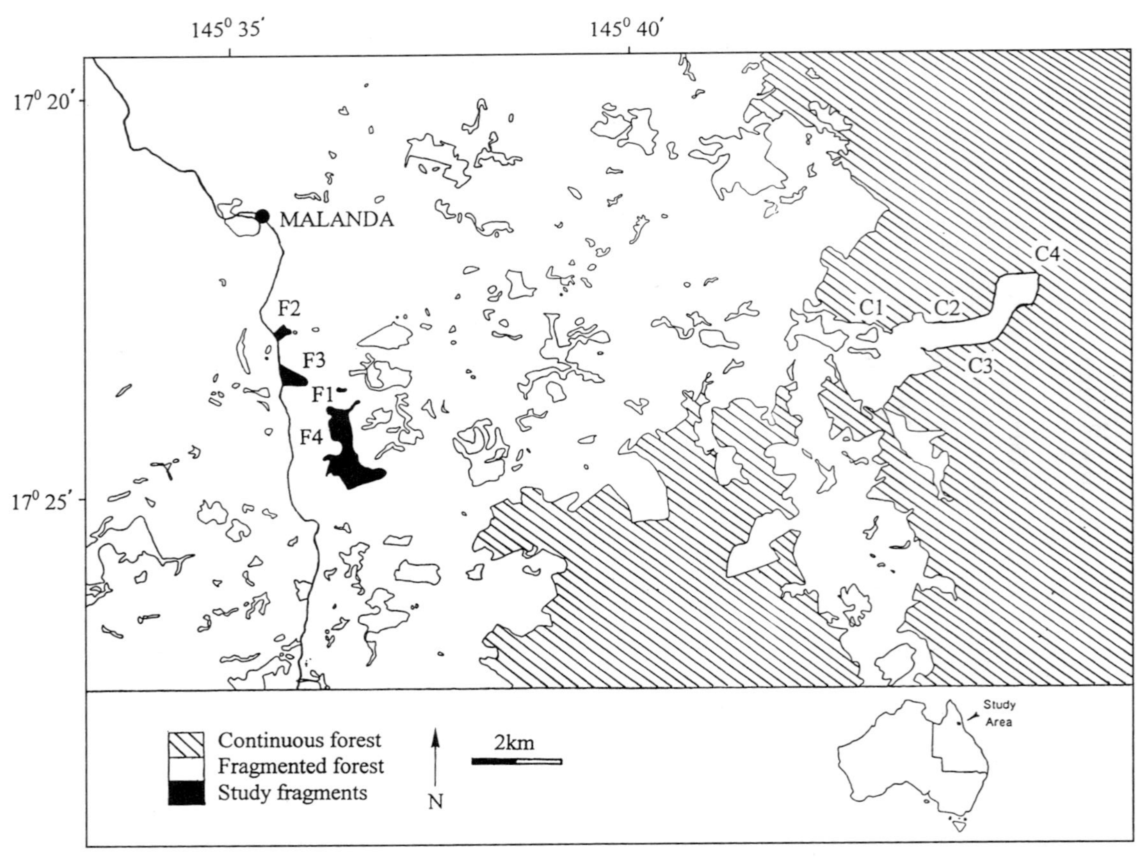

Figure 19.1. Map of the study area, showing the distribution of continuous and fragmented forest and the locations of the study sites.

Table 19.1 Predation and dispersal characteristics of the propagules of thirteen large-seeded target tree species.

		Palatability[a]	
Class and species	Endocarp	Flesh	Seed
Class A	Hard	Rat-kangaroo	White-tailed rat
Athertonia diversifolia (Proteaceae)			
Beilschmiedia bancrofti (Lauraceae)			
Endiandra palmerstonii (Lauraceae)			
Class B	Hard	Cassowary, rat-kangaroo, possum, fruit bat, birds	White-tailed rat
Cryptocarya pleurosperma (Lauraceae)			
Endiandra insignis (Lauraceae)			
Pouteria castanosperma (Sapotaceae)			
Class C	Hard	Cassowary, possum, fruit bat, birds	Rodents (not *Uromys*)
Corynocarpus cribbianus (Lauraceae)			
Class D	Soft	Same as class B	Nil
Beilschmiedia tooram (Lauraceae)			
Cryptocarya oblata (Lauraceae)			
Endiandra sankeyana (Lauraceae)			
Class E	Soft	Same as class B	Invertebrates[b]
Syzygium cormiflorum (Myrtaceae)			
Syzygium kuranda (Myrtaceae)			
Class F	Soft	Rat-kangroo, rodents, birds	Chewed by rodents and rat-kangarros, birds
Syzygium gustavioides (Myrtaceae)			

a. The effects of the animals listed are very unequal. In particular, the possums and fruit bats are poorly understood, but are thought to have a relatively minor effect on most species listed. Rat-kangaroo = musky rat-kangaroo (*Hypsiprymnodon moschatus*); possum = coppery brushtail possum (*Trichosaurus vulpecula*); fruit bat = *Pteropus* spp.

b. Rodents eat weevil larvae, not the seed itself.

therefore concluded that our trapping method provided an accurate indication of the number of trappable animals present on each plot.

Seed Removal Experiments

In the first year of the study (1990–1991), we measured removal rates of marked seeds positioned beneath mother trees within each plot, using seven target species that were fruiting at that time. Marks on seeds indicated that white-tailed rats were heavy predators on certain species, but many of these palatable seeds simply disappeared. Nighttime observations revealed that the removed seeds were invariably carried away by white-tailed rats, but their fates remained unknown.

In the following fruiting season (1991–1992), we attempted to ascertain the fates of transported seeds. We glued bobbins of nylon thread to the seed cases of three palatable species *(Athertonia diversifolia, Beilschmiedia bancrofti, Pouteria castanosperma)* and placed them on the forest floor with the end of the thread tied to a small stake. When the seed was carried away, the thread unwound and enabled the seed case to be located later.

To assess seed removal rates in the second year, 600 seeds of each of the seven target species were deployed in a thrice-replicated design at three fragment and three continuous sites. Two replicates were also employed in the 3 ha fragment, but not in its continuous forest counterpart due to limited seed availability. Each replicate consisted of a 4 × 4 grid with 20 m spacing between points. At one random grid point, fifteen seeds were placed on a mini-grid at 20 cm intervals. A single seed was placed at each of the remaining fifteen grid points. Our purpose in this design was to compare predation on clustered seeds (as occur naturally beneath mother trees) with that on dispersed seeds, and to determine whether there was any interaction with forest fragmentation.

In addition, we censused diaspores on the forest floor to assess whether diaspore availability affected seed predation and to obtain additional information on identities of seed predators or dispersers. A strip transect (5 m × 1,000 m long) through each plot was searched monthly from December 1990 to March 1992. Any damage to fruits or seeds was noted and, wherever possible, the agent responsible was identified from tooth or bill marks.

Captive Studies of Rodent Diets

Finally, to supplement field observations on animal diets, we conducted cafeteria-style feeding trials on captive individuals of the five common rodents at our sites, offering them a choice of fruits of the large-seeded trees. We tested eleven white-tailed rats, twelve *Rattus fuscipes,* eight *Melomys cervinipes,* and four each of *M. hadrourus* and *R. leucopus* during a three-day period.

RESULTS

Seed Predation and Tree Dispersal Syndromes

We grouped the thirteen target tree species into six classes depending on which animals attacked or dispersed their fruits or seeds, as gleaned from this study and other sources (table 19.1). Six of the tree species were placed in classes A and B, which are separate from all the others because theirs were the only seeds eaten by white-tailed rats. Frugivores such as the cassowary ingest the fruits of class B species, but not those of class A. Class C is similar to class B in that both the flesh and seeds were eaten, but the mammal eating the seeds was not the white-tailed rat (we failed to identify it). The rate of predation on this seed class was much lower than that inflicted by white-tailed rats (see below).

Class D fruits are ingested by frugivores and their seeds dispersed undamaged; we did not record any predation on this seed class. Class E seeds are also dispersed by frugivores, but heavy seed damage was inflicted by insect larvae as well as by rodents searching for the larvae (rodents did not eat the seeds themselves). Class F diaspores were chewed, but not destroyed, by a variety of animals.

We identified seed types preferred by white-tailed rats by means of the feeding trials, observations of animal behavior in the field, and marks on the attacked fruits and seeds. White-tailed rats were never seen to eat the fleshy pericarp of any target species, although they chewed through the outer flesh of class A and B species in order to attack the seed. They rejected seeds in classes C through E. Other rodents at the study sites ate little or no seed material.

Table 19.2 Mean number of adult and subadult white-tailed rats (excluding juveniles) captured at the study sites during each of nine quarterly sampling periods from March 1990 to March 1992.

	Fragment (ha)				Continuous forest			
	3.0	10	20	97	1	2	3	4
Mean	2.6	5.6	9.0	13.4	9.3	8.1	9.7	17.1
SD[a]	1.4	2.6	4.0	4.7	4.6	4.0	3.8	6.6
Test[b]	A	B	C	D	C	C	C	E

a. SD of arithmetic means increase with the value of the mean. There is no expectation that rat numbers at one site would remain constant over nine quarterly censuses.

b. Sites with the same letter are not significantly different ($P > .05$) as determined by Duncan's multiple range test after a two-way ANOVA on square root-transformed data. The least significant ratio after back-transforming the least significant difference between transformed means was 1.51.

The musky rat-kangaroo *(Hypsiprymnodon moschatus),* a small (ca. 600 g), primitive macropod, scatter-hoards seeds and fruits beneath leaf litter over distances of up to 100 m, and is probably an important influence on regeneration of tree classes C and D. Cassowaries ingest diaspores and defecate clumps of undamaged seeds, which remain on the soil surface until they germinate or are destroyed by faunal, fungal, or bacterial action. If they do germinate, they frequently result in clusters of densely packed seedlings (Stocker and Irvine 1983).

Effect of Forest Fragmentation on Animals

Of the live-trapped animals, only white-tailed rats had a strong effect on seed survival and dispersal, so data presented here are confined to the adults of this species (table 19.2). The average number of adult white-tailed rats captured was significantly lower in the two smallest fragments (3 and 8 ha) than in either the two largest fragments (20 and 97 ha) or the continuous forest sites. Laurance (1994) also found that white-tailed rat densities were reduced in smaller (< 20 ha) forest fragments.

Musky rat-kangaroos were never captured in this study. They were commonly observed at the continuous forest sites, but not in forest fragments. Likewise, Laurance (1994) and Laurance and Laurance (1995) concluded that musky rat-kangaroos were present in continuous forest, but rare or absent in all but the largest (ca. 500 ha) forest fragments they examined.

Cassowaries were present at the continuous forest sites and visited the largest (97 ha) fragment, but were not detected in the other fragments. Detection of these birds is relatively easy because their dung piles are readily observed. Cassowaries are known to be reluctant to cross agricultural land between forested areas (Crome and Moore 1988).

Seed Predation and Dispersal

The data on the rate of disappearance of seeds on the forest floor are taken from the seed-spooling experiment in year two for *Athertonia diversifolia, Beilschmiedia bancrofti,* and *Pouteria castanosperma* and from seeds deployed beneath mother trees in year one for the

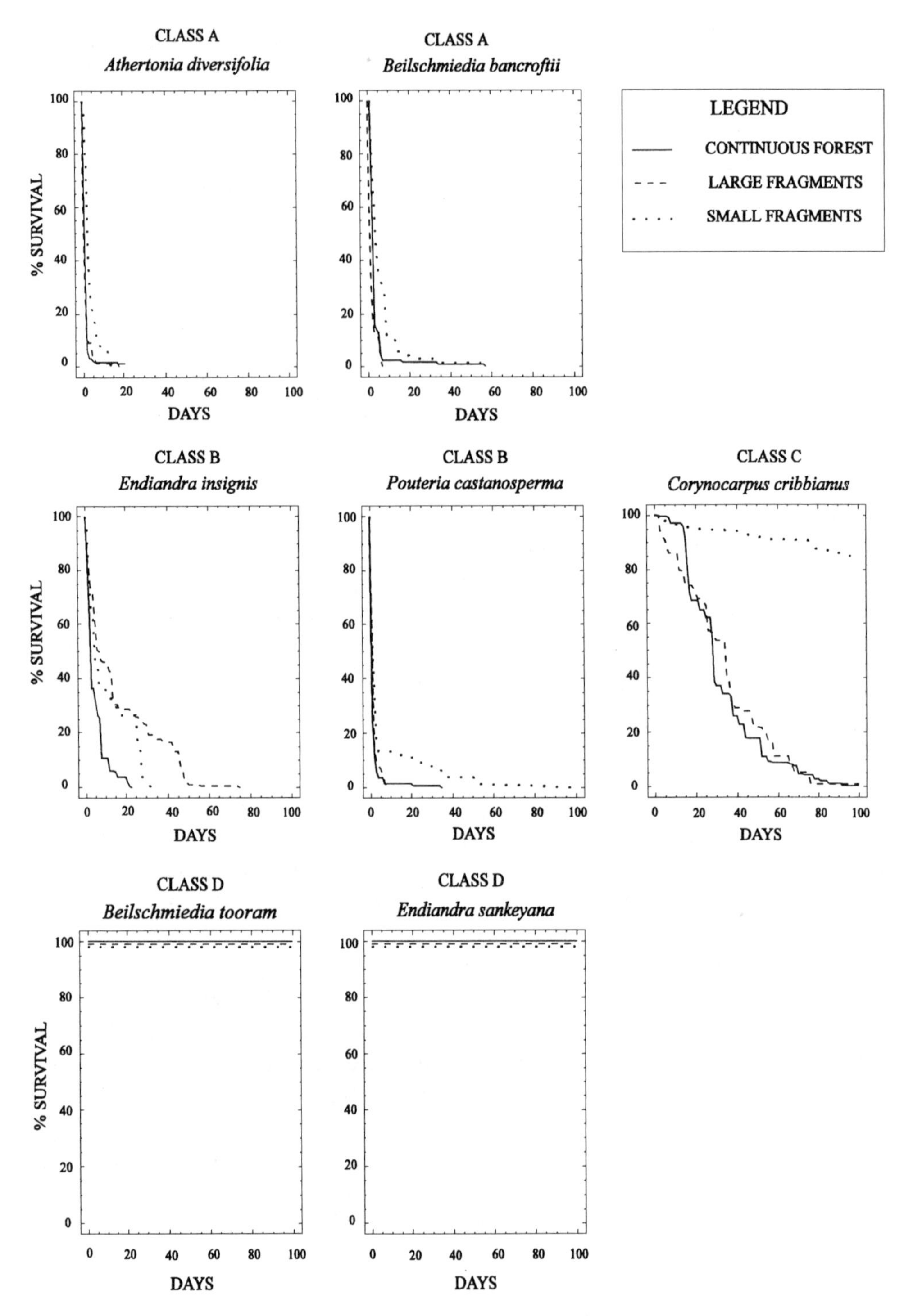

Figure 19.2. Predation rates on seven species of large seeds in continuous rainforest and in large (20 and 97 ha) and small (3 and 8 ha) rainforest fragments. Species are classified as in table 19.1.

Table 19.3 The fates of seeds from three tree species palatable to white-tailed rats (*Uromys caudimaculatus*).

	Fragment (ha)				Continuous forest		
Seed fate	3.0	10	20	97	C1	C2	C3
Original data[a]							
Cached	44	23	3	8	12	25	3
Eaten	26	36	45	33	26	36	37
Unknown	30	41	52	59	62	39	60
Redistributed data[b]							
Cached	64	46	44	44	40	52	46
Eaten	36	54	56	56	60	48	54
G-test	18.8**	0.6	2.0	1.4	7.6*	1.9	0.3

a. Seed fates expressed as a percentage of seeds displayed on the forest floor (n = 270 per site).

b. The likely fates of the same seeds after those of "unknown" fate were redistributed (according to the ratio of seeds known to be eaten or cached after being moved by the rats; this ratio is calculated for each tree species over all treatments, and the results expressed as a percentage of all seeds in the experiment). *G*-tests were used to test for deviations in the ratio of cached : eaten seeds at each site from the overall average from all sites.

$^*P < .01$ $^{**}P < .001$

four other species (fig. 19.2). There was no significant difference in the rate of seed predation between clustered and dispersed seeds, and therefore these data have been pooled.

In continuous forest, all class A and B seeds were removed or eaten by white-tailed rats within a few days, whereas class D and E seeds suffered no predation whatever. The seeds of the single class C species *(Corynocarpus cribbianus)* were removed more slowly than class A and B seeds, and tooth marks on the fruits indicated that the predator was a rodent (but not the white-tailed rat). We did not find sufficient class E and F fruits to assess predation rates. Although cassowaries were present in the continuous forest (Crome and Moore 1988) and are known to eat class D fruits (Stocker and Irvine 1983), none of the class D fruits monitored in year one were removed by any agency (fig. 19.2).

For class A and B species, there was a tendency for total predation of seeds to be achieved later in the two smallest fragments than in continuous forest and the two largest fragment sites, but this difference was not significant (fig. 19.2).

When spooled seeds of three species palatable to white-tailed rats *(A. diversifolia, B. bancrofti* and *P. castanosperma)* were studied, the seeds were either eaten in situ or moved. Some removed seeds were traced to caches in the ground. The remainder had an unknown fate because the rats removed the bobbins. There were large differences between sites in terms of the percentage of cached seeds (table 19.3). These differences did not appear to be associated with fragmentation or fragment size, but sites with few records of seed caching tended to have high numbers of seeds of unknown fate.

The incidence of caching was probably underestimated in some cases because the white-tailed rats at several sites usually removed the bobbins before caching seeds. To compensate for this, we redistributed the "unknown" category in proportion to the ratio of "removed and eaten" to "removed and cached" seeds of known fate of each species.

Table 19.4 Adult density and juvenile : adult ratios of trees with seeds in palatability classes A (*Athertonia diversifolia, Beilschmiedia bancrofti*), B (*Endiandra insignis, Pouteria castanosperma*), and D (*Beilschmiedia tooram, Endiandra sankeyana*).

	Fragments		Continuous forest	
Class	Adults/ha	Juvenile : adult ratio	Adults/ha	Juvenile : adult ratio
A	5	0.3	5	0.1
B	3	4.5	3	8.8
D	6	47.1	13	60.3

This manipulation yielded relatively consistent cache rates across all study sites (table 19.3), and these rates did not differ consistently between fragmented and continuous forest.

Overall, about half of the 1,800 seeds employed in the experiment were cached, but all the seeds whose fate was known (971) were eventually eaten (fig. 19.2). Most cached seeds were removed and eaten within a few days, but a few survived for several weeks. The longest survivor was an *A. diversifolia* seed that was dug up and eaten after being buried for 125 days. In contrast, the longest survival time for an uncached seed was 9 days.

Tree Regeneration

The ratio of juvenile to mature trees was used as an indication of post-fragmentation regeneration rates for different palatability classes (table 19.4). Only juveniles with a stem diameter of less than 0.5 cm were used to ensure that they had become established since fragmentation (although juvenile trees can remain in this size class for decades [J. H. Connell, pers. comm.], it is probable that the great majority were established after fragmentation occurred some forty years previously). Mature trees were defined as those size classes that produced full crops of fruit (> 20 cm stem diameter, for most species), determined from our records of fruiting trees during the fruit census.

There was no significant difference in juvenile:adult ratios between continuous and fragmented forest for any tree species. The ratio of juveniles to adults was an order of magnitude higher for the unpalatable class D species than for the class A and B species, which have seeds palatable to white-tailed rats. This difference highlights the considerable effect that seed predation can have on regeneration potential. Surprisingly, however, this disparity was not reflected in the density of adult trees; class A, B, and D species all occurred at similar densities (table 19.4).

DISCUSSION

Rats and Seed Predation in Fragmented Forests

White-tailed rats can have a strong negative effect on certain types of seeds that they find palatable (see fig. 19.1). This intense predation is probably the cause of the much lower

Table 19.5 Germination times (days) for seeds examined in this study.

Species	Germination time
Class A	
Athertonia diversifolia	300–350
Beilschmiedia bancrofti	140–3,285
Class B	
Endiandra insignis	35–100
Pouteria castanosperma	26–50
Class C	
Beilschmiedia tooram	18–22
Endiandra sankeyana	25–100

Sources: Hyland 1989; A. Irvine, pers. obs.

ratio of juvenile to adult plants in class A and B species than in those whose seeds are not palatable to white-tailed rats.

At the outset, one would suppose that seed predation could threaten the local extinction of trees when the result is that the smallest juvenile stage is represented by less than one plant per adult (compared with about fifty juveniles per adult in species with unpalatable seeds; see table 19.4). These same seed predators, however, may enhance regeneration of trees with palatable seeds because they also disperse and bury seeds, which is conducive to seed germination and seedling establishment. Although most buried seeds persisted for only a few days before being recovered and eaten, a few survived for more than 100 days. It is reasonable to suppose that some buried seeds do remain undetected long enough for germination to occur (table 19.5), if only because we recorded juvenile stages of the palatable species. A similar situation has been described for a palm (Smythe 1989) and a large-seeded South American tree (Forget 1990).

In circumstances such as these, in which a single dominant seed predator and disperser exercises such a critical influence on tree regeneration, any substantial change in the density of the predator could have a marked effect on tree populations (Howe 1989; Osunkoya 1994). Our finding that white-tailed rats had much lower densities in small fragments (see table 19.2) is therefore significant. However, we were unable to detect any differences in plant demography between small and large fragments and continuous forest. The apparent reason was that, even in the small fragments, white-tailed rats consumed all the palatable seeds. They also buried a similar percentage of available seeds on all the plots, regardless of their abundance.

It is not known whether white-tailed rat populations ever crash, thereby reducing predation pressure on palatable seeds. They did not do so during this study. Moreover, the consistently low density of juvenile trees at our study sites suggests that the effects of white-tailed rats on seed dispersal and predation have not been disrupted for several decades. If white-tailed rat populations were more extinction-prone in fragments, then the effects on the pattern and rate of tree regeneration could be profound. However, white-tailed rats appear to be mobile within the matrix surrounding fragments (Laurance 1994), and the high rate of turnover of adult rats found in this study suggests considerable movement of animals into and out of fragments. Thus, if a fragment population should disappear, it is likely that the fragment would soon be recolonized from nearby source populations.

Cassowaries and Seed Dispersal

The effects of fragmentation are even more difficult to assess for tree species that are mainly dependent upon cassowaries for seed dispersal. The pattern of seed dispersal by cassowaries has not been described, but can be inferred from what is known of the bird's movement patterns (Crome and Moore 1988, 1990). Seed dispersal is likely to be mostly localized within a few hectares, but with occasional long-distance movements. In any event, only a small fraction of the total seed population is actually dispersed by cassowaries, which are territorial (Crome and Moore 1990) and eat only fresh fruit (Lott et al. 1995). Both the remaining seeds and those in cassowary dung are then subject to dispersal by other animals and by overland water flow. From our comparative study of predation rates on clustered and dispersed seeds, we would expect seeds in cassowary dung to be attacked at a high rate, comparable to those beneath the mother trees. Consequently, the effect of an absence of cassowaries in forest fragments on seed dispersal and tree regeneration is likely to be minor, but extremely difficult to quantify.

CONCLUSIONS

Overall, it seems likely that regeneration of large-seeded trees in this region is affected by forest fragmentation, principally because of their strong dependence on animal vectors for seed dispersal. In general, seed-dispersing animals are affected by the size of a fragment, edge effects on forest structure and floristics, the quality of the matrix, and disturbances such as hunting, wood gathering, and access by domestic animals (see Clark and Clark 1984; Howe, Schupp, and Westley 1985; Saunders, Hobbs, and Margules 1991). In the absence of animal dispersers, some seeds will be dispersed only by gravity or water, leading to novel and highly clumped dispersion patterns. Others may fail to regenerate because they are transported by animals into unsuitable matrix habitat.

This study, however, did not detect any effects of forest fragmentation on the density of juvenile trees in northern Queensland. We conclude that change in populations of long-lived trees is likely to be a slow process extending over many decades or centuries. Our ability to predict changes over such a long time frame for such complex systems is weak (Crome 1994; Crome, chap. 31).

Determination of current management priorities for forest fragments and the surrounding agricultural matrix is more likely to be influenced by organisms that show a rapid response to habitat change and management—usually fauna—than by the long-term responses of rainforest trees (see Corlett and Turner, chap. 22). A recent case in point has been the urgent protection of remaining forest fragments on the coastal plain of northern Queensland, which occurred not because of the threat to trees posed by large-scale clearing for sugarcane, but because some fragments were found to sustain the rare mahogany glider (*Petaurus gracilis:* Queensland Department of Environment and Heritage 1995).

GENERAL IMPLICATIONS

1. When rainforest is reduced to fragments in northern Queensland, the behavior and densities of rainforest animals likely to be important predators and dispersers of large seeds are greatly altered.
2. We were, however, unable to detect any significant response to fragmentation in terms of the ratio of juvenile to adult trees.
3. We conclude that the study of higher-order interactions in tropical forests will often be difficult, especially when the organisms studied (long-lived trees) respond very slowly to land use change. Studies of higher-order interactions are extremely demanding of resources and have a low probability of success.
4. Conservation and management initiatives for rainforests require human motivation before they will be adopted. It is unlikely that long-lived trees, which appear to respond slowly to fragmentation and are difficult to study, will provide a sufficient stimulus. Fauna are likely to have more appeal; because many species are dramatically and rapidly affected by fragmentation, they are likely to provide less equivocal data on which to base management decisions.

ACKNOWLEDGMENTS

We would like to thank Andrew Dennis for sharing with us his knowledge of frugivory and seed predation in the local rainforests, particularly with respect to the musky rat-kangaroo. Mervyn Thomas provided statistical advice. We are also grateful to Joe Connell, Bill Laurance, John Winter, Jack Putz, David Westcott, Segun Osunkoya, Peter Green, and Andrew Dennis for their useful comments on earlier drafts of this chapter.

20

Dispersal and the Dynamics of Genetic Structure in Fragmented Tropical Tree Populations

John D. Nason, Preston R. Aldrich, and J. L. Hamrick

NATURAL habitats are being destroyed worldwide at an alarming rate. Human encroachment into forested regions is particularly rapid in the Tropics, where forests are being cut at an annual rate of 100,000 to 200,000 km^2 (Katzman and Cale 1990; see also Whitmore, chap. 1). While the total land area contained in large, continuous tracts of tropical forest is diminishing, remnant forest fragments are rapidly increasing in number as a result of incomplete deforestation. Such fragments form a mosaic of small (often 1–5 ha) habitat islands embedded in a human-modified matrix in which abiotic and biotic processes are dramatically altered (Harris 1984; Bierregaard et al. 1992). As a result, the within-fragment and among-fragment dynamics of populations will largely determine which tropical forest species can maintain themselves in disturbed landscapes.

Many tropical tree species may be particularly vulnerable to this landscape transformation due to both their own low densities and the disruption of their pollen and seed vector associations. Accordingly, the eventual representation of these tree species by relatively few individuals in any one fragment will lead to small effective population sizes unless sufficient pollen and seed movement occurs among fragments. Because animals are the primary dispersal agents of tropical tree pollen and seeds (Wheelwright 1988; Bawa 1990; Levey, Moermond, and Denslow 1994), the level of genetic and demographic connectedness among forest patches will ultimately depend on vector abundance and behavior. If vector dispersal is curtailed by fragmentation, then remnant forest populations will become genetically and demographically independent (Lande and Barrowclough 1987; Templeton et al. 1990). Over successive generations, genetic variation within individual patches will decrease, and variation among patches will increase, due to random drift. Subsequent extinction of individual patch populations would therefore represent a loss of genetic variation at the species level (Templeton et al. 1990; Hamrick and Nason 1996).

It is currently uncertain, however, whether trees in tropical forest fragments retain a functional level of genetic connectivity with the surrounding landscape. For example, habitat corridors may provide avenues for gene dispersal between patches and may also contain individuals that can contribute to the local dynamics of adjoining patches. Similarly, the surrounding matrix may contain free-standing remnant trees that are important foci of seed deposition (e.g., Guevara and Laborde 1993) and that direct pollen and seeds to nearby patches (Levin 1995). Further, some vectors simply may not perceive environmental heterogeneity on the same scale or quality as humans and may move freely between patches. In general, very little empirical evidence exists regarding the importance of these factors in determining the dynamics of tropical tree populations in a fragmented environment.

This chapter summarizes ecological and genetic data on patterns of pollen and seed dispersal in tropical trees. We begin by considering the interaction of dispersal, spatial scale, and genetic structure in a fragmented landscape. We then review the literature on dispersal and genetic structure in continuous tropical forests, both because the trajectories that fragmented populations may take will depend on conditions prior to fragmentation, and because available information has been obtained principally in continuous forests. In this section, as in subsequent ones, we elaborate the similarities and differences of pollen versus seed dispersal. Next, we consider how preexisting spatial genetic variation influences the organization of genetic variation within and among recently fragmented populations. We then develop predictions regarding short-term and long-term cumulative effects of fragmentation in a metapopulation framework. Finally, we summarize the importance of dispersal to the conservation of tropical areas and the design of nature reserves.

DISPERSAL AND GENETIC STRUCTURE

Genetic structure refers to the distribution of genetic variation in space and time. It arises through a complex interaction of historical and evolutionary factors, key among which is dispersal. Dispersal among established populations, or gene flow, reduces genetic differentiation among populations and may introduce novel genetic variation to the recipient populations (Wright 1931). If two populations exchange genes at a high rate, however, so that a significant amount of the recruitment in one is dependent upon the other, then they are perhaps better considered a single population. In this sense, dispersal not only influences genetic structure, but also defines the limits of the population itself.

Wright (1943, 1946) quantified the relationship between pollen and seed dispersal and the spatial dimension of a population in his neighborhood model, formulated for the extreme case of a continuous, uniformly distributed species. The area occupied by a neighborhood is defined as that region from which the parents of centrally located progeny may be treated as if chosen at random, and is estimated as a linear function of the variances in pollen and seed dispersal distances (Crawford 1984). The size of a neighborhood or the number of individuals within it is determined both by the extent of dispersal and by local population density. In theory, a neighborhood of N individuals is expected to respond to the effects of genetic drift at the same rate as an effective population of the same size (N_e). Neighborhoods smaller than twenty individuals should experience substantial genetic drift, while drift should be moderate with two hundred individuals and virtually absent when numbers exceed one thousand (Crawford 1984). Thus, dispersal distances that are large with respect to the separation of conspecifics will increase the neighborhood size, decreasing both the rate of genetic drift and the amount of spatial genetic differentiation among neighborhoods.

The neighborhood concept is particularly useful for understanding the relationship between dispersal and population density and the effects of these factors on spatial genetic structure. Despite the low overall density of many tropical tree species in continuous forests, substantial variation in density occurs within and among species. Most taxa are clumped at some spatial scale, for example, in conjunction with spatial heterogeneity in the environment (e.g., Denslow 1985; Hubbell and Foster 1986; Clark 1994; Hartshorn and Hammel 1994). Furthermore, fragmentation can result in dramatic changes in plant density. Based on theoretical predictions of evolutionary models and on empirical data, largely from temperate systems, pollinator and frugivore dispersal distances would be expected to decrease with increasing host plant density (e.g., Levin and Kerster 1969; Beattie 1976; Manasse and Howe 1983; Ellstrand, Devlin, and Marshall 1990; House 1992; Stacy et al. 1996), while pollen and seed dispersal distances in abiotically dispersed species should not. This response of animal vectors to the spatial distribution of reproductive individuals should strongly influence the spatial dimension of realized genetic neighborhoods: adult plant density should be negatively correlated with the area of the neighborhood. Further, gene flow between neighborhoods depends not on the physical distance between them, but rather on the number of intervening neighborhoods. As a result, density also affects the spatial scale at which a population may differentiate in response to microgeographic selection or genetic drift. If neighborhood areas are small with respect to the spatial scale of environmental heterogeneity, then gene frequencies may vary between neighborhoods as a result of local selection pressures. If, in contrast, the numbers of individuals within neighborhoods are sufficiently small, genetic drift can overwhelm the effects of weak selection.

DISPERSAL AND GENETIC STRUCTURE IN CONTINUOUS TROPICAL FORESTS

In plants, estimates of the total genetic diversity maintained within a species or population have primarily been based on allozyme analyses. Compared with annual, and in particular, autogamous angiosperms, woody species are generally characterized by high amounts of genetic variation (Hamrick and Godt 1989). Although relatively few tropical tree species have been studied using allozyme markers (the majority in Central America), as a group they exhibit levels of genetic diversity comparable to those found in temperate angiosperm and gymnosperm tree species (Hamrick, Godt, and Sherman-Broyles 1992; Hamrick 1994). Moreover, though the genetic structures associated with ecological traits such as pollen and seed dispersal syndrome, growth form, longevity, and successional status (Loveless and Hamrick 1984; Hamrick and Godt 1989) and patterns of geographic distribution (Moran and Hopper 1987; Hamrick and Godt 1989; Hamrick, Godt, and Sherman-Broyles 1992) are perhaps better understood for temperate species, levels of genetic diversity vary substantially among tropical tree species (e.g., Hamrick and Loveless 1986; Moran, Muona, and Bell 1989; Nason, Herre, and Hamrick, in press). For example, genetic diversity increases significantly with geographic distribution (Loveless

1992) and with local abundance within an area (Hamrick and Murawski 1991; but see Nason, Herre, and Hamrick, in press). Furthermore, although there is significantly less genetic variation in shrubs than in trees, this is not true for shrub species with wider geographic distributions (Loveless 1992). Levels of genetic diversity are also lower in early successional species than in late successional species, but these differences are not significant, based on the limited number of species studied (Loveless 1992).

Detailed studies examining genetic structure at various spatial scales are few, and those studies that are available are often conducted at differing spatial scales. As a result, comparisons among species and locations are difficult, and the neighborhoods of tropical tree species are only poorly understood. Fine-scale genetic structure has been shown to develop around parents in seedling cohorts, particularly in low-density species, and may be important in the founding of fragment populations (Hamrick, Murawski, and Nason 1993). This structure, however, tends to decay with time (presumably owing to random mortality). In general, low levels of genetic differentiation over distances of up to a few kilometers (e.g., Heywood and Fleming 1986; Hamrick and Loveless 1989; Eguiarte, Pérez-Nasser, and Pinero 1992; Alvarez-Buylla and Garay 1994; Loiselle et al. 1995) suggest that large neighborhood numbers and areas are typical of many tropical tree species. Although numerous factors are responsible for these patterns, the movement of pollen and seeds is an important determinant of the development of genetic structure over a wide range of spatial scales.

POLLEN DISPERSAL

Numerous studies have been conducted on the pollination biology of tropical trees. Particularly important to understanding the development of genetic structure are studies evaluating mating systems and patterns of pollen movement. The mating system involves patterns of parentage within a population and establishes the genotypic composition of the progeny cohort, while pollen dispersal distance determines the spatial scale at which this part of the life cycle influences genetic structure.

Mating Systems

The mating system is one of the primary determinants of genetic structure in plant populations and of their susceptibility to loss of genetic diversity and inbreeding depression following fragmentation. The low adult population density and apparent hyperdispersion of many tropical tree species led early researchers to question whether insect pollinators could move among widely spaced conspecific trees to effect outcrossing. This led to the suggestion that self-pollination should be the prevalent form of reproduction in tropical forests (Corner 1954; Baker 1959; Federov 1966). Although the relative importance of localized selection, genetic drift, and gene flow in the evolution of population genetic structure was much debated, it was generally believed that individual species possessed relatively low amounts of genetic variation partitioned among, as opposed to within, local subpopulations (Federov 1966).

A growing body of empirical data, however, indicates that self-fertilization is not a common feature of tropical forest communities and that most tropical trees have mechanisms that promote or ensure outcrossing. For example, dioecy, an obligate form of outcrossing, is common in lowland Neotropical forests, with up to 23% of tree species having this sexual system (reviewed in Bawa 1990; Kress and Beach 1994). Although proportionately more tree species are self-compatible in montane forests, where low temperatures may limit the activity of many insect pollinators (Sobrevila and Arroyo 1982; Tanner 1982), cross-pollination experiments in lowland forests have demonstrated that hermaphroditic tropical trees are generally highly outcrossing and that many species are self-incompatible (e.g., Ashton 1969; Bawa 1974; Chan 1981; Lack and Kevan 1984; Bawa, Perry, and Beach 1985; Kevan and Lack 1985; Bawa 1990). Allozyme analyses of mating systems have helped to confirm the prevalence of outcross fertilization as well as providing estimates of the proportions of selfed and outcrossed progeny. Nineteen of twenty-two tropical angiosperm tree species examined to date have estimated outcrossing rates of greater than 80%, with only three species appearing to have mixed mating systems incorporating comparable levels of self and outcross reproduction (Murawski et al. 1990; Murawski and Hamrick 1992; Murawski, Dayanandan, and Bawa 1994).

While many features of reproductive systems are intrinsic to individual species, there is some evidence that they may also be associated with the local density of flowering conspecifics. For example, Bawa, Perry, and Beach (1985) found two of three rare or uncommon hermaphroditic rainforest tree species they examined to be self-compatible, suggesting that the evolution of the incompatibility system may respond to long-term patterns of a species' abundance. In another example, House (1992) found that pollen-limited seed set increased with nearest-neighbor distance in the three spatially clumped dioecious species she studied in tropical Australia. In contrast, genetic analyses of mating systems in other species with lower population densities have found high rates of outcrossing (O'Malley and Bawa 1987; O'Malley et al. 1988; Nason, Herre, and Hamrick, in press), indicating, indirectly, that pollinators may effectively disperse long distances between flowering trees. Other studies specifically examining variation in the mating system relative to population density also indicate the potential for complex responses by pollinators to host plant structure. For example, although allozyme analyses of the mating system were found to be unaffected by variation in population density in the locally abundant, beetle-pollinated understory palm *Astrocaryum mexicanum* (Aracaeae) at Los Tuxtlas in Mexico (Eguiarte, Pérez-Nasser, and Pinero 1992), the much lower-density, self-compatible, bat and hawkmoth-pollinated canopy emergent *Cavanillesia platanifolia* (Bombacaceae) has lower outcrossing rates when surrounded by fewer co-flowering conspecifics (Murawski et al. 1990; Hamrick and Murawski 1991; Murawski and Hamrick 1992). Similarly, a low-density Panamanian population of the bat-pollinated, pantropical canopy emergent *Ceiba pentandra* (Bombacaceae: Murawski and Hamrick 1992) possesses a mixed mating system, with estimates for individual trees ranging from complete selfing to complete outcrossing. Though it occurs at much higher densities, the large-bee-pollinated canopy tree *Shorea trapezifolia* (Dipterocarpaceae) in Sri Lanka also possesses a mixed mating system, with significant variation in outcrossing rates found among individuals and between years (Murawski, Dayanandan, and Bawa 1994). These examples demonstrate the potentially dual nature of even large, volant pollinators in promoting both outcrossing and self-fertilization in the same population.

Pollen Dispersal Distances

Rather than increased selfing rates, low plant population densities may be associated with expanded pollinator foraging ranges and dispersal distances. In fact, studies of pollinator movement in continuous tropical forests suggest that many tropical pollinators are capable of moving long distances between flowering conspecifics (e.g., Heithaus, Fleming, and Opler 1975; Linhart 1973; Haber and Frankie 1989). For example, using fluorescent dyes, Webb and Bawa (1983) demonstrated pollen movement of up to 225 m (mean 37.5 m) in the low-density, hummingbird-pollinated shrub *Malvaviscus arboreus* (Malvaceae) and up to 31 m (mean 6.47 m) in a high-density population of the butterfly-pollinated herb *Cnidoscolus urens* (Euphorbiaceae). Pollen movements of 600 m were observed in the butterfly-pollinated vine *Psiguria warscewiczii* (Cucurbitaceae: Murawski 1987), and Janzen (1971) found that certain euglossine bees may forage over distances of several kilometers. Moreover, in a disturbed area of tropical dry forest in Guanacaste Province, Costa Rica, Frankie, Opler, and Bawa (1976) recorded individuals of eight species of bees moving between the flowering crowns of *Andira inermis* (Fabaceae) trees located 0.8 km apart. Nonetheless, while even relatively isolated trees receive substantial outcross pollen, reductions in plant population density may negatively affect the total amount of pollen received by individual trees. For example, local reductions in the numbers of flowering males resulted in significant decreases in pollination success and fruit set in three dioecious tree species in tropical Australia (House 1992, 1993).

Over the last decade, statistical and genetic techniques that were developed to address questions of paternity in humans have been "borrowed" to quantify spatial patterns of mating and thus effective pollinator and pollen dispersal distances in natural plant populations. Employing allozyme markers, a small but growing number of "paternity analysis" studies have documented substantial long-distance pollen flow as well as density-dependent effects in continuous forest populations. For example, although Boshier, Chase, and Bawa (1995) identified pollen movement over distances of up to 280 m in a high-density Costa Rican population of the moth-pollinated canopy tree *Cordia alliodora* (Boraginaceae), they also found evidence for an increase in localized mating among genetically correlated individuals associated with higher-density patches of flowering trees. In general, Hamrick and Murawski (1990) have suggested that the effective breeding unit for locally common tropical tree species could be on the order of 25–50 ha. The majority of tropical tree species, however, occur as uncommon, widely dispersed individuals (Davis and Richards 1933a,b; Black, Dobzhansky, and Pavan 1950; Richards 1952). For these species, a much larger area may be necessary to encompass interbreeding individuals. Indeed, though mating among near neighbors was common for individuals occurring in small, isolated clusters of *Spondias mombin* (Anacardiaceae), a small-bee-pollinated species studied in a mature forest on Barro Colorado Island (BCI) in central Panama, substantial pollen flow was observed over distances of at least 300–

350 m (Stacy et al. 1996). The same study also found that in *Calophyllum longifolium* (Clusiaceae), 38% of pollen moved at least 210 m, more than twice the mean distance between nearest neighbors, and that for two trees of *Turpinia occidentalis* (Staphyleaceae), at least 27.5% of the progeny resulted from pollen flow from a minimum of 235 m away. Similarly, paternity analyses indicate more than 25% pollen flow over distances of 500 m in two bee-pollinated leguminous canopy tree species, *Platypodium elegans* and *Tachigali versicolor* (Fabaceae), which occur at densities well below one adult individual per hectare on BCI (Hamrick and Murawski 1990).

Finally, in an unprecedented example of long-distance pollen dispersal, patterns of paternity in two rare strangler fig species (*Ficus* spp., Moraceae) on BCI indicate that their minute, species-specific wasp pollinators routinely disperse distances of over 4 km (Nason, Herre, and Hamrick, in press). Based on asynchronous flowering at the population level and adult densities of well below ten individuals per square kilometer, breeding populations of these species are apparently substantially larger than the size of the 15 km^2 island preserve. Similarly large areas are presumably necessary to maintain populations of the pollinators and seed dispersal agents on which these tree species depend.

In general, pollen dispersal may be localized and reproductive success correlated with the density of flowering conspecifics in many tropical trees. This may be particularly true of understory species with unspecialized pollinators. Nevertheless, paternity analyses conducted to date on canopy tree populations clearly demonstrate the potential for long-distance pollen dispersal in several species with very different pollination syndromes. Unfortunately, paternity analysis techniques have yet to be applied to a sufficient number of tropical tree species (or populations) to permit generalizations concerning the effects of factors such as growth form, successional status, or pollination syndrome on the distribution of pollen dispersal distances. However, the studies described above, combined with information on genetic structure, suggest that populations or breeding neighborhoods of many species are characterized by large numbers of potentially interbreeding individuals that, depending on local densities, may occupy tens to hundreds of hectares.

SEED DISPERSAL

While pollen dispersal involves the movement of a single haploid genome, seed dispersal involves the movement of genes derived from both pollen and ovule parents. Thus, on a per unit basis, seed dispersal will more effectively counter differentiation among populations than pollen dispersal. The effect of long-distance seed movement on genetic structure, however, is not simply a function of the rate of dispersal. In contrast to pollen flow, which constitutes migration between established populations, seed flow may result in either migration or the colonization of vacant habitats (Hamrick and Nason 1996). These different fates of pollen and seed flow may have very different effects on genetic structure (Wade and McCauley 1988) and may explain why the breeding system of a species has been found to be significantly correlated with the degree of genetic differentiation among plant populations, while the seed dispersal syndrome has not (Hamrick and Godt 1989). When mechanisms of seed dispersal are generalized into broader categories, however, tropical species with biotic dispersal exhibit significantly less genetic differentiation among subpopulations separated by a few kilometers than do abiotically dispersed species (Loveless 1992). Further, the development of fine-scale spatial genetic structure among juveniles within tropical tree populations also appears to be strongly influenced by biotic and abiotic mechanisms of seed dispersal as well as by the density of fruiting trees (Hamrick, Murawski, and Nason 1993).

Short-Distance Seed Movement

Numerous ecological studies have examined localized seed dispersal in tropical forest stands. Yet, due to the absence of suitable maternally inherited markers, studies utilizing genetic techniques to quantify patterns of seed dispersal are few compared with those examining pollen dispersal. Typically, seed dispersal patterns follow a leptokurtic distribution from the parental source, with numerous seeds falling near the parent and fewer farther away (Howe and Smallwood 1982; Willson 1993). This pattern of seed dispersal produces patches of juveniles (seed and seedling shadows) that share maternal and, perhaps, paternal parents, resulting in half- and full-sib genetic correlations among individuals (half sibs share a common mother, full sibs have both parents in common). The detectability of genetic structure associated with seed shadows will vary, however, depending on the degree to which the seed shadows of different maternal individuals overlap. For example, Hamrick, Murawski, and Nason (1993) found that seedlings and juveniles of *Platypodium elegans,* a low-density, wind-dispersed tree, had a pronounced genetic structure over a scale of approximately 300 m (fig. 20.1A), while *Alseis blackiana* (Rubiaceae), a wind-dispersed tree occurring at a much higher density, had lower genetic correlations manifested over a much smaller spatial scale, about 20 m (fig. 20.1B). Even lower genetic correlations and a smaller genetic patch structure were found in *Swartzia simplex* var. *ochnacea* (Fabaceae), a high-density treelet with large bird-dispersed seeds (fig. 20.1C). In general, while saplings show significant positive genetic correlations over nearest-neighbor distances, this genetic structure is much reduced or absent over the same spatial scales in adults of these three species.

Generally, studies of seed dispersal over fairly limited distances reveal little difference in the slopes of dispersal curves for wind-dispersed versus animal-dispersed tropical trees and shrubs (Willson 1993). Although one might expect that wind dispersal would be less effective under a closed canopy than animal dispersal, a large proportion of animal-dispersed seeds may be dropped beneath the maternal tree due to messy foragers and low foraging intensities. Furthermore, because animals typically forage in a density-dependent fashion (Levey, Moermond, and Denslow 1994), parental plant density could influence dispersal curves. Manasse and Howe (1983) found evidence that high fruiting density in a local neighborhood decreased fruit removal rates in several tropical trees (cf. Sargent 1990). Subsequent gravity dispersal of the remaining fruits should decrease mean dispersal distances and decrease the genetic neighborhood area. Such localized dispersal may not be as common with wind-dispersed species.

While seed dispersal determines the spatial distribution of propagules within tropical tree populations, actual patterns of establishment may be modified by a variety of post-dispersal processes, such as sib competition, predator satiation, or herbivore-, parasite-,

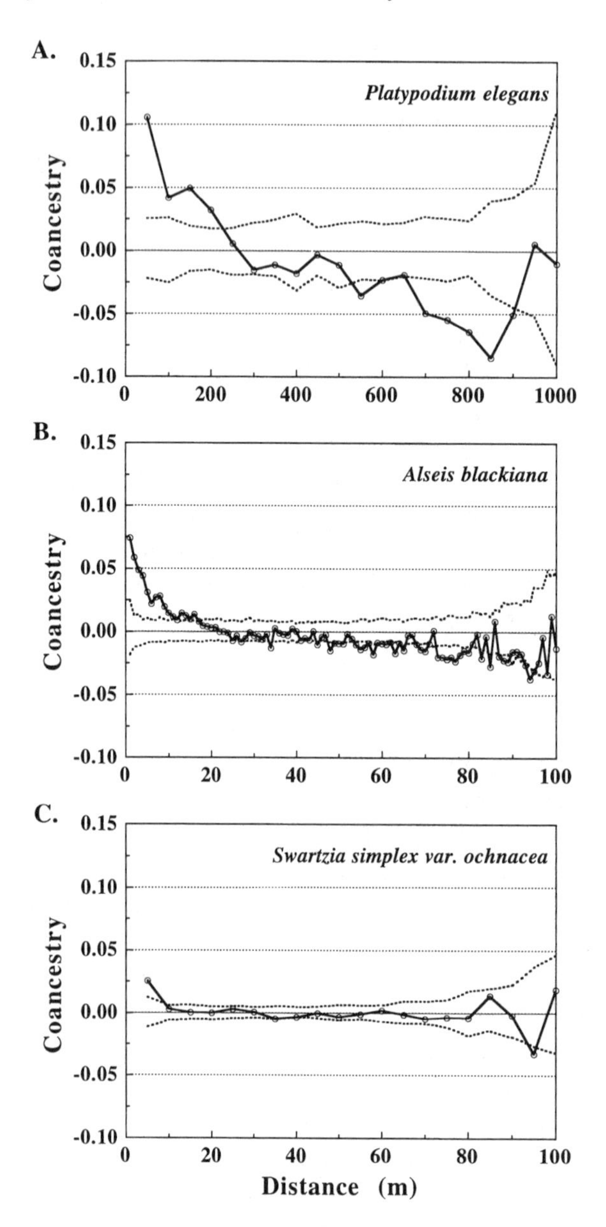

Figure 20.1. Multilocus estimates of kinship, a genetic structure statistic, plotted as a function of the spatial distance between saplings, for three tree species occurring in continuous moist tropical forest on Barro Colorado Island, Panama. The open circles represent mean estimates of kinship for successive distance intervals, with the dashed lines indicating a 95% bootstrap confidence envelope about the null hypothesis of no spatial genetic structure (kinship = 0). *(A)* 1–2 cm diameter at breast height individuals of *Platypodium elegans,* a large-seeded, wind-dispersed canopy tree with low adult densities (< 1/ha). *(B)* 1–2 cm dbh individuals of *Alseis blackiana,* a small-seeded, wind-dispersed canopy tree with high adult densities (> 30/ha). *(C)* Individuals smaller than 1 cm dbh of *Swartzia simplex* var. *ochnacea,* a large-seeded, animal-dispersed treelet with high adult densities (> 100/ha).

or pathogen-mediated density-dependent selection. Acting at both the population and community levels, these ecological processes may increase or decrease effective dispersal distances and, thus, the development of spatial genetic structure. Currently, there is insufficient empirical data to predict the effects of these processes on dispersal and genetic structure in either fragmented or continuous forest stands. However, given the potentially ameliorating effects of gene movement via pollen and random sources of mortality during stand thinning, these postdispersal processes may have relatively little effect on the genetic structure of adult tropical tree populations (Hamrick, Murawski, and Nason 1993).

Long-Distance Seed Movement

The tail of the leptokurtic distribution of seed dispersal may allow seeds to migrate into other populations or subpopulations, homogenizing genetic differences among populations. Although dispersal at this scale (i.e., interpopulation migration) is a critical facet of plant evolutionary ecology (Portnoy and Willson 1993), it has not been well studied in tropical trees. Existing studies are generally of the "point-source" type, in which seed movement from a single parent or group of parents is monitored (e.g., Augspurger 1986). While this method provides good estimates of short-range dispersal, it discounts the importance of long-distance moves because some proportion of seeds may leave the study area. This approach is also of limited utility in high-adult-density stands, where seed shadows are likely to overlap, unless seeds can be followed individually by the use of dyes (e.g., Boshier, Chase, and Bawa 1995) or genetic markers (e.g., Hamrick, Murawski, and Nason 1993).

Although the potential for long-distance, animal-mediated seed dispersal is great, actual patterns are difficult to predict. Common frugivores such as bats and birds can fly long distances and often have extensive foraging ranges. The foraging and seed deposition behavior of animals, however, introduces vagaries into the dispersal process that are not as pronounced in abiotic modes. Foraging and seed deposition patterns can lead to correlations among dispersed progeny that contribute to fine-scale genetic structure within populations and reduce the effectiveness of migration in countering genetic differentiation among populations (Levin 1988; Hamrick and Nason 1996). For example, seeds that are packaged and dispersed together, as in the multiseeded infructescences of figs or the genus *Piper* (Piperaceae), will consist of some combination of full and half sibs. Bats may carry sibs within a fruit from the parent to a feeding roost (Fleming 1981), so that although the potential for long-distance movement exists, recruited seedlings may consist primarily of patches of genetically correlated individuals. Similar correlations may arise if frugivores forage on single-seeded fruits preferentially within an individual crown or among genetically related individuals in a clump, then carry the seeds to a roost or a perch, as do some birds. These forms of correlated dispersal, though a potential form of migration, may actually enhance the development of spatial genetic structure (Whitlock and McCauley 1990).

DISPERSAL AND GENETIC STRUCTURE IN TROPICAL FOREST FRAGMENTS

The immediate effects of fragmentation on the genetic composition of tropical forests will be determined by the spatial scale and patterning of fragmentation relative to that of preexisting genetic and demographic structure. Over subsequent generations, changes in genetic composition will depend upon the presence and behavior of pollen and seed vectors and the relative occurrence of migration versus extinction and recolonization.

Immediate Effects of Fragmentation on Genetic Structure

The genetic structure of a forest will be altered by the initial loss and fragmentation of habitat. Aside from that resulting in local extinction, the degree of this change will depend on the way in which fragments are sampled from the landscape. The consequences of this sampling process can be considered in terms of their effects on the absolute number of individuals contained within a fragment and the spatial scale of fragmentation relative to that of preexisting genetic structure.

For tropical tree species exhibiting little or no spatial genetic structure over a landscape, the principal determinant of the rate of loss of variation within fragments will be their effective population sizes. Intuitively, the greater the number of individuals lost through deforestation, the more severe the genetic bottleneck. Nei, Maruyama, and Chakraborty (1975) have shown that the size of the bottleneck has a greater effect on the maintenance of rare alleles, while the temporal duration of the bottleneck has more influence on overall levels of genetic diversity. For example, a tropical tree species occurring at a density of two adults per hectare might have two hundred adults in a 100 ha tract of forest. If deforestation removes 90% of these individuals, the probability of retaining heterozygosity is nearly 100% for a locus with two alleles beginning at equal frequencies. Not until the initial frequency of one allele falls below .02 does the probability of fixation of the more common allele exceed 50%. The importance of the loss of novel or rare alleles should not be discounted, however, since some loci, such as those that determine self-incompatibility (Wright 1965) or pathogen resistance (Kinloch and Littlefield 1977), typically segregate for numerous low-frequency alleles that may be critical to the long-term viability of a population.

If spatial genetic structure exists in a species prior to fragmentation, it becomes necessary to consider not only the number of individuals in a fragment, but also the spatial scales of fragmentation and preexisting genetic structure. Localized pollen and seed dispersal decreases effective population numbers and contributes to the formation of genetic neighborhoods within continuous forest stands independent of microenvironmental variation. Levels of genetic variation within and among isolated fragments will be determined by the size of the fragments relative to the spatial scale of the preexisting genetic structure. Subsequent to fragmentation, the distribution of genetic variation within and among remnant patch populations will depend on the spatial scale of fragmentation relative to the area of preexisting neighborhoods. If fragments contain several neighborhoods, then genetic diversity will largely be partitioned within rather than among fragments, whereas if fragments are smaller than neighborhood areas, then the opposite will be true (fig. 20.2).

Allozyme analyses of pollen dispersal patterns and the spatial genetic structure of continuous forest populations suggest that neighborhoods of tropical tree species are often larger than many forest fragments. Alternatively, genetic structure may be associated with features of the environment, such as slope, soil type, or water availability, resulting in clines in gene frequencies (fig. 20.3). If surviving fragments are not randomly distributed relative to the environmental structure, then adaptive genetic variation will be lost at a rate that is higher than that expected for neutral variation. Such a scenario could have

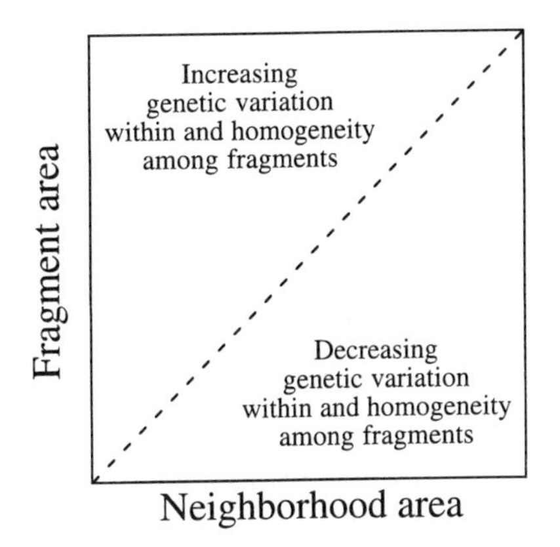

Figure 20.2. Theoretical relationship between neighborhood area in continuous forest and the effect of fragment area on the preservation of total genetic variation. Above the dashed diagonal, fragments are larger than a neighborhood and can contain several neighborhoods; below the diagonal, fragments fail to encompass preexisting neighborhoods.

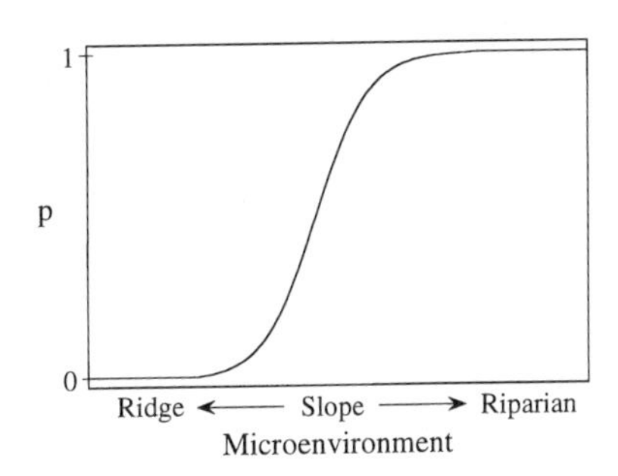

Figure 20.3. A hypothetical cline in the frequency of an adaptively significant allele (p) relative to microenvironmental variation in a continuously forested area. If the pattern of fragmentation does not effectively encompass the range of environmental variation, then adaptive genetic variation will be lost both within and among fragments. Such losses may occur, for example, when remnant forests are restricted to riparian or ridgetop locations as a result of human-associated land use patterns.

important fitness consequences and could hinder reforestation efforts. In general, fragmentation at a given scale will have different implications for species differing in the spatial dimensions of their genetic structure.

Short-term Effects of Fragmentation on Genetic Structure

Over time, alterations in dispersal dynamics in a fragmented landscape will influence the level of patch autonomy and the importance of stochastic processes. At equilibrium between mutation and genetic drift, the island model of population genetic structure (Wright 1931) predicts that four or more migrants per generation, drawn at random from surrounding subpopulations (breeding units), should counteract differentiation among subpopulations due to genetic drift. The number of migrants needed is the product of the effective size of the subpopulation (N_e) and the rate of gene flow (m). Thus, when a subpopulation is large, genetic drift can be countered by migration rates of a few percent or lower. In small subpopulations of, say, twenty individuals, however, migration rates on the order of 20% or higher are necessary to counteract loss of genetic diversity due to drift. Even higher rates may be necessary if the immigrants are genetically correlated, as when they originate from a small number of source trees or neighboring subpopulations (Whitlock and McCauley 1990; Hamrick and Nason 1996). While measures of genetic variability such as heterozygosity and quantitative genetic variation are less strongly affected by founder effects (Lande 1980), computer simulations indicate that in isolated subpopulations of nine and twenty-five individuals (representative of the population sizes of many tropical trees in fragments), genetic drift is the primary factor determining the loss of genetic variation (Lacy 1987). Rapid increases in the population growth rates of pioneer or other light-loving species in newly formed fragments and at fragment margins may alleviate the genetic consequences of bottlenecks in population size for these species, but regeneration of primary forest species, a far more diverse community, is less likely to be enhanced by changes in the biotic and abiotic environment associated with fragmentation. These predictions not only stress the importance of substantial pollen and seed movement between forest fragments, but also emphasize that a large proportion of successful recruits must result from immigration.

In the only study providing direct estimates of rates of pollen flow into fragmented populations of tropical trees (Nason, Herre, and Hamrick, in press), individual fig trees isolated on narrow peninsulas and small islands in Gatun Lake, Panama, had high fruit set, indicating substantial fig wasp-mediated pollen flow from surrounding forested areas. Further, genetic analyses indicated that seeds on these trees were fathered by numerous pollen donors. However, studies of other species with less specialized pollination systems have indicated density-dependent reductions in both the quantity and quality of pollen received by individuals in fragmented populations. Levin (1995) reviewed data on seed set in individuals of temperate, self-incompatible species that were spatially isolated from surrounding populations, and concluded that although seed set typically occurs, there is a 25–50% reduction on a per flower basis relative to individuals in intact populations. Similarly, Aizen and Feinsinger (1994a) found reductions in seed set associated with fragmentation for several bee-pollinated woody species occurring in an Argentine dry forest, suggesting that reproduction may be pollen limited in smaller fragments. In both studies, however, it is not clear to what degree the observed changes in pollination and reproductive success can be attributed to the direct effects of fragmentation on plant density as opposed to its more general effects on the environment.

The effect of tropical forest fragmentation on progeny fitness has been examined for the canopy tree *Spondias mombin* in central Panama. In a low-density population occurring in undisturbed primary forest on BCI, Stacy et al. (1996) found this small-bee-pollinated species to be highly outcrossing (mean outcrossing rate $t = 0.989 \pm 0.16$), with pollen dispersal occurring over distances of up to at least 300 m. Using a chi-square test, we (J. D. N. and J. L. H.) compared reproduction in three continuous forest *Spondias* populations with populations on small islands located in Gatun Lake. Populations occurring on BCI and closely adjacent Orchid Island (15 km^2 in combined area) are considered to be located in continuous forest of differing successional status. The island fragments range from approximately 0.5 to 4 ha in area and are located from less than 100 m to approximately 500 m from the nearest continuous forest. Vegetational structure on these islands varies from open and savanna-like to closed-canopy primary successional forest. Although there were no apparent differences in the sizes of the indehiscent, fibrous pericarps, germination in the greenhouse was significantly reduced for seeds from islands and primary successional continuous forests relative to those from older continuous forest populations (table 20.1). Genetic analyses of the mating system and rates of pollen gene

Table 20.1 Reductions in the viability of *Spondias mombin* seeds collected from continuous forest of differing successional stages and from island fragment populations as measured relative to an old forest population (FDP) occurring on Barro Colorado Island.

Population	Forest type	*N*	Reduction in seed viability relative to FDP population (%)
Continuous forest			
FDP	Old forest	2,331	
OI	Secondary successional	581	7.1[n.s]
LC	Primary successional	847	26.1*
One large island fragment (ca. 4 ha)			
DL (30 adults)	Primary successional	2,969	37.1*
Six small island fragments (0.5–1.0 ha)			
I (24 adults each)	Primary successional	779	38.5*

Note: In general, the germination rate of seeds from island populations was more than 35% lower than that from the FDP population.

n.s. Chi-square test of reduction in seed viability not significant.

*Chi-square test, $P < .001$.

flow are in progress, but preliminary results suggest that the reductions in seed viability may have resulted from inbreeding depression due to increased rates of selfing or biparental inbreeding in the spatially isolated populations.

Historically, large populations of outcrossing species have been expected to maintain high genetic load via the accumulation of deleterious recessive alleles (Wallace 1968). Given little evidence of strong fine-scale spatial genetic structure in the adults of tropical tree species (Hamrick, Murawski, and Nason 1993; Loiselle et al. 1995), fragment populations at their inception should exhibit little change from their source populations in the expression of genetic load. However, even in dioecious or self-incompatible species, biparental inbreeding within fragments over successive generations should increase inbreeding depression at a rate inversely proportional to the effective population size (Crow and Kimura 1970). For *Spondias mombin,* biparental inbreeding may be a factor in reducing seed viability in the isolated fragments because these small populations are nearly ninety years old, a period of time over which a number of generations of this early successional species may have transpired.

Further indications of the potential effect of fragmentation on pollen dispersal come from studies of pollinator assemblages. Research in Amazonian fragments indicates that breaks in the forest of as little as 80 m can diminish interfragment movements for some insects and birds (Bierregaard et al. 1992).

Fragmentation also creates edge effects that are often characterized by alterations in the composition of insect communities, with light-loving species tending to replace understory species as one nears the edge of a fragment (e.g., butterfly communities: Brown and Hutchings, chap. 7). The vegetation surrounding a fragment may also be influential in determining pollinator abundance and variety, and may serve as a source of non-native pollinators (e.g., Aizen and Feinsinger 1994b), which could further disrupt native plant-insect associations.

No genetic data are available on fragmentation and seed dispersal in tropical trees, but a number of studies have examined the behavior of different dispersal agents in fragmented landscapes (see also Harrington et al., chap. 19; Restrepo, Renjifo, and Marples, chap. 12). Charles-Dominique (1986) and Foster, Arce, and Wachter (1986) found that bird-dispersed and nonvolant mammal-dispersed taxa have more abrupt declines in seed movement across forest-pasture ecotones than do wind- or bat-dispersed taxa. These findings agree with the observations that wind dispersal may actually be facilitated by fragmentation (Fore et al. 1992) and that bats are known to travel long distances and traverse diverse habitats during foraging (Heithaus and Fleming 1978). In contrast, Guevara and Laborde (1993) found the seeds of mature forest tree species to be dispersed by birds over 300 m from seed sources. Similarly, Estrada et al. (1993) found that agricultural sites are heavily used by birds and mammals, and captured numerous birds and bats in pastures that were carrying seeds of forest canopy species. Additionally, Rolstad (1991) found that some tropical bird species use numerous fragments as part of their foraging ranges. Nevertheless, the high frequency of terra firme, understory bird taxa in Amazonia that are delimited by rivers (Capparella 1990) suggests that understory frugivorous birds, a class of seed dispersers important to many tropical woody species, may be physically or behaviorally incapable of movement through areas either naturally or artificially clear of forest cover.

In general, tropical tree populations should be coincidental with patch area only insofar as their pollen and seed vectors respond to the scale of patch structure and to changes in host density resulting from fragmentation. Therefore, it may be heuristic to consider "population units" aside from the physical landscape structure when examining the potential genetic consequences of fragmentation. While spatially structured genetic models, such as the island (Wright 1931) and stepping-stone (Kimura and Weiss 1964) models, can provide useful insights, they assume an equilibrium relationship between the effects of gene flow and genetic drift—a relationship that is unlikely to apply to contemporary fragments given their largely recent origins and the longevity of tropical trees. Furthermore, by arbitrarily treating remnant forest patches as "populations," these models confound population area with patch area. In fact, the actual genetic neighborhood, which should be closer to an actual "population unit," may extend well beyond the limits of a fragment, or alternatively, may consist only of individuals within the fragment and a few adjacent individuals outside. The use of genetic markers to determine parentage should clarify the actual population structure of tropical tree species in fragmented landscapes.

Long-Term Effects of Fragmentation on Genetic Structure

In landscapes in which gene movement between forested patches is low, the long-term effects of fragmentation will be to reduce genetic variation and individual heterozygosity within patches and to increase genetic variation among patches via genetic drift. Fragment-level extinctions or the destruction of individual fragments will further reduce the genetic diversity of species throughout the disturbed area and may result in the loss of novel genes as well as locally adapted gene complexes (Hamrick and Nason 1996). This potential for long-term erosion of genetic variability can be countered only by natural long-distance pollen or seed immigration, natural recolonization, or restoration efforts. It is therefore important to conserve both relict fragments and suitable habitat for pollinators and frugivores. Indeed, animal-pollinated temperate tree species characterized by small population sizes similar to those of many fragmented tropical tree populations have been shown to maintain levels of genetic diversity comparable to those of their more abundant and widespread congeners (e.g., Moran and Hopper 1987).

Extinction and recolonization will enhance genetic differentiation among populations if colonists originate from a single source population or if colonists are drawn at random from surviving populations, but at a rate less than half that of migration (Wade and McCauley 1988). Differentiation will be further enhanced if dispersal events are correlated, such as via multiseeded fruits, or if surviving populations are monomorphic for different combinations of alleles as a result of genetic drift (Whitlock and McCauley 1990; Hamrick and Nason 1996). Empirical studies of these processes are not available for tropical trees, due in part to their longevity and to the difficulty of documenting extinction in large patches. Moreover, for seed dispersal to be a colonization, rather than a migration, event, by definition the seed must fall outside the boundaries of any population in the area. Identification of these events requires a knowledge of the local neighborhood structure, because a currently unoccupied patch may, nonetheless, fall well within the boundaries of a widely dispersing species that simply is not occupying that particular space at that time.

In general, changes in the spatial genetic structure of a species over time are strongly influenced by patterns of gene flow and colonization. Therefore, models constructed to predict the spatial distribution of genetic variation in tropical tree species occurring in a fragmented landscape must incorporate the effects of disturbance on the dispersal and density-dependent foraging behavior of pollinators and seed dispersers. While certain classes of pollinators are known to forage over large areas and even to migrate substantial distances, the effectiveness of any such pollinators at maintaining reproductive success and genetic connectivity between forest fragments is virtually unknown. This stresses the need for ecological and genetic studies to compare pre- and post-fragmentation pollination and reproductive success, as well as neighborhood structure, in tree species with different population densities, breeding systems, and degree of pollinator specificity. Similarly, additional information is needed on the direct effects on seed dispersal of different seed dispersers and their responses to disturbance. Furthermore, because different animal guilds act as pollen vectors and seed vectors, the pollination and seed dispersal biologies of a species must be considered jointly if any attempt to predict the short-term and long-term consequences of fragmentation on genetic structure is to be successful.

GENERAL IMPLICATIONS

1. Analyses of spatial patterns of mating and genetic structure in continuous tropical forests indicate that the neighborhood areas of individual tree species may be on the order of

tens to hundreds of hectares, larger than the sizes of forest remnants in many disturbed landscapes.

2. If pollinator and frugivore movement is restricted by disturbance so that tree populations are localized to individual fragments, genetic variation should be partitioned primarily among as opposed to within fragments. However, if fragmented landscapes represent a metapopulation of interacting remnant subpopulations, then, depending on genetic correlations among migrants and the rate of migration versus local extinction and recolonization, genetic variation should be partitioned primarily within, as opposed to among, fragments.
3. Unfortunately, it is not presently clear how changes in the size and degree of isolation of fragments affect rates and spatial patterns of plant dispersal. Moreover, it is not well understood how plant characteristics such as successional status, growth form, environmental requirements, and dispersal syndrome (e.g., bat, bird, insect, wind) affect the number of individuals of a species within newly formed fragments as well as subsequent rates of migration.
4. Further ecological and genetic studies are needed before we can predict the genetic effects of fragmentation on surviving tree species. Given the dispersal capabilities of tropical animal vectors, however, many tree species may persist as metapopulations of great potential importance to reforestation and conservation programs.
5. Research results highlight the potential importance of relict fragments and other forest remnants as components of tree demes and reservoirs of genetic variation, and also as "stepping stones" for maintaining pollinator and frugivore populations.

ACKNOWLEDGMENTS

We thank V. Apsit and E. Stacy for contributing to discussions of the demographic and genetic consequences of habitat fragmentation in tropical forests.

21

Plant Dispersal in Fragmented Landscapes: A Field Study of Woody Colonization in Rainforest Remnants of the Mascarene Archipelago

Christophe Thébaud and Dominique Strasberg

UNDERSTANDING how habitat fragmentation affects ecological processes that involve animal and plant populations—and how this may influence species and habitat persistence in the long term—is a major challenge for conservation biologists (Simberloff and Abele 1982; Simberloff 1988; Saunders, Hobbs, and Margules 1991; Harris and Silva-Lopez 1992; Hanski 1994). Recently, numerous empirical studies of habitat fragmentation have addressed the effects of loss of original habitat, reduction of patch size and associated edge effects, and increasing isolation of habitat patches on the properties of populations inhabiting fragmented landscapes (e.g., Robinson et al. 1992; Margules, Milkovits, and Smith 1994; McCoy and Mushinsky 1994; Holt, Robinson, and Gaines 1995; Murcia 1995; Malcolm, chap. 14). However, the mechanisms that determine persistence or extinction in fragmented landscapes are still poorly understood (but see Hanski, Pakkala et al. 1995; Hanski, Poyry et al. 1995).

Clearly, the effects of habitat fragmentation vary among higher taxa and even among closely related species (Bierregaard et al. 1992; Aizen and Feinsinger 1994a; Margules, Milkovits, and Smith 1994). A landscape that appears fragmented for one species may not be so for another (e.g., Diffendorfer, Gaines, and Holt 1995). For a given species, population responses to fragmentation depend not only on the relative size of habitat fragments, but also on the spatial scale at which the fragments are arrayed, and how fragmentation mediates dispersal success across the landscape (Doak, Marino, and Kareiva 1992; Fahrig and Merriam 1994). The extent to which populations of species inhabiting fragments are arranged in a metapopulation, and the degree to which that metapopulation favors species persistence, appear to be largely determined by colonizing ability (Harrison 1989; Ebenhard 1991; Holt 1993). For a particular spatial arrangement of habitat patches, fragmentation could thus have a greater effect on species with lower colonization rates (Laurance 1991a; Lawton 1995).

Fragments are generally not embedded in a matrix of completely unsuitable habitats, particularly in forest regions (but see Uhl 1988). Although it has been rarely mentioned in the literature (but see Thomas 1994; Lamb et al., chap. 24), expansion of fragment vegetation through colonization of the surrounding matrix can be important for species persistence at a regional scale. For example, it may result in increases in local population sizes, and distances between fragments may be reduced, facilitating interfragment movements. This process, however, may ultimately depend on the extent to which the matrix surrounding fragments is colonized by new suites of organisms from the regional species pool, such as pioneer and exotic species, that can exploit the new habitat conditions caused by fragmentation. Such species can reduce the ability of local inhabitants to colonize intervening areas, either directly (e.g., via competition or predation) or indirectly (e.g., via disease transmission).

Although the importance of internal versus external dynamics of habitat fragments is still debated (Saunders, Hobbs, and Margules 1991), it seems clear that we must study the rate and pattern of colonization outside habitat fragments and how they differ among organisms if we are to understand the mechanisms by which species are affected by fragmentation in the long term (Ebenhard 1991; Lawton 1995; Pickett and Cadenasso 1995). There have been several attempts to quantify movements among fragments in animals (e.g., Harrison 1989; Diffendorfer, Gaines, and Holt 1995; Villard, Merriam, and Maurer 1995). In contrast, we know virtually nothing about how plant species move around fragmented landscapes (but see Matlack 1994a for a notable exception), despite the central role played by dispersal and establishment limitation in many terrestrial plant communities (Harper 1977; Grubb 1986; Huston and Smith 1987).

In this chapter, we report on woody plant colonization in a naturally fragmented landscape in lowland tropical rainforest on the island of La Réunion (Mascarene Archipelago). The purpose of this study was to assess how quickly plant species inhabiting forest fragments colonize intervening areas and to explore the mechanisms that may underlie the observed patterns.

METHODS

The Island of La Réunion

La Réunion, located in the southwestern Indian Ocean, 800 km east of Madagascar, is the largest island (2,512 km^2) of the Mascarene Archipelago, which also comprises Mauritius and Rodrigue. The island probably formed above a fixed hot spot and consists of two coalescing shield volcanoes, extinct Piton des Neiges (3,069 m above sea level [ASL]) and Piton de La Fournaise (2,560 m ASL), which is one of the most active volcanoes on earth, with 153 eruptions recorded during the past 350 years (Bachelery 1981). The tropical climate is dominated in winter by southeast trade winds and in summer by the western Indian cyclone system (Cadet 1977).

Like other tropical archipelagoes (Carlquist 1974; Vitousek 1988; Primack 1993), the Mascarene Archipelago as a whole has been greatly affected by human disturbance since it was first colonized in the early sixteenth century (Cheke 1987a). Habitat destruction has been tremendous: virtually all the original forest covering Rodrigue has been destroyed, while a mere 5% of the original cover has been left in Mauritius (Strahm 1994). In contrast, on the island of La Réunion, 57,000 ha of primary forest still remains, constituting about 25% of the estimated original extent (Strasberg 1994). However, as a result of deforestation and rugged topography, La Réunion's remaining forest is severely

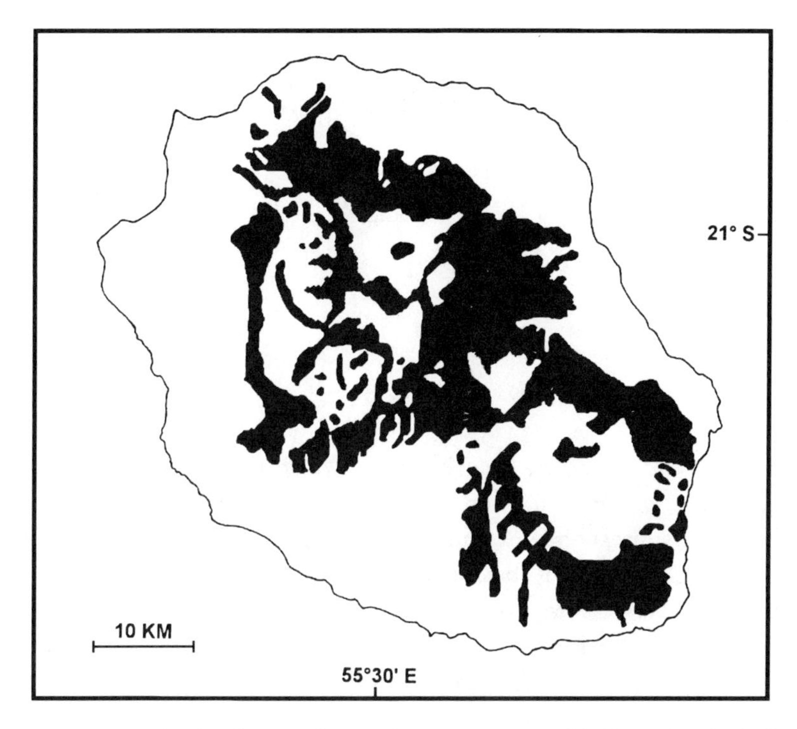

Figure 21.1. The distribution of native forest remnants (in black) on the island of La Réunion. Forest is thought to have covered most of the island, except for mountaintops, before human colonization.

fragmented (fig. 21.1), with large tracts found only above 500 m elevation, and no more than 1% of the lowland forest remaining (fig. 21.2). Ironically, most lowland forest remnants are located on the slopes of the active volcano, where they often take the form of forest islands embedded in a matrix of lava flows of various ages (Cadet 1977; Strasberg 1994). This close juxtaposition of primary forest fragments and lava flows that are geologically similar but differ in age provides an ideal system in which to determine the rates and patterns of plant colonization from fragments into intervening areas.

Study Area

The study was conducted at Grand Brûlé, on the lower southeastern flank of Piton de La Fournaise volcano (21°16′ S, 55°48′ E) in the lowland rainforest zone (elevation 0–400 m: Strasberg 1994). Mean annual rainfall varies from 4,000 to 5,000 mm, and annual mean temperature is near 22°C (Cadet 1977). Typically, forest remnants across the study area vary in size from 0.5 to 15 ha, and are embedded in a landscape matrix of recent lava flows, most of whose ages range from 5 to 300 years (Strasberg 1994). Descriptions of the vegetation are presented by Cadet (1977) and Strasberg (1994, 1996).

The forest remnant flora is particularly diverse and luxuriant, with *Doratoxylon apetalum* (Sapindaceae), *Labourdonnaisia callophylloides* (Sapotaceae), *Molinea alternifolia*

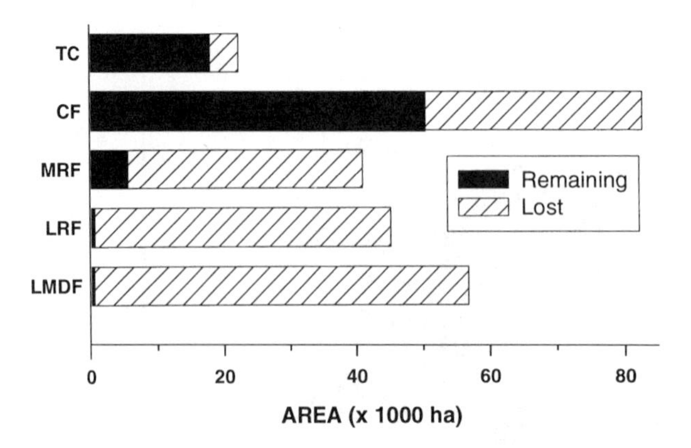

Figure 21.2. Changes in forest cover on the island of La Réunion since human colonization. Abbreviations: LMDF, lowland and montane semi-dry forest; LRF, lowland rainforest; MRF, montane rainforest; CF, cloud forest; TC, tropical-alpine vegetation. (Adapted from Strasberg 1994.)

(Sapindaceae), *Nuxia verticillata* (Loganiaceae), *Ocotea obtusata* (Lauraceae), *Polyscias repanda* (Araliaceae), and *Syzygium borbonicum* (Myrtaceae) being the most common tree species. The understory is often dominated by shrubs such as *Chassalia corallioides* (Rubiaceae), *Gaertnera vaginata* (Rubiaceae), and the tree fern *Cyathea borbonica,* while the canopy supports a very diverse assemblage of epiphytes, mostly orchids and ferns. The main characteristics of the forest structure include a lack of emergent trees, high densities of trees with small crowns, and a scarcity of vines (Strasberg 1996). Moreover, most woody species can become established on young lava (Cadet 1977).

In the last three centuries, people have introduced numerous plant species to La Réunion, which have now spread throughout the landscape, with lava flows serving as invasion corridors (Macdonald et al. 1991). The most prominent exotic plants established in the system are *Ardisia crenata* (Myrsinaceae), *Boehmeria penduliflora* (Urticaceae), *Rubus alceifolius* (Rosaceae), *Psidium cattleianum* (Myrtaceae), and *Casuarina equisetifolia* (Casuarinaceae) (Macdonald et al. 1991).

Data Collection and Analysis

From a walk-through survey, we selected three small forest remnants, ranging in size from 2.5 to 12 ha, that showed little or no sign of human activity and were clearly bounded by recent (<100 years) lava flows of greater than 150 m width. The three remnants were located between 100 and 300 m ASL and were no more than 500 m apart. They shared the same aspect and general topography, and also had a similar, shallow layer of soil derived from volcanic ash (andisols; <10 cm deep) directly over basalt (Raunet 1991; D. Strasberg, pers. obs.). The selection of these remnants minimized physical, climatic, and soil gradients that might influence plant distributions. Ages of bordering lava flows ranged from 15 to 91 years at the time of the study, and all lava flows consisted of the same basalt. Two remnants were contiguous with the same 15-year-old lava flow, but one of them was also bordered on its opposite side by a 48-year-old flow. Therefore, we used four sites in this study: sites A and B along a 15-year-old flow, site C along a 48-year-old flow, and site D along a 91-year-old flow, with sites A and C sharing a common forest remnant.

Transects 60 m long, extending 50 m onto the lava flow and 10 m into the forest remnant, were laid out perpendicularly to the forest-flow boundary. Three transects were established at 20 m intervals along the forest edge at each site except site D, where we established only two transects. Plots of 2 × 2 m were centered at 3 m intervals along each transect. In each plot, we counted all woody plants taller than 0.25 m and recorded species identity.

Prior to analyses of colonization rates, richness in native species known to occur in lowland rainforest remnants was computed on a per plot basis. We then examined how species richness varied with distance from the forest edge to ensure that our study was performed at the appropriate spatial scale. This was done by performing tests of linear trends in species richness over the distance gradient using a bootstrap procedure (SAS Institute 1992).

For each species in our study, we calculated the colonization rate (m/year) for each transect as the distance (m) from the forest edge to the farthest individual of the species on the lava flow divided by the age (in years) of the lava flow. When a species was present in the forest remnant only, we considered the colonization rate to be zero. When a species was absent from a transect, rates were reported as missing values.

We scored each species for dispersal mode (wind-dispersed or vertebrate-dispersed), growth form (tree, shrub, or vine), and vegetative propagation (absent or present) to determine whether these traits were associated with variation in species' colonization rates. The effects of site, transect within a site, dispersal mode, growth form, and vegetative propagation on colonization rates were analyzed with an analysis of variance (SAS Institute 1990). For vertebrate-dispersed species, we further examined the effect of fruit size on colonization rates. Fruit lengths were obtained from a large database on Mascarene fleshy-fruited plants that we are currently constructing.

RESULTS

Species Richness as a Function of Distance from Remnants

Twenty-eight species occurring in forest remnants had colonized lava flows (table 21.1). In contrast, only three forest species *(Coffea mauritiana, Ficus lateriflora,* and *Pandanus purpurescens)* were restricted to remnants and never colonized intervening areas. Four additional species *(Blechnum tabulare, Mussaenda arcuata, Senecio ambavilla,* and *Stoebe passerinoides)* were present only on lava flows and were eliminated from the analyses because all are pioneer species, never found in remnants.

Colonization declined with distance from the forest remnant at all sites (all $P < .008$ after adjusting for multiple tests; fig. 21.3). In addition, species richness decreased by 80–90% over a distance of 50 m at all but the oldest site (D), where the decline in richness was only 30–40%. This finding indicates that a 50 m transect was appropriate to capture most species that had colonized lava flows from our forest remnants.

Factors Affecting Colonization Rates

Differences in colonization rates among species that successfully colonized lava flows were substantial, as rates ranged from 0.01 to 2.07 m/year (see table 21.1). About 40% of the species had very low rates (< 0.1 m/year), while only two tree species, the wind-dispersed

Table 21.1 Woody species and mean colonization rates outside forest remnants at Grand Brûlé, La Réunion, Mascarene Archipelago.

Species	Rate (m/yr)	Growth[a]	Dispersal[b]	Fruit length
Agauria salicifolia	2.07 (11)	T	W	
Sideroxylon borbonicum	1.09 (7)	T	F	11.9
Nuxia verticillata	0.78 (11)	T	W	
Danais fragrans	0.64 (11)	V	W	
Aphloia thaeiformis	0.53 (11)	T	F	11.1
Doratoxylon apetalum	0.47 (11)	T	F	17.9
Antirhea borbonica	0.44 (11)	T	F	11.6
Polyscias repanda	0.33 (11)	T	F	6.7
Piper pyrifolium	0.26 (4)	V	F	8.0
Geniostoma borbonicum	0.25 (9)	S	F	17.6
Acanthophoenix rubra	0.24 (2)	T	F	9.4
Pittosporum senacia	0.23 (8)	S	F	7.2
Cordyline mauritiana	0.23 (1)	S	F	5.1
Weinmannia tinctoria	0.22 (3)	T	W	
Smilax anceps	0.20 (6)	V	F	9.0
Ficus reflexa	0.17 (4)	T	F	9.3
Molinaea alternifolia	0.17 (11)	T	F	12.7
Gaertnera vaginata	0.12 (11)	S	F	14.5
Ocotea obtusata	0.07 (9)	T	F	18.9
Memecylon confusum	0.06 (5)	S	F	6.8
Labourdonnaisia calophylloides	0.06 (6)	T	F	29.6
Callophyllum tacamahaca	0.06 (5)	T	F	48.8
Cyathea borbonica	0.04 (10)	T	W	
Casearia coriacea	0.04 (5)	T	F	25.6
Syzygium sp.	0.03 (8)	T	F	10.6
Ficus mauritiana	0.03 (4)	T	F	45.1
Cnestis glabra	0.02 (5)	V	F	22.9
Chassalia coralloides	0.02 (8)	S	F	10.3
Xylopia richardii	0.01 (4)	T	F	40.1
Coffea mauritiana	0.00 (1)	S	F	21.2
Ficus lateriflora	0.00 (1)	T	F	16.7
Pandanus purpurescens	0.00 (4)	S	F	55.2

Note: The number of transects in which a species was present is given in parentheses.

a. Growth form: (T) Tree, (S) Shrub, (V) Vine.

b. Dispersal mode: (W) Wind-dispersed, (F) Fleshy fruit, seeds dispersed by vertebrates.

Agauria salicifolia and the fleshy-fruited *Sideroxylon borbonicum,* had colonization rates greater than 1 m/year. Variation in colonization rates was best explained by differences in dispersal mode and growth form (table 21.2). Colonization rates were on average about 3.5 times higher in wind-dispersed species than in species producing fleshy fruits dispersed by vertebrates (fig. 21.4B). Comparisons of least-squares means (results not shown, but see fig. 21.5) revealed, however, that the difference between dispersal modes was more pronounced on younger (sites A and B) than on older sites (site D). This difference accounted for much of the variation in colonization rates among sites (table 21.2; fig. 21.4A). Trees colonized significantly faster than shrubs (fig. 21.4C), but

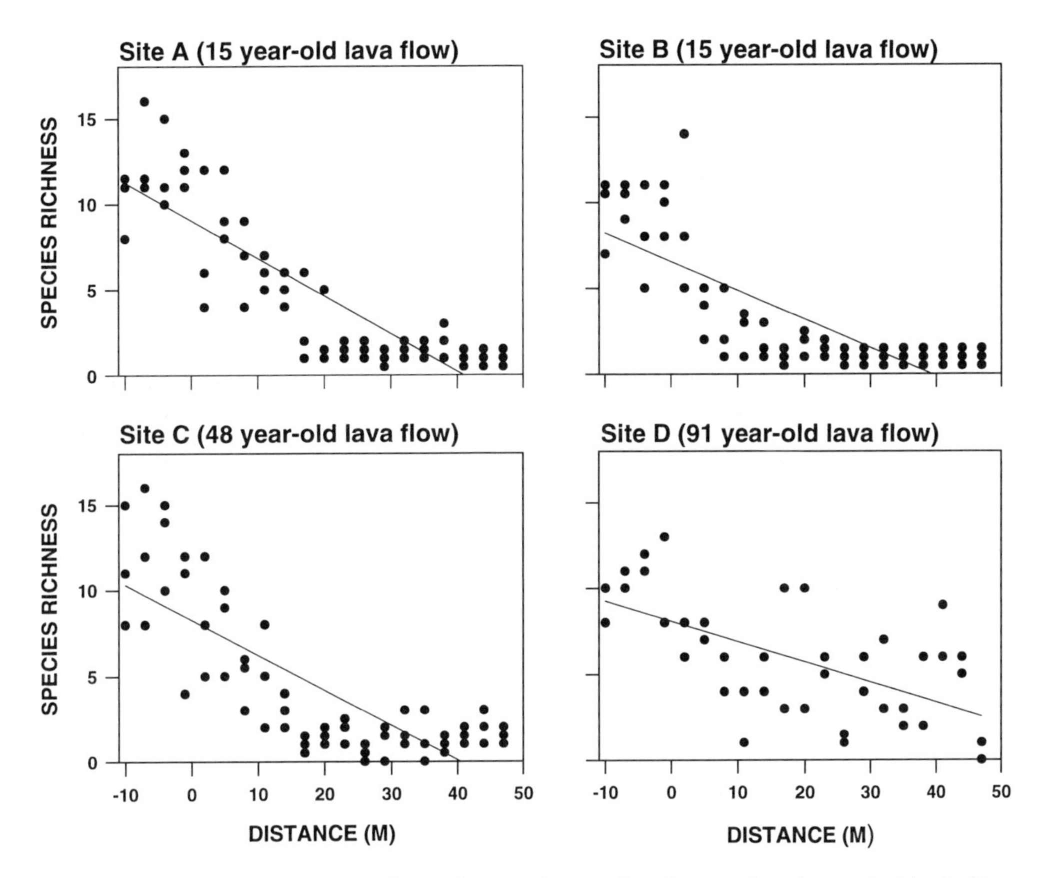

Figure 21.3. Species richness on lava flows relative to distance from forest at four sites on the island of La Réunion. Negative distance values indicate plots inside forest remnants; positive distance values indicate distance into lava flows. Regression lines are shown for the sake of clarity only.

Table 21.2 Analysis of variance of colonization rates of forest remnant species into contiguous lava flows.

Source	df	SS	*F*	*P*
Site	3	9.44	11.64	.0001
Transect (site)	7	2.37	1.25	.2765
Dispersal mode	1	15.34	56.76	.0001
Growth form	2	4.68	8.65	.0002
Vegetative propagation	1	0.01	0.04	.8344
Site * dispersal mode	3	8.57	10.57	.0001
Site * growth form	3	3.45	4.25	.0062
Error	197	53.25		

Note: Colonization rates were calculated as the distance from the forest edge to the farthest individual of the species on the lava flow, divided by age of the lava flow. Partial (Type III) sums of squares were used to construct *F*-tests.

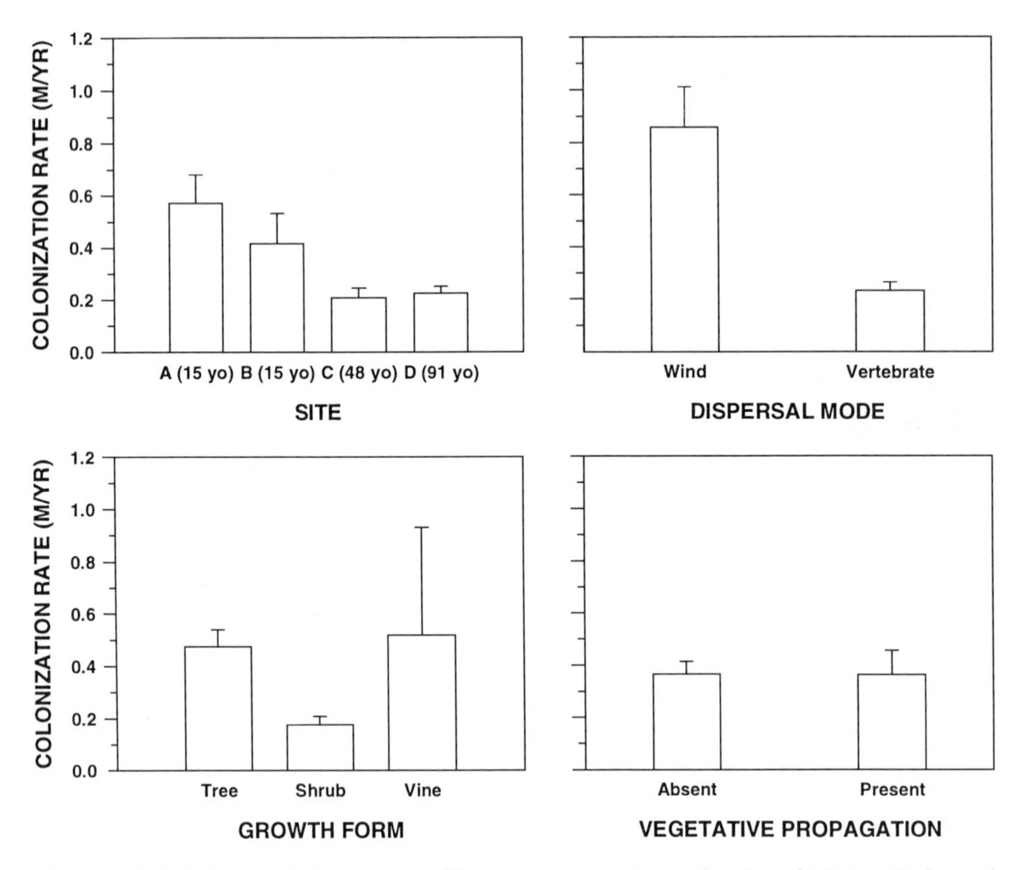

Figure 21.4. Variation in colonization rates of forest remnant species as a function of *(A)* site, *(B)* dispersal mode, *(C)* growth form, and *(D)* presence or absence of vegetative propagation. Data shown are the mean ± 1 SE.

differences between growth forms were statistically significant only for the two youngest sites (A, B: $P < .0001$ and $P < .004$, respectively), with trees and shrubs showing similar rates of colonization (0.22 m/year) at the oldest sites (C, D). In contrast, the presence or absence of vegetative propagation had no effect on colonization rates (table 21.2; fig. 21.4D).

Fruit Size and Colonization Rate

For vertebrate-dispersed species, there was a significant negative relationship between rates of colonization and fruit size ($P < .002$, $r_s = -.24$, Spearman rank correlation; fig. 21.6). As a check on the validity of the analysis, we excluded the three species restricted to the forest remnant because they may fail to colonize lava flows owing to a lack of suitable shaded habitat, and we used mean colonization rates instead of raw values. With these adjustments, however, colonization rates and fruit size were still significantly negatively correlated ($P < .005$, $r_s = -.53$, Spearman rank correlation).

DISCUSSION

Colonization Rates

Our study illustrates the process by which vegetation in forest fragments can expand through a gradual spread of species into intervening areas. However, most colonization rates reported here were low (< 1 m/year), and as distance from the remnant increased, there were consistent and marked declines in the number of species colonizing the matrix, even on the oldest (91 years old) flows. This observation makes clear that woody plant species move slowly across this particular fragmented landscape.

What limits the rate of plant colonization outside forest remnants? In our study system, all colonizing plants must have been recruited from the seed rain because no seed bank exists in the new lava substrate. Availability of suitable microsites may limit seedling establishment on lava flows (Drake and Mueller-Dombois 1993). At Grand Brûlé, germination and seedling establishment are often restricted to cracks or other microtopo-

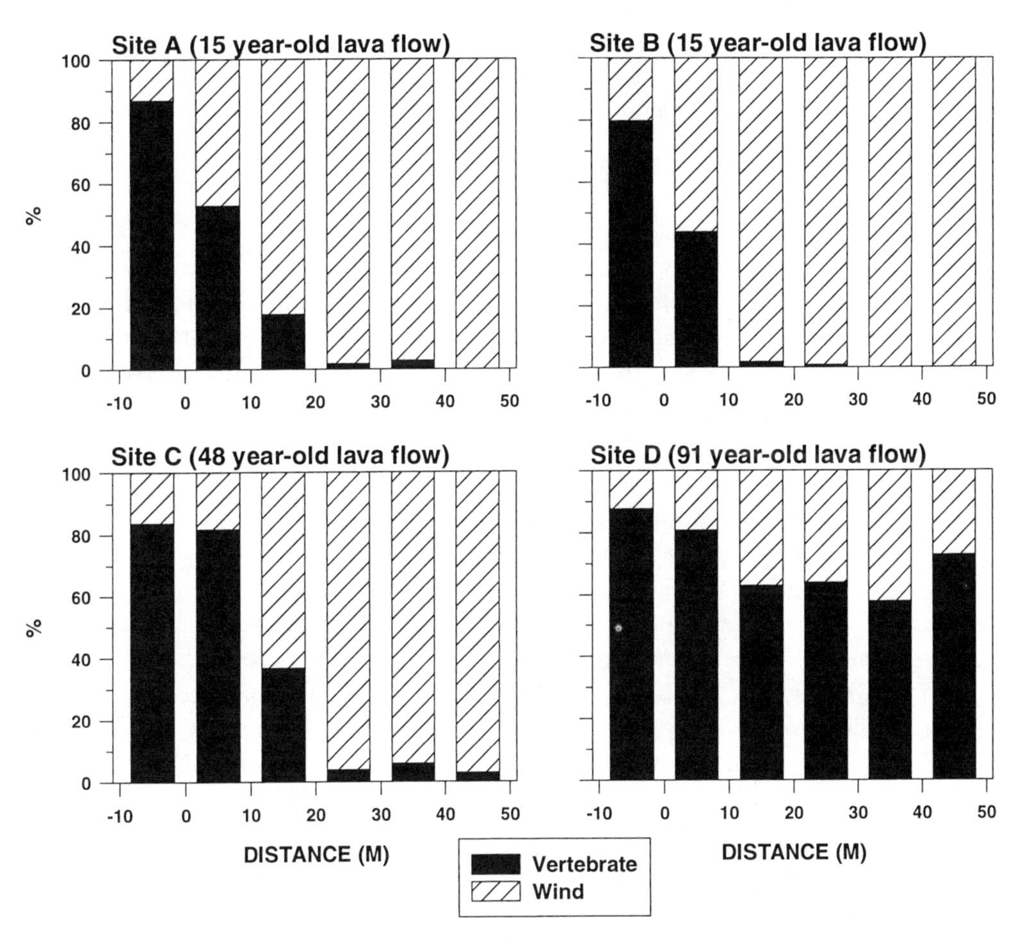

Figure 21.5. Changes in prevalence of dispersal mode with distance from forest remnants at four sites on the island of La Réunion. Data are mean percentages based on species densities.

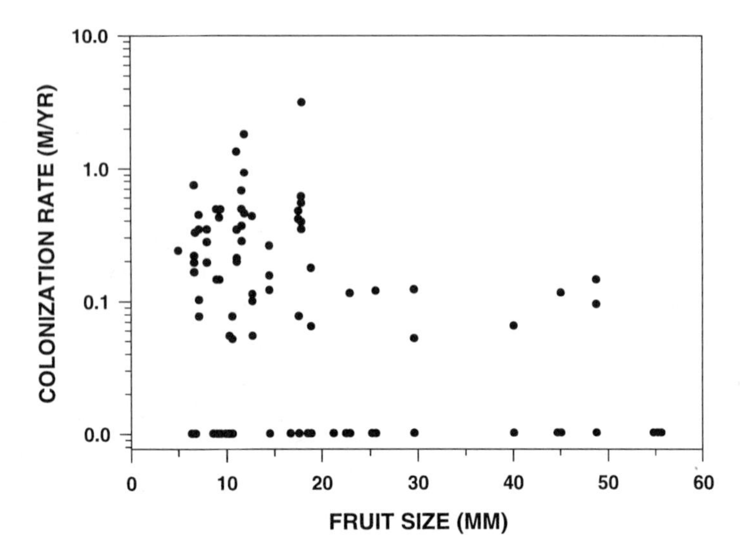

Figure 21.6. Rates of colonization of fleshy-fruited forest remnant species in relation to fruit size. Data shown are mean rates per species per site where the species was present.

graphic features occurring at relatively low densities (D. Strasberg and C. Thébaud, pers. obs.). Lava is chemically uniform, but structurally variable (Bachelery 1981). The structure of lava, however, never changes in a systematic way across flows (Strasberg 1994), and crack density is unlikely to be related to distance from forest fragments. In addition, there are no obvious changes in substrate conditions due to successional processes along the length of the observed gradients, at least at the temporal scale of this study (D. Strasberg and C. Thébaud, pers. obs.). This is probably related to the low average stem densities on the flows, even close to forest edges (< 5 saplings or small trees per m^2: Strasberg 1994). It thus appears that substrate variation can hardly explain the pattern of colonization limitation observed in this study.

Plant establishment may also be limited if the matrix surrounding the forest remnants is colonized by new suites of species that exploit the microsites available for establishment, thereby reducing the carrying capacity of the environment. Macdonald et al. (1991) suggested that this might be the case in lowland rainforest remnants of La Réunion, owing to extensive invasions of recent lava flows by exotic plant species. We observed, however, that, despite a gradual increase in the proportional abundance of exotic species with distance from forest remnants, native forest species remain dominant on the flows, and we failed to detect any negative correlation between the local densities of native and exotic species. This finding leads us to conclude that colonization of the lava matrix must be limited by seed dispersal in rainforest remnants on La Réunion. In other words, distance appears to be the major impediment to colonization for most woody species inhabiting those remnants.

Differences among Species

Colonization rates varied among species, suggesting that species differ in their ability to move over fragmented landscapes. Differential colonization rates among plant species have been attributed to their demographic attributes and dispersal abilities (Harper 1977). In our study, colonization rate was most strongly related to dispersal mode and, to a lesser extent, to growth form.

Trees showed a higher rate of colonization than shrubs, but the effect was moderate and was significant only on the younger flows. There are two possible explanations for differences in colonization dynamics between trees and shrubs. First, plant height may play an important role in determining seed fall patterns around fruiting individuals, although such an effect is likely for wind-dispersed, but not vertebrate-dispersed, species (Willson 1993). Second, individual trees may have greater seed production than individual shrubs, but this difference may be counterbalanced by differences in local abundance, as shrubs are more abundant in these forests (Strasberg 1996).

More importantly, wind-dispersed species were much better colonizers than fleshy-fruited species dispersed by vertebrates. In other systems, the pattern is usually reversed (Willson 1993; Matlack 1994a; Whittaker and Jones 1994). Vertebrate-dispersed seed shadows often show no distance effect because animals that carry seeds are capable of long-range movement, and fleshy-fruited species are generally consumed by a variety of vertebrate species differing in their foraging behavior (e.g., Willson and Crome 1989). At Grand Brûlé, most vertebrate-dispersed species exhibit very low colonization rates, with only one species *(Sideroxylon borbonicum)* having a rate greater than 1 m/year. Colonization rate also was negatively correlated with fruit size (see fig. 21.6); rates of colonization in species having fruits longer than 20 mm (eight of twenty-seven vertebrate-dispersed species) never exceeded 0.1 m/year.

Loss of Disperser Faunas

Why did vertebrate-dispersed species have unusually low colonization rates? One possible explanation is that vertebrate-dispersed plants have strong habitat preferences and become established only under forest cover. Most of the woody species on La Réunion, however, are to some extent niche generalists, capable of using a broad range of successional phases (Cadet 1977). In addition, most species involved in this study did indeed colonize lava flows, although at slow rates (see table 21.1).

Alternatively, it is possible that plant-disperser interactions have been altered by changes in disperser assemblages owing to habitat fragmentation and extinctions of animal species on the island (Cox et al. 1991). Hence, very low colonization rates may reflect either a complete absence of dispersers or the loss of efficient dispersers, such as flying foxes and birds, that can travel long distances before defecating seeds. Although the Mascarene Archipelago gained its reputation from the early extinction of the Mauritius dodo, it has in fact suffered many additional animal extinctions (Cheke 1987a). In the last three hundred years, an estimated 58% of the land vertebrates have gone extinct, with La

Réunion itself losing twenty-seven of forty-two species, including two flying foxes, two fruit pigeons, three or four parrots, one tortoise, and two skinks (Strasberg 1994). Among extant species, only three small (< 60 g) native birds *(Hypsipetes borbonicus, Zosterops borbonicus,* and *Z. olivacea)* and an introduced turtledove *(Streptopelia picturata)* are potential seed dispersers, and only one of these *(H. borbonicus)* is likely to play a significant role in seed dispersal because the others are mostly insectivorous (*Zosterops* spp.) or seed-eaters (*S. picturata*) (Barre and Barau 1982; Cheke 1987b; C. G. Jones 1987; C. Thébaud and D. Strasberg, pers. obs.). Such a dramatic loss of vertebrate species offers a compelling explanation for the low colonization rates of many, if not most, fleshy-fruited species in this study.

General Implications

1. Theoretical metapopulation models have recently emphasized the importance of dispersal as a process that may promote persistence of species at a regional scale. Unfortunately, quantifying dispersal in fragmented landscapes is no easy task. Our results indicate, however, that knowledge of how species recolonize fragmented landscapes, and why they differ in their colonization ability, can be essential if we are to understand how habitat fragmentation contributes to declines of species.
2. Most woody species from the lowland rainforest remnants of the island of La Réunion colonize the landscape very slowly and only over short distances. The degree to which colonization rates are similar in upland forest remnants on La Réunion remains an open question, although we suspect that the patterns are not dramatically different.
3. Different plant species may respond very differently to fragmentation. We found that fleshy-fruited woody species dispersed by vertebrates had remarkably low rates of colonization. This finding probably relates to the extinction of most seed dispersers on the island of La Réunion, and may well reflect the inability of many plants to adjust to disperser loss (Cox et al. 1991; Leigh et al. 1993). Given the high incidence of fleshy-fruitedness in the woody flora of La Réunion, these findings cast some doubt on the long-term survival prospects of many species, including those that are still locally abundant in forest remnants.

ACKNOWLEDGMENTS

We wish to thank Jimmy Turpin for invaluable help during fieldwork, Michel Borderes, of the Office National des Forêts, for permitting work at Grand Brûlé, and Joël Dupont, who provided unpublished maps and botanical expertise. We are grateful to Dan Simberloff for his continuous encouragement and comments on an early draft of this chapter. Comments by Illka Hanski, Andrew Graham, Bill Laurance, and an anonymous reviewer greatly improved the chapter. Logistic support for this research was provided by Jacques Figier. This study was supported by French Ministère de l'Environment grant SRETIE-91206 and a predoctoral fellowship from Conseil Régional de La Réunion to D. Strasberg. Manuscript preparation was supported by the core grant to the Natural Environment Research Council (NERC) Center for Population Biology.

22

Long-Term Survival in Tropical Forest Remnants in Singapore and Hong Kong

Richard T. Corlett and I. M. Turner

ALTHOUGH the rate of tropical deforestation has accelerated tremendously in recent decades, the process itself is not new. Large areas of forest were cleared in many parts of the Tropics hundreds of years ago (Whitmore 1990). These past deforestation events provide an opportunity to study long-term processes that cannot be studied experimentally, including secondary succession over long periods (Corlett 1995) and the fate of species in isolated forest fragments (Turner and Corlett 1996). Inevitably, however, the greater chronological depth of long-term studies is paid for by a loss in precision: there can be no pre-fragmentation baseline studies, no detailed impact histories, and no reliable controls.

The experimental approach to forest fragmentation is exemplified by the Biological Dynamics of Forest Fragments Project, which involves fragments that have been isolated for up to fifteen years (Bierregaard et al. 1992; Kapos et al., chap. 3; Brown and Hutchings, chap. 7; Tocher, Gascon, and Zimmerman, chap. 9; Bierregaard and Stouffer, chap. 10; Malcolm, chap. 14). Longer-term investigations have had to make use of rainforest fragments that were isolated by deforestation (e.g., Estrada et al. 1993; Laurance 1994) or flooding (e.g., Leigh et al. 1993) well before the start of the study. Many studies have examined changes in fragments created in the twentieth century, but older fragments have so far received little attention, in part, presumably, because well-documented fragmentation histories from earlier periods of deforestation are rare. We need to know the long-term effects of fragmentation now, however, so that we can respond effectively to the present deforestation crisis. How large must rainforest remnants be to preserve a significant fraction of the original flora and fauna, and for how long can they do so? Which species are most vulnerable over the course of decades or centuries? Is a new equilibrium eventually reached, or does species loss continue? Such questions cannot be answered by experimental studies.

We present here studies of fragmented forests in two parts of tropical Asia: Singapore, which probably has the best-documented deforestation history in Southeast Asia (Corlett 1992a,b), and Hong Kong, with a considerably longer but more poorly documented history of human impact (Dudgeon and Corlett 1994; Corlett 1997). In Singapore, most

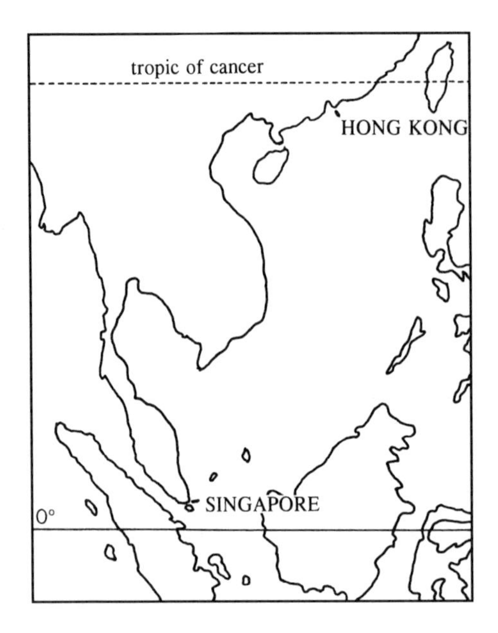

Figure 22.1. Map of Southeast Asia showing the locations of Singapore and Hong Kong.

Table 22.1 The most species-rich vascular plant families in the forest floras of Singapore and Hong Kong (rank order in parentheses).

Family	Singapore	Hong Kong
Pteridophytes	130 (2)	135 (1)
Annonaceae	61 (5)	10
Cyperaceae	11	24 (9)
Euphorbiaceae	102 (4)	33 (5)
Fagaceae	21	33 (5)
Lauraceae	41 (10)	30 (7)
Leguminosae	40	40 (3)
Melastomataceae	43 (9)	9
Moraceae	58 (6)	26 (8)
Myrsinaceae	20	22 (10)
Myrtaceae	46 (8)	8
Orchidaceae	161 (1)	67 (2)
Palmae	49 (7)	8
Rubiaceae	118 (3)	35 (4)

existing fragments have been isolated for at least 150 years, while in Hong Kong, fragmentation was completed more than 350 years ago.

STUDY AREAS

Singapore and Hong Kong are both densely populated, tropical, Asian city-states. Until the last few centuries they would have been linked to each other by continuous forest cover, broken only by rivers and shallow marine straits less than 1 km wide (fig. 22.1). Their floras and faunas are similar at the family and generic levels (table 22.1), although some equatorial plant families and genera are absent from Hong Kong. They differ considerably, however, in both their physical environments and their histories.

The Republic of Singapore is an independent state of 2.8 million people at the southern tip of the Malay Peninsula, 137 km north of the equator. It consists of the island of Singapore (574 km^2) and more than fifty smaller islands. The main island has a mean elevation of only 15.1 m and a maximum, at Bukit Timah Hill, of 162 m. The central part of the island, which includes all the surviving primary forest remnants, is underlain by granite. Singapore has a virtually aseasonal equatorial climate (fig. 22.2). The mean annual rainfall in the center of the island is 2,200–2,400 mm, and no month has a mean rainfall of less than 140 mm. Singapore is outside the typhoon belt, and strong winds are rare.

Hong Kong is a British Dependent Territory on the South China coast, 130 km south of the tropic of Cancer, with a population of 6.0 million. Hong Kong consists of a section of the Chinese mainland (Kowloon and the New Territories, 782 km^2) and numerous islands, the largest of which are Lantau Island (142 km^2) and Hong Kong Island (78 km^2). The total land area is 1,076 km^2. The topography is extremely rugged, and there is little natural flat land. The highest point is at Tai Mo Shan (957 m) in the New Territories. The commonest rock types are volcanic tuffs (which support most of the remaining forest) and granites. Hong Kong has a monsoonal climate with both rainfall and temperature seasonal (fig. 22.2). The January mean (15.8° C) and absolute minimum (0° C in January, 1893) temperatures are both exceptionally low for the lowland Tropics. Mean annual rainfall over most of the Territory (and all the forest) is in the range of 1,600–3,000 mm, of which more than 70% typically falls between May and September. However, winter low temperatures reduce the effect of the dry season, and the majority of tree species are evergreen. Typhoons are common along the South China coast, and there have been several direct hits on Hong Kong this century.

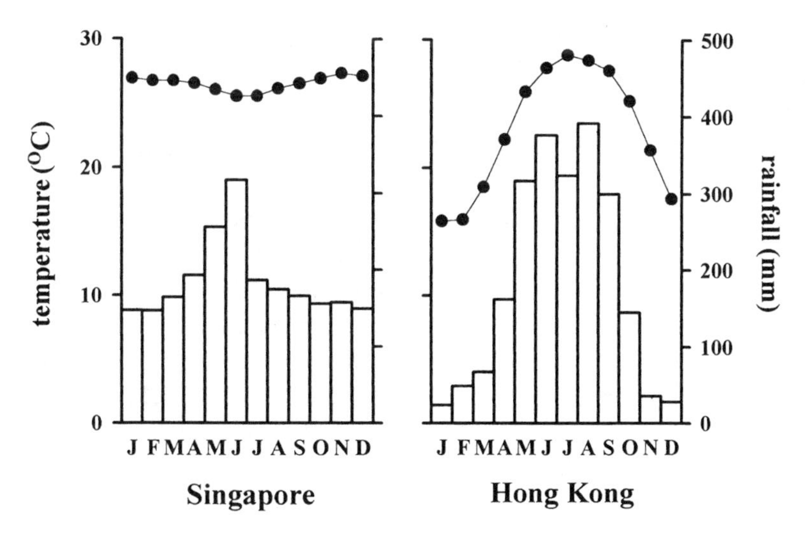

Figure 22.2. Long-term monthly mean temperature and rainfall in Singapore and Hong Kong.

DEFORESTATION

Singapore

Were it not for human activities, Singapore would be almost entirely covered in forest. The main forest type would be lowland evergreen rainforest. The family Dipterocarpaceae would dominate the emergent layer, as it characteristically does in the western Malesian forests (Whitmore 1990). There is no evidence for significant forest clearance in the interior of the island before 1800, although coastal settlements have existed for centuries (Corlett 1992a,b). Most of the deforestation occurred in the period 1819–1900, after the foundation of the British colony led to a rapid rise in population. Most initial clearance of primary forest was for the cultivation of gambier *(Uncaria gambir),* the extract of which was used medicinally or in tanning and dyeing. Gambier grows best on soil newly cleared of forest, and its harvest requires a roughly equal area of forest to provide firewood for boiling the gambier leaves (Jackson 1965). The gambier growers moved on when the soil and firewood supply were exhausted, after 15–25 years, and abandoned plantations were invaded by dense grass *(Imperata cylindrica)* or scrubby forest. Often the land was eventually used again for other crops.

Cantley (1884, 8) describes the results of this "reckless, migratory cultivation":

> Such Crown forests as remain uncut are widely distributed in isolated patches over the island. These forest patches or clumps are of various sizes, from half an acre or so to about twenty-five acres [10 ha], and of no particular shape; their distance from each other may average a quarter of a mile [0.4 km], though often exceeding a mile [1.6 km]. The interspace is generally waste grassland which supports, as a rule, only strongly-growing grass known locally as "lalang" *(Imperata cylindrica).*

Cantley estimated the total area of forest on Crown land as 2,000 ha, with a similar area on private land, and noted that all large trees of timber quality and other valuable species had been removed from most patches. Many of these forest fragments were cleared over the next sixty years. From 1884, most of the larger patches were included in forest reserves, but most of these were eventually abandoned (Corlett 1992b). Only the central part of the island (ca. 1,600 ha) has received continuous protection to the present day, and primary forest remnants are now confined to this area. Here, the cessation of cultivation and the control of grassland fires has permitted the growth of secondary forest (Corlett 1991), restoring links between many of the fragments. The construction of reservoirs, roads, and recreational and military facilities, however, has subsequently created new dispersal barriers between the fragments. Today, remnant patches of primary forest (in the sense of Corlett 1994), up to 45 ha in size and totalling about 200 ha, are embedded in 1,400 ha of secondary forest thirty to a hundred years old, which is in turn divided into two large and several smaller patches (fig. 22.3). Most of the primary forest remnants have received little disturbance in the last fifty years, but military training and recreational activities have disturbed much of the secondary forest to some extent.

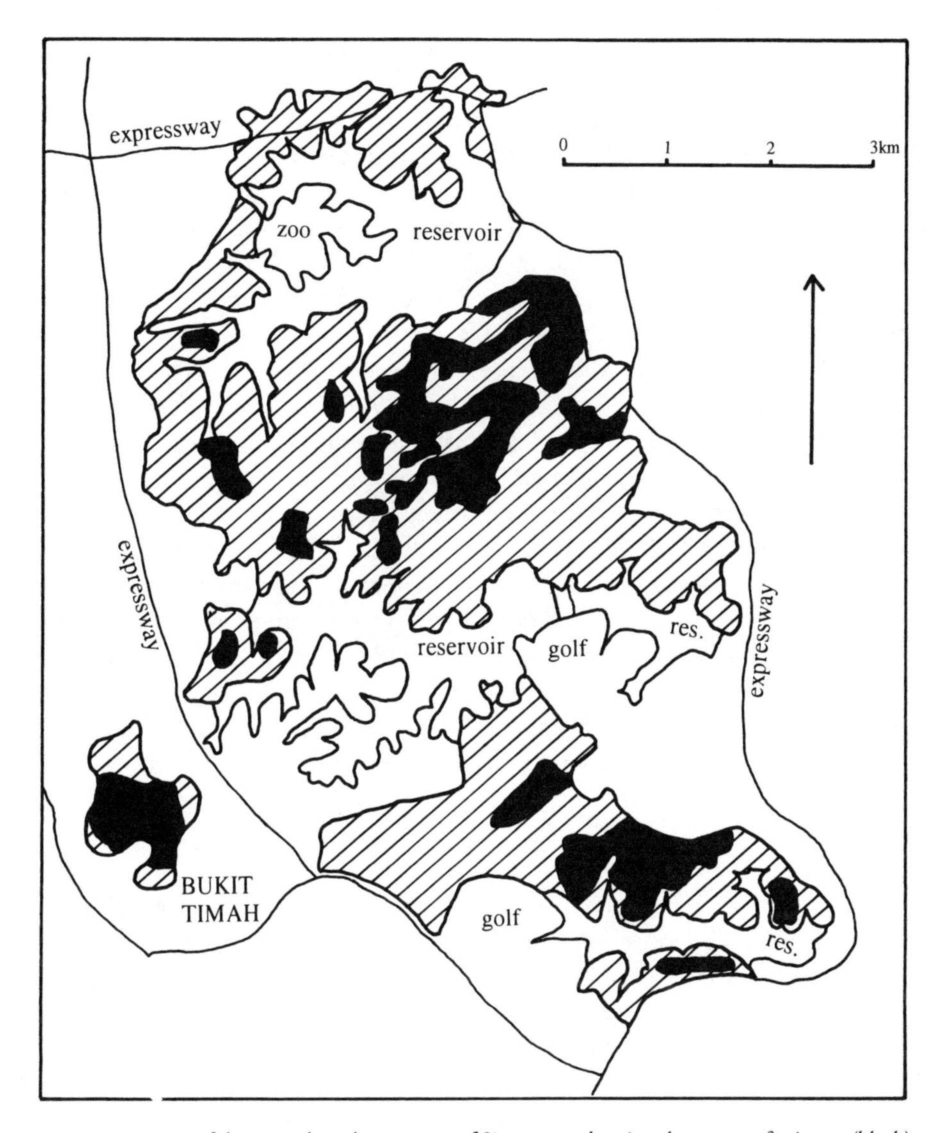

Figure 22.3. Map of the central catchment area of Singapore, showing the extent of primary (black) and secondary (diagonal lines) forest. The map is based on aerial photographs and probably exaggerates the extent and contiguity of primary forest. Minor roads and small openings within forested areas are not shown.

Hong Kong

Hong Kong would also be covered in forest if not for recent human activities (Dudgeon and Corlett 1994). Judging by the surviving flora, the forest at all altitudes would be dominated by members of the Lauraceae and Fagaceae, but below about 400 m altitude, the families Euphorbiaceae, Rubiaceae, and Moraceae would also be prominent, together with representatives of other largely tropical genera. Above this altitude the flora has a more subtropical (or tropical montane) aspect.

There is archaeological evidence for human settlement in Hong Kong at least 6,000 years ago (Meachem 1994), and although population densities appear to have been low before the sixteenth century, the susceptibility of secondary grasslands to dry season fires would have made deforestation a cumulative process once it started. From the seventeenth century on, both Chinese and European accounts of the region describe or imply a largely deforested landscape, but we have found no reliable earlier records (Dudgeon and Corlett 1994; Corlett 1997).

Early descriptions of Hong Kong all use words such as "barren," "bleak," and "bare" to describe the landscape (Dudgeon and Corlett 1994). All flat land was cultivated, and the hills were covered in grass, with a few scattered shrubs and stunted pine trees. Forest patches were preserved near villages for reasons of *feng shui* ("winds and waters"), the traditional Chinese system for determining favorable locations for settlements. Most were less than 10 ha in area, but the largest exceeded 150 ha. All the *feng shui* woods that survive today appear to be secondary, having probably originated from a combination of planting and secondary succession, but descriptions by nineteenth-century botanists suggest that some of the larger woods present then may have included remnants of the primeval forest (Corlett 1997). At higher altitudes there were numerous tiny forest patches—at most a few hectares in extent—in deep ravines and similar sites protected by topography from fire and cutting. In total, by the mid-nineteenth century, forest seems to have covered 2–5% of what is today the Territory of Hong Kong, with only a fraction of that being primary forest.

Britain took over Hong Kong Island in 1841, and the New Territories were leased from China in 1898. Large-scale reforestation started in the 1880s, but no protection was given to the existing forest. All the larger patches recorded in the nineteenth century had disappeared by the 1930s, although the total forest cover at this time (largely plantations of the native conifer *Pinus massoniana*) was probably the highest it had been for several centuries. The demand for fuelwood and timber during the Japanese occupation (1941–1945) caused the destruction of almost all the plantations and other accessible forest patches. Aerial photographs of the Territory from 1945 show forest again confined to *feng shui* woods and scattered small patches in topographically protected sites.

Since World War II, rural depopulation and increased protection of the countryside for water catchments, recreation, and conservation have allowed forest succession to proceed in some areas. Secondary forest now covers about 9% of the Territory, with a similar area in tall shrubland and an additional 5% under plantations, largely of exotic species (Ashworth et al. 1993). The largest forest patches are 150–200 ha in area, but most are less than 25 ha. There are now no forest patches that are entirely primary, but many include a topographically protected core area containing poorly dispersed species not found in secondary forest and some individual trees that appear to be very old, suggesting that such areas were never entirely cleared. In addition, some of the surviving *feng shui* woods are probably several hundred years old. Grassland is still the major nonforest vegetation type, but is being replaced by species-rich shrublands in fire-protected areas (Ashworth et al. 1993). Less than 20% of the territory is currently urbanized.

METHODS

The major focus of this chapter is botanical, with information on the vertebrates included both for comparison and because of the implications of vertebrate extinctions for the survival of these floras. The faunal information has been compiled from the literature and from the unpublished observations of many local naturalists. The botanical information comes from parallel projects investigating the taxonomy, ecology, and conservation status of the floras of Singapore (Turner 1994; I. M. Turner et al. 1994) and Hong Kong (Corlett 1997). As a result of Hong Kong's larger area and much greater topographic complexity, the current status of its flora is less well known than that of Singapore's, but both floras are well studied by tropical standards.

All recorded vascular plant species were classified by life form (tree, shrub, herb, liana, vine, epiphyte, and strangler) and assigned to one of three major habitat types (coastal, inland forest, and noncoastal open habitats: Turner 1994). This classification is relatively straightforward in Singapore, where primary forest remnants retain most of their original flora and structure. In Hong Kong, however, species recorded as forest epiphytes in regional floras survive as lithophytes in damp, treeless ravines, and potential tree species persist in shrub form in regularly burned areas. Both life form and habitat assignments were therefore sometimes made on the basis of potential rather than actual status. This is numerically significant only for the forest epiphyte category, for which results should therefore be interpreted with caution.

RESULTS

Singapore's Forest Fauna

Nineteenth-century records of Singapore's fauna are poor, and the first reliable accounts come from after the major period of deforestation (Corlett 1992b). At least half (70 spp.) of the resident inland forest bird species have become extinct since records began, with 31 species disappearing before 1940 and 39 since then (Lim 1992). These forest bird extinctions include all of the pheasants (Phasianidae: 3 spp.), hornbills (Bucerotidae: 3 spp.), trogons (Trogonidae: 2 spp.), broadbills (Eurylaimidae: 5 spp.) and kingfishers (Alcedinidae: 3 spp.) and most of the barbets (Capitonidae: 4/5), woodpeckers (Picidae: 8/11), babblers (Muscicapidae: 7/13), and bulbuls (Pycnonotidae: 5/10). At least nine mammal species are also extinct, mostly since 1940: the slow loris *(Nycticebus coucang),* tiger *(Panthera tigris),* leopard *(Panthera pardus),* leopard cat *(Prionailurus [Felis] bengalensis),* wild pig *(Sus scrofa)* (which, however, still survives on the island of Pulau Tekong), greater mouse deer *(Tragulus napu),* barking deer *(Muntiacus muntjak),* sambar deer *(Cervus unicolor),* and porcupine *(Hystrix brachyura).* The nineteenth-century status of six other carnivore species that are absent today is uncertain, as is the current status of several species of bats and rodents that were reliably recorded in the past. There have been at least twenty freshwater fish extinctions, mostly in the past sixty years (Lim and Ng 1990). No amphibians or reptiles are known to have become extinct, but the reptiles in particular have been inadequately studied.

The surviving forest bird and mammal faunas are dominated by a few species that are very abundant, probably more so than in the extensive primary forests of Malaysia, although there are no quantitative records for comparison. Many of the dominant vertebrates, such as the plantain squirrel *(Callosciurus notatus)* and striped tit-babbler *(Macronous gularis),* are highly adaptable species that also thrive in a variety of secondary vegetation types, including gardens. The majority of forest vertebrates, in contrast, are rare; several bird and mammal species apparently have populations of fewer than twenty individuals (Lim 1992). Many of the earlier bird and mammal extinctions were undoubtedly caused or accelerated by hunting, but the extinction rate—approximately one vertebrate species per year—has shown no sign of decreasing in recent decades, when hunting has been largely prevented. No exotic vertebrate has become established in areas covered by closed, native forest, although exotic birds, rodents, and freshwater fishes are prominent in the open-country biota (Corlett 1992b).

Singapore's Forest Flora

The total recorded vascular plant flora of Singapore's inland forests contains 1,674 species (Turner 1994). In comparison with other tropical rainforest areas, Singapore's forest flora is rich in tree species, but relatively poor in climbers and epiphytes (table 22.2: Turner 1994). The pteridophytes and the three most species-rich angiosperm families (Orchidaceae, Rubiaceae, and Euphorbiaceae) account for 30% of the flora (see table 22.1). An estimated 1,196 species (71% of the recorded forest flora) survive today (I. M. Turner et al. 1994). The loss of species has not been uniform across life forms and families. More than half the forest epiphytes have been lost, whereas three-quarters of the tree and shrub species have persisted. Of the major families, the Orchidaceae has suffered the highest extinction rate (86%), followed by the Rubiaceae (44%) and the Meliaceae (42%).

Only one exotic species (*Clidemia hirta,* Melastomataceae) is commonly found in the interior of major forest areas (Corlett 1988). In an isolated 4 ha primary forest fragment in the Singapore Botanic Gardens, however, regeneration is being smothered by exotic lianas, including *Dioscorea sansibarensis, Tanaecium jaroba,* and *Thunbergia grandiflora* (Turner and Tan 1992). The distinction between primary forest remnants and even the oldest secondary forest is still clear, both in aerial photographs and on the ground. The secondary forest has a lower stature, a more even canopy, a poorly developed under-

Table 22.2 Percentages of species with different life forms in the forest floras of Singapore and Hong Kong.

	Singapore	Hong Kong
Trees	45.8	31.7
Shrubs	10.5	12.6
Herbs	12.5	30.2
Lianas	12.5	18.1
Vines	3.8	1.1
Epiphytes	15.0	6.4

story, and much lower species diversity (Corlett 1991; Turner et al. 1997). Dipterocarps and other taxa with large wind-dispersed seeds are entirely absent, and those with large animal-dispersed fruits are rare. Most of the primary forest fragments have boundaries with secondary forest, but even where the primary forest margins are exposed, obvious influences on their structure and species composition do not usually extend more than 50 m into the forest (see also Kapos et al., chap. 3; Turton and Freiburger, chap. 4; Laurance, chap. 6). The interiors of the largest fragments are superficially indistinguishable from the extensive primary forests of the Malaysian mainland.

Hong Kong's Forest Fauna

The recorded nineteenth-century vertebrate fauna of Hong Kong included no species completely dependent on forest. The rhesus macaque *(Macaca mulatta)* was the only primate, and there were no squirrels or large deer and no forest specialist birds. We can only speculate about the composition of the primeval forest fauna, but historical evidence suggests that it would have been much richer than any surviving in the Hong Kong region today. There were gibbons (*Hylobates* sp.) in the region until the seventeenth century, although not necessarily in Hong Kong, and both elephants and rhinoceros are mentioned in earlier Chinese records (Goodyer 1992; Dudgeon and Corlett 1994). There have been few recorded vertebrate extinctions in Hong Kong since the nineteenth century, and none of these species was dependent on forest. The known mammalian extinctions are all carnivores persecuted by people: the dhole *(Cuon alpinus),* fox *(Vulpes vulpes hoole),* tiger *(Panthera tigris),* leopard *(Panthera pardus),* and large Indian civet *(Viverra zibetha).* The only documented bird extinction is the ring-necked pheasant *(Phasianus colchicus),* which occupied fire-maintained grassland.

Two different subspecies of the tree squirrel *Callosciurus erythraeus* were introduced to Hong Kong Island and the New Territories in recent decades and are now well established, although they live at much higher densities in urban fringe habitats than in the forest (Dudgeon and Corlett 1994). In the past fifty years, at least seven species of forest birds have also become established of their own accord or from escaped or released caged birds (H. K. Kwok, pers. comm.). Most of these were probably part of the primeval forest fauna and so cannot be considered exotics.

Hong Kong's Forest Flora

The total forest flora recorded from Hong Kong since 1841 contains 950 species. As in Singapore, this flora is dominated by woody species, but herbs and climbers are proportionately more abundant, and trees and epiphytes less so (table 22.2). Also as in Singapore, the pteridophytes and the three most abundant angiosperm families (Orchidaceae, Leguminosae, and Rubiaceae) account for 30% of the flora. Six of the ten most diverse families are the same as in Singapore, but the Annonaceae, Melastomataceae, Myrtaceae, and Palmae are impoverished in comparison (see table 22.1). More than a third of the recorded flora has not been reported in the last thirty years, but recent surveys of inaccessible sites suggest that, in most cases, this finding reflects inadequate research rather than extinction.

No recognizably exotic plant species are found in most forest areas, except for planted individuals, but, given Hong Kong's long history of human impact, it is possible that some introduced species have become so well naturalized that they are considered part of the native flora. The pinewood nematode *(Bursaphelenchus xylophilus),* which was probably introduced from North America several decades ago, has virtually eliminated the native pine *(Pinus massoniana),* an important pioneer species (Dudgeon and Corlett 1994). The distinction between primary forest and old secondary forest in Hong Kong has been obscured by centuries of human impact, but areas of definite secondary forest, developed since 1945 on previously treeless sites, have a very uniform flora, dominated by species of *Persea* (Lauraceae). Members of the Fagaceae—the most species-rich tree family in the local flora—are absent from secondary forest.

DISCUSSION

Differences in Faunal and Floral Extinctions

Singapore and Hong Kong are extreme examples of tropical deforestation and fragmentation. In both areas, more than 98% of the original forest was cleared, and the surviving fragments were highly disturbed and isolated within a matrix of fire-maintained grassland. Although we cannot be certain of the composition of the original flora and fauna in either area, regional comparisons place upper bounds on the pre-fragmentation diversity.

There have been many faunal extinctions in both Singapore and Hong Kong. Among birds—the best-studied group—at least 50% of the forest specialists have become extinct in Singapore, while apparently none have survived in Hong Kong. Although most vertebrate species seem to have survived the initial period of active deforestation in Singapore, extinction rates, at least among birds, mammals, and freshwater fishes, have remained high subsequently, despite the control of hunting and protection of the remaining forest. Today, Singapore's forest vertebrate fauna is dominated by a few abundant species, and the remaining species are rare. The situation in Hong Kong, where the only surviving vertebrates are those that can utilize the nonforest matrix, can be seen as the logical endpoint of this trend. The success of several forest specialist vertebrates that have colonized or been introduced into Hong Kong in the last fifty years, however, suggests that the remaining forest could support more vertebrate species now that hunting has largely stopped.

Plant extinctions, in contrast, have been far less frequent than expected based on the massive loss of forest area, and a surprisingly rich forest-dependent flora has persisted in the tiny surviving fragments. Even if we assume pre-fragmentation forest floras of 2,000 vascular plant species in both areas—which is more than we believe likely—then about 60% of the forest flora has survived in Singapore and about 47% in Hong Kong, far more than would be predicted by the species-area relationship (I. M. Turner et al. 1994; see also Andersen, Thornhill, and Koopowitz, chap. 18). In Singapore, most plant extinctions since the nineteenth century are likely to have resulted from the loss of forest fragments outside the protected central area. Compared with the vertebrate fauna, plant species losses from surviving fragments seem to have been relatively less severe, except in the smallest remnant. The 4 ha Singapore Botanic Gardens' Jungle, which is not buffered by secondary forest, has lost 51% of its native plant species since the 1890s (Turner et al. 1996).

The remarkably low long-term extinction rate estimated for Hong Kong's forest flora (ca. 47%) may not be representative of the Tropics as a whole. Not only does Hong Kong's rugged topography provide numerous protected sites where forest plants can persist, but exposure to frequent typhoons and occasional frosts may have selected for a greater resilience to major disturbance than is usual in tropical floras.

Extinctions in Singapore's forest flora have not been uniform; epiphytes and orchids (most of which are epiphytic) are the most vulnerable. It is tempting to interpret the relative impoverishment of both these groups in the Hong Kong flora as a continuation of this trend, but in the absence of a pre-fragmentation baseline flora it is impossible to separate the influence of fragmentation from that of climate. The disproportionate loss of epiphytes in Singapore may simply reflect their shorter generation times, although it could also be explained by changes in canopy microclimate: satellite-derived estimates of forest canopy temperature show that a peripheral forested zone at least a kilometer wide is 0.5–1.0°C warmer than the central part of Singapore's current forest area (Nichol 1994).

Floristic diversities in both Singapore and Hong Kong are probably still considerably above equilibrium; many species in both Singapore and Hong Kong are known only from single sites or very few individuals, and thus are likely to join the ranks of extirpated species sooner or later. Moreover, wherever plant populations are very small or highly fragmented, loss of within-species genetic diversity is expected, although we have no data on this from our study areas.

Possible Mechanisms of Extinction

How can the great difference in the persistence of plant and vertebrate species be explained? The simplest explanation is that many large vertebrates have succumbed to hunting pressure, while few plant species have suffered an equivalent degree of exploitation. More generally, most plants are long-lived compared with vertebrates and thus pass less frequently through the vulnerable regeneration stage. Tree species that have not reproduced successfully since fragmentation may still survive in Singapore, and conceivably even in Hong Kong. Furthermore, an individual plant is sensitive to changes only in its immediate vicinity, while most warm-blooded vertebrates require large areas of suitable habitat for individual survival.

Forest plants often require animal vectors for transporting their seeds and pollen (see Nason, Aldrich, and Hamrick, chap. 20). Many forest plants in Singapore and Hong Kong are reproducing successfully, however, suggesting that plant-animal relationships in the rainforest are not as specialized as is often suggested. There is certainly no evidence for the rapid extinction vortex due to loss of large vertebrates that some have predicted (Howe 1984).

Additional evidence for the long-term survival of diverse rainforest floras in fragmented landscapes comes from a study by Meave and Kellman (1994) of small riparian forest patches (1–160 ha in area) in the savanna region of Belize. They demonstrated that these small patches, which may have existed throughout the Holocene, support a flora comparable in species composition and diversity to that of continuous forests in Central America. Similarly, Winter (1988) suggests that the diverse flora of Australian subtropical rainforests survived late Quaternary dry periods in fragments that were too small to support rainforest-specialist mammals. However, the similarities between natural and anthropogenic rainforest fragments require further, critical examination before too much emphasis is placed on these findings (Turner and Corlett 1996).

Forest Regeneration

Secondary forest develops rapidly on almost any fire-protected site in Singapore and Hong Kong, despite long-term soil degradation. However, these secondary forests accrete plant species slowly and selectively (see also Lamb et al., chap. 24). One major filter is the composition and behavior of the disperser fauna (Corlett 1995; Thébaud and Strasberg, chap. 21). Faunal extinctions have disproportionately eliminated large frugivores, and in both Singapore and Hong Kong, most dispersal of forest seeds into early successional vegetation is done by small and medium-sized open-country birds, which feed in the forest canopy but do not enter the forest interior (Corlett 1991, 1996). The common bird species have gape limits of 10–15 mm—smaller than many forest fruits. Only civets provide efficient dispersal for larger fruits and seeds across open areas (Dudgeon and Corlett 1994), and they are rare in Singapore. The widespread and abundant long-tailed macaques *(Macaca fascicularis)* in Singapore and the more localized rhesus macaques *(M. mulatta)* in Hong Kong are also potential seed dispersers that readily enter nonforest vegetation. They are inefficient dispersers, however, and rarely move large seeds over long distances (Corlett and Lucas 1990). In Singapore, entire crops of large-seeded fruits of species unattractive to macaques, such as most Myristicaceae, may be dropped under the parent tree, while in Hong Kong, all species of Fagaceae meet the same fate. Wind dispersal of seeds is relatively unimportant in the forest floras as a whole, and most of the wind-dispersed canopy trees in Singapore, such as the dipterocarps, have such large fruits that dispersal distances are small even under ideal conditions.

Seed predators may also impede both regeneration and the colonization of new sites. Although there is no comparative data, the elevated abundance of two squirrels, *Callosciurus notatus* and *Sundasciurus tenuis,* in Singapore may have resulted in exceptionally high seed predation. Recent dipterocarp fruiting in the Botanic Gardens' Jungle failed to recruit any seedlings, apparently because the squirrels attack freshly fallen and newly germinated fruits.

Invasion of existing forest fragments by exotic plants and animals has not been a major problem in either Singapore or Hong Kong, although numerous exotics are established in disturbed open habitats in both areas (Corlett 1988, 1992b,c; Turner and Tan 1992; Ng, Chou, and Lam 1993; Dudgeon and Corlett 1994). Whether this represents some form of community-level resistance to invasion by rainforests (see Laurance, chap. 6) or simply the tendency of invading exotics to require open or disturbed habitats, or both, is an important question. If the latter is true, vigilance against potentially invasive exotics should be increased.

Conservation of Forest Floras

Our results show that extinction from small rainforest fragments is rapid for vertebrates, but surprisingly slow for vascular plants. This does not mean that fragmentation is not a serious problem for rainforest floras (see Andersen, Thornhill, and Koopowitz, chap. 18). Hundreds of plant species have already become locally extinct in both Singapore and Hong Kong, and these taxa are becoming increasingly prone to global extinction as the remaining forests in the region suffer the same fate. Moreover, there is no evidence from either area that a new equilibrium has been reached; much of the surviving forest flora in both areas exists as tiny populations, which are vulnerable to both stochastic processes and environmental change. The slow rate of floristic extinction does not provide a solution to the fragmentation problem, but only a breathing space, during which a solution must be found (Turner and Corlett 1996). It is clear that any long-term solution will involve the resynthesis of diverse, continuous forests from the surviving fragments (Lamb et al., chap. 24). How this can be done in areas with degraded soils and impoverished faunas (see Thébaud and Strasberg, chap. 21) is perhaps the most important question raised by the results of this study.

GENERAL IMPLICATIONS

1. In Singapore and Hong Kong, forest-dependent vertebrate species are lost rapidly from isolated fragments. Although this loss is undoubtedly hastened by hunting and other post-fragmentation factors, it continues even in well-protected areas.
2. The vascular plant flora apparently declines much more slowly than the vertebrate fauna, and a substantial fraction of the original flora can persist even after several hundred years in unprotected forest fragments. There is, however, no sign of a new equilibrium being reached in Singapore and Hong Kong, and extinction still continues. Extinction rates vary among life forms and families, with epiphytes and orchids being most vulnerable in Singapore.
3. Even on degraded soils, secondary forest develops rapidly on sites protected from fire and tree cutting. However, secondary forest accretes species slowly and selectively, and is floristically impoverished even after a hundred years. Natural succession alone appears unlikely to restore the rainforest from surviving fragments in degraded landscapes with impoverished disperser faunas.
4. The persistence of many species in primary forest fragments means that the conservation of these sites should receive highest priority, particularly in areas that no longer have extensive tracts of forest. Protection and management of isolated fragments will probably reduce the extinction rate, but the only long-term solution will be the resynthesis of diverse, continuous forest.

Section V
Restoration and Management of Fragmented Landscapes

INTRODUCTION
Management of Tropical Landscapes

In the Tropics, as elsewhere, land management encompasses a diverse range of activities. In this section the emphasis is on practical methods, tools, and technologies that can contribute to landscape management and restoration, especially in terms of reducing or mitigating the effects of forest fragmentation.

At the outset, it is important to emphasize that active habitat management is usually a time-consuming and expensive activity. Resource management agencies and organizations commonly operate on very limited and inadequate budgets. For this reason, there is a great need to target management projects carefully and to expend scarce resources wisely. Applied research can contribute to this process by highlighting the most essential management and conservation goals and by developing or evaluating promising new methods or technologies.

The first two chapters in this section emphasize the restoration of degraded lands or forest remnants. Nature reserves rarely comprise more than a small fraction of any biogeographic region, and thus habitats outside reserves are almost always important for the persistence of some species, populations, and ecological processes. Moreover, the type and spatial configuration of matrix habitats surrounding fragments can strongly affect the dynamics and composition of fragment biotas (see chaps. 9, 10, 13, 14, 20, and 32). Because the matrix is so important, reforestation initiatives are likely to play an ever-increasing role in tropical landscape management.

Habitat Restoration

In chapter 23, Virgílio Viana, André Tabanez, and João Luis Batista assess the ecological consequences of fragmentation and techniques for forest restoration in the Atlantic moist forests of Brazil. These forests constitute a global biodiversity "hotspot" in the sense that they have extremely high levels of endemism and have experienced massive clearing and fragmentation. Viana's team found that small forest remnants in this region are highly prone to a range of physical and anthropogenic disturbances, and often become increasingly degraded over time. Because these small remnants are ecologically "nonsustainable," the authors have devised and tested a series of management practices (mainly involving the control of heavy vines and experimental tree plantings) to improve forest structure and tree regeneration. In general, their results suggest that restoration practices can help to reverse the ecological degradation of small forest remnants. They also discuss the economic and social realities involved in implementing such programs.

Chapter 24, by David Lamb and his colleagues, provides a thorough and very practical explanation of strategies for the restoration of degraded tropical landscapes. Drawing on case studies from northern Queensland and Puerto Rico, the authors present an integrated discussion of the ecological, practical, and socioeconomic factors that affect reforestation initiatives in the Tropics. For example, because reforestation is expensive and time-consuming, they discuss measures for maximizing the attractiveness of reforestation projects (such as integrating timber production and reforestation) or increasing their efficiency (such as planting tree species that are highly attractive to native frugivores, which then transport many new plant species into reforestation plots). They also discuss the most appropriate locations for reforestation activities, such as faunal corridors, nature reserves, and habitats for endangered species.

Remote Sensing and GIS

The last two chapters in this section focus on the use of remote sensing technology and geographic information systems (GIS) in landscape management. In chapter 25, Elizabeth Kramer uses remote sensing and GIS to monitor changes in forest cover and composition at Guanacaste National Park in Costa Rica. The Guanacaste area encompasses some of the last remnants of tropical dry forest (monsoonal forest) in Central America, and the Costa Rican government has gone to considerable lengths to increase and restore the remaining Guanacaste forests. Kramer discusses both the strengths and limitations of remote sensing technology for monitoring and managing forest recovery. She also applies an innovative software package, Fragstats, to quantify landscape patterns and changes over time.

Chapter 26, by Virginia Dale and Scott Pearson, focuses on the state of Rondônia in southeastern Amazonia, a region that has experienced remarkably rapid human colonization and deforestation over the past two decades. Dale and Pearson use remote sensing to describe ongoing patterns of forest clearing, then build a simulation model to predict the pattern and scale of deforestation in the future. An important contribution of their modeling approach is that they can contrast the effects of different land use practices in the region. For example, they demonstrate dramatic differences between a "worst case" scenario (typified by the type of rampant forest loss associated with the Trans-Amazon Highway) and a "best case" scenario (the result of sustainable agroforestry methods practiced by some Rondônian farmers) in terms of deforestation rates and patterns. This type of information can be an invaluable tool for making informed land use decisions in rapidly developing regions.

SYNTHESIS

Collectively, these chapters contain some important lessons for practical landscape management. For example, habitat restoration clearly has an important role to play in conservation and wildlife management, especially in regions such as the Atlantic coastal forests of Brazil (chap. 23) and the tropical dry forests of Central America (chap. 25), which have been heavily degraded by past land uses. Under these circumstances, restoration measures can improve the viability of small forest remnants (chap. 23), expand key wildlife habitats (chap. 25), and increase landscape connectivity by revegetating stream margins (chap. 24).

Although it can be relatively expensive and time-consuming, the efficiency of reforestation can be improved by exploiting natural regeneration mechanisms. For example, forests can regenerate on denuded lands quite rapidly if the land is protected from fire, grazing, and encroachment, and if small forest patches have been retained as sources of forest seeds, mycorrhizae, and seed dispersers (chap. 25). Natural successional processes can also be accelerated by planting tree species that are attractive to a variety of frugivorous birds and bats, which will then transport new forest species into the area (chap. 24).

For private landowners, financial incentives will often be necessary to justify reforestation. Such incentives can be provided by planting combinations of trees that include timber or crop species as well as those attractive to wildlife (chap. 24). Even intensively managed areas such as timber plantations can harbor a surprisingly diverse array of native plants if natural regeneration is permitted in the understory (chap. 24). Sensitivity to the values and perspectives of landowners is an essential prerequisite for promoting forest rehabilitation on private lands (chap. 23).

Increasingly, remote sensing and GIS offer powerful tools for assessing patterns and change in forest cover over large areas. These technologies have myriad applications, from measuring the rate and patterning of deforestation (chap. 26) to assessing successional and phenological changes in vegetation over time (chap. 25). The spatial resolution and effectiveness of these technologies are improving rapidly, and one obvious implication of this is that researchers wishing to assess temporal changes in landscapes must often compare older, less precise images with newer, higher-resolution ones (chap. 25). This problem highlights an immediate need for establishing accurate baseline imagery for rapidly developing regions, using the best available technology, for comparison with future imagery.

There are now over sixty metrics that can be used to quantify landscape patterns (chaps. 25 and 26). One of the more promising applications of remote sensing and GIS involves integrating landscape metric and species-based approaches so that the potential effects of fragmentation on target species can be predicted over large areas. Studies in Amazonia, for example, suggest that the responses of different animal species to fragmentation at the landscape and regional scales can be predicted from knowledge of their area requirements and gap-crossing abilities (chap. 26 and references therein).

Finally, it should be emphasized that there is still enormous scope for *proactive* habitat management in the Tropics. Rapid forest conversion is a fact of life in many tropical regions, but the ecological effects of forest loss and fragmentation can be partly mitigated by legislation that seeks to control the deforestation process. For example, legal guidelines that prohibit clearing of vegetation along streams and watercourses can help to maintain ecosystem connectivity and reduce erosion and stream sedimentation. Likewise, measures that prohibit clearing on steep slopes promote topsoil and watershed protection and retain some forest remnants. There probably has been far too little attention paid to active management of the deforestation process in rapidly developing regions.

23

Dynamics and Restoration of Forest Fragments in the Brazilian Atlantic Moist Forest

Virgílio M. Viana, André A. J. Tabanez, and João Luis F. Batista

FOREST fragmentation is a widespread phenomenon that is almost invariably associated with frontier expansion in both tropical and temperate regions (Harris 1984). Tropical forest fragmentation has received increasing attention over the past decade due to high deforestation rates and consequent fragmentation effects in tropical regions (Blockhuss et al. 1992; Whitmore, chap. 1).

From a conservation perspective, forest fragmentation is particularly problematic in areas with extensive and rapid deforestation, high endemism, and where protected area networks underrepresent natural landscape heterogeneity. This is clearly the case in the Atlantic moist forests of Brazil, the *Mata Atlântica.* In this region, where some of the highest levels of plant diversity in the world have been recorded (Mori, Boom, and Prance 1981), remaining forest cover is often less than 5% of the total pre-settlement forest area (CIMA 1991). More importantly, most forest remnants are privately owned by farmers, and their fate is highly dependent upon the attitudes of those farmers and local communities.

Ecosystem fragmentation is directly linked to land use dynamics in rural and urban areas, which is driven by economic, social, cultural, institutional, and technological factors (Viana 1995; Dale and Pearson, chap. 26). Conservation of forest fragments in Brazilian Atlantic moist forests will have to be based on (1) an understanding of the factors driving land use dynamics and development, (2) an evaluation of the ecological consequences of fragmentation, and (3) development of technologies for restoration and sustainable management. While there is a growing body of published literature on the effects of fragmentation (Laurance et al., chap. 32), there is a particular scarcity of research on restoration technologies for tropical forest fragments (Janzen 1988b; Lamb et al., chap. 24).

In this chapter, we focus on the ecological consequences of fragmentation and technologies for forest restoration in the Atlantic moist forests of Brazil. We address part of a general question: are small fragments of these forests self-sustainable? Our hypothesis is that these forest fragments are often not self-sustainable and require restoration efforts to maintain their ecological functions and biodiversity in the long run. This hypothesis is based on the understanding that edge effects (Lovejoy et al. 1986; Kapos 1989; Laurance

1991b; Waldorff and Viana 1993; Didham, chap. 5; Laurance et al., chap. 32), vine colonization following logging and fire disturbances (Baur 1964; Uhl and Buschbacher 1985; Putz 1992; Veríssimo et al. 1992; Tabanez 1995), and reduced tree population sizes (Soulé 1987; Viana, Tabanez, and Aguirre 1992) all contribute to a progressive ecological deterioration of forest fragments.

If forest fragments are undergoing degradation, our predictions are that (1) tree recruitment will be lower than mortality rates, (2) edge effects will increase over time, (3) populations of several tree species will be small, and (4) there will be poor forest structure dominated by low-diversity eco-units (here defined as patches of forest at distinct stages of succession or degradation, *sensu* Oldeman 1983). If forest restoration practices are to be effective, they should facilitate successional processes, decreasing the dominance of vines and increasing tree canopy cover.

We begin by briefly describing the deforestation situation in the Atlantic moist forest, then present a case study of a forest fragment that has been studied as a part of the Biology and Management of Forest Fragments Project carried out by the University of São Paulo. We conclude by discussing the implications of current research for the conservation of forest fragments in the region.

THE FRAGMENTATION SCENE

The Atlantic moist forests of eastern Brazil *(Mata Atlântica)* are a unique set of ecosystems, with very high diversity and endemism, that once covered about 12% of the country (1,100,000 km^2: SOS Mata Atlântica & INPE 1993) in sixteen Brazilian states. These forests are among the most threatened tropical ecosystems in the world, largely due to conversion of forested lands to agriculture and pasture (Fonseca 1985). Contemporary deforestation in the Atlantic forests began with the settlement of Portuguese colonists along the coast in the sixteenth century and progressed as the agricultural frontier—mainly coffee plantations—moved toward the interior (Viana and Tabanez 1996).

The level of deforestation varies among different states (SOS Mata Atlântica & INPE 1993). Several states, particularly those in northeastern Brazil, have very little forest cover remaining. This is the case in Alagoas, for example, which had most (99%) of its original forest cover replaced by agriculture and pasture (table 23.1). In contrast, other states, such as Santa Catarina (69% deforested), maintain much higher amounts of forest cover.

Historically, annual deforestation rates have also varied throughout the Brazilian Atlantic moist forests. Rates in the state of São Paulo, for example, reached their maximum between 1920 and 1935 (fig. 23.1A). In the states of Paraná, Minas Gerais, and Espírito Santo, the peaks of deforestation rates occurred in the periods of 1937–1955, 1947–1953, and 1975–1980, respectively (figs. 23.1B, C, and D). These different histories of deforestation are important to our understanding of present dynamics and future restoration needs of forest fragments.

These historical differences in deforestation rates across the region mean that different fragments have been isolated and populations of plants and animals have had their effective population sizes reduced over different periods of time. For example, many forest fragments in the Paraíba Valley, between the states of São Paulo and Rio de Janeiro,

Table 23.1 Forest cover in the Brazilian Atlantic rainforest (area in km^2).

State	Original cover	Present cover	% remaining
Sergipe	12,000	85	1
Pernambuco	20,000	394	2
Alagoas	13,000	367	3
Bahia	140,000	7,446	5
Minas Gerais	250,000	13,300	5
Rio Grande do Sul	170,000	11,282	7
Paraíba	6,000	560	9
Espírito Santo	45,500	4,587	10
Rio de Janeiro	44,000	5,001	11
Goiás	50,000	6,000	12
Mato Grosso do Sul	80,000	10,816	14
São Paulo	201,000	32,210	16
Paraná	180,000	34,336	19
Ceará	10,000	2,000	20
Rio Grande do Norte	2,600	594	21
Santa Catarina	77,000	23,730	31
Total	1,300,000	152,702	12

Source: CIMA 1991.

became isolated over a hundred years ago. A contrasting situation can be found in forest fragments in the lower Rio Doce Valley in Espírito Santo, most of which became isolated in the last few decades. Plant and animal populations in fragments in the Paraíba Valley have had more time to respond to the effects of fragmentation than those in the Rio Doce Valley. It is reasonable to hypothesize that changes in the structure and dynamics of plant and animal populations should vary between regions with different histories of deforestation. These differences need to be taken into account when interpreting ecological data and designing conservation strategies.

Deforestation levels also vary significantly within subregions and even within individual states. In general, deforestation levels are higher in landscapes with topographies favorable for agricultural mechanization, soils with higher fertility, and sites closer to roads and markets. Forest remnants are thus not always a good sample of the original forest ecosystems (Viana and Tabanez 1996). Landscape-level analyses, combining geomorphological information with forest cover data, are needed to identify forest types that are more or less represented in remnants so as to design sound conservation strategies.

Most remnants of the Atlantic moist forests are on private property and are thus highly vulnerable to continued disturbances. For example, protected reserves account for less than 1% of the plateau forest region of São Paulo State (northwest of the coastal mountains). These privately owned remnants often provide critical habitat for endangered species that are not protected in existing reserves (Viana and Tabanez 1996). Innovative approaches that incorporate private landowners as key factors in a long-term conservation strategy are needed to deal with the current threats to remnants of Brazilian Atlantic moist forest (Viana 1995).

Remnants of the Atlantic moist forests are typically small, isolated, and highly dis-

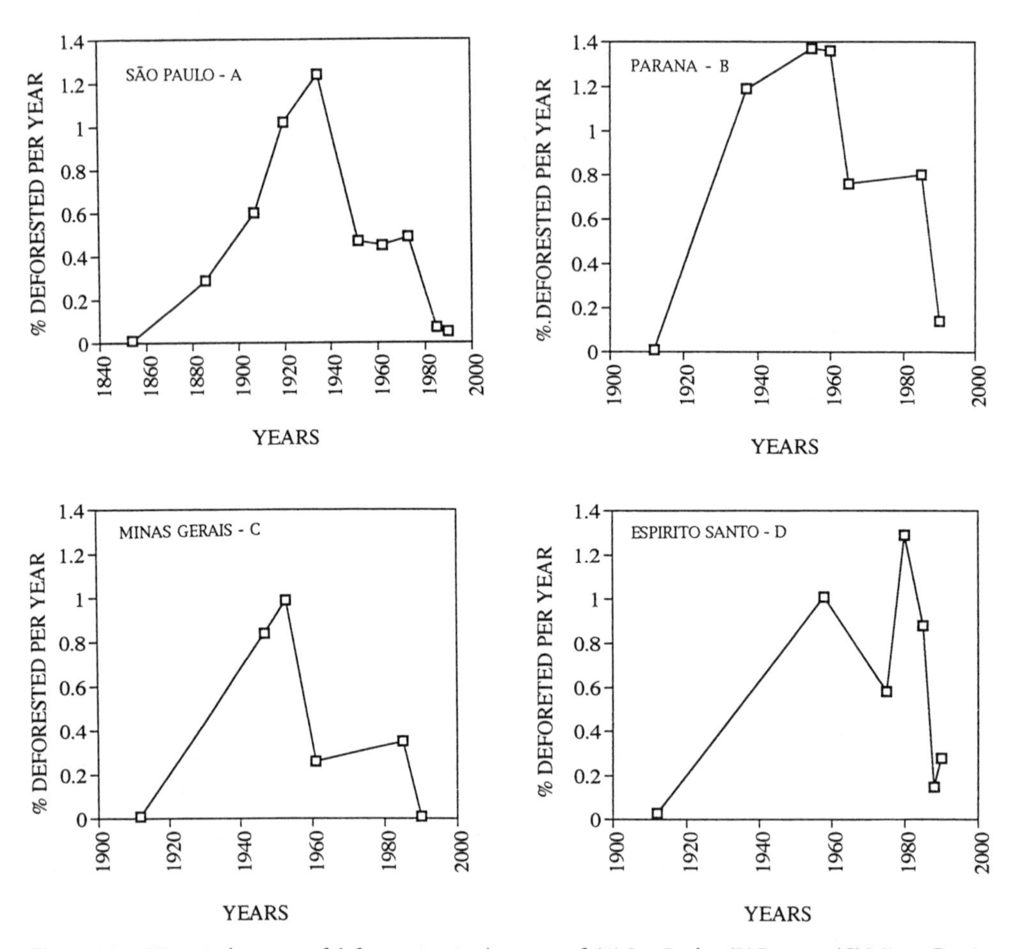

Figure 23.1. Historical process of deforestation in the states of *(A)* São Paulo, *(B)* Paraná, *(C)* Minas Gerais, and *(D)* Espírito Santo. (Data from SOS Mata Atlântica and INPE 1993.)

turbed. For example, in the Piracicaba region, in the state of São Paulo, the remaining forest cover is made up of old growth (0.81%) and secondary forests (1.23%), distributed among 29 and 73 fragments respectively (fig. 23.2). Small forest fragments (<50 ha) constitute 89.9% of remaining old-growth patches and 87.1% of secondary forest patches (DEPRN 1991).

The effectiveness with which forest fragments can provide "ecosystem functions" for the Atlantic forest region will depend on their long-term sustainability. If forest fragments are not self-sustaining, restoration and conservation measures will be essential. Therefore, an understanding of the factors that lead to forest degradation, as well as the development of alternatives for forest restoration, is vital to the development of a conservation strategy for the region (Viana, Tabanez, and Aguirre 1992). We must also improve our understanding of the biology of endangered species and how they respond to habitat fragmentation (Brown and Brown 1992; Bernardes, Machado, and Rylands 1990). A third key element in developing conservation strategies is understanding the decision-

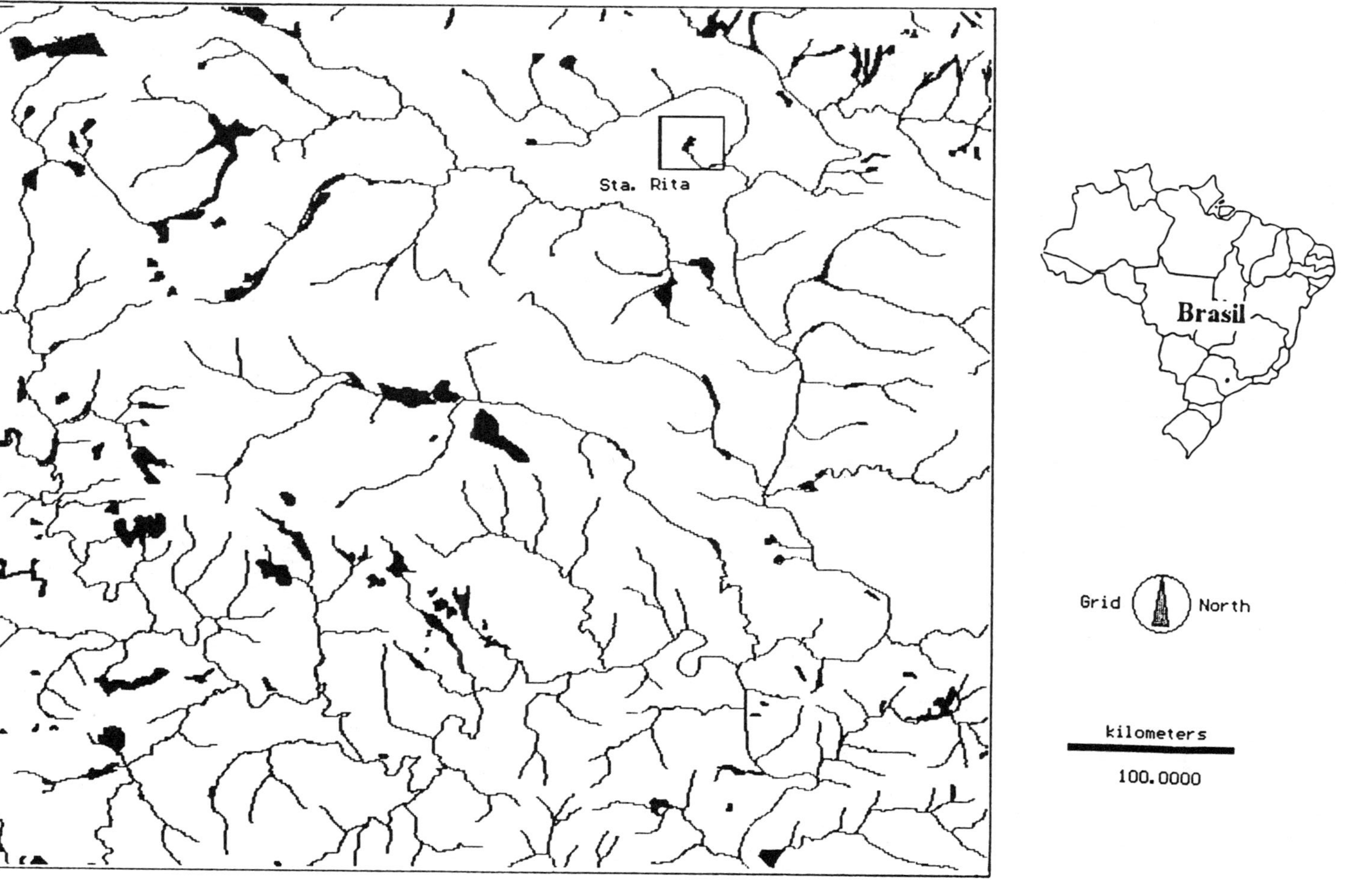

Figure 23.2. Fragmented forest landscape in Piracicaba, São Paulo, Brazil, showing the location of the Santa Rita forest fragment.

making process of landowners and other stakeholders (Viana 1995). Finally, we must take into consideration the fact that deforestation is still going on at an alarming rate, given the very small fraction that remains (SOS Mata Atlântica & INPE 1993). We must improve current policies and develop environmental education campaigns to reverse the current situation.

The present situation offers some reasons for hope. Both government institutions and nongovernmental organizations have improved their monitoring structures, and environmental awareness is increasing rapidly. However, given their highly fragmented situation, Brazil's Atlantic forests, especially in the plateau region, require special attention from the scientific and environmental communities.

CASE STUDY: THE SANTA RITA FOREST

The study site is located in the central part of the state of São Paulo, on the western slopes of the Atlantic moist forest (see fig. 23.2). The forest, characterized as "plateau forest," is semi-deciduous, with high diversity, and is dominated by emergent trees up to 55 m in height. The soil, a mixture of ultisols and alfisols, is fertile and well drained. The climate is seasonal (mesothermic with dry winter, classified as CWA in the Köppen system of classification), characterized by a pronounced dry season coinciding with the austral winter (average rainfall is 1,257 mm/year). The six drier months each average less than 100 mm of rainfall.

The Santa Rita fragment is 9.5 ha in area and relatively well protected, with no records of logging. The former owner of the forest practiced intensive agriculture in the remainder of the area, and chose not to plant sugarcane around the fragment in order to protect it from fires. Recently the area was sold, and the new owner planted sugarcane all the way to the edge of the forest. The result was a fire in mid-1994, after we had collected the data analyzed in this chapter. There is still some illegal hunting going on, although this is rather limited today.

METHODS

The Santa Rita fragment, like other plateau forests, can be seen as a mosaic of patches with similar structures and successional characteristics, or eco-units. Four types of eco-units have been identified in primary forest fragments in the study area: low forest (or low *capoeira*), high forest (or high *capoeira*), bamboo-dominated forest, and mature forest (or *mata madura*). The first three eco-units are degraded patches, but are not the result of recent human activities. Rather, they are patches that were degraded either by human activities when isolation occurred in the beginning of this century, or by edge effects and liana infestation. The low forest eco-units are 2–5 m high, with low plant diversity and basal area (<10 m^2/ha), dominated by vines and climbers, while the high forest eco-units are 5–15 m high, with higher plant diversity and basal area. The mature forest and bamboo-dominated eco-units have high values for both basal area and canopy height (Viana and Tabanez 1996; Tabanez, 1995).

The study was carried out in three 10 m wide transects that cut across the fragment (fig. 23.3). A total of 7,671 m^2 (about 8% of the fragment area) was surveyed. In the

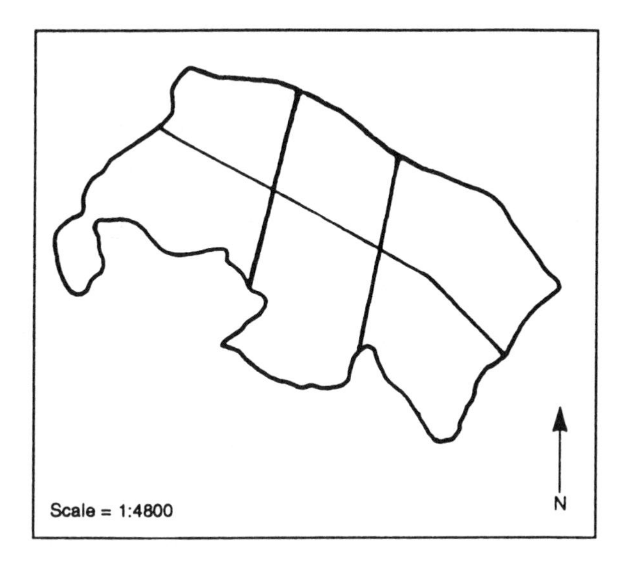

Figure 23.3. The Santa Rita forest fragment, showing the location of transects.

inventory, all trees greater than 5 cm diameter at breast height (DBH) were measured, tagged, and identified. We surveyed the plot in 1991 and again in 1994. Mortality and recruitment (ingrowth of plants greater than 5 cm DBH) values were obtained from these inventories. All plants were identified, and specimens were deposited at the Herbarium of the Escola Superior de Agricultura Luiz de Queiroz (ESALQ) of the University of São Paulo.

A forest restoration experiment was conducted in the nearby Usina Capuava fragment, which has a similar forest structure. The experiment included four treatments and a control. One treatment involved only vine cutting, and three involved vine cutting combined with plantings of three different mixtures of tree species. All treatments were applied to low forest eco-units (Tabanez 1995).

We used two approaches to analyze aspect and edge effects on the variables we studied. First, we used generalized linear models with a Poisson distribution for the meristic response variables (number of trees alive, number of dead trees, and number of ingrowth, or recruited, trees). Second, for basal area measurements (m^2/ha), which are continuous, we used classic analysis of variance (ANOVA). Edge distance was treated as a series of discrete classes, not as a continuous variable, considering that response variables (e.g., basal area, number of trees) were obtained in 10 m long plots along transects. This allowed the use of linear models (generalized or classic ANOVA) to analyze response variables with respect to distance from edge. In all models, the interaction of aspect and edge distance was tested before the main effects. The main effect of distance from edge, when significant (5% level), was tested as orthogonal polynomials. Due to lack of normality, the difference in basal area among eco-units was tested with a Kruskal-Wallis test.

RESULTS AND DISCUSSION

As discussed below, our results appear to confirm some of our predictions and accord with studies in other fragments in the region (Viana, Tabanez, and Aguirre 1992; Viana

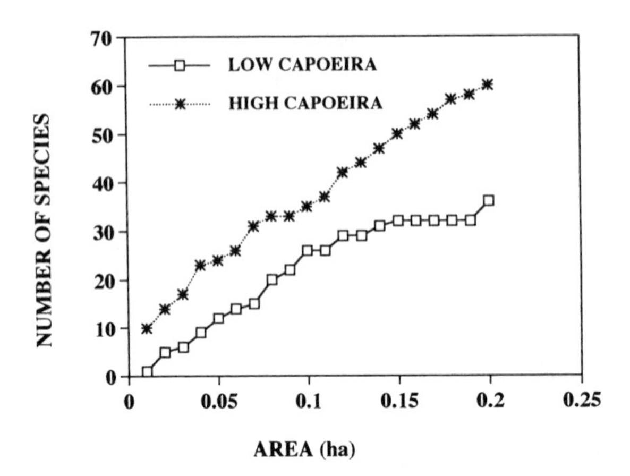

Figure 23.4. Species-area relationships for trees in the low and high secondary forest *(capoeira)* eco-units.

and Tabanez 1996). Nevertheless, because the results reported here were generated from only one fragment, they should be interpreted with caution and not automatically generalized to all fragments in the area.

Forest Dynamics

The Santa Rita fragment has a relatively high tree diversity (Shannon-Weaver index = 3.84, with 118 species identified in 0.7 ha, or 8% of the fragment surveyed), with a highly variable structure. The forest is dominated by *Astronium graveolens* Jacq. (Anacardiaceae) and *Securinega guaraiuva* Kuhlm. (Euphorbiaceae). In comparison with other fragments currently being studied in the region, this is a relatively diverse and well-structured forest (Viana and Tabanez 1996). The fragment was dominated by low-diversity eco-units: low forest (26.1% of the sampled area) and high forest (60.4%) were abundant, while mature forest (6.2%) and bamboo-dominated forest (7.3%) were rare. Diversity indices (Shannon-Weaver) for these four eco-units were 1.16, 2.02, 4.11, and 4.03, respectively. The species-area curves for the low and high forest eco-units exhibited marked differences (fig. 23.4).

Contrary to our prediction, overall tree recruitment in the Santa Rita fragment (106 trees/ha) was higher than tree mortality (63 trees/ha). This means that the number of trees increased during the study period. However, recruitment and mortality varied among eco-units (table 23.2). In both high and low forest eco-units, recruitment was higher than mortality. In contrast, mortality was higher than recruitment in mature and bamboo eco-units. This difference is important because the highest mortality occurred in eco-units with the highest diversity, while the highest regeneration occurred in those with the second lowest diversity. A comparison of species losses among eco-units, however, did not indicate a clear trend (table 23.3). Some eco-units appeared to be losing species, but it is too early to draw conclusions about this point.

In accord with our predictions, the number of live trees increased significantly ($P = .004$; fig. 23.5A) with distance from the forest edge, while the number of dead trees decreased (P = .012) with distance from the edge (fig. 23.5B). The number of newly recruited trees did not vary significantly with distance from the edge (P = 0.49; fig. 23.5C; all χ^2). Basal area of live trees also did not vary significantly with distance from the edge (P = .268; fig. 23.5D). Similarly, basal area of dead trees and basal area of newly recruited trees did not vary significantly with distance from forest edge (P = .47 and P = .932, respectively; all ANOVAs).

These results indicate an ongoing process of forest degradation near the forest margin. Forest edges have lower densities of live trees and higher mortality rates than the forest interior. The fact that tree basal area did not vary with distance from the forest edge seems to be the result of a few large trees surviving near edges. Because recruitment did not vary significantly with distance from the forest edge, the current process of edge effects appears to be more related to factors affecting tree mortality than to those affecting tree recruitment. However, while the relation between distance from edge and density of live trees was significant (P = .001), this relation was not significant for tree mortality (P = .534, χ^2). Tree mortality seems to have a bimodal pattern, which fits a third-order model (P = .0003, χ^2; fig. 23.5B).

These results indicate the complexity of edge effects in the plateau fragments we studied. This is especially true of the bimodal pattern in tree mortality rates (see also Didham, chap. 5). Edge effects often seem to be nonlinear processes with complex patterns at varying spatial and temporal scales. Long-term studies are needed to further evaluate the dynamics of edges in these fragmented ecosystems.

There was a significant (P = .0003) effect of aspect on the number of live trees. East-facing and west-facing edges had the highest number of live trees, and north-facing edges

Table 23.2 Mortality and recruitment of trees in different eco-units in the Santa Rita forest fragment during the 1991–1994 study period.

Eco-unit	% of Area	Recruitment (no./ha)	Mortality (no./ha)
Low second growth	26.6	79	49
High second growth	59.8	124	62
Old-growth forest	6.2	88	123
Bamboo forest	7.4	88	123
Total	100	106	63

Table 23.3 Changes in species numbers from 1991 to 1994 in different forest eco-units in the Santa Rita fragment.

	Number of species		
Eco-unit	1991	1994	% Change
Low second growth	32	30	−6.0
High second growth	106	105	−0.9
Old-growth forest	26	26	0
Bamboo forest	30	29	−3.3

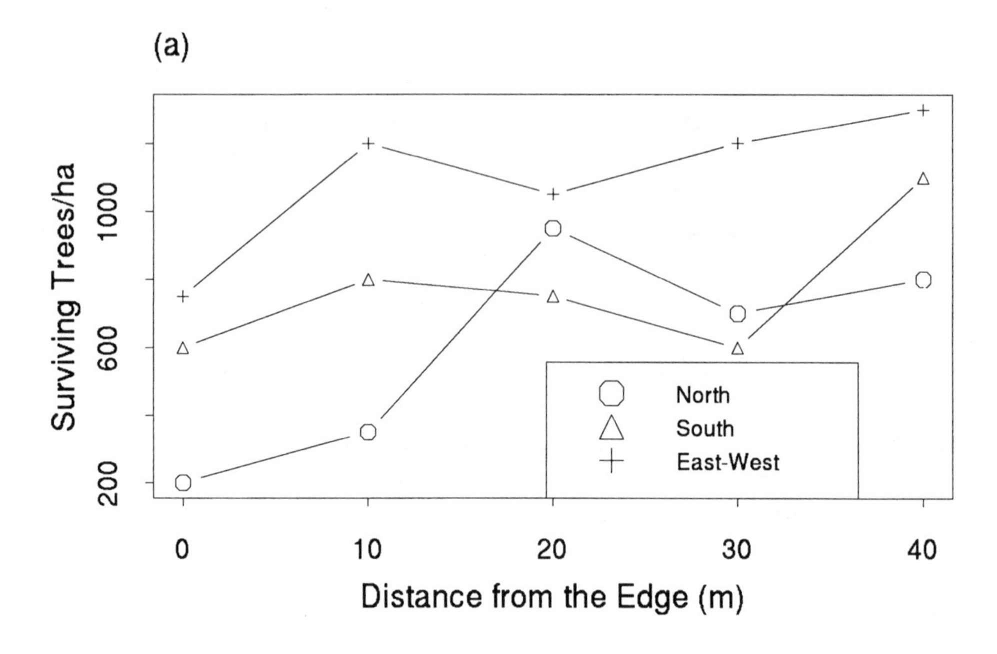

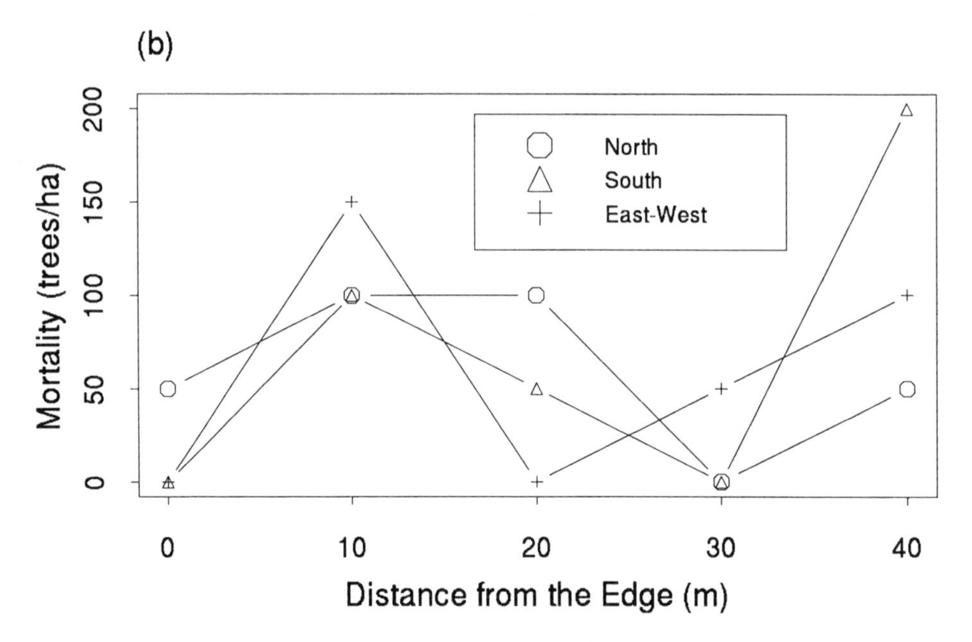

Figure 23.5. Effects of distance from the forest edge on *(a)* density of live trees, *(b)* density of dead trees, *(c)* density of ingrowth, and *(d)* basal area of live trees.

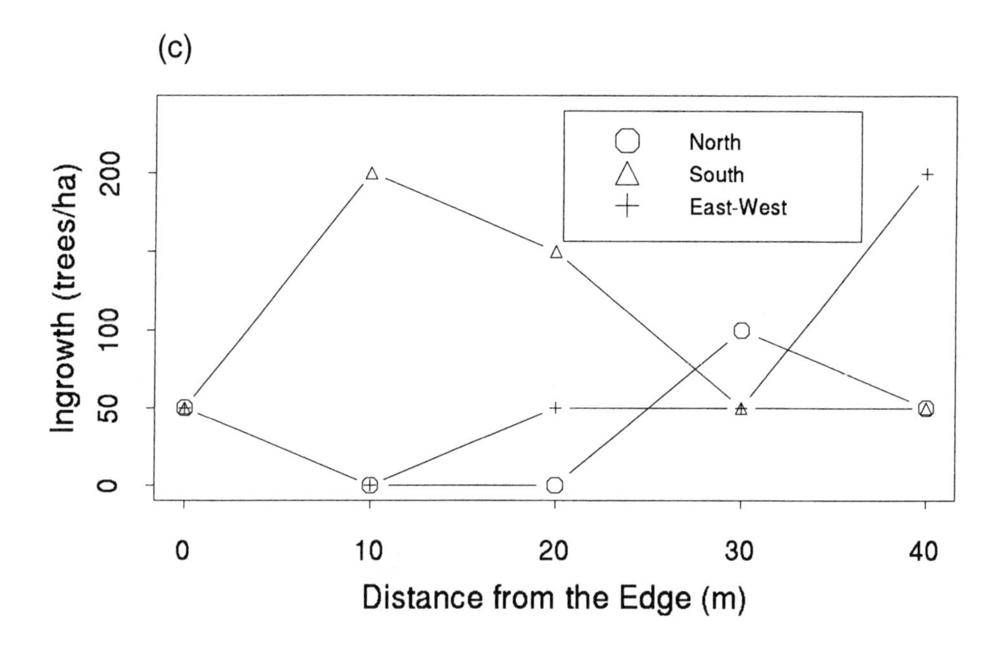

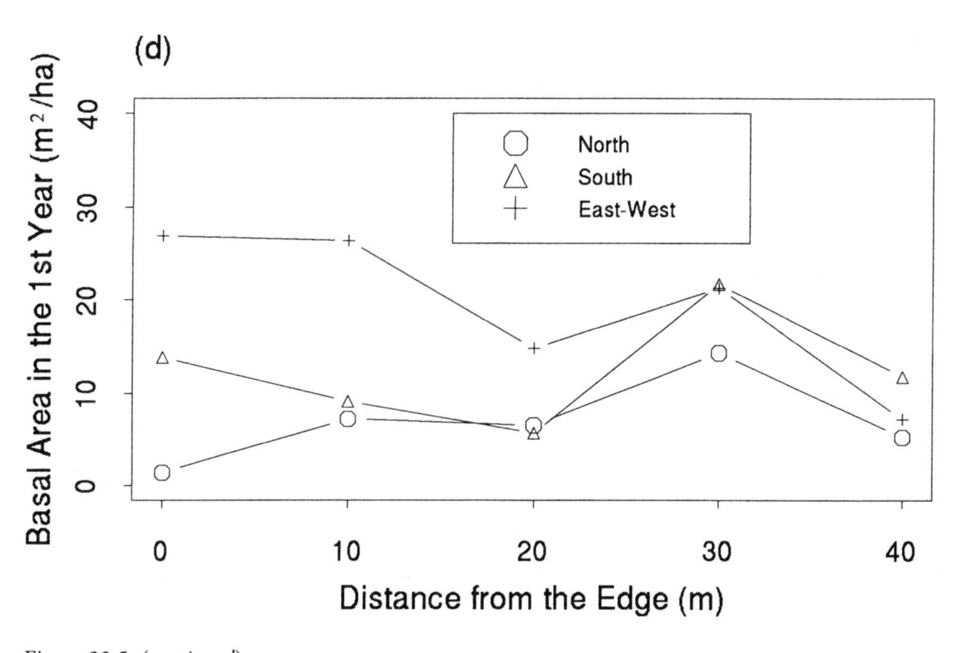

Figure 23.5. (*continued*)

Table 23.4 Changes in population size over a three-year period for tree species in the Santa Rita fragment.

Population size	% Increase			% Decrease		
	< 15	15–30	> 30	< 15	15–30	> 30
< 15	2	1	1	5	1	4
15–30	3	0	0	3	1	0
30–45	2	0	0	1	0	0
45–60	1	0	0	1	0	0
> 60	0	0	0	0	0	0
Total	8	1	1	10	2	4

Note: Tree species are divided ito five classes based on their estimated total population size in the fragment. Of 118 tree species present in the fragment, 82 had no changes in population size during the study period.

the lowest (fig. 23.5A). Basal area of live trees also varied significantly with aspect (P = .0270; fig. 23.5D). The number of newly recruited trees, however, did not vary significantly with aspect (P = .47; fig. 23.5C), nor did the number of dead trees (P = .93, fig. 23.5B; all χ^2).

These results demonstrate that forest structure varies with aspect, as indicated by the lower number of live trees and lower basal area near northern edges. In the Southern Hemisphere, northern edges receive more sunlight than southern edges. Also, at our study site, northern edges receive prevailing winds during drier months, while southern edges receive prevailing winds during the humid summer season (ESALQ, meteorological data). The effect of aspect on forest edges is likely to be an important component in the dynamics of fragments in the Atlantic moist forest, especially in its subtropical regions (see also Turton and Freiburger, chap. 4).

Declining Tree Populations

Tree populations in fragments are not only small, but in several cases are decreasing in size (table 23.4). Populations of sixteen species declined over the study period. Such declines are particularly problematic for species that already have small population sizes. Ten species that had estimated population sizes of fewer than fifteen individuals experienced further reductions (table 23.4). Several tree populations appear to have no regeneration, and the remaining individuals may be "living dead," that is, alive but reproductively nonfunctional. *Cedrela fissilis* (Meliaceae), a gap opportunist canopy tree that was heavily exploited in the region for its highly valuable timber, but supposedly not logged in this fragment, had no individuals smaller than 21 cm DBH. *Chorisia speciosa* (Bombacaceae), another gap opportunist species, had no individuals smaller than 13 cm DBH.

Very little is known about the processes leading to differential mortality and recruitment rates of forest tree species in fragments in the Atlantic moist forest, and much more research is needed on this theme. We hypothesize that high mortality rates are associated with (1) short life cycles; (2) anthropogenic pressures, especially for those species that have high commercial value and are selectively harvested, such as palm heart *(Euterpe edulis);* and (3) high vulnerability to windthrows due to mechanical (wood resistance) or architectural (root depth) characteristics (see Laurance 1991b; Laurance, chap. 6).

We also hypothesize that low recruitment is correlated with dependence on animal pollinators and seed dispersers that are themselves vulnerable to fragmentation (see Nason, Aldrich, and Hamrick, chap. 20; Thébaud and Strasberg, chap. 21) and poorly adapted to the forest environments that dominate forest fragments. The problem of maintaining species with low population sizes is further compounded in highly fragmented landscapes where forest patches are distant from each other. Low-density (rare) species also have a greater chance of being absent in nearby patches due to sampling effects.

Forest Restoration Practices

Restoration practices aimed at facilitating successional processes in low forest eco-units produced the expected changes in forest structure (Tabanez, 1995). Basal area and crown cover were most strongly affected by vine control and certain combinations of experimental plantings. Long-term monitoring of these treatments is needed to test whether or not improved forest structure will result in increased levels of tree diversity, as we hypothesize.

The costs of restoration practices for low forest eco-units in the fragment interior were on the order of U.S.$1,050.00/ha, and they would be applicable to about 1.7 ha of low forest eco-units in the Santa Rita forest fragment. Considering that this fragment contains several different eco-units, each with different restoration needs, a plan to restore the whole fragment entails an overall cost of U.S.$2,718.50 (table 23.5). This plan includes vine control in high forest and enrichment plantings with vine control in low forest eco-units. The application of these treatments is simple, and the overall cost is not out of reach of local forest owners (typically sugar mills).

There is a need for further research on the silvicultural and ecological aspects of forest restoration in order to reduce its costs and make it accessible to a wider range of landowners. Agroforestry systems are a promising strategy for reducing restoration costs, especially in forest edges. Part of the restoration costs would be offset by revenues from transient agroforestry crops (corn, beans, pumpkins). These revenues could pay for vine

Table 23.5 Estimated costs of restoration practices for the Santa Rita forest fragment.

Eco-unit	Area (ha)	Treatments[a]	Cost (US$/ha)
Low secondary forest			
Edge	0.63	Agroforestry	1,574.00
Interior	1.71	Enrichment	1,050.0
High secondary forest	5.38	Vine control	94.50
Old-growth forest	0.54	—	—
Bamboo forest	0.66	—	—
Total	8.92		2,718.50

a. Restoration of the fragment also involves the control of hunting and fire, which are not included here.

control during the critical phase of establishment of young trees, which is a major component of the restoration costs. Furthermore, many crops (e.g., corn) could serve as food sources for animal populations in forest fragments (e.g., primates). Because these practices would last for only two to three years, they would have minimal negative effects on forest ecosystem characteristics.

In summary, forest fragments in the plateau region of the Brazilian Atlantic moist forest are unlikely to be self-sustainable. The situation described for the Santa Rita fragment is representative of a significant part of the remaining forest cover of the Atlantic forests, especially in the plateau region, where fragments are smaller, more isolated, and more disturbed than those along the coastal hillsides (Viana 1990; Viana, Tabanez, and Aguirre 1992; Viana and Tabanez 1996). There is an urgent need for both applied research and conservation initiatives.

GENERAL IMPLICATIONS

1. Forest fragments are not simply isolated patches of primary forest. Most fragments have a complex history of disturbance events—mainly fire, logging, and hunting—and the effects of isolation are usually exacerbated by these disturbances. In many cases, fragment history seems to be the key factor determining forest structure and dynamics.
2. Fragments comprise a mosaic of eco-units with varying ecological characteristics. Assessments of the conservation value and restoration needs of a fragment should consider the relative proportions of these eco-units.
3. Edge effects seem to be a continuing phenomenon, even in old fragments. Edge effects are complex and require detailed analysis for individual fragments.
4. Conservation of tree diversity should target selected species that seem most vulnerable to fragmentation. These include species that *(a)* depend on animal pollinators and seed dispersers that are themselves vulnerable to fragmentation (this problem is exacerbated in cases in which there are obligate, specialized mutualisms); *(b)* have short life cycles or low survival rates (vegetative propagation techniques should be explored for ex situ genetic conservation of such species: see Andersen, Thornhill, and Koopowitz, chap. 18); *(c)* are poorly adapted to the secondary forest environment that dominates small, highly disturbed forest fragments; *(d)* have small and isolated populations, especially those that naturally occur at low densities; *(e)* have high vulnerability to windthrows due to their mechanical or architectural characteristics; and *(f)* are subject to high anthropogenic pressures, especially those that have high commercial value and are selectively harvested (e.g., *Euterpe edulis*).
5. Restoration of Atlantic moist forest fragments appears to be feasible. Such efforts need to be based on both the ecological characteristics of the fragment and the socioeconomic and cultural perceptions of the farmers that own the fragments. Further research is needed to lower the costs of restoration and to develop a series of restoration alternatives that suit the varying socioeconomic and cultural perspectives of landowners. The ultimate test of a "good" restoration practice is its adoption by farmers, not simply publication in scientific journals.

ACKNOWLEDGMENTS

We would like to acknowledge the financial support of the Boticário Foundation and FAPESP; the institutional support of IPE and IPEF; the generous collaboration of the owners of the Santa Rita forest fragment; the students and researchers of the Laboratory of Tropical Forestry at ESALQ; Ricardo Rodrigues for his assistance in botanical identifications; William Laurance, Richard Bierregaard, and Rod Keenan, who reviewed earlier versions of this chapter; Henrique Nascimento, who assisted in data analysis; and Lili Viana, for her support in preparing the figures and revising one of the final versions of the chapter.

24
Rejoining Habitat Remnants: Restoring Degraded Rainforest Lands

David Lamb, John Parrotta, Rod Keenan, and Nigel Tucker

In the closing years of the twentieth century, the preservation of the world's declining biological diversity has emerged as a major conservation challenge. Land managers and policy makers in most developing countries have to deal with the immediate problems of rural poverty and an increasing need for food and living space on the one hand, and the demands of the world community to conserve biological diversity on the other. In many Western industrial countries, present levels of social and economic development have been achieved through the capitalization of natural resources, and it is hardly surprising that this is seen as a model for others to follow. The model is appropriate, however, only if one accepts the accompanying environmental costs, and while some less-developed countries are apparently willing to do so, many are concerned about the consequences such development may have for their biological resources (Kahn and McDonald, chap. 2).

The problem of preserving biological diversity is perhaps most acute in landscapes occupied by moist tropical forests, both because of their great biological richness and apparent fragility (but see Lugo 1995) and because they are being subjected to rapid land use changes. Large areas of forested land are being cleared for food production, but failure to recognize the importance of tightly linked forest nutrient cycles in maintaining soil productivity has resulted in declining crop production (Jordan 1993), and as a result many cleared lands are being abandoned in a degraded form. Settlers and indigenous farmers are thus forced to exploit new lands, and consequently there are fewer areas of undisturbed forest available for conservation reserves. In addition, many of the remaining undisturbed areas are now isolated fragments, often unrepresentative of the diversity of ecosystems once present in the landscape.

The tropical conservation problem results from a complex mix of economics, politics, and biology. From a biological viewpoint, the challenge is to develop sustainable land management practices that satisfy the needs of the local human population and contribute to the conservation of biological diversity, while simultaneously developing and maintaining a viable and representative system of nature reserves. In this chapter, we highlight ways of dealing with degraded lands to restore their ecological function, to reduce the effects of fragmentation on native fauna and flora, and in some cases, to produce economic benefits for local communities.

RESTORATION GOALS

Three alternative goals when overcoming land degradation in a previously forested area might be

1. *Restoration,* attempting to restore a forested area to its presumed original condition. This new forest would contain the same complement of species and would have the same general structure as the original forest.
2. *Rehabilitation,* in which mostly native species are used in the new forest, but some exotics are included because they are economically or ecologically necessary to overcome the degradation.
3. *Reclamation,* using only exotic species to overcome degradation. Again, these may be chosen for economic or ecological reasons.

These alternatives are similar in that they all aim to re-create productive and functional forest ecosystems on an otherwise degraded landscape, but they differ in the extent to which they seek to reestablish the original forest biodiversity.

Restoration, as defined above, is obviously the preferred option if preservation of biological diversity is the prime objective. However, it is also likely to be the most expensive option if large areas are to be treated, and its feasibility is debatable for several reasons.

1. The richness of tropical forests makes it impossible to plant or directly seed anything more than a small subset of the plant species once present. Moreover, many forest plants require animal mutualists to reproduce, and ensuring that these animals also recolonize degraded sites is problematic.
2. Biotic composition and relative population sizes of different plant species are rarely known for tropical forests. These difficulties are amplified when a site has been degraded and only remnants are left to indicate the likely original condition.
3. Forest clearing for agriculture commonly results in the most fertile land on gentle topography being cleared first. Forest remnants are therefore often unrepresentative of the original forest types in cleared areas, and while some plant and animal species may be found in both situations, many are likely to have disappeared. Even where representative forest remnants persist on more fertile sites, the size of these remnants is often too small to sustain the original biota or to provide adequate sources of colonists for degraded sites.
4. The widespread presence of weed or pest species makes them difficult to eradicate and can hinder the re-creation of original communities.
5. While it may be possible to approximate the original forest composition at a particular site, finer-scale spatial variation may be impossible to specify (except in general terms) because we lack knowledge of site and habitat requirements for the majority of species. Community composition is also highly dynamic, so that aiming at a preferred species composition is "a game with a moving target whose trajectory cannot be accurately predicted" (Simberloff 1990b, 44).
6. Small differences in factors such as climate or population sizes of planted species at the time of establishment may lead to different outcomes. For example, the presence of a particular mycorrhizal symbiont, or the order in which species enter the new ecosystem, can affect the nature and diversity of the ecosystem (Gilpin 1987a; Grange, Brown, and Sinclair 1993).

Add to these problems the stochastic nature of rainforest succession (e.g., Webb, Tracey, and Williams 1972), and it is clear that true restoration is practically impossible to define precisely, much less implement. This does not mean, however, that a more modest goal of reestablishing a representative species-rich community of plants and animals is not achievable. Whether this community has the same species richness and relative population sizes may not matter. In circumstances in which regional levels of biological diversity are plummeting, even modest levels of "restoration" are worth attempting.

WHERE TO RESTORE?

Studies of island biogeography suggest a generalized relationship between species richness and island area of the form

$$\log S = \log C + Z \log A$$

where S is species richness, A is area, C is a constant giving the number of species when A has a value of 1, and Z is also a constant, usually assuming a value of 0.7 or less (Williamson 1981). The actual value of Z (the slope of the log species-log area relationship) varies with the biota and locality in question. The predicted species loss caused by deforestation, which leads to the creation of forest "islands," may therefore be high or low depending on the factors that influence Z, such as the species' vagility and the hostility of the surrounding habitat (Lugo, Parrotta, and Brown 1993). If Z is low, then restoration of massive areas might be needed to significantly increase regional biodiversity where only fragments of an original forest area remain. Such restoration is expensive and often difficult to achieve for social and political reasons.

The situation may not be as difficult as this simple analysis suggests, however. Most forest remnants are not "islands" in a biologically sterile "sea," and many species are not distributed at random across a landscape, but are concentrated in certain key locations. Small increases in the sizes of these key habitats may allow the persistence of larger populations of certain species.

Potential target areas for forest restoration (table 24.1) include the following.

1. *Habitats of particular species.* Because restoration seeks to prevent further species loss, an obvious focus for activity is the habitats of endangered species, especially local endemics (see Andersen, Thornhill, and Koopowitz, chap. 18). It may be possible to define the habitat requirements of these species and attempt to re-create conditions for their reproduction and expansion, although lack of baseline knowledge of habitat conditions or variation in the distribution and density of target species may make this difficult. Artificially increasing the local population size of an endangered species may be a short-term solution, but its longer-term survival will probably require restoration of more extensive areas of forest.

2. *Streamsides.* Riparian ecosystems are productive and comparatively species-rich. Many species are restricted to these habitats, while others range more widely but still depend on these systems (Remsen and Parker 1983; Crome, Isaacs, and Moore 1995).

Table 24.1 Preferred target areas for forest restoration.

Location	Comment
Habitats of particular species	Where populations of particular species have been reduced to critical levels
Streamsides	Often local centers of species richness and important wildlife habitats during drier periods
Degraded lands within existing reserves	To enhance the value of existing conservation areas
Edges of forest remnants	To consolidate boundaries and reduce adverse changes at the edge or within forest fragments
Corridors linking forest remnants	To facilitate species movement between isolated fragments
Islands	Islands are often the only places where exotic pests or predators can be excluded
Within the general degraded land matrix between forest remnants	Wherever it is economically, socially, or politically possible to attempt restoration

Restoration of riparian habitats could allow populations of these latter species to increase and enhance their capacity to colonize other areas. Revegetation of streamside areas also benefits water quality, stabilizes soil surfaces, and reduces erosion.

3. *Degraded areas within and around conservation reserves.* Creation of national parks and nature reserves is often difficult and sometimes controversial (Green, Paine, and McNeely 1991). Costs of managing reserves can also be comparatively high for many tropical countries. It makes sense, therefore, to ensure that degraded areas within these reserves are managed with the aim of optimizing the conservation benefits they provide. The same is true where degraded lands available for restoration are located on the boundaries of nature reserves or forest remnants. In this situation, restoration may aim to consolidate irregular boundaries and reduce the adverse ecological changes that occur at edges and often penetrate some distance into forest remnants (Sisk and Margules 1993; Laurance et al., chap. 32).

4. *Corridors.* Vegetation corridors that link together several forest remnants have been widely discussed as targets for restoration efforts (e.g., Saunders and Hobbs 1991). Corridors may have a particular value in linking remnant areas of forest at different altitudes to facilitate movements of elevational migrants, which are common in the Tropics (Crome 1975). Enabling species to move between remnants may also enhance opportunities for genetic or demographic interchange or facilitate seasonal movements. Much remains to be learned about the most appropriate design for such corridors, such as width, length (and their interaction: Harrison 1992), location, and how edge effects might limit their value. The utility of corridors has been questioned by those who argue that there is little direct evidence that they facilitate species movements in the way that corridor advocates suggest (e.g., Simberloff 1992). Most of the debate has concerned only animal species, and there is currently little field evidence from which to draw definitive conclusions. Irrespective of this debate, it seems likely that corridors beyond a certain minimum width will have value as new habitat even if they have limited utility for increasing animal or plant movement.

5. *Islands.* Degraded island ecosystems may be particularly attractive target areas for

restoration because they are one of the few locations in which pest species can be completely eradicated with a greatly reduced possibility of re-invasion (Downs and Ballantine 1993).

6. *In the matrix between remnants.* Restoration of matrix areas can be used to create new habitat between existing fragments. The size and potential conservation benefit of this habitat will depend on the restoration effort, the landscape pattern of current fragmentation, and the presence of species that can colonize the new habitat. This strategy also offers the advantage of further increasing the biological heterogeneity of the "sea" between remnant forest "islands."

HOW TO RESTORE DEGRADED RAINFOREST LANDS

The first step in any approach to restoration is to alter land use practices responsible for land degradation. In conservation reserves, the management objective is generally to remove human influences to as large an extent as practically possible and assist natural processes that will restore vegetation to a natural condition. In other areas, such as streamsides, corridors, and the general matrix of agricultural areas between remnant vegetation, eliminating human influences is not possible, but destructive patterns of grazing, indiscriminate use of fire, and inappropriate cropping must be altered before systems for reestablishing natural vegetation can be applied.

Four possible methods for restoring natural vegetation are natural regeneration, direct seeding, accelerating natural succession through planting seedlings, and incorporation of restoration goals in timber production programs. These methods will be discussed below, and case studies from Queensland and Puerto Rico will be used to demonstrate the application of the latter two methods.

Natural Regeneration

Restoration through natural regeneration involves protecting the site from further disturbances and allowing natural successional processes to restore a forest community. There are many examples in which natural seed dispersal has led to reforestation of large areas. In the New England region of the United States, forest has reestablished itself over large areas of land cleared for agriculture in the late eighteenth century (Williams 1992). Clearing reached a peak in 1820–1880, when more than 80% of the land was deforested. The present forests are comparatively young, and compositional trends can still be related to past land use (Matlack 1994a). However, the magnitude of the reforestation process in the absence of direct replanting is significant.

Similar patterns of reforestation have occurred in various areas now occupied by rainforest. Jones (1956) reported that large areas of forest in Nigeria appear to have regenerated following extensive agricultural clearing more than two hundred years earlier. Rainforests in Venezuela and Colombia have recovered following widespread wildfires about two hundred fifty years ago (Saldarriaga and West 1986), and there is evidence for natural recovery of rainforests and other tropical forest formations in Puerto Rico during the last fifty years following centuries of intensive agriculture (Birdsey and Weaver 1982).

Recovery times for such natural succession are often very long. Where degradation is severe, limitations to the process may include low availability of propagules (seeds or root stock), seed or seedling predation, loss of dispersal agents, nonavailability of suitable microhabitats for plant establishment, low soil nutrient availability, absence of fungal or bacterial root symbionts that facilitate nutrient uptake, seasonal drought, competition from grasses and ferns, and fire (Uhl and Jordan 1984; Uhl 1987, 1988; Uhl, Buschbacher, and Serrão 1988; Nepstad, Uhl, and Serrão 1991; Parrotta 1992, 1993a; Thébaud and Strasberg, chap. 21). If this approach is consciously applied as a restoration practice, land uses must be restricted to those compatible with restoration objectives.

Direct Seeding

Reestablishment of natural forest can be assisted by broadcast sowing of seed of certain key species that modify the environment and facilitate the reestablishment of other species. This strategy requires some form of weed control and a relatively large supply of seed from appropriate species. Direct seeding has been shown to be more efficient than planting seedlings when establishment rates of greater than 1% of sown seeds can be achieved (C. Van der Woude, pers. comm.). There are many examples of direct seeding being used to revegetate bare soil at mine sites, but few cases in which direct seeding has been used successfully to restore degraded forest lands. Sun, Dickenson, and Bragg (1995) presented results of direct seeding of a rainforest pioneer *(Alphitonia petrei)* in degraded streamside areas in the Wet Tropics of northern Queensland, in combination with herbicide application. In this case, up to 28% of sown seeds became established. If this technique can be tested at a wider variety of sites and proven to be operationally efficient, it may be particularly useful for steep or otherwise inaccessible areas. The difficulty of acquiring seeds of rainforest species in sufficient quantities, however, may limit this option.

Accelerating Natural Succession by Planting Seedlings

Carefully established plantings of particular tree species can also accelerate natural succession. Once a forest structure is established, the site may become suitable once more for forest-dwelling wildlife. Plant species suited to such restoration programs (table 24.2) include those that can tolerate and ameliorate adverse site conditions and those that act

Table 24.2 Priority species for use in planting programs for forest restoration.

Species type	Comment
Nitrogen-fixing species and those able to improve soil fertility	Such species can reduce the need for costly fertilizer amendments.
Species able to grow rapidly and exclude weeds	These species will help create appropriate microclimatic conditions.
Species attractive to frugivores	These species encourage seed dispersal to the site via frugivores.
Mutualistic species able to sustain wildlife at otherwise stressful times of the year	These species may help maintain wildlife populations.
Poor dispersers (e.g., large fruits)	Such species may not colonize the site otherwise.
Rare or threatened species	Plantings will increase local population sizes.

to accelerate the colonization process by attracting seed-dispersing animals (Lugo 1988b; Lugo et al. 1993; Parrotta 1992, 1993a,b).

The "framework species" method developed in northern Queensland (Goosem and Tucker 1995) uses suites of local tree species that promote more rapid natural succession. It has the greatest potential in areas where substantial forest tracts and remnants still exist. These framework species have a number of important features: individuals can grow in close proximity (i.e., they are "gregarious"); they cast sufficient shade to eliminate grasses and weeds and thus rapidly capture the planted site; and they have small to medium-sized fruits that are attractive to a wide range of avian and mammalian frugivores, the major seed dispersers in local closed forests (Goosem and Young 1989).

Pioneer species constitute 30% of framework species. Pioneers rapidly provide food for frugivores and arboreal folivores (J. W. Winter, pers. comm.), create perches and other forest structures, and through self-thinning and short lifespans, quickly create ground habitats such as fallen logs. Figs (*Ficus* spp.) are also important because they can serve as a major food and perching resource to many local frugivores and some folivores (Crome 1975). Other framework species are from later successional stages but become regular, abundant fruit producers within three to six years after establishment. Species that are unlikely to colonize the site quickly because of large diaspore size, are threatened or endangered, or can provide particular food or habitat for endangered wildlife (e.g., *Flindersia* spp.: Laurance 1990) may also be used.

The planted area should be close enough to nearby forest that the arrival of seed and fruit eaters is likely. In northern Queensland, some fruit pigeons and flocks of gregarious frugivores such as metallic starlings *(Aplonis metallica)* and yellow orioles *(Oriolus flavocinctus)* can easily cross about 500 m of open grassland. Other species may have different tolerances. Whether the primary factor in trees attracting frugivores is fruit reward or perch site is uncertain. McDonnell and Stiles (1983) have shown that increased structural complexity alone can enhance seed deposition in planted areas. Their work suggests that if this applies more widely, framework species might be chosen on the basis of their perching value rather than fruit type. If a complete species complement is to be achieved in the restored area, the planting program will also need to provide for the establishment of species with poor natural dispersal abilities, such as those with heavy seeds that rely on wind dispersal.

Inclusion of Restoration Goals in Timber Production Programs

In much of the Tropics it is rarely possible to plant large areas purely for habitat restoration. However, commercial timber and fuelwood plantations on cleared or degraded forest sites may be utilized to meet conservation goals. Native species with high timber value can be planted in mixtures. Alternatively, monocultures of native species carefully selected to match specific site conditions can be established, yielding a mosaic of different species over a particular landscape. The initial species diversity in either situation will not be as high as in the framework species method described above, but the trade-off is that production benefits allow these methods to be applied over larger areas of degraded landscape. This approach (as will be described further below) may therefore yield higher diversity benefits in the longer term.

Alternatively, plantations of exotic species can be used to suppress grasses and weeds and create appropriate understory microclimatic conditions for natural successional processes to increase local biodiversity (e.g., Mathur and Soni 1983; Mitra and Sheldon 1993; Lugo 1992; Parrotta 1992, 1993a,b).

CASE STUDIES

Restoration Efforts in Northern Queensland, Australia

In Australia, clearing of natural vegetation and subsequent land degradation are major environmental problems that have resulted in extensive destruction and fragmentation of forest and woodland habitats. The Australian Tropics have not escaped such effects. Recognition of these problems has recently resulted in government initiatives aimed at mobilizing community support to restore forest on cleared or degraded areas in northern Queensland. Two such initiatives are the Community Nature Conservation Program (CNCP) and the Community Rainforest Reforestation Program (CRRP).

The Atherton Tableland is a focus for both programs. It is an undulating plateau ranging from 700 to 1,000 m in altitude (17° and 17°35′ S), separated from the coast by a narrow coastal plain and a range of granite massifs up to 1,620 m high (fig. 24.1). Annual rainfall on the Tableland proper ranges from 1,400 mm in the north to 3,600 mm in the south, and is strongly seasonal, with a pronounced dry season from July to November. Average temperatures range from a maximum of 28.5°C in December to a minimum of 10.8°C in July (see Laurance, chap. 6, for a more detailed description of the Atherton Tableland).

Before European settlement, the area was mostly rainforest of varying composition, depending on climate and soil type (Tracey 1982, 1986). Clearing began on richer basaltic soils and quickly extended to other areas, but as with most land clearance in the Tropics, there have been many situations in which soil fertility and crop and pasture production declined rapidly following initial clearing (Maggs and Hewett 1993). Significant areas of rainforest in less accessible areas and on poorer soils remained in government ownership, and much of this was logged with varying intensity and quality of management. Logging in state-owned forests ceased following World Heritage listing of the rainforests in 1988.

Case Study 1: Community Nature Conservation Program

The CNCP aims to promote nature conservation outside protected areas by encouraging local communities to help manage their local environment and by providing support to community groups and individuals involved in the restoration of natural plant communities. The CNCP has developed a specialized habitat restoration nursery and works closely with the community group TREAT (Trees for the Evelyn and Atherton Tablelands) to carry out its objectives. Through voluntary nursery and field work and federal

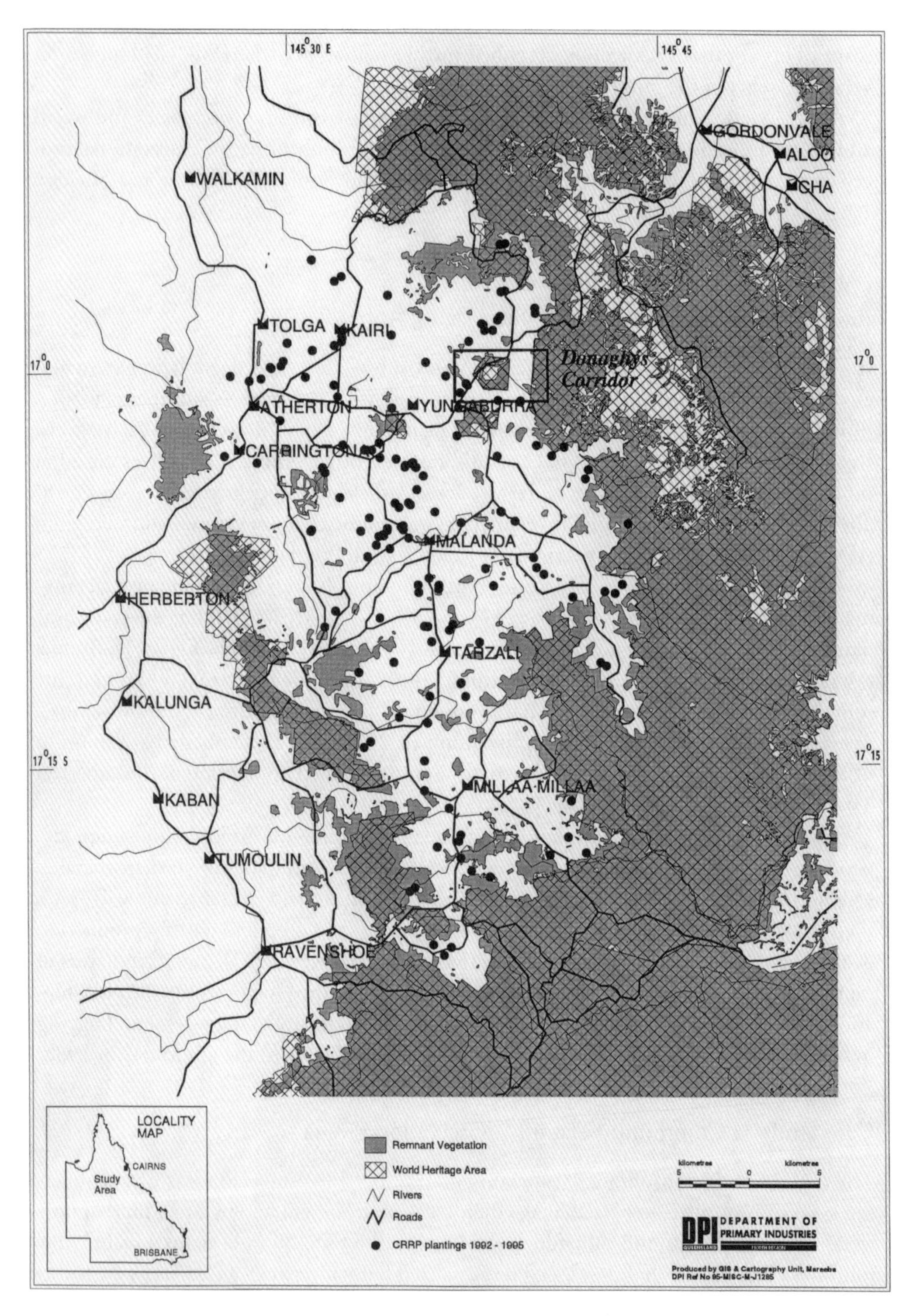

Figure 24.1. Distribution of remaining rainforest in the Atherton Tableland region in northern Queensland. Solid circles indicate location of areas planted under the Community Rainforest Reforestation Program.

and state government support, the CNCP is able to provide free tree seedlings to TREAT members for planting on their properties. Plantings vary from shelterbelts on dairy farms to the repair of riparian zones and expansion of forest remnants. Seedlings are also used by the Queensland Department of Environment and Heritage for restoration projects in and surrounding national parks and other protected areas.

A number of different restoration strategies have been devised for different climatic and edaphic conditions and different scales and intensities of disturbance to natural vegetation. The framework species method described above is commonly used, and specialist legumes such as *Acacia* spp. are used on very degraded soils.

Site preparation involves control of all exotic species prior to planting using knockdown herbicide. All dead material is left to provide mulch cover. Weed control may require two or three herbicide sprayings to exhaust all weeds in the soil seed bank. Plants are randomly established at 1.5 m spacing (about 6,000 stems/ha) during the wet season. This close spacing ensures canopy closure in 12–24 months. An NPK fertilizer is applied at the rate of 100 g per plant at establishment and every month during the wet season thereafter until the trees are two years old. Weeds are excluded by nonresidual herbicide until canopy closure occurs and the area becomes self-maintaining.

Natural colonization following the use of framework species depends on circumstances and time since planting. At one highly degraded site in the coastal lowlands, some 50 plant species (distributed over 45 genera and 31 families) had colonized an area of 200 m^2 by eight years after treatment; these included 36 trees, 3 shrubs, and 11 vine species (Goosem and Tucker 1995). The site was adjacent to an area of riparian rainforest, so dispersal distances were not large.

The framework species method is also being used in the Donaghy's Corridor area (see fig. 24.1) to link a small (500 ha) conservation and recreational reserve, Lake Barrine National Park, with nearby continuous forest. Although the reserve has been isolated for only sixty to eighty years, there have been significant changes to both the small mammal community (Laurance and Laurance 1995) and forest structure (Laurance, chap. 6). A watercourse, Toohey Creek, runs for 2 km over private property between the park and the state forest, and the landowner agreed to revegetation of a linear strip 100 m wide along the creek. Restoration activities began with establishment of fencing and provision of fenced drinking points for cattle. Thirty-eight tree species were selected for planting based on their utility to wildlife and growth rates. Planting began early in 1995 and will be completed over the next three years. Sampling of avian communities, aquatic invertebrates, and mammals is also being undertaken along the corridor route.

Overall, the CNCP has been successful in educating local landowners and community members about ecological processes and the inputs required for successful reestablishment of native rainforest vegetation. It has enabled the development and testing of a number of approaches to forest reestablishment, and its efforts have led to the establishment of about 100 ha of forest in this region. The major limitations of this approach are its expense and the small number of private landowners willing to commit large areas for conservation purposes.

Case Study 2: The Community Rainforest Reforestation Program

Even before logging stopped in Queensland's northern state-owned rainforests, there had been calls for the development of a timber industry based on plantations on private property (Gilmour and Reilly 1970; Tracey 1986). In the 1950s and 1960s, the Queensland government established about 3,000 ha of plantations of native hoop pine *(Araucaria cunninghamii)* and exotic Caribbean pine *(Pinus caribbea)* on land purchased from economically marginal dairy farmers. Various investigations identified substantial areas of land on the Atherton Tableland that had been cleared, but were more suited to catchment protection or production forestry than agriculture (e.g., Gilmour and Reilly 1970). In 1992, federal and state governments agreed to fund the CRRP for four years with the following objectives:

1. Develop a private plantation resource base for a sustainable timber industry in northern Queensland, with a major emphasis on native rainforest species
2. Address problems of land degradation in the Wet Tropics area following extensive inappropriate clearing
3. Provide for improved water quality by establishing vegetative buffers along rivers and streams
4. Train a workforce to support the long-term practice of rainforest plantation establishment (QDPI-Forest Service 1994).

Landowners are recruited for the CRRP by extension staff through advertisements and at public meetings. After agreeing to participate, landowners are expected to prepare the site for planting and protect planted seedlings from livestock. In 1994–1995 a small fee (A$100/ha) was introduced to help cover costs and ensure landowner commitment. The CRRP provides all seedlings, and they are planted by work crews of supervisors and trainees. Grass and weeds are usually slashed before planting, and the area is disc-plowed or ripped by farm tractor. Newly sprouting vegetation is sprayed with knock-down and pre-emergent herbicide. On steep sites, where machinery cannot operate, the area is sprayed with a knock-down herbicide two or three times.

Species are chosen for particular sites based on the knowledge of local rangers and extension officers, the landowner's goals, and the availability of planting stock. The common planting pattern is a mixture of pioneer and later successional species; for example, a row of eucalypts or other pioneer species and a row of rainforest trees consisting of three to eight species. Spacing is generally 5 m between rows and 2 m along the row (1,000 stems/ha). The wide inter-row spacing allows access by a farm tractor for slashing of grass and weeds. Planting takes place following the onset of monsoon rains and can continue until June, depending on the locality. Survival is monitored, and replanting is undertaken wherever necessary. Ongoing weed control is essential for successful establishment and rapid growth, and trees are kept free of weeds by the use of herbicides for two or three years or until crown closure is reached. Subsequent management of the plantation is the responsibility of the landowner.

In the study area, 139,800 seedlings have so far been planted on seventy-six individual properties, for a total planted area of 206 ha (over 1,500 ha has been planted in the entire

Wet Tropics region). Across the Wet Tropics, 129 different species have been planted, although half of the plantings were of 8 species: *Acacia mangium, Agathis robusta, Araucaria cunninghamii, Elaeocarpus angustifolius, Eucalyptus cloeziana, E. grandis, E. pellita,* and *Flindersia brayleyana.*

Establishment success has been relatively high because of the close attention paid to establishment and maintenance. Early height growth has varied considerably among species and site types. While growth has been relatively good, no one species grew better than others on all site types. This variation suggests that there may be considerable scope for developing a regionally diverse plantation system even if plantation monocultures are planted at particular localities.

Some problems encountered in the program (as compared with a traditional timber plantation or restoration project on government-controlled land) have included (1) getting landowners to carry out required activities in a timely fashion; (2) growing sufficient seedlings of appropriate species when the area of different site types to be planted is not known until a few months prior to planting; and (3) the additional cost associated with maintaining plantations of species with different growth rates. Despite these problems, the CRRP has been very successful in reintroducing a relatively high variety of native species into the Wet Tropics landscape as part of a timber production-oriented initiative. Evidence from older plantation trials in the region suggests that such plantations will eventually acquire a diversity of native species in the understory. The program is helping to demonstrate that mixed-species plantings are valuable for the wider range of goods and services they provide, despite their marginally higher cost (Keenan, Lamb, and Sexton, 1995). The value of these plantations for native fauna, particularly in overcoming problems caused by habitat fragmentation, will be determined through a monitoring program to be implemented over the next few years.

Case Study 3: Coastal Tropical Forest Rehabilitation in Puerto Rico

In Puerto Rico, studies have been conducted during the past decade to test the hypothesis that forest plantations like those described above can surmount barriers to natural regeneration and facilitate the restoration of indigenous forest flora on degraded sites. The study area is the University of Puerto Rico's Toa Baja experimental station on the northern (Atlantic) coast of Puerto Rico (18°27′ N, 66°10′ W). Annual precipitation averages 1,600 mm and is moderately seasonal. Formerly supporting coastal dune forest vegetation, this site has been subject to frequent and often severe disturbances during the past century, including forest clearing, cattle grazing, cultivation, topsoil removal, and periodic fires. The surrounding landscape is presently dominated by residential housing and commercial and industrial facilities. Secondary forest is restricted to roadsides and a 10 ha abandoned wooded pasture adjacent to the southern edge of the study site, which supports about twenty native and naturalized tree species. At the time the experimental plantation was established, the site was dominated by grasses (principally *Panicum maximum* and *Tricholaena repens*), along with some twenty species of herbs and vines; woody plants were absent (Parrotta 1993b).

Natural regeneration of secondary forest tree and shrub species was monitored in experimental plantations and unplanted (though protected) control plots established in 1984 and 1989 (Parrotta 1993a, 1995). The older plantation area includes nine tree plots planted with *Albizia lebbek* at a range of densities and four control plots. The younger plantation area consists of twenty-one plots, three each of the following seven species or species combinations: (1) *Casuarina equisetifolia;* (2) *Eucalyptus robusta;* (3) *Leucaena leucocephala;* (4) *C. equisetifolia* and *E. robusta;* (5) *C. equisetifolia* and *L. leucocephala;* (6) *E. robusta* and *L. leucocephala;* and (7) unplanted, protected controls. All of these species are commercially valuable, and all are exotics. Plots received no irrigation or fertilizer. Planted plots were manually hoed during the first six months to control weed competition, and all plots were protected from disturbances, particularly grazing and fire.

Regeneration studies were initiated at 4.5 years after planting, when plots had closed canopies and well-developed forest floors. In the *A. lebbek* and associated control plots, all woody seedlings of secondary forest species taller than 5 cm were identified, mapped, measured, and tagged in 16 m^2 permanent quadrats. In the plots planted in 1989, all secondary forest species found within each plot were identified and recorded. Understory light conditions and forest floor (litter) depth and dry mass were measured at selected intervals during the study period.

Temporal Patterns of Understory Colonization

In both experiments, ground cover during the first two years was generally very sparse relative to the unplanted controls, but included most grasses and herbaceous species present on the site prior to plantation establishment. Regeneration was not noted within plantation areas until the third year after planting. This was the case despite apparently regular seed inputs by visiting frugivorous bird and bat species, the virtual elimination of understory competition by grasses, and the creation of favorable light, temperature, and humidity conditions from the second year onward. This delay appears to be due to biotic requirements for the breaking of seed dormancy and facilitating germination, rather than constraints of the physical environment. Similar observations of soil fauna and microflora succession in Costa Rican plantations suggest that delays in natural regeneration by native forest species (despite abundant seed rain and suitable physical conditions) may be linked to the rate of development of soil microflora that perform functions required for germination (R. Fisher, pers. comm.).

After 3 years, there was a rapid increase in both seedling density and species richness, and at 4.5 years, plantation plots collectively supported a fairly rich flora of native and naturalized trees and shrubs (fig. 24.2). While species richness and density of colonizing secondary forest species tended to peak after about 6 years, basal area of colonists continued to increase, and comprised 15–32% of the total stand basal area at 8.7 years. In the *A. lebbek* plantations, naturally colonizing trees had begun to occupy codominant or dominant crown positions after 5 years. In contrast, no woody seedlings were found in either the control plots or the other unplanted, protected grassland.

In the 1989 plots, the overwhelming majority of colonizing species were produced from seeds dispersed by frugivorous birds and bats (fig. 24.3) and derived from parent trees located 50–300 m from the plantation. Only three wind-dispersed species were

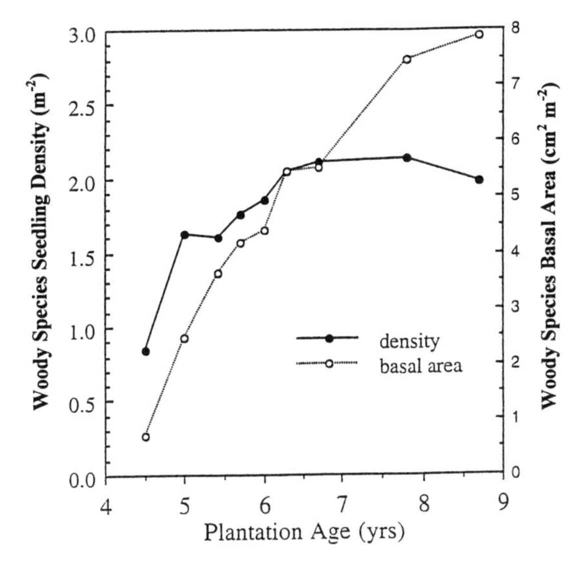

Figure 24.2. Mean density and basal area of colonizing trees and shrubs in *Albizia lebbek* plantation stands at Toa Baja, Puerto Rico. The total numbers of species found in the 192 m² areas sampled in each age class were 9, 14, 18, 20, and 19 at 4.5, 5.0, 6.0, 6.7, and 8.7 years of age, respectively.

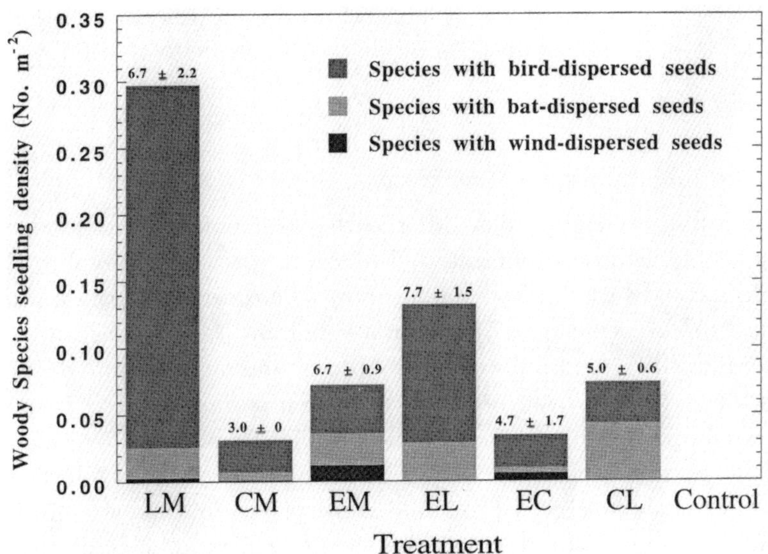

Figure 24.3. Seedling density of understory colonizing tree and shrub species 4.5 years after plantation establishment. Numbers above bars indicate the mean (±SE) number of species for each treatment, based on surveys of three 289 m² plots per treatment. LM = *Leucaena leucocephala;* CM = *Casuarina equisetifolia;* EM = *Eucalyptus robusta;* EL = *E. robusta* and *L. leucocephala* (1 : 1); EC = *E. robusta* and *C. equisetifolia* (1 : 1); CL = *C. equisetifolia* and *L. leucocephala* (1 : 1); Control = unplanted, protected grassland.

found in the plantation: *Cedrela odorata, Lonchocarpus latifolius,* and *Spathodea campanulata.* The absence of these colonizing species in control plots, where their seeds also fall, is probably explained by competition with grasses or unsuitable microclimatic conditions (McClanahan and Wolfe 1987, 1993; Nepstad, Uhl, and Serrão 1991; Guevara, Purata, and Van der Maarel 1992; Parrotta 1992, 1993a).

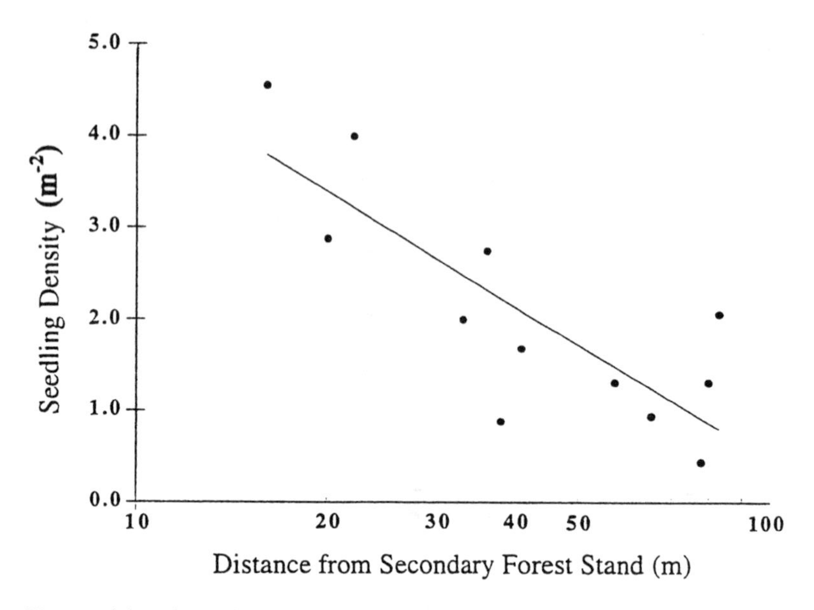

Figure 24.4. Relationship between density of colonizing seedlings and saplings in quadrats under *Albizia lebbek* stands and distance from the nearest secondary forest ($r^2 = .66$, $P < .01$).

Factors Influencing Understory Colonization

The major factors influencing colonization of the plantation understories were distance from seed source, light environment, and overstory species composition. Understory seedling densities and species distributions showed considerable variation with location within the *A. lebbek* plantations. There was a significant ($P < .01$) negative correlation between seedling density (for all colonizing species) and distance from the nearest possible seed source (fig. 24.4), suggesting that proximity to seed sources is an important factor controlling the rate of secondary forest species establishment in plantations (see Thebaud and Strasberg, chap. 21). The relative importance of seed sources external to the plantation would be expected to diminish over time, however, as individuals established within the plantation become reproductively mature, although external seed sources would continue to be important for increasing floristic diversity. Several species, such as *Bourreria succulenta, Citharexylum fruticosum, Cordia polycephala,* and *Schinus terebinthifolius* began flowering and fruiting in the *A. lebbek* plantations by year 5 or 6.

To quantify the light environment at each site, light measurements were made over several days between 1000 and 1300 hours at ten to twenty points. These results were averaged to provide a light value for each site. Measurements were carried out using a hand-held light meter at a height of 1 m. Among *A. lebbek* plantation quadrats, average light intensities at 5.5 years ranged from 2% to 29% of those in the open, and these variations were negatively correlated with the density of colonizing secondary forest species (fig. 24.5). A similar trend was observed in the younger plots. It is unlikely that light intensity had any direct influence on seed germination or subsequent survival, as most species that became established in the plantations can also do so in full sunlight. However, quadrats with higher understory light levels tended to occur on sites with coarser, sandy soils low in organic matter. These sites had lower tree growth and leaf area, and were more likely to be dominated by grasses than were quadrats with lower light intensities. Thus, the relative scarcity of seedlings in the more open quadrats could be due to a combination of increased root competition with herbaceous flora (principally grasses), less favorable soil moisture conditions affected by soil texture and organic matter content, and lower soil nutrient availability (Parrotta 1993a).

Overstory composition appears to exert a strong influence on understory species richness and colonization rates (see fig. 24.3). *L. leucocephala* appears to provide the most favorable conditions for understory colonization by secondary forest species, followed by *E. robusta. C. equisetifolia* supported relatively little new regeneration. In mixed-species treatments, understory species richness and seedling densities were generally intermediate between the values from plots of single species (Parrotta 1995). Significant negative correlations ($P < .05$) were found between understory species richness and forest floor depth, and between seedling density, forest floor depth, and dry mass (fig. 24.6). Fewer species and lower densities occurred in the understories of plots with the greatest forest floor depth (i.e., those with *C. equisetifolia* comprising 50–100% of the overstory), indicating that increased litter accumulation acts as a progressively more severe barrier to regeneration by small-seeded (bird-dispersed) woody species.

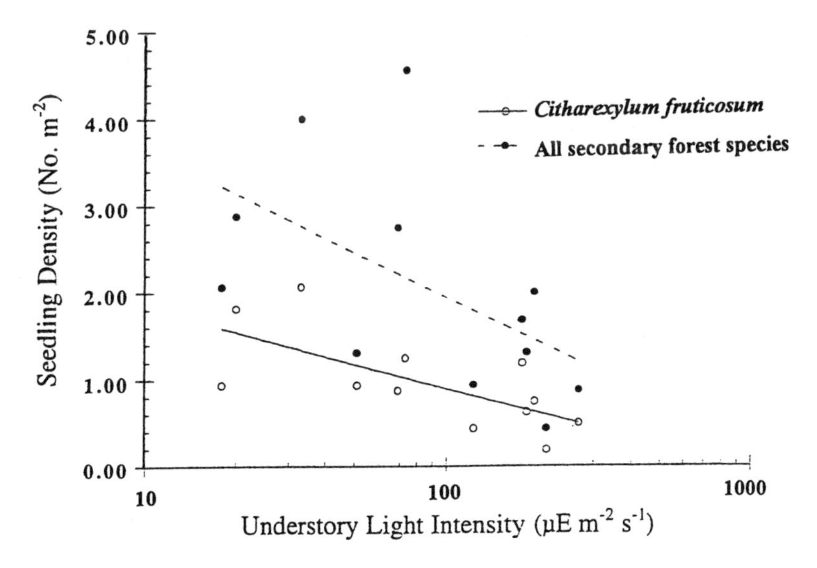

Figure 24.5. Relationship between seedling density and understory light intensity in quadrats in 6.7-year-old *Albizia lebbek* plantations. The understory plants are dominated by *Citharexylum fruticosum* ($r^2 = .48$, $P < .05$; all secondary forest species: $r^2 = .31$, $P < .10$). Correlation coefficients and significance levels are for regressions of the general form: $Y = A - B \log_{10} X$.

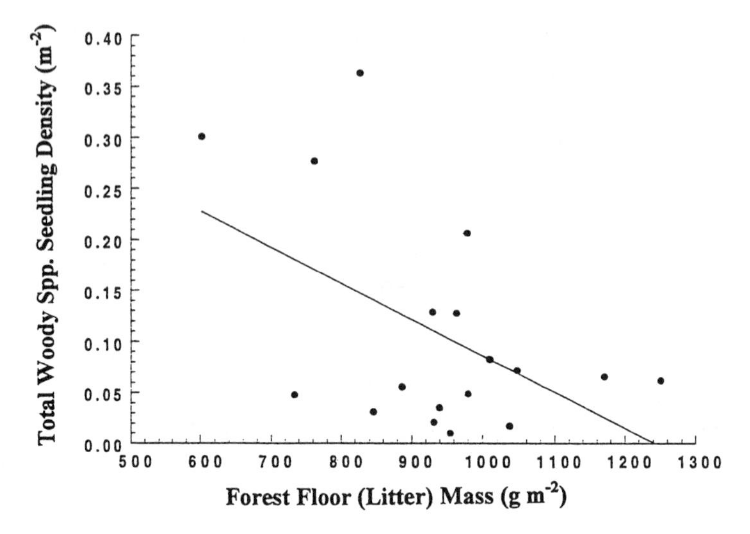

Figure 24.6. Relationship between seedling density of colonizing tree and shrub species and litter mass in single- and mixed-species plantations of *Casuarina equisetifolia, Eucalyptus robusta,* and *Leucaena leucocephala* at 4.5 years of age ($r^2 = .26$, $P < .05$).

DISCUSSION

Three different approaches to the restoration of forest habitats in tropical landscapes were described in the case studies. In case study 1, a variety of species with desirable characteristics for accelerating successional processes (termed "framework species") were planted at high stocking rates with intensive follow-up maintenance, purely for habitat restoration. Case study 2 also involved planting and intensive maintenance, but used a subset of native plants with potential commercial uses. Case study 3 showed that the difference in the two approaches may not be as great as it might initially appear.

Each approach has certain advantages and disadvantages. The framework species method of case study 1 is the most amenable to reestablishing poorly dispersed, rare, or endangered species by including these in the planting mix. The costs of seedling production and labor are high, however, and are not balanced by any potential commercial benefits. Use of this method may thus be restricted to relatively small areas of landscape, except in societies with the financial resources or community motivation to apply it over large areas.

Case study 3 demonstrates that the moderating influence of commercial tree plantations on the understory environment can facilitate reestablishment of native species, potentially increasing biodiversity as well as providing commercial benefits. This method may allow restoration of much larger areas of degraded land. A number of key questions remain unresolved, however. Does the gain in biodiversity reduce the commercial success of the plantation? Does the developing understory inhibit the growth of overstory trees, or is the overstory effectively free of competitive effects once canopy closure is achieved?

Finally, how much biological diversity can be retained when the commercial crop trees are harvested and the area is replanted?

The conservation benefits of such an approach will ultimately depend on the management system used. Where native species with timber value become reestablished in exotic plantations, these may, over time, become the dominant species in the plantations, or they may be incorporated into the production system along with the originally planted species. Choice of plantation species may also strongly influence natural recruitment of native species through factors such as the amount of litter accumulated on the plantation floor.

The use of native species in a timber production program (case study 2) is a flexible alternative that can be directed more or less toward nature conservation depending on the interests of the landowner and the socioeconomic circumstances at the time of harvesting. More complex plantation forestry of this type may become important where societies require a wider range of outputs (Kanowski and Savill 1992; Lamb and Tomlinson 1994). The disadvantages of this approach are that the potential conservation values and timber production benefits are largely unknown, in part because of our limited understanding of the growth rates and wood quality of native species grown in plantation systems.

Biological Factors Affecting the Design of Restoration Strategies

In all three case studies, the process of further species recruitment depends on the design and management of the planting program. Relevant considerations include the location of plantings in relation to primary and secondary forest stands, tree species selection, and planting density (Parrotta 1993a, 1995).

Proximity to seed-bearing trees and the habitat requirements of seed-dispersing birds and bats are key factors influencing development of floristic diversity in restoration plantings or plantations. The importance of perching or roosting sites for seed inputs into open habitats such as grasslands and mined lands has been demonstrated in several recent studies (Guevara, Purata, and Van der Maarel 1986; McClanahan and Wolfe 1987, 1993; Guevara et al. 1992). Where distances to intact forest remnants are large, more complex and creative strategies may be required to maximize the potential for successional processes. "Archipelagoes" of replanted "islands," containing species of particular utility to seed-dispersing wildlife (e.g., framework species; see above), could be created within a "sea" of commercial timber plantations or other kinds of land use. The choice of timber plantation species would influence the recruitment rate of native plants, but the presence of patches of attractant species could facilitate the regeneration process where the distance to potential seed sources is large.

Planting density is important because recruitment of secondary forest species depends on seed inputs, the maintenance of sufficient canopy cover to reduce weed and grass competition, and sufficient habitat diversity for birds and other seed dispersers. If early economic returns are required, harvests should be designed to ensure at least partial canopy closure until the colonizing forest species assume a more dominant structural role within the system.

Socioeconomic Factors Affecting the Design of Restoration Strategies

Sociological context is key to any restoration program (see Viana, Tabanez, and Batista, chap. 23). Land degradation usually has its roots in a particular socioeconomic condition, and policies to address this condition must be implemented to prevent further degradation. Scientists and conservationists often assume that the benefits of restoration are self-evident, but they may not always be apparent to local communities, particularly if they have been gaining some economic benefit from degradative former land uses (Fortmann and Bruce 1988; Gilmour and Fisher 1991; Morrison and Bass 1992). It is therefore advantageous to involve the local community in any restoration program, and this has been a key feature of case studies 1 and 2. This involvement has led to far more effective restoration than if only institutional resources were employed. Involving local people, however, may shift the major objective of a project from optimizing biological diversity to finding a balance between biodiversity and economic production. Where rates of forest loss and land degradation are not being matched by restoration efforts, such a compromise should be willingly accepted.

Perhaps a more difficult issue is whether plantations established for timber production can be diverted to conservation purposes. This will obviously depend on the original reason for plantation establishment. Managers of industrial plantations that required a large capital investment (e.g., a pulp mill) are generally committed to intensive management to satisfy investment criteria and are often not in a position to reduce timber outputs, lengthen rotations, or change harvesting systems to allow a diversity of species to persist during the production cycle. Even in a landscape committed to intensive fiber production, however, there are usually inaccessible areas, streamside zones, naturally poor soils, and other land types that could be used to provide for conservation of regional diversity. Land use systems can also be modified to provide for structural, genetic, and species diversity while managing for fiber production (Bunnell 1990; Franklin 1993).

While there are no universally applicable techniques for tropical forest rehabilitation and restoration, we feel that the case studies presented here illustrate practical approaches worthy of consideration in restoration projects. Further ecological and silvicultural research is necessary to refine these methods and demonstrate their benefits for biological conservation. Research needs include the identification of key barriers to natural forest succession at specific sites; the definition of site conditions under which alternative restoration methods can yield favorable results; assessment of the value of plantations as corridors to facilitate genetic interchange; further investigation of methods for accelerating succession in particular directions; and the development of planting designs and stand management practices that will result in productive, species-rich secondary forests on severely degraded lands.

GENERAL IMPLICATIONS

1. Restoration implies an attempt to reestablish original forest communities in a landscape. This is a difficult task in the Tropics because of the exceptional species richness of tropical forests and our generally poor understanding of their composition and dynamics.

2. There are large areas of degraded forest lands in the Tropics in need of restoration. Large-scale restoration is unlikely to occur, however, because of economic, social, and political constraints. Under these conditions, priority should be placed on restoration of *(a)* the habitats of endangered species, *(b)* corridors linking vegetation remnants along rivers and streams, and *(c)* areas within and around existing conservation reserves. Restored areas need not necessarily be large if they are carefully planned to facilitate movement and genetic exchange between subpopulations of wildlife and plants.
3. All restoration efforts require the prevention of further clearing, grazing, or burning. Revegetation can then occur naturally or be facilitated by planting. The choice of planted species is crucial. Where restoration is being attempted, the planted species should be "framework species" that attract seed dispersers (birds or bats), which in turn bring other plant species into the area.
4. Where socioeconomic factors preclude restoration, commercial considerations will determine the species used in plantations. These considerations need not preclude some enhancement of the landscape's biodiversity, because careful matching of species to specific sites may yield a mosaic of different plantation monocultures. Moreover, inaccessible areas, streamsides, and other small conservation areas can be planted with a range of native species to create greater landscape diversity. Alternatively, mixed-species plantations can be established.
5. Natural dispersal and regeneration is likely to enhance the biodiversity of the understory in both mixed-species plantations and mosaics of monocultures. Because this process could facilitate increased biological diversity over large areas of tropical landscapes, researchers and plantation managers should work together to devise methods to achieve and maintain the biological potential of timber plantations.

ACKNOWLEDGMENTS

We would like to thank Terry Webb and Shonad Philips at the Department of Primary Industries in Mareeba for preparation of Figure 24.1, and Bill Laurance, Steve Goosem, and anonymous referees for their constructively critical comments on the manuscript.

25

Measuring Landscape Changes in Remnant Tropical Dry Forests

Elizabeth A. Kramer

Tropical dry forest is the most endangered habitat in Central America, currently reduced to less than 1% of its original extent (Janzen 1988a). In Costa Rica, the entire area of dry forests was cut over by 1961 (Sader and Joyce 1988). These figures illustrate the urgent need for conservation of remaining dry forest fragments and the importance of establishing techniques for restoration of this habitat.

Unfortunately, much of the literature about tropical ecosystems and their management pertains to either rainforest or savanna systems (Murphy and Lugo 1986); dry forests have been studied much less frequently. Tropical dry forests occur in frost-free areas where the mean annual temperature is higher than 17°C and where mean annual rainfall ranges from 250 to 2,000 mm. An additional feature of tropical dry forest areas is that mean annual potential evapotranspiration (PET) exceeds mean annual precipitation (Murphy and Lugo 1986), whereas in wet tropical forest areas the reverse is true (Ewel 1977). Because of this difference, the factors controlling secondary succession in dry forests differ from those in wet forests. Dry forests are strongly water-limited, with harsh environmental conditions occurring over at least six months of the year, while wet forests are mostly nutrient-limited, with species establishment strongly influenced by competition (Ewel 1977). These differences affect the rates and mechanisms of secondary forest establishment.

The effect of forest fragmentation on biodiversity is the primary concern of many studies of forest remnants (Saunders, Hobbs, and Margules 1991). Fragmentation commonly results in small forest remnants embedded in a matrix of agriculture, secondary vegetation, and degraded land. Few studies, however, have examined the interactions of remnant forest patches with the matrix, or vice versa (Janzen 1988b; but see Thébaud and Strasberg, chap. 21).

This chapter describes the use of geographic information systems (GIS) and remote sensing to aid in the management of forest recovery. The general goal of this study is to measure, and to predict, how vegetation responds to changing land management practices at the landscape scale. In the tropical dry forest areas within Guanacaste National Park in northwestern Costa Rica, land use has changed from extensive agriculture to habitat protection and management for forest recovery. Specific management options

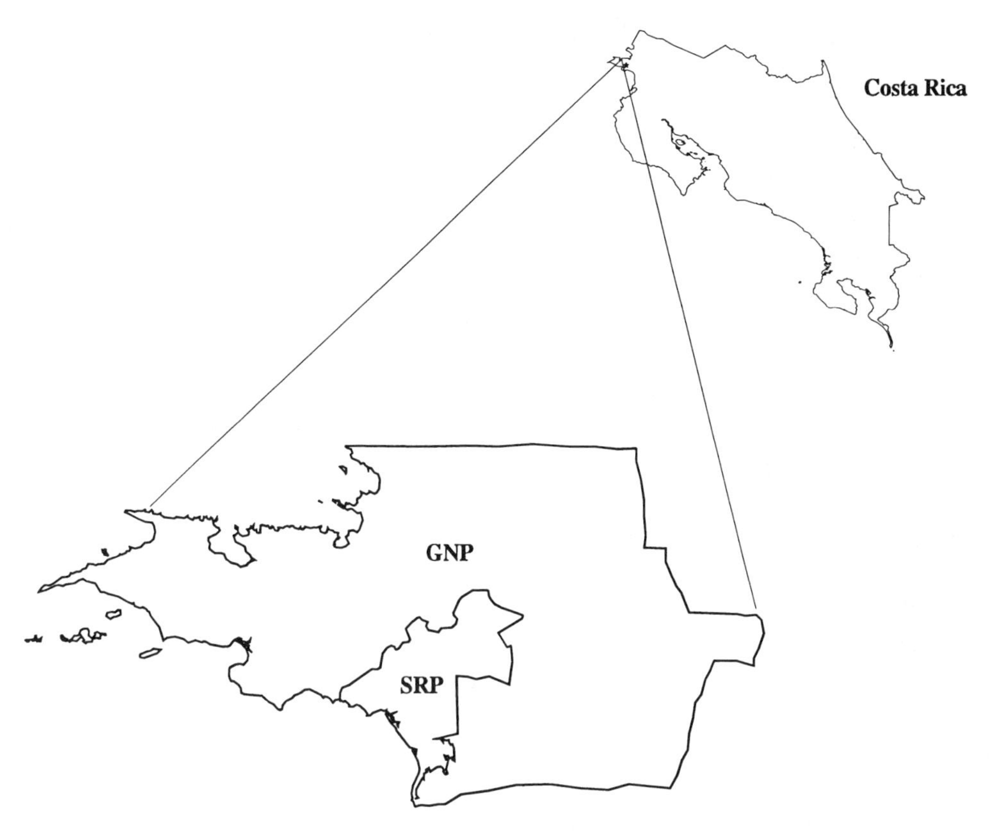

Figure 25.1. Map of Costa Rica, showing the locations of Guanacaste National Park (GNP) and Santa Rosa Park (SRP).

range from allowing natural patch spread and coalescence (see Thébaud and Strasberg, chap. 21) to intensive forest plantings. The results presented in this chapter focus on the development of vegetation databases and measurements of vegetation change over an eight-year period, from 1979 to 1985.

GUANACASTE NATIONAL PARK

The Costa Rican government established Guanacaste National Park (GNP) in the mid-1980s. The park encompasses an area of about 700 km² in northwestern Costa Rica (fig. 25.1). It covers a diversity of ecosystems within an area ranging from the Pacific coast of the Santa Elena Peninsula east toward the peaks of Vulcan Orosi and Cacao, spanning an elevational gradient from sea level to 1,900 m. The area encompasses six of the Holdridge life zones: tropical dry forest, premontane forest, premontane moist forest, premontane wet forest, premontane rainforest, and lower montane rainforest (Holdridge et al. 1971); this study focuses on the tropical dry forest and a portion of the premontane forest. The region has a pronounced dry season that lasts for about six months. Wet season rainfall is about 1,500 mm per year.

The predominant land use prior to establishment of the park was pasture for cattle production, with pastures being largely composed of introduced African grasses (principally jaragua, *Hyparrhenia rufa*) grown in monoculture. Annual burning during the dry season is used to maintain these grasses. Most of the original vegetation has been manipulated, leaving small patches of forest surrounded by agricultural land. In areas where annual burning has ceased, woody vegetation does become established, although natural succession is slower in dry forests than in wet forests. As in all types of natural succession, however, the rate of floristic change at any specific location is limited by the availability of seed sources (Janzen 1988b; Moran et al. 1994).

GNP incorporates three previously existing parks and the surrounding agricultural areas, which are predominantly cattle pasture. Of these, Santa Rosa Park (SRP) is the largest, at 108 km^2. Established in 1977, it is principally dry forest, and is the main focus of this study. At the time of its establishment, SRP also comprised a mosaic of remnant forest patches and degraded agricultural land.

My main goal was to use remotely sensed images and GIS to characterize and measure changes in the vegetation of SRP from 1979 to 1985, and to compare them with changes occurring within the entire GNP area before the establishment of Guanacaste National Park. This strategy permitted me to compare the effects of varying management reimes on vegetation dynamics over large areas, and to assess the interactions of vegetation patches among themselves and with the physical environment.

METHODS

Image Analysis

Remotely sensed images are a cost-effective way to acquire land cover information for a large geographic area (see Elachi 1987; Asrar 1989; Hobbs and Mooney 1990; Wickland 1991; Lillesand and Keifer 1994 for reviews). Remote sensing instruments typically measure electromagnetic radiation reflected and emitted by the earth's surface, which can be collected at multiple scales and times (Quattrochi and Pelletier 1991; Wickland 1991). Because each image is a snapshot in time, by comparing images we can measure temporal change occurring at a particular site.

Satellite images provide information at a scale appropriate for mapping landscape-level vegetation patterns. Table 25.1 provides a comparison of two commonly available image types, Landsat MSS and Landsat TM. Landsat MSS has an 80 m × 80 m ground spatial resolution and is composed of four spectral bands, while Landsat TM has a 30 m × 30 m ground spatial resolution and is composed of seven spectral bands. In this study I analyzed a Landsat MSS scene from 1979 and a Landsat TM scene from 1985. Both scenes were acquired in the dry season when deciduous vegetation was dormant. This timing facilitated vegetation classification and minimized differences in physiological states of the vegetation.

Vegetation classes were chosen to minimize overlap in forest type. Janzen (1988a) defined and described sixteen distinct communities for the entire area of GNP. Due to the minimum resolution of the MSS imagery used in this study, Janzen's classification was simplified to four distinct vegetation classes.

Table 25.1 Landsat sensors, wavelength sensitivities, and ground spatial resolution.

Sensor	Satellite mission[a]	Sensitivity (μm)	Ground spatial resolution (m)
Multispectral scanner (MSS)	1–5	0.5–0.6	79/82[b]
		0.6–0.7	79/82
		0.7–0.8	79/82
		0.8–1.1	79/82
	3	10.4–12.6	240
Thematic mapper (TM)	4, 5	0.45–0.52	30
		0.52–0.60	30
		0.63–0.69	30
		0.76–0.90	30
		1.55–1.75	30
		10.4–12.5	120
		2.08–2.35	30

Source: Lillesand and Keifer 1994.

a. Refers to the satellite number, i.e., mission 1 is the same as Landsat 1.

b. Landsat 1–3: 82 m; Landsat 4–5: 79 m.

Evergreen forests. These are distinctly evergreen during the dry season, and include evergreen canyon forests and evergreen oak forests.

Mixed forests. These are predominantly deciduous but have an evergreen component that is visible during the dry season, and include Janzen's dry forests and alluvial semideciduous bottomland forests.

Deciduous. This class encompasses both forest and nonforested scrub areas without an evergreen component, and includes Janzen's strongly deciduous hillside forests, early successional forests, and abandoned pastures.

Pasture. This class includes pastures and agricultural fields.

Image classification is the science and art of extracting information from raw images (Lillesand and Kiefer 1994). Classification techniques fall into two categories, supervised and unsupervised. Supervised classification involves a person assigning raw pixels to a priori classes. "Training sets" are representative samples of each class that are chosen from the image. These representative training sets are then used by the computer to place the rest of the image into previously defined classes. By contrast, an unsupervised classification involves aggregating pixels from the image into spectral groupings called clusters. The clusters are compared with an image and assigned to a vegetation class. For this study, an unsupervised classification was performed using the ISODATA (Iterative Self-Organizing Data Analysis Technique) algorithm in ERDAS Imagine 8.2 (ERDAS 1994). ISODATA uses the minimum spectral distance to cluster pixels into natural spectral signature classes (ERDAS 1994). The clusters are then assigned to the vegetation classes.

Because the images used were collected while a majority of the vegetation was leafless, underlying soils affected the spectral properties of the data. Therefore, I transformed the images using a tassel-cap transformation (Crist and Kauth 1986). The transformation compressed the original data sets down to three bands, representing brightness, greenness, and wetness. The brightness value is a weighted sum of all of the bands and strongly corresponds to soil reflectance. The greenness value is the orthogonal of the brightness value and strongly corresponds to the amount of green vegetation in the scene. The wetness value relates to canopy and soil moisture (Lillesand and Keifer 1994).

The transformed images were then run through the ISODATA routine to produce 256 spectral clusters. The 256 clusters represent the number of individual classes that can be displayed using an eight-bit graphics monitor. Each of the ISODATA clusters was then placed in one of the four previously described vegetation classes. Class placement was based upon decision rules determined by the pixel's brightness, greenness, and wetness values. Pixels with high greenness and wetness values were assigned to the evergreen class, while those with high brightness and low greenness and wetness values were assigned to pasture. Pixels falling in between these two classes were assigned to mixed or deciduous classes based upon the ratio of the three values. Pixels with higher greenness and wetness were assigned to mixed forest, and those with higher brightness and wetness to deciduous forest. A series of aerial photos taken in February 1988, with 10 m × 10 m ground resolution, was used to aid in class assignment and to assess the accuracy of classification. Because the photos were used to assist in classification, a statistical accuracy assessment could not be performed on the final land cover maps. The vegetation maps for Santa Rosa Park (plate 1) for 1979 were based on images with a 65 m × 65 m resolution, whereas those for 1985 had a 30 m × 30 m resolution.

Landscape Analysis

The detection of vegetation change involved simply overlaying the 1985 land cover map on the 1979 land cover map. A change algorithm was then applied to determine areas of change or no change. Transition and retention frequencies were calculated for each vegetation class combination. Before the images were overlaid, they were resampled to a common pixel size of 65 m × 65 m. The resampling technique used a nearest-neighbor algorithm.

Overlay techniques are useful for determining trends in landscape change, but additional information on patch spatial distributions and shapes is necessary to further reveal why and how the landscape is changing. Spatial pattern analysis was performed using Fragstats software (McGarigal and Marks 1994). Fragstats calculates landscape metrics on three scales: individual patch, vegetation class, and the entire landscape. Metrics fall into a number of classes, including area, patch density, patch size, edge, shape, core area, nearest-neighbor, diversity, contagion, and interspersion (table 25.2). Because SRP was experiencing a different management regime than GNP at the time of the study, results for the two parks are reported separately. Fragstats analyses were run on the entire area of Guanacaste National Park. The Santa Rosa Park area was clipped from the regional area and statistics were derived separately for this region.

Measurements of length, area, and frequency are standard output for most GIS. These data are commonly used by resource managers to inventory resources. Area measures also provide important information for ecologists. Patch size has often been used as the basis for modeling species richness and species distribution patterns. Patch size affects the internal functions of a patch; for example, microclimatic variables, such as internal

Table 25.2 Metrics computed by Fragstats software.

Scale	Acronym	Metric (units)
Area metrics		
Patch	AREA	Area (ha)
Patch	LSIM	Landscape similarity index (%)
Class	CA	Class area (ha)
Class	PLAND	Precentage of landscape (%)
Class/landscape	TA	Total landscape area (ha)
Class/landscape	LPI	Largest patch index (%)
Patch density, patch size, and variability metrics		
Class/landscape	NP	Number of patches
Class/landscape	PD	Patch density (no./100 ha)
Class/landscape	MPS	Mean patch size (ha)
Class/landscape	PSSD	Patch size standard deviation (ha)
Class/landscape	PSCV	Patch size coefficient of variation (%)
Edge metrics		
Patch	PERIM	Perimeter (m)
Class/landscape	TE	Total edge (m)
Class/landscape	ED	Edge density (m/ha)
Shape metrics		
Patch	SHAPE	Shape index
Patch	FRACT	Fractal dimension
Class/landscape	LSI	Landscape shape index
Class/landscape	MSI	Mean shape index
Class/landscape	AWMSI	Area-weighted mean shape index
Class/landscape	DLFD	Double log fractal dimension
Class/landscape	MPFD	Mean patch fractal dimension
Class/landscape	AWMPFD	Area-weighted mean fractal dimension
Nearest-neighbor metrics		
Patch	NEAR	Nearest-neighbor distance (m)
Patch	PROXIM	Proximity index
Class/landscape	MNN	Mean nearest-neighbor distance (m)
Class/landscape	NNSD	Nearest-neighbor standard deviation (m)
Class/landscape	NNCV	Nearest-neighbor coefficient of variation (%)
Class/landscape	MPI	Mean proximity index
Diversity metrics		
Landscape	SHDI	Shannon diversity index
Landscape	SIDI	Simpson diversity index
Landscape	MSIDI	Modified Simpson diversity index
Landscape	PR	Patch richness (no.)
Landscape	PRD	Patch richness density (no./100 ha)
Landscape	RPR	Relative patch richness (%)
Landscape	SHEI	Shannon evenness index
Landscape	SIEI	Simpson evenness index
Landscape	MSIEI	Modified Simpson evenness index
Contagion and interspersion metrics		
Class/landscape	IJI	Interspersion and juxtaposition index (%)
Landscape	CONTAG	Contagion index (%)

Source: McGarigal and Marks 1994.

patch temperature and relative humidity, will change with patch size as well as shape Differences in patch area may also affect processes such as nutrient cycling rates and the size of propagule pools for plant colonization (Forman and Godron 1986).

Ecologists also use shape and edge measures for landscape analysis. Shape usually is expressed as a ratio of edge to area or as a deviation from a standard shape. Patch shape may influence a number of interpatch processes, including changes in vegetation composition, brood parasitism, and the susceptibility of forests to disturbance (Saunders, Hobbs, and Margules 1991; McGarigal and Marks 1994; Laurance, chap. 6).

Edge metrics can be quantified using fractals (Gardner et al. 1987; Krummel et al. 1987; Milne 1988; O'Neill et al. 1988; Turner and Ruscher 1988; Milne 1991). Fractals are commonly used because they can be applied to spatial features over a large number of scales. The term *fractal* was introduced by Mandelbrot to refer to temporal or spatial phenomena that are continuous but not differentiable, and that exhibit partial correlations over many scales (Burrough 1981). Fractal dimensions *(D)* quantify the complexity of a polygon by relating perimeter *(P)* to area *(A)* such that *P* approximates $(A^{D})^{1/2}$ (McGarigal and Marks 1994). A number of studies suggest that human-influenced landscapes show simpler patterns, and thus lower fractal dimensions, than those not heavily affected by humans (Krummel et al. 1987; O'Neill et al. 1988).

Because the degree of patch isolation plays a critical role in island biogeographic theory (MacArthur and Wilson 1967), nearest-neighbor and proximity metrics are probably the most familiar to conservation biologists. Nearest-neighbor metrics are measured from the edge of the focal patch to the edge of the nearest patch of the same vegetation type (McGarigal and Marks 1994). The proximity index is a second measure of patch isolation, calculated as the sum of areas of all patches of the same vegetation type falling within a fixed distance of the focal patch, divided by the square of the mean distance between the focal patch and the other patches (Gustafson and Parker 1992). For this study, the search radius chosen was 195 m (3 pixels) for the MSS imagery and 180 m (6 pixels) for the TM. These distances approximate the limit of seed dispersal by wind, and thus are biologically meaningful (Janzen 1988b).

Contagion and interspersion metrics are related to nearest-neighbor metrics in that they quantify the patch configuration of the entire landscape (McGarigal and Marks 1994). Contagion is a measure developed from percolation theory and neutral models (Gardner et al. 1987; Gardner and O'Neill 1991; Gustafson and Parker 1992) that essentially quantifies the patterning of landscape components. It is a measure of clumping or randomness of patches within the landscape: the more clumped the landscape is, the closer the contagion value will be to 1.0. Contagion indices compare the degree of similarity of individual pixels to the surrounding pixels. Thus, landscapes with large patches have a higher contagion value because adjacent pixels are more likely to be of the same type (McGarigal and Marks 1994).

The interspersion and juxtaposition index uses patch adjacencies, rather than pixel adjacencies, to measure landscape similarity (McGarigal and Marks 1994). Because it works at the patch level rather than the pixel level, it is less sensitive than the contagion index to the resolution of the original land cover map. The interspersion and juxtaposition index measures how well the landscape is "mixed." Landscapes that are not well mixed (more contiguous) have a lower index than those that are well mixed (more fragmented).

Finally, although diversity measures are commonly associated with community studies, diversity exists at many levels of biological organization, including landscape patterns. Two components of diversity are commonly measured: (1) richness (species number) and (2) evenness (the relative distribution of individuals: Romme 1982). A large number of diversity indices have been developed by ecologists. Fragstats calculates the Shannon index, which is more sensitive to richness than evenness, and the Simpson index, which is less sensitive to richness and thus provides information on the distribution of common landscape elements and their respective evenness indices (Magurran 1988; McGarigal and Marks 1994). Evenness indices isolate the evenness component of the diversity index, based upon the proportion of the diversity index that is reduced by the uneven distribution of landscape components (McGarigal and Marks 1994).

RESULTS

There was an overall decrease in the total area of GNP maintained as pasture from 1979 to 1985 (table 25.3). Pasture declined by 28% (from 7,432 ha in 1979 to 5,371 ha in 1985). Pasture within the boundaries of SRP showed a more dramatic reduction, declin-

Table 25.3 A transition matrix showing changes in vegetation classes for Santa Rosa Park and Guanacaste National Park in Costa Rica.

	1985				
1979	Evergreen forest	Mixed forest	Deciduous forest	Pasture	Total
Santa Rosa Park					
Evergreen forest	1,434	374	396	65	2,269
Mixed forest	1,366	585	560	110	2,510
Deciduous	606	331	384	78	1,399
Pasture	485	249	455	119	1,308
Total	3,891	1,540	1,795	371	

	1985				
1979	Evergreen forest	Mixed forest	Deciduous forest	Pasture	Total
Guanacaste National Park					
Evergreen forest	2,833	769	897	428	4,926
Mixed forest	2,688	1,107	1,316	852	5,963
Deciduous	1,326	671	1,031	1,029	4,057
Pasture	1,718	740	1,911	3,063	7,433
Total	8,565	3,286	5,155	5,372	

Note: Values shown are in hectares. The stub heads at the left represent the vegetation classes for the 1979 land cover, while the column heads across the top represent those for the 1985 land cover. Reading across the table provides the transition in hectares from 1979 to 1985. The diagonals represent retetion values.

Table 25.4 Patch area, size, and other spatial indices for Santa Rosa Park and Guanacaste National Park.

	Class area (ha)	Percentage of landscape	Largest patch index	Number of patches	Patch density (no./100 ha)	Mean patch size (ha)	Patch size SD (ha)	Patch size coefficient of variation (%)
Santa Rosa Park: 1979								
Evergreen forest	2,300.93	29.71	7.21	248	3.2	9.28	48.14	518.91
Mixed forest	2,656.26	34.29	9.09	237	3.06	11.21	53.23	474.96
Deciduous	1,422.14	18.36	0.82	441	5.69	3.22	5.93	184.01
Pasture	1,366.37	17.64	1.4	229	2.96	5.97	13.78	230.9
Santa Rosa Park: 1985								
Evergreen forest	3,944.16	50.06	29.73	159	2.02	24.81	208.36	839.95
Mixed forest	1,665.45	21.14	1.14	559	7.10	2.98	7.05	236.56
Deciduous	1,891.35	24.01	9.56	263	3.34	7.19	47.87	665.69
Pasture	377.82	4.8	0.56	108	1.37	3.5	6.25	178.73
Guanacaste National Park: 1979								
Evergreen forest	4,982.54	21.93	5.72	555	2.44	8.98	66.19	737.31
Mixed forest	6,035.41	26.56	4.02	683	3.01	8.84	44.24	500.69
Deciduous	4,145.15	18.24	0.55	1235	5.43	3.36	6.8	202.74
Pasture	7,560.21	33.27	19.28	426	1.87	17.75	212.97	1,200.02
Guanacaste National Park: 1985								
Evergreen forest	8,842.59	38.29	26.06	463	2.00	19.1	280.32	1,467.76
Mixed forest	3,460.41	14.98	0.89	1232	5.33	2.81	8.4	298.9
Deciduous	5,349.96	23.17	4.36	947	4.10	5.65	35.05	620.36
Pasture	5,441.85	23.56	6.10	404	1.75	13.47	80.83	60.1

ing by 72% over the same period (from 1,308 ha to 371 ha). Most pasture was converted to deciduous or evergreen forest vegetation types. For both GNP and SRP, evergreen forest showed the largest increase in area; these forests developed from areas that were formerly pastures as well as mixed or deciduous forests.

In SRP, the distribution of vegetation classes changed dramatically from 1979 to 1985 (table 25.4). There was a large increase in evergreen forest and a commensurate decrease in pasture. The increase in evergreen forest was not caused by an increase in the number of patches, but rather by expansion and coalescence of existing evergreen patches. Changes in the GNP area were not as dramatic. Most of the changes in GNP evergreen forest were a reflection of changes in SRP, which is embedded within GNP.

Values for the three shape indices (landscape shape index, mean shape index, and area-weighted mean shape index) reflect the fact that the overall vegetation within the classes diverges from a regular shape (table 25.5). Both landscapes were influenced by the expansion of evergreen forest patches. Pasture area in both landscapes exhibited declining shape indices, indicating a move toward a more regular shape. Thus, agricultural areas are progressively becoming more compact throughout the region. This finding is confirmed by the double log fractal dimension figures for each of the vegetation classes in both landscapes (table 25.5).

The landscape statistic for mean nearest-neighbor distance indicates that for both GNP and SRP, there was a slight decrease in distance between like patches (table 25.6). This trend was consistent for all of the individual vegetation classes except pasture in

Table 25.5 Shape and fractal indices for Santa Rosa Park and Guanacaste National Park.

	Landscape shape index	Mean shape index	Area-weighted mean shape index	Double log fractal dimension
Santa Rosa Park: 1979				
Evergreen forest	17.4	1.58	5.22	1.62
Mixed forest	25.04	1.88	8.93	1.72
Deciduous	18.87	1.60	2.72	1.74
Pasture	13.89	1.58	3.05	1.59
Santa Rosa Park: 1985				
Evergreen forest	30.69	2.11	24.52	1.63
Mixed forest	28.73	2.29	4.36	1.73
Deciduous	18.24	2.01	7.51	1.56
Pasture	6.19	1.70	2.46	1.48
Guanacaste National Park: 1979				
Evergreen forest	20.25	1.56	6.65	1.61
Mixed forest	32.11	1.80	7.77	1.71
Deciduous	28.52	1.61	2.88	1.76
Pasture	24.02	1.64	13.05	1.59
Guanacaste National Park: 1985				
Evergreen forest	40.38	2.05	33.54	1.59
Mixed forest	33.69	2.23	4.55	1.73
Deciduous	36.27	2.22	7.05	1.66
Pasture	20.07	1.95	6.52	1.49

Table 25.6 Nearest-neighbor, interspersion and juxtaposition, and contagion indices for Santa Rosa Park and Guanacaste National Park.

Year	Mean nearest-neighbor distance (m) by vegetation class	Mean nearest-neighbor distance (m) by landscape	Interspersion and juxtaposition index	Contagion index
Santa Rosa Park: 1979		94	86.84	13.24
Evergreen forest	103.46			
Mixed forest	77.27			
Deciduous	84.08			
Pasture	120			
Santa Rosa Park: 1985		61.7	82.94	25.5
Evergreen forest	49.01			
Mixed forest	46.16			
Deciduous	72.82			
Pasture	134.01			
Guanacaste National Park: 1979		92.4	86.84	13.24
Evergreen forest	117.08			
Mixed forest	80.50			
Deciduous	81.82			
Pasture	110.01			
Guanacaste National Park: 1985		61.9	67.57	32.25
Evergreen forest	54.38			
Mixed forest	59.92			
Deciduous	56.97			
Pasture	88.04			

Note: The latter two indices have no units.

Table 25.7 Diversity and evenness indices for habitat mosaics within Santa Rosa Park and Guanacaste National Park.

	Santa Rosa		Guanacaste	
	1979	1985	1979	1985
Shannon diversity index	1.34	1.16	1.36	1.33
Simpson diversity index	0.73	0.64	0.74	0.72
Modified Simpson diversity index	1.34	1.04	1.34	1.28
Shannon evenness index	0.97	0.84	0.98	0.96
Simpson evenness index	0.97	0.86	0.98	0.96
Modified Simpson evenness index	0.94	0.75	0.96	0.92

Santa Rosa Park, which exhibited increasing distance between like patches as it declined in area.

As pasture decreased within SRP, there was a commensurate decline in the habitat diversity of the area. In contrast, there was very little overall change within GNP over the same period (table 25.7). As expected, evenness values declined for SRP and remained nearly constant for GNP.

DISCUSSION

Landscape Changes within Santa Rosa Park

The results of this study indicate that changes in management and land use practices can rapidly alter landscape patterning. Traditionally, agriculture in the Guanacaste region maintained a particular type of landscape pattern. This pattern was dramatically altered when land use was changed to emphasize forest management and conservation within Santa Rosa Park. A cessation of intensive agricultural practices within SRP allowed for the natural reestablishment of woody vegetation in former agricultural areas. Evergreen forests became more prevalent, and mixed forests and deciduous vegetation also increased. Existing forest patches increased in size and complexity, while pasture areas within the park decreased. Those pasture areas that have persisted are smaller and less complex patches that are maintained by park managers as fire buffers and for research.

The results of the overlay technique were useful as qualitative data, but provided no quantitative information. Difficulties arose in pixel-matching the 1979 and 1985 images. A number of factors prevented exact overlaying, including differences in original pixel resolution and the lack of good ground control information. In addition, the topographic maps used to geo-reference the data did not include many of the smaller rivers and roads found in the imagery. However, the trends detected with the overlay technique did agree with those observed in the landscape pattern evaluation.

Describing landscape patterns is the first step toward understanding the complexity of interactions within the landscape. Once the pattern is established, researchers can begin to explore the changes caused by various ecological and physical processes (Turner 1989; Ripple, Bradshaw, and Spies 1991; Bridgewater 1993). By examining how succession occurs in tropical dry forests, investigators can begin to explain many of the observed changes in pattern.

Climate and Succession in Tropical Dry Forests

A number of studies have assessed small-scale changes in successional plots in dry forests (Opler, Baker, and Frankie 1977; Ewel 1980). These studies indicate that because of the harsh dry season, local tree establishment is more likely to result from vegetative propagation than from seed germination. Thus, the greatest reduction in pasture is likely to result from expansion of forest vegetation along pasture edges. One would also expect to see an increase in stems within deciduous forest patches, where existing parent trees are available to reproduce vegetatively.

Succession in dry forests is slower than that in wet forests (Ewel 1980). A close look at the processes involved, however, shows that dry forest succession occurs very rapidly in the wet season, but average gains in patches are usually offset by declines in the dry season. Thus, measurable increases in forest cover occur only when wet season gains offset dry season losses.

Seasonal rainfall patterns strongly influence successional changes in tropical dry forest. For example, total annual rainfall in GNP varied dramatically between 1980 and 1985 (from a minimum of 917 mm in 1983 to a maximum of 2,241 mm in 1984: Janzen 1988b). Periods of low rainfall can kill seedlings and saplings with shallow root systems, thereby inhibiting successional processes. Regenerating deciduous and mixed forest areas with high proportions of newly established vegetation are therefore highly susceptible to periodic droughts caused by El Niño events.

Tree phenological responses also reflect seasonal rainfall patterns. Although similarities between tropical and temperate deciduous forests are often noted, there are considerable differences that affect the ability of trees to retain their foliage. Flushing and senescence behavior is less predictable in tropical dry forests. Reich and Borchert (1984) found that trees at more mesic sites either held their leaves longer, or shed them and then flushed new leaves more quickly, than those at xeric sites. Variation in leaf fall may also be due to differences in soil moisture and root depth. Trees found in riparian areas, for example, show leaf fall responses more similar to those in wet forests. These differences may account for the substantial increases in evergreen vegetation in SRP from 1979 to 1985, which mainly occurred along river corridors and in wetland areas.

Limitations of Remote Sensing and GIS

One would expect that in the future, landscape change in the Guanacaste National Park area will occur through the expansion and coalescence of patches of various forest types. To accurately measure these fine-scale changes, data should be collected at the appropriate spatial scale. MSS images can measure abrupt habitat changes, such as the transition from pasture to forest, but TM images are is more appropriate for measuring more subtle changes between different forest types.

A word of caution is necessary. A considerable amount of error can be propagated from remotely sensed data as a result of misclassification and problems with geo-referencing. Errors in classification arise from differences in sensors, in phenological and physiological states of the vegetation, and in atmospheric conditions at the time of image acquisition. Images from older satellites are usually poorer in quality than newer images. However, the use of older images may be unavoidable when researchers are attempting to establish historical trends in vegetation change.

Problems also arise in the matching of images for overlays. In many tropical systems, for example, it is difficult to find multiple ground control points because the canopy appears homogeneous and continuous. In many instances ground control information comes from maps that have a large component of error. In this study, inaccuracy may also have been caused by differences in the spatial resolution of MSS and TM images. Landsat TM pixel resolution is more appropriate for this type of study, but TM was unavailable before the early 1980s, and thus MSS images were used for the 1979 data. This approach can effectively reveal qualitative vegetation trends, but the results may be quantitatively imprecise.

GENERAL IMPLICATIONS

1. Geographic information systems are powerful tools that allow the user to expand his or her view beyond the isolated forest patch or nature reserve. The use of GIS is becoming

increasingly important because forest remnants do not exist in isolation. Physical and biological factors affecting a patch are controlled by the type of vegetation adjoining the forest edge as well as the shape, size, and position of the patch within the surrounding landscape.

2. GIS and remote sensing can increase our understanding of the influence of biophysical factors on forest remnants. These tools can be used to quantify landscape patterning, guide our research questions, and develop frameworks for modeling. By taking a more integrated approach to studying forest remnants and fragmented landscapes, ecologists will increase the relevance of their findings to conservation management.
3. Many studies of forest fragmentation continue to focus on island biogeographic theory as an explanation for population declines, and largely ignore changes in physical factors. Such approaches often fail to take into account the influence of the surrounding vegetation and biophysical conditions on remnant populations. Conservation ecologists should begin to expand their research focus from assessing the effects of fragmentation on local populations to devising strategies for maintaining remnants and fragmented landscapes for biodiversity.

ACKNOWLEDGMENTS

This work was partially funded by a scholarship from the Pew Charitable Trust and a grant from the E. P. Odum Foundation. I also thank J. C. Luvall of NASA for providing data and technical support, and C. R. Carroll, E. Box, W. F. Laurance, and two anonymous referees for invaluable comments on earlier drafts of this chapter.

26

Quantifying Habitat Fragmentation Due to Land Use Change in Amazonia

Virginia H. Dale and Scott M. Pearson

RECENT land clearing in Brazil is occurring faster than in any other place on earth (Skole and Tucker 1993). The concomitant fragmentation of tropical forest habitats is degrading the biological diversity of the region. The socioeconomic drivers of land use decisions resulting in forest loss and the effects of those decisions can be understood by developing models of interactions between the drivers and their effects (Turner, Meyer, and Skole 1994). Models that link socioeconomic processes to ecological effects offer such a tool.

In this chapter, we compare the patterns of forest loss recorded by satellite imagery with simulations of deforestation created by a computer model of three alternative land use management strategies. We use spatial indices to compare maps generated from satellite imagery and model projections. The comparison illustrates both the characteristics of these spatial indices and the spatial features of linked socioeconomic and ecological models. Our analyses will help us to understand, predict, and monitor the effects of land use on patterns of loss and fragmentation of tropical forests.

CAUSES OF HABITAT FRAGMENTATION

Habitat fragmentation is recognized as one of the major threats to ecosystems. Although spatial heterogeneity is a natural phenomenon, human activities are altering natural landscapes by changing the abundance and spatial pattern of habitats. Habitat fragmentation occurs when a habitat or land cover type is subdivided either by a natural disturbance (e.g., fires or windthrows) or by human activities (e.g., roads, transmission lines, or agriculture).

The major anthropogenic drivers of land use changes are population, affluence, technology, political economy, political structure, and attitudes and values (Turner, Moss, and Skole 1993). The importance of these factors varies with the situation and the spatial scale of analysis (see Kahn and McDonald, chap. 2). Human population growth can be considered an ultimate cause for most land use changes; however, local demographics, as well as per capita consumption and its variability, modify the effects of population. Three major proximate causes of land use change are economic exploitation of natural resources (logging, mining, hydroelectric power, etc.), population expansion (urbanization and colonization), and the expansion of agriculture (permanent agriculture and shifting cultivation).

Disturbance and land management practices interact with the existing landscape pattern to dramatically affect fragmentation and the related risk of species loss (Gardner, King, and Dale 1994). The most vulnerable species typically require large areas of contiguous habitat, or are restricted to specific habitat types. Land management practices that increase the degree of landscape fragmentation cause a change in the competitive balance between species, further exacerbating threats to species diversity (Gardner, King, and Dale 1994).

INDICES OF SPATIAL PATTERN

Over sixty indices have been developed to quantify landscape pattern (O'Neill et al. 1988; Turner and Gardner 1991; Baker and Cai 1992; Kramer, chap. 25). These indices have become valuable tools for landscape ecologists, allowing researchers to select measures that quantify characteristics of landscapes directly related to ecosystem and population processes.

A combination of different measures is required to provide a holistic description of the abundance and spatial pattern of cover types. These indices, however, should be used with caution, and with an understanding of their statistical properties. For example, the dominance index, derived from information theory, is mathematically related to the Shannon diversity index and the proportions of different land cover types on a map (Gustafson and Parker 1992). Dominance and diversity measure different aspects of the statistical distribution of cover types, but the two measures are often correlated. The index of contagion measures the probability that adjacent cells are of the same land cover type, which is useful for measuring the interspersion of cover types.

A suite of simple measurements can be sufficient for initially describing spatial pattern in habitat maps. Four general types of indices are useful: area of habitat, frequency distribution of patch sizes, measures of patch shape, and length of edge between habitat types. Calculating one index of each type for major habitats provides an interpretable set of quantitative descriptors of landscape pattern.

The area of each habitat type is basic information for most landscape-level analyses. A frequency distribution of patch sizes can be used to examine the connectivity or fragmentation of habitat (i.e., whether it is aggregated into a few large patches or distributed among many small patches). The data necessary for constructing this distribution also contain useful information for comparing maps, such as the absolute and relative sizes of the largest patch (the ecological characteristics of the landscape may, to a large extent, be dictated by the characteristics of the largest patch).

A measure of patch shape, such as edge-to-area ratio, also provides information on whether patches are compact and simple or elongated and complex. Patches with elongated and complex shapes (i.e., high edge-to-area ratio) may serve as dispersal corridors but experience extensive edge effects. The length of edge between different land cover types is useful for predicting habitat availability for species that either prefer or avoid certain types of ecotones. Gustafson and Parker (1992) provide a measure of patch proximity that is useful for estimating the isolation of particular patches. This suite of landscape pattern measures provides information on many ecologically relevant characteristics of habitat maps.

Although remotely sensed data and the spatial indices derived from them together provide a means to describe landscape-level patterns, no one set of measurements can adequately describe landscape patterns for all species or processes. Hence, the significance of landscape patterns must be evaluated from the perspective of particular species and ecological processes. This organism- or process-based perspective (Addicott et al. 1987; Wiens 1989) should be incorporated into definitions of habitat patches and used to select appropriate spatial indices. For example, measures of patch shape and edge type would be useful for evaluating landscape suitability for species requiring forest interior habitat (see Laurance, chap. 6). Dale, Pearson, et al. (1994) and Pearson et al. (1996) provide examples of landscape evaluation from organism-based perspectives. In this chapter, we compare the four general spatial indices derived from remotely sensed data with the same set of indices generated from model projections of land use change in the Brazilian Amazon.

LAND USE CHANGE IN RONDÔNIA

The Brazilian state of Rondônia, located in the southern Amazon Basin (fig. 26.1), is dominated by mature Neotropical forests. In recent decades, government initiatives produced an extensive network of roads that opened the interior forest areas to colonization; an eighteenfold increase in the total length of roads occurred between 1979 and 1988 (Frohn, Dale, and Jimenez 1990). Land for small family farms was parceled into lots averaging 1 km^2 in size (Leite and Furley 1985). The original colonization plan was that, over time, only half of the area of these large lots would be cleared by the farmers. In fact, only about 3 ha are cleared from a lot in any one year due to the difficulty of felling the trees.

Colonists, using slash-and-burn techniques to clear the forest for agriculture, produce a dynamic mosaic of agricultural fields, pasture, regrowth, and mature forest, with most of the clearing originating along roads. On the family-owned farms in central Rondônia, human encroachment into continuous forest typically results in a landscape mosaic of deforested areas and isolated forest fragments on the order of 10–100 ha in area (fig. 26.1). Between 1978 and 1988, 17,717 km^2 of Rondônia's forest was cleared, and an additional 1,417 km^2 of forest was isolated from the contiguous forest into small (<1 km^2) tracts (Skole and Tucker 1993).

A MODEL OF LAND USE CHANGE IN THE AMAZON

A Dynamic Ecological Land-Use Tenure Analysis (DELTA) model has been developed for central Rondônia that describes the effects of small-farmer settlement on deforestation (Southworth, Dale, and O'Neill 1991; Dale, O'Neill et al. 1993; Dale, Southworth et al. 1993; Dale, Pearson et al. 1994). The basic insight underlying the model is that deforestation is a socioeconomic process with ecological implications that may, in turn, affect socioeconomic processes. The model is designed to improve our understanding of the socioeconomic factors that drive the transition from continuous forest to habitat mosaics.

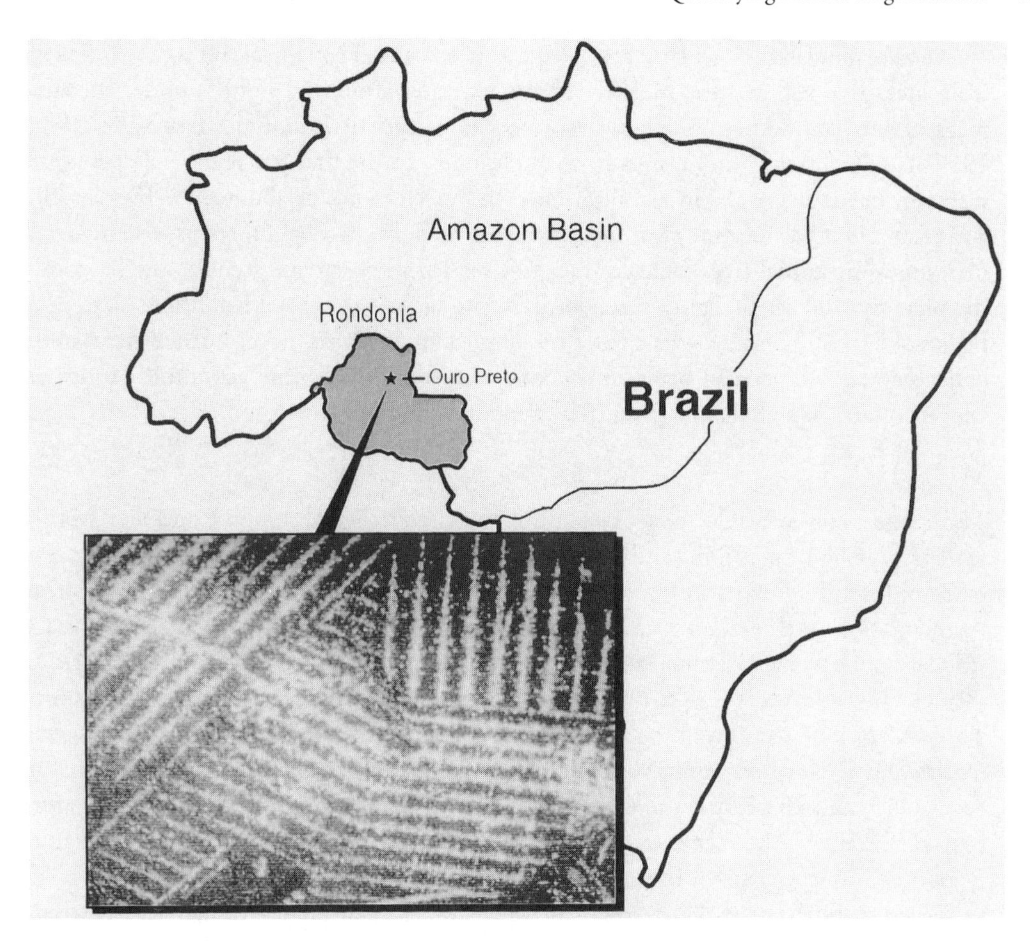

Figure 26.1. Map of Rondônia, Brazil, and Landsat TM image of the study area in central Rondônia. White areas are cleared of forest, while dark areas retain forest. The regular road patterns are clearly distinguishable.

To ensure that the resulting (aggregate) land use patterns appropriately reflect the human settlement process, the model tracks both the history of each farm lot and the history of colonists on the land. The tracking of each lot also allows DELTA to serve as a prescriptive tool by which settlement policy options can be evaluated. For the case study presented here, 294 lots are modeled on the basis of digitized surveys of the area. Each lot is divided into four areas (for a total of 1,176 pixels).

DELTA model simulations suggest that different land management strategies result in unique land cover patterns (Dale, Southworth et al. 1993). Using the model to simulate different land management scenarios permits evaluation of the causes of specific land cover changes. Land use activities that are typical for colonists in Rondônia involve rapid clearing of the forest and almost complete deforestation within eighteen years (Leite and Furley 1985; Coy 1987; Dale and Pedlowski 1992). Model projections for the typical case compare well with changes over time in the patterns of land cover depicted in a complete Landsat image of the region.

We also simulated worst-case and best-case scenarios. The worst-case scenario (taken from areas that suffered the most devastating deforestation by colonists under the auspices of the Trans-Amazon Highway project northeast of Rondônia: Fearnside 1980, 1984, 1986; Moran 1981) results in total clearance in the first ten years. The best-case scenario, however, results in stabilization of forest clearance at about 40% by year 20. This scenario involves some clearing, but no burning, of the virgin forest; it also involves planting of perennial trees. The conditions used for the best-case scenario are based on the practices of three of eighty-seven farmers interviewed in central Rondônia (Dale and Pedlowski 1992). These few farmers grew a diversity of crops, never burned their land, had no cows, and yet had been on the land longer and had more economic profit per year than farmers practicing agricultural methods typical of the region.

METHODS

Three study sites were selected from a remotely sensed image of central Rondônia, Brazil, to represent high, medium, and low levels of development; these sites characterized a range of forest fragmentation (highly fragmented landscape to continuous forest). Forest fragmentation on these sites was compared with projections from the DELTA model simulating the worst-case, typical, and best-case scenarios. Forest cover was mapped from a 1986 Thematic Mapper (TM) image covering the central part of the state of Rondônia. Three 2,025 km^2 sections of the satellite image were selected to represent areas subjected to different levels of economic development. Model comparisons are presented only for years 5, 10, and 20 because there were few changes in the amount of deforestation after year 20 for any of the scenarios. On the basis of when clearing was initiated in central Rondônia, we estimate that the 1986 TM image approximates model year 8.

To compare the spatial pattern of deforestation in central Rondônia with patterns projected by the simulation model, we quantified the area of suitable habitat (forest) and the frequency distribution of habitat patch sizes. The first step in the analysis was to tally the area of suitable habitat and identify distinct patches of habitat in each study area. Percentage of forest, number of patches, and minimum patch size were calculated for the model projections and for the three sections of the Landsat images. In addition, the area of the largest forest patch divided by the total forest area in each map (LC/LCmax) was generated to provide an index of habitat fragmentation (Pearson et al. 1996). Habitat in a landscape is highly fragmented when LC/LCmax is close to 0, and well connected when LC/LCmax is close to 1.0.

RESULTS

The percentage of forest cover is a useful metric for comparing model scenarios, showing changes over time, and illustrating the effects of development (fig. 26.2). The percentage of forest declines both with increases in economic development, as depicted in the Landsat image, and over time in the model scenarios. In addition, less forest survives in the worst-case and typical model scenarios than in the best case. A simple comparison between the model runs and the Landsat image suggests that the percentage of forest cover is similar for the best-case scenario at model year 5 and the site under medium levels of

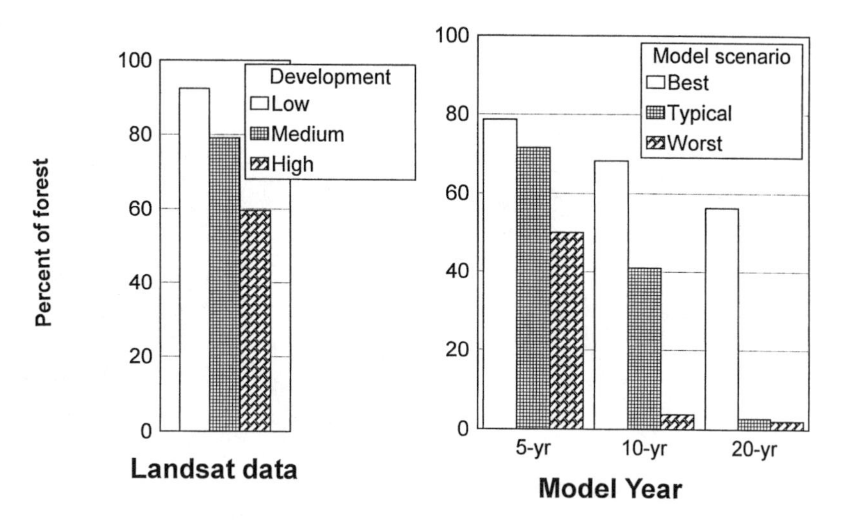

Figure 26.2. Percentage of forest from a 1986 Landsat image of Rondônia at sites experiencing three levels of development and from best-case, typical, and worst-case model scenarios.

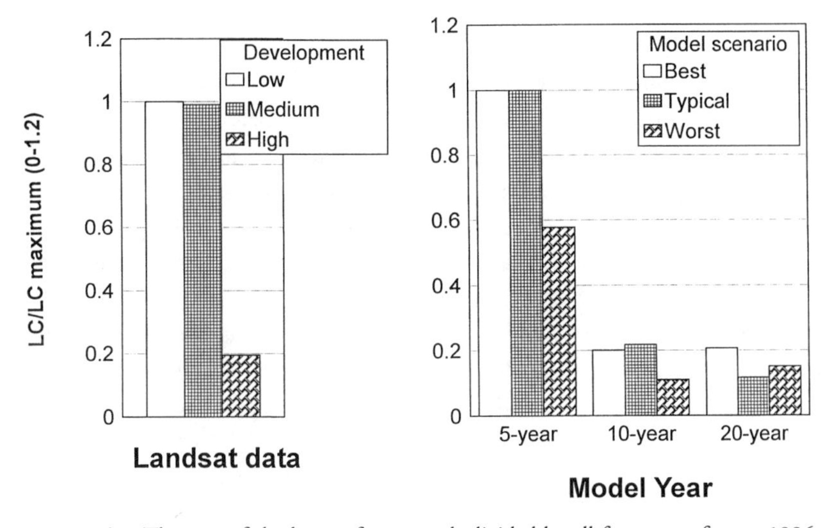

Figure 26.3. The area of the largest forest patch divided by all forest area from a 1986 Landsat image of Rondônia at sites experiencing three levels of development and from best-case, typical, and worst-case model scenarios.

development. The worst-case scenario at model year 5, however, results in less forest than exists at the high-development site at year 8. Thereafter, little forest remains in the worst-case scenario.

The area of the largest forest patch divided by total forest area (LC/LCmax) is lowest for the high-development site and worst-case scenario (fig. 26.3). The ratio is consistently low for model years 10 and 20. Interestingly, the ratio for the worst-case scenario is

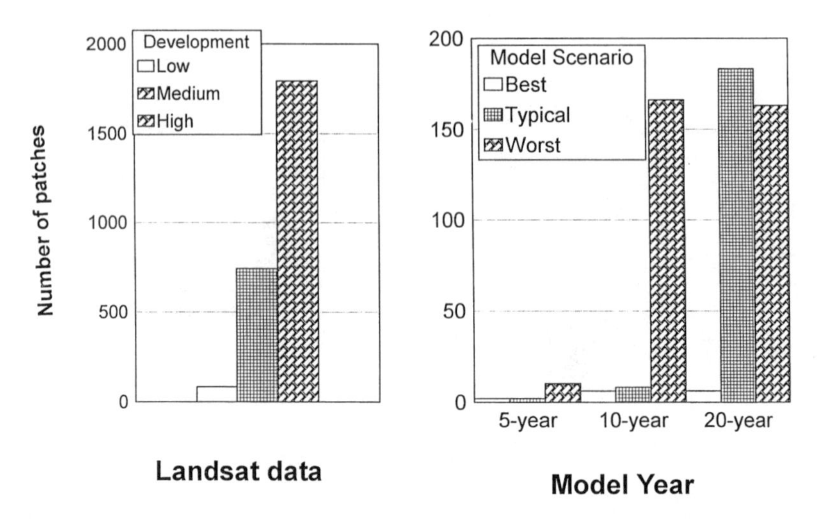

Figure 26.4. Number of forest patches from a 1986 Landsat image of Rondônia at sites experiencing three levels of development and from best-case, typical, and worst-case model scenarios.

slightly higher for model year 20 than for year 10. The increase probably reflects the decline in fragmentation as the landscape becomes dominated by cleared land. During the interval between years 10 and 20, the complete elimination of several small patches results in a landscape with less habitat and fewer habitat patches.

There were relatively few forest patches present at the low-development site and for all the model scenarios at year 5, the best and typical cases at year 10, and the best case at year 20. The concordance between the remote sensing image and the model is consistent, but the absolute number of patches differs because the number of patches increases from 744 to 1,792 between the Landsat sites with medium and high levels of development (fig. 26.4). Furthermore, in the simulation, deforestation is specified to occur adjacent to roads, so the deforested areas are juxtaposed. Thus, the model predicts that deforestation will progress so that areas near roads are deforested before those farther from roads.

For a given percentage of forested area, the Landsat imagery and DELTA model estimate different numbers of patches and a different minimum patch size (fig. 26.5). The Landsat scene has a minimum patch size of only 90 m^2 because of its fine resolution (90 m^2 pixels), whereas the DELTA patch sizes vary depending on the sizes of landowner lots (which were obtained from blueprints for the region); the average lot size of 1 km^2 implies that the mean pixel size (one-quarter of a lot) would be 0.25 km^2. Hence, the DELTA simulations cannot match the fine resolution of Landsat imagery. Nevertheless, coarse-scale projections are useful for simulating the timing and spatial pattern of deforestation over broad regions. For example, the calculation of minimum patch sizes reveals that only in the best-case scenario do patches greater than 33,000 km^2 exist.

DISCUSSION

Spatial Indices and Fragmentation

Habitat fragmentation is inherently complicated and may best be depicted by a combination of spatial indices. LC/LCmax shows the extent to which most of the habitat is aggregated into one (or a few) large patches and thus conveys the degree of connectedness within the landscape. The number of patches depicts the absolute amount of fragmentation. The minimum patch size is probably most important for describing fragmentation effects on those species that require a minimum area of contiguous habitat.

Together, these spatial indices for the low-, medium-, and high-development areas of the Landsat scene and the three model scenarios illustrate the present conditions and potential future of Rondônia. The 1986 Landsat image demonstrates that the fragmentation of forested land depends on the degree of development. Relative to the other areas, the highly developed sites had a lower ratio of maximum patch size to total habitat area, many more patches, and a smaller minimum patch size. We believe that the highly fragmented forest of the developed area would be unlikely to support many of the forest species typical of the Brazilian Amazon (Bierregaard and Lovejoy 1989). Losses of biodiversity are common following fragmentation of tropical forests (Laurance 1990; Newmark 1991). In contrast, the low-development area depicted in the Landsat scene had more forest, a higher LC/LCmax ratio, fewer patches, and larger patches, which probably supported more native species.

Based on comparison with the Landsat image, the DELTA model represents the loss of forest area reasonably well, but due to its coarse spatial resolution, it underestimates the degree of habitat fragmentation. Nevertheless, this model provides a glimpse of the

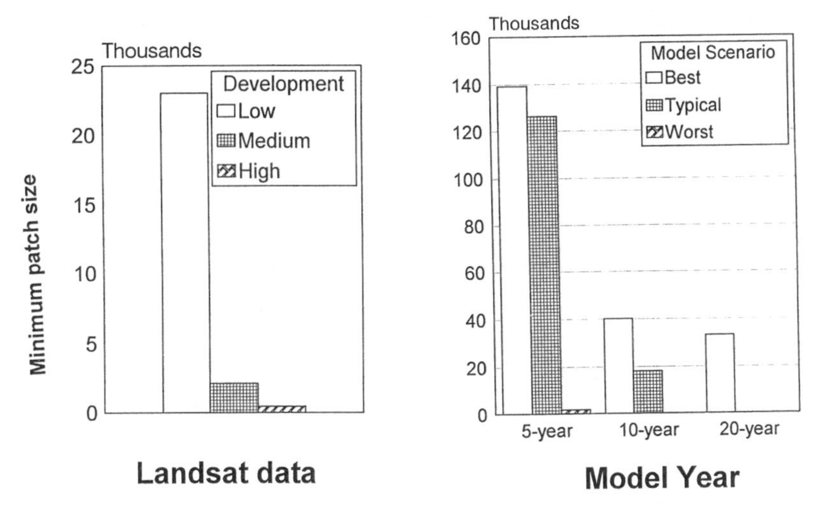

Figure 26.5. Minimum patch size from a 1986 Landsat image of Rondônia at sites experiencing three levels of development and from best-case, typical, and worst-case model scenarios.

range of future conditions that will occur in Amazonia. Its projections show that, as more and more land is cleared for farming, both absolute forest area and the size of forest patches will decline. The ratio of the largest patch size to total habitat area and the minimum patch size both decline rapidly by model year 10. The number of patches increases slightly over time, particularly for the worst-case scenario. These changes portend bleak conditions for species that require a large area of continuous forest, and those with little ability to move between forest fragments.

Fragmentation and Forest-Dependent Species

Interpretation of fragmentation patterns requires consideration of the distribution of forest on the landscape as well as the characteristics of forest species present and their ability to cope with fragmentation. For example, Offerman et al. (1995) recently reviewed attributes of native Amazonian species in relation to different-sized forest patches. When a species that requires large land areas (such as large carnivores or elevational migrants) is present where contiguous patches do not occur, the species' ability to cross gaps between habitat fragments becomes critical for its survival (Bierregaard and Lovejoy 1989; Bierregaard et al. 1992).

Future analyses need to integrate the landscape-metric and species-based approaches so that the potential effects of fragmentation can be assessed at landscape and regional scales (Bierregaard and Dale 1996). Dale, Pearson et al. (1994) have taken the first step in such an analysis by relating gap-crossing ability and area requirements for nine animal groups to projected patterns of deforestation. They found that animals with gap-crossing abilities proportional to their home range size respond similarly to fragmentation, regardless of their taxonomic affiliation. In contrast, species that require large areas but cross only small gaps are more adversely affected by forest fragmentation.

The analyses in this chapter suggest that our model of fragmentation patterns in Rondônia does reflect actual conditions in the central Brazilian Amazon. Thus, the unregulated development of the Rondônia region could dramatically degrade its value for nature conservation. More studies of the effect of forest fragmentation in the Brazilian Amazon are needed to understand, and possibly mitigate, the ongoing ecosystem degradation and threats to biodiversity.

GENERAL IMPLICATIONS

1. Forest fragmentation is complicated and may best be depicted by a combination of spatial indices. The correct interpretation of spatial indices must consider the characteristics of the data used to generate those indices. For example, indices are affected by the number and size of pixels and the distribution across space of land cover types.
2. The spatial resolution of maps greatly affects estimates of minimum patch size and the number of patches. Thus, patches provide a useful landscape metric only if they are small relative to the size of the map.
3. For characterizing extreme conditions (e.g., either intact or highly fragmented forests), the ratio of the largest forest patch to the total forest area may be the most appropriate index.
4. The percentage of area in particular land cover types remains an appropriate way to depict

the overall effects of changes in land use patterns. In Rondônia, for example, the degree of forest fragmentation reflects the level of agricultural development.

5. Future analyses of the effects of deforestation on diversity need to consider how changes in the spatial configuration of forest remnants affect species survival.

ACKNOWLEDGMENTS

Bob Frohn and Woods Hole Research Laboratory provided a land cover map of Rondônia. Reviews of an earlier draft of the chapter by Rob Bierregaard, Michael Kane, Bill Laurance, Bob O'Neill, and Andrew Smith are greatly appreciated. This research received support from the Office of Health and Environmental Research, U.S. Department of Energy under contract no. DE-AC05-84OR21400 with Lockheed Martin Energy Systems, Inc. This chapter is publication number 4,435 of the Environmental Sciences Division of Oak Ridge National Laboratory.

Section VI

Site Selection and Design of Tropical Nature Reserves

INTRODUCTION

At current rates of forest loss, most tropical nations will be drastically modified by human activities within a few decades. Many regions will be dominated by mosaics of agriculture, human settlements, and infrastructure in which forests persist only as small, isolated remnants. Other areas will be set aside as nature reserves, but these are likely to be limited in extent and isolated from other such reserves.

Between these two extremes, yet other areas will be managed as production forests—for timber or non-timber products or as reserves for indigenous peoples, in which some development is likely. Because nature reserves per se will comprise but a small fraction of each nation's land area, semi-natural production forests will clearly play a major role in nature conservation. There is obviously great scope for socioeconomic and ecological research into local strategies, such as Integrated Conservation and Development Projects (ICDPs), that are intended to incorporate both sustainable development and nature conservation.

In the rush toward land development in many tropical regions, one of the major challenges facing biologists and resource managers is the need to identify rapidly areas that most merit protection as nature reserves. In nations such as the Philippines, Guatemala, and Madagascar, in which forest loss has already been extensive, this involves making triage-like evaluations of existing forest remnants. In other nations, such as Brazil and Papua New Guinea, there is still scope for designing reserves, in the sense of influencing factors such as reserve location, size, and shape.

Unfortunately, the distribution of biodiversity in tropical ecosystems is usually very complex, and is far from understood. Many tropical forests exhibit high beta and gamma diversity, so that there is a rapid turnover in species composition as one moves across the landscape. There is often strong local differentiation of vegetation types, species, ecotypes, and populations, and even considerable cryptic variation at the genetic level. Tropical ecosystems probably have the highest proportion of locally endemic species of any biome. These patterns have been shaped and driven by a complex interplay of ecological and historical processes, and the net result is that a limited number of nature reserves will undoubtedly fail to capture more than some fraction of a region's biological diversity.

RESERVE DESIGN IN THE TROPICS

The four chapters in this section all focus on the challenge of identifying key areas for nature conservation. In chapter 27, Andrew Smith assesses the rate of deforestation and the adequacy of existing nature reserves in western Madagascar. The island of Madagascar is clearly a global biodiversity "hotspot" because of its combination of extraordinary endemism and appalling extent of deforestation. In western Madagascar, primary forest cover is now limited to less than 3% of the land area. Smith uses a geographic information system (GIS) and available biophysical data to identify key habitats for lemurs, which are the main "flagship" species for conservation in Madagascar. He demonstrates that forests within 2 to 4 kilometers of villages and roads are at greatest risk of clearing and degradation. He concludes by urging the designation of four new nature reserves that represent very poorly conserved forest types, and for which future protection seems feasible. His chapter makes compelling reading for anyone interested in Madagascar and wildlife conservation.

The latter three chapters employ phylogenetic or biogeographic approaches to identify centers of tropical biodiversity. Chapter 28, by Craig Moritz and his colleagues, examines the distribution of genetic variation among selected rainforest-dependent vertebrates in eastern Australia and New Guinea. The authors demonstrate that the genetic differences among disjunct populations of the same species are generally much greater than would be expected to result from late Pleistocene forest contractions. Their results also reveal that many currently recognized Australasian species arose earlier than the Pleistocene, when the distribution of rainforests was very different from that of today. Their findings highlight the highly dynamic distributions of Australasian rainforests in the past, and demonstrate that contemporary nature reserves and reserve networks must be designed to accommodate the effects of predicted future climatic change.

In chapter 29, James Patton and his co-workers decry the gross inadequacies in our current knowledge of the distribution of Neotropical biodiversity, using their extensive phylogenetic and biogeographic work on South American small mammals for illustration. They demonstrate that even the most basic species inventories are lacking for all but a handful of localities. They also highlight gaping holes in our knowledge of the taxonomy, ecology, and distributional limits of many species. Finally, they demonstrate how phylogeographic analyses can be used to identify regional patterns of diversification and uniqueness, using data on small mammals from lowland forests. The key implication of their work is that, for the bulk of Neotropical taxa, present knowledge of biogeographic patterns is so patchy and coarse that it seriously compromises efforts to identify optimal locations for nature reserves.

In chapter 30, Jon Fjeldså and Carsten Rahbek assess the efficacy of the current protected areas network in South America for conserving bird diversity. They identify a number of areas that are inadequately protected, especially in the Brazilian Atlantic moist forests, in western Ecuador, and in parts of the Andes. Surprisingly, the Amazon Basin emerges as a lower conservation priority, mainly because most bird species in the region have relatively large geographic ranges. Fjeldså and Rahbek also contrast the distributions of "new endemic" and "old endemic" species, and argue that many existing reserves are in the wrong locations to conserve key centers of avian endemism. This thought-provoking chapter provides many insights into the historical processes that shaped biogeographic patterns in South America, and is a must-read for anyone interested in Neotropical ecology and conservation.

SYNTHESIS

The spatial distribution of biodiversity can be assessed in various ways, and the chapters in this section illustrate some strategies that can be used to identify areas of high conservation significance. One method, commonly termed "gap analysis," is to examine the distribution of vegetation types or selected animal groups in relation to existing reserve networks. Poorly protected habitats can then be identified and prioritized for conservation. This approach was useful for identifying the most poorly conserved lemur habitats in western Madagascar (chap. 27) and unprotected centers of avian endemism in South America (chap. 30).

Another general method is phylogeography: documenting geographic patterns of genetic divergence in selected taxa, often using the techniques of molecular genetics. The idea underlying this approach is that genetic patterns in species or species groups reflect the historical distribution of their habitats. The apparent contraction of some rainforests into discrete refugia during Pleistocene glacial periods, for example, would have subdivided populations of many forest species, leading to genetic divergence among their populations. Today, such differences may be either apparent or cryptic, detectable only at the genetic level. From a conservation perspective, the power and utility of phylogeography is that it provides insights into both the historical processes that produced current biogeographic patterns and the spatial distribution of genetic variation within and among species. One application of such data is the identification of regions containing many highly differentiated—and thus more unique—taxa, which should merit high conservation significance. In this volume, phylogeographic analyses are used to reveal dynamic changes in the past distribution of rainforests in Australia and New Guinea (chap. 28) and marked genetic differences between small mammal assemblages in Amazonia and southeastern Brazil (chap. 29).

Interestingly, three of the chapters in this section (28, 29, and 30) arrive at similar conclusions regarding the role of late Pleistocene forest contractions in shaping modern biogeographic patterns. Unlike some earlier authors, these authors do not see Pleistocene refugia as dynamic "species pumps" (centers of rapid incipient speciation), simply because most tropical species appear to be much older than that. Rather, the Pleistocene "sifted" species and shaped the distributions of more ancient taxa, thereby contributing to the complex distribution of tropical biodiversity today. Obviously this is not merely an academic argument, because in nations such as Brazil, entire networks of nature reserves have been located to correspond to apparent locations of Pleistocene refugia.

There are at least four additional concerns regarding the use of Pleistocene refugia as foci for nature reserves. First, the putative locations of refugia are probably influenced by sampling artifacts (Nelson et al. 1990). Second, at least in Brazil, a reserve system based on refugia would fail to protect many existing vegetation types (e.g., Fearnside and Ferraz 1995). Third, historical barriers within the Amazon basin—such as rivers and ancient tectonic ridges—appear to have created complex patterns of faunal endemism not necessarily associated with Pleistocene refugia (chap. 29). Finally, analyses of Neotropical floras suggest that many endemic species are associated with marginal habitats and montane areas that often are not contained within known refugia locations (Gentry 1986). These arguments do not necessarily invalidate the use of refugial locations for nature reserves, but they suggest that an overly simplistic approach to reserve design could be dangerous.

27

Deforestation, Fragmentation, and Reserve Design in Western Madagascar

Andrew P. Smith

MADAGASCAR is recognized as one of the richest and most endangered ecosystems on earth (Mittermeier et al. 1992; Sussman, Green, and Sussman 1994). Its large size, equable climate, extensive tropical rainforest cover, and long isolation from other large landmasses have given rise to a diverse biota with an exceptional level of endemism. The flora includes an estimated eight thousand endemic species of vascular plants, including bizarre, cactuslike species in the endemic family Didiereaceae (Phillipson 1994; fig. 27.1). Plant species diversity equals or rivals that of the Cape Province of South Africa, generally acknowledged to be the highest in the world (Lebrun 1960). An equally unique fauna is dominated by twenty-eight species of endemic lemurs (fig. 27.2), primitive primates widely recognized to be the "flagship" taxa for biodiversity conservation in Madagascar (Mittermeier et al. 1992).

Concern for Madagascar's unique biodiversity stems from an exceptional rate of forest clearing, fragmentation, and land degradation for subsistence agriculture and fuel production. At least 79% of the land surface no longer has a significant primary vegetation cover, and this percentage is increasing annually (Jenkins 1987). A recent assessment, using 1 km resolution satellite imagery, estimated total forest cover to be only 11% of the island area (Nelson and Horning 1993). Forest clearing rates averaged 110,000 ha per year in eastern Madagascar between 1950 and 1985 (Green and Sussman 1990).

The forests of Madagascar have poor regenerative powers due to the scarcity or absence of secondary colonizing species (Jenkins 1987). Large-scale deforestation is therefore essentially an irreversible process except on forest margins.

The remaining forests of Madagascar are subject to a diverse range of pressures. Principal among them are the high growth rate of the human population (which doubles every 22 years: Daume 1990), low per capita annual income (U.S.$230), and high level of subsistence agriculture. Approximately 78% of the population is rural, deriving a subsistence income from grazing or agriculture (Gade 1985). The degradation of productive land as a result of weed invasion, loss of fertility, and erosion maintains pressure on surviving forest remnants.

The equivalent of 1–3% of remaining forest cover is harvested annually for charcoal, Madagascar's principal domestic energy source, mostly from plantations in the north and

Figure 27.1. Southern spiny scrub forest (or dry deciduous thicket), here shown with emergent cactuslike *Alluaudia ascendens* and *Adansonia za* (baobab tree), has the highest proportion of endemic species among all Malagasy forests.

east, but also from natural thorn-scrub forest in the south and west (Pollock 1986; Sussman, Green, and Sussman 1994). Grazing is widespread in secondary forests and savannas, and regular, uncontrolled fires that are started to promote fresh fodder destroy forest margins, prevent forest regeneration, and promote erosion. The savannas of central Madagascar have been severely degraded by erosion, which creates characteristic erosion gullies, or *lavaka,* on hillsides (fig. 27.3). Vertebrates are also hunted for food within forest remnants (Ganzhorn et al. 1990; Smith, Horning, and Moore, in press).

Despite several centuries of scientific observation and collection on Madagascar, even the most basic information on distribution, abundance, threats, and conservation status is lacking for most flora and fauna species. More than 50% of plant species are known from fewer than five locations (G. Schatz, pers. comm.) and can be regarded as "rare" or "poorly known" using conventional criteria for assessment of conservation priorities. Fourteen lemur species have become extinct since the arrival of humans about 2,000 years ago (Tattersall 1982), and more than half of the surviving species are considered endangered or vulnerable (Harcourt and Thornback 1990; Mittermeier et al. 1992).

Efforts to conserve biodiversity in Madagascar have focused on the designation and management of a network of nature reserves referred to as Protected Areas, which cover about 1.05 million ha, or 1.8% of the land surface (Jenkins 1987; Smith, Moore, and Horning 1991). However, there has been no systematic attempt to evaluate the adequacy of these Protected Areas for biodiversity conservation. An objective approach to reserve

Figure 27.2. The crowned lemur *(Lemur coronatus),* a species confined to remnant forests in the northern tip of Madagascar.

Figure 27.3. Eroded savanna landscape typical of the central plateau of Madagascar.

design has been constrained by lack of spatial information on the distribution and abundance of flora and fauna, as well as by the size, shape, biodiversity, condition, and history of forest remnants.

Recognizing the urgent need for collection, modeling, and mapping of this information, the World Wide Fund for Nature (International) commissioned a feasibility study to investigate the need for a Geographic Information System (GIS)-based biodiversity planning service in Madagascar. This study concluded that existing landscape information was so scarce that establishment of a centralized GIS capability for information management would need to be coupled with a program to collect new biological, socioeconomic, and biophysical data (Smith et al. 1990). A pilot study evaluating techniques for collection and spatial analysis of biophysical data using GIS was carried out in a large forest remnant north of Morondava in western Madagascar (Smith, Moore, and Horning 1991).

Classification of remnants as Protected Areas is no guarantee of protection in Madagascar. The Malagasy government lacks the resources for enforcement of forest protection legislation and for adequate staffing and management of nature reserves. Most reserves are understaffed and are threatened by ongoing agricultural encroachment, poaching, burning, and progressive degradation. Intensive reserve management is largely confined to a select number of Integrated Conservation and Development Projects (ICDPs), funded primarily by foreign aid. ICDPs seek to halt forest clearing and degradation by focusing on development of sustainable agricultural practices in buffer zones surrounding reserves and through financial compensation for loss of land. This compensation takes the form of employment of local people as guides and guardians. In 1990, foreign aid expenditures on ICDP and related projects exceeded $4.5 million.

A centralized GIS capability has now been established in Madagascar to coordinate the collection and analysis of biodiversity data generated by ICDPs, but a systematic, national survey of biophysical and socioeconomic data has yet to be initiated. In the interim, the Malagasy government is seeking to identify priority areas for establishment of new ICDPs (J. M. Dufis, pers. comm.) based on the limited available data.

The study described in this chapter contributes to the identification of priority areas for reserve location and management in western Madagascar. It uses updated national and regional GIS databases established for the biodiversity planning service feasibility and pilot studies, and places particular emphasis on the use of lemurs as an indicator group. Lemurs are an ideal indicator group for biodiversity planning in Madagascar because they include some of the most endangered fauna on the island, their ecology and distribution are better known than those of most other vertebrate taxa, they are forest dependent, they are likely to play an important role in seed dispersal and regeneration of rainforest plants (Smith and Ganzhorn 1996), and they have a central role in the ecotourism industry that is being developed as a means of providing alternative employment and stabilizing development in buffer zones around Protected Areas.

The particular aims of this study were to (1) describe patterns of clearing and fragmentation in the Western Domain of Madagascar; (2) evaluate the adequacy of existing nature reserves for protection of representative examples of environmental zones and lemur habitats; and (3) identify priority areas for location of new reserves and ICDPs.

STUDY AREA

This study is confines to an analysis of deforestation and fragmentation in the Western phytogeographic domain of Madagascar. White (1983) classified the forests of Madagascar into four floristic domains: Eastern, Western, Central, and Southern. The eastern region is dominated by evergreen rainforests, which occur on a Precambrian metamorphic basement in high rainfall areas east of the Central Divide (plate 2A). This domain extends to the northwestern coast in the Sambirano region, and includes high-elevation outliers in the north (Montagne d'Ambre) and south (Isalo Massif). The Western Domain is dominated by dry deciduous rainforests, which occur on sedimentary substrates in low to moderate rainfall regions with a pronounced winter dry season of seven or more months. This forest type intergrades with evergreen rainforest in the north (Sambirano) as rainfall increases and with dry forests and scrubs to the south as rainfall decreases (plate 2A). The southern division is dominated by a regionally endemic xeromorphic vegetation adapted to low rainfall and periodic drought. The Central Domain is dominated by savannas with remnant patches of fire-resistant, evergreen taipa *(Uapaca bojeri)* forests in rocky areas and gullies protected from fire. The dominant vegetation type in this domain before the arrival of humans about 1,500 years ago has not been determined with certainty, but is likely to have comprised small patches of fire-tolerant scrubs and savannas in a matrix of western deciduous forest and sclerophyllous, high-elevation evergreen rainforest. Present-day savannas are thought to be a product of anthropogenic burning and Pleistocene climatic fluctuations (Humbert 1927; Burney 1987). During the last Ice Age in the Northern Hemisphere, Madagascar was at times so dry and cool that grassland and dry montane shrubland dominated by Ericaceae spread over vast areas of the central highlands, down to perhaps 1,000 m or lower (D. Burney, pers comm.).

Geology of the Western Domain

Three principal geologies occur in Madagascar: Precambrian metamorphics and lateritic mantles, Permian to Recent sediments and alluvium, and Cretaceous and Tertiary extrusive volcanics (see fig. 27.5A). A sedimentary series including marine and terrestrial deposits extends the length of the west coast in a zone 30–200 km wide. An extensive superficial marine and terrestrial sandy-calcareous carapace overlies these sediments in much of the subcoastal region. The eastern two-thirds of Madagascar and the higher-elevation areas of the west are dominated by an entirely metamorphic Precambrian basement. Extensive patches of Upper Cretaceous volcanics and limited occurrences of late Tertiary volcanics provide significant patches of fertile soils throughout the island.

Climate of the Western Domain

Temperatures are fairly uniform at low elevations, rising from a mean annual average of 23°C in the south to 27°C in the north, but decreasing significantly with elevation to approximately 16°C in the central plateau (Jenkins 1987). Rainfall decreases to the west and south, and as mean annual rainfall declines, the length of the dry season increases. On the subequatorial eastern slopes of the northeast, annual rainfall exceeds 1,500 mm, and no distinct dry season is apparent. This climatic zone extends to the west coast in the Sambirano region of northwestern Madagascar. In the extreme north and on the western slopes of the central plateau, rainfall is high, but there is a pronounced five- to six-month dry season. In the western lowlands, rainfall varies from 500 mm in the south to more than 1,500 mm in the north, and periods of twelve to eighteen months without rain may occur in the south (Jenkins 1987).

Forests of the Western Domain

The principal forest type considered in this study is Western dry forest (fig. 27.4), a seasonally dry (monsoon) rainforest characterized by a high proportion of species that shed their leaves during the winter dry season. Western dry forests vary in structure and floristic composition with climate, substrate, and disturbance history. Four subtypes of Western dry forest were recognized in this study: Western dry (primary) forests; secondary forests; karst forests; and riparian forests. The proportion of evergreen and semi-evergreen species in Western dry forests is higher in the north, on fertile (basaltic) soils, and in moist, sheltered microclimates provided by limestone outcrops (karst forests) and major drainage basins. Forests dominated by baobabs and bottle-tree forms are more prevalent on the less fertile Pliocene sandy-calcareous carapaces common in the sub-coastal zone of the central west. Western dry forest occurs at lower elevations (below 500 m) in the north, where it grades into evergreen rainforests in the Sambirano region. In the south it grades into sclerophyllous dry deciduous thicket (spiny scrub) as rainfall declines. Western dry forest probably once covered most of the Western Domain to an elevation of 500–1,000 m.

Most Western dry forest has been destroyed by clearing and fire and converted to savanna grasslands or mosaics of savanna and secondary forest. Savannas are maintained by frequent burning to promote cattle fodder, but may support a few species of scattered fire-tolerant palms, baobabs, and alien shrubs. Western dry forest does not appear well adapted to recovery after clearing and fire. Disturbed areas are rapidly invaded by alien plants (Jenkins 1987; Phillipson 1994), and there are few endemic pioneer species that initiate natural cycles of recovery. Riparian forests (moist gully and gallery forests) persist as small isolates or linear patches in moist drainage lines protected from fire.

METHODS

Two raster (grid-based) GIS databases were used in this study: a Western Domain subset of a national database (updated from Smith, Moore, and Horning 1991 to include 1950s and 1990 vegetation cover) that covers the whole of Madagascar in a 4 km^2 grid; and a regional database that covers the largest remnant of Western dry forest (Smith, Moore, and Horning 1991; Smith, Horning, and Moore, in press) on the west coast of Madagascar in a 1 ha grid (plate 2B). Mapped environmental attributes in the national and

Figure 27.4. An example of western dry forest in the Kirindy Forest north of Morondava.

Table 27.1 Mapped environmental attributes stored in the national and regional GIS databases used in this study.

National GIS database
Elevation (0–200 m, 200–500 m, and thereafter at 500 m intervals)
Geology (grouped into metamorphics; sedimentary and alluvium; and Cretaceous and Tertiary extrusive volcanics (after Besaire 1969)
Vegetation cover (after Faramalala 1988)
Vegetation type and cover (after Humbert and Cours Darne 1965)
Major roads (after Faramalala 1988)
Protected areas (after Nicoll and Langrand 1989)
Lemur range maps (after Tattersall 1982 and unpublished)
Forest cover (forest, nonforest, and cloud after Nelson and Horning 1993)
Regional (Morondava) GIS database
Disturbance (primary, secondary, and alienated forest; after Duccene, Schroff, and Narson 1988; Smith, Moore, and Horning 1991)
Lemur abundance and diversity (after Smith, Horning, and Moore, in press)
Disturbance risk (after Smith, Horning, and Moore, in press)
Distance to village (km)
Distance to stream (km)
Geology (after Besaire 1969)
Elevation (5 m intervals)
Predicted rainfall

regional GIS databases are listed in table 27.1. Environmental attributes were digitized using a variety of different GIS systems, including ERMS, IDRISI, ARC-INFO, and SPANS, but all were transferred to ERMS (Ferrier 1988) for this analysis.

Environmental Zones

The Western phytogeographic domain of Madagascar was stratified into twelve environmental zones on the basis of unique combinations of geological substrate (three classes) and elevation (four classes) (figure 27.5). Elevation was used as a substitute for climate in the absence of temperature and rainfall data. Environmental zones were used as surrogates for natural community distributions in the absence of more comprehensive biological survey data. Each zone is assumed to support different floral and faunal communities because of differences in soil structure, fertility, and climate. This approach is useful for defining potential biogeographic subregions in countries such as Madagascar where spatial biological data are limited or unavailable. This classification should be considered preliminary until more comprehensive data are available to evaluate associations between environmental zones and biodiversity.

Estimation of Deforestation Rates

Post-human forest cover in western Madagascar has been mapped by Humbert and Cours Darne (1965), Faramalala (1988), and Nelson and Horning (1993). These maps provide a time sequence from which deforestation rates can be estimated. However, due to significant differences in resolution of mapping and vegetation classification methods, these coverages cannot be directly compared without first making some corrections for methodological differences. Humbert and Cours Darne produced the most detailed classification, based on high-resolution aerial photography from 1950. Despite the passage of time, this remains the most comprehensive vegetation mapping for most areas of Madagascar with remaining natural forest cover. Faramalala used Landsat MSS 80 m resolution photo products from 1972–1976 to measure forest cover and broad categories of forest type. Nelson and Horning used very low resolution AVHRR 1 km images from 1990 to classify vegetation as forest, nonforest, or areas obscured by cloud.

Faramalala (1988) classified extensive areas of vegetation surrounding Montagne d'Ambre, in the north, as riparian, in contrast to Humbert and Cours Darne (1965), who classified most of this area as savanna. Limited site inspection of this region by the author showed it to be dominated by highly altered savannas and alien scrubs, with oc-

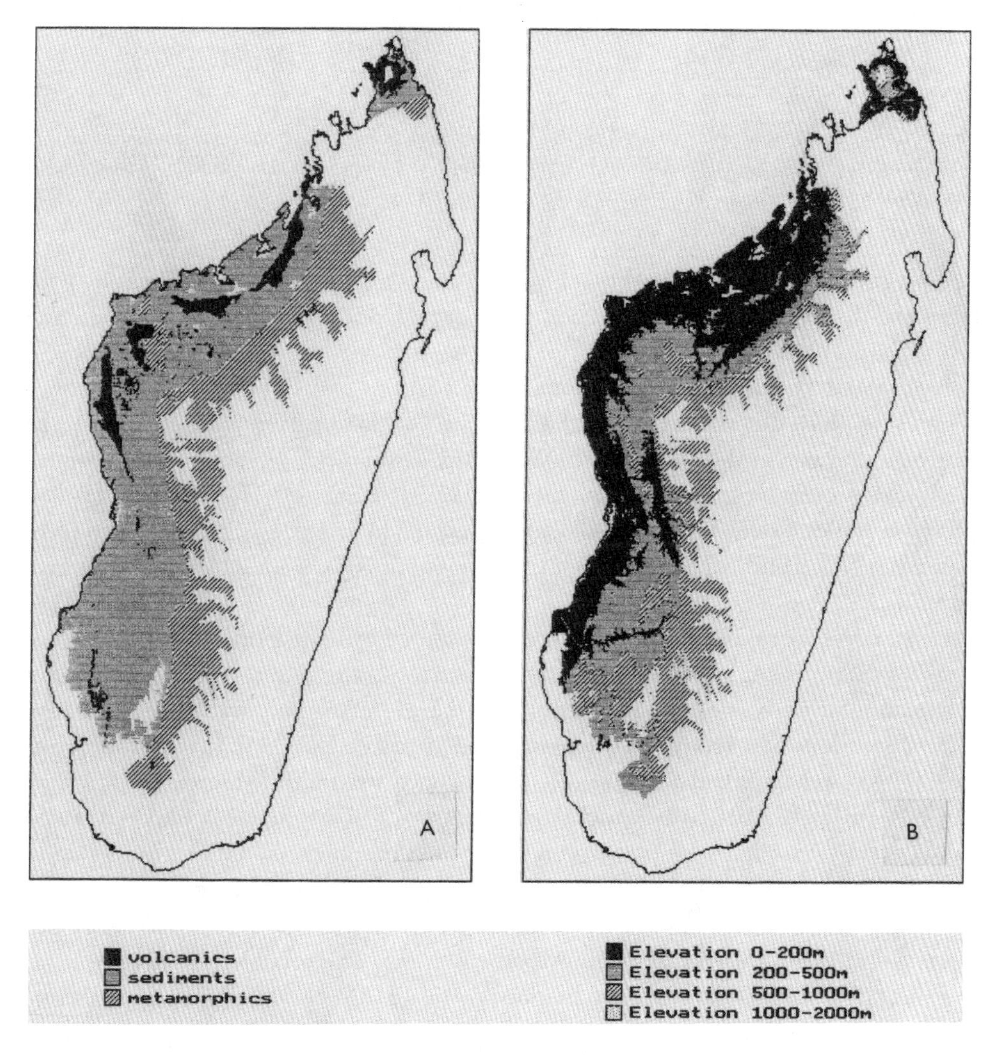

Figure 27.5. Environments of Madagascar. *(A)* Geology. *(B)* Elevation.

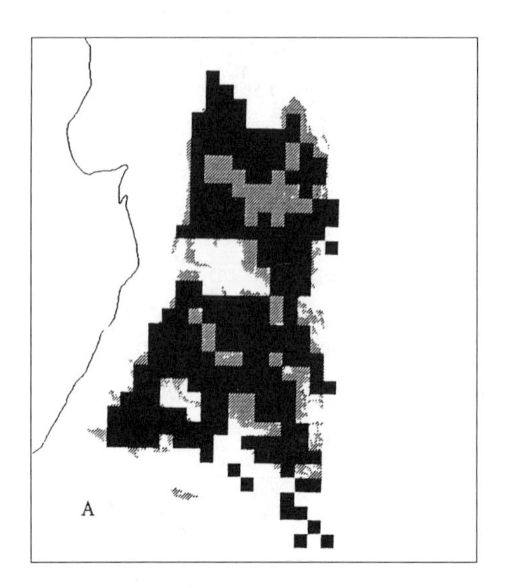

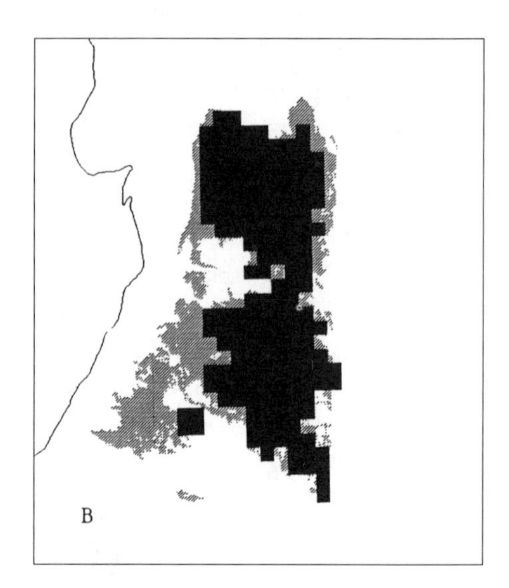

Figure 27.6. Comparison of forest cover in the Kirindy forest remnant (see plate 2B) as determined by *(A)* Faramalala (1988) and *(B)* Nelson and Horning (1993) (black areas) with actual primary forest cover as determined by Smith, Moore, and Horning (1991) (shaded areas).

casional patches of semi-natural vegetation. For the purpose of this study, areas mapped as riparian by Faramalala were assumed to be degraded, consistent with Humbert and Cours Darne.

Nelson and Horning (1993) reported forest cover in two significant areas classified as secondary forest by Faramalala (1988), one in Ankarafantsika Nature Reserve in the northwest and one in the central portion of a large remnant of Western dry forest north of Morondava (in the area covered by the regional GIS database). Descriptions of Ankarafantsika indicate that this area is still largely covered in primary forest vegetation (Jenkins 1987; Nicoll and Langrand 1989), consistent with Nelson and Horning rather than Faramalala.

Forest cover north of Morondava in the area covered by the regional GIS database was compared with recent mapping from Landsat imagery and ground-truthing carried out by the author (Smith, Moore, and Horning 1991; Smith, Horning, and Moore, in press). The distributions of primary forest reported by Faramalala (1988) and Nelson and Horning (1993) were overlaid with forest cover mapped by Smith, Moore, and Horning (1991) to identify patterns of misclassification (fig. 27.6). Faramalala underestimated forest cover by 18% due to apparent misclassification of some seasonal stream vegetation as secondary forest. Site inspection of these areas revealed them to be low-quality, low-diversity forest with the structural features of secondary vegetation (Smith, Moore, and Horning 1991; Smith, Horning, and Moore, in press). The origins and ecology of this vegetation type are not known, but are possibly related to severe, seasonal flood disturbance. Nelson and Horning underestimated forest cover by 14%, but the area excluded by them consisted of low-quality (lower density and lower canopy height) vegetation in a region of low rainfall and high exposure to hunting and harvesting disturbance in the southwest of the area. The sections of forest excluded by Faramalala and by Nelson and Horning were found by Smith, Horning, and Moore (in press) to be of low quality, and supported a lower diversity and abundance of lemur species than the surrounding areas of forest. This comparison suggests that Nelson and Horning have identified remaining areas of high-quality primary forest, while Faramalala has identified both high- and low-quality primary forest and has possibly misclassified some occurrences of low-quality Western dry primary forest as secondary forest.

Faramalala (1988) identified significant occurrences of primary and secondary forest south of Morondava that are absent from Nelson and Horning's (1993) forest cover map, indicating that the forest has either been cleared or misclassified by one or both authors. In the absence of published accounts of vegetation in this region, it was assumed for the purpose of this study to have been cleared or degraded, consistent with Nelson and Horning.

A discrepancy is also apparent between all three historical coverages in the area and distribution of forest on limestone karst formations. Karst forests are of great importance to conservation because the presence of rugged limestone gorges, rills, and pinnacle karst formations *(tsingy)* has provided some protection against clearing. Faramalala (1988) classifies some karst forest recognized by Humbert and Cours Darne (1965) as Western dry primary forest (e.g., in Bemaraha Nature Reserve), while Nelson and Horning (1993) do not recognize significant occurrences of karst forest in the Ankarana and Analamera Nature Reserves in northern Madagascar. These areas support semi-evergreen forests in relatively inaccessible gorges as well as degraded and low-quality forests on exposed limestone plateaus and in peripheral areas (Fowler et al. 1989; Hawkins, Chapman, and Ganzhorn 1990). However, the actual area of vegetation on karst formations is considerably less (< 50%) than the area of karst due to an abundance of exposed rock, which may account for its poor detection by AVHRR imagery.

The preceding discrepancies were taken into account by incorporating the following correction factors: (1) the total forest cover estimate for 1972–1976 (after Faramalala 1988) was calculated after excluding riparian and secondary forest and increasing primary forest cover by 18%; and (2) total forest cover in 1990 (after Nelson and Horning 1993) was calculated after increasing karst forest cover in Ankarana and Analamera Reserves to the levels mapped by Faramalala and increasing all other forest cover values by 14%.

A simplified current vegetation map for the whole of Madagascar was derived by overlaying Nelson and Horning's (1993) forest cover and Faramalala's (1988) karst forest cover on amalgamated vegetation types from Humbert and Cours Darne (1965) (plate 2A).

DEFORESTATION AND FOREST DEGRADATION

Deforestation Rates

The original, pre-human vegetation of Madagascar is thought to have consisted almost entirely of forest. Evergreen rainforests occurred east of the central highlands, and high mountain shrubland and taipa forest, interspersed with patches of natural grassland above 2,000 m, covered the central plateau. Western dry forest covered the entire western slopes

Table 27.2 Estimates of changes in forest cover in the Western Domain of Madagascar from the time of human settlement (approximately A.D. 500) to 1990.

Period	Forest type	Forest area (km²)	% original cover	Deforestation rate (ha/year)[a]	Map source
A.D.500	All	251,660	—	—	—
1950	Primary	25,052	9.9	15,600	Humbert and
	Karst	6,476	2.5		Cours Darne
	Riparian	368	0.2		1965
	Total	31,528	12.5		
1974	Primary	12,440	4.9	106,000	Faramalala
	Secondary	7,408	2.9		1988
	Karst	1,972	0.8		
	Riparian	4,240	1.7		
	Total	16,651[b]	6.6		
1990	All Forest	5,656	2.3	62,500	Nelson and
	Northern Karst[c]	676	0.2		Horning
	Total[a]	7,123	2.8		1993

a. Includes increase of 14% in all forest cover.
b. Excluding riparian and secondary forest and increasing primary forest by 18%.
c. Area of karst forest in Ankarana and Analamera after Faramalala (1988).

and plains (Western phytogeographic region), and dry forests and scrubs covered the semiarid south. The widely scattered riparian forest remnants on the western slopes of the central divide indicate the likely extent of original forest cover. These moist forest remnants persist throughout the Western Domain as tiny patches, mostly too small to map, in moist, sheltered gullies protected from fire.

Deforestation rates in the Western phytogeographic domain of Madagascar were estimated by comparing the percentage of forest cover remaining in 1950 (after Humbert and Cours Darne 1965) with that in 1972–1976 (after Faramalala 1988) and 1990 (after Nelson and Horning 1993), after correcting for differences in methodology as described above. The extent of Western dry forest in Madagascar is estimated to have declined from 12.5% in 1950 to 6.6% in 1974 and 2.8% in 1990, giving average deforestation rates of 61,000 ha per year between 1950 and 1990 (table 27.2, fig. 27.7). This rate can be compared with the estimate of 111,000 ha per year for evergreen rainforest clearing in eastern Madagascar between 1950 and 1985 (Green and Sussman 1990).

Causes of Deforestation and Forest Degradation

The principal cause of deforestation in the Western Domain is slash-and-burn agriculture *(tavy)* carried out by subsistence farmers. Forest is cut and then burned some months later after drying out (fig. 27.8). The cleared land is cultivated for dry-land rice, maize, manioc, and other crops for one or two years before being abandoned due to weed invasion and reduced fertility. Abandoned lands may be recleared at intervals of more than a decade or burned and converted to low-productivity grassland. Annual burning of grassland to promote the growth of nutritious green shoots for zebu cattle prevents shrub and forest regeneration and gradually erodes forest margins. Slash-and-burn agriculture is driven by a combination of rapid population growth and the need to replace abandoned weed-infested and low-fertility croplands.

The regional GIS database encompasses the largest patch of remnant Western dry forest in western Madagascar, which is located between the Andranomena and Tsiribihina Rivers, north of Morondava. Patterns and causes of deforestation in this region were evaluated by superimposing a randomly located, regular grid of fifty-six sample points on the area and modeling the probability of occurrence of primary versus secondary (or cleared) forest at grid points as a function of attributes stored in the regional GIS database, including distance to roads, distance to streams, distance to permanent water, dis-

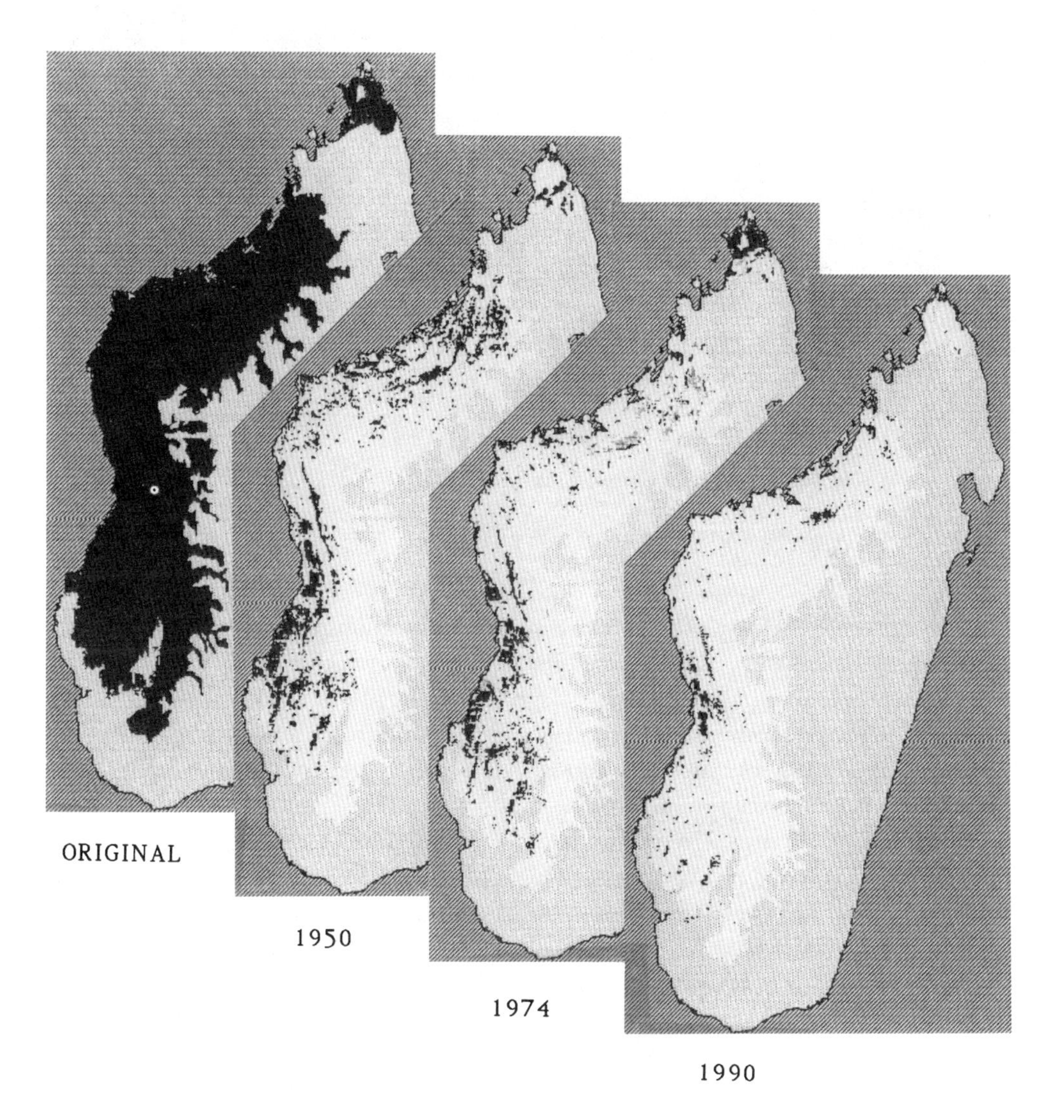

Figure 27.7. Forest cover changes in the Western Domain of Madagascar, 1950–1990.

Figure 27.8. Slash-and-burn clearing of primary Western dry forest occurring east of Morondava in 1990.

tance to the nearest village, geology, elevation, and minimum rainfall. The resulting decision tree model (KnowledgeSeeker 1990; fig. 27.9) shows that the best predictor of deforestation is proximity to existing villages ($P < .001$, $\chi^2 = 20$, df $= 1$). Ninety percent of sites more than 4 km from villages supported primary forest, and 67% of sites within 4 km of villages supported secondary or cleared forest. Within areas more than 4 km from villages, proximity to a road was also a significant predictor of deforestation ($P < .005$, $\chi^2 = 10$, df $= 1$). All sites more than 2 km from roads and more than 4 km from villages supported primary forest, while only 67% of sites within 2 km of a road and more than 4 km from a village supported primary forest. Villages are generally located close to permanent watercourses on alluvial flats.

Forests that have not been cleared may be subject to modification and degradation by timber harvesting, fuel harvesting, and hunting. A number of dedicated timber production forests (Forêt Classée) are located in Western Madagascar. None, however, is suitable for sustainable forestry operations because of low productivity. Existing harvesting operations, which target only a few commercial tree species, will have a short lifespan (<10 years). Probably the most detrimental effect of timber harvesting is the establishment of roads, which facilitate illegal forest use for slash-and-burn agriculture and hunting (Smith, Horning, and Moore, in press).

Hunting is likely to have a significant effect on faunal abundance in remnant primary and secondary forests close to villages. Smith, Horning, and Moore (in press) found a strong linear increase in lemur abundance in primary forests with increasing distance from villages. This effect, which was apparent up to 8 km into secondary forest, was attributed primarily to hunting. In some areas, fauna may derive limited protection from hunting under local taboos *(faddy)*. A *faddy* against hunting of lemurs still had force in Ankarana Reserve in northern Madagascar in 1990 (Hawkins, Chapman, and Ganzhorn 1990).

The preceding findings indicate that substantial buffer zones (4–8 km) are required around forest remnants to protect core areas from hunting, slash-and-burn agriculture, and associated anthropogenic edge effects, particularly in the vicinity of roads and villages. If fragments are approximately circular in shape, they would need to exceed 20,000 ha in order to have a core area more than 8 km from a forest edge, and they would need to exceed 5,000 ha to have a core area more than 4 km from a forest edge.

FOREST FRAGMENTATION

Prior to human occupation of Madagascar, animal movement and genetic exchange in Western dry forests was restricted only by permanent rivers and major expanses of different habitat types, such as humid evergreen forest and semiarid thorn scrub. This natu-

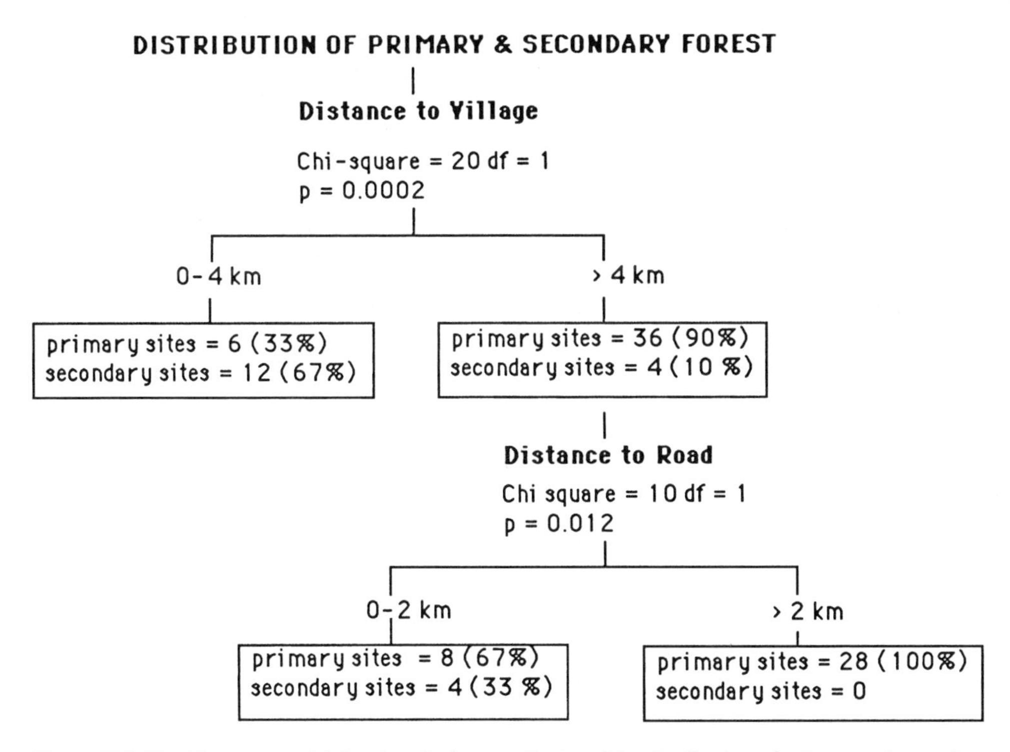

Figure 27.9. Decision tree model showing the best predictors of the distribution of primary and secondary forest in the Morondava regional database.

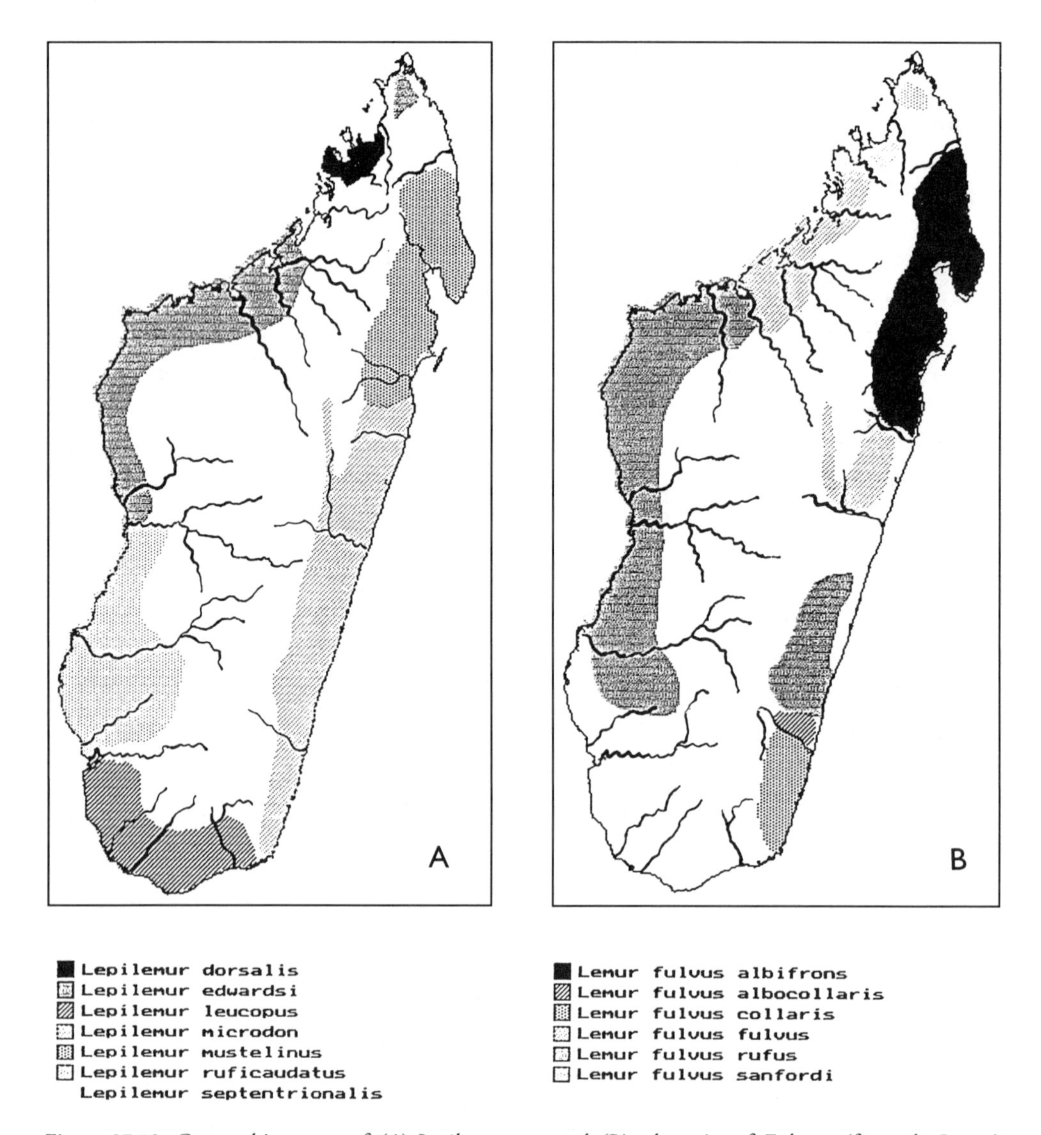

Figure 27.10. Geographic ranges of *(A) Lepilemur* spp. and *(B)* subspecies of *Eulemur* (formerly *Lemur*) *fulvus,* showing coincidence between distribution boundaries and locations of major rivers.

ral barrier effect is illustrated by the distribution patterns of the ecologically homologous subspecies of *Eulemur fulvus* and species in the genus *Lepilemur* (fig. 27.10). These patterns are likely to have developed during the late Tertiary or Pleistocene when expansion of high-elevation grasslands and wooded savannas prevented dispersal around the headwaters of major western and some eastern streams. A major extension of eastern humid evergreen forests into the Sambirano region of northwestern Madagascar separates the Western dry forests of northern Madagascar from those of the west and south. The dry forests in these two regions support distinctly different lemur communities (Smith and Ganzhorn 1996).

Anthropogenic clearing and fragmentation has superseded natural environmental variation as the principal cause of barrier formation and faunal population isolation in Madagascar over the past 1,500 years. Once continuous expanses of Western dry forests between major rivers are now represented by collections of small, highly dispersed fragments (fig. 27.11). Sedentary faunal populations within these fragments are increasingly

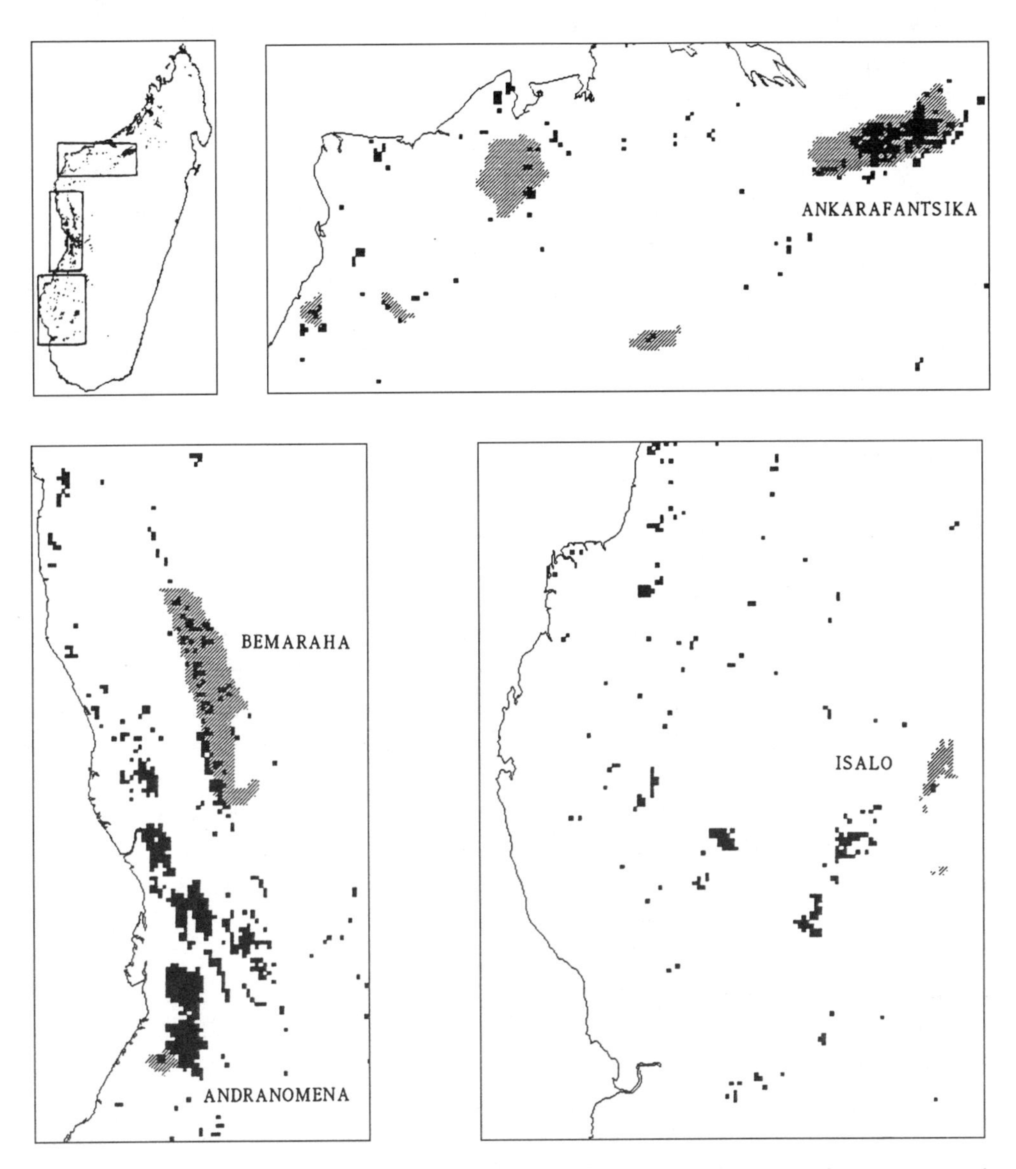

Figure 27.11. Patterns of forest fragment size and distribution in the Western Domain of Madagascar. Each square represents an area of 2 km^2. Shaded areas indicate nature reserves.

vulnerable to extinction due to anthropogenic and natural disturbance, inbreeding, and stochastic demographic events. The risk of faunal extinction will be determined primarily by fragment size and the risk of slash-and-burn agriculture, fire, cyclones, and other disturbances. The fragments of greatest importance to conservation are likely to be those of large size (>5,000 ha) in locations with a low risk of future disturbance. Excessive declines in monsoon forest can be attributed to their greater susceptibility to destruction by fire and their more limited powers of regeneration, particularly in Madagascar. Smith and Ganzhorn (1996) attribute the poor recovery of rainforests in Madagascar to an unusually high dependence on arboreal mammals (lemurs) for dispersal of seeds, and the poor powers of dispersal of lemurs between forest fragments.

The number and size distribution of forest fragments in western Madagascar has changed dramatically over time (table 27.3, fig. 27.12). The proportion of small fragments has increased, and the number of large fragments has decreased, over the past forty-five years. In 1950, five fragments exceeded 102,000 ha in size, while none did so by 1974. By 1990, only two fragments exceeded 51,000 ha, one on sandy-calcareous carapace north of Morondava and one on a rugged, elevated sandy plateau surrounding the Ankarafantsika Nature Reserve. These two patches made up 29% of the total remaining high-quality forest cover. Small patches of less than 600 ha made up the next highest percentage (19%) of remaining forest cover.

RESERVE ADEQUACY

Effects of Deforestation and Fragmentation on Lemur Habitats

Fifteen species and four subspecies of lemurs occur entirely or predominantly within the Western Domain (table 27.4). The effect of deforestation on the area of habitat potentially available to each of these species (habitat range) was estimated by overlaying broad lemur distributions (after Tattersall 1982) on forest cover for the whole of Madagascar. Total potential habitat areas ranged from a low of 2,000 ha for *Propithecus tattersalli* to a high of 2,090,000 ha for *Cheirogaleus medius* (table 27.4). These values do not take into account the effects of fragmentation and isolation, microhabitat variation, and the likely absence of lemur populations from many small isolated habitat patches.

The susceptibility of lemurs to local extinction in small, isolated fragments could be best evaluated by conducting surveys of lemur occurrence in fragments of increasing size throughout their range. The patch size at which lemur frequency of occurrence begins to plateau, or attain 100%, would provide an indication of minimum viable habitat size. Because no studies of this type have been carried out in Madagascar, minimum viable patch sizes must be estimated using less rigorous, indirect means. One indirect measure, minimum patch size with a core area free of hunting disturbance (20,000 ha), has already been considered. Another approach, based on conservation genetics theory, aims to estimate habitat patch sizes necessary to sustain lemur populations above some theoretical limit, below which there is some risk of inbreeding and loss of genetic heterozygosity and allelic diversity. This study uses two theoretical limits, 500 and 5,000 individuals, which span the range generally advocated for long-term maintenance of genetic diversity in vertebrate populations (Schonewald-Cox et al. 1983).

Plates

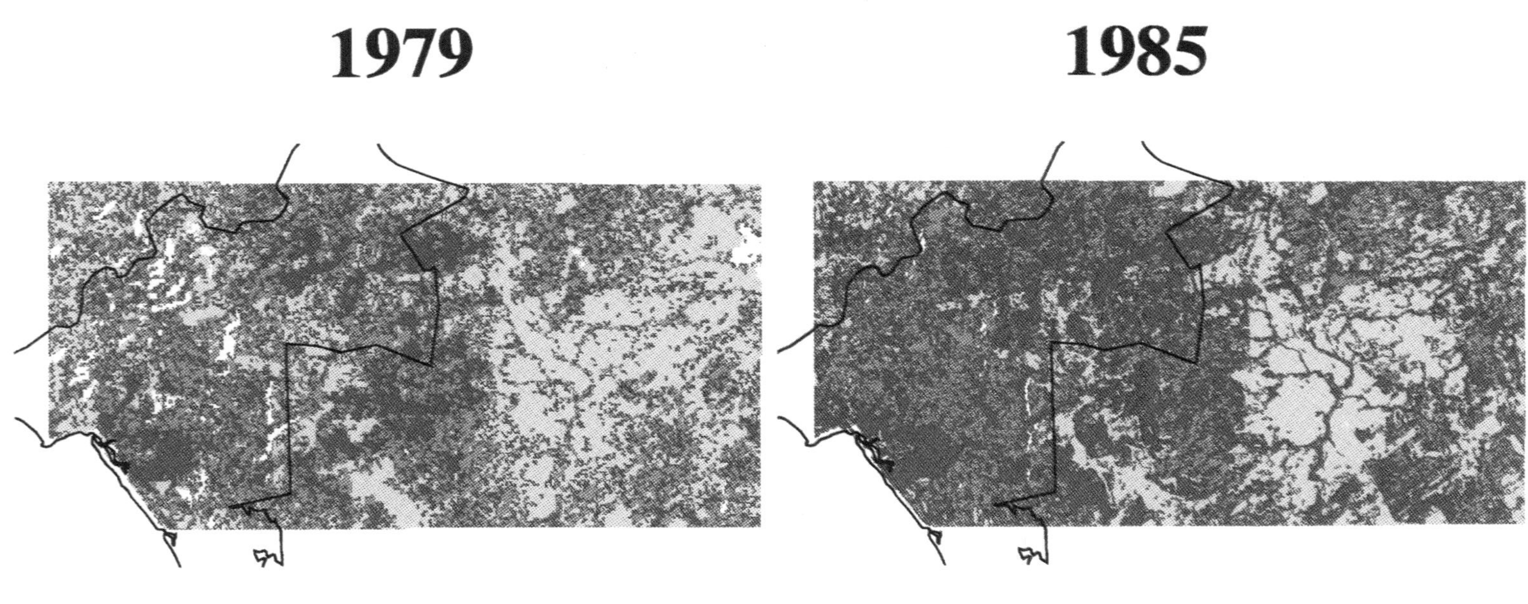

Plate 1. Land cover maps of the dry tropical forest area in Guanacaste National Park and Santa Rosa Park (the area within the outline is Santa Rosa Park), showing the expansion of forest vegetation from 1979 to 1985. Both images were taken during the dry season.

A

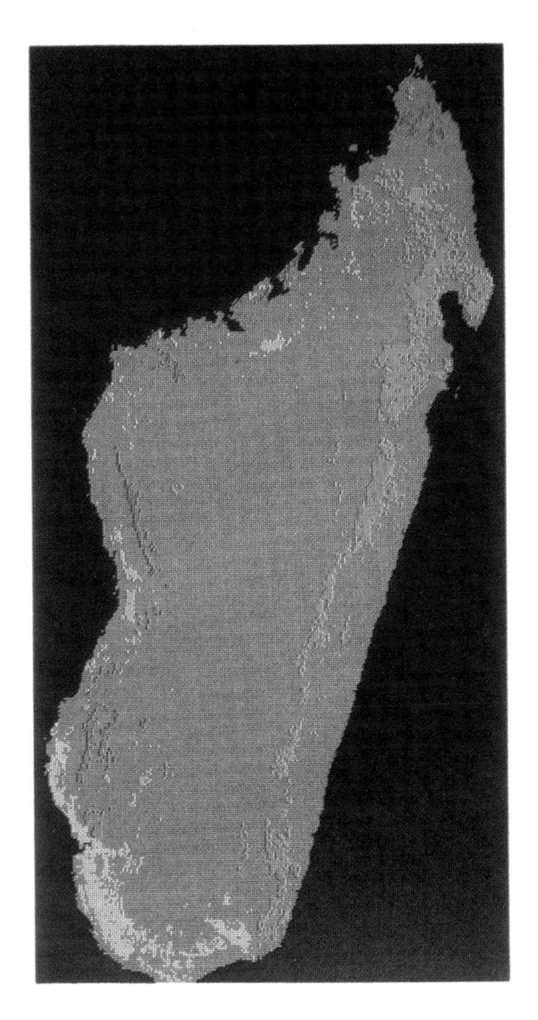

Plate 2. (A) Forest cover in Madagascar in 1990, showing Eastern wet rainforest (dark green); Western dry rainforest (light green); karst (Western dry) forest (blue); spiny scrub (yellow); and savanna grasslands and other vegetation (including plantation, mangrove, littoral, aquatic, and riparian gallery forests) (red). (B) Landsat image of Kirindy Forest in the Morondava region in 1990, showing the largest remaining remnant of Western dry forest in Madagascar (shades of red); savanna and highly altered forest (light yellow and green); and salt flats and aquatic habitat (blue).

B

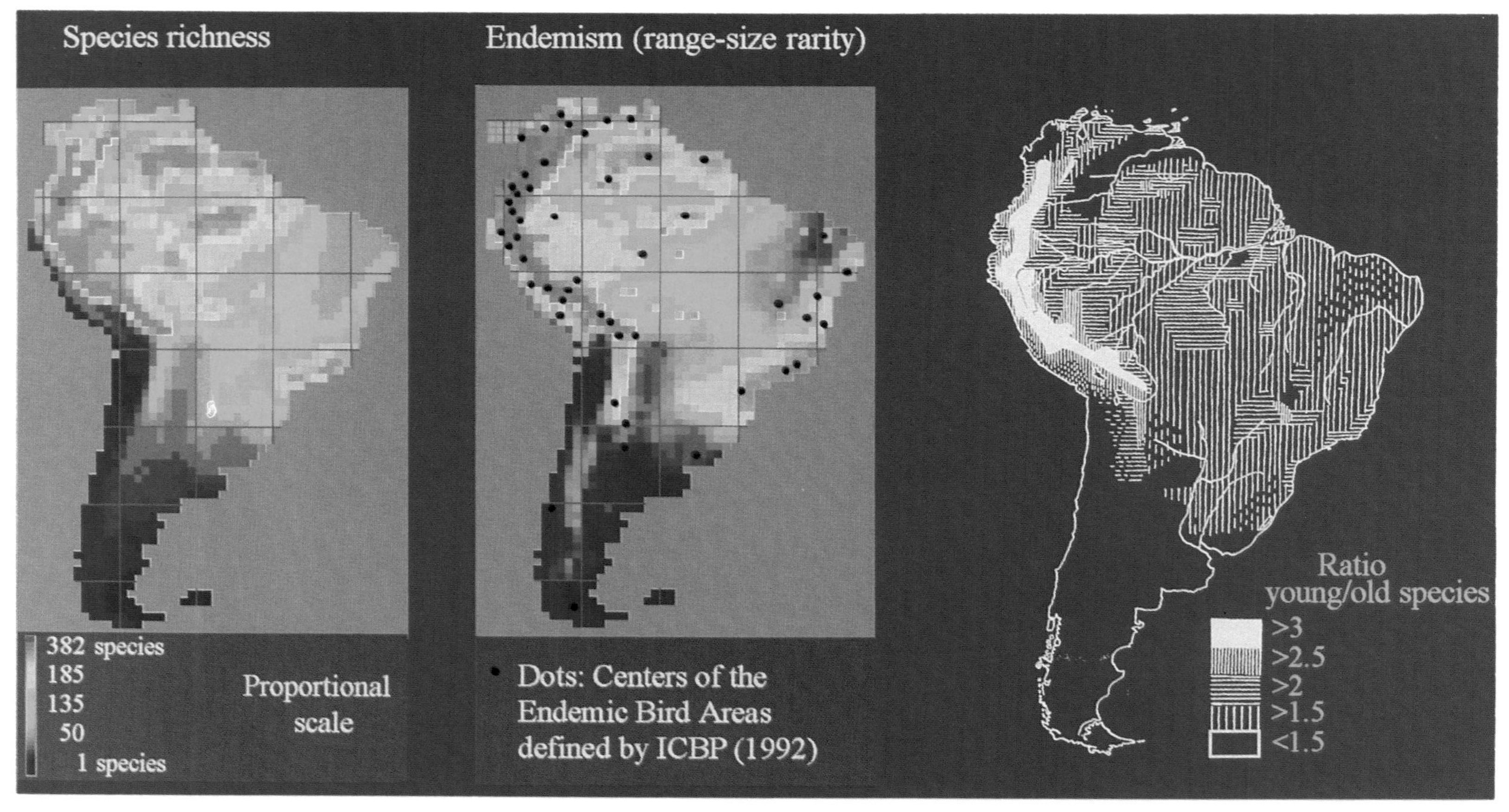

Plate 3. Species richness and endemism of South American forest birds: (**A**) total species richness for the seven taxonomic groups analyzed in this chapter (1,114 species in total); (**B**) endemism, showing increasing range-size rarity as defined by the WORLDMAP software (dots indicate centers of the Endemic Bird Areas defined by ICBP [1992]); (**C**) geographic variation in the ratio between number of species representing strong radiations during the last 5–7 million years and species representing single lineages since before this time, based on DNA divergence data (see Fjeldså 1994; interrupted lines indicate that species density was too low for reliable calculations).

Table 27.3 Changes over time in the number of forest patches in fragment size classes, and the percentage of total forest cover in fragment size classes, in the Western Domain of Madagascar.

Size class	1	2	3	4	5	6	7	8	9	10
$4\ km^2$ units	1	2	3–4	5–8	9–16	17–32	33–64	65–128	129–256	> 256
Number of fragments in size class										
1950	380	124	108	80	42	29	17	16	5	5
1974	204	80	71	47	26	12	9	5	2	0
1990	258	60	42	17	8	3	1	1	2	0
Percentage of total forest area in fragment size class										
1950	4.8	3.0	4.7	6.5	6.5	8.8	10.3	19.2	12.0	24
1974	7.2	5.6	8.7	10.7	11.4	10.3	15.4	17.0	13.6	0
1990	19.3	9.0	11.0	8.2	7.5	5.4	3.6	7.3	29.0	0

Note: Fragment size classes are shown on a $\log_2$ scale with fragment size range expressed in units of $4\ km^2$.

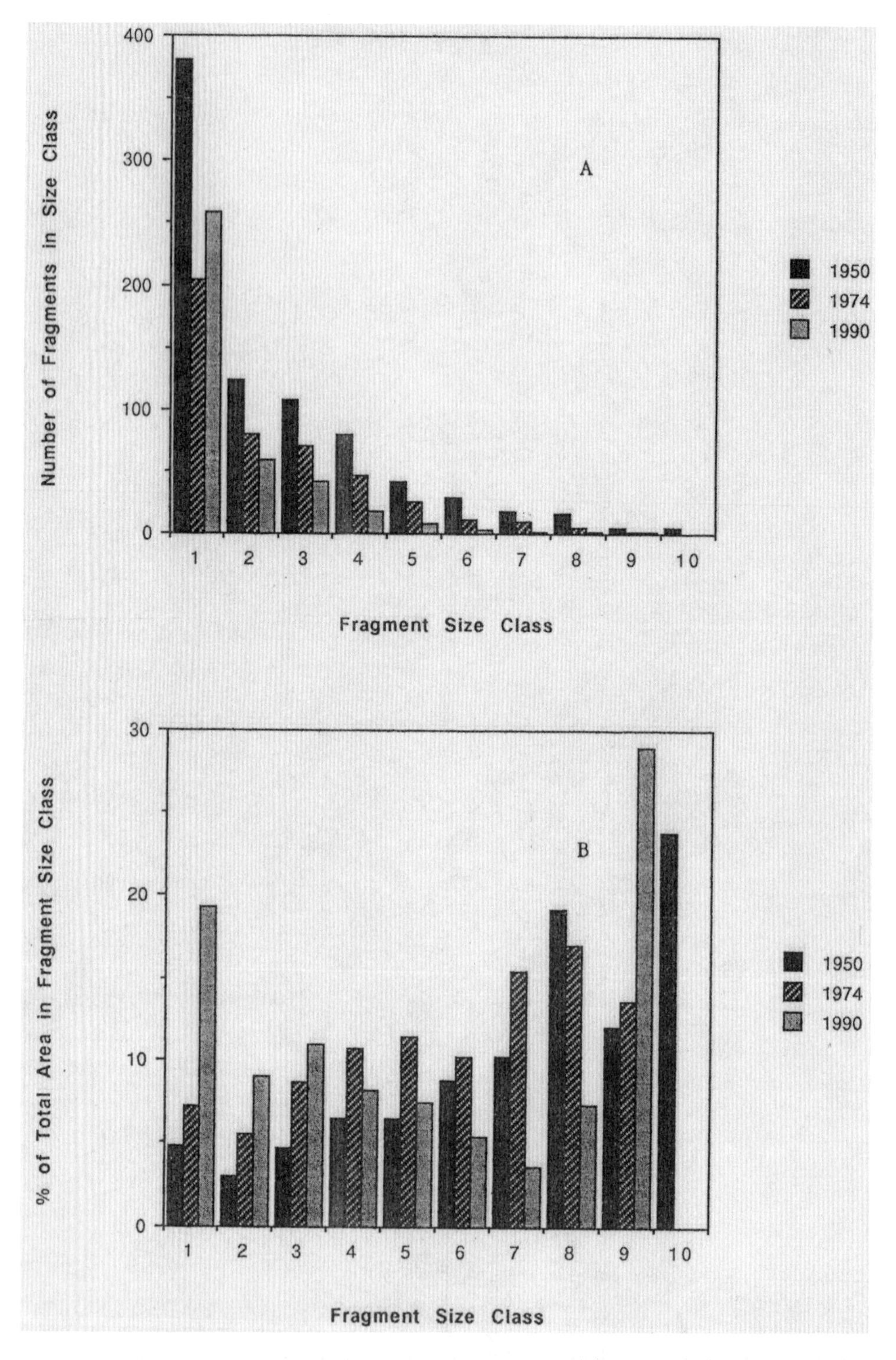

Figure 27.12. Changes over time in *(A)* the number of fragments in different patch size classes and *(B)* the percentage of total remaining forest area in each size class. (See table 27.3 for a description of the size classes used.)

Table 27.4 Lemurs occurring wholly or mainly within the Western Domain, showing total area of potential habitat throughout Madagascar and an estimate of minimum habitat area requirements (in units of 1,000 ha, or 10 km^2) necessary to maintain populations of 500 and 5,000 individuals.

			Habitat requirements (1,000 ha) for population of	
Species	Density (no./ha)	Habitat area	500 individuals	5,000 individuals
Microcebus murinus	0.42	2,078	1.2	11.9
Mirza coquereli	0.20	261	2.5	25.0
Cheirogaleus medius	0.46	2,090	1.1	10.9
Phaner furcifer	1.50	562	0.3	3.3
Lepilemur edwardsii	0.57	274	0.9	8.8
Lepilemur ruficaudatus	1.40	526	0.4	3.6
Lepilemur septentrionalis	0.60	34	0.8	8.3
Eulemur coronatus	0.77	41	0.6	6.5
Eulemur fulvus sanfordi	1.25	42	0.4	4.0
Eulemur f. fulvus	1.70	729	0.3	2.9
Eulemur f. rufus	0.90	623	0.6	5.5
Eulemur mongoz	0.40	125	1.3	12.5
Hapalemur griseus occidentalis	0.40	39	1.3	12.5
Avahi laniger occidentalis	0.67	15	0.7	7.5
Propithecus v. verreauxi	0.80	2,030	0.6	6.3
Propithecus v. deckeni	0.40	148	1.3	12.5
Propithecus v. coquereli	0.60	117	0.8	8.3
Propithecus diadema perrieri	0.20	59	2.5	25.0
Propithecus tattersalli	0.40	2	1.3	12.5

Minimum fragment sizes necessary to support populations of 500 and 5,000 individuals were estimated for each lemur species by multiplying these limits by estimates of average lemur density in Western dry forest (table 27.4). Density estimates for each lemur species were obtained from reviews in Harcourt and Thornback (1990) and more recent estimates in Smith, Horning, and Moore (in press). A density of 0.4 animals/ha was assigned to four species or subspecies *(E. mongoz, H. griseus occidentalis, P. verreauxi deckeni,* and *P. tattersalli)* for which no density estimates were available. Assuming that all remaining forest is equally suitable for lemurs, fragment sizes required to retain populations of 5,000 individuals of the rarest species *(M. coquereli* and *P. diadema perrieri)* exceed 25,000 ha. The number of fragments of this size in the Western Domain has declined from sixteen in 1950 to seven in 1974 and five in 1990. However, not all of these large patches fall within the range of each lemur species.

The information in table 27.4 can be used to estimate lemur population sizes and identify priority areas for conservation of each species. Worked examples are provided for two species, *M. coquereli* and *L. edwardsii.* Based on the distribution of Western dry forest patches within the known range of *M. coquereli* and the minimum patch size of more than 25,000 ha necessary to support long-term viable populations of more than 5,000 individuals (table 27.4), we find that only two such patches remain within the species'

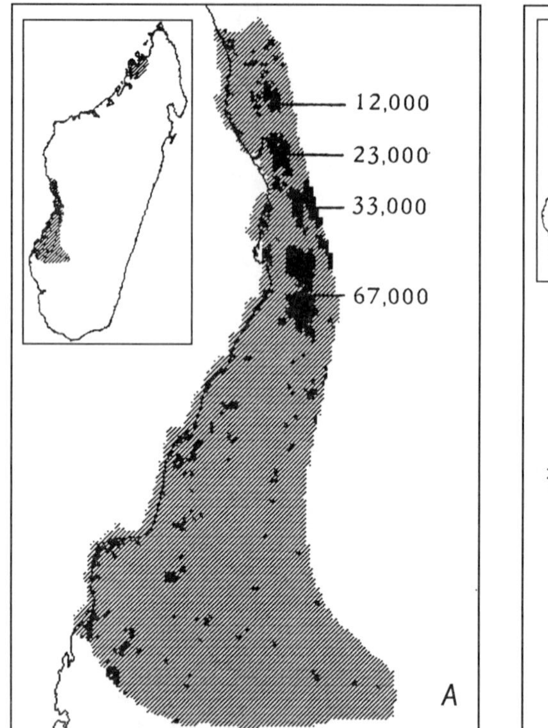

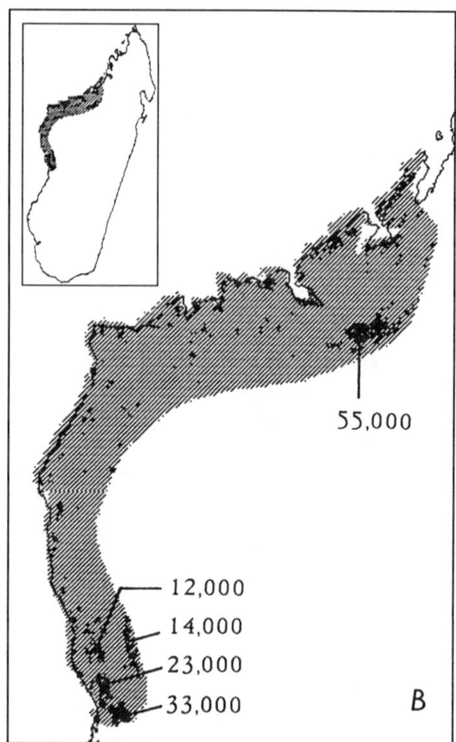

Figure 27.13. Overlay of the geographic ranges of *(A) Mirza coquereli* and *(B) Lepilemur edwardsii* (shaded areas) with the distribution of forest fragments (black areas) to show the actual occurrence of remaining potential habitat. Numbers indicate remnant patch sizes (ha).

range in western Madagascar (fig. 27.13A). The largest patch (67,000 ha) is located north of Morondava between the Andranomena and Tsiribihina Rivers in the region surveyed by the author (Smith, Moore, and Horning 1991; Smith, Horning, and Moore, in press), and an additional significant habitat patch occurs immediately north of the Tsiribihina River. Neither of these patches is effectively protected by existing nature reserves (cross-hatched areas in fig. 27.11). If we assume that all fragments with the potential to support *M. coquereli* populations of more than 500 individuals (those larger than 2,500 ha) are occupied and viable in the short term, the number and area of such fragments can be counted and used to estimate potential short-term population size. Within the range of *M. coquereli,* 442 4 km^2 grid cells occur in fragments made up of seven or more cells, giving a total habitat of 176,800 ha. This area of habitat is sufficient to support a population of approximately 35,000 individuals.

L. edwardsii has a higher density and therefore much smaller theoretical minimum patch sizes than *M. coquereli*: 8,800 ha for maintenance of populations of 5,000 individuals and 900 ha for 500 individuals. The largest habitat patch (55,000 ha) within the range of *L. edwardsii* occurs in the north of its range, where it falls within the Ankarafantsika Nature Reserve (fig. 27.13B). Three additional unprotected habitat patches of greater than 8,800 ha occur north of the Tsiribihina River, and two patches each of

Table 27.5 The total area and percentage of original forest cover remaining in each of the twelve environmental zones of western Madagascar.

Environmental zone	Original cover (km²)	1990 cover (km²)	% cover remaining	Representation in protected areas (km²)
Sediments				
1. 0–200 m	77,788	3,696	4.7	368
2. 200–500 m	54,632	992	1.8	672
3. 500–1,000 m	17,848	508	2.8	0
4. > 1,000 m	885	8	0.9	0
Metamorphics				
5. 0–200 m	8,352	108	1.3	0
6. 200–500 m	25,100	144	0.6	0.4
7. 500–1,000 m	50,884	640	1.3	0
8. > 1,000 m	3,548	12	0.3	0
Volcanics				
9. 0–200 m	12,944	96	0.7	16
10. 200–500 m	5,264	36	0.7	0
11. 500–1,000 m	1,750	24	1.4	0
12. > 1,000 m	656	4	0.6	0

7,000 ha occur nearby in the Bemaraha Nature Reserve. Although forest in the Bemaraha Reserve is mapped as fragmented and discontinuous due to its dispersal among karst formations, it is likely that most of these fragments are connected to some degree by small corridors of forest; therefore, the effective patch size within the reserve is likely to exceed 14,000 ha. The total area of habitat in patches of more than 900 ha available within the range of this species is 211,000 ha. This is sufficient to support approximately 120,000 individuals. Approximately one-third of this habitat occurs within existing nature reserves.

Representation of Environmental Zones in Fragments and Nature Reserves

The proportion of forest remaining in each of the twelve environmental zones was determined by overlaying forest cover (Nelson and Horning 1993) on the environmental zones. Results indicate that deforestation has been severe in all zones, but disproportionately higher on volcanic and metamorphic substrates (table 27.5). Low-elevation sediments, with 4.7% cover remaining, have been least affected. This discrepancy can be attributed to differences in soil fertility and vegetation susceptibility to fire on different substrates. The most extensive fragments of surviving low-elevation sedimentary forest occur north of Morondava on Pliocene sandy carapace. This region lacks the fertility and water-holding capacity to provide a productive arable substrate, in contrast to surrounding river alluvium and richer basaltic soils. The next largest surviving remnant, in Ankarafantsika Nature Reserve, occurs on infertile sandy soils on an elevated, dissected plateau, where it is afforded additional protection by rugged terrain (Jenkins 1987). Other significant expanses of forest on low-elevation and medium-elevation sediments, such as those in Bemaraha, Ankarana, and Analamera Nature Reserves, are largely protected by the rugged relief of limestone karst formations. Accessible vegetation on fertile sedimentary, volcanic, and metamorphic substrates has been almost entirely cleared or destroyed by burning.

Significant areas of medium- to high-elevation Western dry forest occur on sedimentary substrates in the southwest. None is protected, and little is known about their current condition or status. An example of high-elevation diverse Western dry forest, which includes evergreen, deciduous, and fire-resistant savanna elements, occurs in the elevated, dissected sandstone Isalo massif and nature reserve. This area is considered an outlier of the Central Domain and was not included in the Western Domain GIS database for this study, but could possibly be included as a significant example of high-elevation Western dry forest. The forests in this area are patchy and degraded by frequent fires (Jenkins 1987).

Very little forest remains on volcanic substrates in the Western Domain, and most of the surviving patches are small (<500 ha). Volcanic substrates in the west and south have been almost entirely cleared. The best opportunity for protection of vegetation and habitat on this substrate occurs in the north, on the Montagne d'Ambre volcanic massif. This region has been regarded as a humid, evergreen outlier of the Eastern evergreen forest domain, but it supports many unique species of plants and animals and could more appropriately be considered a distinct biogeographic subregion in its own right, comparable to the Sambirano region, which it adjoins to the southwest. Evergreen forests on the wet summit of Montagne d'Ambre are located within a nature reserve, and some remnants of deciduous Western dry forest occur at lower elevations within the boundaries of the Ankarana Nature Reserve. More detailed vegetation and habitat mapping of this highly diverse and important region is necessary to establish opportunities for location and management of new reserves.

Little vegetation remains on metamorphic basement rocks, and it is not known with certainty what vegetation type originally predominated in these zones. The lateritic clay soils of these zones may have supported a vegetation type more susceptible to degradation by fire and erosion than those occurring on the gray soils of the sedimentary zones. The only significant occurrence of vegetation in this environmental zone in existing reserves is a 4,000 ha patch of Western dry forest in Bora Nature Reserve (see fig. 27.14). This reserve is currently small and little known. Smaller patches of remnant forest surround the area, so it is possible that the reserve could be expanded and managed to maintain a larger, more viable forest remnant.

When we overlay the locations of existing nature reserves on the distribution of forest cover, we find that the overall representation of different environmental zones in existing nature reserves is poor (see fig. 27.14). (Primary and secondary forest cover in 1974 [Faramalala 1988] was included as well as Nelson and Horning's 1993 estimate of 1990 forest cover because of the possibility that Nelson and Horning underestimated the occurrence of some low-quality vegetation cover in the semiarid southwest and on karst formations in the north.) Reserves are almost totally lacking on metamorphic and volcanic substrates, even though such areas support the most luxuriant examples of Western dry forest (Jenkins 1987). It is evident that in Madagascar, as in many other countries, reserves have been primarily located in areas unsuitable—or least suitable—for

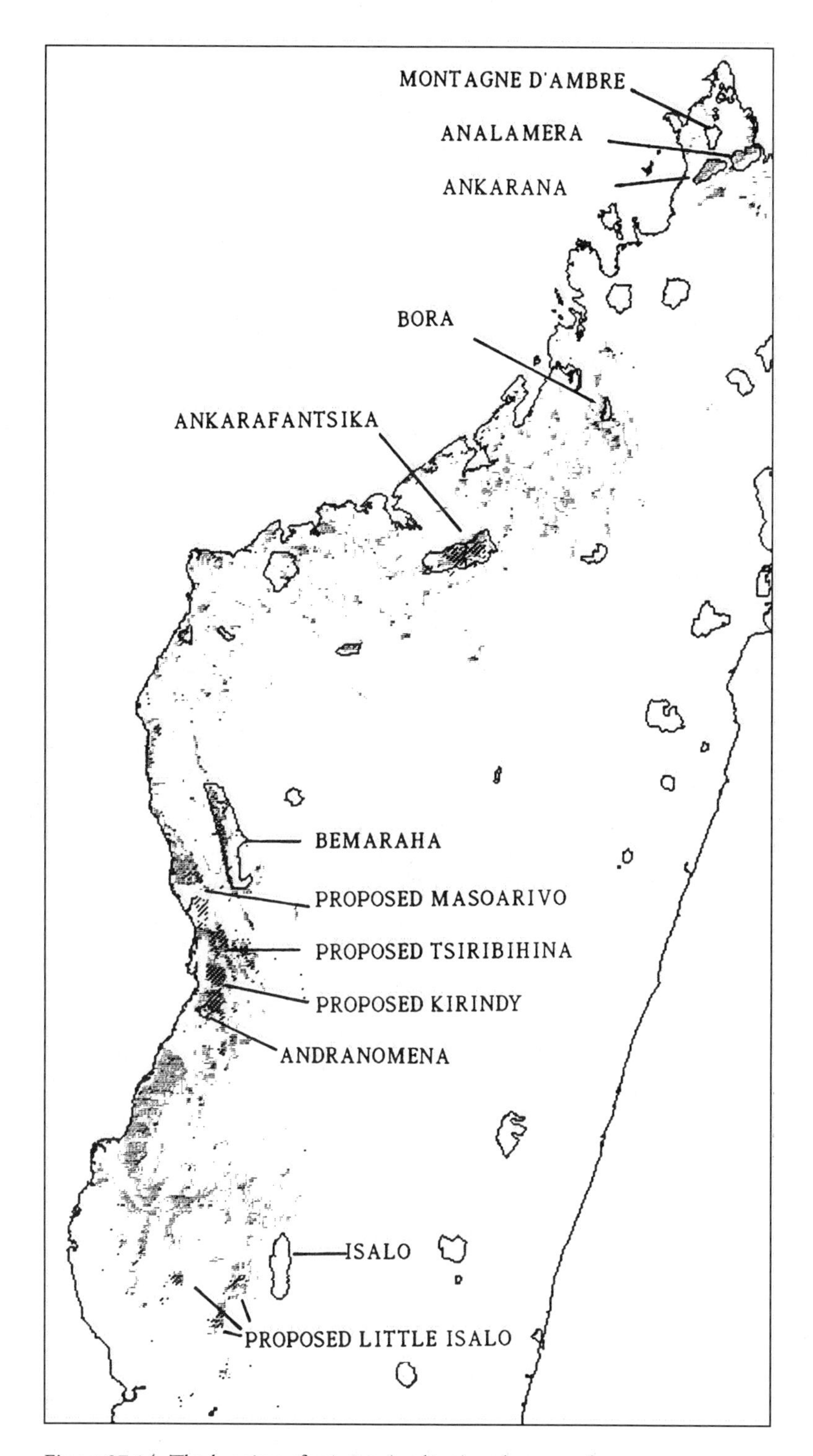

Figure 27.14. The location of existing (outlines) and proposed new nature reserves in the Western Domain overlaid on the distribution of primary and secondary forest cover in 1974 (shaded: after Faramalala 1988) and forest cover in 1990 (hatched black: after Nelson and Horning 1993). Only those reserves referred to in the text are labeled.

alternative use and development. This strategy is likely to have caused the extinction of many species and the loss of some of the best examples of Western dry forest.

Reserve Selection and Management Priorities

The biodiversity and endemism of Madagascar is so rich that almost all surviving remnants are likely to be of considerable biological importance and value (Jenkins 1987). A strong case can be made for protection of all surviving forest remnants, particularly in the Western Domain of Madagascar, where only 2.8% of high-quality primary forest survives and less than 22% of surviving forest is located within nature reserves. Clearing has been more severe in the Western Domain than in the Eastern and Southern Domains, and existing remnants are less well protected by reserves. The establishment of four new large reserves or reserve networks and the expansion of some existing reserves is recommended to protect most of the remaining fragments of Western dry forest larger than 5,000 ha.

A new Protected Area, "Kirindy," is recommended to protect the largest remaining fragment of Western dry forest (see fig. 27.14). An ICDP is also essential in this location to prevent ongoing illegal clearing and degradation of existing remnants by hunting and road building. An existing small nature reserve is located in the southwestern corner of this fragment, but more than 50% of the reserve has been cleared or degraded since 1950, and the remainder is at risk of destruction due to its proximity to roads and villages (Smith, Horning, and Moore, in press). A more suitable location for the reserve has been identified in the northwestern section of the fragment (Smith, Horning, and Moore, in press). Overlaying of biodiversity and disturbance risk maps has revealed this area to have a high biodiversity and low risk of disturbance. The southern section of the fragment is subject to low-intensity harvesting of commercial timber. While this may not affect biodiversity directly, the road-building activities necessary for timber extraction increase the risk of future clearing and hunting disturbance. Sufficient background biophysical information is available for this site to justify the immediate initiation of an ICDP.

Another new reserve, "Tsiribihina," is recommended to protect remaining large fragments between the Tsiribihina and Manambolo Rivers, and a further nearby reserve, "Masoarivo," is recommended to protect remnants north of the Manambolo River and inland from the township of Masoarivo (see fig. 27.14). Because little is known about the current status of biodiversity in and socioeconomic threats to these two regions, both should be the subject of conservation evaluation surveys prior to management.

Another new reserve, "Little Isalo" (see fig. 27.14), incorporating three medium-sized forest remnants, is recommended to protect examples of Western dry forest on medium- to high-elevation (500–1,000 m) sedimentary substrates. Because little information is available about these sites, they will also need to be evaluated prior to management.

Existing reserves for which more information is required as a precursor to possible expansion, restoration, and more active management include Bora Nature Reserve, for protection of the only significant examples of Western dry forest on metamorphic substrates, and all forest remnants in the northern tip of Madagascar, including the existing Montagne d'Ambre, Ankarana, and Analamera reserves. As well as supporting most of the remaining Western dry forest on volcanic substrates, the northern region is so diverse in geology, topography, and climate and so rich in endemic plant and animal species that it deserves more detailed evaluation as a separate biogeographic domain in its own right.

GENERAL IMPLICATIONS

1. This study highlights the importance of monitoring and analyzing patterns of vegetation cover change in developing countries as a foundation for biodiversity conservation.
2. This study also demonstrates that, in Madagascar, seasonally dry (monsoon) forests have declined more severely, and are under greater threat, than tropical wet forests. Recent conservation efforts in Madagascar have focused most heavily on protection of wet forests in the north and east, leaving western dry forests relatively ignored and underprotected. Similar patterns have occurred in other tropical regions, such as Central America, Southeast Asia, and Australia.
3. There is clearly a need to focus global conservation priorities on the protection of limited remaining areas of seasonally dry rainforest in both developing and developed countries (see also Kramer, chap. 25).
4. This study highlights the futility of locating protected areas in regions where there is a high risk of future disturbance. This is particularly true in developing countries, where clearing and hunting by subsistence farmers are difficult to control and concepts of land tenure are poorly developed and enforced. Where there is a high risk of disturbance, biodiversity patterns alone should not determine the pattern of reserve location, unless the technology and resources are available to mitigate future threats.
5. This and previous studies have shown that GIS can be used to predict and map both biodiversity and disturbance risk at both national and regional scales. This information can be readily incorporated into conservation evaluation processes for identifying priority areas for conservation and management.

ACKNOWLEDGMENTS

I wish to thank the following individuals and organizations for their valuable contributions to this study: N. Horning (WWF-Madagascar) for assisting with the establishment of early (1990) national and regional GIS databases and supplying AVHRR 1990 forest cover in ERMS format; Kate MacGregor and Nick Rollings (Department of Ecosystem Management) for digitizing the Humbert and Cours Darne vegetation cover; S. Ferrier (NSW National Parks and Wildlife Service) for assisting with the establishment of the national GIS database and reviewing this chapter; D. Moore (Department of Ecosystem Management) for helping to establish the national and regional GIS databases; K. Glander, W. Laurance, and R. O. Bierregaard, Jr., for reviewing this chapter; NSW National Parks and Wildlife Service for supplying ERMS GIS software; and M. Nicoll and S. O'Conner of WWF Madagascar for their foresight in initiating GIS studies in Madagascar and providing organizational and logistic support for field surveys.

28

Molecular Perspectives on Historical Fragmentation of Australian Tropical and Subtropical Rainforests: Implications for Conservation

Craig Moritz, Leo Joseph, Michael Cunningham, and Chris Schneider

There is increasing concern over the effects of recent fragmentation of rainforests in Australia and elsewhere (Bierregaard et al. 1992; Laurance 1994; Whitmore, chap. 1). This concern has resulted in concentrated efforts to identify the effects of habitat fragmentation at genetic, species, and community levels for management purposes. However, the recent anthropogenic loss and isolation of habitat can also be viewed as the latest in a long history of changes in rainforest distributions, stretching back to the Miocene and beyond. Contrary to the previous view that rainforests developed in stable climates and landforms (Prance 1982b), it is now clear from several lines of evidence that many areas now supporting rainforest have had a turbulent history, particularly in the late Tertiary and Quaternary (Flenley 1979; Terborgh 1992a). Therefore, we have two scales of fragmentation to consider: recent anthropogenic clearing and long-term natural processes.

A historical perspective is relevant to the issue of current anthropogenic rainforest fragmentation in several ways. First, understanding the responses of rainforests to previous changes in climate should put us in a better position to interpret and, perhaps, predict the effects of anticipated global warming (Walker 1990; Nix 1991; Hopkins et al. 1993). Second, historical changes in rainforest distribution are likely to be major determinants of the current distributions of genetic and species diversity. Historical contractions of rainforest areas are presumed to have created disjunct refugia within which a subset of rainforest-dependent species were protected from extinction, and between which there was the potential for allopatric divergence and, perhaps, speciation (Haffer 1969, 1993; Vanzolini and Williams 1970; Prance 1982b; Mayr and O'Hara 1986; but see Endler 1982 for an alternative interpretation). In the context of reserve design, it is important to understand both the current distribution of diversity and the processes that gave rise to it (Brooks, Mayden, and McLennan 1992). An appreciation of how individual species have responded to historical changes in rainforest distribution might also help us to identify taxa that are either sensitive or resilient to recent anthropogenic fragmentation of rainforest (Balmford 1996), although this remains to be demonstrated.

This chapter reviews studies of molecular variation within and between vertebrate species occurring in subtropical and tropical Australian rainforests in the context of other evidence for changes in rainforest distribution. The goal of these molecular studies is to provide insights into current and historical population processes (e.g., gene flow, population size: Slatkin 1987; Avise 1994; Moritz 1995) and the process of speciation among rainforest taxa. Uniquely, molecular comparisons also provide a (crude) estimate of the timing of these events (e.g., Knowlton et al. 1993). With this information, we can identify historically isolated populations or communities that, as reservoirs of genetic diversity, warrant conservation efforts (Avise 1992; Moritz 1994a). An alternative, and perhaps conflicting, use of these studies is to identify and conserve areas with concentrations of young, actively speciating taxa (e.g., Fjeldså 1994; Fjeldså and Rahbek, chap. 30).

NATURAL FRAGMENTATION ON A LARGE SCALE

The tropical and subtropical rainforests of eastern Australia occur in several disjunct patches (fig. 28.1) on the relatively mesic coasts of Queensland and northern New South Wales (NSW; reviewed by Webb and Tracey 1981; Adams 1992). The three major rainforest areas are referred to here as the "Wet Tropics" (Cooktown to Townsville, 15–19° S), "central Queensland" (20–22° S), and "southern Queensland" (including nearby rainforests in northern NSW; 27–32° S). There are also some significant rainforest areas in Cape York, north of the Wet Tropics (Kikkawa, Montieth, and Ingram 1981). These rainforest areas include the mesic components of a chain of high and relatively cool montane islands surrounded by warmer or drier environments, termed the "mesotherm archipelago" by Nix (1991), which includes the central mountains of Papua New Guinea. The eastern Australian rainforests are widely regarded as the remnants of vegetation that covered much of the continent during the early Miocene, but contracted as the climate became more seasonal, cooler, and drier from the mid-Tertiary onward (reviewed by Kemp 1981; Truswell 1993). Information on the spatial and temporal dynamics of this contraction is sparse because of the small number of sites with appropriate fossil or pollen records.

There are substantial differences in the numbers of rainforest-dependent vertebrates between the three major tropical-subtropical rainforest areas in eastern Australia. Winter (1988) commented on the lack of rainforest specialist mammals in the southern Queensland forests as compared with the Wet Tropics and on the presence of more ecotonal species in the former area, although the difference is not so marked for reptiles and amphibians (table 28.1; see also Covacevich and McDonald 1991). Winter suggested that the absence of rainforest specialist mammals in the southern areas could be due in part to insufficient size of late Pleistocene refugia, compared with more substantial refugia in the Wet Tropics (see below). In principle, this hypothesis could be tested by comparing levels of within-population genetic variation in rainforest-dependent species, such as certain reptiles and frogs, that survived rainforest contractions in both areas. Following from the assumption that refuge area (and thus historical population size) was lower in the central and southern rainforests, we would predict that surviving populations should have lower sequence diversity than ecologically similar species in the Wet Tropics.

The relationships among congeneric vertebrate species from the three areas are critical to understanding the history of natural fragmentation, yet remain to be determined. The

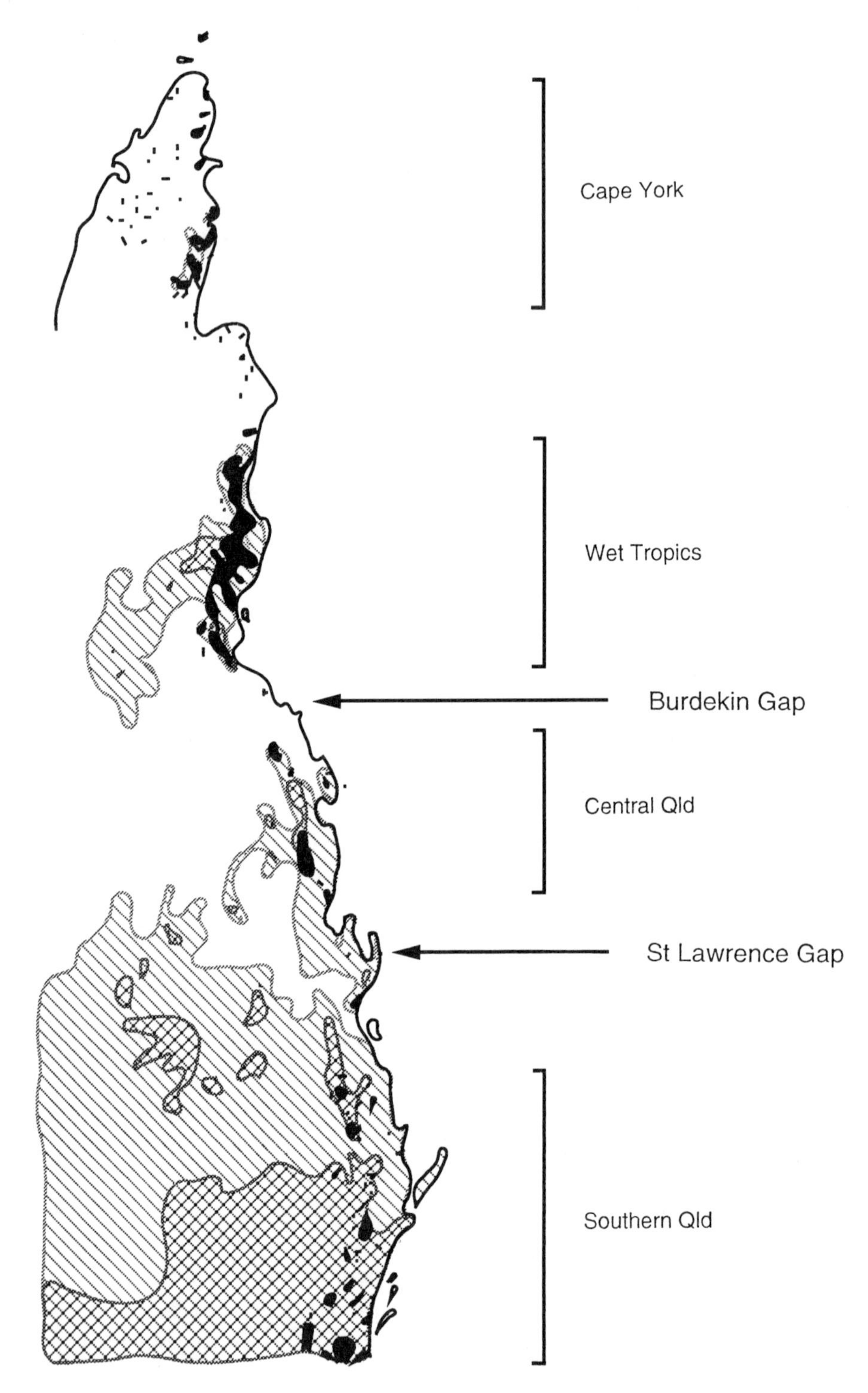

Figure 28.1. Distribution of tropical and subtropical rainforests in eastern Australia (black areas; adapted from Webb and Tracey 1981) overlain with the distribution of mesotherm climates (adapted from Nix 1991). Cross-hatching indicates mesotherm-dominant climates; diagonal striping indicates transitional mesotherm/megatherm climates. Arrows indicate previously recognized warm, dry corridors known as the Burdekin and St Lawrence Gaps.

Table 28.1 Numbers of rainforest specialist mammals and herpetofauna in major areas of eastern Australian subtropical and tropical rainforest.

		Rainforest specialist species		
Unit	Area[a]	Mammals[b]	Reptiles[c]	Frogs[c]
Cape York	260,260	5	6	4
Wet tropics	791,300	13	26	22[d]
Central Queensland	172,556	0	15	4
Southern Queensland—Northern NSW	436,444	0	18	10

a. Estimates of rainforest area (ha) were obtained from Webb and Tracey 1981 except for Central Queensland, obtained from G. Lake, Queensland Forest Service. The Southern Queensland—Northern NSW figure was obtained by subtraction of the Central Queensland estimate from the combined estimate of Webb and Tracey.

b. Winter 1988. c. Modified from Covacevich and McDonald 1993.

d. A large proportion of amphibians restricted to the Wet Tropics rainforests are terrestrial-breeding microhylids (e.g., *Sphenophryne* and *Cophixalis*), which are absent from the central and southern areas.

recent description of new rainforest-restricted species from the central Queensland region (e.g., Couper, Covacevich, and Moritz 1993) has substantially increased the known diversity and endemicity of the vertebrate fauna, but it remains to be seen whether there are any consistent area relationships. Among the widely distributed rainforest specialist reptile and amphibian species that occur in Central Queensland, five are shared with the Wet Tropics and six with the Southern Queensland subtropical forests (Covacevich and McDonald 1993). The depauperate avian fauna of the Central Queensland rainforests has variously been associated with the Wet Tropics (Schodde and Calaby 1972) or southern (Cracraft 1991) faunas. Some of the endemic bird species from the central forests appear to have phylogenetic affinities with southern rainforest taxa, while others are associated with northern rainforest taxa (reviewed by Joseph, Moritz, and Hugall 1993).

There have been few molecular comparisons of rainforest species distributed across the three areas. Mitochondrial DNA (mtDNA) was compared for small samples of three widely distributed, rainforest-inhabiting bird species, the yellow-throated scrubwren *(Sericornis citreogularis),* large-billed scrubwren *(Sericornis magnirostris),* and eastern whipbird *(Psophodes olivaceus),* as well as a habitat generalist, the white-browed scrubwren *(Sericornis frontalis)* (table 28.2: Joseph, Moritz, and Hugall 1993; Joseph and Moritz 1994). The three rainforest species showed substantial divergence between the southern forests and the Wet Tropics, whereas the generalist white-browed scrubwren was variable, but showed no geographic structure. For the large-billed scrubwren and whipbird, the central Queensland populations were genetically similar to those of the southern forests and divergent from the Wet Tropics populations. These preliminary data implicate rainforest contraction to either side of the dry and warm "Burdekin Gap" (see fig. 28.1) as a significant historical vicariance event. By comparison, the current warm, dry corridor separating the southern and central rainforest areas does not appear to have greatly inhibited historical gene flow in these widespread rainforest species.

Analyses of other widely distributed rainforest species or sister species are needed to test the generality of this result. The available information on levels of mtDNA sequence diver-

Table 28.2 Measures of mean mtDNA sequence difference (%, for cytochrome-*b*) within and between populations of birds from three major rainforest areas.

Species	*N*	SQ	C	NQ	S-NQ	C-SQ	C-NQ
Scrubwrens							
Large-billed	11	0.3	0.0	0.3	3.7	0.0	3.0
White-browed	9	1.6	0.0	1.5	0.0	0.0	0.1
Whipbird[a]	6	1.1	—	1.0	0.5	0.4	0.9

Source: Joseph, Moritz, and Hugall 1993; Joseph 1994.

Note: SQ = southern Queensland; C = central Queensland; NQ = northern Queensland (Wet Tropics). The values between areas are corrected for within-area variation. *N* represents the number of sequences compared.

a. Only one individual from central Queensland was analyzed.

gence among related species of leaf-tail geckos *(Phyllurus, Saltuarius)* and *Sphenomorphus*-group skinks from the different rainforest areas indicate very large sequence divergences (>15%; C. Moritz and C. Schneider, unpub.) and, accordingly, ancient separations.

RAINFOREST VICARIANCE WITHIN THE WET TROPICS

The rainforests of the Wet Tropics occupy an area of about 800,000 ha, spread across lowland and disjunct montane habitats (fig. 28.2A). Several lines of evidence suggest that there have been dramatic fluctuations in the nature and distribution of these rainforests during the Quaternary (ca. 1.5 million years B.P. to present).

First, long-term pollen records from several sites on the Atherton Tableland dating as far back as 190,000 years B.P. (Kershaw 1986) and from offshore deposits dating back to the early Pleistocene, 1.4 million years B.P. (Kershaw 1994), have been analyzed. Detailed records from the Tableland sites indicate alternating periods of cool gymnosperm-dominated forests, dry fire-adapted sclerophyllous vegetation, and warm-wet angiosperm-dominated rainforests. Under the cool and dry conditions of the late Pleistocene (18,000 years B.P.), the vegetation was dominated by sclerophyll woodlands. The current angiosperm rainforests are less than 10,000 years old and are floristically different from the rainforests of the preceding interglacial period. Based on the long-term record, Kershaw (1994) speculated that the wetter rainforest types could have existed as isolates surrounded by drier Araucarian rainforests for much of the Quaternary.

Second, Webb and Tracey (1981) identified a series of putative Pleistocene refugia (fig. 28.2B), based in part on present concentrations of narrow endemics and primitive angiosperms.

Third, Nix (1991) modeled the geographic distribution of climates suitable for the current rainforest types under various scenarios. During the drier, cooler period presumed to have occurred 18,000 years B.P., the appropriate climates were restricted relative to the present, with a major disjunction between the Atherton and Carbine Tablelands (fig. 28.2B).

Fourth, Hopkins et al. (1993) have identified extensive late Pleistocene and Holocene charcoal deposits, even within the putative refugia suggested by Webb and Tracey (1981). This finding suggests that the refugia, if they existed, were deeply dissected by fire-prone sclerophyllous vegetation. An alternative view is that substantial refugia did not exist;

rather, rainforest contracted to numerous local microrefugia and subsequently expanded from these (Stocker and Unwin 1989; Hopkins et al. 1993). Another significant process on the Atherton Tableland is basalt flows, some of which are less than 10,000 years old and which inevitably affected the distribution of rainforests and their fauna (Willmott and Stevenson 1989).

Genetic Consequences of Historical Rainforest Fragmentation

If the Wet Tropics rainforests contracted to geographically isolated refuges for substantial periods, then this should be reflected by geographic patterns of genetic variation that are congruent among similarly affected species. To test this hypothesis, we are examining patterns of mtDNA phylogeography in rainforest-restricted species that are distributed across the montane rainforest blocks. Historically isolated populations should have closely re-

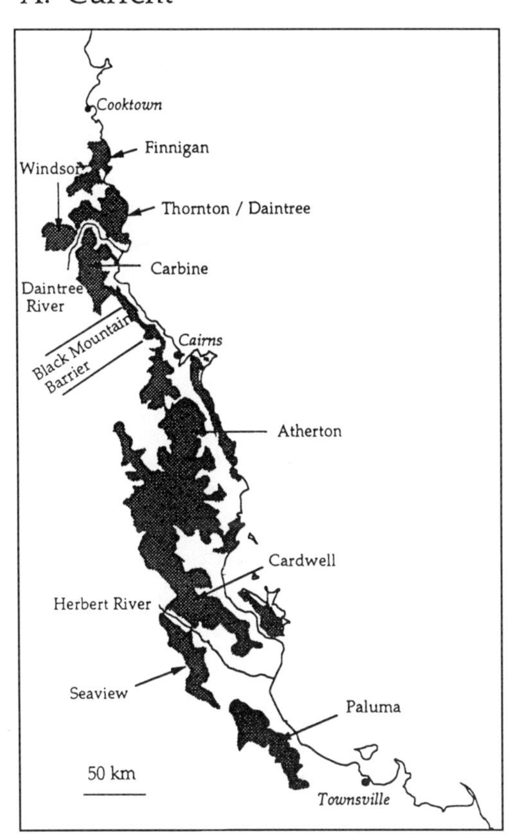

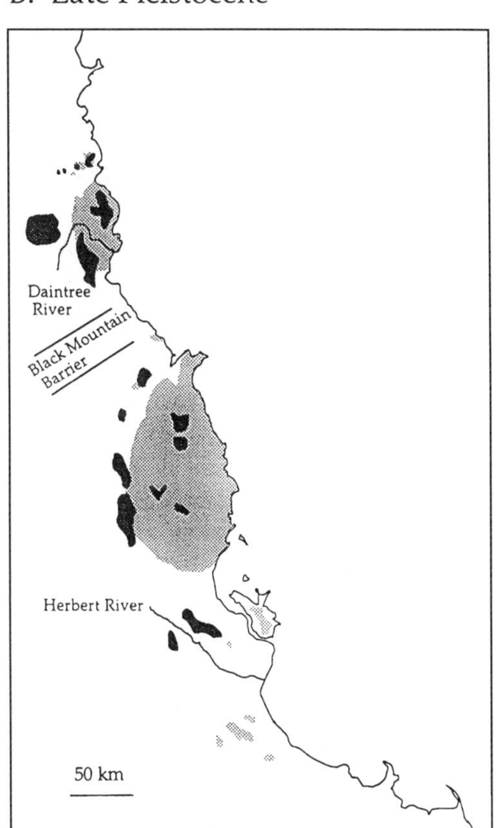

Figure 28.2. Distribution of rainforests in the Wet Tropics. *(A)* Current distribution (dark shading), with major rainforest blocks and the "Black Mountain Barrier" labeled. *(B)* Distribution of refuge areas (dark shading) predicted by Webb and Tracey (1981), overlain with the areas predicted to have had climates suitable for the current rainforest types 18,000 years B.P. (light stippling; from Nix 1991).

Table 28.3 Average sequence difference among all sampled individuals (%, from cytochrome-*b*) of rainforest-dwelling vertebrate species from the Wet Tropics; and the net sequence difference between populations to the south and north of the BMB, corrected for within-area variation.

Species	*N*	Mean sequence difference (%)	Net divergence across BMB (%)
Prickly skink[a]	44	3.9	6.1[b]
Chowchilla[a]	11	1.2	2.4[b]
Grey-headed robin[a]	26	0.9	1.4[b]
Yellow-throated scrubwren	25	0.6	0.4[b]
Large-billed scrubwren	17	0.5	0.2
Atherton scrubwren[a]	23	0.2	0.0

Source: Joseph, Moritz, and Hugall 1995.

a. Endemic species.

b. Sequence variation in these species was significantly structured across the BMB.

lated alleles, distinct from those in other isolates, and this pattern should be repeated across similarly affected species (Avise et al. 1987; Avise and Ball 1990).

A comparison of one skink and five bird species from the Wet Tropics revealed striking congruence in their geographic patterns of mtDNA variation, but also an order of magnitude difference in the amount of sequence variation (Joseph, Moritz, and Hugall 1995). Four of the six species tested showed strong differentiation to either side of the "Black Mountain Barrier" (BMB; see fig. 28.2A), and for at least two—the grey-headed robin *(Poecilodryas albispecularis)* and prickly skink *(Gnypetoscincus queenslandiae)* (see Moritz, Joseph, and Adams 1993)—the genetic divergence was greater than expected from isolation by distance within either area. However, the level of net sequence divergence between the northern and southern populations of species with significant geographic variation varied from 6% in the prickly skink to 0.4% in the yellow-throated scrubwren (table 28.3). The prickly skink, grey-headed robin, and chowchilla *(Orthonyx spaldingii)* also showed strong phylogeographic structure, with alleles from the northern populations being closely related, as were those from the southern populations. Preliminary data from rainforest geckos *(Carphodactylus laevis, Saltuarius cornutus)* and stream-dwelling frogs *(Litoria nannotis, L. genimaculata)* indicate levels of sequence divergence across the BMB similar to that in the prickly skink, but genetic breaks at additional sites are evident in some taxa (C. Schneider and M. Cunningham, unpub.).

The two exceptions to the above pattern were the large-billed and Atherton *(Sericornis keri)* scrubwrens (table 28.3). The former is explicable, as the large-billed scrubwren is reported to be less rainforest-restricted than its yellow-throated congener (Crome 1990). The lack of geographic structure and reduced sequence variation in the Atherton scrubwren was unexpected, given that this rainforest specialist is thought to occur in disjunct high-altitude localities (Nix and Switzer 1991). One possible explanation is that the species was reduced to a single area during the glacial maximum and recolonized the other area during the relatively cool, wet period from 7,500 to 5,000 years B.P. (see Nix 1991). Another possibility is that an advantageous mtDNA variant has recently swept though the species, eliminating information on population history (e.g., Rand, Dorfsman, and Kann 1994).

Excluding the Atherton scrubwren, the amount of genetic difference across the BMB decreases with estimates of the ability of species to move between rainforest isolates. This finding, the geographic congruence among species of the pattern of intraspecific mtDNA variation, and the location of a common genetic break at the site predicted to have been a climate-induced gap (fig. 28.2B; Nix 1991), combine to implicate climate-induced rainforest contraction as the cause of these patterns. This provides strong evidence for the role of allopatric divergence among rainforest refugia, a central plank of the Haffer (1969) proposal.

The Timing of Intraspecific Divergences

The segment of cytochrome-*b* sequenced for the species described above evolves at approximately 1–2% per million years in mammals. Rates appear to vary by about fivefold among vertebrates, with large-bodied and cold-blooded species being slower (Martin and Palumbi 1993; Rand 1994). The substantial sequence differences in all but the scrubwrens are incompatible with late Pleistocene divergences; rather, the vicariance events seem much earlier. One possibility is that there was a single, major vicariance event, perhaps during the late Tertiary, that isolated the populations, and the more vagile bird species decreased the level of sequence difference through subsequent occasional gene flow. Alternatively, there may have been a succession of increasingly severe isolation events at the BMB from the late Tertiary onward, with the early, relatively mild events affecting the more sedentary reptiles and amphibians and the later, more severe events (e.g., 18,000 years B.P.; fig. 28.2B) affecting the more vagile bird species. Some of the variation could also be due to differences in rates of mtDNA evolution, but, if anything, the cold-blooded reptiles and amphibians are expected to have a slower rate (Martin and Palumbi 1993), which would exacerbate these differences.

Interspecific Comparisons

Although most of the vicariance events affecting within-species variation predate the mid- to late Pleistocene, the speciation events are older still. The available mtDNA sequence comparisons among rainforest species from the Wet Tropics and their congeners indicate large sequence divergences. For ringtail possums, all interspecific comparisons but one yield sequence divergences greater than 12%, and the species from the Wet Tropics are interspersed among those from other areas, especially Papua New Guinea (PNG: fig. 28.3A; see also Baverstock et al. 1990). This pattern suggests old, possibly Miocene, divergences among these species and is consistent with findings based on DNA hybridization (Springer et al. 1992). The exception is the *Pseudocheirus herbertensis-cinereus* complex, with a sequence difference of about 2%; these two species, distributed on either side of the BMB, remain a candidate for Pleistocene divergence. Tree frogs of the *Litoria eucnemis* complex from the Wet Tropics, Cape York, and PNG show sequence divergences (at the more slowly evolving 16S rRNA gene) ranging from 4% to 12%

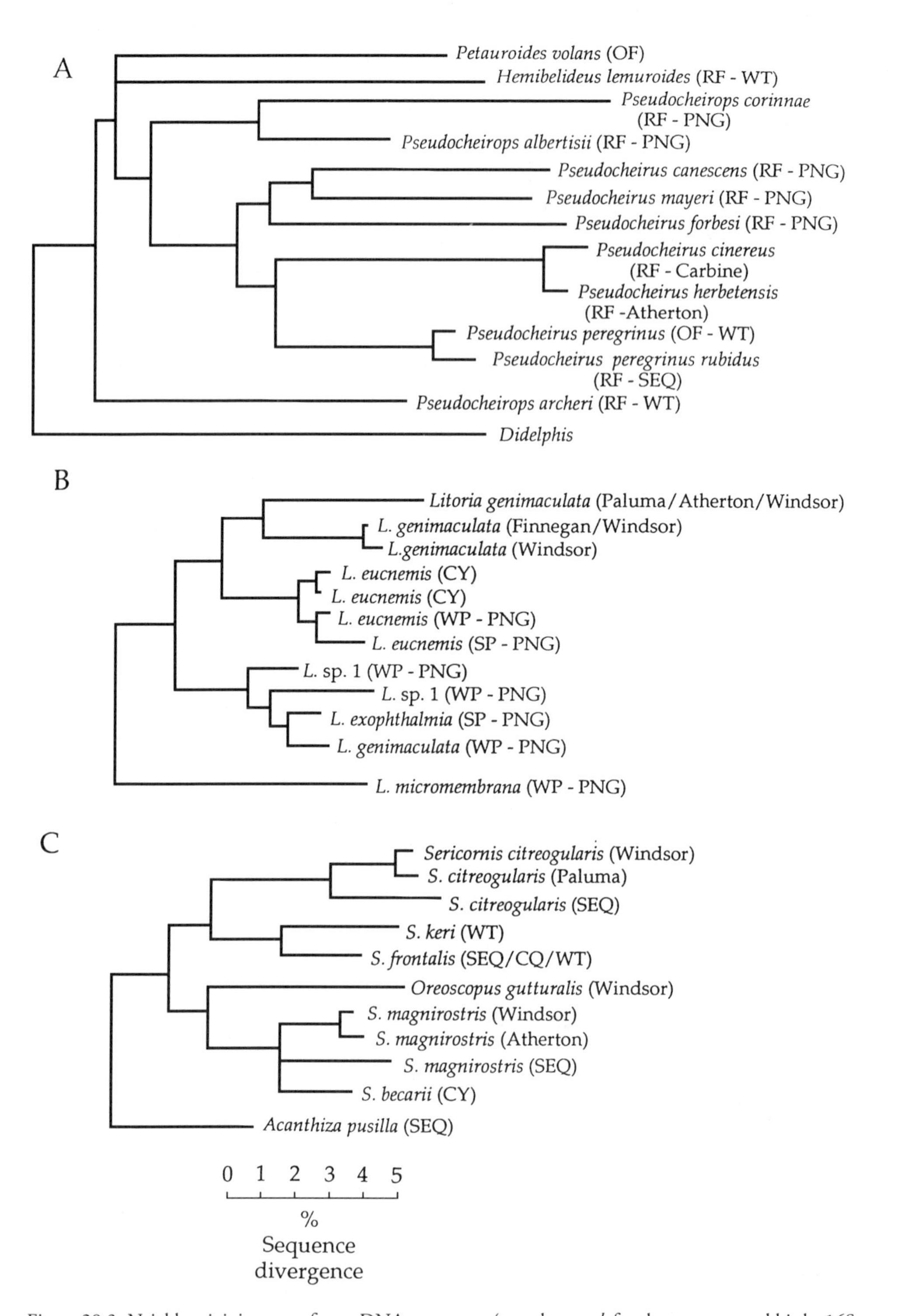

Figure 28.3. Neighbor-joining trees for mtDNA sequences (cytochrome-*b* for the possums and birds, 16S rRNA for the frogs) from rainforest vertebrate species from the Wet Tropics and related species from elsewhere. The branch lengths are scaled according to estimates of sequence divergence. *(A)* Various species of ringtail possums and the greater glider *(Petauroides volans),* rooted with the new world opossum *Didelphis.* *(B)* Different populations and species of tree frogs in the *Litoria eucnemis* complex from the Wet Tropics,

among recognized species, again with taxa from the Wet Tropics and elsewhere interleaved (fig. 28.3B). Finally, the Wet Tropics scrubwrens show interspecific cytochrome-*b* divergences of more than 5%, with more widely distributed species located within the clade (fig. 28.3C; see also Christidis, Schodde, and Baverstock 1988; Joseph and Moritz 1993).

We conclude from the above that intraspecific divergence, and probably also speciation, has occurred in rainforest vertebrates through natural processes of rainforest fragmentation, but at different spatial and temporal scales. In the taxa studied so far, isolation of geographic populations appears to predate the Pleistocene in several species, particularly the amphibians and reptiles, with the BMB being a consistent point of vicariance. By contrast, the majority of speciation events appear to be much older, probably dating back to the Pliocene or mid- to late Miocene, when the rainforest had a quite different and broader distribution (Truswell 1993). Deep sister species relationships between rainforest taxa from PNG and the Wet Tropics are a consistent feature, although some shallow divergences between Cape York and PNG populations suggest more recent dispersal (e.g., *L. eucnemis,* fig. 28.3B; see also Kikkawa, Montieth, and Ingram 1981). Thus, whereas the current distribution of intraspecific diversity is dominated by ongoing vicariance patterns, the distribution of recognized species is shaped by longer-term processes, including differential extinctions and ecological processes.

LATE PLEISTOCENE REFUGIA REVISITED

Even allowing for known variation in rates of molecular evolution, it is difficult to reconcile the above molecular data and similar evidence from Neotropical systems (e.g., Heyer and Maxson 1982; Silva and Patton 1993; Patton et al., chap. 29; Fjeldså and Rahbek, chap. 30) with the assertion that *late* Pleistocene refugia had a major role in speciation and generation of diversity within species (e.g., Haffer 1969; Vanzolini and Williams 1970; Whitmore and Prance 1987; cf. Haffer 1993; Bush 1994). Rather, the relatively severe late Pleistocene fluctuations may have acted primarily as species filters, causing many rainforest species to go extinct or diversify into a broader range of habitats (Winter 1988).

The effects of late Pleistocene refugia and subsequent expansion from them may be detectable from finer-scale patterns of geographic genetic variation. On the Atherton Tableland, populations of the prickly skink from the base of one putative refuge, Mt. Bartle Frere, have mtDNA alleles phylogenetically distinct from those in the central and southern Tableland (Joseph, Moritz, and Hugall 1995), and preliminary data from the gecko *Carphodactylus laevis* (C. Schneider, unpub.) suggest a similar pattern.

Cape York, and Papua New Guinea (PNG), rooted with a PNG species from outside the complex (note that the 16S rRNA evolves at about half the rate of cytochrome-*b*). *L.* sp. is an undescribed species from PNG (S. Richards, pers. comm.). *(C)* Species of scrubwrens from the Wet Tropics and elsewhere, rooted with the buff-rumped thornbill *(Acanthiza reguloides). S. frontalis* occurs around the southern coast to Western Australia and inhabits a wide variety of habitats. OF = open forest; RF = rainforest; WT = Wet Tropics; SEQ = Southeastern Queensland; CP = Chimbu Province, PNG; WP = Western Province, PNG; CY = Cape York Peninsula; CQ = Central Queensland.

It is difficult to use gene phylogenies to test hypotheses about historical biogeography in the late Pleistocene and Holocene because of the small amount of sequence divergence expected over this period. One fruitful approach, however, may be to contrast allelic genealogies from inside and outside putative refuges for a number of rainforest-dependent species, testing for the genetic signature of historical population expansion in the latter locations (Slatkin and Hudson 1991; Nee, Holmes, and Harvey 1995). Such analyses offer some hope of merging molecular approaches with more traditional paleoecological approaches (e.g., palynology: Kershaw 1994; charcoal dating: Hopkins et al. 1993) to study the historical dynamics of rainforest expansion and contraction.

IMPLICATIONS FOR CONSERVATION

> For all the concern about gene pools, conservationists on the whole have found the slow evolutionary process something too awe-inspiring to provide a future for; we hope that by preserving what we have now, the emergence of new species will continue to occur. Perhaps a more positive approach will soon grow from molecular biology. (Walker 1990, 30)

Measuring and Mapping Biodiversity

The majority of the rainforest species so far examined here and in Neotropical studies (Patton et al., chap. 29) show deep genetic divergences and strong geographic structure among populations. These findings suggest that descriptions of biodiversity based on distributions of vegetation types or endemic species may not adequately represent the underlying evolutionary diversity of the biota (see also Fjeldså 1994; Fjeldså and Rahbek, chap. 30). In the case of the Wet Tropics fauna, several species show reciprocal monophyly of mtDNA on either side of the BMB. For the prickly skink, the divergence in mtDNA is corroborated by divergence at allozyme loci (Moritz, Joseph, and Adams 1993), so that populations from the two areas would be regarded as separate "Evolutionarily Significant Units" (ESUs, *sensu* Moritz 1994b; Vogler and DeSalle 1994) for the purposes of assigning conservation value. It is likely that many of the other species we examined also include multiple ESUs. This species-by-species approach to describing evolutionary diversity may be of little practical benefit, however, given the vast number of species still to be examined.

A more practical approach is to identify geographic areas within which populations of a high proportion of species have been evolving independently from conspecific populations in other areas (Avise 1992). For the Wet Tropics rainforests, the patterns of sequence divergence suggest that the rainforest areas on either side of the BMB should be accorded separate priority for management because each contains an evolutionarily divergent community. As additional species, perhaps of lower vagility (e.g., flightless beetles, snails), are examined, it is possible that other rainforest blocks within each of the two areas will be found to be distinct. This approach is hierarchical, wherein areas separated by major zones of divergence in a number of taxa are defined first and assigned highest conservation priority. Subsets containing ESUs for less vagile taxa are then identified from within these major areas.

Historical Perspectives and Maintenance of Processes

An equally important contribution of molecular data is the refinement of our understanding of the processes of diversification, such as the relative roles of refugia, ecological gradients, and riverine barriers (Patton and Smith 1992; Patton, da Silva, and Malcolm 1994; Joseph, Moritz, and Hugall 1995). Improved understanding of these processes and how they operate in different areas is essential if we are to design a reserve system that both protects existing diversity and allows for future evolution (Erwin 1991; Fjeldså 1994). In particular, it is important to maintain both major centers of intraspecific diversity and the ephemeral connections between them—the former as reservoirs of genetic variation, and the latter to allow for whatever genetic interactions occur among the different evolutionary lineages.

A key conclusion from genetic studies of rainforest vertebrates from the Wet Tropics is that historical changes in climate have strongly affected the ranges and connectivity of populations through their effects on rainforest distribution. Species found to have been substantially affected by these past events might also be expected to be sensitive to future global warming. For these in particular, we must implement a management system that allows for geographic shifts in preferred habitats as climatic conditions change in the short to medium term. This conclusion reinforces earlier calls for the location of reserves across environmental gradients and the provision of appropriate corridors and buffer zones (e.g., Walker 1990).

The design of refuges also needs to take into account the viability of the component species within the rainforest mosaic (Witting, McCarthy, and Loeschcke 1994). To what extent can the sensitivity of individual species or guilds to recent anthropogenic fragmentation be inferred from comparative molecular data? The viability of populations within fragments is determined by a number of interacting ecological factors (e.g., edge effects, microhabitat quality, resource availability) that cannot be addressed though molecular studies. However, with specific reference to fragmented mammal populations on the Atherton Tableland, Laurance (1991a, 1994) concluded that the major determinant of viability is the ability of species to use the matrix, in this case open pastures and intermittent riparian vegetation.

One possibility, still to be evaluated, is that the extent of molecular divergence across historical barriers may be a useful predictor of the current ability of species to traverse the matrix between rainforest patches and thus maintain viable metapopulations; that is, species showing large divergences should be more prone to isolation within anthropogenic fragments. This idea makes two major assumptions: (1) that animals responded to the habitat separating historical isolates as they will to that surrounding anthropogenic fragments; and (2) that the habitat requirements of species have remained similar through time. Neither of these assumptions is likely to be strictly true, but this idea remains a hypothesis worth testing.

GENERAL IMPLICATIONS

1. Molecular comparisons among populations and species of rainforest fauna are revealing high levels of genetic diversity and deep genetic divisions usually not apparent from exter-

nal phenotypes. These observations suggest that an understanding of geographic patterns of historical vicariance will improve the ability of conservation managers to design reserve systems that encompass evolutionary as well as species diversity.

2. The major contribution of these studies, however, is to improve our understanding of the temporal and spatial *processes* that shaped the current distribution of diversity. An approach to reserve design that ignores these ongoing processes is clearly inadequate to ensure the long-term viability of the system, given ongoing extinctions and the increasing likelihood of significant global climatic change.

ACKNOWLEDGMENTS

The research described in this chapter has been supported by grants from the Wet Tropics Management Authority, the Australian Research Council, and the Cooperative Research Centre for Tropical Rainforest Ecology and Management. We are grateful to Andrew Hugall, Jiro Kikkawa, Marcia Lara, Keith McDonald, and John Winter for extensive discussions, and to Andrew Hugall and Anita Heideman for assistance with preparation of figures. The clarity of presentation was improved through comments on the manuscript from Peter Baverstock, Bill Laurance, and Jim Patton.

29

Diversity, Differentiation, and the Historical Biogeography of Nonvolant Small Mammals of the Neotropical Forests

James L. Patton, Maria Nazareth F. da Silva, Márcia C. Lara, and Meika A. Mustrangi

"These days, it's either high diversity or lots of endemics. That's what sells reserves."
(Mythical campside conversation between the late Al Gentry and the late Ted Parker, two of the world's most knowledgeable Neotropical biologists: Foster 1995, 12).

As true as this statement might be, it is equally true that our present knowledge of both diversity and endemicity of most rainforest mammals is inadequate to identify, much less sell, biological reserves in the Neotropical region, except in the most global sense. Of the approximately 4,600 species of living mammals in the world (Wilson and Reeder 1993), some 1,150 are known from the Neotropics (Patterson 1994), and about 80% of these are endemic to that region (Cole, Reeder, and Wilson 1994). Most of these endemics are small-bodied, secretive, and nocturnal—bats, marsupials, and rodents.

While authors disagree as to the relative species richness of tropical versus temperate zones within South America (Chesser and Hackett 1992; Mares 1992; Pimm and Gittleman 1992), few doubt that Neotropical forests sustain some of the richest mammalian communities in the world. The country of Peru alone contains over 10% of the world's mammalian diversity (462 species: Pacheco et al. 1995), the vast majority of which is found in tropical forest communities. Unfortunately, however, knowledge of actual species richness in communities within Neotropical forests is extremely limited, except perhaps for primates and the large-bodied ungulates and carnivores.

As we will discuss below, previously unknown species are being revealed by careful and extensive fieldwork in poorly sampled areas. The application of modern technical and analytical methods is also delimiting unappreciated levels of diversity, even in those areas that are relatively well known biologically. Of equal importance is that knowledge of distributional limits (both ecological and geographic), population demography, and basic life history parameters—or of virtually all aspects of general natural history—is woefully inadequate for most species inhabiting Neotropical forests. As a consequence, at this most critical time, when maximal information is needed to make difficult but necessary triage decisions, to design reserves, or to construct effective recovery plans for species already on the edge of extinction, we are effectively unable to provide it for the majority of South American mammals.

We were originally asked to contribute to this volume in the general area of historical vicariance and centers of mammalian endemism in South America. In the review that follows, we argue that a discussion of this topic, while of considerable interest to a broad community of both scholars and planners, is not possible with present information on mammals, except at the grossest geographic scale. To some, this statement might be surprising, given the fact that mammalian diversity is relatively limited in comparison to many other animal groups and because of our inherent interest in taxa close to ourselves. The type of analysis and presentation given by Fjeldså and Rahbek (chap. 30) for Neotropical birds, however, is something that we mammalogists can only hope for in the future.

Consequently, herein we explore, in a limited way, the existing database on the systematics, distribution, population parameters, and other attributes of the mammalian denizens of Neotropical lowland forests, with an emphasis on what we know and what must be understood before generalizations can be achieved. We focus on small-bodied terrestrial and arboreal marsupials and rodents and do not treat large-bodied ungulates or their predators, nor do we concern ourselves with primates or the highly diverse Neotropical microchiropteran bats. We point out that, despite over four hundred years of study of Neotropical mammals (for a historical review of mammalian exploration in South America, see Hershkovitz 1987), we have a poor understanding of important yet elemental components of the systematics, biogeography, and population biology of small mammals. These elements include, but certainly are not limited to, the nature of species boundaries and the distributional limits of species, areas of endemism and even variations in local species richness, and local persistence and population structure.

We end this chapter with recommendations for further work. These are efforts we feel are necessary if correct estimates of mammalian species diversity are to be made, if the regional uniqueness of faunas is to be recognized, and if the phylogenetic uniqueness of given taxa is to be appreciated. We recognize that time is severely limited for the requisite geographic surveys, systematic revisions, and longitudinal population studies. As field biologists with active programs in the Neotropical forests, we are all too cognizant of the extent to which native communities have disappeared, have been altered by exotics, or are in imminent danger of loss. We also emphasize that the identification of critical habitats and species, as well as specific geographic areas, for preservation is not a simple task. The criteria for selection are numerous and extremely diverse. They involve concepts ranging from ecological keystone species to phylogenetic uniqueness and the preservation of evolutionary process (e.g., Smith, Bruford, and Wayne 1993; Greene 1994; Moritz 1994b). Our remarks are intended both to provide an overview of our present knowledge and to point in appropriate directions for further activity, in the areas of both scientific research and political discussion.

MEASURES OF LOCAL SPECIES RICHNESS

Emmons (1984) provided the first comparative data for diversity estimates of nonvolant Neotropical forest mammals for seven sites across the Amazon Basin. She noted differences in species richness, with higher levels in western Amazonia than in the Guiana

Shield region of Brazil, and suggested that environmental factors such as soil fertility and undergrowth density affect local mammal diversity. Her estimates were based largely on visual transect censuses supplemented by trapping—a comprehensive methodology, but one that cannot provide complete inventories. Indeed, the number of localities within the roughly 9,000,000 km^2 of lowland South American forests for which reasonably complete inventories of mammalian species are available can be counted on the fingers of two hands. For example, faunal lists for nonvolant mammals have been published for only four areas in eastern Peru: the Manu Biosphere Reserve, with 70 species recorded in the vicinity of Cocha Cashu on the Rio Manu (Terborgh, Fitzpatrick, and Emmons 1984; Janson and Emmons 1990; Pacheco et al. 1993); Reserva Cusco Amazonico on the Rio Madre de Dios, with 56 species recorded (Woodman et al. 1991, 1995); Panguana Biological Station in the Rio Pachitea drainage, with 59 species (Hutterer et al. 1995); and the Aguaruna Indigenous Area on the Rio Cenepa in the Alto Rio Marañón drainage, with 61 species (Patton, Berlin, and Berlin 1982). Additional studies or inventories have been completed, or are in progress, in other parts of Amazonia. When published, these will add importantly to our understanding of diversity patterns within the basin, and collectively will provide tests of Emmons's (1984) thesis.

In comparison to baseline inventory work, there is an even greater paucity of ecological or longitudinal population studies for most wet-forest small mammals, although this is also being rectified at a number of sites in Amazonia (see, for example, Charles-Dominique et al. 1981; Janson and Emmons 1990; Malcolm 1988, 1990, chap. 14). Two of us (J. L. P. and M. N. F. S., along with J. R. Malcolm and C. A. Peres) recently completed a year-long survey of the nonvolant mammalian fauna of the Rio Juruá Basin in western Amazonian Brazil. When the study is published, this will represent one of the most thoroughly documented areas in the lowland Amazonian forest for both species richness and community structure. Fifty-two species of marsupials and rodents alone were recorded in our survey. At least seven of those species were new to science at the time of their initial collection (e.g., Patton and Silva 1995; M. N. F. da Silva 1995).

For the *Mata Atlântica,* great effort has been expended over the past decade by several Brazilian research teams to survey this highly diverse and very poorly known fauna. Again, however, these endeavors are still ongoing, with only limited results published to date (e.g., Fonseca and Kierulff 1989; Stallings 1989), and the available database remains inadequate. For example, the biological reserve of Fazenda Intervales in São Paulo state was the subject of intensive surveys for small mammals from 1988 to 1990, with twenty-three species of marsupials and rodents recorded (Manço et al. 1991). Limited sampling by two of us (M. A. M. and J. L. P.) in less than a month in 1992 and 1994 yielded at least five additional species at this site, an increase of nearly 20% in recognized species richness. Moreover, recent taxonomic revisions have doubled the recognized species diversity in some of the more common taxa of the *Mata Atlântica* (e.g., the murid rodent *Delomys* [Voss 1993] and the murine opossum *Marmosops* [Mustrangi and Patton, 1997]), and have reorganized others into geographically coherent phylogenetic clades (the echimyid rodent *Trinomys* [Lara 1994]).

THE ADEQUACY OF CURRENT TAXONOMY

Generalizations about both alpha and beta diversity are dependent upon the quality of available taxonomies. Yet for the majority of wet-forest, nonvolant small mammals, current taxonomy may substantially underestimate true species diversity. Moreover, accurately mapped species ranges are lacking for most taxa, as are hypotheses of phylogenetic relationships. These three components (species boundaries, geographic ranges, and phylogenetic relationships) are all equally important aspects of diversity, and together define the base for an eventual understanding of the historical development of that diversity.

Few taxonomic revisions have been completed, yet such revisionary work both establishes species boundaries and records the geographic and ecological distributions of recognized species. Most available revisions were published in the early and middle parts of the twentieth century, during a period in which taxa were "lumped" in a wholesale fashion, usually without documented reason. For example, in the 1950s, P. Hershkovitz, regarded by many as the dean of Neotropical mammalogy, revised several wet-forest murid rodent genera. For the arboreal rice rat *Oecomys,* he consolidated some twenty-five names into two species, the large-bodied *O. concolor* and the smaller *O. bicolor* (Hershkovitz 1960). This genus is currently being restudied by G. G. Musser and M. D. Carleton, and their preliminary list includes not two, but at least thirteen species (Musser and Carleton 1993). This latter view is supported by our own surveys on the Rio Juruá, where five species of *Oecomys* are readily recognizable by morphological, cytological, and biochemical characters. Similarly, Hershkovitz (1960) suggested the synonymy of some twenty taxa of terrestrial rice rats under the species *Oryzomys capito,* one of the more common murid rodents of the lowland forests. This same opinion was expressed by Cabrera (1961) in his influential checklist of South American mammals. Both views were adopted by most authors until karyotypic data indicated that this concept of *O. capito* included four distinct species that were broadly sympatric throughout the western Amazon (Gardner and Patton 1976).

New species are also being discovered at an impressive rate (Patterson 1994). Recently, many thorough small mammal surveys of Neotropical forest sites in which vouchers have been preserved and studied have uncovered previously unrecognized taxa (e.g., Pacheco 1991; Hershkovitz 1994; Woodman et al. 1995; Patton and Silva 1995). As mentioned above, at least seven of the fifty-two species (13%) of nonvolant small mammals trapped along the length of the Rio Juruá in western Amazonian Brazil were new to science. In each case, these are true additions to estimates of overall diversity, not simply the elevation of known subspecies to specific rank.

For some important groups, thorough field surveys have revealed a substantial increase in local species richness. Without such knowledge, any local diversity figure is an underestimate to an unknown degree. For example, the most common rodents in Neotropical forests are typically species of the terrestrial spiny rat genus *Proechimys* (family Echimyidae). These medium-sized to large rats (300 to nearly 1,000 g) are very abundant, very diverse, and one of the most taxonomically difficult groups of nonvolant mammals. The most recent review (Patton 1987) identified a series of species groups, but was unable to delineate species boundaries or geographic ranges for most recognized taxa. Nevertheless, three to four species could often be recognized in western Amazonian community assemblages, and these overlapped broadly in their use of both space and food resources (Emmons 1982). More recent surveys, using nontraditional molecular and cytological characters in addition to morphological features, have radically changed our appreciation of the total species diversity within this genus as well as the geographic and ecological limits of individual species. For example, M. N. F. da Silva (1995) identified eight species of *Proechimys* along the length of the Rio Juruá in western Brazil, recognizing four as new to science. Three of these were shown to have distributions extending well beyond the Rio Juruá, into southern Peru and adjacent Bolivia, into northern Peru, and into central Brazil. Some of these species are apparently limited to seasonally flooded forests—*várzea* along "whitewater" rivers or *igapó* in "blackwater" regions—while others of similar morphology replace one another geographically, and yet others co-inhabit the nonflooded upland (terra firme) forest at single localities across large geographic regions. The known number of true species within this diverse genus will only increase as thorough field surveys, complemented by analyses of morphological and biochemical data, continue.

THE IMPORTANCE OF PHYLOGENETIC RECONSTRUCTION

Phylogenetic hypotheses provide us with the means to understand the processes of evolutionary diversification—to test the veracity of explicit models (e.g., Patton and Smith 1992; Patton, Silva, and Malcolm 1994) and to identify "evolutionarily significant units" for conservation (Moritz 1994b). The histories of geographic areas can also be deduced from the phylogenetic concordance of the organisms occurring within them (Rosen 1978). As important as phylogenetic assessments might be, there are few hypotheses of relationship for Neotropical mammals, particularly within and between genera of marsupials and rodents. This should not be surprising given the lack of revisionary studies and the uncritical delineation of species boundaries.

Our own efforts with respect to phylogenetic analysis of Neotropical mammals have centered on a few taxa of didelphid marsupials and murid and echimyid rodents. At the moment these analyses lack the necessary geographic samples and are insufficient in numbers of taxa to suggest a comprehensive understanding of the dynamic history of the Neotropical mammalian biota. Nevertheless, we provide here a limited discussion bearing on several important historical biogeographic questions: (1) How unique are the mammalian taxa of the *Mata Atlântica* with respect to their Amazonian relatives? (2) Is there a consistent pattern of relationship among geographic areas within the Amazonian region, as suggested by authors working with other organisms? and (3) Do mammalian examples offer support for any of the hypotheses developed previously for the diversification of tropical biotas? Included here would be the riverine hypothesis first elaborated by Alfred Russel Wallace in 1849 (Wallace 1849; substantiated by Ayres and Clutton-Brock 1992), Haffer's (1969) and Vanzolini and Williams' (1970) Pleistocene refuge hypothesis, and other more recently developed ideas (such as those of Tuomisto et al. [1995] and Räsänen et al. [1995]).

We have used six generic or infrageneric groups of small, common mammals to investigate the concordance of area relationships within Neotropical lowland forests, using Cracraft's (1985, 1988; Cracraft and Prum 1988) hypothesis of centers of avifaunal diversity as a model (fig. 29.1). The taxa examined were species of the didelphid marsupials *Philander, Didelphis, Metachirus,* and *Micoureus* and two species complexes of the murid rodent genus *Oryzomys.* Each of these genera is broadly distributed within lowland Neotropical forests, and current taxonomies (Gardner 1993; Musser and Carleton 1993) suggest either single species or sets of closely related species within each. The data used for phylogenetic reconstruction are comparative sequences of the mitochondrial cytochrome-*b* gene, and the concordance among gene, organism, and area cladograms is being assessed by the methods of Page (1993, 1994). Complete analyses of these data will be presented elsewhere, but the important points are summarized here.

Three important patterns of sequence diversity were found in each of the six marsupial and rodent taxa examined. First, sequences within geographic regions are quite similar, with levels of difference maximally 2–3% between different localities. Second, all sequences within regions are typically monophyletic with respect to those of other regions. Third, divergence levels between sequence clades of the different geographic regions can be quite high, above 12% in the case of *Philander, Micoureus,* and both *Oryzomys* complexes. In each case, the samples from the southeastern Brazilian coastal forests are the most divergent phylogenetically (fig. 29.1). The geographic uniqueness of Atlantic forest representatives of taxa that are otherwise common and widespread mirrors an overall pattern of endemicity in the *Mata Atlântica;* for example, some fourteen genera of primates and rodents are unique to this region (Voss 1993). Both the high degree of endemicity and the sequence divergence (above 12%) evidenced between clades suggest that the *Mata Atlântica* has been isolated from Amazonia for a considerable period of time.

Our sampling points within the Amazon Basin are limited, but there is some consensus for cladistic relationships among the broad geographic areas illustrated in fig. 29.1. For example, a northeastern-southeastern Amazonian regional affinity is supported by analyses for *Micoureus, Philander,* and *Oryzomys capito;* data for *Didelphis, Metachirus,* and *O. macconnelli* are incomplete in this regard. Similarly, a northwestern-southwestern area relationship is supported for both *Micoureus* and *Oryzomys capito.* In general, the sharpest division within Amazonia lies between western and eastern areas for each taxon, with sequence divergence levels ranging from 8% to 12%. Consequently, the concordance of these geographic relationships and their temporal depth (based on sequence divergences) suggest a similar, if not common, underlying history for these regions. These findings also suggest that the trans-Amazonian differences in species diversity identified by Emmons (1984) might have a historical, as well as a present-day ecological, basis.

The depth of divergence across the entire 3,000 km west-to-east axis of Amazonia is mirrored by less deep, but still striking, divergences within more restricted geographic areas. For example, the Rio Juruá basin, one of the major river basins within southwestern Amazonia, is divisible into sharply defined upper and lower regions by the concordance of haplotype clades of four different genera of arboreal echimyid rodents, with sequence divergence averaging 6% in each (Silva and Patton 1993). This analysis has now been extended to include other genera (the echimyid *Proechimys* and the murids

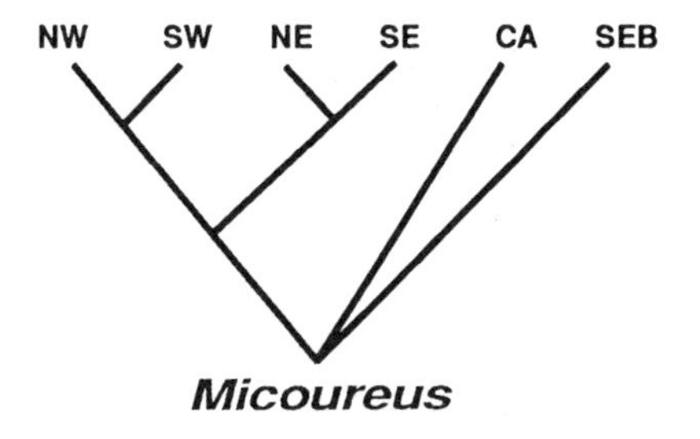

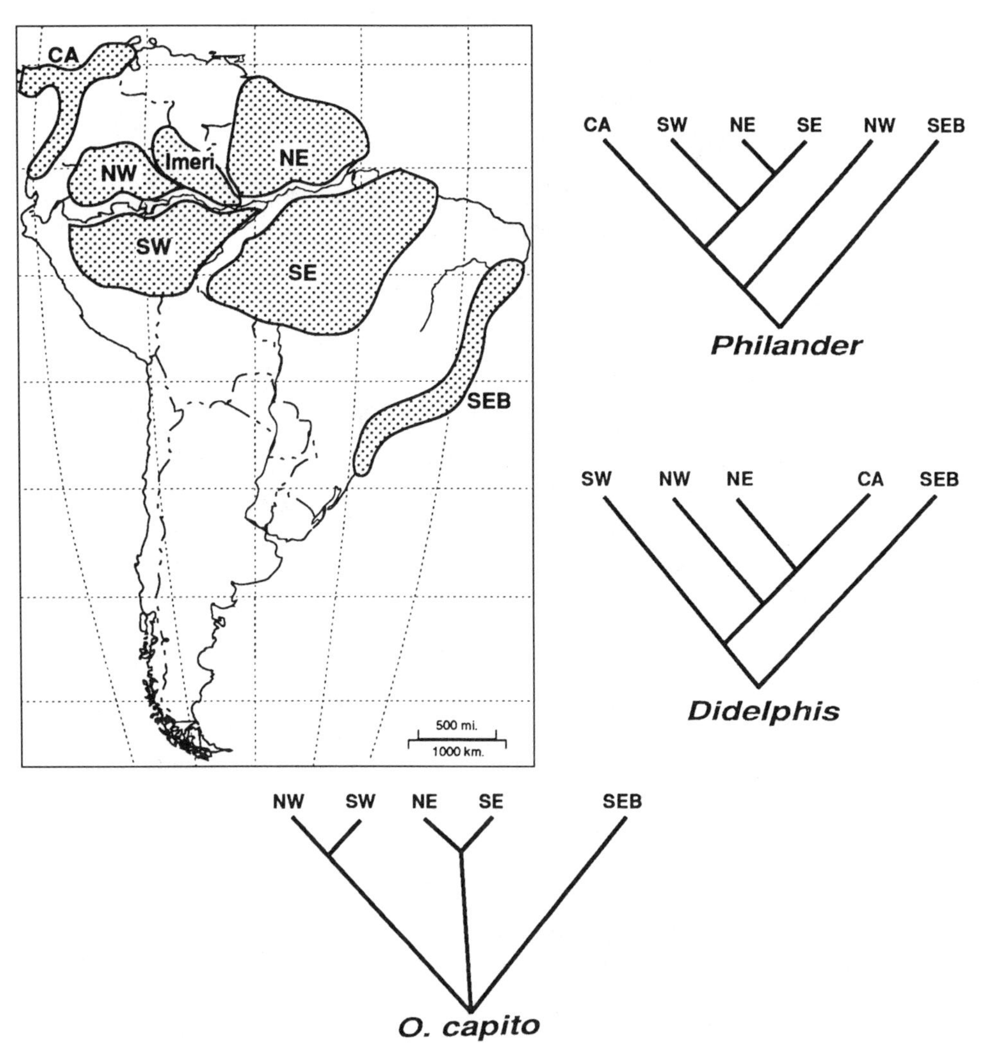

Figure 29.1. Centers of endemism hypothesized for lowland Neotropical forest birds (shaded areas on map: after Cracraft 1985, 1988; Cracraft and Prum 1988). The cladograms illustrate area relationships for samples of four Neotropical small mammals from the avian centers of endemism: the didelphid marsupials *Didelphis, Philander,* and *Micoureus* and the murid rodent *Oryzomys capito* complex. Each of these cladograms is derived from phylogenetic analysis of haplotypes of the cytochrome-*b* mitochondrial gene (from 630 to 801 base pairs of DNA sequence).

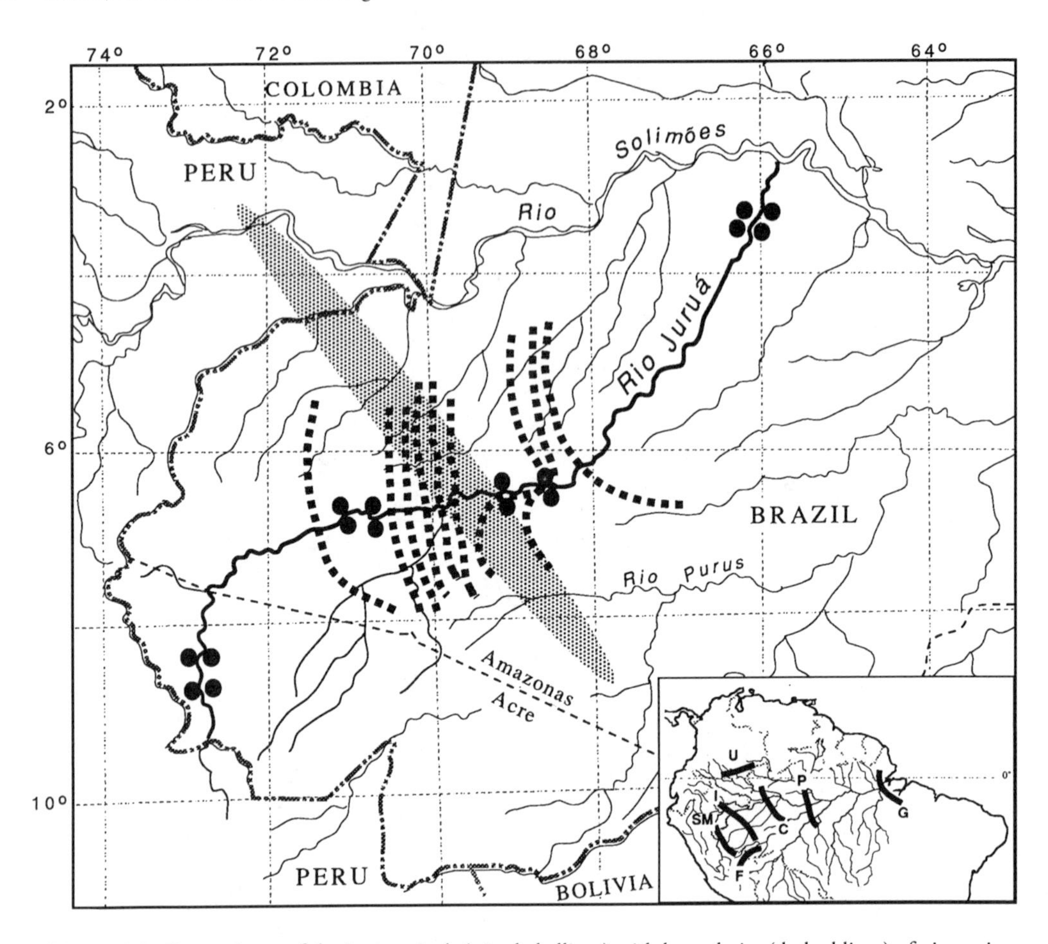

Figure 29.2. Concordance of the Iquitos Arch (stippled ellipse) with boundaries (dashed lines) of nine pairs of sister species or haplotype clades for taxa of echimyid *(Mesomys, Isothrix, Makalata, Dactylomys,* and *Proechimys)* and murid *(Neacomys* and *Oecomys)* rodents in the central Rio Juruá basin of the western Amazon (in part from Silva and Patton 1993 and M. N. F. da Silva 1995). Sample localities along the river are indicated by solid circles. *Inset:* Major structures and geomorphological zones of the Amazon basin, from west to east: SM = Serra do Moa arch, F = Fitzcarrald arch, I = Iquitos arch, U = Uaupés arch, C = Carauarí arch, P = Purús arch, and G = Gurupá arch (from Räsänen, Salo, and Kalliola 1987, 1992).

Neacomys and *Oecomys*), and the same pattern of abrupt transition between phylogenetic haplotype clades, with divergence levels in the neighborhood of 6% or higher, has been found repeatedly (fig. 29.2; M. N. F. da Silva 1995). Interestingly, the point of transition between clades coincides with a geological structure known as the Iquitos Arch, the presumptive tectonic product of Andean uplift that marks the boundary between the pre-Pleistocene Acre and Central Amazon fluvial deposition systems (Räsänen, Salo, and Kalliola 1987; Räsänen et al. 1992).

The extent and similarity of observed sequence divergence between geographic representatives of what have almost universally been considered single species have been unexpected findings. Divergence levels of 10% or more argue for much more ancient cladistic events than expected, certainly older than the commonly invoked forest refuge formation of the late Pleistocene (Haffer 1969; but see Haffer 1993). Indeed, the dissection of the present Amazon Basin by major structural arches (inset, fig. 29.2) resulting from Andean uplift suggests a common vicariant history that might underlie diversification patterns for organisms in general across Amazonia (see also Fjeldså and Rahbek, chap. 30). If this hypothesis is verified by subsequent analyses, then both lineage diversification and the resultant geographic patterns of species diversity of nonvolant small mammals may have a deeper history in the Amazon than previously thought. As such, the positions of paleobasins may signal present-day centers of endemism and diversity more strongly than those of the putative Pleistocene refugia suggested by other researchers (review in Brown 1987b). This perspective reinforces the recent view that Amazonian speciation, and thus geographic patterns of species diversity, cannot be explained entirely by any single model of vicariance or climatic change (Bush 1994). Rather, it is more likely that pre-Quaternary events established the major regional divisions of species complexes, with subsequent speciation or redistribution of existing species due to refuge formation (Haffer 1969), floodplain dynamics (Salo 1988b), and/or ecological heterogeneity (Tuomisto et al. 1995) superimposed upon this earlier pattern.

PROSPECTUS: CURRENT AND FUTURE DIRECTIONS

Patterson (1994, 85) closed his recent paper on Neotropical mammalian diversity by stating, "The question of 'how many species of Neotropical mammals are there?' can only be answered with far more study, both in the field and in the lab." While we agree completely, it remains clear that we are on the verge of a quantum leap in our knowledge of small mammals of Neotropical forests. A significant number of scholars, especially in Latin American countries, have become active in field and laboratory research within the last decade, and the efforts of their scholarship are beginning to bear fruit. Moreover, substantial synthetic efforts, such as that by Emmons and Feer (1990), have provided explicit direction and provoked focused field efforts, if only by pointing directly to areas (geographic and conceptual) where information is both missing and needed. We hope that the remarks above, and those that follow, will have a similar impact.

To our mind, there are at least two clear directions that must be incorporated into continuing, and future, research protocols. First, intensive and extensive local surveys should be initiated immediately. These surveys should follow specified protocols that adequately sample both terrestrial and arboreal components of communities (e.g., Malcolm 1991a) and target geographic regions of biogeographic importance (e.g., the transition areas across the structural arches identified in fig. 29.2). Direct observational diurnal and nocturnal transect surveys (like those of Emmons 1984) are important, but of limited value in areas where both the recognition and identification of species are often difficult. Consequently, as emphasized by Emmons and Feer (1990), properly prepared museum voucher specimens are still required. This is particularly true of the small-bodied marsupials and rodents, for which age and local variation is extensive. Without vouchers, species identifications will remain suspect, new taxa cannot be recognized, and critical revisionary work, even in relatively well known groups, cannot be accomplished. It is particularly important that conservationists, and especially agencies with regulatory authority, value systematic research and the inventories upon which that research is based. Such studies will yield much of the necessary data for effective management decisions.

Second, long-term demographic studies at many sites are required for understanding critical components of species habitat ranges and population dynamics as these relate to local persistence and structure. We lack basic information for most small-bodied species, including estimates of local abundance, seasonal and longer-term changes in density, habitat usage, individual movement patterns, food habits and the importance of individual species as seed predators or dispersers, and so forth. Assessments of habitat use, for example, and particularly knowledge of how individual species respond to secondary and disturbed habitats, are prerequisites for developing informative predictions of long-term population viability in the face of habitat fragmentation (see Malcolm, chap. 14). These studies should also include the examination of genetic variability using currently available molecular methods. We lack anything but the most rudimentary appreciation of the genetic structure of populations, including everything relating to how, and why, genetic variation is apportioned within and among local populations (see Patton, Silva, and Malcolm 1996). Each of these issues is important, not only for assessing the roles that individual species play in complex ecosystems (Eisenberg 1990), but also for understanding the species themselves as evolving entities, their immediately past history, and their long-term future. In the long run, our ability to conserve the mammalian evolutionary heritage of the lowland Neotropical forests rests on the compilation of this very diverse information.

GENERAL IMPLICATIONS

1. A major aim of this chapter has been to emphasize the inadequacy of our knowledge of small mammal species in Neotropical wet forests and thus to encourage the continuation, or initiation, of species inventories with their resultant systematic and ecological analyses. We cannot preserve what we don't know—either taxa that are as yet unrecognized by the scientific community or those that are unknown as members of local faunas.
2. Although we have very general notions about a west-to-east gradient in local species abundances across Amazonia, our knowledge of community structure is too poor to assess the adequacy of existing reserves as foci for the long-term preservation of Neotropical fauna. In the main, we also lack sufficient information to suggest the locations of additional areas for protection.
3. Nevertheless, it seems clear that the Atlantic coastal forests of Brazil contain many phylogenetically divergent taxa, and that protection of the existing remnant patches in this very threatened region should continue to receive high priority (see Viana, Tabanez, and Batista, chap. 23).
4. Within Amazonia, congruent patterns of taxon diversification suggest an important role for vicariance between paleodrainages in the upper Amazon Basin. If future studies show similar patterns elsewhere, these geologically defined paleobasins will represent good targets for conservation emphasis.
5. Finally, the magnitudes of molecular sequence divergence we have recorded among lineages of quite different taxa are far greater than expected for the vicariant events of late Pleistocene refuge formation. Rather, our data suggest that pre-Quaternary events estab-

lished the basic levels of small mammal diversity, which were modified subsequently by the climatic and ecological events of the Quaternary. Thus, Pleistocene refuges are perhaps a better model for explaining current distribution patterns than for identifying centers of diversification.

ACKNOWLEDGMENTS

Our thanks go to those individuals who have helped us to accumulate an understanding of small mammal diversity in South America, especially Al Gardner, Louise Emmons, Rob Voss, Jay Malcolm, and Carlos Peres. Bill Laurance and Craig Moritz invited us to participate in this volume and waited more than patiently for us to write this chapter; we hope we have not done them too much of an injustice. We also thank our colleagues Albert Ditchfield, Liz Hadley, Eileen Lacey, Manuel Ruedi, and Peg Smith for helping to shape our ideas. Finally, we thank those agencies that have provided educational and research support: fellowship support from the Conselho Nacional de Desenvolvimento Científico e Tecnológico (CNPq) to M. C. L. and M. N. F. S. and from the Fundação de Amparo à Pesquisa do Estado de São Paulo (FAPESP) to M. A. M., and financial assistance for field and laboratory research from the Museum of Vertebrate Zoology and grants from Wildlife Conservation International, the National Geographic Society, and the National Science Foundation.

30

Species Richness and Endemism in South American Birds: Implications for the Design of Networks of Nature Reserves

Jon Fjeldså and Carsten Rahbek

In conservation biology the focus has recently shifted from individual species to entire biomes. The global strategy for conserving biodiversity (McNeely et al. 1990) assigns high priority to "areas with particular species richness and endemism." The inherent conflicts between these two criteria are noteworthy. Geographic patterns of species richness may reflect current carrying capacity, determined by energy and habitat flux, while endemism to a much larger extent reflects evolutionary events (Fjeldså 1994, 1995). Past histories are relevant to current initiatives to conserve species and sustainably manage natural resources because many such histories reflect intrinsic natural properties of specific areas, which may influence biological diversification as well as human livelihood.

Computer programs have recently been developed for automated priority area analyses that integrate information about how the phylogenetic tree of a certain group is represented in different areas (see reviews in Williams and Humphries 1994; Williams, Gaston, and Humphries 1994). The computer software WORLDMAP (Williams 1994) is designed to identify areas with the highest taxic diversity (numbers of species and higher taxa) and networks of sites with complementary sets of species. However, the lack of assumptions about processes underlying the geographic patterns makes it unclear whether this method of optimizing reserve locations is an adequate tool for minimizing extinction (Fjeldså 1994). Species that evolved under special local conditions may require management strategies different from those that survived in a dynamic landscape by responding opportunistically to shifting conditions.

Other pragmatic approaches to reserve design that were developed without knowledge of underlying environmental processes were the identification of botanical "hotspots" (Myers 1990) and Endemic Bird Areas (ICBP 1992). These studies identified areas where many species face strong extinction risks simply because of exceptional local concentrations of unique species. However, Brooks, Mayden, and McLennan (1992) and Erwin (1991) propose that highest priority should be given to areas important for the process of biological diversification. Anticipating a contemporary mass extinction of species, they want to identify the best places for biological reconstruction.

In this chapter, the WORLDMAP software is used to compare distributional data for 1,114 species of Neotropical birds of primarily forest-dependent families with the distribution of existing protected areas. This comparison has been performed for the entire South American continent, and is supplemented with a detailed study of the area we believe to be the most important for evolution of new bird species: the Andes. We describe a balanced strategy for reserve selection that considers inherent differences between areas that are important for (1) the differentiation of new species and (2) the survival of old ones. We emphasize the relationship between endemism and the ecoclimatic properties of centers of endemism, and identify major gaps in the existing protected areas system. Unfortunately, the data sets now available for taxonomic groups other than birds are of a poorer quality, and it is therefore not yet possible to document the universality of our results.

METHODS AND MATERIALS

Analytical Approach

WORLDMAP (version 3.18/3.19) is a PC-based graphical tool originally designed for fast, interactive assessment of priority areas for conserving biodiversity (Williams 1994). The program accommodates data on distributions of large numbers of species. Relative differences in endemism can be expressed in an objective way by adding up the rarity scores for each species (i.e., the inverse of the number of grids with records) per grid cell (see Usher 1986; Williams 1994 for details). These range-size rarity scores do not represent a perfect measure of faunistic uniqueness, as they reflect numbers of endemic species as well as numbers of widespread species. However, this bias will not affect the conclusions drawn in this chapter.

Among several options for identifying conservation priorities contained in WORLDMAP, we used the principles of gap analysis, employed as a nationwide conservation evaluation program in the United States and elsewhere (Scott et al. 1993). By comparing distributions of large numbers of species with protected area networks, this analysis locates gaps in the networks and identifies a fully representative reserve network. Some of the assumptions and limitations of gap analysis were summarized by Butterfield, Csuti, and Scott (1994). A limitation of the present continent-wide analysis is that only a rather crude geographic resolution is possible (1° grid baseline map). When such maps are superimposed in WORLDMAP, a very high beta (between habitat) diversity may result where the 1° grid units span major ecotones or altitudinal gradients, especially along the Andes. However, we regard it as being outside the scope of this chapter to analyze how much of the variation in species richness can be explained from environmental amplitudes within the individual grid units.

Information about existing protected areas was mainly derived from IUCN (1992). These protected areas were established over several decades, mainly on an ad hoc basis, and they differ greatly in extent and legal status and in how well the biological resources are managed. Many areas were established to support indigenous people, others mainly because of cultural heritage or scenic landscapes. Although good species lists exist for some protected areas, in most cases all we can do is assume that species inhabiting a certain grid unit are present inside protected parts of this unit. We did not find it feasible to examine in detail which of the individual reserves are managed well enough to maintain biodiversity. Instead, we based our analysis on the assumption that species are well protected if they occur in grid units with (1) three nominally different reserves, (2) formal protection of at least one-third of the unit's area, or (3) effective protection of the biologically most unique parts of the unit. We identified 253 such grid units, which we hereafter refer to as the *existing network of well-protected areas.* We also make the assumption that a species is "safe" if it is represented in at least three of the well-protected grid units.

New areas proposed for biodiversity management consist of irreplaceable grids, new flexible grids, and flexible grids from ties (see Williams 1994 for details). The first two categories (irreplaceable and new flexible grids) make up the minimum set of areas needed to achieve the conservation goal (i.e., a network of biodiversity-managed grids that includes all species in at least three grids). Irreplaceable grids are defined as those containing species whose total range falls within one to three grids. New flexible grids are areas that could be exchanged for other areas, although this may require larger sets of areas than the minimum set. Flexible areas from ties are the most likely alternatives to identified flexible grids if biodiversity management is impossible within the latter areas due to conflicts with other interests (e.g., development interests).

Database on Neotropical Birds

For continent-wide bird distributions, we used the breeding ranges of baseline maps prepared by R. S. Ridgely and W. L. Brown and kindly placed at our disposal by the Academy of Natural Sciences of Philadelphia. These maps show verified records and an estimated range outline for each species. Although they had already been revised by a number of persons, we checked the maps again, and included data gathered by Fjeldså and Krabbe (1990) and Collar et al. (1992), as well as those on restricted-range species in the databases of BirdLife International, and a database on central Brazilian birds compiled by J. M. Cardoso da Silva. When adjusting the maps, we assumed that species were continuously present between collecting points unless existing habitat maps or an absence of records from well-studied sites suggested distribution gaps.

Our study focused on forest birds. The definition of forest birds, however, is not straightforward. Rainforest biomes may well include species of nonforest habitat. Many birds that inhabit forest habitat can also live at edges and in woodland thickets, or colonize such habitats in part of their range. J. M. C. da Silva (1995b) has demonstrated that 393 of the 837 breeding birds of the Brazilian Cerrado, a savanna biome, are in fact forest dependent and are associated with gallery forests or forests on patches of nutrient-rich soils. Because of such problems of definition, we found it most meaningful to analyze large taxonomic groups that have radiated mainly in rainforest biomes, irrespective of whether some of the species have expanded into other habitats. Because this chapter was prepared before all South American bird distributions had been digitized, we used data only for hummingbirds (Trochilidae); trogons (Trogonidae), kingfishers (Alcedinidae), and motmots (Momotidae) (viz., all Coraciiformes); jacamars (Galbulidae), puffbirds (Bucconidae), barbets (Capitonidae), toucans (Ramphastidae), and woodpeckers (Picidae) (viz., all Piciformes); cotingas (Cotingidae); manakins (Pipridae); tyrant flycatchers (Tyrannidae, including the fluvicoline radiation, which has adapted to open land); and tanagers (Thraupinae; including *Tersinia*). These families and orders were analyzed individually, and finally all 1,114 species were lumped together. Examination of the not yet fully processed data on some other large forest-adapted groups, such as antbirds (Formicariidae) and wrens (Troglodytidae), indicates that adding those groups would not substantially alter the results of our analysis.

The geographic scale used in the continent-wide analysis is too coarse to assess whether a species identified as "safe" actually lives within the boundaries of biodiversity management zones. More detailed biogeographic patterns are shown below for the region that probably has the highest degree of local differentiation of species, namely, the montane (temperate zone) forest and mist vegetation in the Andes of Ecuador, Peru, and Bolivia. This latter analysis included taxa that had a distribution of less than 50,000 km^2, and employed the full data set of ICBP (now BirdLife International), with additional information gathered by teams from the Zoological Museum of Copenhagen. In this analysis our operational concept of endemism differed from that used by ICBP in that we accepted as separate units well-marked local forms (evolutionary species currently classified as subspecies of a biological species; see Fjeldså 1995 for taxa included). We also examined populations and taxa that can be regarded as geographic *relicts,* in the sense that they comprise widely disjunct populations or are separated by a significant geographic gap from related taxa (Cronk 1992).

To discuss our results in relation to processes underlying the biogeographic patterns, we review below the results of recent studies (Fjeldså 1992, 1994, 1995; Fjeldså and Lovett, in press). An important new methodology was the use of an enormous data set of interspecific DNA hybridizations (Sibley and Ahlquist 1990) for a preliminary differentiation of groups that diversified rapidly during the last few million years (Pliocene/Pleistocene) from those representing older phylogenetic branches. Comparisons of the biogeographic patterns of these groups can be used to identify "evolutionary fronts," as well as those areas where species (of potentially diverse origins) simply accumulated over long time spans. An ecoclimatic comparison of different areas can then be used to characterize those that are important for biological diversification.

RESULTS

Continent-Wide Distributions and Conservation Status of Bird Species

Hummingbirds. The hummingbirds (fig. 30.1) show a strong association with the humid parts of the tropical Andes region. This is particularly clear when endemism is considered, as the majority of species inhabiting the Amazon lowlands and the savanna woodlands are widespread (this applies quite generally in birds, as reflected by the very low number of Amazonian and Cerrado birds meeting ICBP's [1992] criteria of endemism). The current network of well-protected areas includes 220 species (89.3% of all species). A minimum of 15 additional grids are needed to add the last 23 species, and 51 additional grids are needed if all species are to be included in at least three well-protected

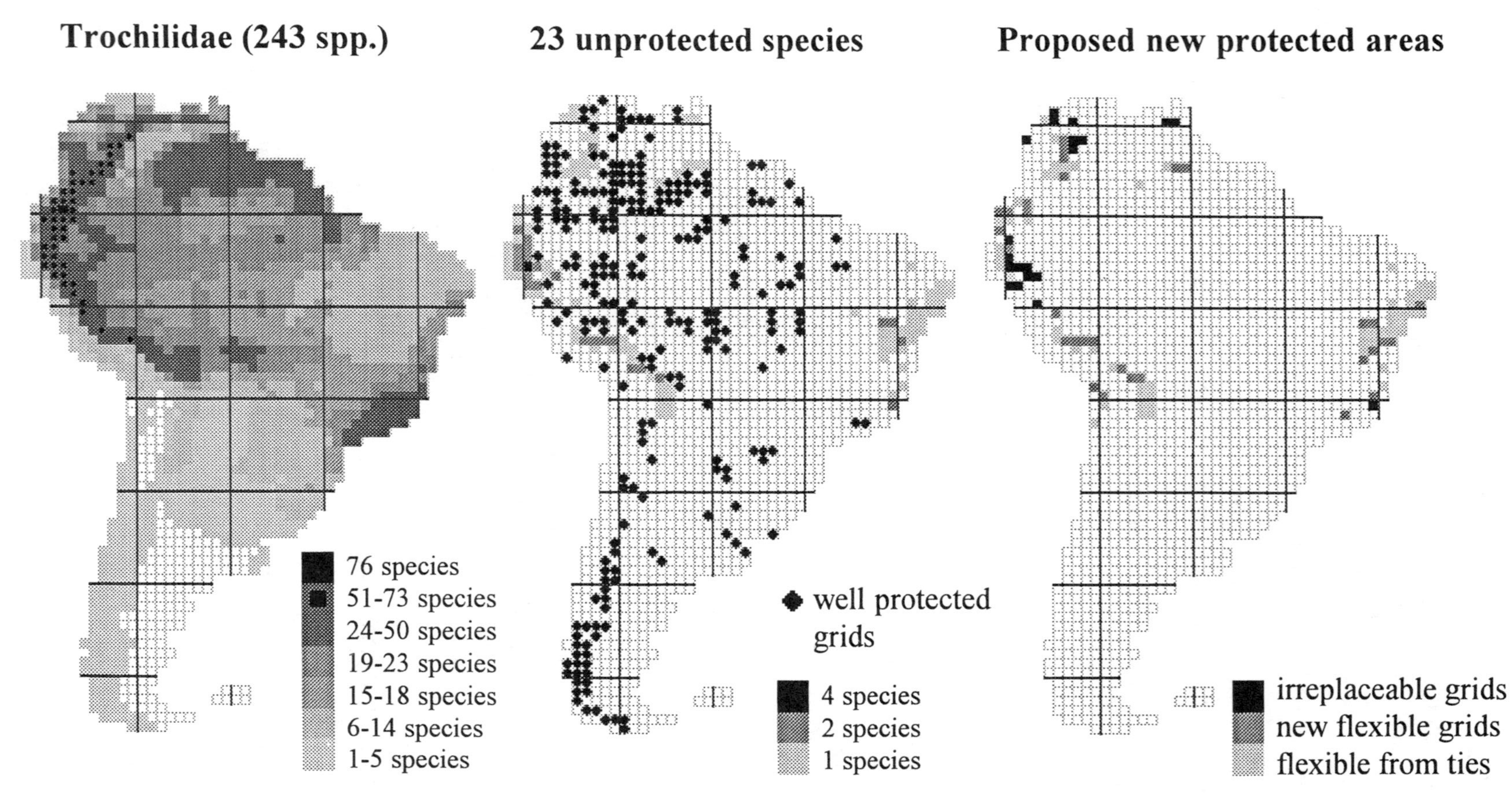

Figure 30.1. *(A)* Species richness for hummingbirds (Trochilidae); *(B)* distributions of 23 species not covered by the current network of well-protected areas; *(C)* new protected areas needed to include all species in at least three well-protected grids.

areas. Although the total number of optional grids (fig. 30.1C) is 111, there are only twenty irreplaceable grids, located in Sucre, Venezuela; in the Perija Mountains on the Venezuelan/Colombian border; on the Caribbean coast and in the eastern Andes of Colombia; in the western Andes in Ecuador into northern Peru; and one grid on the southern Espírito Santo coast of Brazil (an irreplaceable grid in northwestern Colombia is an artifact caused by introgression of a Central American species).

Coraciiform and Piciform Birds. Both the Coraciiformes and the Piciformes are best represented in lowland rainforest (fig. 30.2), with the Piciformes showing elevated species richness in grid units intersected by major rivers in the Peruvian/Ecuadorian portion and below Manaus in the Amazon Basin, reflecting displacement of closely related species across the rivers. Both groups are also well represented in gallery forests, thickets, and woodlands in the savanna regions (the genus *Colaptes* has even adapted to open land). Few genera are adapted to montane forest. Because few species have restricted distributions, all of the Coraciiformes are within at least three protected grids. Among the Piciformes, 6 species were not included in the existing network of well-protected areas, and a minimum of 23 additional areas need protection to include all species in at least three well-protected grids. Seven irreplaceable grid cells (fig. 30.2C) occur in four locations: the Orinoco Delta; Amazonas/San Martín; sub-Andean forest in Puno, Peru; and southern Rondônia in Brazil.

Cotingas and Manakins. Cotingas (fig. 30.3) are widespread in lowland and montane rainforests, but a concentration of species occurs along the eastern Andean slope. The manakins (not illustrated) are widespread in tropical lowland and submontane forest, with a distinctive concentration of species near escarpments north and south of the Amazon lowlands. Both groups, particularly the manakins, show increased species richness in grids intersected by major branches of the Amazon River. The existing network of well-protected areas leaves 6 of the 60 cotingas, but none of the 52 manakins, unprotected. At least 15 additional grids need protection to include all species of cotingas in at least three well-protected grids; seven irreplaceable grids (fig. 30.3C) occur near Rio de Janeiro and in Andean treeline habitat. For manakins, one additional site is needed in the Venezuelan Tepuis and another farther inland from the Brazilian Atlantic forests to have all species represented in at least three grids.

Tyrant Flycatchers. Tyrant flycatchers (fig. 30.4) show a wide distribution, with some subgroups adapted to open terrain. In contrast to other groups, their peak diversity is on the wet eastern slope of Cordillera Real in southeastern Peru. Fifteen species are not covered by the existing network of well-protected areas. Some extensive shaded areas in figure 30.4C refer to only four species of open land birds: *Polystictus superciliaris* in the Brazilian highlands, *Pseudocolopteryx dinelli* and *Xolmis salinarum* in the Chaco, and *Neoxolmis rufiventris* on Patagonian steppes. A minimum of 45 additional areas need protection to include all species in at least three well-protected grids. Ten irreplaceable grids are distributed as follows: one in the Andes of northern Ecuador; four in northern Peru; one in the Andes of central Peru; and four in Brazil, east of Rio Madeira, in Alagoas, and in Caiapó in Mato Grosso.

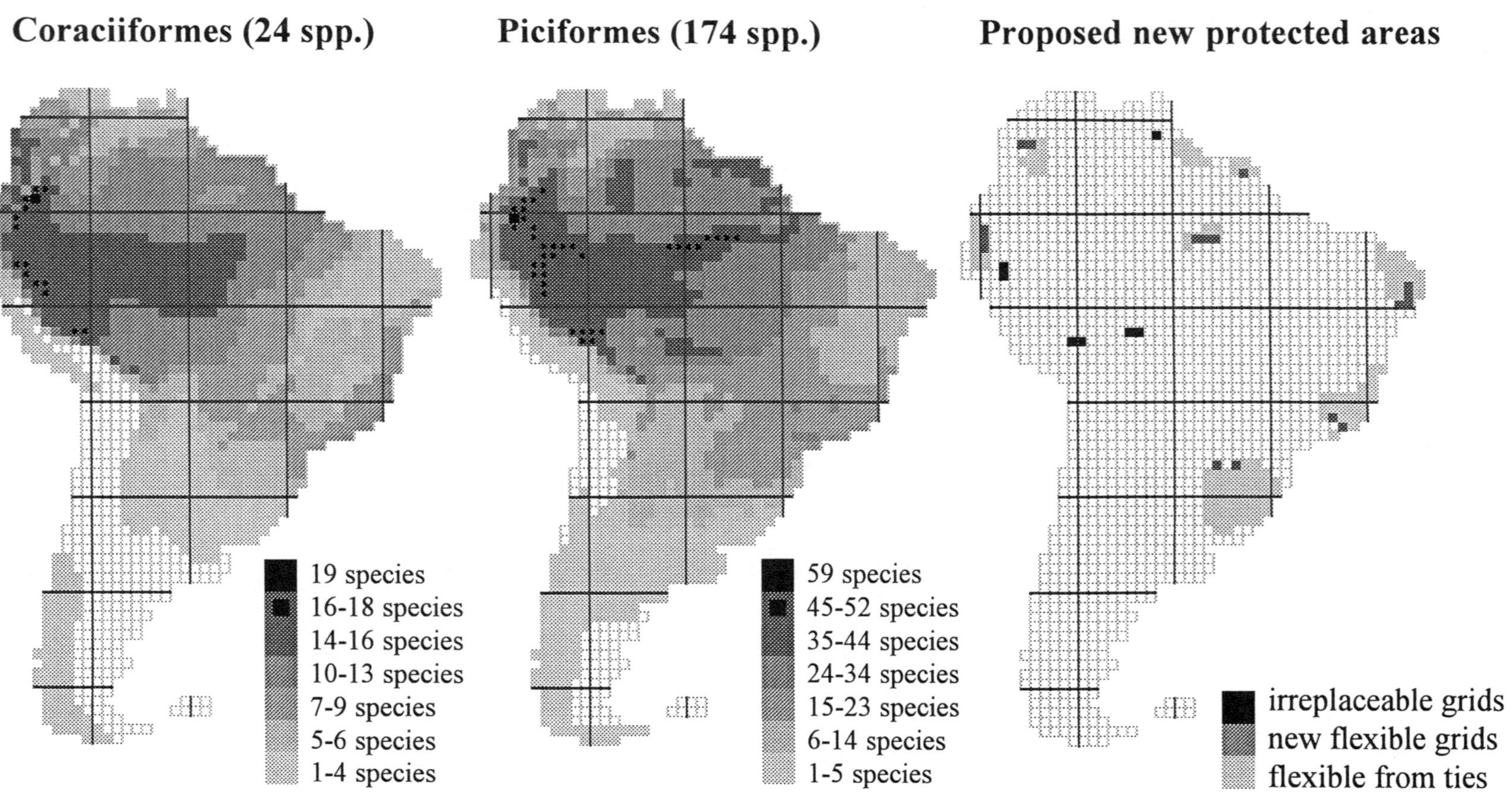

Figure 30.2. Species richness for *(A)* trogons, kingfishers, and motmots (Coraciiformes); *(B)* species richness for jacamars, puffbirds, barbets, toucans, and woodpeckers (Piciformes); *(C)* new protected areas needed to include all 198 species in at least three well-protected grids.

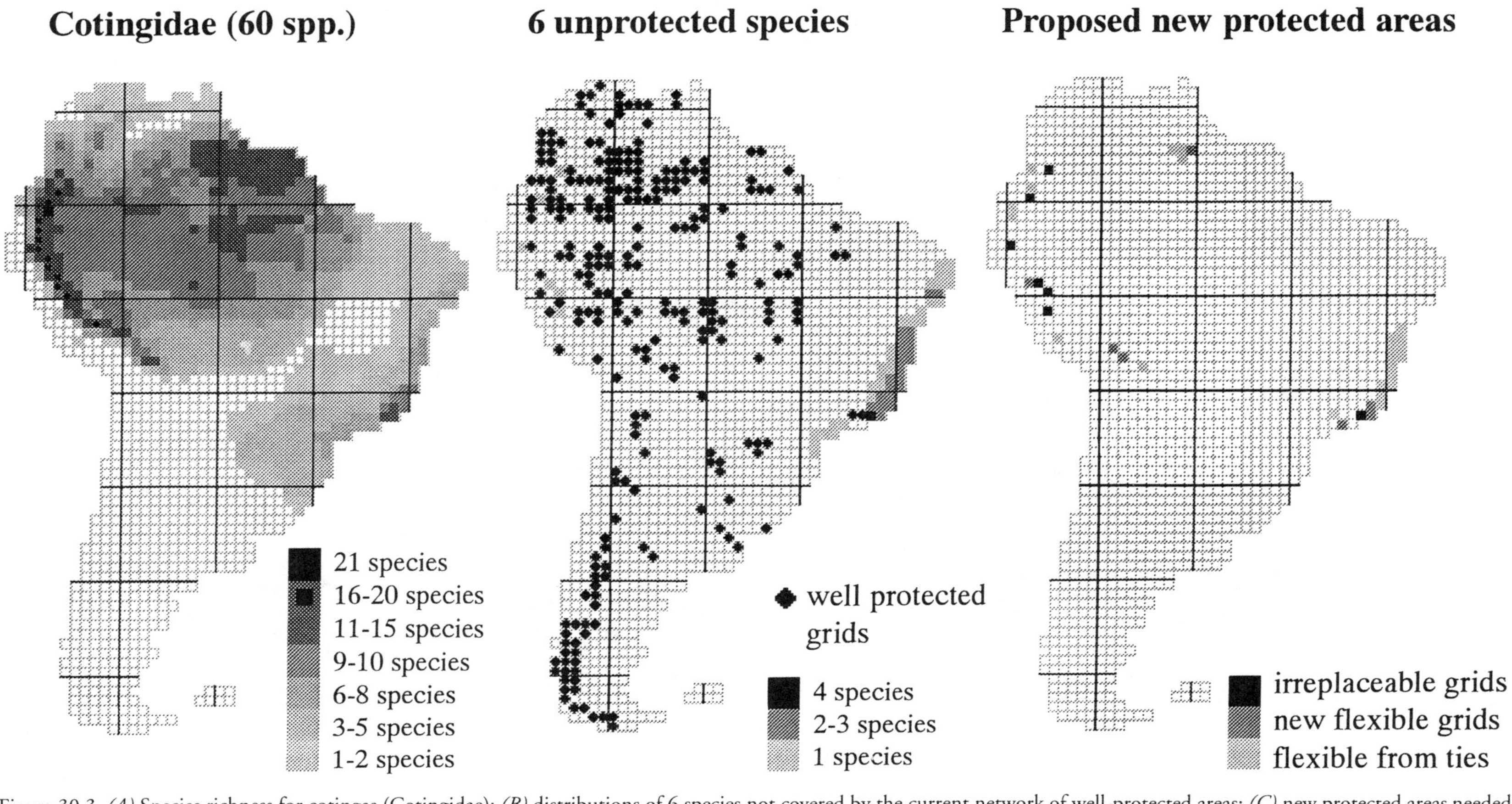

Figure 30.3. *(A)* Species richness for cotingas (Cotingidae); *(B)* distributions of 6 species not covered by the current network of well-protected areas; *(C)* new protected areas needed to include all cotingas in at least three well-protected grids.

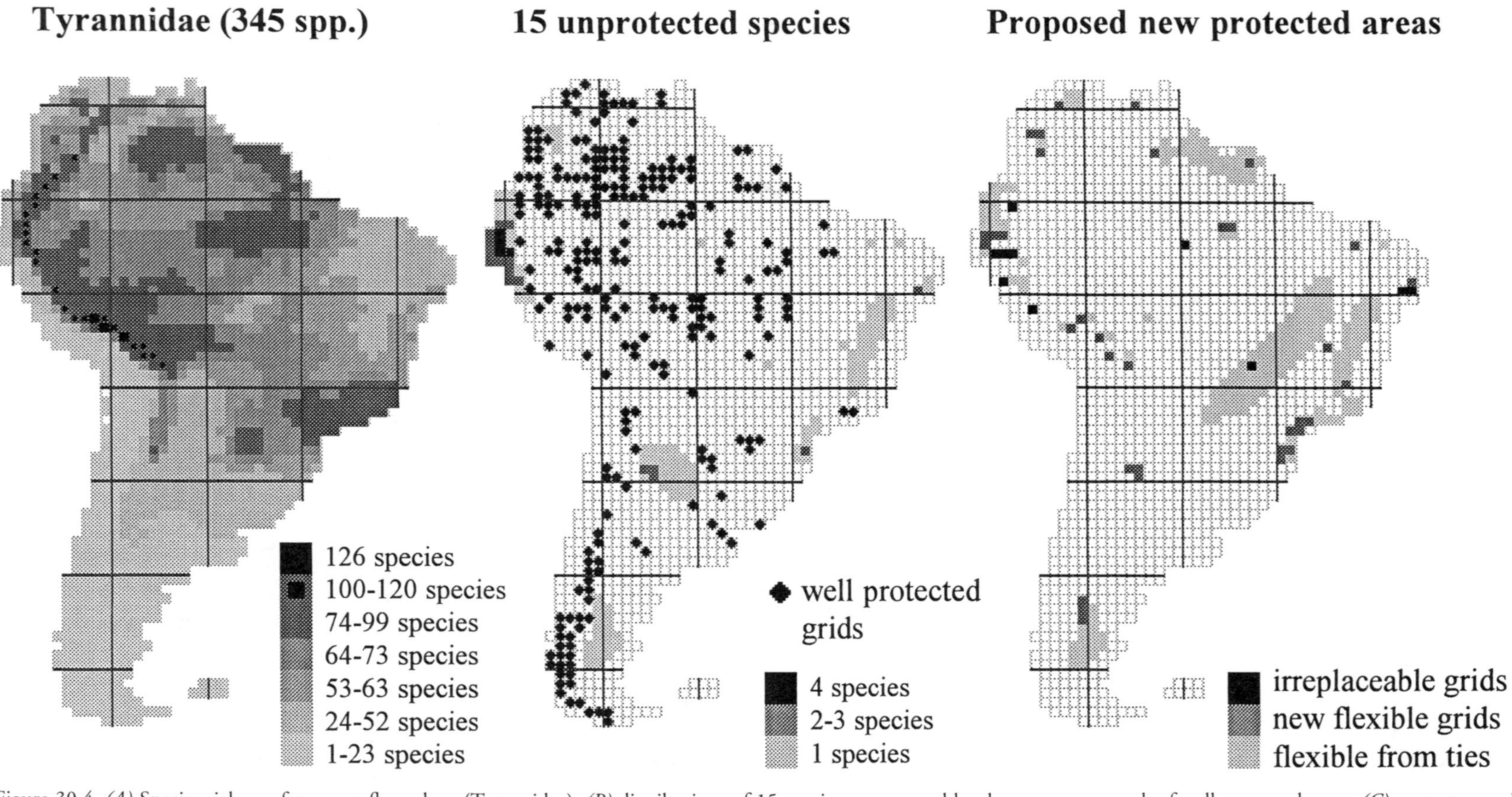

Figure 30.4. *(A)* Species richness for tyrant flycatchers (Tyrannidae); *(B)* distributions of 15 species not covered by the current network of well-protected areas; *(C)* new protected areas needed to include all species in at least three well-protected grids.

Tanagers. Tanagers (fig. 30.5) exhibit high species richness in rainforests and also fairly high numbers in forest and woodland mosaics in the Brazilian highlands. Most species are members of mixed feeding flocks that move around in the forest canopy, which may explain their tendency to have wide distributions. Some smaller distributions, however, are found in the Andes and in the Brazilian Atlantic forests. Ten species are not covered by the existing network of well-protected areas. A minimum of 37 additional grids need protection to include all species in at least three well-protected areas, with nineteen irreplaceable grids located in Sucre, Monagas, Cordillera de Mérida, and the Perija Mountains of Venezuela; in western Colombia; at the Ecuadorian/Peruvian border, around the upper Huallaga River, and in Cuzco, Peru; and south of Rio Madeira, in the north of the Pantanal, and in southern Espírito Santo, Brazil.

The Adequacy of the Protected Areas Network in the Andean "Species Factory"

The tropical Andean region is a key area for evolution of new bird species in South America. Twenty-eight Endemic Bird Areas (EBAs) in this region contain 4 to 44 strictly confined species each (ICBP 1992). Analyses at a finer scale than the continent-wide analyses above show that typical montane (temperate zone) forest birds representing recent radiations are aggregated into several EBAs (fig. 30.6A). To illustrate the degree of complementarity of different EBAs, separate symbols were used for adjacent sets of species. The resulting map (fig. 30.6A) clearly illustrates the degree of aggregation of those neoendemic species that make up EBAs. It should be added that additional EBAs exist in the foothill zone, such as in the submontane rainforests of northwestern Ecuador, on subtropical ridges north and south of the Marañón depression in northern Peru (mainly the same species are present on either side of this assumed barrier), and in the Bolivian foothills. Most of the representative species of the Tumbes center of endemism (for names of centers, see Fjeldså 1995) of southwestern Ecuador and adjacent Peru inhabit the low foothills, and other areas of endemism occur in the Peruvian coastal desert. Species with relict distributions (fig. 30.6B) also cluster in specific areas, giving a pattern very similar to that in figure 30.6A.

Clearly, the planning of current protected areas (shaded areas in figure 30.6, which include those in the sub-Andean zone) could not have been based on knowledge of the geographic distribution of endemic species. Fortunately, many areas along the eastern Andean slopes are virtually uninhabited by humans because of steep terrain and inhospitable climate. However, the treeline zone, which is particularly important for the differentiation of species, nearly everywhere is strongly modified by burning for grazing. Thirty of the 176 species included in figure 30.6 may not exist inside any formally protected area.

In Bolivia, a large protected zone near the Cochabamba-Santa Cruz boundary (Amboró, Sajta Ichilo, El Pirai, and Río Grande Masicurí; altogether 12,000 km^2) covers continuous forest from treeline to foothills. However, it is situated on the ecotone between two EBAs, and therefore protects marginal populations rather than core areas for restricted-range species (see Fjeldså 1993 regarding the efficiency of placing biodiversity conservation zones in the centers of endemic bird areas). The planning of natural

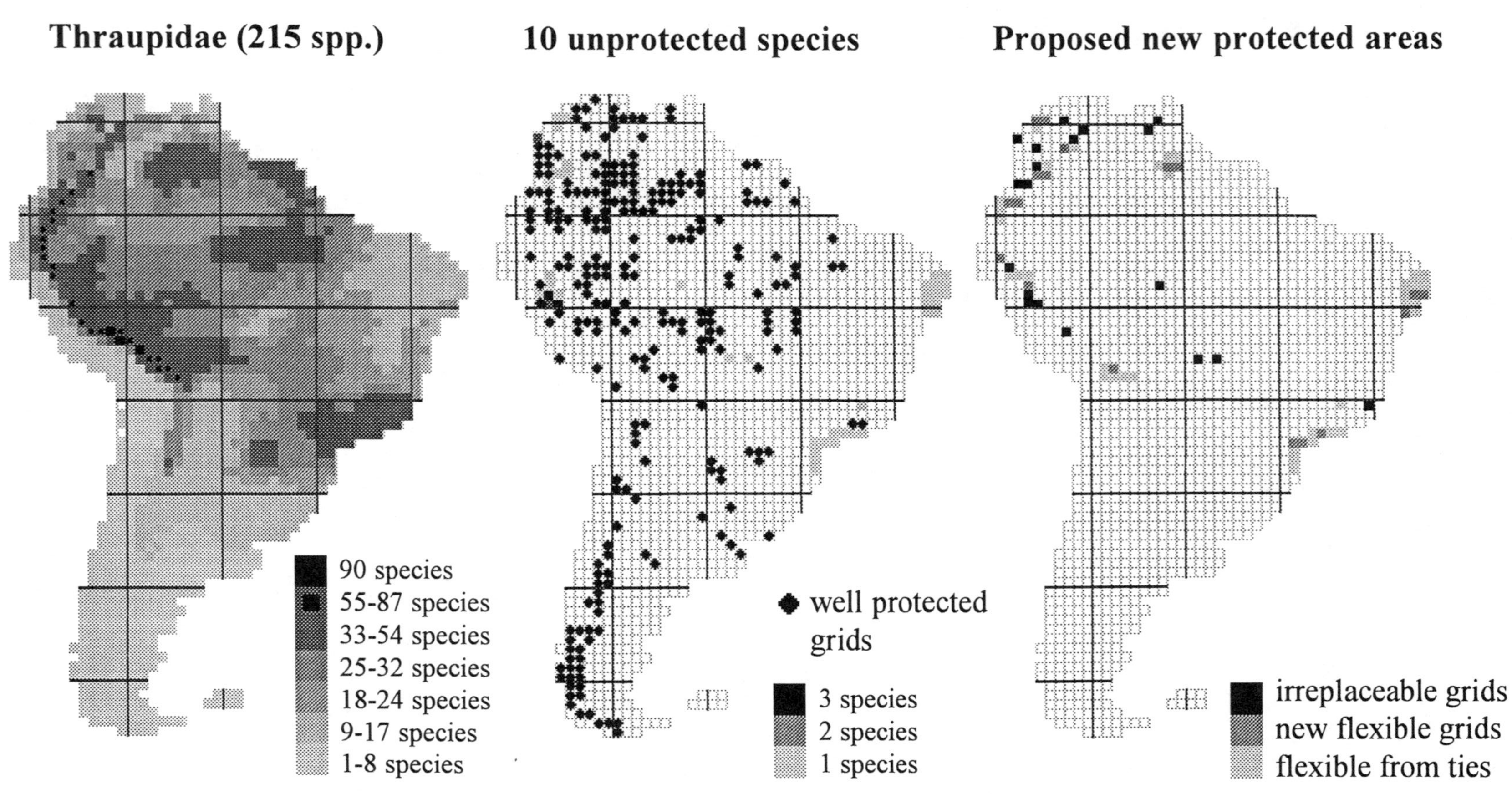

Figure 30.5. *(A)* Species richness for tanagers (Thraupinae); *(B)* distributions of 10 species not covered by the current network of well-protected areas; *(C)* new protected areas needed to include all species in at least three well-protected grids.

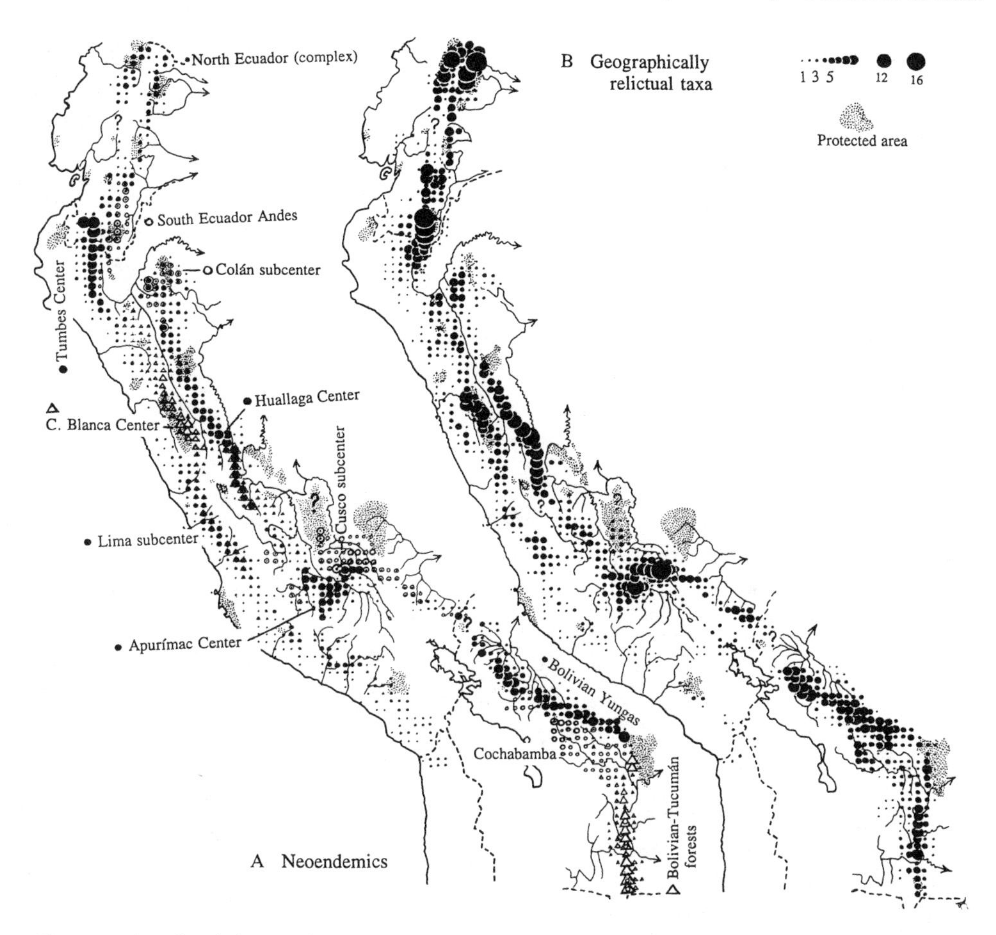

Figure 30.6. A detailed view of endemism in the Andes of Ecuador, Peru, and Bolivia, compared with the distribution of protected areas (shaded). *(A)* Variation in numbers of bird species of montane rainforest and mist vegetation that have restricted ranges (less than ca. 50,000 km^2) and are part of vicariance patterns in groups considered to represent strong Pleistocene radiations (separate symbols are used for adjacent sets of endemic species). *(B)* Distribution of species with relict distribution ranges. (Redrawn from Fjeldså 1995, in which the species are subdivided on additional maps and all 176 taxa used are listed.)

resource management is still in its initial stage in the "Valles" of southern Bolivia, with the Tariquía National Reserve (2,469 km^2) being the only fully established protected area. Because of accelerating habitat degradation, biodiversity management zones are urgently needed in the large tracts of semi-evergreen forest that still exist in Montes Chapeados and in the unique montane woodland patches in Cochabamba.

In Peru, Huascarán National Park (3,400 km^2) combines wild mountain scenery and aggregates of endemic species. Only two small protected areas exist (but do not cover the biologically most unique habitats) in the Cuzco and Apurímac centers of endemism. The large Río Abiseo (2,745 km^2) and Manu (15,328 km^2) National Parks cover a wide spectrum of habitats from treeline to lowlands, but are unfortunately placed between, rather than inside, montane centers of endemism (the Manu park, however, extends into a lowland area with several endemic species). The protected cloud forest habitats on the Pacific slope of northern Peru mainly contain widespread species, but the tiny and biologically unique cloud forests in the Lima subcenter have no protection at all. The low-altitude areas of endemism (upper Marañón Valley, coastal Peru) have minimal protection.

In Ecuador, the protected areas system covers the majority of species in the main Andean chains. The Podocarpus National Park (1,462 km^2) includes all endemic birds found along the continental divide in the south (Rahbek et al. 1995). However, there is a catastrophic lack of areas designated for biodiversity management in the isolated mountain ridges and subtropical foothills in the southwest, as described by Best (1992). Fortunately, the Ecuadorian dry forest type is protected in adjacent Peru, in Cerros de Amotape and El Angolo (altogether 1,563 km^2).

Combined Species Richness

Plates 3A and 3B combine all groups from figures 30.1–30.5. The gradual increase in species richness along the Amazon River from the lowest part to the base of the Andes is apparent. When carefully corrected for area effects, the regional species richness of South American land birds peaks at 500–1,000 m elevation (Rahbek 1995*a*,b). Endemicity (plate 3B) is decidedly highest in the Andes region, reflecting the mosaic of interlocking Endemic Bird Areas (the centers of all EBAs are marked with dots in plate 3B; see ICBP 1992; Long et al. 1996). High levels of endemism are also found in the isolated lowland areas to the north, in the Guianas, and in southeastern Brazil, but the levels are relatively low in most of the Amazon Basin. The enormous total species richness of the Amazon Basin is therefore partly a consequence of its huge extent. Although closely related species often replace each other in different parts of the Amazon Basin, the individual species forming such vicariant patterns usually have large ranges compared with Andean local endemics, and these ranges often encompass several protected areas.

DISCUSSION

Characteristics of Species-Rich Areas and Those Representing Evolutionary Fronts

The pattern of branch lengths in DNA-based phylogenies (Fjeldså 1994, based on data in Sibley and Ahlquist 1990) reveals that the extraordinary avian species richness of South America evolved abruptly since the Late Miocene period. When biogeographic and DNA divergence data are integrated (see Fjeldså 1994 for methods), it is apparent that much of the avian speciation in the last few million years has occurred in the tropical Andes region (plate 3C). These mountain ranges were uplifted in the Late Miocene, blocking the earlier outlet of the Amazon River into the Pacific. Tectonic changes east of the Andes might also have provided opportunities for speciation, for example, by creating hydrologically unstable lowland plains that isolated populations adapted to well-drained habitats (Hanagarth 1993; Patton, Silva, and Malcolm 1994; J. M. C. da Silva 1995a). However, it is apparent that the species richness of the hydrologically dynamic lowland plains (Orinoco Basin, lower midsection and Ucayali fluvial mosaics of the Amazon, and Chaco) is caused mainly by the persistence of old (and widespread) species, rather than by the differentiation of new ones (plate 3C). The traditional view, that the diversity of species in the lowland rainforest arose as a consequence of isolation caused by cyclic climatic/vegetational changes during the Pleistocene (see Whitmore and Prance 1987 for reviews), is contradicted by the simple fact that most species there are older than the Pleistocene (Amorim 1991; Fjeldså 1995).

Salo (1988a), Kalliola, Puhakka, and Danjoy (1993), and Tuomisto (1994) have thoroughly documented how the complex tectonic history of the Amazon Basin and the continuous horizontal migration of river meanders within the floodplains have maintained a dynamic mosaic of different successional stages, each with different biological communities. Climatic changes may compound this disturbance regime; whether or not historical climatic-vegetational changes were direct causes of speciation, they may at least have increased the rate of habitat turnover. In addition, the Orinoco and most of the Amazon lowlands are affected by occasional drought years caused by the El Niño-Southern Oscillation (Philander 1989). Although these droughts are brief, their relative frequency, erratic periodicity, sudden onset, and severe effects on the biota (Meggers 1994) may affect life history adaptations such as ecological tolerance and dispersal potential.

Although differentiation of new species has several causes, we emphasize here the potential role of specific stable places within the highly dynamic Andean landscape mosaic, where the ecological conditions remained relatively unchanged over time. Because of dramatic mountain formation, the region as a whole is characterized by great ecoclimatic variation. However, the local aggregates of relict forms (plate 3B) suggest that specific sites provided permanently suitable ecoclimatic conditions for the species living there, at least in the late Pleistocene (see Fjeldså 1992, 1995). A correlation between distributions of relict forms (fig. 30.6B) and endemism resulting from vicariant distributions of closely related species (fig. 30.6A) would seem to indicate that new species arose mainly by isolation on those sites that provided the most stable conditions during periods of changing global climate.

Alternatively, the aggregates of endemics in figure 30.6 might be caused by a high local "species carrying capacity." If this were true, a correlation between endemism and total species richness would also be expected. As a test, the avifaunas of twenty-eight well-studied montane forest areas along the eastern Andes slope were compared (Fjeldså 1995). Spearman rank tests did not demonstrate a correlation between the number of endemics and total species richness, but a good correlation was found between the number of neoendemic species (the product of Pliocene or Pleistocene radiations), which are part of a vicariance pattern, and the number of species with relict distributions (r_s = .788, $P < .001$). The densities of neoendemic species were also positively correlated with those of only the oldest (Tertiary) relict taxa ($r_s = .557$, $P < .01$). These results suggest that aggregates of endemics are not due to local peaks in current "species carrying capacity," but rather reflect patterns of long-term ecoclimatic stability.

Certain centers of endemism extend right across deep valleys with an arid climate on the bottom (fig. 30.6; see Fjeldså 1995 for details). For example, gaps in the montane rainforest associated with the upper Marañón and Huallaga Rivers cut right through the sedimentary series, with no tectonic faults, indicating that they were formed by erosion intersecting the eastern Andean ridge throughout the uplift. Thus, the extension of some centers of endemism across such gaps suggests that the endemism is associated with special ecoclimatic conditions on both sides of the gaps, rather than with the gaps per se. Species displacements across such gaps (O'Neill 1992) may be a secondary phenomenon caused by a low chance of establishment by jump-dispersal into an area already occupied by closely related species (this may also be the case for displacements across major tributaries and sections of the Amazon). In plants as well as birds, the speciation process in the high Andes seems to be associated with isolation of relict populations in places with special local conditions (Kessler 1995).

A characteristic feature of the assumed stable sites is persistent rain and fog (Fjeldså 1995). Similarly, a global analysis by Long (1994) demonstrated a widespread correlation between endemicity of birds and cloud forest habitat. According to Graves (1985, 1988), the range disjunctions and differentiation of local populations of Andean birds are highest in the narrow treeline zone. Treeline species may be sensitive to climatic changes because changes in fog persistence on certain Andean slopes interact with the narrow range of the treeline zone.

Forests on mountain ridges with persistent fog may "comb" moisture out of the atmosphere, even when it does not rain (see Bruijnzeel 1990 for review). There is some evidence that tiny forest patches, once established, can maintain a stable local climate that makes them self-sustaining (Kerfoot 1968; Pócs 1974). In this way, habitat patches with an evergreen aspect can be maintained locally in semiarid regions.

The special importance of northwestern Peru and southwestern Ecuador is determined by the "collision" of the cold Peruvian Current, originating in the south, and the warm Equatorial Current (Best 1992). This phenomenon causes sustained cloudiness on the sub-Andean ridges (especially in Piura and Lambayeque; see Salati 1985, fig. 25), where the cloud forests form permanent water catchment areas for the lowlands of southwestern Ecuador. The position of this meteorological front is static because it is determined by the equator rather than by shifting climates. On the arid Peruvian Pacific slope, atmospheric inversion causes fairly stable mist zones, with favorable conditions for cloud forest development at 2,600–3,200 m. The ecoclimatic conditions for cloud forest formation may have been most persistent in the Lima Province because of the steep, 6,000 m high altitudinal gradient (see Salati 1985, fig. 25).

Conservation Priorities in Relation to Species Richness and Endemism

A simple summing up of species richness would assign conservation priority to lowland rainforests with a high degree of habitat flux, and in particular to submontane forests, where lowland and highland biotas overlap. A large proportion of this diversity, however, is made up of widespread species. These species often maintain themselves at very low population densities, or are patchily distributed, but may respond efficiently to opportunities that arise locally in the dynamic habitat mosaic in which they live. In this case, site-oriented conservation efforts may not be the best strategy for minimizing extinction. When WORLDMAP compares the summation of species distributions with a preselected network of well-protected areas, attention is shifted to those 60 (5%) of the 1,114 bird species in figures 30.1–30.5 that fall outside the network, mainly because of their restricted distributions. This implies a shift in focus from the core of the large rainforest blocks to a few places at the periphery with exceptional species richness, or to some exclusive areas right outside this boundary. It is noteworthy that not a single irreplaceable new site was suggested within the Amazon floodplains, although a few sites were identified at scarps and headwaters near the southern fringe of the basin and in the foothill zone in Peru. Thus, the many large protected areas that already exist in the Amazon Basin, mainly in Brazil near its borders with Venezuela, Colombia, Peru, and Bolivia, include nearly the total avian diversity of this region.

Very few irreplaceable and new flexible grids were identified in the Brazilian highlands, despite very little protection there. The forest birds recorded in this region are mainly "intruders" from the Amazon and Atlantic rainforests that have followed gallery forests into the region (J. M. C. da Silva 1995b).

A critical situation is found in the Atlantic forests of Brazil because of their considerable endemism, a severe loss of forest in the past, and a grossly insufficient number of protected areas (see also Viana, Tabanez, and Batista, chap. 23). The Alagoan Atlantic Slope, with 10 endemic birds, nearly lacks protected areas, and the large (1,000 km^2) Serra de Bocaina National Park near Rio de Janeiro does not cover the full assortment of endemics of the southeastern Brazilian lowlands and mountains (53 and 19 restricted bird species, respectively: Long et al. 1996).

Other serious gaps contain exclusive local habitats in Venezuela and in the tropical Andes region. Because of their high endemism, biodiversity management is needed in nearly all grid units in the Andean zone, from southern Ecuador to 12° S in Peru and farther south along the eastern slope (figs. 30.1–30.6). A particularly critical situation exists in southwestern Ecuador. Altogether, the gaps identified correspond very well with the critical and urgent priorities of ICBP (1992), except that we found the Guajiran lowlands, Santa Marta Mountains, and Upper Rio Negro white-sand forests to be well covered by the predefined network of protected areas.

Focusing on centers of endemism, Fjeldså (1992, 1993) found that populations of 59% of all land birds and 66.7% of the threatened and near-threatened avian species inhabiting the zone above 3,500 m in Peru and Bolivia exist within three 100 km^2 areas with complementary sets of endemic species. Fortunately, many endangered Andean birds appear to maintain viable populations in very small and isolated habitat fragments (see Whitmore, chap. 1). Where species have evolved by divergent evolution of small, local isolates, as may often be the case in the Andean habitat mosaic, inbreeding depression may not present as much of a problem as it could have, had the historical population sizes been large (see various papers in Loeschcke, Tomiuk, and Jain 1994).

GENERAL IMPLICATIONS

1. In continental biotas, local aggregates of endemic species exist mainly in ecotonal areas and regions with great topographic contrasts and locally persistent fog. In the tropical zone, it is mainly in these same regions that human pressure on the environment is most intense. For example, Balmford and Long (1994) demonstrated a worldwide correlation between the diversity of restricted-range forest birds and the mean annual deforestation

rate. The most extreme rate of forest fragmentation (ca. 6% annually) is found in southeastern Brazil, which also has as many as 10 restricted-range species per 1° map grid (Balmford and Long 1994). Ecuador and Bolivia also exhibit comparable patterns. In many parts of the Andes, the mist zones have been intensively cultivated for millennia. Thus, the most serious gaps in the network of protected areas (see figs. 30.1–30.6) are exactly those zones that have the poorest natural resource management.

2. We propose that top priority be given to the remaining patches of evergreen forest in southwestern Ecuador, to specific areas in the Andes of Peru and Bolivia, and to Brazilian Atlantic forests. These areas all combine pronounced endemism, high deforestation rates, and severe gaps in the network of well-protected areas. The assumed causal connection between aggregates of endemic species and special ecoclimatic factors is an additional reason for concentrating management effort in these areas.
3. We do not question the uniqueness of the enormous tracts of Amazonian and Guianan rainforests, nor their value for maintaining biodiversity and climatic stability. However, because few species in these regions have restricted ranges, and because of the complex role of disturbance regimes in maintaining regional biodiversity, precise goals for concentrated conservation efforts are difficult to identify. We believe that international agreements on forests and climate, renegotiation of the General Agreement on Tariffs and Trade, and national development policies may have a stronger effect on these ecosystems than site-oriented actions. The top priorities for conserving the biodiversity of these ecosystems would be actions on the political level (see Kahn and McDonald, chap. 2).

ACKNOWLEDGMENTS

This analysis was possible only because of long-term financial support from the Danish Natural Science Research Council (currently grant no. 11-0390) and good collaboration with a number of people. P. Williams kindly provided the WORLDMAP software and did a great extra service by programming according to our specific wishes. For provision of distributional data we thank R. Ridgely and the Academy of Natural Sciences of Philadelphia, BirdLife International, N. Krabbe, S. Maÿer, and J. M. Cardoso da Silva. For many stimulating discussions and comments on the manuscript, we thank P. Arctander, N. Burgess, G. R. Graves, N. Krabbe, J. Lovett, J. M. Cardoso da Silva, J. L. Patton, R. O. Bierregaard, Jr., and W. F. Laurance.

Section VII

Summary and New Perspectives

INTRODUCTION

This final section presents syntheses of important concepts discussed in the volume's various chapters and also forwards some challenging new perspectives. The last two chapters have been carefully cross-referenced so that readers can easily refer to particular chapters for further discussion.

A CRITICAL LOOK AT FRAGMENTATION RESEARCH

In chapter 31, Francis Crome presents a thoughtful—and thought-provoking—critique of tropical forest fragmentation research. In his lively essay, Crome challenges researchers to think deeply about their subjects and to grapple with fundamental conceptual and methodological issues that plague not just fragmentation research, but much of contemporary science. For example, he attacks the paradigm of hypothetico-deductivism, the short-term nature of most current research, and the slavish and overly simplistic use of significance tests. He also highlights the risks of seeking generalizations, especially when one is dealing with real-life landscapes. Finally, he considers the challenges of assessing ecosystem connectivity and the efficacy of faunal corridors. Even if one disagrees with parts of Crome's thesis, one's science will probably be better for having considered it.

SYNTHESIS OF KEY CONCEPTS

Chapter 32, compiled by William Laurance and fourteen co-authors, canvasses a wide range of concepts concerning the study, management, and conservation of fragmented tropical landscapes. The ideas are presented in outline form to ensure that the concepts are easily accessible. One of the chapter's main themes is ecological changes in fragmented forests; issues such as the kinetics of extinction, species invasions, edge effects, hyperabundance, and higher-order effects are reviewed. Another key theme is forest management; important topics include the conservation of biodiversity "hotspots," biological and social values of forest remnants, habitat restoration, and the economics of deforestation.

PRIORITIES FOR FUTURE RESEARCH

Chapter 33, compiled by Richard Bierregaard and fourteen co-authors, highlights essential priorities for future research. Like the penultimate chapter, it is presented in outline format and canvasses a large number of issues. Among these are the need for improved estimates of species extinction rates, additional research on the efficacy and design of faunal corridors, and a much sharper focus on assessing the importance of matrix habitats. Management-related issues are also highlighted, including the need for greater dialogue among researchers and resource managers, and for the development of methods for harvesting timber and non-timber resources while incurring minimal ecological damage to forests.

31

Researching Tropical Forest Fragmentation: Shall We Keep On Doing What We're Doing?

Francis H. J. Crome

Understanding and managing fragmentation is a significant component of conservation biology, which itself is an applied ecological discipline. As such, there needs to be some confidence in the theoretical bases and methods of fragmentation biology. As practitioners, we can sometimes lull ourselves into believing that our theory and methods are fine, but this is not necessarily the case. Fragmentation biology is a challenging subject, and this book, arguably the best compilation of tropical fragmentation studies yet produced, highlights many of the challenges we face. This chapter extends the challenge a little further by exploring the possibility that there are fundamental philosophical and practical difficulties in researching tropical forest fragmentation.

There has been controversy for some time, emanating from a small but lucid group of thinkers (e.g., Haila 1986; Simberloff and Cox 1987; Simberloff 1994, 1995), about the practical application of the various theories that are central to fragmentation biology. Moreover, the number of studies in the Tropics is sadly small. Most fragmentation studies have been done in the temperate zone, although there are some notable exceptions (such as this volume, Lovejoy et al. 1984, 1986; Laurance 1989, 1990, 1991a, and Bierregaard et al. 1992.)

I must stress at the outset that I am discussing only the problems of doing research to manage already fragmented systems, not the problems of how science could provide guidelines for "safely" fragmenting intact forest.

I will draw on work on the Atherton Tableland, a fragmented landscape of agriculture and upland forest remnants in Queensland, for examples. Our team studied the characteristics of rainforest trees that make them more or less able to survive in fragments (Crome 1994), and how wildlife uses fragmentary stream corridors (Crome, Isaacs, and Moore 1995). For the first project we attempted to determine whether fragmentation was affecting selected large-fruited trees by studying their regeneration and population biology in four fragments and four matched sites in continuous forest. We chose large-fruited species for which the only specialized disperser is the southern cassowary *(Casuarius casuarius),* which has been shot out of the fragments. Having thus lost their major disperser, the large-fruited species could be disadvantaged in fragments (see Harrington et al., chap. 19).

THE PROBLEM OF COMPLEXITY

One of the most difficult aspects of fragmentation studies is the size of the subject. At the landscape level, fragmentation has the potential to affect regional and local climate, hydrology, animal and plant survival, animal movements and behavior, plant dispersal, regional population dynamics of species, pollination systems, the evolution of the local biota, and, of course, the way people live. Some of the factors that can influence individual patches of forest are changes to internal ecological processes, nutrient runoff, the penetration of pests, weeds, and other exotic species, changed internal disturbance rates, changes to the local climate brought about by fragmentation, altered hydrological regimes, and edge effects (see Saunders, Hobbs, and Margules 1991).

I would like to focus on two particular ways in which this complexity can present the researcher with difficulties: one is the lack of knowledge about the intact system, and the second is the management requirement for at least some degree of comprehensiveness in our knowledge.

Knowledge of the Intact System

Fragmentation is a system-level phenomenon, and assessing its effect presupposes that we know how the intact system operates; otherwise no effect can be estimated. One could try framing questions in clever ways to get around the problem—for example, "How do we maintain ecosystem processes as they are?"—but it is impossible to avoid the requirement of some knowledge of the intact system. Such understanding, however, is often rudimentary, and in the case of tropical forests, even simple species inventories are grossly incomplete. Thus, the researcher is forced to split resources and study both the process and how it is affected, and runs the risk of doing neither well. In the case of our Atherton Tableland work, we had no knowledge of seed predation and dispersal dynamics in intact forest. This forced us to divide our resources between fragments and intact forest, thereby halving the effort in fragments and leaving us with small sample sizes.

Comprehensiveness

There needs to be a minimum amount of knowledge before one can say anything sensible about an entire system, yet researchers usually operate in only a few areas of expertise. This is fine for academic ecology, but most of us do not have the luxury of operating totally theoretically. We daily face the social and political demands, generated by ourselves as much as anybody, that we do "useful" research.

In the face of this pressure, many of us offer management advice of some sort. In giving this advice, however, we are faced with the dilemma of knowing that a multitude of things are happening, but not knowing how they are happening within that specific landscape. Just identifying which effects may be dominating is a problem.

We need to be reasonably aware of our system before we can even approach designing studies, let alone making management suggestions. We have to be conscious of the ramifications of being right or wrong, and of the options foregone elsewhere when time and resources are committed to managing a particular landscape or series of fragments. We have to be conscious of all those things we are not covering, and aware of the possible negative consequences of the management options we may suggest.

Sometimes we are lucky enough to have lived in a landscape for a long time and developed a reasonable local knowledge base. This experience can assist in avoiding major pitfalls, but the dilemma of conservation biologists and applied ecologists giving advice with limited information is the norm. Obviously one can't do everything, so one should concentrate on specific areas, research them well, and be willing to give a "best guess" if needed.

Conservation biology is a "crisis discipline" (Soulé 1991) wherein educated guessing is not unreasonable. Unfortunately, the breadth of fields that fragmentation covers and the comparative scarcity of experienced field people make it a particular problem area in which serious mistakes can be made as a result of too narrow a focus and the incentive to "be seen to be useful." The popularity of corridors is an example. Their importance in theoretical analyses of metapopulation dynamics (Gotelli 1991; Pulliam and Danielson 1991) has made them a central theme of fragmentation biology, and their utility is widely accepted (see review in Bennett 1990). Intuitively, they seem to be useful and a good thing to have.

Almost lone voices in the wilderness, Simberloff and Cox (1987) and Simberloff (1994) have leveled keen criticism at the concept, both because of the possibility of negative effects and because of the foregone options that investment in corridors would engender (see also Noss 1987). Given a broad enough perspective, purchases of corridors can be seen to be counterproductive despite the vociferous support corridors receive (Simberloff 1994).

Compounding this problem is the short-term nature of most current research. In the case of our tree work, how do we interpret measurements taken over a particular two- to four-year period in the life of trees that live for several hundred? Despite that problem, we felt that we generated information pertinent to whether or not particular tree species need to have their populations augmented to survive, whether fragments should have cassowaries reintroduced, whether manipulation of seed predator populations is needed, and whether planting corridors would be useful to small mammals and tree survival.

However, if we incorrectly say trees do not need augmenting, the populations will die out. If we incorrectly say they require a tree planting program, we will be planting them at the expense of other species. Reintroducing cassowaries, or any other large animal, is an expensive, protracted, and difficult process, and will divert funds from other operations. How good is our data to make such suggestions?

Arguably the biggest problem is foregone options. We worked on a few species out of a huge tree flora, just fifteen out of seven hundred. Even extending our results to all seventy large-fruited species, we still cover only 10% of the tree species. We worked only on trees and rodents. What about the more difficult species, such as frogs, quolls (marsupial carnivores), fungi, lichens, and so forth? Dan Simberloff (pers. comm.) makes the point that we must concentrate on species whose loss may have particular ecological significance. But even when we do this by choosing supposed keystone species, such as cassowaries, conservation biologists should be asked the question (or ask themselves): "Is the organism (or group) you have used as the basis of your recommendations the most important one to use, and what are the consequences of focusing on it?"

THE PROBLEM OF GENERALITY AND THEORY

Generalization and prediction are characteristics that separate "science" from natural history (Peters 1980). In fragmentation studies, however, generalizations are particularly difficult to make. Each landscape appears unique in its pattern of fragmentation, the activities in the surrounding matrix, the composition of the biota, and the relative importance of the various effects of fragmentation. Each situation described in this book is different. In northern Queensland, the rainforests support a far simpler and more generalized fauna than those in, for example, Malaysia. Fjeldså and Rahbek's Amazonian forests (chap. 30) have at least two to three times as many bird species as those on the Atherton Tableland. Northern Queensland lacks the large mammals found in the forests of the United States.

There are different ranges of problems in each landscape. In Western Australia, the problem is sheer habitat loss and soil salinization. In northern Queensland, it is probably expansion of the human population. In the U.S. Pacific Northwest, it is timber harvesting (e.g., Dixon and Juelson 1987; Simberloff 1987). In this immense diversity of situations, patterns, and problems, how useful can any generalities really be—even at the scientific, let alone the practical, level? Can we apply results from Brazilian Atlantic forests to Queensland mangroves other than at the most generalized level—small fragments have fewer species than large ones, corridors may or may not be useful, animal movement is important, weeds may or may not be invading (and if so, this may or may not be important), fire regimes may have changed, and edge effects can occur.

Obviously, we do have valuable scientific generalizations about fragmentation, which, if broadly applicable, could save some research effort. But generalizations give us only a shopping list of possible problems that may be occurring in a specific landscape. Take, for example, edge effects. Edge effects, like corridors, have become a management issue and are generally perceived as a problem (Kroodsma 1984; Schonewald-Cox and Bayless 1986; Harris 1988; Reese and Ratti 1988; Saunders and Hobbs 1991). Applying generalizations about edge effects to the Atherton Tableland would generate concerns about erosion of forest edges and invasion of weeds. Laurance (1991b) has measured a distance effect on the occurrence of early successional vegetation within patches. However, aerial photographs dating from World War II show that forest edges are expanding into cleared areas unless prevented from doing so. In this case, how do we use our generalizations about edge effects? If our forest edges are expanding, do we need to bother about edge effects at all?

Of course, it is important to attempt to use generalities and to think about issues generated by other studies. But is it really useful for site management? Will it enable us to reduce our research effort? It did not do so in our studies. We have not been able to take anything from other landscapes at face value, and have always been in the situation of having to know something about most things in our landscape.

To a certain extent this problem is rooted in fundamental philosophical issues. The simple mechanistic approach to nature, which holds that phenomena are attributable to a limited set of natural laws that science can elucidate, should be obsolete by now, but still appears to be widely accepted (Suppe 1977 and Hacking 1983; cited in Haila, Saunders, and Hobbs 1993). The workings of the natural world are not representable by a set of tidy principles that a limited range of simple hypotheses and tests will uncover. A good example is the lack of success in using deterministic biological and economic principles in fisheries management (Constable 1991 and references therein).

Our problem is that it is easy to slip into the belief that ecosystems are machines that can be completely described and their behavior predicted. They cannot. Ecosystems have been described as middle number systems—one cannot use either a small number approach (dealing with things individually) or a large number approach (dealing with things as an average) with such systems (O'Neill, Hunsaker, and Levine 1992). The futility of our narrow, mechanistic approach to ecosystems is perhaps best summed up by O'Neill, Hunsaker, and Levine (1992, 1446):

> One of the most interesting features of middle number systems is the uncompromising reaction of mankind in general and scientists in particular. "Give me a bit more time, a bigger computer, and a lot more money and I will crack the problem. I simply don't understand enough yet." There is no way to demonstrate definitively the error of this credo, but it remains a statement of blind faith. The statement is founded on an undying faith in the fundamental orderliness of the universe and the near-infinite capacities of the human mind to grasp that orderliness. Heisenberg's uncertainty principle, the stochastic nature of quantum mechanics, chaotic analysis, and the human experience of several centuries have done little or nothing to shake our confidence.

Generalizations, formalized as ecological theories, cannot be regarded as simple, neat, restricted ideas that can be tested once and for all, but must rather be viewed as broad intellectual fields indicating directions and possibilities. Haila (1989) gives a good example in the history of the theory of island biogeography. The theory can be interpreted narrowly, in the sense of being a restrictive, testable theory, and as an explanatory model that explains species numbers on islands resulting from an equilibrium between colonization and extinction. Alternatively, it can be viewed more broadly as a program that draws attention to the dynamic nature of island communities and focuses research effort on the mechanisms that determine the colonization and extinction processes in specific situations. According to Haila, the narrow view has prevailed, and the theory, at least in its application, is seen as the study of species-area regressions with equilibrium as their explanation. The theory "ossified into a simple formula that began to suppress creative thinking" (Haila 1989, 237; see also Levins and Lewontin 1980; Simberloff 1980).

Haila (1989) further developed this broader appreciation of what ecological theories are by suggesting that they are myths (see also Boulding 1980; Ghilarov 1992). A myth is not a falsehood, but rather a powerful guide to ways of thinking and understanding. It is not a pejorative term. It is a way of explaining or grasping concepts or things that are too complex to explain in words, or too difficult to directly grasp with our intellects. An analogy is that myths are signposts to what really is, but are not real in themselves.

The problem with not recognizing ecological theories as myths is that we can become attached to them and actually believe them, failing in the process to recognize them as symbols and signposts. Worse still, we can try to use them for "management."

It is this narrow view of theory and our attachment to it that has generated the corridor problem. Rather than being regarded as myths that indicate direction and that need to be contemplated within the broader operations of specific landscapes, untested conceptions of corridor utility are being peddled as being true (see above). The same seems to be happening with metapopulation dynamics. It is a nice theory, but has fast become a conservation "tool." Management agencies can now hire people with a basic knowledge of how to run metapopulation dynamics software to run analyses for endangered species management (pers. obs.)! It is reassuring to see that Simberloff (1995) has turned his attention to the real-world usefulness of metapopulation dynamics theory.

This discussion of theory and myths is not a trivial philosophical aside. Our attitudes toward generalization and theory can have profound practical outcomes, especially when we become attached to a theory, believe in its reality, and vigorously go out to manage nature according to it. When we regard theories as tight, real entities and devote ourselves to their analysis, we can limit our horizons and, worse, attempt to make the world fit them. A lot of ecological discussion is not about nature, but about theories, generalizations, or models supposed to represent nature.

THE PROBLEM OF ANALYTICAL METHODS

Tame, Toy Problems versus Wicked, Real Problems

Hypothetico-deductivism and experimentation are the methodological ideals ecologists aspire to. Although we use a range of methods in ecology, reductionism and experimentation, set up in such a way as to allow hypothesis testing in a null framework, and analyzed using Classical statistical techniques, seems to be considered the most elegant or scientific way, if not the only acceptable way, of carrying out research. This approach, however, may not always be appropriate for studies of fragmentation—the subject is too complicated.

As a brief reminder, the hypothetico-deductive procedure operates as follows: (1) speculation; (2) formal hypothesis formation; (3) deduction-prediction; (4) data gathering; (5) data-hypothesis evaluation; and (6) explanation/speculation (= 1). Under this scheme, falsification, but never verification, is possible. According to Haila (1982, 260, referring to Fretwell 1972), the speculative character of hypotheses "is often emphasized by maintaining that former experience or data have no role in the invention of explanatory hypotheses; matching up hypotheses and data can only be undertaken afterward by testing specific predictions of the hypotheses." This is a logically restrictive use of hypotheses to explore nature. Haila gave three examples of synthetic and ecologically real hypotheses that are too complicated for the use of hypothetico-deductivism in that they attempt to integrate the effects of several factors on community functioning.

Miller (1993) was more exacting in his critique of the use of analytical, reductionist science in natural resource decision making. He talked about "tame" and "wicked" problems, tame ones being circumscribed, well known, and simple, and wicked ones being complicated, poorly known, and messy. Reductionist science is good with the former but inappropriate for the latter, where more holistic thinking is necessary. Miller (1993) gave a good real-world example of a spraying program that had been established in New Brunswick pine forests to control the spruce budworm. A major spraying program to control outbreaks of this pest was the focus of a public campaign to end spraying. Public concern was first manifested when spraying was implicated in an unusually high occurrence of a rare children's disease, Reye's syndrome. The authorities took a reductionist, analytical approach by setting up expert panels to determine whether there was scientific proof of harm, a prerequisite for the authorities to stop spraying. The panels myopically sifted through evidence linking the syndrome with spraying without realizing the impossibility of teasing out the subtleties of such an issue. When no simple causal links between spraying and the disease were found, the spraying program continued. Had the panels or the authorities taken a nonreductionist, holistic approach, they would have looked at the entire problem and seen that health was a subsidiary issue to the real issues of a lack of diversity in the New Brunswick economy, overdependence on the pine forests, and inappropriate forest policy. Fixing these big issues would have probably meant that there was no need for spraying. Instead, the reductionist approach de-focused from the fundamental problems, which presumably remain unaddressed.

Sprague and Sprague (1976, quoted in Preston and Bedford 1988) put a more general cast on the issue. They define "real" problems as important, large, complex, and multidimensional issues that are inaccessible to manipulation by an empiricist. "Toy" problems are simple, small, well structured, and controllable.

Sprague and Sprague (1976) use the same terms in relation to research. Real research is well-bounded research with high scientific rigor, with specific protocols and approaches applied simply and directly to discrete cause-and-effect situations. Toy research is simple observation (but see below) and may not even involve quantification. Real research is best for toy problems, and toy research is often all one can do on real problems. The trick is not to mismatch research and problem type. Never do toy research on toy problems, and recognize the extraordinary limitations in attempting real research on real problems.

Fragmentation is a wicked, real problem. It is all but impossible to set up higher-level field experiments—at the level of patches versus mainland, corridors versus no corridors, this type of patch versus that type of patch. Even if the well-known, but often ignored, problems of design and pseudoreplication (Hurlbert 1984) can be overcome, it is not often that an optimum experimental design on paper can be translated to the field because the necessary situations do not exist. Anyway, time and resource management considerations usually preclude adequate sampling.

In an attempt to encourage researchers to come to grips with the problem of deciding whether corridors are useful or not, Nicholls and Margules (1991) have clearly shown what would be required to set up experiments to investigate, within the hypothetico-deductive paradigm, the utility of corridors in the Western Australian Wheatbelt. Their suggestion was to select replicated sets of patches that are either connected to source areas or isolated, then systematically defaunate subsets of them and test the null hypothesis that reinvasion of connected defaunated patches is no faster than reinvasion of isolated ones. The design appears perfect, but is impossible to put into practice. First, it is unlikely to be affordable—it would probably be cheaper to simply buy the remaining patches and plant corridors. Second, the moral and ethical issues with defaunating would be insuperable. (They could not be avoided by using invertebrates because the problem fauna is vertebrates.) Finally, the array of patches needed to do the experiment simply do not exist in the desired pattern (Nicholls and Margules 1991).

Even if these problems could be overcome in another landscape, it is unlikely that the experiment could be run long enough to get an outcome. How long does one wait—20 years? 50 years? Such periods are ecologically, but not experimentally, reasonable. And finally, which species does one choose? To get a broad picture of the utility of corridors would require a cascade of experiments studying a range of species and corridor types and positions as well as the potential negative effects of corridors. Ultimately, the question of whether corridors are useful or not is so unspecific as to be a non-question. The answer is, "it depends."

Statistics Are Not Recipes

Given that, despite all of the above, we wish to go ahead with an experiment in some form, we have to face the issue of how to analyze it and make a decision about our hypotheses. This, in turn, presupposes that we are quite clear as to what our hypotheses are and that we understand the difference between our scientific and statistical hypotheses—the two are often confused (see Simberloff 1990a; Underwood 1990).

The commonest way to approach the hypothetico-deductive/falsification system is by statistical testing in a null framework, and the commonest types of tests are significance tests. We set things up as yes or no questions and use significance tests to help us decide yes or no. Significance testing is one of our tried and true tools, yet it rarely addresses the scientific hypothesis, and is only sometimes useful for applied problems.

Significance testing is a tool we have borrowed from statistics, but, like all borrowed tools, it is imperfect, often fails to work properly, and we usually abuse it. To examine this problem further, however, it is useful to remind ourselves of the controversies that exist between different schools of statistics (Efron 1978). These controversies are based on differing philosophies of statistical inference. Generally, ecologists are focused on the procedural problems of Classical statistics (e.g., Hurlbert 1984; Simberloff 1990a; Underwood 1990) and appear to be unaware of these controversies or the fundamentals of inference.

The problem is that there is no general consensus among theoretical statisticians about what, if anything, constitutes a sound framework for making inferences. Consequently, several schools of statistics have developed, of which the Classical and Bayesian are probably the largest. There is also diversity within these schools, with many shades of Bayesian statistics and a number of different approaches to Classical statistics.

The Classical (or Frequentist) school has provided the most commonly used tool in ecology, the significance test. As a demonstration of this, I took an issue of *Ecology* at random (December 1993) and found that seventeen of the twenty-five papers in the issue relied almost entirely on Neymann-Pearson style (see below) significance testing. The Classical school adheres to the familiar Frequentist concept of probability. Probability is conceived as a limiting relative frequency in some long-run (hypothetically infinite) replication of a chance system. For example, if we toss a coin repeatedly, we can record the relative frequency (proportion) of heads and plot this against the sample size (number of tosses). It is assumed that as the sample size tends to infinity, the observed proportion of heads tends to a fixed constant—the probability of obtaining a head. Here, probability is a physical attribute of a system—it is not conditioned on the data, and there is no place for subjectivity in the process.

In this view, probability has no meaning in relation to a particular, single event. Probability statements may not be made about specific hypotheses, particular estimates, intervals, or tests, but only about the long-run behavior of the procedures that are used.

The Classical school asserts that a particular parameter of interest has a fixed and knowable value that can be objectively estimated. With the parameter fixed, the data are treated as variables and the likelihood of getting the data, given the parameter, is determined. This then leads to a reassessment of the value of the parameter.

The Classical school is further divisible into the Neymann-Pearson and Fisherian traditions, which differ in their approaches to hypothesis testing. The Neymann-Pearson tradition holds that a significance test is a decision procedure and that the interpretation of a nonsignificant result depends upon the test's power. The Fisherian tradition, on the other hand, does not consider the test a decision procedure and considers power illogical. To the Fisherian, a nonsignificant test result is uninterpretable.

The Bayesian school takes a fundamentally different view of the process of inference. In the Bayesian school, subjectivity is fundamental. It says that there is no such thing as objectivity—that all we ever have are beliefs, and we use data to change them. It uses conditional or subjective probability. Probability is conceived as the subjective assessment of the likelihood of an event, for example, will it rain today? Hence, it is a measure of one's personal degree of uncertainty about an event. It is conditional in that it depends upon the existing data; for example, given the values of these particular bird counts, what is the probability that this fragment has bird populations that are 25% smaller than those in intact forest?

The Bayesian school does not consider that a parameter has a particular objective value; it asserts, rather, that we can only form beliefs about what that value is likely to be depending upon the data we have. It treats the data as fixed and the parameter as the variable. The procedure is to formally accept an initial value (or estimate, or a distribution of belief about the location of a parameter), combine the data with this, and generate a modified belief based on the data. It is a learning process that treats subjectivity formally. The mathematics involved are difficult, and there are relatively few ecological studies published using these methods (e.g., Gazey and Staley 1986; Crome, Thomas, and Moore 1996). However, for those who can accept their philosophical foundations, Bayesian methods have the advantage of allowing plain English statements to be made about individual situations—a luxury denied users of Classical statistics.

Returning now to the Atherton Tableland, we wanted to find out whether being in a fragment reduced a large-fruited tree's ability to regenerate. We started out by setting up a hypothesis that went as follows: "We know dispersers are gone from fragments, therefore seeds of large-fruited trees should be collecting beneath parents, where high seed predation rates would be expected; thus fewer seeds should be escaping predators and, consequently, there should be fewer seedlings per adult in fragments as compared with intact forest." From this we were able to generate further hypotheses and predictions about predation rates, seed deposition, and seedling occurrence. We generated the prediction, for example, that there should be lower seedling densities in fragment populations than in intact forest.

To "test" this prediction, we could have counted seedlings and compared the results using significance tests. However, this process gets a little complicated. Following the logic of falsification, we should say, "We can never prove that there are fewer seedlings. But we can refute the alternative hypothesis that there are as many or more seedlings. We will therefore test this other hypothesis, which we call the null hypothesis, on the basis that refuting it will leave the only logical alternative, that is, the original hypothesis that there are fewer seedlings." We then act as if the null hypothesis were true and generate a distribution for a chosen test statistic (e.g., Student's *t,* chi-square), which gives us our statistical null distribution. We then find out whereabouts in this null distribution our value of the test statistic lies, and make a decision about "significance" based on arbitrary rules about this position (5%, 1%, etc.). If we decide that the word "significant" can be used, we reject the null hypothesis and implicitly accept the alternative original one.

Now suppose we reject the null hypothesis and accept that there are lower seedling densities in patches than at mainland sites. We think we have an important result, but in fact we don't—because what does it matter if they are different? What does the difference matter in terms of tree survival or population dynamics or management decisions?

The problem is that there are inevitably going to be differences, but what *level* of difference is necessary to initiate changes in a scientific hypothesis and in the way one mentally models the system? In performing a significance test, we are testing whether the population means or variances are identical (i.e., whether the difference between them is exactly zero). Trivially and obviously this is not so, since no two populations can ever be identical. If we gather enough data, we can inevitably reject the null hypothesis because we are really testing the universally false statement that "these two things are identical."

Hence, setting up a significance test switches the object of falsification from the original hypothesis to the logically opposite null. Increasing sample size should act to put the original hypothesis through harder and harder tests, but instead—because it leads almost inevitably to rejection of the null—it makes it easier for the original hypothesis. It has even been suggested (Binder 1963, in Oakes 1986) that sample sizes can be too large for sound inference!

This appears quite absurd, but is only so because we wanted a simple decision about whether two things (seedling densities) differed or not. Simply saying that they should be different does not really enhance our understanding, because one needs to know how big a difference is relevant to one's theory. The statement that really needs testing is "Is the difference between these two things important to me or not?" This can be done only by having a theory well developed enough so that the size of difference between seedling populations that would lead to abandoning or modifying the theory can be determined. This size difference can then be tested for.

One can try to get around this problem by saying that it is first necessary to find out whether there is a difference—finding out how big it is and whether it matters is the next stage. That is only avoiding the issue.

Statistical Power

The problem of the size of the effect that we are interested in brings us to another issue: the power of a significance test. As indicated above, power is not a concept that is used in the Fisherian school. Its rules are simple and draconian: a significant result allows rejection of the null hypothesis; a nonsignificant result allows us to do nothing. The result cannot be interpreted except to say "Repeat the experiment or get more data" (presumably until you get a significant result!). This is a difficult position to be in. Most of us know the dangers of interpreting nonsignificant results, but in practice it is very common for them to be interpreted. Even if we obey the rules and do not formally interpret them, the act of publishing significant and nonsignificant results together and discussing only the significant ones makes implications by omission.

Power is, however, central to Neymann-Pearson-style significance testing, which is briefly learned in biometrics courses, then promptly filed in the "never to be used" archives of our minds. The calculation of power supposedly allows us to judge the degree of support that the test gives to the two hypotheses. A nonsignificant effect can be interpreted if the power is known. For example, if the Type II error in our seedling test were 0.6, we could say that, although there was a nonsignificant result suggesting that seedlings are at the same density in fragments and in intact forest, no conclusion can be drawn because of the low power of the test. (In an earlier paper [Crome 1994] I gave this example and an incorrect interpretation of the power statement—it is easy to do.)

However, although nonsignificant results are often interpreted, or implications made about them, Type II errors are hardly ever calculated, despite the fact that they must be in order to do this. This is a general problem for conservation biology, where both Type I and Type II errors need to be calculated, but where all we usually do is attempt to minimize Type I errors. Given that we have failed to reject a null hypothesis, the natural and commonest tendency, even if it is not done explicitly, is to behave as if there is no difference, even if intellectually we understand that we cannot behave in such a way.

Why do we continue to ignore Type II error? Unless one is a Fisherian and really obeys the strictures about interpretation, it is methodologically compulsory to set both error rates. Perhaps part of the reason is the difficulty of doing so—but also, perhaps, it is because we would often discover just how weak our significance tests actually are.

Additional Challenges

There are other problems with significance testing—for example, controversy about the real meaning of a *P* value, or indeed, whether it has any meaning at all (Berger and Sellke 1987); the use of tail areas; the bizarre behavior that these methods, and those of other schools, sometimes show (Efron 1978; Oakes 1986; Berger and Berry 1988); and the problems of analyzing accumulating data (Berger and Berry 1988).

Most of us are aware of, or occasionally reminded of (Hurlbert 1984), the superficial problems that lead to misuse of significance testing (pseudoreplication, nonindependence, inadequate sample sizes, non-normal data, etc.). Few of us, however, are aware of these more fundamental aspects of the tests themselves and the often convoluted interpretational difficulties associated with them. Perhaps the average user of significance tests, without knowing it, smears him- or herself over the three major statistical schools, and disobeys the rules of each: we act as if Fisherian and often believe we are; attempt Neymann-Pearson approaches, but never properly; and interpret tests in a Bayesian fashion.

The difficulties of statistical inference make misinterpretation easy. How well understood are the convoluted ways in which one has to think to be true to the Classical methods (see Oakes 1986)? We all know, for example, that we cannot interpret a 95% confidence interval to mean that "there is a 95% chance that the true value of what we are trying to measure lies between *x* and *y*." This is a Bayesian, not a Classical, interpretation. Neither does it mean that "if I did this experiment a large number of times, then in 95% of cases the true value would lie between *x* and *y*." What it says, really, is something about the method rather than the specific situation. It means, "If I did this test a large number of times, then in 95% of cases, the test would give me a confidence interval within which the true value lies. In my specific case, the true value is or is not between *x* and *y*; however, because my method usually gives me the right answer, I will in this case choose to believe that the value is between *x* and *y*." It requires considerable clear thinking.

In theory, hypothesis testing is not about decision making—decision theory is available for that. In practice, however, statistics and hypothesis testing cannot wriggle out of the patently obvious fact that they can be used, and often are used, as decision-making procedures (pers. obs., and see Good 1978; Moore 1978).

Studies of fragmentation would benefit from a broadening of analytical approaches. Bayesian methods could be used more commonly. Where a Classical framework is preferred, hypotheses and experiments are perhaps better approached using estimation and confidence interval methods rather than significance testing. The former have the added advantage of forcing deep thinking about the hypothesis.

Objectivity

The above would not be a problem if hypothetico-deductivism and Classical statistics were not so dominant and did not have such power to mold research methods. This power is founded in part on the false belief that these methods produce objectivity. They do not. Objectivity eludes scientists as it does everybody else, but we can fool ourselves into believing that, because we are scientists and because science opines to be objective, we are therefore objective. Our science culture determines not only how we interpret and approach things, but also how we set ourselves up as observers, and indeed, what we observe (see Miller 1993 for further discussion). Even Classical statistical methods are subjective (Good 1978; Berger and Berry 1988).

Classical methods are fine in themselves. Our problem is that we use them to dictate field operations because we believe in their objectivity. Worse, we use them as blunt instruments against each other, and particularly against our students. Field operations are often designed so that a significance test can be carried out because significance tests are usually demanded by journals. Proper statistical vetting is essential, but my experience is that the mechanisms of the tests often take precedence over the philosophy and logic. What is the point of demanding that we do significance tests if significance tests themselves are problematical things to do?

THE PROBLEM OF CONNECTIVITY

Fragments do not exist in isolation. A central tenet of landscape ecology and conservation biology is that there is connectivity between fragments. The entire landscape acts in concert to support the biota and maintain ecological systems. This is intuitively obvious and well documented for vertebrates, particularly mammals and birds (Arnold, Weeldenberg, and Steven 1991; Saunders and De Reibera 1991; Andrén 1994). Date, Ford, and Recher (1991) have shown how rainforest fragments act as stepping stones for frugivorous pigeons moving between major forest blocks. In general, it is well known that rainforest frugivores may suffer periodic and sometimes unpredictable and catastrophic food shortages, during which they depend upon scattered and rare food resources. At such times individual patches can become very important (Foster 1982; Howe 1984; Crome 1991a; Crome and Benntruperbäumer 1991).

Altitudinal migration of birds may connect upland forest blocks with lowland rainforest remnants. McClure (1974) has shown how complex local migration systems of rainforest birds can be. If such movements involve pollinators and seed dispersers, and if remnant patches support them at one end of their migration, then loss of the patches will have dynamic ramifications throughout the entire system. Biogeographic regions may be connected by migration. The more one looks, the more the connections anastomose across the landscape, across states, across nations.

Although we know the connectivity is there, our need to perform rigorous science sometimes forces us into a schizophrenic type of behavior. We are aware that patches are not independent, yet, because we need statistical independence for our "tests," we usually behave as if they are. This is a particular issue in studies of mobile species such as birds, mammals, and many invertebrates. Perhaps in many of these studies it would be best not to use statistics at all.

We came to this decision in our studies of vertebrate use of riverine remnants. We could see individuals going from patch to patch! The analytical issues involved were not trivial. Biological and statistical independence are different things, but the former has great bearing on the latter. Lack of statistical independence can be dealt with by developing complex modeling procedures, but we decided it was not worth it. All students of fragmentation and landscape ecology have a problem here, but the problem doesn't lie with us or with the landscape. It lies with the set of analytical tools, developed mostly for agricultural and industrial situations, that we insist on using. We hammer our round landscape into the square hole of our methods.

Connectivity is particularly important in view of the common minimalist attitudes in our society. A "minimalist" attitude is the belief that there is some exact, definable minimum conservation effort that will ensure the survival of species or systems. "What is the minimum number of reserves to preserve this species?" "What is the minimum number of fragments?" "What is the minimum size of a fragment?" "What is the minimum area you need to preserve in national parks?"

In the context of conservation in fragmented landscapes, the minimalist approach is to determine what bits of habitat can be safely got rid of or left to disappear. However, the richness of biodiversity and connectivity will always frustrate efforts to answer such questions. One could probably determine which bits are the most important (such as the big bits), but even then, you need to know the risks of losing something. Let us say that certain tree species occupy certain types of forest patches. Frogs occupy other types of patches, land snails yet other types. If you pile species upon species and then consider connectivity, you may end up concluding that every bit is useful and important.

Faced with the problem of interdependence and connectivity, one must conclude that getting rid of patches in an already highly fragmented landscape such as the Atherton Tableland involves unknown risks to the surviving biota. One could argue that, for most fragmented tropical landscapes, data on biodiversity distribution and connectivity are nonexistent, and therefore speculation about a patch's importance is illegitimate.

The importance of connectivity is such that it has to be given as much emphasis as biodiversity distribution, but the technological difficulties of measuring it are far more severe. Static surveys can be done given enough resources, but getting good data on animal and plant propagule movements is hard, time-consuming, and expensive. Given the difficulty of the connectivity problem, if one wishes to minimize the environmental risk, then one should abandon any "minimalist" view and regard all the bits and pieces of the forest in the landscape as important. The bottom line is that we have the minimum for conservation now.

This may sound impractical and an abrogation of responsibility. After all, a minimalist attitude would be that "you're going to lose some of them anyway, so if you don't want to be involved in the decision making, we'll do it without you." It is not such an abrogation, however. It means a redirection of energy away from reductionist studies of fragments to issues of policy, governance, and human consciousness. A further implication of taking such a stand is a shift in emphasis from the fragments to the management of the matrix in which they are embedded (e.g., Bierregaard and Stouffer, chap. 10). If the biota in the fragmented landscape is to persist, then management of the matrix becomes all-important.

THE PROBLEM OF BALANCE

Closely related to minimalism is the issue of balance. We all like to be reasonable, and we seek (although we do not necessarily find) balance in most things. In issues of landscape and tropical forest conservation, there are going to be trade-offs and negotiations to reach some sort of balance. But such negotiation processes must have some end. A stand has to be taken somewhere; otherwise there will be continuous cycles of "balancing"—each cycle resulting in further erosion of the remaining forest until it is completely lost. At a workshop on landscape rehabilitation at Tammin in the western Australian Wheatbelt in October 1991 (pers. obs.), the participants were told that the area had been 93% cleared for wheat, and it was suggested that a moratorium be placed on clearing. This was met with cries of "You want everything! For goodness sake, you need a balance." As ecologists, we must be able to say what we think is an end-point for forest clearing.

THE PROBLEM OF INTERMEDIATE AND OTHER PEOPLE'S REALITIES

As applied ecologists, we have to be realists. The world's societies are going through a difficult period. Materialism, population growth, and political and economic rationalism still dominate. In the words of Chris Margules (pers. comm.), "A particular cultural orientation holds sway—one that views nature as a provider of products, not as a medium for the persistence of human beings. We have to change this cultural orientation. In the meantime, land is being cleared and we have to use all means at our disposal to stop it. Our long-term task is to change this cultural orientation. Our short-term task is to deal with the reality of this particular cultural orientation to save what we can. Hopefully, this cultural orientation will be a brief historical aberration."

From the point of view of conservation biologists, our economics and politics are realities at a certain level only. They are, in another sense, deep illusions. As ecologists and human beings, we can see the craziness in much of our present reality. We see our life-support systems being degraded before our eyes, and wonder at the "reality" that our present economic systems can actually measure "benefit" in it.

Because we see landscape degradation as ultimately more important than preservation of individual lifestyles, we perceive our reality as being deeper. When, therefore, we are urged to "be realistic" about a political or social situation, we are being told to be realistic at an intermediate level, comfortable for our antagonist. When urged to be realistic, it is hard to stop halfway in the process when "realistically" we are in a very serious situation in this world.

Each person's reality is another's illusion. As conservation biologists, we know we have a fundamental reality—to care for the earth—but we need to understand that the realities of others are just as fundamental to them. Such was the case with the plains reserves of Burma. These lowland reserves had been set up by the colonial forest department as part of its policy to maintain timber supplies. By the late nineteenth and early twentieth century, with increasing landlessness and the closing of the agricultural frontier, these reserves were under severe pressure, and access to them became a central demand of the Burman nationalist movement. These reserves became symbols of colonialism. Forest offenses rapidly increased from the early 1920s to 1930s, including attacks on forestry officers and the murder of one. Calls for deforesting the reserves became a catch phrase. Politicians and the elite, the major beneficiaries of deforestation, vied with each other for the honor of deforesting the reserves (Bryant 1994). The reserves were decimated. To an ecologist, the need for lowland forest reserves is so fundamental that the craziness of deforesting is transparent. The Burmese reality, however, was that Burmans owned Burma, and they had fundamental rights to their forests.

Reality is created by thought. Our societies are built on thoughts and dreams, and the more powerful the thoughts, the more chance they have of becoming reality. Change, from the liberation of nations to the establishment of multibillion-dollar resource projects, comes about by people with powerful thoughts creating new realities. It does not come about by people succumbing to urges to be "realistic," yielding to intermediate reality, or giving up. The message is simple—be as realistic as possible by recognizing where your and others' reality really lies, and what the bottom line is. Above all, don't give up.

SUGGESTIONS

Only you, the practitioner, have the right to decide whether any of the above has relevance to your work or not. Nonetheless, the following exhortations are at least relevant:

1. Don't accept the above and try to prove it wrong.
2. Try holistic approaches.
3. Recognize that the real product of our research is not the research, but ourselves. Our research and experience help mold us, and we are the real products we offer society.
4. Recognize the importance of high-quality natural history studies. By natural history, I mean the sort of descriptive studies characterized by deep thinking and the application of logic that were common in the nineteenth century. Such studies are rarely funded today, yet they would provide the kind of background on species' behavior and distributions that would make quantitative research much better, and would also provide at least some of the information necessary for decision making. Often the most useful information we can bring to bear on problems is our natural history knowledge. We should make more use of it.
5. Be deeply aware of the history of your landscape—make historical researches as important as your biological ones.
6. Live in your landscape—don't just visit it. Things happen over long periods that you will never see otherwise.
7. Enjoy and play with theory, but don't take it seriously. It is never a substitute for good observation and common sense.
8. Be suspicious of all but the most obvious generalities. Completely disbelieve the obvious ones. Be particularly wary of generalities that have gone straight from theory to management suggestions without a good intermediate data-gathering phase.
9. Operate across a range of intellectual areas.
10. Recognize the short time scale ahead of us and act accordingly. Recognize the time scale of your research in relation to this, and if your research will take too long, then redesign it.
11. The simplest reality we can come to terms with is that the people who inhabit our study landscape and own the fragments we study are the most powerful force acting upon them, yet we tend to ignore them in favor of examining "natural" forces such as edges, climate, and so forth. In practice, people have to be considered the most important and powerful functional part of the system.
12. We must accept that ecological data is often very rough and we can't get much out of

it, yet it is what we have to deal with. It is important, therefore, that we diversify our approaches.

13. Always ask whether hypothetico-deductivism can be applied to the ideas or theories one has, or whether the situations and ideas are too complex.
14. Don't misuse statistics. Just because statistics are usually misused, misinterpreted, or misunderstood doesn't mean they are bad methods, and it doesn't mean you should do the same.
15. Read up on the fundamental philosophies and differences of the various schools of statistics and the intricacies of induction.
16. When developing a hypothesis, think about it deeply. Work toward understanding the idea. Do not have the use of a statistical test in mind at all. Decide under what circumstances you would abandon your hypothesis. Think up some predictions, pretend you have the results of a test, then say, "So what?" Deep and clear thinking and logic may obviate much data collecting and analysis. When the theory is well understood, the decisions about appropriate methods and analysis can be made. Put thinking ahead of the statistics.
17. Be dictated to by the requirements of a statistical procedure, and use that procedure only when you thoroughly understand the philosophy of inference underlying that procedure.
18. Use statistical procedures from a range of schools and strictly adhere to their respective methods and interpretation. For example, do a Fisherian significance test properly and interpret it properly. Then set up a formal Neymann-Pearson test and interpret it formally (this means setting both Type I and II error rates beforehand, among other things). Then do an estimation procedure. Then switch hats and do a Bayesian analysis. Take the results of all four, noting their different behavior, and come to your conclusion. Good analysis and interpretation are as important as the fieldwork, so allot adequate time and resources to both.
19. Never force anybody, particularly your students, to perform particular statistical procedures unless you understand them yourself.
20. Develop a relationship and partnership with a good biometrician and give up significance testing. This doesn't mean giving up publishing. Be bold!
21. For conservation biology, formally assess the practical consequences of what you are doing and consider this in the initial development of a study. Do a "what if?" thought procedure. Ask yourself, "If this effect I am studying is a real one, can anything be done about it? Is this a practical red herring? Does it really matter? Is the solution a simple one that is harmless and a good thing to do anyway?" If so, it may be cheaper and easier to institute the solution than to do research to find out whether the problem exists.
22. Recognize that someone might take your advice based on your research. Try to estimate what the costs will be if you are wrong.
23. Put as much, if not more, emphasis on the matrix as on the fragments in your research.

AUTHOR'S NOTE

This chapter was produced in response to the editors' request that a revamping of my original essay (Crome 1994) be included in this volume. The original essay has been partially rewritten and updated, but the chapter retains much of the original structure and wording.

32

Tropical Forest Fragmentation: Synthesis of a Diverse and Dynamic Discipline

William F. Laurance, Richard O. Bierregaard, Jr., Claude Gascon, Raphael K. Didham, Andrew P. Smith, Antony J. Lynam, Virgílio M. Viana, Thomas E. Lovejoy, Kathryn E. Sieving, Jack W. Sites, Jr., Mark Andersen, Mandy D. Tocher, Elizabeth A. Kramer, Carla Restrepo, and Craig Moritz

CONSENSUS can be difficult to reach in any scientific discipline. In the study and management of fragmented tropical landscapes, this task is complicated by several factors: the recent and explosive growth of the field, the widely varying approaches of different investigators, the diversity of land use histories, and the enormous natural variation inherent in the tropical forest biome.

Given these challenges, it is hardly surprising that we initially approached our attempt at a general synthesis with more than a little trepidation. Fortunately, many of this volume's contributing authors were able to attend a symposium and workshop on tropical forest fragmentation as part of the 1995 annual meeting of the Ecological Society of America in Snowbird, Utah. During the workshop we discussed and debated a wide range of issues, eventually producing a draft of the following synthesis. We discovered that there was much we agreed upon.

The goal of this penultimate chapter is to survey ideas pertaining to the ecology and management of fragmented tropical landscapes. Our principal question is: What do we currently know—or think we know—about fragmented tropical forests? Obviously, a single chapter like this one could never be exhaustive, but we have attempted to highlight important concepts, then direct the interested reader to pertinent chapters in the volume and some strategic references elsewhere.

The chapter is divided into three sections. The first focuses on the vulnerability of tropical biotas to fragmentation, while the second details relevant additional features of tropical ecosystems. The final section focuses on the management and conservation of fragmented landscapes. Each section is divided into a series of themes or concepts, and each of these is supported by a number of specific points. While our emphasis is clearly on tropical forests, we believe that many of the principles described below apply to fragmented ecosystems in general.

I. VULNERABILITY OF TROPICAL FOREST BIOTAS

1. *In general, tropical forest biotas are highly vulnerable to habitat fragmentation* because they contain
   A. Greater species richness than any other terrestrial ecosystem (Myers 1984).

B. Many species with restricted or patchy distributions (Diamond 1980; Andersen, Thornhill, and Koopowitz, chap. 18), which often are missing or poorly represented in fragments due to the sampling effect (Wilcox 1980; Lynam 1995).
C. Many rare species, which typically have small population sizes in fragments and thus are especially vulnerable to local extinction (see Andersen, Thornhill, and Koopowitz, chap. 18; Bierregaard and Stouffer, chap. 10; Brown and Hutchings, chap. 7; Viana, Tabanez, and Batista, chap. 23).
D. Many ecological and habitat specialists, which often avoid modified habitats and may be sensitive to ecological changes in fragments such as edge effects (e.g., Laurance 1990, 1991a, 1994, chap. 6; Quintela 1985). (Nevertheless, many organisms traditionally considered primary forest species actually use regrowth forest, especially if it adjoins primary forest [see Bierregaard and Stouffer, chap. 10; Malcolm, chap. 14; Tocher, Gascon, and Zimmerman, chap. 9]).
E. Many species in coevolved interdependencies—such as plant-pollinator, host-parasite, and mimicry relationships—which are likely to be vulnerable to higher-order effects or "ripple effect" extinctions in fragments (L. E. Gilbert 1980).
F. Finally, immigration appears to be important in maintaining local populations of many rare trees (Hubbell and Foster 1986; Nason, Aldrich, and Hamrick, chap. 20) and animals (Laurance 1991a) in tropical forests, and fragmentation often drastically impedes local immigration rates (Didham, chap. 5).

2. *Species vary greatly in their responses to habitat fragmentation.* The most vulnerable species are often those that
   A. Avoid or rarely use matrix habitats surrounding fragments, or are intolerant of habitat changes inside fragments (Bierregaard and Stouffer, chap. 10; Didham, chap. 5; Malcolm, chap. 14; Warburton, chap. 13).
   B. Have large area requirements, such as top carnivores and large-bodied species (Soulé, Wilcox, and Holtby 1979; Schaller and Crawshaw 1980; Newmark 1987).
   C. Are vulnerable to hunting, harvesting, or other forms of exploitation (Redford and Robinson 1987; Smith, chap. 27; Viana, Tabanez, and Batista, chap. 23).
   D. Occur at low population densities (however, locally common species also can be highly vulnerable to fragmentation: Laurance 1990, 1991a; Bierregaard and Stouffer, chap. 10; Didham, chap. 5; Lynam, chap. 15; Sieving and Karr, chap. 11).
   E. Have unstable or highly variable populations (Karr 1982a), or are dependent on such species because of strong or obligate ecological relationships (L. E. Gilbert 1980; Lovejoy et al. 1986).
   F. Have limited dispersal abilities (Laurance 1990, 1991a).
   G. Have low fecundity (Sieving and Karr, chap. 11).

3. *Kinetics of extinction.* Within the same fragment, the time to extinction varies considerably between different taxa.
   A. Some species, such as long-lived trees, are likely to be functionally extinct in fragments well before their populations have actually disappeared. These spe-

cies—fancifully termed the "living dead" (Janzen 1986b)—will be doomed to local or global extinction unless they can be preserved by active management (Harrington et al., chap. 19).

B. Sensitive vertebrates often disappear more rapidly from fragments than most vascular plants, and apparently have greater areal requirements for long-term survival (Corlett and Turner, chap. 22).

C. The time to extinction may be roughly scaled by generation time (Diamond 1984; Pimm, Jones, and Diamond 1988). However, other factors, such as habitat specificity, natural density, and trophic level, are probably also important.

D. Being short-lived and sensitive to fine-scale environmental variation, invertebrates may respond very rapidly to ecological changes in forest fragments (Brown and Hutchings, chap. 7; Didham, chap. 5; Weishampel, Shugart, and Westman, chap. 8).

4. *Habitat fragmentation is invariably destructive to natural biotas* and results in
   A. Habitat loss and alteration.
   B. Increased edge effects.
   C. Increased isolation of forest populations.
   D. An influx of homeless individuals from destroyed or altered habitats (Bierregaard and Lovejoy 1988).
   E. Invasions of forest remnants by exotic species or those favoring disturbed habitats (Lynam, chap. 15; Tocher, Gascon, and Zimmerman, chap. 9).

The following may also occur:

   F. The area becomes more accessible to hunting and other forms of exploitation (Redford and Robinson 1987; Peres and Terborgh 1995; Smith, chap. 27).
   G. Disturbance regimes in forest remnants, such as the frequency of windthrow and fire, may be altered (Laurance, chap. 6; Lynam, chap. 15; Viana, Tabanez, and Batista, chap. 23).
   H. Regional climatic changes may occur (Salati et al. 1979).

5. *Biotas on oceanic islands are often highly vulnerable to habitat modification* despite their limited species richness because they have
   A. Many endemic species (Elton 1958).
   B. Many species with small geographic ranges (Diamond 1980).
   C. Low resistance to invasions of exotic species and pathogens (Elton 1958; Atkinson 1985; Vitousek 1988).
   D. Low rates of immigration, reducing the chance that species that disappear will be able to recolonize the island, or that immigrants will bolster small island populations.

II. FEATURES OF FRAGMENTED TROPICAL FORESTS

6. *Stability and resilience of tropical forests*
   A. Until quite recently, tropical forests were regarded as being stable over geological time scales. However, evidence from biogeographic and geomorphological re-

search increasingly points to the fact that tropical forests have been highly dynamic in their distribution (Haffer 1969; Webb and Tracey 1981; Prance 1982a; Moritz et al., chap. 28). For example, there appear to have been dramatic changes in forest cover during the Pleistocene and early Holocene in many tropical regions worldwide (Prance 1982a; Heaney 1991).

B. Many tropical forests are also highly dynamic over ecological time scales. Physical and climatic perturbations such as river meanders, landslides, windstorms, and fires often create mosaics of patches of differing sizes and disturbance histories (Pickett and Thompson 1978; Foster 1980; Salo et al. 1986).

C. Despite the dynamic nature of many tropical forests, the current loss and fragmentation of forests is occurring at a scale that is unprecedented in geological or ecological history, and clearly is a phenomenon for which many tropical species are poorly adapted. However, ecosystems in areas that have been subjected to region-wide disturbance over geological time periods may prove to be relatively resilient (Brown and Brown 1992; Balmford 1996).

7. *Species invasions.* Our knowledge of invasions in tropical forests is limited, but the following generalities may apply.

   A. Undisturbed habitats are very rarely invaded, and fragmentation clearly predisposes habitats to invasions of exotics and nonforest species (Brown and Hutchings, chap. 7; Malcolm, chap. 14; Laurance, chap. 6; Lynam, chap. 15; Tocher, Gascon, and Zimmerman, chap. 9).

   B. Little is known about faunal invasions in tropical forests. In Thailand, *Rattus rattus* readily invades fragmented monsoonal forests (Lynam, chap. 15). In Queensland, several exotic and non-rainforest rodents can invade fragmented rainforests, but are far less abundant (<2%) than native rodents in fragments (Laurance 1994).

   C. Unless small or disturbed, evergreen forests are quite resistant to plant invasions (Humphries and Stanton 1992; Viana, Tabanez, and Batista, chap. 23). Gap-colonizing exotics often occur in evergreen forest fragments, but usually have an ephemeral existence, disappearing from gaps as canopy closure progresses. Nevertheless, exotic vines and climbers can have an important effect along forest edges (Laurance, chap. 6; Viana, Tabanez, and Batista, chap. 23).

   D. Deciduous and semi-deciduous forests (e.g., monsoonal and tropical dry forests) appear far more prone to weed invasions than evergreen forests, probably because many annual weeds can germinate, become established, and reproduce before seasonal canopy closure occurs (Janzen 1983; Hopkins, Tracey, and Graham 1990; Fensham, Fairfax, and Cannell 1994; Laurance, chap. 6).

8. *Hyperabundant populations of some animal and plant species appear to be a general feature of fragmented systems* because

   A. Some species in fragments respond positively to forest edges or surrounding matrix habitats (Brown and Hutchings, chap. 7; Didham, chap. 5; Laurance, chap. 6; Lynam, chap. 15; Malcolm, chap. 14; Tocher, Gascon, and Zimmerman, chap. 9).

B. Large carnivores often disappear from fragmented landscapes, and their absence can lead to elevated abundances of small and medium-sized omnivores in fragments—a phenomenon termed "mesopredator release" (Soulé et al. 1988; Terborgh 1992b; Terborgh et al., chap. 17).
C. The loss of species in fragments may result in reduced interspecific competition, allowing persisting species to achieve unusually high densities (analogous to density compensation on islands: MacArthur, Diamond, and Karr 1972).
D. A hostile matrix can cause "frustrated dispersal" for some species in fragments, resulting in elevated population densities (Adler and Levins 1994; Terborgh et al., chap. 17).
E. The invasion of remnants by exotic mammals with high reproductive rates (e.g., rabbits, rats, mice) can lead to elevated abundances of predators and the subsequent decline of secondary prey species—a phenomenon termed "hyperpredation" (Smith and Quinn 1996).

9. *Although edge effect phenomena are extremely diverse, several trends appear relatively consistent across different studies and regions.*
   A. Measurable microclimatic changes are generally limited to a zone within 15–60 m of forest edges (Kapos 1989; Williams-Linera 1990b; Kapos et al., chap. 3; Turton and Freiburger, chap. 4).
   B. For many physical phenomena, a reasonable assumption for the maximum penetration of edge effects is about 100 m (Didham, chap. 5). However, some physical edge effect phenomena, such as wind disturbance, can occur over larger spatial scales (200–500 m: Laurance 1991b, chap. 6).
   C. Insects often increase in diversity and abundance near edges, usually as a result of elevated numbers of edge and generalist species (Brown and Hutchings, chap. 7; Didham, chap. 5).
   D. The composition of faunal communities and functional groups usually changes as one crosses forest edges (Quintela 1985; Didham, chap. 5).
   E. Some animals, such as certain insects, birds, and small mammals, are "edge avoiders," often becoming uncommon within 50–100 m of forest edges (Quintela 1985; Lynam 1995; Laurance, chap. 6).
   F. The frequency of treefall gaps may increase in forest remnants as a result of increased wind shear forces (Lovejoy et al. 1986; Leigh et al. 1993; Laurance, chap. 6) and elevated tree mortality near edges (Kapos et al., chap. 3).
   G. Edge effects may exhibit considerable spatial and temporal variation, and some edge phenomena may not change monotonically as a function of edge distance (Murcia 1995; Didham, chap. 5).
   H. The intensity of predation on nesting birds and seeds may increase near edges and in fragments (Harrington et al., chap. 19). This increase may be driven by mesopredator release (Sieving and Karr, chap. 11; Terborgh et al., chap. 17) or by an influx of generalist predators from surrounding matrix habitats (Wilcove 1985; Gibbs 1991).
   I. Subsistence hunting in rainforests may have a negative effect on faunal popula-

tions many kilometers from forest boundaries, necessitating substantial buffer zones around protected areas (e.g., Smith, Horning, and Moore, in press).

J. Shifts in boundary locations and post-fragmentation disturbances can complicate studies of edge effects (Crome, chap. 31; Viana, Tabanez, and Batista, chap. 23).

K. The study of edge effects is still in its infancy, and the precautionary principle dictates that estimates of the maximum penetration of edge effects into forests (fig. 32.1) should be doubled for management purposes. This is not to imply, however, that small, edge-dominated forest remnants lack conservation value for wildlife (Gascon 1993; Turner and Corlett 1996).

10. *Population and molecular genetics.* Although only limited work has been conducted in tropical forests, the following patterns may be typical.

A. Molecular comparisons reveal deep genetic divisions among rainforest faunal populations that are not apparent from studies of phenotypes. These findings suggest that an understanding of geographic patterns of historical vicariance is relevant to the design of nature reserves and to efforts to preserve genetic diversity (Fjeldså and Rahbek, chap. 30; Moritz et al., chap. 28; Patton et al., chap. 29).

B. The effects of recent (<50 years old) habitat fragmentation on genetic diversity may be minor relative to the effects of historical (i.e., Pleistocene) vicariance (Moritz et al., chap. 28).

C. At present, there is little evidence to suggest that a loss of heterozygosity or allelic diversity is commonly a driving force of local extinction in fragmented populations (Lande 1988; Pimm, Jones, and Diamond 1988; Sieving and Karr, chap. 11). However, inbreeding, by reducing fecundity or survivorship, may render small populations more extinction-prone, especially if the affected population exhibits a low growth rate (Mills and Smouse 1994).

D. Genetic neighborhoods of many tropical tree species are often on the order of ten to several hundred hectares. This is larger than the sizes of many forest fragments (Nason, Aldrich, and Hamrick, chap. 20).

11. *Higher-order interactions in tropical forests.*

A. Although tropical forests are regarded as the most ecologically complex of all land communities, the number of clearly documented cases of higher-order effects in fragmented forests is limited. Some examples include the loss of army ant-following birds in Amazonian forest fragments (Bierregaard and Lovejoy 1989; Harper 1989); the decline of some insectivorous bird species as an apparent result of mesopredator release on Barro Colorado Island, Panama (Sieving 1992; Sieving and Karr, chap. 11); and changes in tree communities on tiny man-made islands in Lake Gatun, Panama, following the disappearance of large seed-eating rodents (Putz, Leigh, and Wright 1990).

B. A considerable number of tropical plants have obligate relationships with one or a few species of animal pollinators. Because fragmentation can strongly alter pollinator assemblages (Powell and Powell 1987; Aizen and Feinsinger 1994b), the fitness and reproduction of such plants may be reduced in forest fragments

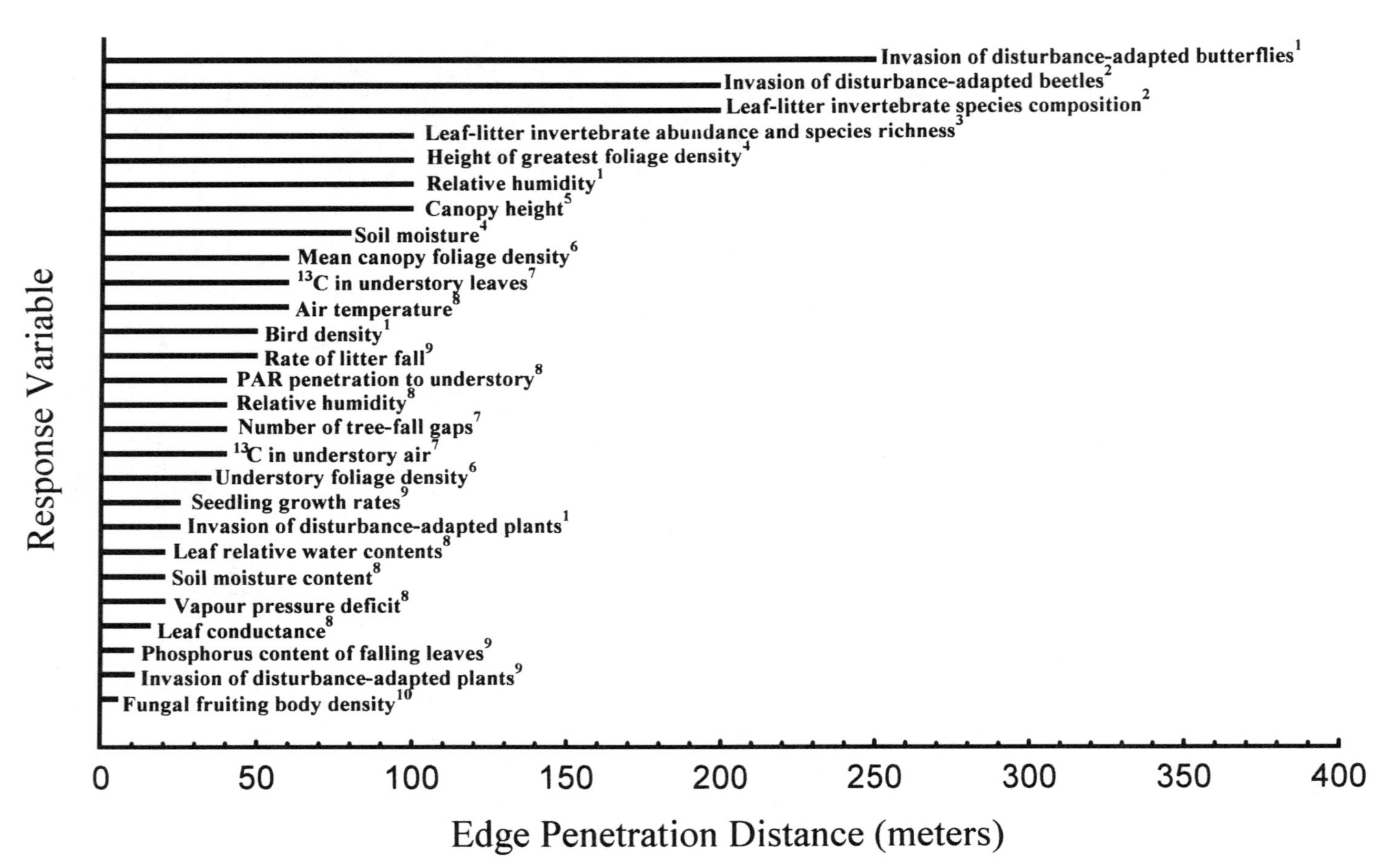

Figure 32.1. Penetration distances of various edge effects into forest remnants, measured as part of the Biological Dynamics of Forest Fragments Project in central Amazonia. Superscripted numbers indicate data sources: 1 = Lovejoy et al. 1986; 2 = Didham, in press; 3 = Didham, chap. 5; 4 = Camargo and Kapos 1995; 5 = Camargo 1993; 6 = Malcolm 1994; 7 = Kapos et al. 1993; 8 = Kapos 1989; 9 = Bierregaard et al. 1992; 10 = R. K. Didham, pers. obs.). PAR = photosynthetically active radiation.

(Linhart and Feinsinger 1980; Aizen and Feinsinger 1994a; Nason, Aldrich, and Hamrick, chap. 20).

C. Ecological processes such as nutrient cycling can also be altered as a result of higher-order effects (Didham, chap. 5). For example, Klein (1989) demonstrated that organic decomposition rates were reduced in Amazonian forest fragments as a result of the loss of specialized dung and carrion beetles.

D. The nature, direction, and magnitude of higher-order effects may be extremely difficult to predict, given the inherent complexities of ecological interactions and the nonlinear nature of many interactions (Crome, chap. 31; Laurance, chap. 6). This realization reinforces the need for caution when undertaking management decisions.

III. MANAGEMENT AND CONSERVATION OF TROPICAL FORESTS

12. *Deforestation is a highly nonrandom process.* The following patterns may be typical.
    A. The most vulnerable and degraded forest types are usually a highly nonrandom subset of all forest areas. Areas that are accessible and productive for agriculture are most prone to clearing (Dale and Pearson, chap. 26).
    B. Forest remnants are usually a nonrandom subset of local habitat and topographic types (Smith, chap. 27). Remnants commonly persist in steep, inaccessible, or low-productivity areas and in areas protected from fire.
    C. Considerable effort should be focused on identifying, classifying, and conserving rare or poorly protected forest types within each nation or biogeographic region (Fearnside and Ferraz 1995; Fjeldså and Rahbek, chap. 30; Kramer, chap. 25; Patton et al., chap. 29; Smith, chap. 27).
    D. Tropical dry forests (monsoon rainforests) have been devastated in many regions of the world. In Central America, for example, 98% of the tropical dry forests have been destroyed (Kramer, chap. 25). In Madagascar, primary monsoon rainforest has been reduced by more than 97% (Smith, chap. 27), while in Australia, over 80% of the monsoon rainforest has been lost (Auslig 1990). Monsoon rainforests are under threat because they are susceptible to fire (Lynam, chap. 15; Smith, chap. 27), may be better suited for conversion to agriculture than evergreen rainforests, and have limited powers of regeneration (Kramer, chap. 25).
    E. Large expanses of continuous forest are becoming increasingly rare in the Tropics, especially in Central America, Africa, and Asia (McCloskey 1993; Whitmore, chap. 1). In peninsular Thailand, for example, less than 20% of the land area is covered by tropical forests. Lowland rainforest constitutes only 1–5% of the total forest area, and less than half of this is included within existing protected areas (Brockelman and Baimai 1993). Most remaining forest tracts in Thailand are smaller than 125 km^2 and are internally fragmented and degraded (Lynam, chap. 15).

13. *Even small forest remnants may have important biological, economic, and social values,* such as
   A. Harboring populations of locally endemic species. For example, some plant, invertebrate, and small vertebrate taxa apparently can persist for considerable periods even in small forest remnants (Gascon 1993; Turner and Corlett 1996; Corlett and Turner, chap. 22; but see also Didham, chap. 5).
   B. Acting as "stepping stones" for faunal or floral dispersal or migration, thereby increasing ecosystem connectivity (Kozakiewicz and Szaki 1995; Powell and Bjork 1995; Crome, chap. 31; Nason, Aldrich, and Hamrick, chap. 20).
   C. Providing a source of immigrants to bolster declining populations in nearby forest remnants—termed the rescue effect (Brown and Kodric-Brown 1977)—or colonists to reestablish former populations that have disappeared (consistent with metapopulation dynamics theory: Fahrig and Merriam 1985; Lidicker 1995; Nason, Aldrich, and Hamrick, chap. 20).
   D. Providing food or shelter for migratory species, such as birds, bats, and moths, that exhibit pronounced elevational movements during different seasons of the year (Janzen 1986a; Loiselle and Blake 1992).
   E. Conserving local populations of trees and their propagules, which are crucial for subsequent reforestation efforts (Lamb et al., chap. 24; Nason, Aldrich, and Hamrick, chap. 20).
   F. Providing windbreaks, stabilizing soils, ameliorating flooding, and other "natural ecosystem services" (Gentry and Lopez-Parodi 1980).
   G. Providing a source of game and other non-timber products to local communities (Fearnside 1989; Peters, Gentry, and Mendelsohn 1989; Balick and Mendelsohn 1992; Schelhas and Greenberg 1996).
   H. Providing readily accessible areas for recreation and education (WTMA 1995).
   I. On a regional level, objective criteria can be used to score and rank existing forest remnants to determine conservation priorities (table 32.1; Smith, chap. 27). Such systems can be valuable in GIS applications.

Table 32.1 Example of a simple ranking system that can be used to assess the relative values of existing forest remnants for biological conservation

	Conservation value		
Criteria	High	Medium	Low
Representation of the habitat in reserves	< 1% in reserves	1–10% in reserves	> 10% in reserves
Endemic species	> 1 present	1 present	None
Disturbance	Pristine	Modified	Degraded
Matrix type	Forest or mixed	Agroforest	Agriculture
Isolation from other forest	< 100 m	100–1,000 m	> 1,000 m
Connectivity	Linked for most species	Linked for many species	Poorly linked
Size	> 300 ha	3–300 ha	< 3 ha
Shape	Roughly circular	Intermediate shape	Irregular
Habitat diversity	High (> 2 habitats)	Medium (2 habitats)	Low (1 habitat)

14. *Conservation of biological "hotspots."*
    A. Many tropical regions exhibit an alarming combination of high biological endemism, severe deforestation, and inadequate protected area systems. Some examples include Madagascar (Smith, chap. 27), the montane forests of West Africa, the Chocó region of Colombia, New Caledonia, parts of the Philippines, and many lowland forests in Asia (Myers 1986, 1988; Lynam, chap. 15).
    B. In South America, local aggregates of endemic birds appear to be concentrated in ecotonal areas, regions with great topographic contrasts, and areas with locally persistent fog. Unfortunately, human pressure on these areas is often great. A number of biological "hotspots" are under severe threat in South America, including southwestern Ecuador, the Andes of Peru, Colombia, and Bolivia, and the Atlantic coastal forests of Brazil. All of these areas have high levels of endemism, severe past and present deforestation, and major gaps in the protected areas network (Fjeldså and Rahbek, chap. 30).
    C. Major centers of diversity and endemism for birds (Fjeldså and Rahbek, chap. 30) and plants (Gentry 1986) in South America often do not correspond closely with the locations of putative Pleistocene refugia. Hence, a conservation strategy that focuses mainly on presumptive refugial locations will be inadequate for biological conservation.
    D. One of the greatest challenges in tropical biology is the conservation of locally endemic species (Gentry 1986; Andersen, Thornhill, and Koopowitz, chap. 18). Local endemics can be effectively conserved only by networks of multiple reserves, stratified across major environmental gradients. In many regions, however, such measures are impossible because of prior deforestation. In such cases it becomes vital to conserve existing forest remnants, which may harbor relict populations of local endemics. Some species may also be conserved by ex situ conservation methods such as botanical gardens and gene banks (Andersen, Thornhill, and Koopowitz, chap. 18).
15. *Many nature reserves are, or will become, internally fragmented by roads, highways, and other linear clearings.* Such clearings can have important effects on reserve communities and should be avoided wherever possible. Ecological changes associated with linear clearings include the following.
    A. Some forest vertebrates avoid even small (<20 m wide) clearings, and thus their populations may become fragmented by narrow linear barriers (Bierregaard and Lovejoy 1988; Burnett 1992; Brockelman and Srikosamatara 1993; Goosem, chap. 16).
    B. There may be pronounced microclimatic changes and other physical edge effects associated within linear clearings (Goosem, chap. 16).
    C. Roads and other linear clearings facilitate invasions of non-rainforest species (Goosem, chap. 16) and hunters into reserves (Arbhabhirama et al. 1988; Brockelman and Baimai 1993; Peres and Terborgh 1995; Smith, chap. 27).
    D. The effects of habitat fragmentation on forest biotas may be exacerbated if the remnant or reserve is also internally fragmented (Goosem, chap. 16).

E. Roads and highways with high traffic volumes in forest reserves can cause considerable mortality of amphibians, lizards, snakes, small mammals, and birds (Goosem, chap. 16).

F. Roads and associated traffic disturbances can cause substantial "halo effects." Some birds and larger mammals, for example, become much less abundant in the vicinity of roads and hiking trails (Griffiths and van Schaik 1993; Goosem, chap. 16).

16. *Restoration and rehabilitation of degraded landscapes.*

A. Reforestation projects attempt to reestablish the original vegetation in an area. This is very difficult to do in the Tropics because the original composition of the forest often is unknown, and there generally is only a limited knowledge of the factors that influence the dynamics and trajectories of succession (Lamb et al., chap. 24).

B. Although secondary forest develops rapidly in many tropical areas, plant species richness often accumulates quite slowly. Even regrowth over a century old usually does not contain all the species present in primary forest (Corlett and Turner, chap. 22).

C. Successional processes alone may not lead to the recovery of a natural forest structure. This is especially true in areas subject to invasions of aggressive weedy species, and where natural disperser faunas have declined or disappeared locally (Corlett and Turner, chap. 22; Lamb et al., chap. 24; Thébaud and Strasberg, chap. 21).

D. Reforestation is expensive, and therefore projects must be carefully targeted. Examples of prime targets for reforestation include habitats of endangered species, faunal corridors, and areas within and around existing nature reserves (Lamb et al., chap. 24; Viana, Tabanez, and Batista, chap. 23).

E. Fires are highly disruptive to natural rainforest successional processes and may result in the conversion of burnt areas to dry forest (Viana, Tabanez, and Batista, chap. 23), bamboo thickets (Lynam, chap. 15), or other degraded forest types (Uhl and Buschbacher 1985; Uhl 1987).

F. On private lands, forest restoration should be based on both ecological principles and the socioeconomic and cultural perceptions of landowners. Researchers should aim at providing a "menu" of restoration options to landowners and should strive to incorporate the landowner's perspectives in the design, implementation, and evaluation of reforestation programs (Viana, Tabanez, and Batista, chap. 23).

G. Technologies for tropical reforestation aiming at both environmental protection and wood production have improved significantly in recent years (Lamb et al., chap. 24). In the Atlantic coastal forests of Brazil, for example, the cost of mixed plantings of native trees has declined from about U.S.$7,000/ha to U.S.$1,000/ha, mainly as a result of greater efficiencies in tree production and reforestation (Viana, Tabanez, and Batista, chap. 23).

Table 32.2 A comparison of matrix habitats from a wildlife conservation perspective.

			Protection from			Natural	
Matrix type	Gene flow	Wildlife habitat	Climatic extremes	Exotic species	Fire	environmental services[a]	Total score
Fully protected forest	4	4	4	4	4	4	24
Non-timber harvest only	4	3	4	4	4	4	23
Low-intensity selective logging[b]	4	3	4	3	3	4	21
Traditional forest management[c]	3	3	4	3	3	4	20
Medium-high intensity logging	3	3	3	3	1	3	16
Mature regrowth forest	3	2	3	2	2	3	15
Low-diversity agroforestry	2	2	2	2	2	2	12
Plantation forests	1.5	1.5	3	2	3	3	14
Water	0	0	0	4	4	1	9
Row crops	1	1	0	0	1	0	3
Cattle pastures	1	1	0	0	0	1	3

Note: Each habitat type was subjectively scored by a panel of fifteen researchers for six key attributes, and the scores were totaled to yield a hierarchy from most (high scores) to least favorable (low scores).

a. Such as flood amelioration, erosion control, and functioning as a carbon sink.

b. Harvest rate of < 1 tree/ha.

c. Refers to low-density populations of traditional swidden agriculturalists and hunters such as the Yanomami Indians of northern Amazonia.

17. *Managing the matrix.*
    A. Ecologists are becoming increasingly aware that the composition of matrix habitats surrounding fragments often plays a crucial role in determining the structure and dynamics of fragment biotas (Laurance 1990, 1991a; Bierregaard et al. 1992; Bierregaard and Stouffer, chap. 10; Malcolm, chap. 14; Tocher, Gascon, and Zimmerman, chap. 9; Warburton, chap. 13).
    B. From an ecological perspective, the most favorable matrix is usually the one in which the habitat structure, floristic composition, and microclimate is most similar to that of primary forest.
    C. Matrix habitats can be ranked for key attributes affecting forest fragments, such as their capacity to allow movement and gene flow for forest species, provide wildlife habitat, and ensure physical protection from climatic extremes, fire, and exotic species. A semi-objective scoring of these attributes creates a hierarchy of matrix habitat types ranging from most to least preferable (table 32.2).
18. *Economics of tropical deforestation.*
    A. Microeconomic factors, such as poorly defined timber leasing agreements, a failure to define property rights for landholders, and inappropriate government policies, may strongly influence regional and national rates of deforestation (Frohn, Dale, and Jimenez 1990; Dale, Pearson et al. 1994). However, macroeconomic conditions may also be important determinants of deforestation, independent of microeconomic factors (Kahn and McDonald, chap. 2).
    B. Because macroeconomic factors appear to contribute to deforestation, mechanisms must be developed to ensure that wealthy nations compensate developing

nations for preserving their environmental assets. Debt-for-nature swaps and carbon-offset projects are a good beginning, but far more comprehensive policies are direly needed (Kahn and McDonald, chap. 2).

C. Population growth affects deforestation rates by exacerbating both microeconomic and macroeconomic pressures, typically leading to lower per capita income, increased consumption needs, and greater exploitation of natural resources (Kahn and McDonald, chap. 2).

D. Most nations will retain only relatively small areas of forest for purely conservation purposes. Consequently, managed, multiple-use forests will play an essential role in biodiversity conservation. Retention of managed forests is to be *greatly preferred* over land uses that result in the large-scale loss or fragmentation of forest cover (Whitmore, chap. 1). International lending and aid agencies should consider this point carefully when assessing the viability and sustainability of proposed development projects (cf. Norse 1987).

33

Key Priorities for the Study of Fragmented Tropical Ecosystems

Richard O. Bierregaard, Jr., William F. Laurance, Jack W. Sites, Jr., Antony J. Lynam, Raphael K. Didham, Mark Andersen, Claude Gascon, Mandy D. Tocher, Andrew P. Smith, Virgílio M. Viana, Thomas E. Lovejoy, Kathryn E. Sieving, Elizabeth A. Kramer, Carla Restrepo, and Craig Moritz

The fate of tropical forests and their biotas depends on a bewilderingly complex suite of factors. Historically, changes in global climatic patterns have probably had the strongest effect on the distribution of the world's tropical forests. Within the past half-century, however, a growing number of anthropogenic factors have overwhelmed climate as prime determinants of the extent and nature of these ecosystems.

In reality, factors such as international trade relationships and treaties, national development policies, laws and their enforcement, and the local consciousness and education of residents and colonists all interact to hasten or slow the incursion of humans into the tropical forests of the world. One need only compare the current activities of such conservation organizations as Conservation International or the World Wide Fund for Nature with their target species-oriented programs of just two decades ago to see that these are indeed the arenas in which many conservationists now feel that they can act most effectively. Prospecting for biomedically active compounds in Costa Rica, ethnobotany, "debt-for-nature swaps," ecotourism, and the empowerment of local citizens' groups now highlight the annual reports of the major conservation foundations.

The essential first step toward slowing the rapid erosion of tropical ecosystems is to create conditions in which conservation is desired. Only when diverse sociological, political, and economic factors are aligned in such a way that tropical forests are seen by a majority—or a sufficiently vocal and convincing minority—as worthy of preservation and management, and resources are consequently made available to these ends, will managers and conservation biologists be able to begin effective stewardship of the tropical landscape (e.g., Bodmer 1994).

In theory, three approaches to conservation management are possible: preserving what we have now, planning the nature of the landscape that development will leave behind, and managing what is left. Ideally, we should set aside vast tracts of forest in national parks of no less than 2 million hectares each (Thiollay 1989; Terborgh 1992b). Even under the most optimistic scenarios, however, human populations will continue to expand into tropical forests, seeking material and economic gain or simply a place to live. Costa Rica, the tropical country with the most progressive, conservation-oriented government in the world, has committed to establishing only 25% of its countryside in protected areas. Because we have no reason to believe that other countries will protect more than a small percentage of their land in large reserves, we can safely predict that most of the world's remaining tropical forests will eventually become mosaics of forest remnants surrounded by modified habitats (see Franklin 1993). Accepting this reality, as resource managers, we would like to design the size and configuration of the remnants before development begins (Laurance and Gascon, in press), but this is rarely an option. By far the most probable scenario is that we will often have to manage an existing landscape mosaic created by colonization or economic exploitation that proceeded with little or no regard for its conservation implications.

For managers of existing landscape mosaics, resources, in the broadest sense, are always likely to be limiting. Such resources include not only money, but also information, trained personnel, and the genetic substrate required for species reintroductions, reforestation, and so forth. Priorities must be set, and inevitably compromises will have to be made, in determining how scarce resources are channeled. Hence, we need to have clear goals. If preserving local species richness per se is the main objective, a landscape mosaic of pasture, second growth, fragments, and a large tract of primary forest might fulfill the goal better for some species (e.g., frogs) than a single larger tract of primary continuous forest. Second-growth species, however, tend to be widespread and are rarely threatened on a regional scale. If preservation of a unique biota and its myriad interactions is the goal, a larger tract of forest, undisturbed by clearing, will almost certainly be called for.

Whether we are managing an existing landscape or enjoy the luxury of planning development before the fact, successful conservation depends upon our understanding of the physical and biological changes that occur in the aftermath of habitat fragmentation, and our ability to mitigate them. This book represents but a sample of the international effort to fathom the process of forest fragmentation at the species and ecosystem levels. In the preceding chapter (Laurance et al., chap. 32), we have attempted to distill some of the hard-won understanding we have already achieved down to a modest number of generalizations. We have also pointed out where and why generalizations seem impossible.

In this chapter, we outline unanswered questions and whole avenues of research that we feel are needed to fill critical gaps in our current knowledge base. In a provocative essay, Crome (chap. 31) addresses many methodological and philosophical challenges confronting students of habitat fragmentation. It is our hope that if researchers address the following questions with the intellectual and statistical rigor that Crome advocates, we will be better able to maximize the compatibility of development and conservation in the tropical landscape.

In broadest terms, our research needs range from the most basic information regarding the natural history of tropical forest species to knowledge of how the global marketplace influences local deforestation rates. At the species level, we need to know how species live in undisturbed forests so that we can understand why some are sensitive to fragmentation while other, closely related species are not. We need to understand higher-order interactions between species. Such baseline biological data are necessary, but not sufficient, to manage a fragmented landscape. On regional and global scales, we must understand better the socioeconomic factors that drive deforestation itself. As Crome (1994, chap. 31) and

Pickett and Cadenasso (1995) have strongly asserted, we cannot study tropical forest systems without also considering the species that most threatens their existence: *Homo sapiens.*

We offer the following menu of questions and issues as a challenge to those who would work to study and conserve the richest and most intricate terrestrial ecosystems on earth.

I. METHODOLOGICAL ISSUES

1. *At present, studies of tropical forest fragmentation are highly eclectic, often rendering generalizations tenuous at best.* More integrated studies are needed. Only in the Biological Dynamics of Forest Fragments Project (BDFFP) has there been some level of integration between studies of various taxa in fragmented habitats. Reasons for the highly varied findings of different projects include
   - A. Different methods.
   - B. Different habitats and biotas.
   - C. Different target taxa.
   - D. Different spatial scales.
   - E. Different matrix habitats and land uses surrounding fragments.
   - F. Different fragment ages.
2. *Many fragmentation studies suffer from the confounding effects of edge and isolation.* Rigorously designed studies are needed to separate these effects (Laurance and Yensen 1991; Didham, chap. 5).
3. *We need to explicitly assess the degree to which physical and biotic edge effects change with edge age* (Kapos et al., chap. 3; Didham, chap. 5).
4. *Studies of true islands can provide useful analogues for habitat fragments, but extrapolations between the two should be made with caution because of certain features of islands.*
   - A. For many nonflying species, there is an absence of dispersal between islands.
   - B. On islands, there are no matrix populations of wildlife or of exotic species.
   - C. Climatic effects associated with islands differ from those in altered terrestrial systems.
   - D. Edge effects on islands probably differ from those in mainland situations.
   - E. Anthropogenic land bridge islands offer the best analogues to landscape fragments because they are initially "supersaturated" and are similar in age to most contemporary fragments (see Sieving and Karr, chap. 11; Lynam, chap. 15; Terborgh et al., chap. 17). Pleistocene land bridge islands have had millennia to equilibrate, but may offer insights into the distant future of habitat remnants.
5. *Studies of fragmented systems must not exclude consideration of ongoing human influences.* Local human populations continue to live in and around most fragments, and they may continually alter the surrounding matrix (see Crome, chap. 31). Additionally, both anthropogenic and natural climatic change is likely to alter the distribution of suitable habitat for many species. Species that have limited abilities to migrate (especially plants) may be especially hard-hit by the combination of climatic change and habitat fragmentation (see Peters and Lovejoy 1992; Smith, Smith, and Shugart 1992).

II. SPECIES-LEVEL ENQUIRIES

6. *Fundamental information on the natural history of most species is lacking.* Natural history data will clearly enhance our ability to generate predictive models of the effects of land use practices on native species. Specifically, we need to know
   A. The size of territory or home range required for important or keystone fauna.
   B. Demographic data on annual survivorship and fecundity.
   C. Yearly and seasonal variation in population sizes in natural habitats.
   D. The likelihood that a given species or species assemblage will use and traverse different types of matrix habitat. Work at the BDFFP, for example, has revealed surprising results for birds (Bierregaard and Lovejoy 1989; Stouffer and Bierregaard 1995a,b), frogs (Tocher, Gascon, and Zimmerman, chap. 9), and euglossine bees (Powell and Powell 1987).
   E. For plants, information on obligate links to pollinators and seed dispersers.
   F. Much better data on species distributions in order to improve estimates of expected rates of extinction under different patterns and scales of deforestation (Andersen, Thornhill, and Koopowitz, chap. 18; Patton et al., chap. 29).
7. *What are the characteristics of vulnerable species or taxa?* There is no single "model" or "target" taxon that provides special insights into forest fragmentation. Different taxa respond differently, and there are distinct advantages and disadvantages of studying any particular group (table 33.1).
8. *Are species that are locally vulnerable to extinction also prone to global extinction?* Such species may include
   A. Narrow endemics (Andersen, Thornhill, and Koopowitz, chap. 18).
   B. Island species (Carlquist 1974).
   C. Habitat specialists (Shafer 1990).
   D. Rare species (Terborgh and Winter 1980), which often include large-bodied species, those at high trophic levels, and those that are patchily distributed (which usually are habitat specialists).
   E. Migratory or nomadic species that require different habitats at different stages of the life cycle.
   F. Species with low reproductive rates.
   G. Species with limited dispersal abilities.
   H. Species prone to overhunting or overexploitation.
9. *What contributions can molecular genetics make to the management of fragmented populations?* At the most basic level, biochemical and molecular genetics can assess the relationship between heterozygosity and fitness, levels of inbreeding, and the loss of alleles in small populations and the implications this might have for the adaptive potential of remnant populations (Frankel and Soulé 1981; Allendorf and Leary 1986). Some inroads into these questions for have been made for plant (Menges 1991; Holderegger and Schneller 1994; Young and Merriam 1994; Nason, Aldrich, and Hamrick, chap. 20) and animal (Briscoe et al. 1992) populations, but our understanding of the issues involved is still minimal. Important questions and research

Table 33.1 Advantages and disadvantages of various taxa for studies of tropical forest fragmentation.

Group	Advantages	Disadvantages
Insects	Species-rich Ecologically diverse Functionally important Easily sampled Populations respond rapidly to change Many local endemics	Too species rich—studies focus on small groups so ecological diversity grossly underrepresented Taxonomy and ecology poorly understood Less public appeal Relatively small area needs
Reptiles	Species-rich In some areas, taxonomically well known Often sensitive to changes in microclimate	Ecological data often limited May require relatively small areas of suitable habitat
Birds	Well studied taxonomically Readily sampled Large area requirements Important pollinators and seed dispersal agents	May be less vulnerable than nonflying species because of high vagility Overstudied relative to other groups
Mammals	Large area needs Lower vagility than birds High public appeal (e.g., primates)	Often secretive and difficult to sample Somewhat limited species richness
Trees	Structurally important Many ecological interdependencies Taxonomically poorly known in some areas	Slow responses to fragmentation Hard to identify

needs for fragmented populations (including tropical as well as other species) include the following.

A. Are short-term demographic factors of greater significance than population genetic factors to the immediate management and recovery of small populations (Lande 1988; see also Caro and Laurenson 1994), or is there a direct link between population genetics and demography with short-term conservation relevance (Avise 1995)?

B. The collection of genetic information can be useful for managing genetic diversity (usually at the intraspecific level) for its own value as well as for long-term planning and policy. Key objectives include

   i. Estimating and managing the overall adaptive potential within (Lande 1995) and among populations (Lesica and Allendorf 1995).

   ii. Discovering evolutionarily significant units (ESUs) within species (Moritz 1994b).

   iii. Elucidating phylogenetic histories of taxa with important implications for conservation planning (e.g., Moritz et al., chap. 28; Patton et al., chap. 29; Fjeldså and Rahbek, chap. 30).

iv. Conducting among-species assessments of phylogeographic structure and identifying evolutionarily important biotic regions (Erwin 1991; Brooks, Mayden, and McLennan 1992; Fjeldså and Rahbek, chap. 30) defined by the presence of ESUs of multiple species (Avise 1992; Joseph, Moritz, and Hugall 1995).

C. Molecular ecology (Moritz 1994a,b, 1995), the use of genetic markers to complement ecological and demographic studies (Avise 1995), may contribute in the following ways.

i. Identifying management units (Moritz 1994a) that are genetically distinct enough to be considered demographically independent over ecological time scales (although not sufficiently so to be called ESUs: Avise 1995).

ii. Assessing the structure and connectedness of apparently isolated local populations (see also item 10).

iii. Clarifying the distinction between historical processes and those resulting from recent anthropogenic changes in habitats.

10. *A metapopulation approach may help us to understand the dynamics of populations in natural and fragmented landscapes.*

A. A prerequisite for understanding metapopulation dynamics (Levins 1969) and refining models of their behavior is assessing the level of connectivity between fragmented populations. Two approaches to this question involve

i. Empirical data (e.g., mark-recapture, radiotelemetry) on the rate and extent of animal movements between patches in a landscape matrix.

ii. Data from molecular markers, which, although in their infancy, offer great promise for assessing the structure and connectedness of metapopulations (Hastings and Harrison 1994) and inferring various population processes (Moritz 1995).

B. We need further theoretical developments relating changes in gene flow or demographic trends to the rapid fluctuations in size and connectivity of populations (Moritz 1995) that occur commonly in fragmented tropical habitats (Laurance et al., chap. 32).

III. COMMUNITY-LEVEL ISSUES

11. *The overall amount of extinction in fragmented landscapes is difficult to assess, let alone predict* (Brown and Brown 1992; Heywood and Stuart 1992; Whitmore, chap. 1). Further work on assessing extinction rates at local, regional, and global scales is needed. The following questions are relevant.

A. Do species disappearances and ecological changes in small fragments foreshadow those that will eventually occur in large fragments (see Terborgh et al., chap. 17), or are small and large fragments qualitatively different?

B. Are plants intrinsically better than vertebrate animals at surviving in fragmented habitats (see Corlett and Turner, chap. 22), or will many plant populations persisting in fragments eventually dwindle to extinction because their crucial pollinators or seed dispersers have disappeared (Thébaud and Strasberg, chap. 21; Viana, Tabanez, and Batista, chap. 23)?

C. Are many long-lived plants functionally extinct in fragments—the "living dead" (Janzen 1986b)—or can they persist almost indefinitely?

D. How important are "ecological distortions" in driving local extinctions of species in fragmented forests (see Terborgh et al., chap. 17)?

E. How effective and robust are various models relating extinction rates to levels of deforestation (e.g., Andersen, Thornhill, and Koopowitz, chap. 18), and how similar are the predictions of different models under varying assumptions?

F. Are predicted rates of global species extinctions strongly and linearly correlated with human population size?

12. *What features of fragmented communities influence their vulnerability to invasions of exotic and generalist species?* Some possible factors include

A. Forest type (e.g., evergreen vs. deciduous: Laurance, chap. 6).

B. Matrix type (e.g., agricultural vs. secondary habitats).

C. Fragment size (Brown and Hutchings, chap. 7; Lynam, chap. 15; Smith, chap. 27).

D. Disturbance regimes within the fragment (e.g., storm damage, logging, fire: Laurance, chap. 6; Viana, Tabanez, and Batista, chap. 23).

13. *Higher-order interactions in tropical forests need much further study.* The following questions are relevant.

A. Can higher-order effects be predicted, or will the inherent complexities of interactions and nonlinear relationships often render them unpredictable (Laurance, chap. 6)?

B. Because they support many coevolved mutualisms, are tropical forests more prone to higher-order effects than other kinds of terrestrial communities?

C. Are higher-order effects profitable to study, or are they so inherently complicated and multifactorial in nature that there will be little chance of detecting changes following fragmentation (see Harrington et al., chap. 19) or even planning to avoid changes following fragmentation?

D. Do the "ecological distortions" observed on recent land bridge islands (Terborgh et al., chap. 17) also occur in habitat fragments, or do islands and fragments behave differently after isolation?

14. *Are some regions more vulnerable to fragmentation than others, and if so, can we identify them in advance?* The two hypotheses below suggest that current or historical disturbance regimes may influence the resilience of a region's biota to anthropogenic disturbance. Are they broadly applicable?

A. The historical stability hypothesis: In general, regions where total forest cover has been stable over geological time have high species richness and many ecological and habitat specialists, and thus are highly vulnerable to forest fragmentation (Moritz et al., chap. 28; Fjeldså and Rahbek, chap. 30). Conversely, in regions where total forest cover has been unstable historically, we might expect communities to be more resilient (Balmford 1996).

B. The current disturbance hypothesis: Regions that experience frequent and major natural disturbances (e.g., hurricane and cyclone areas) are more resilient and resistant to habitat loss and fragmentation than those that experience fewer natural disturbances (Lugo 1988a; Corlett and Turner, chap. 22).

IV. RESOURCE MANAGEMENT AND CONSERVATION

15. *There is a need for better communication among resource managers, landscape ecologists, and conservation biologists, and for greater familiarity with each other's literature.*
    A. Landscape ecology and fragmentation research are in many respects going on in parallel worlds using different terms for the same processes—edges versus ecotones, and so forth. Both fields share a similar focus, but there seems to be little sharing of research among the two disciplines.
    B. Scientists should make a concerted effort to address research questions that are of *practical* relevance to resource managers (see Crome, chap. 31). Greater dialogue between researchers and resource managers is needed during the initial, conceptual stages of research projects to ensure that this occurs.
    C. Scientists should make extraordinary efforts to ensure that their research findings become available to real-life managers, conservationists, and the lay public. Merely publishing the results in refereed journals is inadequate. Rather, scientists should also publish their findings in popular venues and in the technical journals read by resource managers, and should exploit a variety of avenues to ensure that their key recommendations for land use are understood and, wherever possible, implemented.
16. *How do resident landowners perceive and interact with forest fragments* (Schelhas 1996)? What strategies are most effective in promoting a sympathetic attitude toward forest conservation among rural landowners (Viana, Tabanez, and Batista, chap. 23), and how are these best implemented?
17. *What are the effects of additional disturbances on forest fragments?* How do additional disturbances such as logging (Whitmore, chap. 1), hunting (Smith, chap. 27), fire (Viana, Tabanez, and Batista, chap. 23), and internal roads (Goosem, chap. 16) exacerbate the ecological effects of forest fragmentation?
18. *A range of predictive models are needed to conserve and effectively manage fragmented landscapes.* These include spatial, socioeconomic, and bioclimatic models at both global and regional scales.
    A. Spatial models linking remote sensing, geographic information systems (GIS), and good scientific data on the responses of species to forest fragmentation are currently being developed (e.g., Dale, Pearson et al. 1994), but need refinement. Such models are required for all regions presently experiencing rapid or large-scale deforestation (e.g., Dale and Pearson, chap. 26; Smith, chap. 27). These models may incorporate empirical data on extinction rates or on the habitat and area requirements of key taxa. Such models can integrate information on the susceptibility of different taxa regionwide with an understanding of the socioeconomic factors that influence patterns of forest colonization and exploitation

(e.g., Dale, Southworth et al. 1993; Offerman et al. 1995; Bierregaard and Dale 1996).

B. GIS models incorporating major biophysical (e.g., vegetation type, geology, soil type) and topographic data layers are direly needed for "gap analysis" (Scott et al. 1987, 1993) and land use planning in many tropical regions (e.g., Fearnside and Ferraz 1995; Fjeldså and Rahbek, chap. 30).

C. Bioclimatic models such as BIOCLIM and DOMAIN can be used to evaluate potential changes in species distributions under various scenarios of future changing climate (so long as suitable regional climate surfaces are available: Nix and Switzer 1991). The interaction of deforestation and global climatic change should be studied because the ability of species to alter their distributions in response to changing climate may be greatly diminished in fragmented landscapes (see Peters and Lovejoy 1992; Smith, Smith, and Shugart 1992).

19. *Are reserve design algorithms useful for developing or improving systems of protected areas in tropical regions?* Researchers (e.g., Margules and Stein 1989) have devised models that can propose minimum sets of reserves needed to preserve known plant communities. An analogous approach should be applicable to any community so long as the baseline biophysical data are available. Reserve design algorithms, however, intrinsically employ a minimalist approach to reserve designation (Crome 1994), and hence may be of limited practical utility. The gap analysis strategy (see Scott et al. 1987, 1993) may be a more ecologically viable method, at least for identifying gaps in existing reserve systems.

20. *The planning and delineation of new reserves must involve consideration of a wide range of issues beyond the simple question of viability of populations of target taxa.* Factors such as indigenous peoples' issues, local attitudes toward conservation, and the costs of protecting preserves will strongly influence the success or failure of protected areas. The following seminal studies have highlighted avenues of research that are ripe for further study.

A. Peres and Terborgh (1995) discussed the economics of managing reserves of different shapes and sizes.

B. Smith (chap. 27) considered the efficacy of reserves in relation to their distance from human settlements.

C. Fearnside and Ferraz (1995) presented a gap analysis for Amazonia and discussed the economic opportunities for establishing additional reserves.

D. Silva and Sites (1995) assessed the role of indigenous-Amerindian reserves in the conservation of Amazonian biodiversity (focusing on squamate reptiles). Similar analyses should be performed for other taxa.

21. *In general, the modified matrix surrounding tropical forest fragments has been poorly studied.* Research is needed to explicitly assess the effects of different matrix habitat types on

A. Animal movements and interfragment dispersal (see Aberg et al. 1995).

B. Weed and exotic animal invasions.

C. Faunal population dynamics in fragments.

D. Species richness of fragment assemblages (see Harris 1984 for some temperate examples).

E. Conservation values of remnants.

22. *Remarkably little work has been focused on the efficacy and design of faunal corridors in the Tropics.* Indeed, even in temperate regions, the debate over whether corridors are effective and desirable is not, and perhaps cannot, be resolved (Noss 1987; Simberloff and Cox 1987; Harris and Scheck 1991; Soulé and Gilpin 1991; Crome, chap. 31). Not so gracefully sidestepping this issue, we suggest the following generalities and associated questions.

A. The concept of "minimum critical corridor width" may be of limited utility because there is probably a monotonic increase in the proportion of primary forest species using a corridor as width increases. It may, however, be possible to identify general thresholds for designing corridors for particular taxa.

B. Corridor width is unlikely to be the only relevant factor. Do factors such as corridor length, topographic position, and aspect also influence corridor effectiveness?

C. The species requiring the widest corridors are often forest interior specialists that are highly vulnerable to fragmentation (W. F. Laurance 1990; S. G. W. Laurance 1996).

D. There may be no generalities for the design of corridor systems—or indeed, for many specific questions in landscape ecology. In each case the answer may be, "It depends on the particular situation" (Crome, chap. 31).

23. *Are forest remnants likely to play a crucial role in reforestation efforts by being key sources of plant propagules (including locally adapted ecotypes) and of native fauna* (Guevara, Purata, and Van der Maarel 1986; McClanahan and Wolfe 1993; Silva, Uhl, and Murray 1996; Corlett and Turner, chap. 22; Viana, Tabanez, and Batista, chap. 23)? Studies of the practical values of forest remnants—even tiny ones—will provide valuable information to regional planners and wildlife managers.

24. *Research is urgently needed to devise ways of harvesting forests without greatly affecting species composition* (Johns 1992a,b; Putz and Pinard 1993; Prabhu 1994; Whitmore, chap. 1; Kahn and McDonald, chap. 2).

V. SOCIOECONOMICS

25. *More empirical work is needed to examine the relationship between macroeconomic conditions and national deforestation rates* (see Kahn and McDonald, chap. 2).

26. *There is an urgent need to develop policies and practices that promote sustainable uses of forest resources.* These can include extractive reserves, the development of sustainable agroforestry and timber-harvesting techniques, the creation of markets for non-timber products, aquaculture, and wild animal husbandry for food production (Kahn and McDonald, chap. 2). These policies and practices will require better understanding of socioeconomics on the local rather than the macroeconomic scale.

27. *There is a great need to develop new, innovative policies that provide mechanisms by which wealthy nations compensate developing nations for preserving their forests* (Kahn and McDonald, chap. 2). The debt-for-nature swap was one such approach. Since

Thomas Lovejoy first proposed this novel scheme in 1984, swaps have become less effective for a variety of reasons (Kahn and McDonald, chap. 2). In recent years, carbon-offset projects have become an increasingly popular way to generate revenue for forest conservation and management in developing nations (see, e.g., Putz and Pinard 1993). Similar linkages and strategies are certainly lurking near the horizon, awaiting the creative insights of scientists or policy makers with a firm grasp of the ecological and socioeconomic underpinnings of the rapidly changing global arena.

Contributors

Preston R. Aldrich
Department of Botany
University of Georgia
2502 Plant Sciences
Athens, Georgia 30602-7271
United States of America

Mark Andersen
Department of Fishery and Wildlife Sciences
New Mexico State University
Las Cruces, New Mexico 88003-0003
United States of America

João Luis F. Batista
Department of Forest Sciences, ESALQ
University of São Paulo
Caixa Postal 9
Piracicaba, São Paulo 13418-900
Brazil

Richard O. Bierregaard, Jr.
Department of Biology
University of North Carolina at Charlotte
Charlotte, North Carolina 28223
United States of America

Keith S. Brown, Jr.
Departamento de Zoologia
Instituto de Biologia
Universidade Estadual de Campinas
C.P. 6109
Campinas, São Paulo 13083-970
Brazil

Ana Rita Bruni
Sociedad Conservacionista Audubon de Venezuela
Apartado 80450
Caracas 1080-A
Venezuela

José Luis Camargo
Projeto Mico-Leão-Dourado
C.P. 113.049
28820-970, Silva Jardim/RJ
Brazil

Richard T. Corlett
Department of Ecology and Biodiversity
University of Hong Kong
Pokfulam Road
Hong Kong

Francis H. J. Crome
P.O. Box 447
Atherton, Queensland 4883
Australia

Michael Cunningham
Cooperative Research Centre for Tropical Rainforest Ecology and Management/ Department of Zoology
University of Queensland
Brisbane, Queensland 4072
Australia

Virginia H. Dale
Environmental Sciences Division
Oak Ridge National Laboratory
P.O. Box 2008
Oak Ridge, Tennessee 37831
United States of America

Raphael K. Didham
Biodiversity Division
Department of Entomology
The Natural History Museum
Cromwell Road
London SW7 8RD
England

Jon Fjeldså
Centre for Tropical Biodiversity
Zoological Museum
University of Copenhagen
Universitetsparken 15
DK 2100 Copenhagen 0
Denmark

Heidi Jo Freiburger
Cornell Laboratory of Ornithology
159 Sapsucker Woods Road
Ithaca, New York 14850
United States of America

Gislene Ganade
School of Biological Sciences
Macquarie University
Sydney NSW 2109
Australia

Claude Gascon
Departamento de Ecologia
Instituto Nacional de Pesquisas da Amazônia
C.P. 478
Manaus, Amazonas 69011-970
Brazil

Miriam Goosem
Wet Tropics Management Authority
P.O. Box 2050
Cairns, Queensland 4870
Australia

J. L. Hamrick
Department of Botany
University of Georgia
2502 Plant Sciences
Athens, Georgia 30602-7271
United States of America

Graham N. Harrington
Division of Wildlife and Ecology
CSIRO Tropical Forest Research Centre
P.O. Box 780
Atherton, Queensland 4883
Australia

Roger W. Hutchings
Department of Entomology
University of Maryland
1300 Symons Hall
College Park, Maryland 20742
United States of America

Anthony K. Irvine
Division of Wildlife and Ecology
CSIRO Tropical Forest Research Centre
P.O. Box 780
Atherton, Queensland 4883
Australia

Leo Joseph
Department of Zoology
University of Queensland
Brisbane, Queensland 4072
Australia

James R. Kahn
Department of Economics
University of Tennessee
Knoxville, Tennessee 37996-0550
United States of America

Valerie Kapos
World Conservation Monitoring Centre
219 Huntingdon Road
Cambridge CB3 0DL
United Kingdom

James R. Karr
Institute for Environmental Studies
Engineering Annex, FM-12
Box 352200
University of Washington
Seattle, Washington 98195
United States of America

Rod Keenan
Queensland Forest Research Institute
Department of Primary Industries
Atherton, Queensland 4883
Australia

Harold Koopowitz
Department of Ecology and Evolutionary Biology
University of California, Irvine
Irvine, California 92717
United States of America

Elizabeth A. Kramer
Institute of Ecology
University of Georgia
Athens, Georgia 30602
United States of America

David Lamb
Cooperative Research Centre for Tropical Rainforest Ecology and Management/ Department of Botany
University of Queensland
Brisbane, Queensland 4072
Australia

Márcia C. Lara
Department of Zoology
University of Queensland
Brisbane, Queensland 4072
Australia

William F. Laurance
Biological Dynamics of Forest Fragments Project
Instituto Nacional de Pesquisas da Amazônia
C. P. 478
Manaus, Amazonus 69011-970
Brazil
and
Cooperative Research Centre for Tropical Rainforest Ecology and Management
CSIRO Tropical Forest Research Centre
P.O. Box 780
Atherton, Queensland 4883
Australia

Lawrence Lopez
Departmento de Zoologia
Universidad Nacional Agraria de La Molina
Casilla 234
La Molina, Lima
Peru

Thomas E. Lovejoy
Division of External Affairs
Smithsonian Institution
Washington, D.C. 20560
United States of America

Antony J. Lynam
Wildlife Conservation Society
2300 Southern Boulevard
Bronx, New York 10460
United States of America

Jay R. Malcolm
Faculty of Forestry
University of Toronto
Toronto, Ontario M5S 3B3
Canada

Paul Marples
Department of Zoology
223 Bartram Hall
University of Florida
Gainesville, Florida 32611
United States of America

Judith A. McDonald
Department of Economics
Rauch Business Center
Lehigh University
621 Taylor Street
Bethlehem, Pennsylvania 18105-3117
United States of America

Les A. Moore
Division of Wildlife and Ecology
CSIRO Tropical Forest Research Centre
P.O. Box 780
Atherton, Queensland 4883
Australia

Craig Moritz
Cooperative Research Centre for Tropical Rainforest Ecology and Management/ Department of Zoology
University of Queensland
St Lucia, Queensland 4067
Australia

Meika A. Mustrangi
Museum of Vertebrate Zoology
3101 Valley Life Sciences Building
University of California
Berkeley, California 94720
United States of America

John D. Nason
Department of Biological Sciences
The University of Iowa
138 Biology Building
Iowa City, Iowa 52242-1324
United States of America

John Parrotta
International Institute of Tropical Forestry
USDA Forest Service, Box 25,000
Rio Piedras, Puerto Rico
United States of America

James L. Patton
Museum of Vertebrate Zoology
3101 Valley Life Sciences Building
University of California
Berkeley, California 94720
United States of America

Scott M. Pearson
Department of Biology
Mars Hill College
Mars Hill, North Carolina 28754
United States of America

Carsten Rahbek
Centre for Tropical Biodiversity
Zoological Museum
University of Copenhagen
Universitetsparken 15
DK 2100 Copenhagen 0
Denmark

Luis Miguel Renjifo
Department of Biology
University of Missouri
80001 Natural Bridge Road
St Louis, Missouri 63121-4499
United States of America

Carla Restrepo
Biological Sciences
Gilbert Building, Room 109
Stanford University
Stanford, California 94305-5020
United States of America

Chris Schneider
Cooperative Research Centre for Tropical Rainforest Ecology and Management/ Department of Zoology
University of Queensland
Brisbane, Queensland 4072
Australia

Herman H. Shugart
Department of Environmental Sciences
University of Virginia
Charlottesville, Virginia 22903
United States of America

Kathryn E. Sieving
Department of Wildlife Ecology and Conservation
University of Florida
303 Newins-Ziegler Hall
Gainesville, Florida 32661
United States of America

Maria Nazareth F. da Silva
INPA/Coordenação de Entomologia
C.P. 478, Alameda Cosme Ferreira
1756 Manaus, Amazonas 69083-000
Brazil

Jack W. Sites, Jr.
Department of Zoology
Brigham Young University
P.O. Box 25255
Provo, Utah 84602-5255
United States of America

Andrew P. Smith
Austeco Pty. Ltd.
84 Brown Street
Armidale, NSW 2350
Australia

Philip C. Stouffer
Department of Biological Sciences
Southeastern Louisiana University
Hammond, Louisiana 70402
United States of America

Dominique Strasberg
Laboratoire de Biologie Végétale
Université de La Réunion
BP 7151
F-97715 Saint-Denis messag Cédex 9
Ile de La Réunion
France

André A. J. Tabanez
Department of Plant Biology
Louisiana State University
502 Life Sciences Building
Baton Rouge, Louisiana 70803
United States of America

José Tello
Department of Biology
University of Missouri
8001 Natural Bridge Road
St Louis, Missouri 63121-4499
United States of America

John Terborgh
Center for Tropical Conservation
Duke University
P.O. Box 90381
Durham, North Carolina 27708-0381
United States of America

Christophe Thébaud
NERC Centre for Population Biology
Imperial College at Silwood Park
Ascot, Berkshire SL5 7PY
England

Alan Thornhill
Department of Ecology and Evolutionary Biology
University of California, Irvine
Irvine, California 92717
United States of America

Mandy D. Tocher
Department of Zoology
University of Canterbury
Christchurch
New Zealand

Nigel Tucker
Queensland Department of Environment
McLeish Road
Lake Eacham, via Yungaburra
4872 Queensland
Australia

I. M. Turner
Department of Botany
National University of Singapore
Lower Kent Ridge Road 0511
Singapore

Stephen M. Turton
Cooperative Research Centre for Tropical Rainforest Ecology and Management/ Department of Tropical Environmental Science and Geography
James Cook University
P.O. Box 6811
Cairns, Queensland 4870
Australia

Virgílio M. Viana
Department of Forest Sciences, ESALQ
University of São Paulo
Caixa Postal 9
Piracicaba, São Paulo 13418-900
Brazil

Elisa Wandelli
Ministero da Agricultura e Reforma Agrária
Centro de Pesquisa Agroflorestal da Amazônia Occidental
Embrapa CPAA
Manaus, AM 69011
Brazil

Neil H. Warburton
P.O. Box 344
Gordonvale
Queensland 4865
Australia

John F. Weishampel
Department of Biology
University of Central Florida
Orlando, Florida 32816
United States of America

Walter E. Westman (deceased)
Applied Science Division
Lawrence Berkeley Laboratory
Berkeley, California 94720
United States of America

T. C. Whitmore
Department of Geography
Cambridge University
Downing Place
Cambridge CB2 3EN
England

Douglas Yu
Museum of Comparative Zoology
Harvard University
Cambridge, Massachusetts 02138
United States of America

Barbara L. Zimmerman
Conservation International
24 Falcon Street
Toronto, Ontario M4S 2P5
Canada

References

Aberg, J., G. Jansson, J. E. Swenson, and P. Angelstam. 1995. The effect of matrix on the occurrence of hazel grouse *(Bonasa bonasia)* in isolated habitat fragments. *Oecologia* 103 : 265–69.

Acevedo, C. I. 1987. Contribución al conocimiento de las relaciones entre colibrís y flores con observaciones de sus ciclos anuales en el bosque altoandino de Iguaque, Boyacá. Thesis, Departamento de Biología, Pontificia Universidad Javeriana, Bogotá, Colombia.

Adams, J. 1985. The definition and interpretation of guild structure in ecological communities. *Journal of Animal Ecology* 54 : 43–60.

Adams, P. 1992. *Australian rainforests.* Oxford University Press, London.

Addicott, J. F., J. M. Aho, M. R. Antolin, D. K. Padilla, J. S. Richardson, and D. A Soluk. 1987. Ecological neighborhoods: Scaling environmental patterns. *Oikos* 49 : 340–46.

Adler, G. H. 1994. Tropical forest fragmentation and isolation promote asynchrony among populations of a frugivorous rodent. *Journal of Animal Ecology* 63 : 903–11.

Adler, G. H., and R. Levins. 1994. The island syndrome in rodent populations. *Quarterly Review of Biology* 69 : 473–90.

Adler, G. H., and J. O. Seamon. 1991. Distribution and abundance of a tropical rodent, the spiny rat, on islands in Panama. *Journal of Tropical Ecology* 7 : 349–60.

Aizen, M. A., and P. Feinsinger. 1994a. Forest fragmentation, pollination, and plant reproduction in a Chaco dry forest, Argentina. *Ecology* 75 : 330–51.

———. 1994b. Habitat fragmentation, native insect pollinators, and feral honey bees in Argentine "Chaco Serrano." *Ecological Applications* 4 : 378–92.

Allendorf, F. W., and R. F. Leary. 1986. Heterozygosity and fitness in natural populations of animals. Pages 57–76 in M. E. Soulé, ed., *Conservation biology: The science of scarcity and diversity.* Sinauer Associates, Sunderland, Mass.

Alvarez, E., L. Balbas, I. Massa, and J. Pacheco. 1986. Aspectos ecológicos del Embalse de Guri. *Interciencia* 11 : 325–33.

Alvarez-Buylla, E. R., and A. A. Garay. 1994. Population genetic structure of *Cecropia obtusifolia,* a tropical pioneer tree species. *Evolution* 48 : 437–53.

Amorim, D. S. 1991. Refuge model simulations: Testing the theory. *Revista Brasileira de Entomologia* 35 : 803–12.

Anderson, S. H., K. Mann, and H. H. Shugart. 1977. The effect of transmission-line corridors on bird populations. *American Midland Naturalist* 97 : 216–21.

Andrade, G. I., ed. 1993. *Carpanta Selva Nublada y Páramo.* Fundación Natura, Santa Fé de Bogotá, Colombia.

Andrade, G. I., and H. Rubio-T. 1994. Sustainable use of the tropical rainforest: Evidence from the avifauna in a shifting-cultivation habitat mosaic in the Colombian Amazon. *Conservation Biology* 8 : 545–54.

Andrén, H. 1994. Effects of habitat fragmentation on birds and mammals in landscapes with different proportions of suitable habitat: A review. *Oikos* 71:355–66.

Andrews, A. 1990. Fragmentation of habitat by roads and utility corridors: A review. *Australian Zoologist* 26: 130–41.

Appanah, S., and F. E. Putz. 1984. Climber abundance in virgin dipterocarp forest and the effect of pre-felling climber abundance on logging damage. *Malaysian Forester* 47:335–42.

Arango, S. 1993. Morfología y comportamiento alimenticio de las aves frugívoras de Carpanta. Pages 127–40 in G. I. Andrade, ed., *Carpanta selva nublada y páramo.* Fundación Natura, Santa Fé de Bogotá, Colombia.

———. 1994. *El papel de las aves dispersoras de semillas en la regeneración de pastizales en el Alto Quindío, Andes Centrales, Colombia.* Fundación Herencia Verde, Cali, Colombia.

Arbhabhirama, A., D. Phantumvanit, J. Elkington, and P. Ingkasuwan. 1988. *Thailand natural resources profile.* Oxford University Press, Oxford.

Arnold, G. W., J. R. Weeldenberg, and D. E. Steven. 1991. Distribution and abundance of two species of kangaroo in remnants of native vegetation in the central wheatbelt of Western Australia, and the role of native vegetation along road verges and fencelines as linkages. Pages 273–80 in D. A. Saunders and R. J. Hobbs, eds., *Nature conservation 2: The role of corridors.* Surrey Beatty, Chipping Norton, N.S.W., Australia.

Ashton, P. S. 1969. Speciation among tropical forest trees: Some deductions in light of recent evidence. *Biological Journal of the Linnean Society* 1:155–96.

Ashworth, J. H., R. T. Corlett, D. Dudgeon, D. S. Melville, and W. S. M. Tang. 1993. *Hong Kong flora and fauna: Computing conservation.* World Wide Fund for Nature-Hong Kong, Hong Kong.

Asrar, G. 1989. *Theory and applications of optical remote sensing.* John Wiley, New York.

Atchley, W. R., and D. Anderson. 1978. Ratios and the statistical analysis of biological data. *Systematic Zoology* 27:71–78.

Atkinson, I. A. E. 1985. The spread of commensal species of *Rattus* to oceanic islands and their effects on island avifaunas. Pages 35–81 in P. J. Moors, ed., *Conservation of island birds.* International Council for Bird Preservation, Cambridge.

Augspurger, C. K. 1986. Morphology and dispersal potential of wind-dispersed diaspores of Neotropical trees. *American Journal of Botany* 73:353–63.

August, P. V. 1983. The role of habitat complexity and heterogeneity in structuring tropical mammal communities. *Ecology* 64:1495–1507.

Auslig. 1990. *Atlas of Australian resources.* Vol. 6, *Vegetation.* Auslig, Canberra, Australia.

Avise, J. C. 1992. Molecular population structure and the biogeographic history of a regional fauna: A case history with lessons for conservation biology. *Oikos* 63:62–76.

———. 1994. *Molecular markers, natural history and evolution.* Chapman and Hall, New York.

Avise, J. C. 1995. Mitochondrial DNA polymorphism and a connection between genetics and demography of relevance to conservation. *Trends in Ecology and Evolution* 9:686–90.

Avise, J. C., J. Arnold, R. M. Ball, E. Bermingham, T. Lamb, J. E. Neigel, C. A. Reeb, and N. C. Saunders. 1987. Intraspecific phylogeography: The mitochondrial DNA bridge between population genetics and systematics. *Annual Review of Ecology and Systematics* 18:489–522.

Avise, J. C., and R. M. Ball. 1990. Principles of genealogical concordance in species concepts and biological taxonomy. *Oxford Surveys of Evolutionary Biology* 7:45–68.

Ayres, J. M., and T. H. Clutton-Brock. 1992. River boundaries and species range size in Amazonian primates. *American Naturalist* 140:531–37.

Bachelery, P. 1981. Le Piton de la Fournaise (Ile de La Réunion), étude volcanologique, structurale, et pétrologique. Ph.D. dissertation, University of Clermont-Ferrand, France.

Baker, H. G. 1959. Reproductive methods as factors in speciation in flowering plants. *Cold Spring Harbor Symposium on Quantitative Biology* 24:177–91.

Baker, R. R. 1984. The dilemma: When and how to go or stay. Pages 279–96 in R. I. Vane-Wright and

P. R. Ackery, eds., *The biology of butterflies.* Royal Entomological Society Symposium 11. Academic Press, London.

Baker, S. L., and Y. Cai. 1992. The role programs for multiscale analysis of landscape structure using the GRASS geographic information system. *Landscape Ecology* 7 : 291 – 302.

Bakowski, C., and M. Kozakiewicz. 1988. The effect of a forest road on bank vole and yellow-necked mouse populations. *Acta Theriologica* 33 : 345 – 53.

Báldi, A., and T. Kisbenedek. 1994. Comparative analysis of edge effects on bird and beetle communities. *Acta Zoologica Academiae Scientarum Hungaricae* 40 : 1 – 14.

Balick, J. M., and R. Mendelsohn. 1992. Assessing the economic value of traditional medicines from tropical rain forests. *Conservation Biology* 6 : 128 – 30.

Balmford, A. 1996. Extinction filters and current resilience: The significance of past selection pressures for conservation biology. *Trends in Ecology and Evolution* 11 : 193 – 96.

Balmford, A., and A. Long. 1994. Avian endemism and forest loss. *Nature* 372 : 623 – 24.

Balslev, H. 1979. *Juncaceae.* Flora of Ecuador, no. 11. Swedish Research Councils Publishing House, Stockholm, Sweden.

Banguero, H. 1993. *La población de Colombia 1938 – 2025. Una visión retrospectiva y prospectiva para el país, los departamentos y sus municipios.* Colección de Edición Previa, Editorial Universidad del Valle, Cali, Colombia.

Barass, A. N. 1985. The effects of highway traffic noise on the phototactic and associated reproductive behaviour of selected anurans. Ph.D. dissertation, Vanderbilt University, Nashville, Tenn.

Barnett, J. L., R. A. How, and W. F. Humphries. 1978. The use of habitat components by small mammals in eastern Australia. *Australian Journal of Ecology* 3 : 277 – 85.

Barney, G. O., ed. 1980. *The Global 2000 Report to the President.* U.S. Council on Environmental Quality, Washington, D.C.

Barre, N., and A. C. Barau. 1982. *Oiseaux de La Réunion.* Imprimerie Cazal, St-Denis, Ile de La Réunion, France.

Bates, H. W. 1862. Contributions to an insect fauna of the Amazon Valley. Lepidoptera: Heliconidae. *Transactions of the Linnean Society of London* 23 : 495 – 566.

———. 1863. *The naturalist on the River Amazons.* J. Murray, London.

Baur, B. N. 1964. *The ecological basis of rainforest management.* Forestry Commission, Sydney, Australia.

Baverstock, P. R., M. Krieg, J. Birrell, and G. M. McKay. 1990. Albumin immunological relationships of Australian marsupials. II. The Pseudocheiridae. *Australian Journal of Zoology* 38 : 519 – 26.

Bawa, K. S. 1974. Breeding systems of tree species of a lowland tropical community. *Evolution* 28 : 85 – 92.

———. 1990. Plant-pollinator interactions in tropical rain forests. *Annual Review of Ecology and Systematics* 21 : 399 – 422.

Bawa, K. S., D. R. Perry, and J. H. Beach. 1985. Reproductive biology of tropical lowland rain forest trees. I. Sexual systems and incompatibility mechanisms. *American Journal of Botany* 72 : 331 – 45.

Beattie, A. J. 1976. Plant dispersion, pollination and gene flow in *Viola. Oecologia* 25 : 291 – 300.

Beccaloni, G. W., and K. J. Gaston. 1995. Predicting the species richness of Neotropical forest butterflies: Ithomiinae (Lepidoptera, Nymphalidae). *Biological Conservation* 71 : 71 – 86.

Becker, P., J. S. Moure, and F. J. A. Peralta. 1991. More about Euglossine bees in Amazonian forest fragments. *Biotropica* 23 : 586 – 91.

Bellingham, P. J., V. Kapos, N. Varty, J. R. Healey, E. V. J. Tanner, D. L. Kelly, J. W. Dalling, L. S. Burns, D. Lee, and G. Sidrak. 1994. Hurricanes need not cause high mortality: The effects of Hurricane Gilbert on forests in Jamaica. *Journal of Tropical Ecology* 8 : 217 – 23.

Belshaw, R., and B. Bolton. 1994. A survey of the leaf litter ant fauna in Ghana, West Africa (Hymenoptera: Formicidae). *Journal of Hymenoptera Research* 3 : 5 – 16.

Bennett, A. F. 1990. *Habitat corridors: Their role in wildlife management and conservation.* Department of Conservation and Environment, Victoria, Australia.

———. 1991. Roads, roadsides and wildlife conservation: A review. Pages 99–117 in D. A. Saunders and R. J. Hobbs, eds., *Nature conservation 2: The role of corridors.* Surrey Beatty, Sydney, Australia.

Bennet, E. L., and J. O. Caldecott. 1981. Unexpected abundance: The trees and wildlife of the Lima Belas Estate forest reserve, near Slim River, Perak. *The Planter* 57:516–19.

Bentrupperbaumer, J. 1988. *Numbers and conservation status of cassowaries in the Mission Beach area following Cyclone Winifred.* Report to Queensland National Parks and Wildlife Service, Brisbane, Australia.

Berger, J. O., and D. A. Berry. 1988. Statistical analysis and the illusion of objectivity. *American Scientist* 76:159–65.

Berger, J. O., and T. Sellke. 1987. Testing a point null hypothesis: The irreconcilability of *P* values and evidence. *Journal of the American Statistical Association* 82:112–39.

Bernardes, A. T., A. B. M. Machado, and A. B. Rylands. 1990. *Fauna brasileira ameaçada de extinção.* Fundação Biodiversitas, Belo Horizonte, Brazil.

Bernardino, F. S., Jr., and G. H. Dalrymple. 1992. Seasonal activity and road mortality of the snakes of the Pa-hay-okee wetlands of the Everglades National Park, USA. *Biological Conservation* 62:71–75.

Besaire, H. 1969. *Carte geologique feulle morondava.* Madagascar Service Geologique, Antananarivo, Madagascar.

Best, B. J., ed. 1992. *The threatened forests of south-west Ecuador.* Biosphere Publications, Leeds, U.K.

Besuchet, C., D. H. Burckhardt, and I. Löbl. 1987. The "Winkler/Moczarski" eclector as an efficient extractor for fungus and litter Coleoptera. *Coleopterist's Bulletin* 41:392–94.

Bierregaard, R. O., Jr. 1990. Avian communities in the understory of Amazonian forest fragments. Pages 333–43 in A. Keast, ed., *Biogeography and ecology of forest bird communities.* Academic Publishing, The Hague, Netherlands.

Bierregaard, R. O., Jr., and V. H. Dale. 1996. Islands in an ever-changing sea: The ecological and socioeconomic dynamics of Amazonian rainforest fragments. Pages 187–204 in J. Schelhas and R. Greenberg, eds., *Forest patches in tropical landscapes.* Island Press, Washington, D.C.

Bierregaard, R. O., Jr., and T. E. Lovejoy. 1988. Birds in Amazonian forest fragments: Effects of insularization. Pages 1564–79 in H. Ouellet, ed., *Acta XIX Congressus Internationalis Ornitholigici,* vol. II. University of Ottawa Press, Ottawa, Ontario, Canada.

———. 1989. Effects of forest fragmentation on Amazonian understory bird communities. *Acta Amazônica* 19:215–41.

Bierregaard, R. O., Jr., T. E. Lovejoy, V. Kapos, A. A. dos Santos, and R. W. Hutchings. 1992. The biological dynamics of tropical rain forest fragments. *Bioscience* 42:859–66.

Bilsborrow, R., and M. Geores. 1994. Population, land-use and the environment in developing countries: What can we learn from cross-national data? Pages 106–33 in K. Brown and D. W. Pearce, eds., *The causes of tropical deforestation: The economic and statistical analysis of factors giving rise to the loss of tropical forests.* University College London Press, London.

Binder, A. 1963. Further considerations on testing the null hypothesis and the strategy and tactics of investigating theoretical models. *Psychological Review* 70:107–15.

Birdsey, R. A., and P. L. Weaver. 1982. *The forest resources of Puerto Rico.* U.S. Department of Agriculture, Forestry Research Bulletin SO-85. Southern Forest Experimental Station, New Orleans, La.

Black, G. A., T. Dobzhansky, and C. Pavan. 1950. Some attempts to estimate species diversity and population density of trees in Amazonian forests. *Botanical Gazette* 111:413–25.

Blake, J. G. 1991. Nested subsets and the distribution of birds on isolated woodlots. *Conservation Biology* 5:58–66.

Blakers, M., S. Davies, and P. Reilly. 1984. *The atlas of Australian birds.* Melbourne University Press, Melbourne, Australia.

Blockhuss, J. L., M. Dillenbeck, J. Sayer, and P. Wegge. 1992. *Conserving biological diversity in managed tropical forests.* International Union for the Conservation of Nature and Natural Resources (IUCN), Gland, Switzerland.

Bodmer, R. E. 1994. Managing wildlife with local communities in the Peruvian Amazon: The case of the Reserva Comunal Tamshiyacu-Tahuayo. Pages 113–34 in D. Western, R. M. Wright, and S. C. Strum, eds., *Natural connections: Case studies in community-based conservation.* Island Press, Washington, D.C.

Boecklen, W. J., and N. J. Gotelli. 1984. Island biogeographic theory and conservation practice: Species-area, or specious-area relationship? *Biological Conservation* 29:63–80.

Bolger, D. T., A. C. Alberts, R. M. Sauvajot, P. Potenza, C. McCalvin, D. Tran, S. Mazzoni, and M. E. Soulé. In press. Effects of habitat fragmentation on native chaparral mammals: Island and edge effects. *Ecological Applications.*

Bolger, D. T., A. C. Alberts, and M. E. Soulé. 1991. Occurrence patterns of bird species in habitat fragments: Sampling, extinction, and nested species subsets. *American Naturalist* 137:155–66.

Bookstein, F. L., B. Chernoff, R. Elder, J. Humphries, G. Smith, and R. Strauss. 1985. *Morphometrics in evolutionary biology.* The Academy of Natural Sciencecs of Philadelphia, Philadelphia, Pa.

Boose, E. R., D. R. Foster, and M. Fluet. 1994. Hurricane impacts to tropical and temperate forest landscapes. *Ecological Monographs* 64:369–400.

Borges, S. H. 1995. Comunidade de aves em dois tipos de vegetação secundária da Amazônia central. M.Sc. thesis, INPA/Universidade do Amazonas, Manaus, Amazonas, Brazil.

Boshier, D. H., M. R. Chase, and K. S. Bawa. 1995. Population genetics of *Cordia alliodora* (Boraginaceae), a Neotropical tree. 3. Gene flow, neighborhood, and population structure. *American Journal of Botany* 82:484–90.

Boulding, K. E. 1980. Science: Our common heritage. *Science* 207:831–36.

Brawn, J. D., J. R. Karr, and J. D. Nichols. 1995. Demography of birds in a Neotropical forest: Effects of allometry, taxonomy, and ecology. *Ecology* 76:41–51.

Brett, D. 1989. Sea birds in the trees. *Ecos* 61:4–8.

Bridgewater, P. B. 1993. Landscape ecology, geographic information systems and nature conservation. Pages 23–36 in R. Haines-Young, D. R. Green, and S. H. Cousins, eds., *Landscape ecology and GIS.* Taylor and Francis, London.

Briscoe, D. A., J. M. Malpica, A. Robertson, G. J. Smith, R. Frankham, R. G. Banks, and J. S. F. Barker. 1992. Rapid loss of genetic variation in large captive populations of *Drosophila* flies: Implications for the genetic management of captive populations. *Conservation Biology* 6:416–25.

Broadbent, J. A., and I. Cranwell. 1979. *Faunal studies for the proposed Mount White-Kariong-Ourimbah sections of the Sydney-Newcastle Freeway F3.* Parts 1 and 2. Environmental and Urban Studies Report No. 45. Macquarie University, Sydney, Australia.

Brockelman, W. Y., and V. Baimai. 1993. *Conservation of biodiversity and protected area management in Thailand.* World Bank/GEF Pre-investment Study, Mahidol University, Bangkok, Thailand.

Brockelman, W. Y., and S. Srikosamatara. 1993. Estimation of density of gibbon groups by use of loud songs. *American Journal of Primatology* 29:93–108.

Brody, A. J., and M. R. Pelton. 1989. Effects of roads on black bear movements in western North Carolina. *Wildlife Society Bulletin* 17:5–10.

Brooks, D. R., R. L. Mayden, and D. A. McLennan. 1992. Phylogeny and biodiversity: Conserving our evolutionary legacy. *Trends in Ecology and Evolution* 7:55–59.

Browder, J. O. 1988. Public policy and deforestation in the Brazilian Amazon. Pages 247–83 in R. Repetto and M. Gillis, eds., *Public policies and the misuse of forest resources.* Cambridge University Press, Cambridge.

Brown, J. H. 1971. Mammals on mountain-tops: Non-equilibrium insular biogeography. *American Naturalist* 105:467–78.

———. 1995. *Macroecology.* The University of Chicago Press, Chicago, Ill.

Brown, J. H., D. W. Davidson, J. C. Munger, and R. S. Inouye. 1986. Experimental community ecology: The desert granivore system. Pages 41–61 in J. Diamond and T. J. Case, eds., *Community ecology.* Harper and Row, New York.

Brown, J. H., and A. Kodric-Brown. 1977. Turnover rates in insular biogeography: Effect of immigration on extinction. *Ecology* 58:445–49.

Brown, K., and D. W. Pearce, eds. 1994. *The causes of tropical deforestation: The economic and statistical analysis of factors giving rise to the loss of tropical forests.* University College London Press, London.

Brown, K. S., Jr. 1972. Maximizing daily butterfly counts. *Journal of the Lepidopterists' Society* 26:183–96.

———. 1979. *Ecologia geográfica e evolução nas florestas neotropicais.* UNICAMP, Campinas, São Paulo, Brazil.

———. 1980. A review of the genus *Hypothyris* Hübner (Nymphalidae), with description of three new subspecies and early stages of *H. daphnis. Journal of the Lepidopterists' Society* 34:152–72.

———. 1984. Species diversity and abundance in Jaru, Rondônia (Brazil). *News of the Lepidopterists' Society* 1984(3): 45–47.

———. 1987a. Biogeografia e conservação das florestas Atlântica e Amazônica brasileiras. Pages 85–92 in SEMA/IWRB/CVRD, *Desenvolvimento e impacto ambiental em áreas de trópico úmido brasileiro: A experiência da CVRD.* CVRD, Rio de Janeiro, Brazil.

———. 1987b. Conclusions, synthesis, and alternative hypotheses. Pages 175–96 in T. C. Whitmore and G. T. Prance, eds., *Biogeography and Quaternary history in tropical America.* Oxford University Press, Oxford.

———. 1991. Conservation of Neotropical environments: Insects as indicators. Pages 349–404 in N. M. Collins and J. A. Thomas, eds., *The conservation of insects and their habitats.* Academic Press, London.

———. 1992. Borboletas da Serra do Japi: Diversidade, habitats, recursos alimentares e variação temporal. Pages 142–86 in L. P. C. Morellato, ed., *História natural da Serra do Japi: Ecologia e preservação de uma área florestal no sudeste do Brasil.* Editora UNICAMP, Campinas, São Paulo, Brazil.

———. In press. The use of insects in the study, inventory, conservation and monitoring of biological diversity in Neotropical habitats, in relation to traditional land use systems. In T. Hirowatari, ed., *Decline and conservation of butterflies in Japan,* III. Lepidopterological Society of Japan, Osaka, Japan.

Brown, K. S., Jr., and W. W. Benson. 1974. Adaptive polymorphism associated with multiple Müllerian mimicry in *Heliconius numata* (Lepidoptera: Nymphalidae). *Biotropica* 6:205–28.

———. 1977. The Heliconians of Brazil (Lepidoptera: Nymphalidae). VII. Evolution in modern Amazonian nonforest islands: *Heliconius hermathena. Biotropica* 8:95–117.

Brown, K. S., Jr., and G. G. Brown. 1992. Habitat alteration and species loss in Brazilian forests. Pages 119–42 in T. C. Whitmore and J. A. Sayer, eds., *Tropical deforestation and species extinction.* Chapman and Hall, London.

Bruijnzeel, L. A. 1990. *Hydrology of moist tropical forests and effects of conversion: A state of knowledge review.* UNESCO International Hydrological Program, Free University, Amsterdam.

Bryant, E. H. 1984. A comparison of electrophoretic and morphometric variability in the face fly, *Musca autumnalis. Evolution* 38:455–58.

Bryant, E. H., S. A. McCommas, and L. M. Combs. 1986. The effect of an experimental bottleneck upon quantitative genetic variation in the housefly. *Genetics* 114:1191–1211.

Bryant, R. L. 1994. Fighting over the forests: Political reform, peasant resistance and the transformation of forest management in late colonial Burma. *Journal of Commonwealth and Comparative Politics* 32: 244–60.

Buechner, M. 1987. Conservation in insular parks: Simulation models of factors affecting the movement of animals across park boundaries. *Biological Conservation* 41:57–76.

Bull, G. A. D., and E. R. C. Reynolds. 1968. Wind turbulence generated by vegetation and its implications. *Forestry* (Suppl.) 41:28–37.

Bunnell, F. H. 1990. Forestry wildlife: W(h)ither the future? Pages 163–76 in A. F. Pearson and D. A. Challenger, eds., *Forests, wild and managed: Differences and consequences.* Students for Forestry Awareness, University of British Columbia, Vancouver, Canada.

Burgess, J. C. 1991. Economic analyses of frontier agricultural expansion and tropical deforestation. M.Sc. thesis, University College, London.

Burgess, R. L., and D. M. Sharpe, eds. 1981. *Forest island dynamics in man-dominated landscapes.* Ecological Studies, vol. 41. Springer-Verlag, New York.

Burghouts, T., G. Ernsting, G. Korthals, and T. D. Vries. 1992. Litterfall, leaf litter decomposition and litter invertebrates in primary and selectively logged dipterocarp forest in Sabah, Malaysia. *Philosophical Transactions of the Royal Society of London B* 335:407–16.

Burnett, S. E. 1989. The effects of a road on the movement behaviour of small rainforest mammals in North Queensland. B.Sc. (Hons.) thesis, James Cook University, Townsville, Australia.

———. 1992. Effects of a rainforest road on movements on small mammals: Mechanisms and implications. *Wildlife Research* 19:95–104.

Burney, D. 1987. Late Holocene vegetation change in central Madagascar. *Quaternary Research* 28:130–43.

Burrough, P. A. 1981. Fractal dimensions of landscapes and other environmental data. *Nature* 294:240–42.

Buse, A., and J. E. G. Good. 1993. The effects of conifer forest design and management on abundance and diversity of rove beetles (Coleoptera: Staphylinidae): Implications for conservation. *Biological Conservation* 64:67–76.

Bush, M. B. 1994. Amazonian speciation: A necessarily complex model. *Journal of Biogeography* 21:5–17.

Butterfield, B. R., B. Csuti, and J. M. Scott. 1994. Modelling vertebrate distribution for gap analysis. Pages 53–68 in R. I. Miller, ed., *Mapping the diversity of nature.* Chapman and Hall, London.

Cabrera, A. 1961. Catálogo de los mamíferos de América del Sur. *Revistas Museo Argentino Ciencias Naturales "Bernardino Rivadavia"* 4:309–732.

Cadet, T. 1977. La végétation de l'Ile de La Réunion: Étude phytoécologique et phytosociologique. Dissertation, University of Aix-Marseille III, France.

Callaghan, C. J. 1983. A study of isolating mechanisms among Neotropical butterflies of the subfamily Riodininae. *Journal of Research on the Lepidoptera* 21:159–76.

Camargo, J. L. C. 1993. Variation in soil moisture and air vapour pressure deficit relative to tropical rain forest edges near Manaus, Brazil. M.Phil. dissertation, University of Cambridge.

Camargo, J. L. C., and V. Kapos. 1995. Complex edge effects on soil moisture and microclimate in central Amazonian forest. *Journal of Tropical Ecology* 11:205–11.

Camargo, M. N. 1979. Sistema de classificação dos solos brasileiros. EMBRAPA-SNLCS, Rio de Janeiro, Brazil.

Campbell, I. C., and T. J. Doeg. 1989. Impact of timber harvesting and production on streams: A review. *Australian Journal of Marine and Freshwater Research* 40:519–39.

Campbell, N. J. H. 1995. Mitochondrial control region variation in Australian mosaic-tailed rats *(Melomys* and *Uromys):* Applications to taxonomy, phylogeography and conservation. Ph.D. dissertation, Southern Cross University, Lismore, New South Wales, Australia.

Cantley, N. 1884. Report on the forests of the Straits Settlements. Singapore Printing Office, Singapore.

Capistrano, A. D. 1994. Tropical forest depletion and the changing macroeconomy, 1967–85. Pages 68–85 in K. Brown and D. W. Pearce, eds., *The causes of tropical deforestation: The economic and statistical analysis of factors giving rise to the loss of tropical forests.* University College London Press, London.

Capparella, A. P. 1990. Neotropical avian diversity and riverine barriers. *Acta XX Congressus Internationalis Ornithologici* 1:307–16.

Carlquist, S. 1974. *Island biology.* Columbia University Press, New York.

Caro, T. M., and M. K. Laurenson. 1994. Ecological and genetic factors in conservation: A cautionary tale. *Science* 263:485–86.

Case, T. J. 1975. Species numbers, density compensation, and colonizing ability of lizards on islands in the Gulf of California. *Ecology* 56:3–18.

———. 1983. Niche overlap and the assembly of island lizard communities. *Oikos* 41:427–33.

Case, T. J., and M. L. Cody. 1987. Testing theories of island biogeography. *American Scientist* 75:402–11.

Case, T. J., M. E. Gilpin, and J. M. Diamond. 1979. Overexploitation, interference competition, and excess density compensation in insular faunas. *American Naturalist* 113:843–54.

Cavelier, J., and A. Etter. 1995. Deforestation of montane forests in Colombia as a result of illegal plantation of opium *(Papaver somniferum).* Pages 541–50 in S. P. Churchill, H. Balslev, E. Forero, and J. L. Luteyn, eds., *Biodiversity and conservation of neotropical montane forests.* Proceedings of the Neotropical Montane Forest Biodiversity and Conservation Symposium, New York Botanical Garden, New York.

Center for Strategic and International Studies and the Massachusetts Institute of Technology. 1995. Sustainable development: Strategies for reconciling environment and economy in the developing world. *The Washington Quarterly* 18:189–224.

Chan, H. T. 1981. Reproductive biology of some Malaysian Dipterocarps. III. Breeding systems. *Maylasian Forestry* 44:28–36.

Chapman, F. M. 1917. The distribution of bird life in Colombia. *Bulletin of the American Museum of Natural History* 33:167–92.

———. 1928. The nesting habits of Wagler's oropendola *(Zarhynchus wagleri)* on Barro Colorado Island. *Bulletin of the American Museum of Natural History* 58:123–66.

———. 1929. *My tropical air castle.* D. Appleton-Century Co., New York.

———. 1935. The courtship of Gould's manakin *(Manacus vitellinus vitellinus)* on Barro Colorado Island. *Bulletin of the American Museum of Natural History* 68:471–525.

———. 1938. *Life in an air castle.* D. Appleton-Century Co., New York.

Charles-Dominique, P. 1983. Ecology and social adaptations in didelphid marsupials: Comparison with eutherians of similar ecology. Pages 395–422 in J. F. Eisenberg and D. G. Kleiman, eds., *Advances in the study of mammalian behavior.* American Society of Mammalogists, Shippensburg, Pa.

———. 1986. Inter-relations between frugivorous vertebrates and pioneer plants: *Cecropia,* birds and bats in French Guyana. Pages 119–36 in A. Estrada and T. H. Fleming, eds., *Frugivores and seed dispersal.* Dr. W. Junk, Dordrecht, Netherlands.

Charles-Dominique, P., M. Atramentowicz, M. Charles-Dominique, H. Gérard, A. Hladik, C. M. Hladik, and M. F. Prévost. 1981. Les mamiféres frugivores arboricoles nocturnes d'une forêt guyanaise: Inter-relations plantes-animaux. *Revue d'Ecologie (Terre et Vie)* 35:341–435.

Chasen, F. N. 1940. A handlist of Malaysian mammals. *Bulletin of the Raffles Museum of Singapore* 15:1–209.

Chauvel, A. 1983. Os latossolos amarelos, álicos, argilosos dentro dos ecosistemas das bacias experimentais do INPA e da região vizinha. *Acta Amazônica* 12 (Supplemento):47–60.

Cheke, A. S. 1987a. An ecological history of the Mascarene Islands, with particular reference to extinctions and introductions of land vertebrates. Pages 5–89 in A. W. Diamond, ed., *Studies of Mascarene Island birds.* Cambridge University Press, Cambridge.

———. 1987b. The surviving native birds of Réunion and Rodrigues. Pages 301–58 in A. W. Diamond, ed., *Studies of Mascarene Island birds.* Cambridge University Press, Cambridge.

Chen, J., J. F. Franklin, and T. A. Spies. 1992. Vegetation responses to edge environments in old-growth Douglas-fir forests. *Ecological Applications* 2:387–96.

———. 1993a. Contrasting microclimates among clearcut edge and interior of old-growth Douglas-fir forest. *Agricultural and Forest Meteorology* 63:219–37.

———. 1993b. An empirical model for predicting diurnal air-temperature gradients from edge into old-growth Douglas-fir forest. *Ecological Modelling* 67:179–98.

Chesser, R. T., and S. J. Hackett. 1992. Mammalian diversity in South America. *Science* 255:1502–4.

Christidis, L., R. Schodde, and P. R. Baverstock. 1988. Genetic and morphological differentiation and phylogeny in the Australo-Papuan scrubwrens (*Sericornis,* Acanthizidae). *Auk* 105:616–29.

CIMA. 1991. *Relatório da Comissão Interministerial sobre desenvolvimento e meio ambiente.* Brasília, Brazil.

Clark, D. A. 1994. Plant demography. Pages 90–105 in L. A. McDade, K. S. Bawa, H. A. Hespenheide, and G. S. Hartshorn, eds., *La Selva: Ecology and natural history of a Neotropical rainforest.* University of Chicago Press, Chicago, Ill.

Clark, D. A., and D. B. Clark. 1984. Spacing dynamics of a tropical rainforest tree: Evaluation of the Janzen-Connell model. *American Naturalist* 124:769–88.

Clark, W. C., D. D. Jones, and C. S. Holling. 1979. Lessons for ecological policy design: A case study of ecosystems management. *Ecological Modelling* 7:1–53.

Clarke, G. M. 1993. Fluctuating asymmetry of invertebrate populations as a biological indicator of environmental quality. *Environmental Pollution* 82:207–11.

———. 1995. Relationships between developmental stability and fitness: Applications for conservation biology. *Conservation Biology* 9:18–24.

Clarke, G. M., and L. J. McKenzie. 1992. Fluctuating asymmetry as a quality control indicator for insect mass rearing processes. *Journal of Economic Entomology* 85:2045–50.

Cloudsley-Thompson, J. L. 1959. Studies in diurnal rhythms. IX. The water relations of some nocturnal tropical arthropods. *Entomologia Experimentalis et Applicata* 13:187–93.

Cody, M. L. 1983. Bird diversity and density in South African forests. *Oecologia* 59:201–15.

Cohn-Haft, M., A. Whittaker, and P. C. Stouffer. In press. A new look at the "species-poor" central Amazon: Updates and corrections to the avifauna north of Manaus, Brazil. *Ornithological Monographs.*

Cole, F. R., D. M. Reeder, and D. E. Wilson. 1994. A synopsis of distribution patterns and the conservation of mammal species. *Journal of Mammalogy* 75:266–76.

Collar, N. J., L. P. Gonzaga, N. Krabbe, A. Madroño Nieto, L. G. Naranjo, T. A. Parker III, and D. C. Wege. 1992. *The ICBP/IUCN Red Data Book.* 3d ed. Part 2, *Threatened birds of the Americas.* International Council for Bird Preservation (ICBP), Cambridge.

Collins, N. M., J. A. Sayer, and T. C. Whitmore. 1991. *The conservation atlas of tropical forests—Asia and the Pacific.* Simon and Schuster, New York.

Connor, E. F., and E. D. McCoy. 1979. The statistics and biology of the species-area relationship. *American Naturalist* 113:791–833.

Constable, A. J. 1991. The role of science in environmental protection. *Australian Journal of Marine and Freshwater Research* 42:527–38.

Corlett, R. T. 1988. The naturalized flora of Singapore. *Journal of Biogeography* 15:657–63.

———. 1991. Plant succession on degraded land in Singapore. *Journal of Tropical Forest Science* 4:151–61.

———. 1992a. The angiosperm flora of Singapore. 1. Introduction. *Gardens Bulletin, Singapore* 44:3–21.

———. 1992b. The ecological transformation of Singapore. *Journal of Biogeography* 19:411–20.

———. 1992c. The naturalized flora of Hong Kong: A comparison with Singapore. *Journal of Biogeography* 19:421–30.

———. 1994. What is secondary "forest?" *Journal of Tropical Ecology* 10:445–47.

———. 1995. Tropical secondary forests. *Progress in Physical Geography* 19:159–72.

———. 1996. Characteristics of vertebrate-dispersed fruits in Hong Kong. *Journal of Tropical Ecology.* In press.

———. 1997. Human impact on the flora of Hong Kong Island. In N. G. Jablonski, D. K. Ferguson, M. J. Singer, M. J. Tooley, and W. W. S. Yim, eds., *Proceedings of the 4th International Conference on the Evolution of the East Asian Environment.* Center of Asian Studies, University of Hong Kong, Hong Kong.

Corlett, R. T., and P. W. Lucas. 1990. Alternative seed handling strategies in primates: Seed spitting by long tailed macaques *(Macaca fascicularis). Oecologia* 82:166–71.

Corner, E. J. H. 1954. The evolution of tropical forests. Pages 34–46 in J. S. Huxley, A. C. Hardy, and E. B. Ford, eds., *Evolution as a process.* Allen and Unwin, London.

Corredor, G. L. 1989. *Estudio comparativo entre la avifauna de un bosque natural y un cafetal tradicional en el Quindio.* Trabajo de Grado, Departamento de Biología, Universidad del Valle, Cali, Colombia.

Couper, P. J., J. A. Covacevich, and C. Moritz. 1993. A review of the leaf-tailed geckos endemic to eastern Australia: A new genus, four new species, and other new data. *Memoirs of the Queensland Museum* 34: 95–124.

Covacevich, J. A., and K. R. McDonald. 1991. Frogs and reptiles of tropical and subtropical eastern Australian rainforests: Distribution patterns and conservation. Pages 281–310 in G. W. Werren and P. Kershaw, eds., *The rainforest legacy.* Australian Government Publishing Service, Canberra, Australia.

———. 1993. Distribution and conservation of frogs and reptiles of Queensland rainforests. *Memoirs of the Queensland Museum* 34:189–99.

Cox, P. A., T. Elmqvist, E. D. Pierson, and W. E. Rainey. 1991. Flying foxes as strong interactors in South Pacific Island ecosystems: A conservation hypothesis. *Conservation Biology* 5:448–54.

Coy, M. 1987. Rondônia: Frente pioneira e programa Polonoroteste. O processo de diferenciação o sócio-econômica na periferia e os limites do planejamento público. *Tübingen Geographische Studien* 95:253–70.

Cracraft, J. 1985. Historical biogeography and patterns of differentiation within the South American avifauna: Areas of endemism. Pages 49–84 in P. A. Buckley, M. S. Foster, E. S. Morton, R. S. Ridgely, and F. G. Buckley, eds., *Neotropical Ornithology.* Ornithological Monographs, no. 36. American Ornithologists Union, Washington, D.C.

———. 1988. Deep-history biogeography: Retrieving the historical pattern of evolving continental biotas. *Systematic Zoology* 37:221–36.

———. 1991. Patterns of diversification within continental biotas: Hierarchical congruence among areas of endemism of Australian vertebrates. *Australian Systematic Botany* 4:211–27.

Cracraft, J., and R. O. Prum. 1988. Patterns and processes of diversification: Speciation and historical congruence in some Neotropical birds. *Evolution* 42:603–20.

Crawford, T. J. 1984. What is a population? Pages 135–73 in B. Shorrocks, ed., *Evolutionary ecology.* Blackwell Scientific Publications, Oxford.

Crist, E. P., and R. J. Kauth. 1986. The tasseled cap de-mystified. *Photogrammetric Engineering and Remote Sensing* 48:243–50.

Croat, T. B. 1978. *Flora of Barro Colorado Island.* Stanford University Press, Stanford, Calif.

Crockett, C. M., and J. F. Eisenberg. 1986. Howlers: Variations in group size and demography. Pages 54–68 in B. B. Smuts, D. L. Cheney, R. M. Seyfarth, R. W. Wrangham, and T. T. Struhsaker, eds., *Primate societies.* University of Chicago Press, Chicago, Ill.

Crome, F. H. J. 1975. The ecology of fruit pigeons in tropical North Queensland. *Australian Wildlife Research* 2:155–85.

———. 1990. Rainforest successions and vertebrates. Pages 53–64 in L. J. Webb and J. Kikkawa, eds., *Australian tropical rainforest—science, values, meaning.* CSIRO, Melbourne, Australia.

———. 1991a. Is silvicultural treatment of rainforests a threat to the conservation of frugivores? Pages 38–40 in I. B. Kayanja and E. L. Edroma, eds., *African Wildlife: Research and Management.* International Council of Scientific Unions, Paris, France.

———. 1991b. Wildlife conservation and rain forest management—examples from north-east Queensland. Pages 407–18 in A. Gomez-Pompa, T. C. Whitmore, and M. Hadley, eds., *Rain forest regeneration and management.* Parthenon Press, Paris, France.

———. 1994. Tropical forest fragmentation: Some conceptual and methodological issues. Pages 61–77 in C. Moritz and J. Kikkawa, eds., *Conservation biology in Australia and Oceania.* Surrey Beatty, Sydney, Australia.

Crome, F. H. J., and J. Benntruperbäumer. 1991. *Management of cassowaries in the fragmented rainforests of north Queensland.* Report to the Endangered Species Program, Australian National Parks and Wildlife Service, Canberra, Australia.

Crome, F. H. J., J. Isaacs, and L. A. Moore. 1995. The utility to birds and mammals of remnant riparian vegetation and associated windbreaks in the tropical Queensland uplands. *Pacific Conservation Biology* 1:328–43.

Crome, F. H. J., and L. A. Moore. 1988. *The southern cassowary in north Queensland—a pilot study.* Vol. 2, *The biology of the Cassowary.* CSIRO Division of Wildlife and Ecology, Canberra, Australia.

———. 1990. Cassowaries in North-eastern Queensland: Report of a survey and a review of their status and conservation and management needs. *Australian Wildlife Research* 17:369–85.

Crome, F. H. J., and G. C. Richards. 1988. Bats and gaps: Microchiropteran community structure in a Queensland rain forest. *Ecology* 69:1960–69.

Crome, F. H. J., M. Thomas, and L. A. Moore. 1996. A novel Bayesian approach to assessing impacts of rainforest logging. *Ecological Applications* 6:1104–23.

Cronk, Q. C. B. 1992. Relict floras of Atlantic islands: Patterns assessed. *Biological Journal of the Linnean Society* 46:91–103.

Crow, J. F., and M. Kimura. 1970. *An introduction to population genetics theory.* Harper and Row, New York.

Crowell, K. L. 1962. Reduced interspecific competition among the birds of Bermuda. *Ecology* 61:194–98.

———. 1986. A comparison of relict versus equilibrium models for insular mammals of the Gulf of Maine. *Biological Journal of the Linnean Society* 28:37–64.

Cuadros, T. 1988. Aspectos ecológicos de la comunidad de aves en un bosque nativo en la Cordillera central en Antioquia, Colombia. *Hornero* 13:8–20.

Cuatrecasas, J. 1958. Aspectos de la vegetación natural. *Revista de la Academia Colombiana de Ciencias Exactas Físicas y Naturales* 10:221–64.

Daily, G. C., and P. R. Ehrlich. 1995. Preservation of biodiversity in small rainforest patches: Rapid evaluations using butterfly trapping. *Biodiversity and Conservation* 4:35–55.

Dale, J. E. 1988. The control of leaf expansion. *Annual Review of Plant Physiology and Plant Molecular Biology* 39:267–95.

Dale, V. H., R. V. O'Neill, M. A. Pedlowski, and F. Southworth. 1993. Causes and effects of land-use change in central Rondônia, Brazil. *Photogrammetric Engineering and Remote Sensing* 59:997–1005.

Dale, V. H., R. V. O'Neill, F. Southworth, and M. A. Pedlowski. 1994. Modeling effects of land management in the Brazilian settlement of Rondônia. *Conservation Biology* 8:196–206.

Dale, V. H., S. M. Pearson, H. L. Offerman, and R. V. O'Neill. 1994. Relating patterns of land-use change to faunal biodiversity in the central Amazon. *Conservation Biology* 8:1027–36.

Dale, V. H., and M. A. Pedlowski. 1992. Farming the forests. *Forum for Applied Research and Public Policy* 7:20–21.

Dale, V. H., F. Southworth, R. V. O'Neill, A. Rosen, and R. Frohn. 1993. Simulating spatial patterns of land-use change in Rondônia, Brazil. Pages 29–56 in R. H. Gardner, ed., *Some mathematical questions in biology.* Lectures in Mathematics and Life Sciences. American Mathematical Society, Providence, R.I.

Date, E. M., H. A. Ford, and H. F. Recher. 1991. Frugivorous pigeons, stepping stones and weeds in northern New South Wales. Pages 241–45 in D. A. Saunders and R. J. Hobbs, eds., *Nature conservation 2: The role of corridors.* Surrey Beatty, Chipping Norton, N.S.W., Australia.

Daume, E. 1990. *Britannica world data, 1990.* University of Chicago, Chicago, Ill.

Davidson, D. W., D. A. Samson, and R. S. Inouye. 1985. Granivory in the Chihuahuan desert: Interactions within and between tropical levels. *Ecology* 66:486–502.

Davis, T. A. W., and P. W. Richards. 1933a. The vegetation of Moraballi Creek, British Guiana: An ecological study of a limited area of tropical rain forest, part I. *Journal of Ecology* 21:350–84.

———. 1933b. The vegetation of Moraballi Creek, British Guiana: An ecological study of a limited area of tropical rain forest, part 2. *Journal of Ecology* 22:106–55.

Dawson, D. G. 1981. Counting birds for a relative measure (index) of density. Pages 12–16 in C. J. Ralph and J. M. Scott, eds., *Estimating numbers of terrestrial birds.* Studies in Avian Biology, no. 6. Cooper Ornithological Society, Lawrence, Kans.

Denslow. J. S. 1985. Disturbance mediated coexistence of species. Pages 307–23 in S. T. A. Pickett and P. S. White, eds., *The ecology of natural disturbance and patch dynamics.* Academic Press, Orlando, Fla.

Department of Main Roads, New South Wales. 1980. Wildlife-highway interrelationships—literature review. Chapter 4 in *Department of Main Roads, N.S.W. Sydney-Newcastle Freeway F3: Mt. White to Ourimbah Section.* Department of Main Roads, Sydney, Australia.

DEPRN. 1991. *Projeto Olho Verde.* Departamento de Proteção aos Recursos Naturais do Estado de São Paulo. São Paulo, Brazil.

Dial, R. 1995. Species-area curves and Koopowitz et al.'s simulations of stochastic extinctions. *Conservation Biology* 9:960–61.

Diamond, J. M. 1972. Biogeographic kinetics: Estimation of relaxation times for avifaunas of southwest Pacific islands. *Proceedings of National Academy of Sciences (USA)* 69:3199–3203.

———. 1975a. Assembly of species communities. Pages 342–444 in M. L. Cody and J. M. Diamond, eds., *Ecology and evolution of communities.* Harvard University Press, Cambridge, Mass.

———. 1975b. The island dilemma: Lessons of modern biogeographic studies for the design of natural reserves. *Biological Conservation* 7 : 129–46.

———. 1980. Patchy distributions of tropical birds. Pages 57–74 in M. E. Soulé and B. A. Wilcox, eds., *Conservation biology: An evolutionary-ecological perspective.* Sinauer Associates, Sunderland, Mass.

———. 1981. Flightlessness and the fear of flying in island species. *Nature* 293 : 507–8.

———. 1984. "Normal" extinctions of isolated populations. Pages 191–246 in M. H. Nitecki, ed., *Extinctions.* University of Chicago Press, Chicago, Ill.

Diamond, J. M., K. D. Bishop, and S. V. Balen. 1987. Bird survival in an isolated Javan woodland: Island or mirror? *Conservation Biology* 1 : 132–42.

Dickman, C. R. 1988. Body size, prey size, and community structure in insectivorous mammals. *Ecology* 69 : 569–80.

Didham, R. K. In press. An overview of invertebrate responses to forest fragmentation. In A. Watt, N. E. Stork, and M. Hunter, eds., *Forests and insects.* Chapman and Hall, London.

Didham, R. K., J. Ghazoul, N. E. Stork, and A. J. Davis. 1996. Insects in fragmented forests: A functional approach. *Trends in Ecology and Evolution* 11: 255–60.

Diffendorfer, J. E., M. S. Gaines, and R. D. Holt. 1995. Habitat fragmentation and movements of three mammals *(Sigmodon, Microtus,* and *Peromyscus). Ecology* 76 : 827–39.

Dixon, K. R., and T. C. Juelson. 1987. The political economy of the Spotted Owl. *Ecology* 68 : 772–76.

Doak, D. F., P. C. Marino, and P. Kareiva. 1992. Spatial scale mediates the influence of habitat fragmentation on dispersal success: Implications for conservation. *Theoretical Population Biology* 41 : 315–36.

Dodd, C. K., Jr. 1990. Effects of habitat fragmentation on a stream-dwelling species, the flattened musk turtle *Stenotherus depressus. Biological Conservation* 54 : 33–45.

Dodson, P. 1978. On the use of ratios in growth studies. *Systematic Zoology* 27 : 62–67.

Downs, D. R., and W. J. Ballantine. 1993. Conservation and restoration of New Zealand island ecosystems. *Trends in Ecology and Evolution* 8 : 452–57.

Drake, D. R., and D. Mueller-Dombois. 1993. Population development of rain forest trees on a chronosequence of Hawaiian lava flows. *Ecology* 74 : 1012–19.

Dressler, R. L. 1990. *The orchids: Natural history and classification.* Harvard University Press, Cambridge, Mass.

Ducenne, H., U. Schroff, and A. Narson. 1988. Carte des végétations: Etablie a partir des scenes SPOT (30.8.86) ET SOJUZ (29-5-87) des observations du SAF-Côte-ouest. SAF, unpublished manuscript.

Dudgeon, D., and R. T. Corlett. 1994. *Hills and streams: An ecology of Hong Kong.* Hong Kong University Press, Hong Kong.

Duelli, P., M. Studer, I. Marchand, and S. Jakob. 1990. Population movements of arthropods between natural and cultivated areas. *Biological Conservation* 54 : 193–207.

Duellman, W. E. 1978. *The biology of equatorial herpetofauna in Amazonian Ecuador.* Miscellaneous Publication of the University of Kansas Museum of Natural History, no. 65. University of Kansas, Lawrence, Kans.

———. 1979. The herpetofauna of the Andes: Patterns of distribution, origin, differentiation, and present communities. Pages 371–445 in W. E. Duellman, ed., *The South American herpetofauna: Its origin, evolution, and dispersal.* Museum of Natural History, monograph 7, University of Kansas, Lawrence, Kans.

Duellman, W. E., and L. Trueb. 1986. *Biology of amphibians.* McGraw-Hill, New York.

Duff, A. B., R. A. Hall, and C. Marsh. 1984. A survey of wildlife in and around a commercial tree plantation in Sabah. *Malaysian Forester* 47 : 197–213.

Dunning, J. B., Jr. 1993. *Handbook of avian body masses.* CRC Press, Boca Raton, Fla.

Dunphy, M. 1988. *Rainforest weeds of the Big Scrub.* Rainforest Rehabilitation Workshop, November 1988, National Parks and Wildlife Service, Wollongbar, New South Wales, Australia.

Eacham Historical Society. 1979. *Eacham Shire: Yesterday and today.* Eacham Shire Historical Society, Herberton, Queensland, Australia.

Eanes, W. F. 1978. Morphological variance and enzyme heterozygosity in the monarch butterfly. *Nature* 276:263–64.

Ebenhard, T. 1991. Colonization in metapopulations: A review of theory and observations. *Biological Journal of the Linnean Society* 42:105–21.

Efron, B. 1978. Controversies in the foundation of statistics. *American Mathematics Monthly* 85:231–46.

Eguiarte, L. E., N. Pérez-Nasser, and D. Pinero. 1992. Genetic structure, outcrossing rate and heterosis in *Astrocaryum mexicanum* (tropical palm): Implications for evolution and conservation. *Heredity* 69: 217–28.

Ehrlich, P., and A. Ehrlich. 1981. *Extinction.* Ballantine Books, New York.

Eisenberg, J. F. 1980. The density and biomass of tropical mammals. Pages 35–56 in M. E. Soulé and B. A. Wilcox, eds., *Conservation biology: An evolutionary-ecological perspective.* Sinauer Associates, Sunderland, Mass.

———. 1990. Neotropical mammal communities. Pages 358–68 in A. W. Gentry, ed., *Four Neotropical rainforests.* Yale University Press, New Haven, Conn.

Eisenmann, E. 1952. Annotated checklist of the birds of Barro Colorado Island, Panama Canal Zone. *Smithsonian Miscellaneous Collections* 117:1–62.

Elachi, C. 1987. *Introduction to the physics and techniques of remote sensing.* John Wiley, New York.

Electricity Generating Authority of Thailand. 1980. Chiew Larn Project: Environmental and ecological investigation. Final Report, vol. II. Electricity Generating Authority of Thailand.

Ellstrand, N. C., B. Devlin, and D. L. Marshall. 1990. Gene flow by pollen into small populations: Data from experimental and natural stands of wild radish. *Proceedings of the National Academy of Sciences USA* 86:9044–47.

Elton, C. S. 1958. *The ecology of invasions by animals and plants.* Chapman and Hall, London.

———. 1975. Conservation and the low population density of invertebrates inside Neotropical rain forest. *Biological Conservation* 7:3–15.

Emlen, J. T., and M. J. DeJong. 1981. The application of song detection threshold distance to census operations. *Studies in Avian Biology* 6:346–52.

Emmons, L. H. 1982. Ecology of *Proechimys* (Rodentia, Echimyidae) in south-eastern Peru. *Tropical Ecology* 23:280–90.

———. 1984. Geographic variation in densities and diversities of non-flying mammals in Amazonia. *Biotropica* 16:210–22.

———. 1988. A field study of ocelots *(Felis pardalis)* in Peru. *Revue d'Ecologie (Terre Vie)* 43:113–57.

Emmons, L. H., and F. Feer. 1990. *Neotropical rainforest mammals: A field guide.* University of Chicago Press, Chicago, Ill.

Endler, J. A. 1982. Problems in distinguishing historical from ecological factors in biogeography. *American Zoologist* 22:441–52.

ERDAS. 1994. *Imagine 8.2.* ERDAS, Atlanta, Ga.

Erwin, T. L. 1991. An evolutionary basis for conservation strategies. *Science* 253:750–52.

Escobar, F. 1994. Excremento, coprófagos y deforestación en bosques de montaña al suroccidente de Colombia. Thesis, Universidad del Valle, Cali, Colombia.

Espinal, L. S., J. Tosi, Jr., E. Montenegro, G. Toro, and D. Díaz. 1977. *Zonas de vida o formaciones vegetales de Colombia.* Instituto Geográfico Agustín Codazzi, Bogotá, Colombia.

Esseen, P. A. 1994. Tree mortality patterns after experimental fragmentation of an old-growth conifer forest. *Biological Conservation* 68:19–29.

Estrada, A., R. Coates-Estrada, D. Meritt, Jr., S. Montiel, and D. Curiel. 1993. Patterns of frugivore species richness and abundance in forest islands and in agricultural habitats at Los Tuxtlas, Mexico. *Vegetatio* 107/108:245–57.

Ewel, J. J. 1977. Differences between wet and dry forest successional tropical ecosystems. *Geo-Eco-Trop.* 1: 103–17.

———. 1980. Tropical succession: Manifold routes to maturity. *Biotropica* 12:2–7.

Faaborg, J. 1979. Qualitative patterns of avian extinction on Neotropical land-bridge islands: Lessons for conservation. *Journal of Applied Ecology* 16:99–107.

Faeth, S. H. 1984. Density compensation in vertebrates and invertebrates: A review and experiment. Pages 491–509 in D. R. Strong, Jr., D. Simberloff, L. G. Abele, and A. B. Thistle, eds., *Ecological communities: Conceptual issues and the evidence.* Princeton University Press, Princeton, N.J.

Fahrig, L., and G. Merriam. 1985. Habitat patch connectivity and population survival. *Ecology* 66:1762–68.

———. 1994. Conservation of fragmented populations. *Conservation Biology* 8:50–59.

Fahrig, L., J. H. Pedlar, S. E. Pope, P. D. Taylor, and J. F. Wegner. 1995. Effect of road traffic on amphibian density. *Biological Conservation* 73:177–82.

FAO (Food and Agriculture Organization of the United Nations)/UNESCO. 1971. *Soil map of the world at 1:5,000,000.* Vol. 1: *Legend.* UNESCO, Paris.

———. 1989. *Classification and mapping of vegetation types in tropical Asia.* FAO, Rome.

———. 1991. *FAO yearbook production 1990.* Vol. 44. FAO Statistics Series no. 99. FAO, Rome.

———. 1992. Third interim report on the state of tropical forests by forest resources assessment 1990 project. FAO, Rome.

———. 1993a. Committee on forestry eleventh session: Report on forest resources assessment 1990. Secretariat note. FAO, Rome.

———. 1993b. Forest resources assessment 1990: Tropical countries. FAO Forestry Paper 112. FAO, Rome.

———. 1993c. Summary of the final report of the forest resources assessment 1990 for the tropical world. FAO, Rome.

———. 1995. Forest resources assessment 1990: Global synthesis. FAO Forestry Paper 124. FAO, Rome.

Faramalala, M. H. 1988. Étude de la végétation de Madagascar a l'aide des donées spatiales. Ph.D. dissertation, L'Université Paul Sabatier de Toulouse (Sciences), France.

Farquhar, G. D., M. H. O'Leary, and J. A. Berry. 1982. On the relationship between carbon isotope discrimination and the intercellular carbon dioxide concentration in leaves. *Australian Journal of Plant Physiology* 9:121–37.

Farris, J. S. 1969. On the cophenetic correlation coefficient. *Systematic Zoology* 18:279–85.

Fearnside, P. M. 1980. Land use allocation of the Transamazon Highway colonists of Brazil and its relation to human carrying capacity. Pages 114–38 in F. B. Scazzochio, ed., *Land, people and planning in contemporary Amazonia.* Occasional Paper no. 3. University of Cambridge Center for Latin American Studies, Cambridge.

———. 1984. Land clearing behavior in small farmer settlement schemes in the Brazilian Amazon and its relation to human carrying capacity. Pages 255–71 in A. C. Chadwick and S. L. Sutton, eds., *Tropical rain forests: Ecology and management.* Blackwell Scientific Publications, Leeds, U.K.[1]

———. 1986. *Human carrying capacity of the Brazilian rainforest.* Columbia University Press, New York.

———. 1989. Extractive reserves in Brazilian Amazonia: An opportunity to maintain tropical rain forest under sustainable use. *Bioscience* 39:387–93.

Fearnside, P. M., and J. Ferraz. 1995. A conservation gap analysis of Brazil's Amazonian vegetation. *Conservation Biology* 9:1134–47.

Federov, A. A. 1966. The structure of the tropical rain forest and speciation in the humid tropics. *Journal of Ecology* 54:1–11.

Feinsinger, P. 1976. Organization of a tropical guild of nectarivorous birds. *Ecological Monographs* 46: 257–91.

Fensham, R. J., R. J. Fairfax, and R. J. Cannell. 1994. The invasion of *Lantana camara* L. in Forty Mile Scrub National Park, north Queensland. *Australian Journal of Ecology* 19:297–305.

Ferrier, S. 1988. Environmental resource mapping system (E-RMS), user's manual for version 1.2. National Parks and Wildlife Service of New South Wales, Sydney, Australia.

Ferris, C. R. 1979. Effects of Interstate 95 on breeding birds in northern Maine. *Journal of Wildlife Management* 43:421–27.

Fitzpatrick, J. W. 1980. Foraging behavior of Neotropical tyrant flycatchers. *Condor* 82:43–57.

Fjeldså, J. 1992. Biogeographic patterns and evolution of the avifauna of relict high-altitude woodlands of the Andes. *Steenstrupia* 18:9–62.

———. 1993. The avifauna of the *Polylepis* woodlands of the Andean highlands: The efficiency of basing conservation priorities on the patterns of endemism. *Bird Conservation International* 3:37–55.

———. 1994. Geographical patterns for relict and young species of birds in Africa and South America and implications for conservation priorities. *Biodiversity and Conservation* 3:207–26.

———. 1995. Geographical patterns of neoendemic and older relict species of Andean forest birds: The significance of ecologically stable areas. Pages 89–102 in S. P. Churchill, H. Balslev, E. Forero, and J. L. Luteyn, eds., *Biodiversity and conservation of neotropical montane forests.* Memoirs of the New York Botanical Garden. Bronx, New York.

Fjeldså, J., and N. Krabbe. 1990. *Birds of the high Andes.* Zoological Museum, University of Copenhagen and Apollo Books, Svendborg, Denmark.

Fjeldså, J., and J. C. Lovett. In press. Geographical patterns of phylogenetic relicts and phylogenetically subordinate species in tropical African forest biota. *Biodiversity and Conservation* (1997).

Fleming, T. H. 1981. Fecundity, fruiting pattern and seed dispersal in *Piper amalgo* (Piperaceae), a bat dispersed tropical shrub. *Oecologia* 51:42–46.

Flenley, J. R. 1979. *The equatorial rainforest: A geological history.* Butterworth, London.

Fonseca, G. A. B. da. 1985. The vanishing Brazilian Atlantic forest. *Biological Conservation* 34:17–34.

———. 1988. Patterns of small mammal species diversity in the Brazilian Atlantic forest. Ph.D. dissertation, University of Florida, Gainesville.

Fonseca, G. A. B. da, and M. C. M. Kierulff. 1989. Biology and natural history of Brazilian Atlantic forest small mammals. *Bulletin Florida State Museum, Biological Sciences* 34:99–152.

Fonseca, G. A. B. da, and J. G. Robinson. 1990. Forest size and structure: Competitive and predatory effects on small mammal communities. *Biological Conservation* 53:265–94.

Ford, J. 1981. Hybridisation and migration in Australia: Populations of the little and rufous-breasted bronze-cuckoos. *Emu* 81:209–22.

Fore, S. A., R. J. Hickey, J. L. Vankat, S. I. Guttman, and R. L. Schaefer. 1992. Genetic structure after forest fragmentation: A landscape ecology perspective on *Acer saccharum. Canadian Journal of Botany* 70: 1659–68.

Forget, P.-M. 1990. Seed dispersal of *Vouacapoua americana* (Caesalpiniaceae) by caviomorph rodents in French Guiana. *Journal of Tropical Ecology* 6:459–68.

Forman, R. T. T., and M. Godron. 1986. *Landscape ecology.* John Wiley and Sons, New York.

Fortmann, L., and J. W. Bruce. 1988. *Whose trees? Proprietary dimensions of forestry.* Westview Press, Boulder, Colo.

Foster, R. B. 1980. Heterogeneity and disturbance in tropical vegetation. Pages 75–92 in M. E. Soulé and B. A. Wilcox, eds., *Conservation biology: An evolutionary-ecological perspective.* Sinauer Associates, Sunderland, Mass.

———. 1982. Famine on Barro Colorado Island. Pages 201–12 in E. G. Leigh, Jr., A. S. Rand, and D. S. Windsor, eds., *The ecology of a tropical forest: Seasonal rhythms and long-term changes.* Smithsonian Institution Press, Washington, D.C.

———. 1995. A campside conversation. *Member's Report,* Spring 1995. Conservation International, Washington, D.C.

Foster, R. B., J. Arce, and T. S. Wachter. 1986. Dispersal and the sequential plant communities in Amazonian Peru floodplain. Pages 357–70 in A. Estrada and T. H. Fleming, eds., *Frugivores and seed dispersal.* Dr. W. Junk, Dordrecht, The Netherlands.

Fowler, K., and M. C. Whitlock. 1994. Fluctuating asymmetry does not increase with moderate inbreeding in *Drosophila melanogaster. Heredity* 73:373–76.

Fowler, S. V., P. Chapman, D. Checkley, S. Hurd, M. McHale, J. E. Ramangason, P. Stewart, and R. Walters.

1989. Survey and management proposals for a tropical deciduous forest reserve at Ankarana in northern Madagascar. *Biological Conservation* 47:297–313.

Frankel, O. H., and M. E. Soulé. 1981. *Conservation and evolution.* Cambridge University Press, New York.

Frankie, G. W., P. A. Opler, and K. S. Bawa. 1976. Foraging behavior of solitary bees: Implications for outcrossing of a Neotropical forest tree species. *Journal of Ecology* 64:1049–57.

Franklin, I. F. 1980. Evolutionary change in small populations. Pages 135–49 in M. E. Soulé and B. A. Wilcox, eds., *Conservation biology: An evolutionary-ecological perspective.* Sinauer Associates, Sunderland, Mass.

Franklin, J. F. 1993. Preserving biodiversity: Species, ecosystems, or landscapes? *Ecological Applications* 3: 202–5.

Franklin, J. F., and R. T. T. Forman. 1987. Creating landscape patterns by forest cutting: Ecological consequences and principles. *Landscape Ecology* 1:5–18.

Frawley, K. J. 1983a. Forestry and land management in northeast Queensland 1859–1960. Ph.D. dissertation, Australian National University, Canberra, Australia.

———. 1983b. A history of forest and land management in Queensland, with particular reference to the north Queensland rainforest. Report to Rainforest Conservation Society of Queensland, Brisbane, Australia.

Fretwell, S. D. 1972. *Populations in a seasonal environment.* Princeton University Press, Princeton, N.J.

Frith, C. B., and D. W. Frith. 1990. Seasonality of litter invertebrate populations in an Australian upland tropical rainforest. *Biotropica* 22:181–90.

Fritschen, L. J. 1985. Characterization of boundary conditions affecting forest environment phenomena. Pages 3–23 in B. A. Hutchinson, and B. B. Hicks, eds., *The forest-atmosphere interaction.* D. Reidel, Boston, Mass.

Frohn, R. C., V. H. Dale, and B. D. Jimenez. 1990. *Colonization, road development and deforestation in the Brazilian Amazon Basin of Rondônia.* ORNL/TM-11470. Oak Ridge National Laboratory, Oak Ridge, Tenn.

Gade, D. W. 1985. Madagascar and non-development. *Focus* 35:14–21.

Galli, A., C. F. Leck, and R. T. T. Forman. 1976. Avian distribution patterns in forest islands of different sizes in central New Jersey. *Auk* 93:356–64.

Ganzhorn, J. U., A. W. Ganzhorn, J. P. Abraham, L. Andriamanarivo, and L. Ramanajatovo. 1990. The impact of selective logging on forest structure and tenrec populations in western Madagascar. *Oecologia* 84:126–33.

Gardner, A. L. 1993. Order Didelphimorphia. Pages 15–23 in D. E. Wilson and D. M. Reeder, eds., *Mammal species of the world, a taxonomic and geographic reference,* 2d ed. Smithsonian Institution Press, Washington, D.C.

Gardner, A. L., and J. L. Patton. 1976. Karyotypic variation in oryzomyine rodents (Cricetinae) with comments on chromosomal evolution in the Neotropical cricetine complex. *Occasional Papers of the Museum of Zoology, Louisiana State University* 49:1–48.

Gardner, R. H., A. W. King, and V. H. Dale. 1994. Interactions between forest harvesting, landscape heterogeneity, and species persistence. Pages 65–75 in D. C. LeMaster and R. J. Sedjo, eds., *Modeling sustainable forest ecosystems.* American Forests, Washington, D.C.

Gardner, R. H., B. T. Milne, M. G. Turner, and R. V. O'Neill. 1987. Neutral models for the analysis of broad-scale landscape pattern. *Landscape Ecology* 3:19–28.

Gardner, R. H., and R. V. O'Neill. 1991. Pattern, process, and predictability: The use of neutral models for landscape analysis. Pages 289–308 in M. G. Turner and R. H. Gardner, eds., *Quantitative methods in landscape ecology.* Springer-Verlag, New York.

Garland, T., and W. G. Bradley. 1984. Effects of a highway on Mojave Desert rodent populations. *American Midland Naturalist* 111:47–56.

Garwood, N., D. P. Janos, and N. Brokaw. 1979. Earthquake-caused landslides: A major disturbance to tropical forests. *Science* 205:997–99.

Gascon, C. 1990. Relative importance of habitat characteristics in the maintenance of a larval anuran assemblage in the Tropics. Ph.D. dissertation, Florida State University, Tallahassee.
———. 1991a. Breeding of *Leptodactylus knudseni*: Responses to rainfall variation. *Copeia* 1991:248–51.
———. 1991b. Population and community level analyses of species occurrences of central Amazonian rainforest tadpoles. *Ecology* 72:1731–46.
———. 1993. Breeding-habitat use by five Amazonian frogs at forest edge. *Biodiversity and Conservation* 2:438–44.
Gaston, K. J., and T. M. Blackburn. 1995. Birds, body size and the threat of extinction. *Philosophical Transactions of the Royal Society of London B* 347:205–12.
Gates, J. E. 1991. Power line corridors, edge effects, and wildlife in forested landscapes in the Central Appalachians. Pages 13–32 in J. E. Rodiek and E. G. Bolen, eds., *Wildlife and habitats in managed landscapes.* Island Press, Washington D.C.
Gates, J. E., and I. W. Gysel. 1978. Avian nest dispersion and fledging success in field-forest ecotones. *Ecology* 59:871–83.
Gauch, H. G., Jr. 1982. *Multivariate analysis in community ecology.* Cambridge University Press, Cambridge.
Gazey, W. J., and Staley, M. J. 1986. Population estimation from mark-recapture experiments using a sequential Bayes algorithm. *Ecology* 67:941–51.
Gentry, A. H. 1983. Dispersal ecology and diversity in Neotropical forest communities. *Sonderband Naturwisseschaften Ver. Hamburg* 7:303–14.
———. 1986. Endemism in tropical versus temperate plant communities. Pages 153–81 in M. E. Soulé, ed., *Conservation biology: The science of scarcity and diversity.* Sinauer Associates, Sunderland, Mass.
———. 1990a. Floristic similarities and differences between southern Central America and upper and central Amazonia. Pages 141–57 in A. H. Gentry, ed., *Four Neotropical rainforests.* Yale University Press, New Haven, Conn.
———, ed. 1990b. *Four Neotropical rainforests.* Yale University Press, New Haven, Conn.
———. 1992a. Diversity and composition of Andean forests of Peru and adjacent countries: Implications for their conservation. *Memorias del Museo de Historia Natural U.N.M.S.M. (Lima)* 21:11–29.
———. 1992b. Tropical forest biodiversity: Distributional patterns and their conservational significance. *Oikos* 63:19–28.
Gentry, A. H., and L. H. Emmons. 1987. Geographical variation in fertility, phenology, and composition of the understory of Neotropical forests. *Biotropica* 19:216–27.
Gentry, A. H., and J. Lopez-Parodi. 1980. Deforestation and increased flooding of the upper Amazon. *Science* 210:1354–55.
George, T. L. 1987. Greater land bird densities on island vs. mainland: Relation to nest predation level. *Ecology* 68:1393–1400.
Getz, L. L., F. R. Cole, and D. L. Gates. 1978. Interstate roadsides as dispersal routes for *Microtus pennsylvanicus. Journal of Mammalogy* 59:208–12.
Ghilarov, A. M. 1992. Ecology, mythology and the organismic way of thinking in limnology. *Trends in Ecology and Evolution* 7:22–25.
Gibbs, J. P. 1991. Avian nest predation in tropical wet forest: An experimental study. *Oikos* 60:1551–61.
Gilbert, F. S. 1980. The equilibrium theory of island biogeography: Fact or fiction? *Journal of Biogeography* 7:209–35.
Gilbert, L. E. 1980. Food web organization and the conservation of Neotropical diversity. Pages 11–33 in M. E. Soulé and B. A. Wilcox, eds., *Conservation biology: An evolutionary-ecological perspective.* Sinauer Associates, Sunderland, Mass.
Gillis, M. 1988a. Indonesia: Public forests, resource management and tropical forests. Pages 43–114 in R. Repetto and M. Gillis, eds., *Public policies and the misuse of forest resources.* Cambridge University Press, Cambridge.
———. 1988b. West Africa: Resource management policies and the tropical forest. Pages 299–342 in

R. Repetto and M. Gillis, eds., *Public policies and the misuse of forest resources.* Cambridge University Press, Cambridge.

Gilmour, D. A., and R. F. Fisher. 1991. *Villagers, forests and foresters: The philosophy, process and practice of community forestry in Nepal.* Sahayogi Press, Kathmandu, Nepal.

Gilmour, D. A., and J. J. Reilly. 1970. Productivity survey of the Atherton Tableland and suggested land use changes. *Journal of the Australian Institute of Agricultural Science* 32:259–72.

Gilpin, M. E. 1987a. Experimental community assembly: Competition, community structure and the order of species introductions. Pages 151–61 in W. R. Jordan, M. E. Gilpin, and J. E. Aber, eds., *Restoration ecology: A synthetic approach to ecological research.* Cambridge University Press, Cambridge.

———. 1987b. Spatial structure and population vulnerability. Pages 125–40 in M. E. Soulé, ed., *Viable populations for conservation.* Cambridge University Press, Cambridge.

Gilpin, M. E., and M. E. Soulé. 1986. Minimum viable populations: Processes of species extinction. Pages 19–34 in M. E. Soulé, ed., *Conservation biology: The science of scarcity and diversity.* Sinauer Associates, Sunderland, Mass.

Glanz, W. E. 1982. The terrestrial mammal fauna of Barro Colorado Island: Censuses and long-term changes. Pages 455–68 in E. G. Leigh, Jr., A. S. Rand, and D. M. Windsor, eds., *The ecology of a tropical forest: Seasonal rhythms and long-term changes.* Smithsonian Institution Press, Washington, D.C.

———. 1990. Neotropical mammal densities: How unusual is the community on Barro Colorado Island, Panama? Pages 287–313 in A. Gentry, ed., *Four Neotropical rainforests.* Yale University Press, New Haven, Conn.

Gliwicz, J. 1984. Population dynamics of the spiny rat *Proechimys semispinosus* on Orchid Island (Panama). *Biotropica* 16:73–78.

Good, I. J. 1978. Alleged objectivity: A threat to the human spirit. *International Statistical Review* 46:65–66.

Goodwin, D. 1976. *Crows of the world.* Comstock Publishers, Ithaca, N.Y.

Goodyer, N. J. 1992. Notes on the land mammals of Hong Kong. *Memoirs of the Hong Kong Natural History Society* 19:71–78.

Goosem, M. W., and H. Marsh. In press. Fragmentation of a small mammal community by a powerline corridor through tropical rainforest. *Wildlife Research.*

Goosem, S. P., and N. I. J. Tucker. 1995. *Repairing the rainforest: Theory and practice of rainforest re-establishment in North Queensland's Wet Tropics.* Wet Tropics Management Authority, Cairns, Australia.

Goosem, S. P., and P. A. R. Young. 1989. Rainforests of the humid Tropics of North Queensland: A review and discussion of population ecology. Internal report, Queensland Department of Environment and Heritage, Brisbane, Australia.

Gotelli, N. J. 1991. Metapopulation models: The rescue effect, the propagule rain, and the core-satellite hypothesis. *American Naturalist* 138:768–76.

Graham, M., and P. Round. 1994. *Thailand's vanishing flora and fauna.* Finance One Public Company Ltd., Bangkok, Thailand.

Grange, A. C., V. R. Brown, and G. S. Sinclair. 1993. Vesicular-arbuscular mycorrhiza: A determinant of plant community structure and early succession. *Functional Ecology* 7:616–22.

Graves, G. R. 1985. Elevational correlates of speciation and intraspecific geographic variation in plumage in Andean forest birds. *Auk* 102:556–79.

———. 1988. Linearity of geographic range and its possible effect on the population structure of Andean birds. *Auk* 105:47–52.

Green, G. M., and R. W. Sussman. 1990. Deforestation history of the eastern rain forests of Madagascar from satellite images. *Science* 248:212–15.

Green, M., J. Paine, and J. McNeely. 1991. The protected area system. Pages 60–67 in M. Collins, J. A. Sayer, and T. C. Whitmore, eds., *The conservation atlas of tropical forests: Asia and the Pacific.* MacMillan, London.

Greene, H. W. 1994. Systematics and natural history, foundations for understanding and conserving biodiversity. *American Zoologist* 34:48–56.

Griffiths, M., and C. P. van Schaik. 1993. The impact of human traffic on the abundance and activity of Sumatran rain forest wildlife. *Conservation Biology* 7:623–26.

Groombridge, B., ed. 1992. *Global biodiversity.* Chapman and Hall, London.

Grubb, P. J. 1977a. Control of forest growth and distribution on wet tropical mountains, with special reference to mineral nutrition. *Annual Review of Ecology and Systematics* 8:83–107.

———. 1977b. The maintenance of species richness in plant communities: The importance of the regeneration niche. *Biological Review* 52:107–45.

———. 1986. Problems posed by sparse patchily distributed species in species-rich communities. Pages 207–25 in J. Diamond and T. J. Case, eds., *Community ecology.* Harper and Row, New York.

Guevara, S., and J. Laborde. 1993. Monitoring seed dispersal at isolated standing trees in tropical pastures: Consequences for local species availability. Pages 319–38 in T. H. Fleming and A. Estrada, eds., *Frugivory and seed dispersal: Ecological and evolutionary aspects.* Kluwer Academic Publishers, Dordrecht, The Netherlands.

Guevara, S., J. Meave, P. Morena-Casosole, and J. Labrode. 1992. Floristic composition and structure of vegetation under isolated trees in neotropical pastures. *Journal of Vegetation Science* 3:655–64.

Guevara, S., S. E. Purata, and E. Van der Maarel. 1986. The role of remnant forest trees in tropical secondary succession. *Vegetatio* 66:77–84.

Guillotin, M. 1982. Rythmes d'activité et régimes alimentaires de *Proechimys cuvieri* et d'*Oryzomys capito velutinus* (Rodentia) en forêt Guyanaise. *Revue d'Ecologie (Terre et Vie)* 36:337–71.

Gullison, R. E., and E. Losos. 1992. The role of foreign debt in deforestation in Latin America. Working paper, Department of Ecology and Evolutionary Biology, Princeton University, Princeton, N.J.

Gunderson, L. H. 1992. Spatial and temporal dynamics in the Everglades ecosystem with implications for water deliveries to Everglades National Park. Ph.D. dissertation, University of Florida, Gainesville.

Gunnarson, B. 1988. Spruce-living spiders and forest decline: The importance of needle-loss. *Biological Conservation* 43:309–19.

Gustafson, E. J., and G. R. Parker. 1992. Relationship between landcover proportion and indices of landscape spatial pattern. *Landscape Ecology* 7:101–10.

Haas, C. A. 1995. Dispersal and use of corridors by birds in wooded patches on an agricultural landscape. *Conservation Biology* 9:845–54.

Haber, W. A., and G. W. Frankie. 1989. A tropical hawkmoth community: Costa Rican dry forest Sphingidae. *Biotropica* 21:155–72.

Hacking, I. 1983. *Representing and intervening: Introductory topics in the philosophy of natural science.* Cambridge University Press, Cambridge.

Haffer, J. 1969. Speciation in Amazonian forest birds. *Science* 165:131–37.

———. 1993. Time's cycle and time's arrow in the history of Amazonia. *Biogeographica* 69:15–45.

Hagan, J. M. III, and D. W. Johnston, eds. 1992. *Ecology and conservation of Neotropical landbird migrants.* Smithsonian Institution Press, Washington, D.C.

Haila, Y. 1982. Hypothetico-deductivism and the competition controversy in ecology. *Annales Zoologici Fennici* 19:255–63.

———. 1986. On the semiotic dimension of ecological theory: The case of island biogeography. *Biological Philosophy* 1:377–87.

———. 1989. Ecology finding evolution finding ecology. *Biological Philosophy* 4:235–44.

Haila, Y., D. Saunders, and R. Hobbs. 1993. What do we presently understand about ecosystem fragmentation? Pages 45–55 in D. A. Saunders, R. J. Hobbs, and P. R. Ehrlich, eds., *Nature conservation 3: Reconstruction of fragmented ecosystems.* Surrey Beatty, Chipping Norton, N.S.W., Australia.

Haines-Young, R., D. R. Green, and S. Cousins. 1993. *Landscape ecology and GIS.* Taylor and Francis, London.

Hallwachs, W. 1986. Agoutis *(Dasyprocta punctata):* The inheritors of Guapinol (*Hymenaea courbaril:* Leguminosae). Pages 285–304 in A. Estrada and T. H. Fleming, eds., *Frugivores and seed dispersal.* Dr. W. Junk, Dordrecht, The Netherlands.

Hamrick, J. L. 1994. Genetic diversity and conservation in tropical forests. Pages 1–9 in R. M. Drysdale, S. E. T. John, and A. C. Yapa, eds., *Proceedings: International symposium on genetic conservation and production of forest tree seed.* ASEAN-Canada Forest Tree Seed Center.

Hamrick, J. L., and M. J. W. Godt. 1989. Allozyme diversity in plant species. Pages 43–63 in A. H. D. Brown, M. T. Clegg, A. L. Kahler, and B. S. Weir, eds., *Plant population genetics, breeding, and genetic resources.* Sinauer Associates, Sunderland, Mass.

Hamrick, J. L., M. J. W. Godt, and S. L. Sherman-Broyles. 1992. Factors influencing levels of genetic diversity in woody plant species. *New Forests* 6:95–124.

Hamrick, J. L., and M. D. Loveless. 1986. Isozyme variation in tropical trees: Procedures and preliminary results. *Biotropica* 18:201–7.

———. 1989. The genetic structure of tropical populations: Associations with reproductive biology. Pages 129–46 in J. H. Bock and Y. B. Linhart, eds., *The evolutionary ecology of plants.* Westview Press, Boulder, Colo.

Hamrick, J. L., and D. A. Murawski. 1990. The breeding structure of tropical tree populations. *Plant Species Biology* 5:157–65.

———. 1991. The effect of the density of flowering individuals on the mating systems of nine tropical tree species. *Heredity* 67:167–74.

Hamrick, J. L., D. A. Murawski, and J. D. Nason. 1993. The influence of seed dispersal mechanisms on the genetic structure of tropical tree populations. Pages 281–98 in T. H. Fleming and A. Estrada, eds., *Frugivory and seed dispersal: Ecological and evolutionary aspects.* Kluwer Academic Publishers, Dordrecht, The Netherlands.

Hamrick, J. L, and J. D. Nason. 1996. Consequences of dispersal in plants. In O. E. Rhodes, Jr., R. K. Chesser, and M. H. Smith, eds., *Population dynamics in ecological space and time.* University of Chicago Press, Chicago, Ill.

Hanagarth, W. 1993. *Acerca de la geoecologia de las sabanas del Beni en el norest de Bolivia.* Instituto de Ecología, La Paz, Bolivia.

Hanski, I. 1986. Population dynamics of shrews on small islands accord with equilibrium model. *Biological Journal of the Linnean Society* 28:23–36.

———. 1994. Patch-occupancy dynamics in fragmented landscapes. *Trends in Ecology and Evolution* 9: 131–35.

Hanski, I., and M. E. Gilpin. 1991. Metapopulation dynamics: Brief history and conceptual domain. *Biological Journal of the Linnean Society* 42:3–16.

Hanski, I., T. Pakkala, M. Kuussaari, and G. Lei. 1995. Metapopulation persistence of an endangered butterfly in a fragmented landscape. *Oikos* 72:21–28.

Hanski, I., J. Poyry, T. Pakkala, and M. Kuussaari. 1995. Multiple equilibria in metapopulation dynamics. *Nature* 377:618–21.

Harcourt, C., and J. Thornback. 1990. *Lemurs of Madagascar and the Comores.* The IUCN Red Data Book, International Union for the Conservation of Nature and Natural Resources (IUCN), Gland, Switzerland.

Harper, J. L. 1977. *Population biology of plants.* Academic Press, London.

Harper, L. H. 1989. Birds and army ants *(Eciton burchelli):* Observations on their ecology in undisturbed forest and isolated reserves. *Acta Amazônica* 19:249–63.

Harris, G. P. 1980. Temporal and spatial scales in phytoplankton ecology: Mechanisms, methods, models, and management. *Canadian Journal of Fisheries and Aquaculture* 37:877–900.

Harris, L. D. 1984. *The fragmented forest: Island biogeography theory and the preservation of biotic diversity.* University of Chicago Press, Chicago, Ill.

———. 1988. Edge effects and conservation of biotic diversity. *Conservation Biology* 2:330–32.

Harris, L. D., and J. Scheck. 1991. From implications to applications: The dispersal corridor principle applied to the conservation of biological diversity. Pages 189–220 in D. A. Saunders and R. J. Hobbs, eds., *Nature conservation 2: The role of corridors.* Surrey Beatty, Sydney, Australia.

Harris, L. D., and G. Silva-Lopez. 1992. Forest fragmentation and the conservation of biological diversity.

Pages 197–237 in P. L. Fiedler and S. K. Jain, eds., *Conservation biology: The theory and practice of nature conservation, preservation, and management.* Chapman and Hall, London and New York.

Harrison, J. L. 1957. Habitat of some Malayan rats. *Proceedings of the Zoological Society of London* 128:1–21.

———. 1958. Range of movement of some Malayan rats. *Journal of Mammalogy* 39:190–206.

Harrison, R. L. 1992. Towards a theory of inter-refuge corridor design. *Conservation Biology* 6:293–95.

Harrison, S. 1989. Long-distance dispersal and colonization in the bay checkspot butterfly. *Ecology* 70: 1236–43.

Hartshorn, G. S., and B. E. Hammel. 1994. Vegetation types and floristic patterns. Pages 73–89 in L. A. McDade, K. S. Bawa, H. A. Hespenheide, and G. S. Hartshorn, eds., *La Selva: Ecology and natural history of a Neotropical rainforest.* University of Chicago Press, Chicago, Ill.

Haslett, J. A. 1994. Community structure and the fractal dimensions of mountain habitats. *Journal of Theoretical Biology* 167:407–11.

Hastings, A., and S. Harrison. 1994. Metapopulation dynamics and genetics. *Annual Review of Ecology and Systematics* 25:167–88.

Hawkins, A. F. A., P. Chapman, and J. U. Ganzhorn. 1990. Vertebrate conservation in Ankarana special reserve, northern Madagascar. *Biological Conservation* 49:83–110.

Hay, M. E. 1994. Species as "noise" in community ecology: Do seaweeds block our view of the kelp forest? *Trends in Ecology and Evolution* 9:414–16.

Heaney, L. R. 1984. Mammalian species richness on islands on the Sunda Shelf, southeast Asia. *Oecologia* 61:11–17.

———. 1986. Biogeography of mammals in SE Asia: Estimates of rates of colonization, extinction and speciation. *Biological Journal of the Linnean Society* 28:99–125.

———. 1991. A synopsis of climatic and vegetational change in Southeast Asia. *Climatic Change* 19:53–61.

Heimerdinger, M. A., and R. C. Leberman. 1966. The comparative efficiency of 30 and 36 mm mesh mist nets. *Bird-Banding* 37:280–85.

Heithaus, E. R., and T. H. Fleming. 1978. Foraging movements of a frugivorous bat, *Carollia perspicillata* (Phyllostomatidae). *Ecology* 48:127–43.

Heithaus, E. R., T. H. Fleming, and P. A. Opler. 1975. Foraging patterns and resource utilization in seven species of bats in a seasonal tropical forest. *Ecology* 56:841–54.

Helle, P., and J. Muona. 1985. Invertebrate numbers in edges between clear-fellings and mature forests in northern Finland. *Silva Fennica* 19:281–94.

Henderson, A., S. P. Churchill, and J. Luteyn. 1991. Neotropical plant diversity. *Nature* 351:21–22.

Hero, J. M. 1990. An illustrated key to tadpoles occurring in central Amazonian rain forest, Amazonia, Brazil. *Amazonia* 1:201–62.

Hershkovitz, P. 1960. Mammals of northern Colombia, preliminary report no. 8: Arboreal rice rats, a systematic revision of the subgenus *Oecomys,* genus *Oryzomys. Proceedings of the United States National Museum* 110:513–68.

———. 1987. A history of the recent mammalogy of the Neotropical Region from 1492 to 1850. *Fieldiana: Zoology,* new series 39:11–98.

Hershkovitz, P. 1994. The description of a new species of South American hocicudo, or long-nose mouse, genus *Oxymycterus* (Signodontinae, Muroidea), with a critical review of the generic content. *Fieldiana: Zoology,* new series 79:1–43.

Hester, A. J., and R. J. Hobbs. 1992. Influence of fire and soil nutrients on native and non-native annuals at remnant vegetation edges in the Western Australian wheatbelt. *Journal of Vegetation Science* 3:101–8.

Heyer, W. R., and L. R. Maxson. 1982. Neotropical frog biogeography: Paradigms and problems. *American Zoologist* 22:397–410.

Heywood, J. S., and T. H. Fleming. 1986. Patterns of allozyme variation in three Costa Rican species of *Piper. Biotropica* 18:208–13.

Heywood, V. H., G. M. Mace, R. M. May, and S. N. Stuart. 1994. Uncertainties in extinction rates. *Nature* 368:105.

Heywood, V. H., and S. N. Stuart. 1992. Species extinctions in tropical forests. Pages 91–117 in T. C. Whitmore and J. A. Sayer, eds., *Tropical deforestation and species extinction.* Chapman and Hall, London.

Hill, M. O. 1979a. DECORANA—A FORTRAN program for detrended correspondence analysis and reciprocal averaging. Cornell University, Ithaca, N.Y.

———. 1979b. TWINSPAN—A FORTRAN program for arranging multivariate data in an ordered two-way table by classification of the individuals and attributes. Ecology and Systematics, Cornell University, Ithaca, N.Y.

Hill, M. O., and H. G. Gauch. 1980. Detrended correspondence analysis, an improved ordination technique. *Vegetatio* 42:47–58.

Hilty, S. L. 1980. Flowering and fruiting periodicity in a premontane rain forest in Pacific Colombia. *Biotropica* 12:292–306.

Hilty, S. L., and W. L. Brown. 1986. *A guide to the birds of Colombia.* Princeton University Press, Princeton, N.J.

Hobbs, R. J. 1993. Effects of landscape fragmentation on ecosystem processes in the Western Australian wheat belt. *Biological Conservation* 64:193–201.

Hobbs, R. J., and H. A. Mooney. 1990. Remote sensing of biosphere functioning. Springer-Verlag, New York.

Hockings, M. 1981. Habitat distribution and species diversity of small mammals in south-east Queensland in relation to vegetation structure. *Australian Wildlife Research* 8:97–108.

Hodl, W. 1990. Reproductive diversity in Amazonian lowland frogs. *Fortschr. D. Zoologie* 38:45–51.

Holderegger, R., and J. J. Schneller. 1994. Are small isolated populations of *Asplenium septentrionale* variable? *Biological Journal of the Linnean Society* 51:377–85.

Holdridge, L. R. 1967. *Life zone ecology.* Rev. ed. Tropical Science Center, San Jose, Costa Rica.

Holdridge, L. R., W. C. Grenke, W. H. Hatheway, T. Liang, and J. A. Tosi, Jr. 1971. *Forest environments in tropical life zones.* Pergamon Press, New York.

Holling, C. S. 1973. Resilience and stability of ecological systems. *Annual Review of Ecology and Systematics* 4:1–23.

———. 1992. Cross-scale morphology, geometry, and dynamics of ecosystems. *Ecological Monographs* 62: 447–502.

Holling, C. S., G. Peterson, P. Marples, J. Sendzimir, K. Redford, L. Gunderson, and D. Lambert. In press. Self-organization in ecosystems: Lump geometries, periodicities, and morphologies. In B. H. Walker and W. L. Steffen, eds., *Global change and terrestrial ecosystems.* Cambridge University Press, Cambridge.

Holloway, J. D., A. H. Kirk-Spriggs, and C. V. Chey. 1992. The response of some rain forest insect groups to logging and conversion to plantation. *Philosophical Transactions of the Royal Society of London B* 335: 425–36.

Holt, R. D. 1993. Ecology at the mesoscale: The influence of regional processes on local communities. Pages 77–88 in R. E. Ricklefs and D. Schluter, eds., *Species diversity in ecological communities: Historical and geographical perspectives.* University of Chicago Press, Chicago, Ill.

Holt, R. D., G. R. Robinson, and M. S. Gaines. 1995. Vegetation dynamics in an experimentally fragmented landscape. *Ecology* 76:1610–24.

Hooghiemstra, H. 1984. Vegetational and climatic history of the high plain of Bogota, Colombia: A continuous record of the last 3.5 million years. Dissertationes Botanicae 79. Cramer, Vaduz, Colombia.

Hopkins, M. S., J. Ash, A. W. Graham, J. Head, and R. K. Hewett. 1993. Charcoal evidence of the spatial extent of the *Eucalyptus* woodland expansions and rainforest contractions in north Queensland during the late Pleistocene. *Journal of Biogeography* 20:59–74.

Hopkins, M. S., J. G. Tracey, and A. W. Graham. 1990. The size and composition of soil seed-banks in the remnant patches of three structural rainforest types in north Queensland. *Australian Journal of Ecology* 15:43–50.

Houghton, R. A. 1991. Tropical deforestation and atmospheric carbon dioxide. *Climatic Change* 19:99–118.

House, S. M. 1992. Population density and fruit set in three dioecious tree species in Australian tropical rain forest. *Journal of Ecology* 80:57–69.

———. 1993. Pollination success in a population of dioecious rain forest trees. *Oecologia* 96:555–61.

Howe, H. E. 1984. Implications of seed dispersal by animals for tropical reserve management. *Biological Conservation* 30:261–81.

———. 1989. Scatter- and clump-dispersal and seedling demography: Hypothesis and implications. *Oecologia* 79:417–26.

Howe, H. E., E. W. Schupp, and L. Westley. 1985. Early consequences of seed dispersal for a Neotropical tree *Virola surinamensis*. *Ecology* 66:781–91.

Howe, H. E., and J. Smallwood. 1982. Ecology of seed dispersal. *Annual Review of Ecology and Systematics* 13:201–28.

Hoyo, J. del, A. Elliott, and I. Sargatal, eds. 1992. *Handbook of the birds of the world*. Lynx Ediciones, Barcelona, Spain.

Hubbell, S. P., and R. B. Foster. 1986. Commonness and rarity in a Neotropical forest: Implications for tropical tree conservation. Pages 205–31 in M. E. Soulé, ed., *Conservation biology: The science of scarcity and diversity*. Sinauer Associates, Sunderland, Mass.

Huber, O. 1986. La vegetacion de la cuenca del Rio Caroni. *Interciencia* 11:301–10.

Hughes, L., M. Dunlop, K. French, M. R. Leishman, B. Rice, L. Rodgerson, and M. Westoby. 1994. Predicting dispersal spectra: A minimal set of hypotheses based on plant attributes. *Journal of Ecology* 82:933–50.

Humbert, H. 1927. Destruction d'une flore insulaire par le feu. Principaux aspects de la végétation à Madagascar. *Mémoires de l'Académie Malagache* 5:1–80.

Humbert, H., and G. Cours Darne. 1965. *Carte internationale du tapis vegetal, Republique Malagache, Madagascar*.

Humphries, S. M., and P. R. Stanton. 1992. Weed invasions in the Wet Tropics of Queensland. Report to Wet Tropics Management Authority, Cairns, Australia.

Hurlbert, S. H. 1984. Pseudoreplication and the design of ecological field experiments. *Ecological Monographs* 54:187–211.

Huston, M., and T. Smith. 1987. Plant succession: Life history and competition. *American Naturalist* 130: 168–98.

Hutchings H. R. W. 1991. Dinâmica de três comunidades de Papilionoidea (Insecta: Lepidoptera) em fragmentos de floresta na Amazônia central. M.Sc. thesis, INPA/Fundação Universidade de Amazonas, Manaus, Amazonas, Brazil.

Hutchison, C. S. 1989. *Geological evolution of South-East Asia*. Clarendon Press, Oxford.

Hutterer, R., M. Verhaagh, J. Diller, and R. Podkoucky. 1995. An inventory of mammals observed at Panguana Biological Station, Amazonian Peru. *Ecotropica* 1:3–20.

Hyland, B. P. M. 1989. A revision of Lauraceae in Australia (excluding *Cassytha*). *Australian Systematic Botany* 2:135–367.

ICBP (International Council for Bird Preservation). 1992. *Putting biodiversity on the map: Global priorities for conservation*. ICBP, Cambridge.

Irving, E. M. 1975. Structural evolution of the northernmost Andes, Colombia. U.S. Geological Survey Professional Paper 846, Washington, D.C.

Isler, M. L., and P. R. Isler. 1987. *The tanagers: Natural history, distribution and identification*. Smithsonian Institution Press, Washington, D.C.

Isotomin, A. V. 1994. Phenotypic diversity in continual and discrete populations of the bank vole in the south taiga. *Zhurnal Obshchei Biologii* 55:477–88.

IUCN (International Union for the Conservation of Nature and Natural Resources). 1992. *Protected areas of the world: A review of national systems*. Vol. 4, *Nearctic and Neotropical*. IUCN, Gland, Switzerland.

Jackson, J. C. 1965. Chinese agricultural pioneering in Singapore and Johore, 1800–1917. *Journal of the Malay Branch of the Royal Asiatic Society* 38:77–105.

Jaenike, J. 1978. Effect of island area on *Drosophila* population densities. *Oecologia* 36:327–32.

Janos, D. P., C. T. Sahley, and L. H. Emmons. 1995. Rodent dispersal of vesicular-arbuscular mycorrhizal fungi in Amazonian Peru. *Ecology* 75:98–104.

Janson, C. H., and L. H. Emmons. 1990. Ecological structure of the nonflying mammal community at Cocha Cashu Biological Station, Manu National Park, Peru. Pages 314–38 in A. W. Gentry, ed., *Four Neotropical rainforests.* Yale University Press, New Haven, Conn.

Janzen, D. H. 1969. Birds and the ant × acacia interaction in Central America. *Evolution* 20:248–75.

———. 1971. Euglossine bees as long-distance pollinators of tropical plants. *Science* 171:203–5.

———. 1983. No park is an island: Increase in interference from outside as park size decreases. *Oikos* 41: 402–10.

———. 1986a. The eternal external threat. Pages 286–303 in M. E. Soulé, ed., *Conservation biology: The science of scarcity and diversity.* Sinauer Associates, Sunderland, Mass.

———. 1986b. The future of tropical biology. *Annual Review of Ecology and Systematics* 17:305–24.

———. 1988a. Guanacaste National Park: Tropical ecological and biocultural restoration. Pages 143–92 in J. Cairns, Jr., ed., *Rehabilitating damaged ecosystems.* CRC Press, Boca Raton, Fla.

———. 1988b. Management of habitat fragments in a tropical dry forest: Growth. *Annals of the Missouri Botanical Garden* 75:105–16.

Jenkins, M. D. 1987. *Madagascar: An environmental profile.* International Union for the Conservation of Nature and Natural Resources (IUCN), Gland, Switzerland.

Jenkins, S. I. 1993. Exotic plants in the rainforest fragments of the Atherton and Evelyn Tablelands, north Queensland. B.Sc. Honors thesis, James Cook University, Townsville, Queensland, Australia.

Jiménez, J. A., K. A. Hughes, G. Alaks, L. Graham, and R. C. Lacy. 1994. An experimental study of inbreeding depression in a natural habitat. *Science* 266:271–73.

Johnels, S. A., and T. C. Cuadros. 1986. Species composition and abundance of bird fauna in a disturbed forest in the Central Andes of Colombia. *Hornero* 12:235–41.

Johns, A. D. 1986. Effects of selective logging on the behavioral ecology of West Malaysian primates. *Ecology* 67:684–94.

———. 1992a. Species conservation in managed tropical forests. Pages 15–54 in T. C. Whitmore and J. A. Sayer, eds., *Tropical deforestation and species extinction.* Chapman and Hall, London.

———. 1992b. Vertebrate responses to selective logging: Implications for the design of logging systems. *Philosophical Transactions of the Royal Society of London B* 335:437–42.

Johns, A. D., and B. G. Johns. 1995. Tropical forests and primates: Long-term co-existence? *Oryx* 29: 205–11.

Johnson, L. 1990. Analyzing spatial and temporal phenomena using geographical information systems: A review of ecological applications. *Landscape Ecology* 4:31–43.

Johnson, R. A. 1954. The behavior of birds attending army ant raids on Barro Colorado Island, Panama Canal Zone. *Proceedings of the Linnean Society, New York* 63–65:41–70.

Johnson, W. C., R. K. Schreiber, and R. L. Burgess. 1979. Diversity of small mammals in a powerline right-of-way and adjacent forest in East Tennessee. *American Midland Naturalist* 101:231–35.

Jones, C. G. 1987. The larger land-birds of Mauritius. Pages 208–300 in A. W. Diamond, ed., *Studies of Mascarene Island birds.* Cambridge University Press, Cambridge.

Jones, E. W. 1956. Ecological studies on the rain forest of southern Nigeria. IV. The plateau forests of the Okamu Forest Reserve. *Journal of Ecology* 44:83–117.

Jones, J. S. 1987. An asymmetrical view of fitness. *Nature* 235:298–99.

Jordan, C. F. 1993. Ecology of tropical forests. Pages 165–98 in L. Pancel, ed., *Tropical forestry handbook.* Springer-Verlag, Berlin.

Jordano, P. 1995. Angiosperm fleshy fruits and seed dispersers: A comparative analysis of adaptation and constraints in plant-animal interactions. *American Naturalist* 145:163–91.

Joseph, L. 1994. A molecular approach to species diversity and evolution in eastern Australian rainforest birds. Ph.D. dissertation, University of Queensland, Brisbane, Australia.

Joseph, L., and C. Moritz. 1993. Phylogeny and historical ecology of eastern Australian scrubwrens *Sericornis* spp: Evidence from mitochondrial DNA. *Molecular Ecology* 2:161–70.

———. 1994. Mitochondrial DNA phylogeography of birds in eastern Australian rainforests: First fragments. *Australian Journal of Zoology* 42:385–403.

Joseph, L., C. Moritz, and A. Hugall. 1993. A mitochondrial DNA perspective on the historical biogeography of middle eastern Queensland rainforest birds. *Memoirs of the Queensland Museum* 34:201–14.

———. 1995. Molecular support for vicariance as a source of diversity in rainforest. *Proceedings of the Royal Society of London B* 260:177–82.

Kahn, J. R., and J. A. McDonald. 1995. Third-world debt and tropical deforestation. *Ecological Economics* 12:107–23.

Kalliola, R., M. Puhakka, and W. Danjoy. 1993. *Amazonia Peruana. Vegetacíon húmeda tropical en el llano subandino.* ONERN, Lima, Peru.

Kanowski, P. J., and P. S. Savill. 1992. Forest plantations: Towards sustainable practice. Pages 21–155 in C. Sargent and S. Bass, eds., *Plantation politics: Forest plantations in development.* Earthscan, London.

Kapos, V. 1989. Effects of isolation on the water status of forest patches in the Brazilian Amazon. *Journal of Tropical Ecology* 5:173–85.

Kapos, V., G. M. Ganade, E. Matsui, and R. L. Victoria. 1993. $\partial^{13}C$ as an indicator of edge effects in tropical rain forest reserves. *Journal of Ecology* 81:425–32.

Karr, J. R. 1971. Structure of avian communities in selected Panama and Illinois habitats. *Ecological Monographs* 41:207–33.

———. 1982a. Avian extinction on Barro Colorado Island, Panama: A reassessment. *American Naturalist* 119:220–39.

———. 1982b. Population variability and extinction in the avifauna of a tropical land-bridge island. *Ecology* 63:1975–78.

———. 1990a. Avian survival rates and the extinction process on Barro Colorado Island, Panama. *Conservation Biology* 4:391–96.

———. 1990b. The avifauna of Barro Colorado Island and the Pipeline Road, Panama. Pages 183–98 in A. Gentry, ed., *Four Neotropical rainforests.* Yale University Press, New Haven, Conn.

———. 1995. Extinctions of birds on Barro Colorado Island. Page 118 in G. K. Meffe and C. R. Carroll, eds., *Principles of conservation biology.* Sinauer Associates, Sunderland, Mass.

Karr, J. R., and K. E. Freemark. 1983. Habitat selection and environmental gradients: Dynamics in the stable tropics. *Ecology* 64:1481–94.

Karr, J. R., S. K. Robinson, J. G. Blake, and R. O. Bierregaard, Jr. 1990. Birds of four Neotropical forests. Pages 237–51 in A. H. Gentry, ed., *Four Neotropical rainforests.* Yale University Press, New Haven, Conn.

Karr, J. R., and R. R. Roth. 1971. Vegetation structure and avian diversity in several New World areas. *American Naturalist* 105:423–35.

Kartawinata, K., S. Riswan, and H. Soedjito. 1980. The floristic change after disturbances in lowland dipterocarp forest in East Kalimantan, Indonesia. Pages 47–54 in J. L. Furtado, ed., *Tropical ecology and development.* International Society of Tropical Ecology, Kuala Lumpur, Malaysia.

Kattan, G. H., H. Alvarez-López, and M. Giraldo. 1994. Forest fragmentation and bird extinctions: San Antonio eighty years later. *Conservation Biology* 8:138–46.

Kattan, G. H., C. Restrepo, and M. Giraldo. 1984. Estructura de un bosque de niebla en la Cordillera Occidental, Valle del Cauca, Colombia. *Cespedesia* 13:23–43.

Katzman, M. T., and W. G. Cale. 1990. Tropical forest preservation using economic incentives. *Bioscience* 40:827–33.

Keenan, R.J., D. Lamb, and G. Sexton. 1995. Experience with mixed species rainforest plantations in North Queensland. *Commonwealth Forestry Review* 75:315–21.

Kemp, E. M. 1981. Tertiary palaeogeography and the evolution of Australian climate. Pages 31–50 in A. Keast, ed., *Ecological biogeography of Australia.* Dr. W. Junk, The Hague, Netherlands.

Kemper, C., and D. T. Bell. 1985. Small mammals and habitat structure in lowland rainforest of Peninsular Malaysia. *Journal of Tropical Ecology* 1:5–22.

Kent, M. 1987. Island biogeography and habitat conservation. *Progress in Physical Geography II* 1:91–102.

Kerfoot, O. 1968. Mist precipitation on vegetation. *Forestry Abstracts* 29:8–20.

Kershaw, A. P. 1986. Climatic change and aboriginal burning in northeast Australia during the last two glacial-interglacial cycles. *Nature* 322:47–49.

———. 1994. Pleistocene vegetation of the humid tropics of northeastern Queensland, Australia. *Palaeogeography, Palaeoclimatology, Palaeoecology* 109:399–412.

Kessler, M. 1995. *Polylepis-Wälder Boliviens: Taxa okologie, verbreitung und geschichte.* J. Cramer, Berlin.

Kevan, P. G., and A. J. Lack. 1985. Pollination in a cryptically dioecious plant *Decaspermum parviflorum* (Lam.) A. J. Scott (Myrtaceae) by pollen collecting bees in Sulawesi, Indonesia. *Biological Journal of the Linnean Society* 25:319–30.

Kikkawa, J., G. B. Montieth, and G. Ingram. 1981. Cape York Peninsula: Major region of faunal interchange. Pages 1695–1742 in A. Keast, ed., *Ecological biogeography of Australia.* Dr. W. Junk, The Hague, Netherlands.

Kimura, M., and G. H. Weiss. 1964. The stepping stone model of population structure and the decrease of genetic correlation with distance. *Genetics* 49:561–76.

King, D. 1978. The effects of roads and open space on the movements of small mammals. B.Sc. (Hons.) thesis, Australian National University, Canberra, Australia.

King, G. C., and W. S. Chapman. 1983. Floristic composition and structure of a rainforest area 25 years after logging. *Australian Journal of Ecology* 5:173–85.

Kinloch, B. B., and J. L. Littlefield. 1977. White pine blister rust: Hypersensitive resistance in sugar pine. *Canadian Journal of Botany* 55:1148–55.

Kitchener, D. J., J. Dell, E. G. Muir, and M. Palmer. 1982. Birds in western Australian wheat belt reserves—Implications for conservation. *Biological Conservation* 22:127–63.

Klein, B. C. 1989. The effects of forest fragmentation on dung and carrion beetle (Scarabaeinae) communities in central Amazonia. *Ecology* 70:1715–25.

Klein, D. R. 1971. Reaction of reindeer to obstructions and disturbances. *Science* 173:393–98.

Kluge, A. G., and W. C. Kerfoot. 1973. The predictability and regularity of character divergence. *American Naturalist* 107:426–42.

KnowledgeSeeker. 1990. *User's Guide, version 2.* FirstMark Technologies, Ottawa, Canada.

Knowlton, N., L. A. Wigt, L. A. Solorzano, D. K. Mills, and E. Bermingham. 1993. Divergence in proteins, mitochondrial DNA, and reproductive compatibility across the isthmus of Panama. *Science* 260: 1629–32.

Koch, L. E. 1983. Morphological characteristics of Australian Scolopendrid centipedes and the taxonomy and distribution of *Scolopendra morsitans* L. *Australian Journal of Zoology* 31:79–91.

———. 1985. The taxonomy of Australian centipedes of the genus *Rhysida* Wood (Chilopoda: Scolopendridae: Otostigminae). *Journal of Natural History* 19:205–14.

Kochummen, K. M., J. V. LaFrankie, and N. Manokaran. 1992. Floristic composition of Pasoh Forest Reserve, a lowland rain forest in Peninsular Malaysia. *Malayan Nature Journal* 45:545–54.

Koopowitz, H. 1992. A stochastic model for the extinction of tropical orchids. *Selbyana* 13:115–22.

Koopowitz, H., M. Andersen, A. D. Thornhill, H. Nguyen, and A. Pham. 1994. Comparison of distributions of terrestrial and epiphytic African orchids: Implications for conservation. Pages 120–24 in A. Pridgeon, ed., *Proceedings of the Fourteenth World Orchid Conference.* HMSO, Edinburgh, Scotland.

Koopowitz, H., A. D. Thornhill, and M. Andersen. 1993. Species distribution profiles of the Neotropical orchids *Masdevallia* and *Dracula* (Pleurothallidinae, Orchidaceae): Implications for conservation. *Biodiversity and Conservation* 2:681–90.

———. 1994. A general stochastic model for the prediction of biodiversity losses based on habitat conversion. *Conservation Biology* 8:425–38.

Kozakiewicz, M., and J. Konopka. 1991. Effect of habitat isolation on genetic divergence of bank vole populations. *Acta Theriologica* 36:363–67.

Kozakiewicz, M., and J. Szaki. 1995. Movements of small mammals in a landscape: Patch restriction or nomadism? Pages 78–94 in W. Z. Lidicker, Jr., ed., *Landscape approaches in mammalian ecology and conservation.* University of Minnesota Press, Minneapolis.

Kremen, C. 1992. Assessing the indicator properties of species assemblages for natural areas monitoring. *Ecological Applications* 2:203–17.

Kress, J., and J. H. Beach. 1994. Flowering plant reproductive systems. Pages 161–82 in L. A. McDade, K. S. Bawa, H. A. Hespenheide, and G. S. Hartshorn, eds., *La Selva: Ecology and natural history of a Neotropical rainforest.* University of Chicago Press, Chicago, Ill.

Kroodsma, R. L. 1982. Edge effect on breeding forest birds along a power-line corridor. *Journal of Applied Ecology* 19:361–70.

———. 1984. Effect of edge on breeding forest bird species. *Wilson Bulletin* 96:426–36.

Kruess, A., and T. Tscharntke. 1994. Habitat fragmentation, species loss, and biological control. *Science* 264:1581–84.

Krummel, J. R., R. H. Gardner, G. Sugihara, R. V. O'Neill, and P. R. Coleman. 1987. Landscape patterns in a disturbed environment. *Oikos* 48:321–24.

Kuhnlein, U., D. Zadworny, Y. Dawe, R. W. Fairfull, and J. S. Gavora. 1990. Assessment of inbreeding by DNA fingerprinting: Development of a calibration curve using defined strains of chickens. *Genetics* 125:161–65.

Lack, A. J., and P. G. Kevan. 1984. On the reproductive biology of a canopy tree, *Syzygium syzygiodes* (Myrtaceae), in a rainforest in Sulawesi, Indonesia. *Biotropica* 16:31–36.

Lacy, R. C. 1987. Loss of genetic diversity from managed populations: Interacting effects of drift, mutation, immigration, selection and population subdivision. *Conservation Biology* 1:143–58.

———. 1992. The effects of inbreeding on isolated populations: Are minimum viable population sizes predictable? Pages 277–96 in P. L. Fiedler and S. K. Jain, eds., *Conservation biology: The theory and practice of nature conservation preservation and management.* Chapman and Hall, New York.

Lamb, D., and M. Tomlinson. 1994. Forest rehabilitation in the Asia-Pacific region: Past lessons and present uncertainties. *Journal of Tropical Forest Science* 7:157–70.

Lambert, F. R. 1992. The consequences of selective logging for Bornean lowland forest birds. *Philosophical Transactions of the Royal Society of London B* 335:443–57.

Lande, R. 1980. Genetic variation and phenotypic evolution during allopatric speciation. *American Naturalist* 116:463–79.

———. 1988. Genetics and demography in biological conservation. *Science* 241:1455–60.

———. 1995. Mutation and conservation. *Conservation Biology* 9:782–91.

Lande, R., and G. F. Barrowclough. 1987. Effective population size, genetic variation, and their use in population management. Pages 87–124 in M. E. Soulé, ed., *Viable populations for conservation.* Cambridge University Press, Cambridge.

Langham, N. 1983. Distribution and ecology of small mammals in three rain forest localities of Peninsular Malaysia with particular references to Kedah Peak. *Biotropica* 15:199–206.

Lanly, J. P. 1982. Tropical forest resources. FAO Forestry Paper 30. FAO, Rome.

Lara, M. C. 1994. Systematics and phylogeography of the spiny rat genus *Trinomys* (Rodentia: Echimyidae). Ph.D. dissertation, University of California, Berkeley.

Laurance, S. G. W. 1996. The utilisation of linear forest remnants by arboreal mammals in tropical Queensland. M.Sc. thesis, University of New England, Armidale, New South Wales, Australia.

Laurance, S. G. W., and W. F. Laurance. 1995. A ground-trapping survey for small mammals in continuous forest and two isolated tropical rainforest reserves. *Memoirs of the Queensland Museum* 38:563–68.

Laurance, W. F. 1989. Ecological impacts of tropical forest fragmentation on nonflying mammals and their habitats. Ph.D. dissertation, University of California, Berkeley.

———. 1990. Comparative responses of five arboreal marsupials to tropical forest fragmentation. *Journal of Mammalogy* 71:641–53.

———. 1991a. Ecological correlates of extinction proneness in Australian tropical rainforest mammals. *Conservation Biology* 5:79–89.

———. 1991b. Edge effects in tropical forest fragments: Application of a model for the design of nature reserves. *Biological Conservation* 57:205–19.

———. 1994. Rainforest fragmentation and the structure of small mammal communities in tropical Queensland. *Biological Conservation* 69:23–32.

Laurance, W. F., J. Garesche, and C. W. Payne. 1993. Avian nest predation in modified and natural habitats in tropical Queensland: An experimental study. *Wildlife Research* 20:711–23.

Laurance, W. F., and C. Gascon. In Press. How to creatively fragment a landscape. *Conservation Biology* (1997).

Laurance, W. F., and J. D. Grant. 1994. Photographic identification of ground-nest predators in Australian tropical rainforest. *Wildlife Research* 21:241–48.

Laurance, W. F., and S. G. W. Laurance. 1996. Responses of five arboreal marsupials to recent selective logging in tropical Australia. *Biotropica* 28:310–22.

Laurance, W. F., and E. Yensen. 1991. Predicting the impacts of edge effects in fragmented habitats. *Biological Conservation* 55:77–92.

Lawton, J. H. 1995. Population dynamic principles. Pages 147–63 in J. H. Lawton and R. M. May, eds., *Extinction rates.* Oxford University Press, Oxford.

Lawton, J. H., and R. M. May, eds. 1995. *Extinction rates.* Oxford University Press, Oxford.

Lawton, J. H., S. Nee, A. J. Letcher, and P. H. Harvey. 1994. Animal distributions: Patterns and processes. Pages 41–58 in P. J. Edwards, R. M. May, and N. R. Webb, eds., *Large-scale ecology and conservation biology.* Blackwell Scientific Publications, Oxford.

Lawton, J. H., and G. L. Wooddroffe. 1991. Habitat and the distribution of water voles: Why are there gaps in a species range? *Journal of Animal Ecology* 60:79–91.

Leary, R. F., and F. W. Allendorf. 1989. Fluctuating asymmetry as an indicator of stress: Implications for conservation biology. *Trends in Ecology and Evolution* 4:214–17.

Lebrun, J. 1960. Sur la richese de la flore de diverses territoires Africanes. *Bulletin des Seances de l'Academie Royal des Sciences Outre-Mer* 6:669–90.

Leck, C. F. 1979. Avian extinctions in an isolated tropical wet-forest preserve, Ecuador. *Auk* 96:343–52.

Leigh, E. G., Jr. 1975. Structure and climate in tropical rain forest. *Annual Review of Ecology and Systematics* 6:67–86.

Leigh, E. G., Jr., A. S. Rand, and D. M. Windsor, eds. 1982. *The ecology of a tropical forest: Seasonal rhythms and long-term changes.* Smithsonian Institution Press, Washington, D.C.

Leigh, E. G., Jr., S. J. Wright, E. A. Herre, and F. E. Putz. 1993. The decline of tree diversity on newly isolated tropical islands: A test of a null hypothesis and some implications. *Evolutionary Ecology* 7:76–102.

Leite, L. L., and P. A. Furley. 1985. Land development in the Brazilian Amazon with particular reference to Rondônia and the Ouro Preto colonization project. Pages 119–40 in J. Hemming, ed., *Change in the Amazon basin.* Manchester University Press, Manchester, U.K.

Lerner, I. M. 1954. *Genetic homeostasis.* John Wiley, New York.

Lescourret, F., and M. Genard. 1994. Habitat, landscape and bird composition in mountain forest fragments. *Journal of Environmental Management* 40:317–28.

Lesica, P., and F. W. Allendorf. 1995. When are peripheral populations valuable for conservation? *Conservation Biology* 9:753–60.

Leung, L. K.-P., C. R. Dickman, and L. A. Moore. 1993. Genetic variation in fragmented populations of an Australian rainforest rodent, *Melomys cervinipes. Pacific Conservation Biology* 1:58–65.

Levenson, J. B. 1981. Woodlots as biogeographic islands in southeastern Wisconsin. Pages 12–39 in R. L. Burgess and D. M. Sharpe, eds., *Forest island dynamics in man-dominated landscapes.* Springer-Verlag, New York.

Levey, D. J., T. C. Moermond, and J. S. Denslow. 1994. Frugivory: An overview. Pages 282–94 in L. A. McDade, K. S. Bawa, H. A. Hespenheide, and G. S. Hartshorn, eds., *La Selva: Ecology and natural history of a Neotropical rainforest.* University of Chicago Press, Chicago, Ill.

Levin, D. A. 1988. Consequences of stochastic elements in plant migration. *American Naturalist* 132: 643–51.

Levin, D. A. 1995. Plant outliers: An ecogenetic perspective. *American Naturalist* 145:109–18.

Levin, D. A., and H. W. Kerster. 1969. The dependence of bee-mediated pollen and gene dispersal upon plant density. *Evolution* 23:560–71.

Levings, S. C., and D. M. Windsor. 1985. Litter arthropod populations in a tropical deciduous forest: Relationships between years and arthropod groups. *Journal of Animal Ecology* 32:157–63.

Levins, R. 1969. Some demographic and genetic consequences of environmental heterogeneity for biological control. *Bulletin of the Entomological Society of America* 15:237–40.

Levins, R., and R. C. Lewontin. 1980. Dialectics and reductionism in ecology. *Syntheses* 43:47–78.

Lewin, R. 1984. Parks: How big is big enough? *Science* 225:611–12.

———. 1986. A mass extinction without asteroids. *Science* 234:14–15.

Lewis, J. G. 1978. Variation in tropical scolopendrid centipedes: Problems for the taxonomist. *Abhandlungen und Verhandlungen des Naturwissen Schaftlichen Vereins in Hamburg* 21:43–50.

———. 1981. *The biology of centipedes.* Cambridge University Press, New York.

Lidicker, W. Z., Jr., ed. 1995. *Landscape approaches in mammalian ecology and conservation.* University of Minnesota Press, Minneapolis.

Lieberman, D., M. Lieberman, R. Peralta, and G. S. Hartshorn. 1985. Mortality patterns and stand turnover rates in a wet tropical forest in Costa Rica. *Journal of Ecology* 73:915–24.

Liggett, A. C., I. F. Harvey, and J. T. Manning. 1993. Fluctuating asymmetry in *Scatophaga stercoraria* L.: Successful males are more symmetrical. *Animal Behavior* 45:1041–43.

Lillesand, T. M., and R. W. Kiefer. 1994. *Remote sensing and image interpretation.* John Wiley, New York.

Lim, K. S. 1992. *Vanishing birds of Singapore.* The Nature Society (Singapore), Singapore.

Lim, K. K. P., and P. K. L. Ng. 1990. *A guide to the freshwater fishes of Singapore.* Singapore Science Center, Singapore.

Lindsey, T. R. 1992. *Encyclopedia of Australian animals: Birds.* Harper Collins, Sydney, Australia.

Linhart, Y. B. 1973. Ecological and behavioral determinants of pollen dispersal in hummingbird-pollinated *Heliconia. American Naturalist* 107:511–23.

Linhart, Y. B., and P. Feinsinger. 1980. Plant-hummingbird interactions: Effects of island size and degree of specialization on pollination. *Journal of Ecology* 68:745–60.

Lipsey, M. W. 1990. *Design sensitivity: Statistical power for experimental research.* Sage Publications, Newbury Park, Calif.

Loeschcke, V., J. Tomiuk, and S. K. Jain, eds. 1994. *Conservation genetics.* Birkhäuser Verlag, Basel, Switzerland.

Loiselle, B. A., and J. G. Blake. 1992. Population variation in a tropical bird community. *Bioscience* 42: 838–45.

Loiselle, B. A., and W. G. Hoppes. 1983. Nest predation in insular and mainland lowland rainforest in Panama. *Condor* 85:93–95.

Loiselle, B. A., V. L. Sork, J. D. Nason, and C. Graham. 1995. Spatial genetic structure of a tropical understory shrub, *Psychotria officinalis* (Rubiaceae). *American Journal of Botany* 82:1420–25.

Lomolino, M. V., J. H. Brown, and R. Davis. 1989. Island biogeography of montane forest mammals in the American Southwest. *Ecology* 70:180–94.

Long, A. 1994. The importance of tropical montane cloud forests for endemic and threatened birds. Pages 79–106 in L. S. Hamilton, J. O. Juvik, and F. N. Scatena, eds., *Tropical montane cloud forests.* Springer-Verlag, New York.

Long, A., M. Crosby, A. J. Stattersfield, and D. C. Wege. 1996. Towards a global map of biodiversity patterns in the distribution of restricted-range birds. *Global Ecology and Biogeography Letters* 5:281–304.

Lott, R. H., G. N. Harrington, A. K. Irvine, and S. McIntyre. 1995. Density-dependent seed predation and plant dispersion of the tropical palm *Normanbya normanbyi*. *Biotropica* 27:87–95.

Lovejoy, T. E. 1975. Bird diversity and abundance in Amazon forest communities. *The Living Bird* 13: 127–91.

———. 1978. Genetic aspects of dwindling populations. Pages 275–80 in S. A. Temple, ed., *Endangered birds: Management techniques for preserving threatened species.* University of Wisconsin Press, Madison.

———. 1980. A projection of species extinctions. Pages 328–31 in G. O. Barney, ed., *The Global 2000 Report to the President.* U.S. Council on Environmental Quality, Washington, D.C.

Lovejoy, T. E., and R. O. Bierregaard, Jr. 1990. Central Amazonian forests and the Minimum Critical Size of Ecosystems Project. Pages 60–71 in A. H. Gentry, ed., *Four Neotropical rainforests.* Yale University Press, New Haven, Conn.

Lovejoy, T. E., R. O. Bierregaard, Jr., A. B. Rylands, J. R. Malcolm, C. E. Quintela, L. H. Harper, K. S. Brown, Jr., A. H. Powell, G. V. N. Powell, H. O. R. Schubart, and M. B. Hays. 1986. Edge and other effects of isolation on Amazon forest fragments. Pages 257–85 in M. E. Soulé, ed., *Conservation biology: The science of scarcity and diversity.* Sinauer Associates, Sunderland, Mass.

Lovejoy, T. E., and D. C. Oren. 1981. Minimum critical size of ecosystems. Pages 7–12 in R. L. Burgess and D. M. Sharp, eds., *Forest island dynamics in man-dominated landscapes.* Springer-Verlag, New York.

Lovejoy, T. E., J. M. Rankin, R. O. Bierregaard, Jr., K. S. Brown, Jr., L. H. Emmons, and M. E. van der Voort. 1984. Ecosystem decay of Amazon forest fragments. Pages 295–325 in M. H. Nitecki, ed., *Extinctions.* University of Chicago Press, Chicago, Ill.

Loveless, M. D. 1992. Isozyme variation in tropical trees: Patterns of genetic organization. *New Forests* 6: 67–94.

Loveless, M. D., and J. L. Hamrick. 1984. Ecological determinants of genetic structure in plant populations. *Annual Review of Ecology and Systematics* 15:65–95.

Lucas, R. M., M. Honzak, G. M. Foody, P. J. Curran, and C. Corves. 1993. Characterizing tropical secondary forests using multi-temporal Landsat sensor imagery. *International Journal of Remote Sensing* 14: 3061–67.

Lugo, A. E. 1988a. Estimating reductions in the diversity of tropical forest species. Pages 58–70 in E. O. Wilson, ed., *Biodiversity.* National Academy Press, Washington, D.C.

———. 1988b. The future of the forest: Ecosystem rehabilitation in the tropics. *Environment* 30:16–20.

———. 1992. Tree plantations for rehabilitating damaged lands in the tropics. Pages 247–55 in M. K. Wali, ed., *Environmental rehabilitation,* vol. 2. SPB Academic Publishing, The Hague, Netherlands.

———. 1995. Management of tropical biodiversity. *Ecological Applications* 5:956–61.

Lugo, A. E., M. Applefield, D. Pool, and R. McDonald. 1983. The impact of Hurricane David on the forests of Dominica. *Canadian Journal of Forest Research* 132:201–11.

Lugo, A. E., J. A. Parrotta, and S. Brown. 1993. Loss in species caused by tropical deforestation and their recovery through management. *Ambio* 22:106–9.

Lwanga, J. F. 1994. The role of seed and seedling predators and browsers in the regeneration of two forest canopy species *(Mimusops bagshawei* and *Strombosia scheffleri)* in Kibale Forest Reserve, Uganda. Ph.D. dissertation, University of Florida, Gainesville.

Lynam, A. J. 1995. Effects of habitat fragmentation on the distributional patterns of small mammals in a tropical forest in Thailand. Ph.D. dissertation, University of California, San Diego.

Lynch, M. 1990. The similarity index and DNA fingerprinting. *Molecular Biology and Evolution* 5:584–97.

Lynch, M., and W. Gabriel. 1990. Mutation load and survival of small populations. *Evolution* 44:1725–37.

MacArthur, R. H. 1969. Patterns of communities in the tropics. *Biological Journal of the Linnean Society* 1: 19–30.

MacArthur, R. H., J. M. Diamond, and J. R. Karr. 1972. Density compensation in island faunas. *Ecology* 53:330–42.

MacArthur, R. H., and E. O. Wilson. 1967. *The theory of island biogeography.* Princeton University Press, Princeton, N.J.

Macdonald, I. A. W., C. Thébaud, W. A. Strahm, and D. Strasberg. 1991. Effects of alien plant invasions on native vegetation remnants on La Réunion (Mascarene Islands, Indian Ocean). *Environmental Conservation* 18:51–61.

MacDonald, J. D. 1973. *Birds of Australia.* Reed Books, Sydney, Australia.

Mack, A. 1993. The sizes of vertebrate-dispersed fruits: A Neotropical-palaeotropical comparison. *American Naturalist* 142:840–56.

Mader, H. J. 1984. Animal habitat isolation by roads and agricultural fields. *Biological Conservation* 29: 81–96.

Madsen, J. 1985. Impact of disturbance on field utilization of pink-footed geese in West Jutland, Denmark. *Biological Conservation* 33:53–63.

Maggs, J., and B. Hewitt. 1993. Organic C and nutrients in surface soil from some primary rainforests, derived grasslands, and secondary rainforests on the Atherton Tableland in North East Queensland. *Australian Journal of Soil Research* 31:343–50.

Magsalay, P., T. Brooks, G. Dutson, and R. Timmins. 1995. Extinction and conservation on Cebu. *Nature* 373:294.

Magurran, A. E. 1988. *Ecological diversity and its measurement.* Princeton University Press, Princeton, N.J.

Main, B. Y. 1981. A comparative account of the biogeography of terrestrial invertebrates in Australia: Some generalities. Pages 1055–79 in A. Keast, ed., *Ecological biogeography of Australia.* Dr. W. Junk, Boston, Mass.

Malcolm, J. R. 1988. Small mammal abundances in isolated and non-isolated primary forest reserves near Manaus, Brazil. *Acta Amazônica* 18:67–83.

———. 1990. Estimation of mammalian densities in continuous forest north of Manaus. Pages 339–57 in A. W. Gentry, ed., *Four Neotropical rainforests.* Yale University Press, New Haven, Conn.

———. 1991a. Comparative abundances of Neotropical small mammals by trap height. *Journal of Mammalogy* 72:188–92.

———. 1991b. The small mammals of Amazonian forest fragments: Patterns and processes. Ph.D. dissertation, University of Florida, Gainesville.

———. 1992. Use of tooth impressions to identify and age live *Proechimys guyannensis* and *P. cuvieri* (Rodentia: Echimyidae). *Journal of Zoology* 227:537–46.

———. 1994. Edge effects in Central Amazonian forest fragments. *Ecology* 75:2438–45.

———. 1995. Forest structure and the abundance and diversity of Neotropical small mammals. Pages 179–97 in M. Lowman and N. M. Nadkarni, eds., *Forest canopies.* Academic Press, New York.

Manasse, R. S., and H. F. Howe. 1983. Competition for dispersal agents among tropical trees: Influences of neighbors. *Oecologia* 59:185–90.

Manço, D. D.-G., E. P. Andriani, F. C. Trematore, R. Gregorin, and S. B. Pereira da Silva. 1991. Levantamento das espécies de mamíferos da Fazenda Intervales, Serra de Paranapiacaba, São Paulo. Unpublished monograph, Universidade de São Paulo, Faculdade de Filosofia, Ciências e Letras de Ribeirão Preto, Brazil.

Manly, B. F. J. 1996. Are there clumps in body-size distributions? *Ecology* 77:81–86.

Mares, M. A. 1992. Neotropical mammals and the myth of Amazonian biodiversity. *Science* 255:976–79.

Margules, C. R. 1992. The Wog Wog Habitat Fragmentation Experiment. *Environmental Conservation* 19:316–25.

Margules, C. R., G. A. Milkovits, and G. T. Smith. 1994. Contrasting effects of habitat fragmentation on the scorpion *Cercophonius squama* and an amphipod. *Ecology* 75:2033–42.

Margules, C. R., and J. L. Stein. 1989. Patterns in the distributions of species and the selection of nature reserves: An example from *Eucalyptus* forests in south-eastern New South Wales. *Biological Conservation* 50:219–38.

Markow, T. A. 1995. Evolutionary ecology and developmental instability. *Annual Review of Entomology* 40:105–20.

Marshall, A. G., and M. D. Swaine. 1992. The Royal Society's South-East Asian rain forest research programme: An introduction. *Philosophical Transactions of the Royal Society of London B* 335:327–30.

Martin, A. P., and S. R. Palumbi. 1993. Body size, metabolic rate, generation time, and the molecular clock. *Proceedings of the National Academy of Science (USA)* 90:4087–91.

Martin, T. E. 1981. Limitation in small habitat islands: Chance or competition? *Auk* 98:715–34.

———. 1985. Selection of second-growth woodlands by frugivorous migrating birds in Panama: An effect of fruit size and plant density? *Journal of Tropical Ecology* 1:157–70.

———. 1992. Breeding productivity considerations: What are the appropriate habitat features for management? Pages 455–73 in J. M. Hagan III and D. W. Johnston, eds., *Ecology and conservation of Neotropical migrant landbirds.* Smithsonian Institution Press, Washington, D.C.

———. 1995. Avian life history evolution in relation to nest sites, nest predation, and food. *Ecological Monographs* 65:101–27.

Martins, M. B. 1989. Invasão de fragmentos florestais por espécies oportunistas de *Drosophila* (Diptera, Drosophilidae). *Acta Amazônica* 19:265–71.

Mason, D. 1995. Effects of a road on the abundance and movements of understory birds in a Venezuelan rain forest. Abstract. Page 56 in Program and Abstracts of the 9th Annual Meeting of the Society for Conservation Biology, Fort Collins, Colo.

Mathur, H. N., and P. Soni. 1983. Comparative accounts of undergrowth under *Eucalyptus* and sal in three different localities of Doon Valley. *Indian Forester* 109:882–90.

Matlack, G. R. 1993. Microenvironment variation within and among forest edge sites in the eastern United States. *Biological Conservation* 66:185–94.

———. 1994a. Plant species migration in a mixed-history forest landscape in eastern North America. *Ecology* 75:1491–1502.

———. 1994b. Vegetation dynamics of the forest edge: Trends in space and successional time. *Journal of Ecology* 82:113–23.

Matthews, C. 1993. Draft management plan for Lake Eacham and Lake Barrine National Parks. Queensland National Parks and Wildlife Service, Cairns, Queensland, Australia.

May, R. M., J. H. Lawton, and N. E. Stork. 1995. Assessing extinction rates. Pages 1–24 in J. H. Lawton and R. M. May, eds., *Extinction Rates.* Oxford University Press, Oxford.

Maynardier, P. G. de, and M. L. Hunter. 1995. The effects of forest roads on amphibian movements in Maine. Abstract. Page 39 in Program and Abstracts of the 9th Annual Meeting of the Society for Conservation Biology, Fort Collins, Colo.

Mayr, E. 1963. *Animal species and evolution.* Harvard University Press, Cambridge, Mass.

Mayr, E., and R. J. O'Hara. 1986. The biogeographic evidence supporting the Pleistocene refuge hypothesis. *Evolution* 40:55–67.

McClanahan, T. R., and R. W. Wolfe. 1987. Dispersal of ornithochorous seeds from forest edges in central Florida. *Vegetatio* 71:107–12.

———. 1993. Accelerating forest succession in a fragmented landscape: The role of birds and perches. *Conservation Biology* 7:279–88.

McCloskey, M. 1993. Note on the fragmentation of primary rainforest. *Ambio* 22:250–51.

McClure, H. E. 1974. *Migration and survival of the birds of Asia.* Applied Scientific Research Corporation of Thailand, Bangkok.

McCoy, E. D., and H. R. Mushinsky. 1994. Effects of fragmentation on the richness of vertebrates in the Florida scrub habitat. *Ecology* 75:446–57.

McCune, B. 1991. *Multivariate analysis on the PC-ORD system.* Holcomb Research Institute, Butler University, Indianapolis, Ind.

McDonnell, M. J., and E. W. Stiles. 1983. The structural complexity of old field vegetation and the recruitment of bird-dispersed plant species. *Oecologia* 56:109–16.

McElroy, A. D., S. Y. Chiu, J. W. Nebgen, A. Aleti, and A. E. Vandegrift. 1975. Water pollution from nonpoint sources. *Water Research* 9:675–81.

McEuen, A. B. 1995. Corridors as conduits and trails as barriers: Daily movement patterns in two species of

Peromyscus. Abstract. Page 56 in Program and Abstracts of the 9th Annual Meeting of the Society for Conservation Biology, Fort Collins, Colo.

McGarigal, K., and B. J. Marks. 1994. *Fragstats: Spatial pattern analysis program for quantifying landscape structure.* Oregon State University, Corvallis, Oreg.

McLellan, B. N., and D. M. Shackleton. 1988. Grizzly bears and resource-extraction industries: Effects of roads on behavior, habitat use and demography. *Journal of Applied Ecology* 25:451–60.

McNaughton, K. G. 1989. Micrometeorology of shelter belts and forest edges. *Philosophical Transactions of the Royal Society of London B* 324:351–68.

McNeely, J. A., K. R. Miller, W. V. Reid, R. A. Mittermeier, and T. B. Werner. 1990. *Conserving the world's biological diversity.* International Union for the Conservation of Nature and Natural Resources (IUCN), Gland, Switzerland.

Meachem, W. 1994. *Archaeological investigations on Chek Lap Kok Island.* Hong Kong Archaeological Society, Hong Kong.

Meave, J., and M. Kellman. 1994. Maintenance of rain forest diversity in riparian forests of tropical savannas: Implications for species conservation during Pleistocene drought. *Journal of Biogeography* 21: 121–35.

Medley, K. E. 1993. Primate conservation along the Tara River, Kenya: An examination of the forest habitat. *Conservation Biology* 7:109–21.

Medway, L. 1978. *The wild mammals of Malaya (Peninsular Malaysia) and Singapore.* Oxford University Press, Kuala Lumpur, Malaysia.

Medway, L., and D. R. Wells. 1976. *The birds of the Malay Peninsula.* Broadwater Press Limited, London.

Meggers, B. J. 1994. Archaeological evidence for the impact of mega-Niño events on Amazonia during the past two millennia. *Climatic Change* 2:321–38.

Mejía, J., H. Meyer, A. Campos, A. Gallego, and A. Velásquez. 1994. El terremoto de Paez, anatomía de un sismo, Parte I: Generalidades y movimientos fuertes. *Revista de la Asociación de Ingenieros del Valle* 72: 14–15.

Menges, E. S. 1991. Seed germination percentage increases with population size in a fragmented prairie species. *Conservation Biology* 2:158–64.

Middleton, J. 1993. The intrusive effects of a powerline clearing on the small mammal community of a tropical rainforest. B.Sc. (Hons.) thesis, James Cook University, Townsville, Australia.

Miller, A. 1993. The role of analytical science in natural resource decision making. *Environmental Management* 17:563–74.

Miller, D. R., J. D. Lin, and Z. N. Lu. 1991. Some effects of surrounding forest canopy architecture on the wind field in small clearings. *Forest Ecology and Management* 45:79–91.

Miller, M. 1991. *Debt and the environment: Converging crisis.* United Nations Publications, New York.

Mills, L. S., and P. E. Smouse. 1994. Demographic consequences of inbreeding in remnant populations. *American Naturalist* 144:412–31.

Milne, B. T. 1988. Measuring fractal dimensions of landscapes. *Applied Mathematics and Computation* 27: 67–79.

———. 1991. Lessons from applying fractal models to landscape patterns. Pages 199–235 in M. G. Turner and R. H. Gardner, eds., *Quantitative methods in landscape ecology.* Springer-Verlag, New York.

Milton, K. 1982. Dietary quality and demographic regulation in a howler monkey population. Pages 273–89 in E. G. Leigh, Jr., A. S. Rand, and D. M. Windsor, eds., *Ecology of a tropical forest: Seasonal rhythms and long-term changes.* Smithsonian Institution Press, Washington, D.C.

Minchin, P. R. 1987. An evaluation of the relative robustness of techniques for ecological ordination. *Vegetatio* 69:89–107.

Ministério de Minas e Energia. 1978a. Projeto RadamBrasil, Folha SA 20 Manaus. Ministério de Minas e Energia, Departamento Nacional de Produção Mineral, Rio de Janeiro, Brazil.

———. 1978b. Projeto RadamBrasil Programa de Integração Nacional: Levantamento de recursos naturais.

Ministério das Minas e Energia, Departmento Nacional da Produção Mineral. Vol. 18. Rio de Janeiro, Brazil.

Mitra, S. S., and F. H. Sheldon. 1993. The use of exotic tree plantations by Bornean lowland forest birds. *Auk* 110:529–40.

Mittermeier, R. A., W. R. Konstant, M. E. Nicoll, and O. Langrand. 1992. *Lemurs of Madagascar, an action plan for their conservation, 1993–1999.* International Union for the Conservation of Nature and Natural Resources (IUCN), Gland, Switzerland.

Møller, A. P. 1983. Damage by rats *Rattus norvegicus* to breeding birds on Danish islands. *Biological Conservation* 25:5–18.

———. 1994. Sexual selection in the barn swallow *(Hirundo rustica).* IV. Patterns of fluctuating asymmetry and selection against asymmetry. *Evolution* 48:658–70.

———. 1995. Patterns of fluctuating asymmetry in sexual ornaments of birds from marginal and central populations. *American Naturalist* 145:316–27.

Moermond, T. C., and J. S. Denslow. 1985. Neotropical avian frugivores: Patterns of behavior, morphology, and nutrition, with consequences on fruit selection. *Neotropical Monographs* 36:865–97.

Mondragón, M. L. 1989. Estructura de la comunidad aviaria en bosques de coníferas y en bosques aledaños de vegetación nativa. Trabajo de Grado, Departamento de Biología, Universidad del Valle, Cali, Colombia.

Moore, P. G. 1978. The mythical threat of Bayesianism. *International Statistical Review* 46:67–73.

Morales, L. C., and S. Gorzula. 1986. The interrelations of the Caroni River Basin ecosystems and hydroelectric power projects. *Interciencia* 11:272–77.

Moran, E. F. 1981. *Developing the Amazon.* Indiana University Press, Bloomington, Ind.

Moran, E. F., E. Brondizio, P. Mausel, and Y. Wu. 1994. Integrating Amazon vegetation, land-use, and satellite data. *Bioscience* 44:329–38.

Moran, G. F., and S. D. Hopper. 1987. Conservation of the genetic resources of rare and widespread eucalypts in remnant vegetation. Pages 151–62 in D. A. Saunders, G. W. Arnold, A. A. Burbidge, and A. J. M. Hopkins, eds., *Nature conservation: The role of remnants of native vegetation.* Surrey Beatty and Sons, Pty Limited, Australia.

Moran, G. F., O. Muona, and J. C. Bell. 1989. Breeding systems and genetic diversity in *Acacia auriculiformis* and *A. crassicarpa. Biotropica* 21:250–56.

Mori, S. A., B. M. Boom, and G. T. Prance. 1981. Distribution patterns and conservation of eastern Brazilian coastal forest species. *Brittonia* 33:233–45.

Moritz, C. 1994a. Applications of mitochondrial DNA analysis in conservation: A critical review. *Molecular Ecology* 3:401–11.

———. 1994b. Defining "Evolutionary Significant Units" for conservation. *Trends in Ecology and Evolution* 9:373–75.

———. 1995. Uses of molecular phylogenies for conservation. *Philosophical Transactions of the Royal Society of London B* 349:113–18.

Moritz, C., L. Joseph, and M. Adams. 1993. Cryptic diversity in an endemic rainforest skink *(Gnypetoscincus queenslandiae). Biodiversity and Conservation* 2:412–25.

Morrison, E., and S. Bass. 1992. What about the people? Pages 92–120 in C. Sargent and S. Bass, eds., *Plantation politics: Forest plantations in development.* Earthscan, London.

Morton, E. S. 1978. Reintroducing recently extirpated birds into a tropical forest preserve. Pages 379–86 in S. A. Temple, ed., *Endangered birds: Management techniques for preserving threatened species.* University of Wisconsin Press, Madison.

Munn, C. A. 1985. Permanent canopy and understory flocks in Amazonia: Species composition and population density. Pages 683–711 in P. A. Buckley, M. S. Foster, E. S. Morton, R. S. Ridgely, and F. G. Buckley, eds., *Neotropical ornithology.* Ornithological Monographs, no. 36. American Ornithologists Union, Washington, D.C.

Munves, J. 1975. Birds of highland clearing in Cundinamarca, Colombia. *Auk* 92:307–21.

Murawski, D. A. 1987. Floral resource variation, pollinator response, and potential pollen flow in *Psiguria warscewiczii*. *Ecology* 68:1273–82.

Murawski, D. A., B. Dayanandan, and K. S. Bawa. 1994. Outcrossing rates of two endemic *Shorea* species from Sri Lankan tropical rain forests. *Biotropica* 26:23–29.

Murawski, D. A., and J. L. Hamrick. 1992. Mating system and phenology of *Ceiba pentandra* (Bombacaceae) in Central Panama. *Journal of Heredity* 83:401–4.

Murawski, D. A., J. L. Hamrick, S. P. Hubbell, and R. B. Foster. 1990. Mating systems of two Bombacaceous trees of a Neotropical moist forest. *Oecologia* 82:501–6.

Murcia, C. 1995. Edge effects in fragmented forests: Implications for conservation. *Trends in Ecology and Evolution* 10:58–62.

Murphy, P. G., and A. E. Lugo. 1986. Ecology of tropical dry forest. *Annual Review of Ecology and Systematics* 17:67–88.

Musser, G. G., and M. D. Carleton. 1993. Family Muridae. Pages 501–755 in D. E. Wilson and D. M. Reeder, eds., *Mammal species of the world, a taxonomic and geographic reference,* 2d ed. Smithsonian Institution Press, Washington, D.C.

Musser, G. G., and C. Newcombe. 1983. Malaysian murids and the giant rat of Sumatra. *Bulletin of the American Museum of Natural History* 174:329–598.

Mustrangi, M. A., and J. L. Patton. 1997. Phylogeography and systematics of the slender mouse opossum *Marmosops* (Marsupialia, Didelphidae). *University of California Publications in Zoology* 130:1–86.

Myers, N. 1984. *The primary source: Tropical forests and our future.* W. W. Norton, New York.

———. 1986. Tropical deforestation and a mega-extinction spasm. Pages 394–409 in M. E. Soulé, ed., *Conservation biology: The science of scarcity and diversity.* Sinauer Associates, Sunderland, Mass.

———. 1988. Tropical forests and their species. Going, going . . . ? Pages 28–35 in E. O. Wilson, ed., *Biodiversity.* National Academy Press, Washington, D.C.

———. 1990. The biodiversity challenge: Expanded hot-spots analysis. *The Environmentalist* 10:243–56.

Nadkarni, N. M., and J. T. Longino. 1990. Invertebrates in canopy and ground organic matter in a Neotropical montane forest, Costa Rica. *Biotropica* 22:286–89.

Nakasathien, S. 1989. Chiew Larn Dam wildlife rescue operation. *Oryx* 23:146–54.

Naranjo, L. G. 1994. Composición y estructura de la avifauna del Parque Regional Natural Ucumarí. Pages 305–28 in J. O. Rangel, ed., *Ucumarí: Un caso típico de la diversidad biótica Andina.* Corporación Autónoma Regional de Risaralda, Colombia.

Nason, J. D., E. A. Herre, and J. L. Hamrick. In press. Paternity analysis of the breeding structure of strangler fig populations: Evidence for substantial long-distance wasp dispersal. *Journal of Biogeography.*

Nee, S., E. C. Holmes, and P. H. Harvey. 1995. Inferring population processes from molecular phylogenies. *Philosophical Transactions of the Royal Society of London B* 349:25–31.

Nei, M., T. Maruyama, and R. Chakraborty. 1975. The bottleneck effect and genetic variability in populations. *Evolution* 29:1–10.

Nelson, B. W., C. A. C. Ferreira, M. F. da Silva, and M. L. Kawalski. 1990. Endemism centers, refugia and botanical collection density in Brazilian Amazonia. *Nature* 345:714–16.

Nelson, B. W., V. Kapos, J. B. Adams, W. J. Oliveira, O. P. G. Braun, and I. L. do Amaral. 1994. Forest disturbance by large blowdowns in the Brazilian Amazon. *Ecology* 75:853–58.

Nelson, R., and N. Horning. 1993. AVHRR-LAC estimates of forest area in Madagascar, 1990. *International Journal of Remote Sensing* 14:1463–75.

Nepstad, D. C., C. Uhl, and A. E. S. Serrão. 1991. Recuperation of a degraded Amazonian landscape: Forest recovery and agricultural restoration. *Ambio* 20:248–55.

Nevo, E. 1978. Genetic variation in natural populations: Patterns and theory. *Theoretical Population Biology* 13:121–77.

Newmark, W. D. 1987. A land-bridge perspective on mammalian extinctions in western North American parks. *Nature* 325:430–32.

———. 1991. Tropical forest fragmentation and the local extinction of understory birds in the Eastern Usambara Mountains, Tanzania. *Conservation Biology* 5:67–78.

Ng, F. S. P. 1983. Ecological principles of tropical rainforest conservation. Pages 359–75 in S. L. Sutton, T. C. Whitmore and A. C. Chadwick, eds., *Tropical rain forests: Ecology and management.* Blackwell Scientific Publications, Oxford.

Ng, F. S. P., and C. M. Low. 1982. Checklist of endemic trees of the Malay peninsula. Malaysian Forestry Department, Research Department Pamphlet 88. Kuala Lumpur.

Ng, P. K. L., L. M. Chou, and T. J. Lam. 1993. The status and impact of introduced freshwater animals in Singapore. *Biological Conservation* 64:19–24.

Nichol, J. E. 1994. An examination of tropical rain forest microclimate using GIS modelling. *Global Ecology and Biogeography Letters* 4:69–78.

Nicholls, A. O., and C. R. Margules. 1991. The design of studies to demonstrate the biological importance of corridors. Pages 49–61 in D. A. Saunders and R. J. Hobbs, eds., *Nature conservation 2: The role of corridors.* Surrey Beatty, Chipping Norton, N.S.W., Australia.

Nicholson, D. I., N. B. Henry, and J. Rudder. 1988. Stand changes in north Queensland rain forests. *Proceedings of the Ecological Society of Australia* 15:61–80.

Nicoll, M. E., and O. Langrand. 1989. *Madagascar: Revue de la conservation et des aires proteges.* World Wide Fund for Nature, Gland, Switzerland.

Nix, H. A. 1991. Biogeography: Patterns and process. Pages 11–39 in H. A. Nix and M. Switzer, eds., *Rainforest animals: Atlas of vertebrates endemic to Australia's Wet Tropics.* Australian Nature Conservation Agency, Canberra, Australia.

Nix, H. A., and M. A. Switzer. 1991. Rainforest animals: Atlas of the vertebrates of Australia's Wet Tropics. Australian National Parks and Wildlife Service, Canberra, Australia.

Norse, E. A. 1987. International lending and the loss of biodiversity. *Conservation Biology* 1:259–60.

Noss, R. F. 1987. Corridors in real landscapes: A reply to Simberloff and Cox. *Conservation Biology* 1: 159–64.

Nowak, R. M. 1991. *Walker's mammals of the world.* 5th ed. John Hopkins University Press, Baltimore, Md.

Oakes, M. 1986. *Statistical inference: A commentary for the social and behavioral sciences.* John Wiley, Chichester, U.K.

Offerman, H. L., V. H. Dale, S. M. Pearson, R. O. Bierregaard, Jr., and R. V. O'Neill. 1995. Effects of forest fragmentation on Neotropical fauna: Current research and data availability. *Environmental Reviews* 3:191–211.

Oldeman, R. A. A. 1983. Tropical rainforest architecture, silvigenesis and diversity. Pages 135–41 in S. L. Sutton, T. C. Whitmore, and A. C. Chadwick, eds., *Tropical rain forests: Ecology and management.* Blackwell Scientific Publications, Oxford.

Olsen, M., and D. Lamb. 1988. Recovery of subtropical rainforest following storm damage. *Proceedings of the Ecological Society of Australia* 15:297–301.

O'Malley, D. M., and K. S. Bawa. 1987. Mating system of a tropical rain forest tree species. *American Journal of Botany* 74:1143–49.

O'Malley, D. M., D. P. Buckley, G. T. Prance, and K. S. Bawa. 1988. Genetics of Brazil nut (*Bertholletia excelsa* Humb. & Bonpl.: Lecythidaceae). 2. Mating system. *Theoretical and Applied Genetics* 76:929–32.

O'Neill, J. P. 1992. A general overview of the montane avifauna of Peru. *Memorias del Museo de Historia Natural (Lima)* 21:47–55.

O'Neill, R. V., C. T. Hunsaker, and D. A. Levine. 1992. Monitoring challenges and innovative ideas. Pages 1443–60 in D. H. Mckenzie, D. E. Hyatt, and V. J. McDonald, eds., *Ecological indicators,* vol. 2. Elsevier, London.

O'Neill, R. V., J. R. Krummel, R. H. Gardner, G. Sugihara, B. Jackson, D. L. DeAngelis, B. T. Milne, M. G.

Turner, B. Zygmunt, S. W. Christensen, V. H. Dale, and R. L. Graham. 1988. Indices of landscape pattern. *Landscape Ecology* 1:153–62.

Opdam, P., D. Van Dorp, and C. J. F. Ter Braak. 1984. The effect of isolation on the number of woodland birds in the Netherlands. *Journal of Biogeography* 11:473–78.

Opler, P. A., H. G. Baker, and G. W. Frankie. 1977. Recovery of tropical lowland forest ecosystems. Pages 379–421 in J. Cairns, Jr., K. L. Dickson, and E. E. Herricks, eds., *Recovery and restoration of damaged ecosystems.* University Press of Virginia, Charlottesville.

Orejuela, J. E., and G. Cantillo, compilers. 1980. *Aves de la Reserva Natural La Planada.* Programa Alegría de Enseñar, Fundación para la Educación Superior-FES, Cali, Colombia.

Orejuela, J. E., R. S. Raitt, and H. Alvarez. 1979. Relaciones ecológicas de las aves en la Reserva Forestal de Yotoco, Valle del Cauca. *Cespedesia* 29/30:7–28.

Osunkoya, O. O. 1994. Post-dispersal survivorship of north Queensland rainforest seeds and fruits: Effects of forest, habitat and species. *Australian Journal of Ecology* 19:52–64.

Osunkoya, O. O., J. E. Ash, M. S. Hopkins, and A. W. Graham. 1994. Influence of seed size and seedling ecological attributes on shade-tolerance of rain-forest tree species in northern Queensland. *Journal of Ecology* 82:149–63.

Otero, L. S., and K. S. Brown, Jr. 1986. Biology and ecology of *Parides ascanius* (Cramer, 1775) (Lep., Papilionidae), a primitive butterfly threatened with extinction. *Atala* 10–12:2–16.

Owen, D. F. 1971. *Tropical butterflies.* Clarendon Press, Oxford.

———. 1975. Estimating the abundance and diversity of butterflies. *Biological Conservation* 8:173–83.

Oxley, D. J., M. B. Fenton, and G. R. Carmody. 1974. The effects of roads on populations of small mammals. *Journal of Applied Ecology* 11:51–59.

Pacheco, S., N. Florencio da Silva, R. Ribon, J. E. Simon, and R. Torres. 1994. Efeito do manejo do cerrado sobre as populações de alguns Tinamidae em Três Marias, Estado de Minas Gerais. *Revista Brasilera Biologia* 54:435–41.

Pacheco, V. 1991. A new species of *Scolomys* (Muridae: Sigmodontinae) from Peru. *Publicaciones del Museo de Historia Natural, Universidad Nacional Mayor de San Marcos, Serie A, Zoologia* 37:1–3.

Pacheco, V., H. de Macedo, E. Vivar, C. Ascorra, R. Arana-Cardó, and S. Solari. 1995. *Lista anotada de los mamíferos Peruanos.* Conservation International Occasional Paper no. 2.

Pacheco, V., B. D. Patterson, J. L. Patton, L. H. Emmons, S. Solari, and C. F. Ascorra. 1993. List of mammal species known to occur in Manu Biosphere Reserve, Peru. *Publicaciones del Museo de Historia Natural, Universidad Nacional Mayor de San Marcos, Serie A, Zoologia* 44:1–12.

Packer, C., A. E. Pusey, H. Rowley, D. A. Gilbert, J. Martenson, and S. J. Brien. 1991. Case study of a population bottleneck: Lions of the N'gorongoro Crater. *Biological Conservation* 5:219–30.

Page, R. D. M. 1993. *Component, version 2.0.* The Natural History Museum, London.

———. 1994. Maps between trees and cladistic analysis of historical associations among genes, organisms, and areas. *Systematic Biology* 43:58–77.

Palik, B. J., and P. G. Murphy. 1990. Disturbance versus edge effects in sugar-maple/beech forest fragments. *Forest Ecology and Management* 32:187–202.

Palmer, R. A., and C. Strobeck. 1986. Fluctuating asymmetry: Measurement, analysis, patterns. *Annual Review of Ecology and Systematics* 17:391–421.

Palo, M. 1994. Population and deforestation. Pages 42–56 in K. Brown and D. W. Pearce, eds., *The causes of tropical deforestation: The economic and statistical analysis of factors giving rise to the loss of tropical forests.* University College London Press, London.

Palo, M., G. Mery, and J. Salmi. 1987. Deforestation in the tropics: Pilot scenarios based on quantitative analyses. Pages 53–106 in M. Palo and J. Salmi, eds., *Deforestation or development in the Third World?* vol. I. Research Bulletin of the Finnish Forestry Institute, Helsinki, Finland.

Panayotou, T., and S. Sungsuwan. 1994. An econometric analysis of the causes of tropical deforestation: The case of Northeast Thailand. Pages 192–210 in K. Brown and D. W. Pearce, eds., *The causes of tropical deforestation: The economic and statistical analysis of factors giving rise to the loss of tropical forests.* University College London Press, London.

Parizek, R. R. 1971. Impacts of highways on the hydrogeologic environment. Pages 151–99 in *Environmental geomorphology: Proceedings of the 1st annual geomorphological symposium.* State University of New York, Birmingham, N.Y.

Park, C. C. 1992. *Tropical rainforests.* Routledge, New York.

Parrotta, J. A. 1992. The role of plantation forests in rehabilitating degraded ecosystems. *Agriculture, Ecosystems and Environment* 41:115–33.

———. 1993a. Assisted recovery of degraded tropical lands: Plantation forests and ecosystem stability. Pages 169–82 in M. Paoletti, W. Foissner, and D. Coleman, eds., *Soil biota, nutrient cycling and farming systems.* Lewis Publishers, Chelsea, U.K.

———. 1993b. Secondary forest regeneration on degraded tropical lands: The role of plantations as "foster ecosystems." Pages 63–73 in H. Leith and M. Lohmann, eds., *Restoration of tropical forest ecosystems.* Dordrecht, The Netherlands.

———. 1995. The influence of overstory composition on understory colonisation by native species in plantations on a degraded tropical site. *Journal of Vegetation Science* 6:627–34.

Parsons, P. A. 1992. Fluctuating asymmetry: A biological monitor of environmental and genomic stress. *Heredity* 68:361–64.

Partridge, I. J. 1994. Will it rain? The effects of the Southern Oscillation and El Niño on Australia. Dept. of Primary Industries, Brisbane, Australia.

Patrick, R. 1973. Effects of channelisation on the aquatic life of streams. Pages 150–54 in *Environmental considerations in planning, design and construction.* Special report no. 138 of the National Academy of Science and National Academy of Engineering. Highway Research Board, Washington, D.C.

Patterson, B. D. 1987. The principle of nested subsets and its implications for biological conservation. *Conservation Biology* 1:323–34.

———. 1990. On the temporal development of nested subset patterns of species composition. *Oikos* 59:330–42.

———. 1994. Accumulating knowledge on the dimensions of biodiversity: Systematic perspectives on Neotropical mammals. *Biodiversity Letters* 2:79–86.

Patterson, B. D., and W. Atmar. 1986. Nested subsets and the structure of insular mammalian faunas and archipelagos. Pages 65–82 in L. R. Heaney and B. D. Patterson, eds., *Island biogeography of mammals.* Academic Press and Linnean Society of London.

Patton, J. L. 1987. Species groups of spiny rats, genus *Proechimys* (Rodentia: Echimyidae). *Fieldiana: Zoology,* new series 39:305–45.

Patton, J. L., O. B. Berlin, and E. A. Berlin. 1982. Aboriginal perspectives of a mammal community in Amazonian Peru: Knowledge and utilization patterns among the Aguaruna Jivaro. Pages 111–28 In M. A. Mares and H. H. Genoways, eds., *Mammalian biology in South America.* Pymatuning Symposium on Ecology, University of Pittsburgh, Pittsburgh, Pa.

Patton, J. L., and M. N. F. da Silva. 1995. A review of the spiny mouse genus *Scolomys* (Rodentia: Muridae: Sigmodontinae) with the description of a new species from the western Amazon of Brazil. *Proceedings of the Biological Society of Washington* 108:319–37.

Patton, J. L., M. N. F. da Silva, and J. R. Malcolm. 1994. Gene genealogy and differentiation among arboreal spiny rats (Rodentia: Echimyidae) of the Amazon Basin: A test of the riverine diversification hypothesis. *Evolution* 48:1314–23.

———. 1996. Hierarchical genetic structure and gene flow in three sympatric species of Amazonian rodents. *Molecular Ecology* 5:229–38.

Patton, J. L., and M. F. Smith. 1992. mtDNA phylogeny of Andean mice: A test of diversification across ecological gradients. *Evolution* 46:174–83.

Pearce, D. W., and J. J. Warford. 1993. *World without end: Economics, environment, and sustainable development.* Oxford University Press, New York.

Pearson, D. L. 1971. Vertical stratification of birds in a tropical dry forest. *Condor* 73:46–55.

Pearson, S. M., M. G. Turner, R. H. Gardner, and R. V. O'Neill. 1996. An organism-based perspective of habitat fragmentation. Pages 77–95 in R. C. Szaro, ed., *Biodiversity in managed landscapes: Theory and practice.* Oxford University Press, New York.

Peres, C. A., and J. W. Terborgh. 1995. Amazonian nature reserves: An analysis of the defensibility status of existing conservation units and design criteria for the future. *Conservation Biology* 9:34–46.

Peters, C. M., A. H. Gentry, and R. Mendelsohn. 1989. Valuation of an Amazonian rainforest. *Nature* 339:655–56.

Peters, R. H. 1980. From natural history to ecology. *Perspectives in Biology and Medicine* 23:191–203.

Peters, R. L., and T. E. Lovejoy, eds. 1992. *Climate change and biological diversity: Proceedings of World Wildlife Funds conference on consequences of global warming for biological diversity.* Yale University Press, New Haven, Conn.

Philander, S. G. 1989. *El Niño and the southern oscillation.* Academic Press, London.

Phillipson, P. B. 1994. Madagascar. Pages 271–81 in S. D. Davis, V. H. Heywood, and A. C. Hamilton, eds., *Centres of plant diversity, a guide and strategy for their conservation.* World Wide Fund for Nature (WWF) and International Union for the Conservation of Nature and Natural Resources (IUCN), Gland, Switzerland.

Pickett, S. T. A., and M. L. Cadenasso. 1995. Landscape ecology: Spatial heterogeneity in ecological systems. *Science* 269:331–34.

Pickett, S. T. A., and J. N. Thompson. 1978. Patch dynamics and the design of nature reserves. *Biological Conservation* 13:27–37.

Pienaar, U. de V. 1968. The ecological significance of roads in a national park. *Koedoe* 11:169–74.

Pimm, S. L., and J. L. Gittleman. 1992. Biological diversity: Where is it? *Science* 255:940.

Pimm, S. L., H. L. Jones, and J. M. Diamond. 1988. On the risk of extinction. *American Naturalist* 132: 757–85.

Pimm, S. L., and A. M. Sugden. 1994. Tropical diversity and global change. *Science* 263:933–34.

Pizzey, G. 1980. *A field guide to the birds of Australia.* Collins, Sydney, Australia.

Pócs, T. 1974. Bioclimatic studies in the Uluguru mountains (Tanzania, East Africa). *Acta Botanica Academiae Scientaricum Hungariae* 20:115–35.

Pollock, J. I. 1986. Primates and conservation priorities in Madagascar. *Oryx* 20:209–16.

Poore, D. 1979. The values of tropical moist forest ecosystems, and the environmental consequences of their removal. *National Parks and Conservation* 17:127–43.

Portnoy, S., and M. F. Willson. 1993. Seed dispersal curves: Behavior of the tail of the distribution. *Evolutionary Ecology* 7:25–44.

Potter, M. A. 1990. Movement of North Island brown kiwi between forest fragments. *New Zealand Journal of Ecology* 14:14–17.

Powell, A. H., and G. V. N. Powell. 1987. Population dynamics of male euglossine bees in Amazonian forest fragments. *Biotropica* 19:176–79.

Powell, G. V. N. 1985. Sociobiology and adaptive significance of interspecific foraging flocks in the Neotropics. Pages 713–32 in P. A. Buckley, M. S. Foster, E. S. Morton, R. S. Ridgely, and F. G. Buckley, eds., *Neotropical ornithology.* Ornithological Monographs, no. 36. American Ornithologists Union, Washington, D.C.

———. 1989. On the possible contribution of mixed species flocks to species richness in Neotropical avifaunas. *Behavioral Ecology and Sociobiology* 24:387–93.

Powell, G. V. N., and R. Bjork. 1995. Implications of intratropical migration on reserve design: A case study using *Pharomachrus mocinno. Conservation Biology* 9:354–62.

Prabhu, R. 1994. Assessing criteria for sustainable forestry. *ITTO Tropical Forest Update* 4:6–8.

Prance, G. T. 1982a. *Biological diversification in the Tropics.* Columbia University Press, New York.

———. 1982b. Forest refuges: Evidence from woody angiosperms. Pages 137–57 in G. T. Prance, ed., *Biological diversification in the tropics.* Columbia University Press, New York.

———. 1990. The floristic composition of the forests of central Amazonian Brazil. Pages 112–40 in A. H. Gentry, ed., *Four Neotropical rainforests.* Yale University Press, New Haven, Conn.

Preston, E. M., and B. L. Bedford. 1988. Evaluating cumulative effects on wetland functions: A conceptual overview and generic framework. *Environmental Management* 12:565–83.

Preston, F. W. 1962. The canonical distributors of commonness and rarity. Parts 1 and 2. *Ecology* 43:185–15, 410–32.

Primack, R. B. 1993. *Essentials of conservation biology.* Sinauer Associates, Sunderland, Mass.

Pulliam, H. R., and B. J. Danielson. 1991. Sources, sinks and habitat selection: A landscape perspective on population dynamics. *American Naturalist* (Supplement) 138:50–66.

Putz, F. E. 1984a. How trees avoid and shed lianas. *Biotropica* 16:19–23.

———. 1984b. The natural history of lianas on Barro Colorado Island, Panama. *Ecology* 65:1713–24.

———. 1992. Silvicultural effects of lianas. Pages 493–501 in F. E. Putz and H. A. Mooney, eds., *The biology of vines.* Cambridge University Press, Cambridge.

Putz, F. E., E. G. Leigh, Jr., and S. J. Wright. 1990. Solitary confinement in Panama. *Garden* 14:18–23.

Putz, F. E., and M. Pinard. 1993. Reduced impact logging as a carbon offset method. *Conservation Biology* 7:755–57.

QDPI-Forest Service. 1994. Community Rainforest Reforestation Program. Annual Report 1993/94. Forest Service, Brisbane, Australia.

Quattrochi, D. A., and R. E. Pelletier. 1991. Remote sensing for analysis of landscapes: An introduction. Pages 50–76 in M. G. Turner and R. H. Gardner, eds., *Quantitative methods in landscape ecology.* Springer-Verlag, New York.

Queensland Department of Environment and Heritage. 1995. Draft conservation plan for the mahogany glider. Queensland Department of Environment and Heritage, Brisbane.

Quintela, C. E. 1985. Forest fragmentation and differential use of natural and man-made edges by understory birds in central Amazonia. M.Sc. thesis, University of Illinois, Chicago, Ill.

Rabinowitz, A. R., and B. G. Nottingham, Jr. 1986. Ecology and behaviour of the jaguar *(Panthera onca)* in Belize, Central America. *Journal of Zoology* 210:149–59.

Rabinowitz, A. R., and S. R. Walker. 1991. The carnivore community in a dry tropical forest mosaic in Huai Kha Khaeng Wildlife Sanctuary, Thailand. *Journal of Tropical Ecology* 7:37–47.

Raguso, R. A., and J. Llorente-B. 1992. The butterflies (Lepidoptera) of the Tuxtlas mountains, Veracruz, Mexico, revisited: Species richness and habitat disturbance. *Journal of Research on the Lepidoptera* 29: 105–33.

Rahbek, C. 1995a. The elevational gradient of species richness: A uniform pattern? *Ecography* 18:200–205.

———. 1995b. The elevational gradient of species richness—with special emphasis on South American tropical landbirds. Ph.D. dissertation, Zoological Museum, University of Copenhagen, Denmark.

Rahbek, C., H. Bloch, M. K. Poulsen, and J. F. Rasmussen. 1995. The avifauna of the Podocarpus National Park—the "Andean jewel in the crown" of Ecuador's protected areas. *Ornitologia Neotropical* 6:113–20.

Ralls, K., and J. D. Ballou. 1983. Extinction: Lessons from zoos. Pages 164–84 in C. M. Schonewald-Cox, S. M. Chambers, B. Macbryde, and W. L. Thomas, eds., *Genetics and conservation: A reference for managing wild animal and plant populations.* Benjamin/Cummings, Menlo Park, Calif.

Ralph, C. J., G. R. Geupel, P. Pyle, T. E. Martin, and D. F. DeSante. 1993. *Handbook of field methods for monitoring landbirds.* General Technical Report PSW-GTR-144, Pacific Southwest Research Station, Forest Service, U.S. Department of Agriculture, Albany, Calif.

Ramage, C. S. 1968. Role of a tropical "maritime continent" in the atmospheric circulation. *Monthly Weather Review* 96:365–70.

Rand, D. M. 1994. Thermal habit, metabolic rate and the evolution of mitochondrial DNA. *Trends in Ecology and Evolution* 9:125–31.

Rand, D. M., M. Dorfsman, and L. M. Kann. 1994. Neutral and non-neutral evolution of *Drosophila* mitochondrial DNA. *Genetics* 138:741–56.

Rangel, J. O., ed. 1994. Ucumarí: Un caso típico de la diversidad biótica Andina. Corporación Autónoma Regional de Risaralda, Colombia.

Rankin-de Mérona, J. M., R. W. Hutchings, and T. E. Lovejoy. 1990. Tree mortality and recruitment over a five-year period in undisturbed upland rainforest of the central Amazon. Pages 573–84 in A. Gentry, ed., *Four Neotropical rainforests.* Yale University Press, New Haven, Conn.

Rankin-de Mérona, J. M., G. T. Prance, R. W. Hutchings, F. M. Silva, W. A. Rodrigues, and M. E. Uehling. 1992. Preliminary results of large-scale tree inventory of upland rain forest in the central Amazon. *Acta Amazônica* 22:493–534.

Räsänen, M. E., A. M. Lina, J. C. R. Santos, and F. R. Negri. 1995. Late Miocene tidal deposits in the Amazonian foreland basin. *Science* 269:386–90.

Räsänen, M. E., R. Neller, J. Salo, and H. Jungner. 1992. Recent and ancient fluvial deposition systems in the Amazonian foreland basin, Peru. *Geological Magazine* 129:293–306.

Räsänen, M. E, J. Salo, and R. Kalliola. 1987. Fluvial perturbance in the western Amazon basin: Regulation by long term sub-Andean tectonics. *Science* 238:1398–1401.

Raunet, M. 1991. *Le milieu physique et les sols de l'Ile de la Réunion.* CIRAD, St-Denis, Ile de La Réunion, France.

Readers' Digest. 1976. *Complete book of Australian birds.* Reader's Digest Pty. Ltd., Sydney, Australia.

Redford, K. H., and J. G. Robinson. 1987. The game of choice: Patterns of Indian and colonist hunting in the Neotropics. *American Anthropologist* 89:650–67.

Reese, K. P., and J. T. Ratti. 1988. Edge effect: A concept under scrutiny. Pages 127–36 in Transactions of the 53rd North American Wildlife and Natural Resources Conference.

Reh, W., and A. Seitz. 1990. The influence of land use on the genetic structure of populations of the common frog *Rana temporia. Biological Conservation* 54:239–50.

Reich, P. B., and R. Borchert. 1984. Water stress and tree phenology in tropical dry forest in the lowlands of Costa Rica. *Journal of Ecology* 72:61–74.

Reichman, O. J., J. H. Benedix, and T. R. Seastedt. 1993. Distinct animal-generated edge effects in a tallgrass prairie community. *Ecology* 74:1281–85.

Reid, W. V. 1992. How many species will there be? Pages 55–74 in T. C. Whitmore and J. A. Sayer, eds., *Tropical deforestation and species extinction.* Chapman and Hall, London.

Remsen, J. V., Jr., and D. A. Good. 1996. Misuse of data from mist-net captures to assess relative abundance in bird populations. *Auk* 113:381–98.

Remsen, J. V., Jr., and T. A. Parker, III. 1983. Contribution of river-created habitats to bird species richness in Amazonia. *Biotropica* 15:223–31.

Rencher, A. C., and F. C. Pun. 1980. Inflation of R^2 in best subset regression. *Technometrics* 22:49–53.

Renjifo, L. M. 1988. Composición y estructura de la comunidad aviaria de bosque andino primario y secundario en la Reserva del Alto Quindío Acaime, Colombia. Thesis, Departamento de Biología, Pontificia Universidad Javeriana, Santa Fé de Bogotá, Colombia.

Renjifo, L. M., G. P. Servat, J. M. Goerck, B. A. Loiselle, and J. G. Blake. In press. Patterns of species composition and endemism in the northern Neotropics: A case for conservation of montane avifaunas. In J. V. Remsen, ed., *Natural history and conservation of Neotropical birds.* Ted Parker Memorial Volume, Ornithological Monographs.

Repetto, R., and M. Gillis, eds. 1988. *Public policies and the misuse of forest resources.* Cambridge University Press, Cambridge.

Restrepo, C. 1990. Ecology and cooperative breeding in a frugivorous bird, *Semnornis ramphastinus.* M.Sc. thesis, University of Florida, Gainesville.

Reville, B. J., J. D. Tranter, and H. D. Yorkston. 1990. Impact of forest clearing on the endangered seabird *Sula abbotti. Biological Conservation* 51:23–38.

Rich, A. C., D. S. Dobkin, and L. J. Niles. 1994. Defining forest fragmentation by corridor width: The influence of narrow forest-dividing corridors on forest-nesting birds in southern New Jersey. *Conservation Biology* 8:1109–21.

Richards, P. W. 1952. *The tropical rain forest.* Cambridge University Press, Cambridge.

Ricklefs, R. E. 1969. An analysis of nesting mortality in birds. *Smithsonian Contributions to Zoology,* no. 9.

Ridgely, R. S., and S. J. C. Gaulin. 1980. The birds of Finca Merenberg, Huila Department, Colombia. *Condor* 82:379–91.

Ridgely, R. S., and G. Tudor. 1989. *The birds of South America.* Vol. 1, *The oscine passerines.* University of Texas Press, Austin.

———. 1994. *The birds of South America.* Vol. 2, *The suboscine passerines.* University of Texas Press, Austin.

Ripple, W. J., G. A. Bradshaw, and T. A. Spies. 1991. Measuring forest landscape patterns in the Cascade Range of Oregon, USA. *Biological Conservation* 57:73–88.

Robbins, R. K., G. Lamas-M., O. H. H. Mielke, D. J. Harvey, and M. M. Casagrande. In press. The species-rich butterfly community at Pakitza, Parque Nacional de Manu, Peru. I. Taxonomic and ecological compositions. In D. E. Wilson, ed., *The biodiversity of Pakitza, Manu National Park, Peru.* Smithsonian Institution Press, Washington, D.C.

Roberts, J., O. M. R. Cabral, and L. F. de Aguiar. 1990. Stomatal and boundary layer conductances in an Amazonian terra firme rain forest. *Journal of Applied Ecology* 27:336–53.

Robinson, G. R., R. D. Holt, M. S. Gaines, S. P. Hamburg, M. L. Johnson, H. S. Fitch, and E. A. Martinko. 1992. Diverse and contrasting effects of habitat fragmentation. *Science* 257:524–26.

Robinson, J. G. 1986. Seasonal variation in use of time and space by the wedge-capped capuchin monkey, *Cebus olivaceus*: Implications for foraging theory. *Smithsonian Contributions to Zoology* 431:1–60.

Robinson, J. G., and K. H. Redford. 1986. Body size, diet, and population density of Neotropical forest mammals. *American Naturalist* 128:665–80.

Robinson, S. K., F. R. Thompson III, T. M. Donovan, D. R. Whitehead, and J. Faaborg. 1995. Regional forest fragmentation and the nesting success of migratory birds. *Science* 267:1987–90.

Rogers, C. A. 1993. Describing landscapes: Indices of structure. M.Sc. thesis, Simon Fraser University, Burnaby, British Columbia, Canada.

Rolstad, J. 1991. Consequences of forest fragmentation for the dynamics of bird populations: Conceptual issues and the evidence. *Biological Journal of the Linnean Society* 42:149–63.

Romme, W. H. 1982. Fire and landscape diversity in subalpine forests of Yellowstone National Park. *Ecological Monographs* 52:199–221.

Rosas, M. L. 1986. Estudio de la estructura de la comunidad de frugívoros en sotobosques del Cañon de Mamarramos en el Santuario de Fauna y Flora de Iguaque, Boyacá. Thesis, Departamento de Biología, Pontificia Universidad Javeriana, Santa Fé de Bogotá, Colombia.

Rosen, D. E. 1978. Vicariant patterns and historical explanation in biogeography. *Systematic Zoology* 27: 159–88.

Rosen, P. C., and C. H. Lowe. 1994. Highway mortality of snakes in the Sonoran Desert of southern Arizona. *Biological Conservation* 68:143–48.

Rost, G. R., and Bailey, J. A. 1979. Distribution of mule deer and elk in relation to roads. *Journal of Wildlife Management* 43:634–41.

Rudd, R. L. 1979. Niche dimension in the bamboo mouse, *Chiropodomys gliroides* (Rodentia: Muridae). *Malayan Nature Journal* 32:347–49.

Rudel, T. 1994. Population, development and tropical deforestation: A cross-national study. Pages 96–105 in K. Brown and D. W. Pearce, eds., *The causes of tropical deforestation: The economic and statistical analysis of factors giving rise to the loss of tropical forests.* University College London Press, London.

Rylands, A. B., and A. Keuroghlian. 1988. Primate populations in continuous forest and forest fragments in central Amazonia. *Acta Amazônica* 18:291–307.

Sader, S. A., and A. T. Joyce. 1988. Deforestation rates and trends in Costa Rica, 1940 to 1983. *Biotropica* 20:11–19.

Salati, E. 1985. Climatology and hydrology of Amazonia. Pages 18–48 in G. T. Prance and T. E. Lovejoy, eds., *Amazonia.* Pergamon Press, Oxford.

Salati, E., A. Dall'Olio, E. Matsui, and J. R. Gat. 1979. Recycling of water in the Amazon Basin: An isotopic study. *Water Resources Research* 15:1250–58.

Salati, E., J. M. Marques, and L. C. B. Molion. 1978. Origem e distribuição das chuvas na Amazônia. *Interciencia* 3:200–205.

Salati, E., and C. Nobre. 1991. Possible climatic impacts of tropical deforestation. *Climatic Change* 19: 177–96.

Salati, E., and P. B. Vose. 1984. Amazon Basin: A system in equilibrium. *Science* 225:129–38.

Saldarriaga, J. G., and D. C. West. 1986. Holocene fires in the northern Amazon basin. *Quaternary Research* 26:358–66.

Salisbury, E. 1974. Seed size and mass in relation to environment. *Proceedings of the Royal Society of London B* 186:83–88.

Salo, J. 1988a. Rainforest diversification in the western Amazon basin: The role of river dynamics. Ph.D. dissertation, Department of Biology, University of Turku, Finland.

———. 1988b. Rainforest diversification in the western Amazon Basin: The role of river dynamics. Report no. 16, University of Turku, Finland.

Salo, J., R. Kalliola, I. Hakkinen, Y. Makinen, P. Niemela, M. Puhakka, and P. D. Coley. 1986. River dynamics and the diversity of Amazon lowland forest. *Nature* 322:254–58.

Sample, V. A., ed. 1994. *Remote sensing and GIS in ecosystem management.* Island Press, Washington, D.C.

Sargent. S. 1990. Neighborhood effect on fruit removal by birds: A field experiment with *Viburnum dentatum* (Caprifoliaceae). *Ecology* 71:1289–98.

SAS Institute, Inc. 1985. *SAS user's guide: Statistics.* Version 5 ed. SAS Institute, Inc., Cary, N.C.

———. 1990. *SAS/STAT guide for personal computers.* Version 6, 4th ed. SAS Institute, Cary, N.C.

———. 1992. *SAS/STAT software: Changes and enhancements.* Version 6, 7th ed. SAS Institute, Cary, N.C.

Saunders, D. A., and C. P. De Reibera. 1991. Values of corridors to avian populations in a fragmented landscape. Pages 221–40 in D. A. Saunders and R. J. Hobbs, eds., *Nature conservation 2: The role of corridors.* Surrey Beatty, Chipping Norton, N.S.W., Australia.

Saunders, D. A., and R. J. Hobbs. 1991. *Nature conservation 2: The role of corridors.* Surrey Beatty, Chipping Norton, N.S.W., Australia.

Saunders, D. A., R. J. Hobbs, and G. W. Arnold. 1993. The Kellerberrin project on fragmented landscapes: A review. *Biological Conservation* 64:185–92.

Saunders, D. A., R. J. Hobbs, and C. R. Margules. 1991. Biological consequences of ecosystem fragmentation: A review. *Conservation Biology* 5:18–32.

Savidge, J. A. 1987. Extinction of an island forest avifauna by an introduced snake. *Ecology* 68:660–68.

Savill, P. S. 1983. Silviculture in windy climates. *Forestry Abstracts* 44:473–88.

Schaller, G. B., and P. G. Crawshaw, Jr. 1980. Movement patterns of jaguar. *Biotropica* 12:161–68.

Schelhas, J. 1996. Land-use choice and forest patches in Costa Rica. Pages 258–84 in J. Schelhas and R. Greenberg, eds., *Forest patches in tropical landscapes.* Island Press, Washington, D.C.

Schelhas, J., and R. Greenberg, eds. 1996. *Forest patches in tropical landscapes.* Island Press, Washington, D.C.

Schodde, R., and J. Calaby. 1972. The biogeography of the Australo-Papuan bird and mammal fauna in relation to Torres Strait. Pages 255–300 in D. Walker, ed., *Bridge and barrier: The cultural history of Torres Strait.* Australian National University, Canberra, Australia.

Schonewald-Cox, C. M., and J. W. Bayless. 1986. The boundary model: A geographical analysis of design and conservation of nature reserves. *Biological Conservation* 38:305–22.

Schonewald-Cox, C. M., J. M. Chambers, B. Macbryde, and W. Thomas. 1983. *Genetics and conservation: A reference manual for managing wild animal and plant populations.* Benjamin Cummings, London.

Schreiber, R. K., and J. H. Graves. 1977. Powerline corridors as possible barriers to movements of small mammals. *American Midland Naturalist* 97:504–8.

Schwarzkopf, L., and A. B. Rylands. 1989. Primate species richness in relation to habitat structure in Amazonian rainforest fragments. *Biological Conservation* 48:1–12.

Scott, M. J., B. Csuti, K. Smith, J. E. Estes, and S. Caicco. 1987. Species richness: A geographical approach to protecting future species diversity. *Bioscience* 37:782–88.

Scott, M. J., F. Davis, B. Csuti, R. Noss, B. Butterfield, C. Groves, H. Anderson, S. Caicco, F. D'Erchia, T. C. Edwards, Jr., J. Ulliman, and R. G. Wright. 1993. *Gap analysis: A geographic approach to the protection of biological diversity.* Wildlife Monographs 123.

Seagle, S. W., and H. H. Shugart. 1985. Faunal richness and turnover on dynamic landscapes: A simulation study. *Journal of Biogeography* 12:499–508.

Shafer, C. L. 1990. *Nature reserves: Island theory and conservation practice.* Smithsonian Institution Press, Washington, D.C.

———. 1995. Values and shortcomings of small reserves. *Bioscience* 45:80–88.

Shaffer, M. L. 1981. Minimum population sizes for species conservation. *Bioscience* 31:131–34.

Shafik, N. 1994. Macroeconomic causes of deforestation: Barking up the wrong tree? Pages 86–95 in K. Brown and D. W. Pearce, eds., *The causes of tropical deforestation: The economic and statistical analysis of factors giving rise to the loss of tropical forests.* University College London Press, London.

Sheperd, A. 1994. The Christmas Island rehabilitation program. *Journal of Tropical Forest Science* 7:18–27.

Sheppard, P. M., J. R. G. Turner, K. S. Brown, Jr., W. W. Benson, and M. C. Singer. 1985. Genetics and the evolution of muellerian mimicry in *Heliconius* butterflies. *Philosophical Transactions of the Royal Society, London B* 308:433–610.

Shmida, A., and M. V. Wilson. 1985. Biological determinants of species diversity. *Journal of Biogeography* 12:1–20.

Shure, D. J., and D. L. Phillips. 1991. Patch size of forest openings and arthropod populations. *Oecologia* 86:325–34.

Sibley, C. G., and J. E. Ahlquist. 1990. *Phylogeny and classification of birds: A study in molecular evolution.* Yale University Press, New Haven, Conn.

Sieving, K. E. 1990. Pheasant cuckoo foraging behavior, with notes on habits and possible social organization in Panama. *Journal of Field Ornithology* 61:41–46.

———. 1991. Differential avian extinction from Barro Colorado Island, Panama: Nest predation and population genetic factors. Ph.D. dissertation, University of Illinois, Urbana-Champaign.

———. 1992. Nest predation and differential insular extinction among selected forest birds of central Panama. *Ecology* 73:2310–28.

Silva, J. M. C. da. 1995a. Biogeographic analysis of the Cerrado avifauna, South America. *Steenstrupia* 21: 49–67.

———. 1995b. Birds of the Cerrado region, South America. *Steenstrupia* 21:69–92.

Silva, J. M. C. da, C. Uhl, and G. Murray. 1996. Plant succession, landscape management, and the ecology of fruit-eating birds in abandoned Amazonian pastures. *Conservation Biology* 10:491–503.

Silva, M. N. F. da. 1995. Systematics and phylogeography of Amazonian spiny rats of the genus *Proechimys* (Rodentia: Echimyidae). Ph.D. dissertation, University of California, Berkeley.

Silva, M. N. F. da, and J. L. Patton. 1993. Amazonian phylogeography: mtDNA sequence variation in arboreal Echimyid rodents (Caviomorpha). *Molecular Phylogenetics and Evolution* 2:243–55.

Silva, N. J. da, and J. W. Sites, Jr. 1995. Patterns of diversity of Neotropical squamate reptile species with emphasis on the Brazilian Amazon and the conservation potential of indigenous reserves. *Conservation Biology* 9:873–901.

Silverman, B. W. 1981. Using kernel density estimates to investigate multimodality. *Journal of the Royal Statistical Society B* 43:97–99.

———. 1986. *Density estimation for statistics and data analysis.* Monographs on Statistics and Applied Probability, 26. Chapman and Hall, London.

Simberloff, D. S. 1976. Experimental zoogeography of islands: Effects of island size. *Ecology* 57:629–48.

———. 1980. A succession of paradigms in ecology: Essentialism to materialism to probabilism. *Syntheses* 43:3–39.

———. 1986. Are we on the verge of a mass extinction in tropical rainforests? Pages 165–80 in D. K. Elliott, ed., *Dynamics of extinction.* Wiley-Interscience, New York.

———. 1987. The Spotted Owl fracas: Mixing academic, applied and political ecology. *Ecology* 68:766–72.

———. 1988. The contribution of population and community biology to conservation science. *Annual Review of Ecology and Systematics* 19:473–511.

———. 1990a. Hypotheses, errors, and statistical assumptions. *Herpetologica* 46:351–57.

———. 1990b. Reconstructing the ambiguous: Can island ecosystems be restored? Pages 37–51 in D. R. Towns, C. H. Daugherty, I. A. E. Atkinson, eds., *Ecological restoration of New Zealand islands.* Department of Conservation, Wellington, New Zealand.

———. 1992. Do species-area curves predict extinction in fragmented forest? Pages 75–86 in T. C. Whitmore and J. A. Sayer, eds., *Tropical deforestation and species extinction.* Chapman and Hall, London.

———. 1994. Fragmentation, corridors and the longleaf pine community. Pages 47–56 in C. Moritz and J. Kikkawa, eds., *Conservation biology in Australia and Oceania.* Surrey Beatty, Chipping Norton, N.S.W., Australia.

———. 1995. Habitat fragmentation and population extinction of birds. *Ibis* 137:105–11.

Simberloff, D. S., and L. G. Abele. 1982. Refuge design and island biogeographic theory: Effects of fragmentation. *American Naturalist* 120:41–50.

Simberloff, D. S., and J. Cox. 1987. Consequences and costs of conservation corridors. *Conservation Biology* 1:63–71.

Simberloff, D. S., J. A. Farr, J. Cox, and D. W. Mehlman. 1992. Movement corridors: Conservation bargains or poor investments? *Conservation Biology* 6:493–504.

Simpson, G. G. 1944. *Tempo and mode in evolution.* Hafner, New York.

Singer, M. C., and P. R. Ehrlich. 1992. Host specialization of satyrine butterflies, and their responses to habitat fragmentation in Trinidad. *Journal of Research on the Lepidoptera* 30:248–56.

Singer, M. C., and L. E. Gilbert. 1978. Ecology of butterflies in the urbs and suburbs. Pages 1–11 in G. W. Frankie and C. S. Koehler, eds., *Perspectives in urban entomology.* Academic Press, New York.

Sisk, T. D., and C. R. Margules. 1993. Habitat edges and restoration: Methods for quantifying edge effects and predicting the result of restoration efforts. Pages 57–69 in D. Saunders, R. J. Hobbs, and P. R. Ehrlich, eds., *Nature conservation 3: Reconstruction of fragmented ecosystems.* Surrey Beatty, Sydney, Australia.

Skole, D., and C. Tucker. 1993. Tropical deforestation and habitat fragmentation in the Amazon: Satellite data from 1978 to 1988. *Science* 260:1905–10.

Skutch, A. F. 1950. The nesting seasons of Central American birds in relation to climate and food supply. *Ibis* 92:185–222.

———. 1960. *Life histories of Central American birds.* Vol. II. Pacific Coast Avifauna Series, no. 34. Cooper Ornithological Society, Berkeley, Calif.

———. 1966. A breeding bird census and nesting success in Central America. *Ibis* 108:1–16.

———. 1969. *Life histories of Central American birds.* Vol. III. Pacific Coast Avifauna Series, no. 35. Cooper Ornithological Society, Berkeley, Calif.

———. 1981. *New studies of tropical American birds.* Publications of the Nuttall Ornithological Club, No. 19.

———. 1985. Clutch size, nesting success, and predation on nests of Neotropical birds, reviewed. Pages 575–95 in P. A. Buckley, M. S. Foster, E. S. Morton, R. S. Ridgely, and F. G. Buckley, eds., *Neotropical Ornithology.* Ornithological Monographs, no. 36. American Ornithologists Union, Washington, D.C.

Slater, P., P. Slater, and R. Slater. 1992. *The Slater field guide to Australian birds.* Kevin Weldon, Adelaide, Australia.

Slatkin, M. 1987. Gene flow and the geographic structure of animal populations. *Science* 236:787–92.

Slatkin, M., and R. R. Hudson. 1991. Pairwise comparisons of mitochondrial DNA sequences in stable and exponentially growing populations. *Genetics* 129:555–62.

Smith, A. P., and J. Ganzhorn. 1996. Convergence in community structure and dietary adaptation in Australian possums and gliders and Malagasy lemurs. *Australian Journal of Ecology* 21:31–46.

Smith, A. P., N. H. Horning, and D. M. Moore. In press. Lemur conservation planning in a region of western Madagascar. *Conservation Biology.*

Smith, A. P., N. H. Horning, S. Olivieri, and L. Andrianifahnana. 1990. Feasibility study for establishment of a biodiversity planning service in Madagascar. A report to WWF-Madagascar. University of New England, Armidale, Australia.

Smith, A. P., D. Moore, and N. Horning. 1991. Madagascar biodiversity planning service pilot study. A report to WWF-Madagascar. University of New England, Armidale, Australia.

Smith, A. P., and D. Quinn. 1996. Patterns and cause of extinction and decline in Australian conilurine rodents. *Biological Conservation* 77:243–68.

Smith, G. C. 1985. Biology and habitat usage of sympatric populations of *Melomys cervinipes* and *M. burtoni. Australian Zoology* 21:307–26.

Smith, N. H. J. 1981. Colonization lessons from a tropical forest. *Science* 214:755–61.

Smith, T. B., M. W. Bruford, and R. K. Wayne. 1993. The preservation of process: The missing element of conservation programs. *Biodiversity Letters* 1:164–67.

Smith, T. M., J. B. Smith, Jr., and H. H. Shugart. 1992. Modeling the response of terrestrial vegetation to climate change in the tropics. Pages 253–68 in J. G. Goldammer, ed., *Tropical forests in transition.* Birkhauser Verlag, Basel, Switzerland.

Smouse, P. E., J. C. Long, and R. R. Sokal. 1986. Multiple regression and correlation extensions of the Mantel test of matrix correspondence. *Systematic Zoology* 35:627–32.

Smythe, N. 1986. Competition and resource partitioning in the guild of Neotropical terrestrial frugivorous mammals. *Annual Review of Ecology and Systematics* 17:169–88.

———. 1989. Seed survival in the palm *Astrocaryum standleyanum*: Evidence for its dependence upon its seed dispersers. *Biotropica* 21:50–56.

Sobrevila, C., and M. T. K. Arroyo. 1982. Breeding systems in a montane tropical cloud forest in Venezuela. *Plant Systematics and Evolution* 140:19–38.

Sokal, R. R. 1986. Phenetic taxonomy: Theory and methods. *Annual Review of Ecology and Systematics* 17:423–42.

Sokal, R. R., and C. A. Braumann. 1980. Significance tests for coefficients of variation and variability profiles. *Systematic Zoology* 29:50–66.

Sokal, R. R., and F. J. Rohlf. 1981. *Biometry.* 2d ed. W. H. Freeman, New York.

———. 1995. *Biometry.* 3d ed. W. H. Freeman, New York.

Somerville, A. 1980. Wind stability: Forest layout and silviculture. *New Zealand Journal of Forest Science* 10:476–501.

Sommer, A. 1976. Attempt at an assessment of the world's tropical forests. *Unasylva* 28:5–25.

SOS Mata Atlântica and INPE. 1993. *Evolução dos remanescentes florestais e ecossistemas associados do domínio da Mata Atlântica.* SOS Mata Atlântica, Instituto de Pesquisas Espaciais, São Paulo, Brazil.

Soulé, M. E. 1971. The variation problem: The gene flow-variation hypothesis. *Taxon* 20:37–50.

———. 1972. Phenetics of natural populations. III. Variation in insular populations of a lizard. *American Naturalist* 106:429–46.

———. 1973. Island lizards: The genetic-phenetic variation correlation. *Nature* 242:191–93.

———. 1979. Heterozygosity and developmental stability: Another look. *Evolution* 33:396–401.

———. 1982. Allometric variation. 1. The theory and some consequences. *American Naturalist* 120:751–64.

———, ed. 1987. *Viable populations for conservation.* Cambridge University Press, Cambridge.

———. 1991. Conservation tactics for a constant crisis. *Science* 253:744–50.

Soulé, M. E., D. T. Bolger, A. C. Alberts, J. Wright, M. Sorice, and S. Hill. 1988. Reconstructed dynamics of rapid extinctions of chaparral-requiring birds in urban habitat islands. *Conservation Biology* 2:75–92.

Soulé, M. E., and M. E. Gilpin. 1991. The theory of wildlife corridor capability. Pages 3–8 in D. A. Saunders and R. J. Hobbs, eds., *Nature conservation 2: The role of corridors.* Surrey Beatty, Sydney, Australia.

Soulé, M. E., and B. R. Stewart. 1970. The niche-variation hypothesis: A test and alternatives. *American Naturalist* 104:85–97.

Soulé, M. E., B. A. Wilcox, and C. Holtby. 1979. Benign neglect: A model of faunal collapse in the game reserves of East Africa. *Biological Conservation* 15:259–72.

Soulé, M. E., and S. Y. Yang. 1973. Genetic variation in side-blotched lizards on islands in the Gulf of California. *Evolution* 27:593–600.

Sousa, W. P. 1984. The role of disturbance in natural communities. *Annual Review of Ecology and Systematics* 15:353–91.

Southgate, D. 1994. Tropical deforestation and agricultural development in Latin America. Pages 134–44 in K. Brown and D. W. Pearce, eds., *The causes of tropical deforestation: The economic and statistical analysis of factors giving rise to the loss of tropical forests.* University College London Press, London.

Southworth, F., V. H. Dale, and R. V. O'Neill. 1991. Contrasting patterns of land use in Rondônia, Brazil: Simulating the effects on carbon release. *International Social Sciences Journal* 130:681–98.

Souza, O. F. F. de, and V. K. Brown. 1994. Effects of habitat fragmentation on Amazonian termite communities. *Journal of Tropical Ecology* 10:197–206.

Sparrow, H. R., T. D. Sisk, P. R. Ehrlich, and D. D. Murphy. 1994. Techniques and guidelines for monitoring Neotropical butterflies. *Conservation Biology* 8:800–809.

Sprague, L. G., and C. R. Sprague. 1976. Management science? *Interfaces* 7:57–62.

Springer, M., G. McKay, K. Aplin, and J. A. W. Kirsch. 1992. Relationships among ringtail possums (Marsupalia: Pseudocheiridae) based on DNA-DNA hybridisation. *Australian Journal of Zoology* 40:423–35.

Stacy, E. A., J. L. Hamrick, J. D. Nason, S. P. Hubbell, R. B. Foster, and R. Condit. 1996. Pollen dispersal in low density populations of three Neotropical tree species. *American Naturalist* 48:275–88.

Stallings, J. R. 1988. Small mammal communities in an eastern Brazilian park. Ph.D. dissertation, University of Florida, Gainesville.

———. 1989. Small mammal inventories in an eastern Brazilian park. *Bulletin Florida State Museum, Biological Sciences* 34:153–200.

Stamps, J. A., M. Buechner, and V. V. Krishnan. 1987. The effects of edge permeability and habitat geometry on emigration from patches of habitat. *American Naturalist* 129:533–52.

Star, J., and J. Estes. 1990. *Geographical information systems: An introduction.* Prentice-Hall, Englewood Cliffs, N.J.

Stiles, F. G. 1975. Ecology, flowering phenology, and hummingbird pollination of some Costa Rican *Heliconia* species. *Ecology* 56:285–302.

———. 1980. The annual cycle in a tropical wet forest hummingbird community. *Ibis* 122:322–43.

———. 1983. Birds: Introduction. Pages 502–29 in D. H. Janzen, ed., *Costa Rican Natural History.* University of Chicago Press, Chicago, Ill.

———. 1985. On the role of birds in the dynamics of Neotropical forests. *International Council for Bird Preservation (ICBP) Technical Publication* 4:49–59.

Stiles, F. G., and A. Skutch. 1989. *A guide to the birds of Costa Rica.* Cornell University Press, Ithaca, N.Y.

Stockard, J., B. Nicholson, and G. Williams. 1985. An assessment of a rainforest regeneration program at Wingham Brush, New South Wales. *Victorian Naturalist* 103:85–91.

Stocker, G. C., and A. K. Irvine. 1983. Seed dispersal by Cassowaries *(Casuarius casuarius)* in North Queensland's rainforests. *Biotropica* 15:170–76.

Stocker, G. C., and G. L. Unwin. 1989. The rain forests of north-eastern Australia—their environment, evolutionary history and dynamics. Pages 241–59 in H. Leith and M. J. A. Werger, eds., *Tropical rain forest ecosystems.* Elsevier, Amsterdam, Netherlands.

Stotz, D. F., and R. O. Bierregaard, Jr. 1989. The birds of the fazendas Porto Alegre, Esteio and Dimona north of Manaus, Amazonas, Brazil. *Revista brasileira de biologia* 49:861–72.

Stotz, D. F., R. O. Bierregaard, Jr., M. Cohn-Haft, P. Peterman, J. Smith, A. Whittaker, and S. V. Wilson. 1992. The status of North American migrants in central Amazonian Brazil. *Condor* 94:608–21.

Stouffer, P. C., and R. O. Bierregaard, Jr. 1993. Spatial and temporal abundance patterns of Ruddy Quail-Doves *(Geotrygon montana)* near Manaus, Brazil. *Condor* 95:896–903.

———. 1995a. Effects of forest fragmentation on understory hummingbirds in Amazonian Brazil. *Conservation Biology* 9:1085–94.

———. 1995b. Use of Amazonian forest fragments by understory insectivorous birds: Effects of fragment size, surrounding secondary vegetation, and time since isolation. *Ecology* 76:2429–45.

Strahm, W. A. 1994. The conservation and restoration of the flora of Mauritius and Rodrigues. Dissertation, University of Reading, U.K.

Strasberg, D. 1994. Dynamique des forêts tropicales de l'Ile de La Réunion. Dissertation, University of Montpellier II, France.

———. 1996. Diversity, size composition and spatial aggregation among trees on a 1-ha rain forest plot at La Réunion. *Biodiversity and Conservation* 5:825–40.

Sun, D., G. Dickenson, and A. Bragg. 1995. Direct seeding of *Alphitonia petrei* for gully revegetation in tropical Northern Australia. *Forest Ecology and Management* 73:249–57.

Suppe, F., ed. 1977. *The structure of scientific theories.* 2d ed. University of Illinois Press, Urbana, Ill.

Sussman, R. W., G. M. Green, and L. K. Sussman. 1994. Satellite imagery, human ecology, anthropology and deforestation in Madagascar. *Human Ecology* 22:333–54.

Sutton, S. L., and N. M. Collins. 1991. Insects and tropical forest conservation. Pages 405–24 in N. M. Collins and J. A. Thomas, eds., *Conservation of insects and their habitats.* 15th Symposium of the Royal Entomological Society of London. Academic Press, London.

Swain, D. P. 1987. A problem with the use of meristic characters to estimate developmental stability. *American Naturalist* 129:761–68.

Swihart, R. K., and N. Slade. 1984. Road crossing in *Sigmodon hispidus* and *Microtus ochrogaster. Journal of Mammalogy* 65:357–60.

Tabanez, A. A. J. 1995. Ecologia e manejo de ecounidades em um fragmento florestal na região de Piracicaba, S.P. M.Sc. thesis, ESALQ, University of São Paulo, Piracicaba, Brazil.

Tanner, E. V. J. 1982. Species diversity and reproductive mechanisms in Jamaica trees. *Biological Journal of the Linnean Society* 18:263–78.

Tattersall, I. 1982. *The primates of Madagascar.* Columbia University Press, New York.

Temple, S. A. 1986. Predicting impacts of habitat fragmentation on forest birds: A comparison of two models. Pages 301–4 in J. Verner, M. L. Morrison, and C. J. Ralph, eds., *Wildlife 2000: Modelling habitat relationships of terrestrial vertebrates.* University of Wisconsin Press, Madison.

Temple, S. A., and J. R. Cary. 1988. Modelling dynamics of habitat-interior bird populations in fragmented landscapes. *Conservation Biology* 2:340–47.

Templeton, A. R., and B. Read. 1984. Factors eliminating inbreeding depression in a captive herd of Speke's gazelle *(Gazella spekei). Zoo Biology* 3:177–99.

Templeton, A. R., K. Shaw, E. Routman, and S. K. Davis. 1990. The genetic consequences of habitat fragmentation. *Annals of the Missouri Botanical Garden* 77:13–27.

Terborgh, J. W. 1971. Distribution on environmental gradients: Theory and a preliminary interpretation of distributional patterns in the avifauna of the Cordillera Vilcabamba, Peru. *Ecology* 52:23–40.

———. 1974. Preservation of natural diversity: The problem of extinction prone species. *Bioscience* 24: 715–22.

———. 1975. Faunal equilibria and the design of wildlife preserves. Pages 369–80 in F. B. Golley and E. Medina, eds., *Tropical ecological systems.* Springer Verlag, New York.

———. 1977. Bird species diversity on an Andean elevational gradient. *Ecology* 58:1007–19.

———. 1988. The big things that run the world—a sequel to E. O. Wilson. *Conservation Biology* 2: 402–3.

———. 1992a. *Diversity and the tropical rainforest.* Scientific American Library, New York.

———. 1992b. Maintenance of diversity in tropical forests. *Biotropica* 24:283–92.

Terborgh, J. W., J. W. Fitzpatrick, and L. H. Emmons. 1984. Annotated checklist of bird and mammal species of Cocha Cashu biological station, Manu National Park, Perú. *Fieldiana: Zoology,* new series 21: 1–29.

Terborgh, J. W., S. K. Robinson, T. A. Parker III, C. A. Munn, and N. Pierpont. 1990. Structure and organization of an Amazonian forest bird community. *Ecological Monographs* 60:213–38.

Terborgh, J. W., and B. Winter. 1980. Some causes of extinction. Pages 119–33 in M. E. Soulé and B. A. Wilcox, eds., *Conservation biology: An evolutionary-ecological perspective.* Sinauer Associates, Sunderland, Mass.

Thiollay, J.-M. 1989. Area requirements for the conservation of rain forest raptors and game birds in French Guiana. *Conservation Biology* 3:128–37.

———. 1992. Influence of selective logging on bird species diversity in a Guianan rain forest. *Conservation Biology* 6:47–63.

Thomas, C. D. 1994. Extinction, colonization, and metapopulations—Environmental tracking by rare species. *Conservation Biology* 8:373–78.

Thurber, J. M., R. O. Peterson, T. D. Drummer, and S. A. Thomas. 1994. Gray wolf response to refuge boundaries and roads in Alaska. *Wildlife Society Bulletin* 22:61–68.

Tilman, D., and J. A. Downing. 1994. Biodiversity and stability in grasslands. *Nature* 367:363–65.

Tracey, J. G. 1982. *The vegetation of the humid tropical region of North Queensland.* CSIRO, Melbourne, Australia.

———. 1986. Trees on the Atherton Tableland: Remnants, regrowth and opportunities for planting. Centre for Resource and Environmental Studies, Working Paper 1986/35, Australian National University, Canberra, Australia.

Truswell, E. 1993. Vegetation changes in the Australian Tertiary in response to climatic and phytogeographic forcing factors. *Australian Systematic Botany* 6:533–57.

Tuomisto, H. 1994. Ecological variation in the rain forests of Peruvian Amazonia: Integrating fern distribution patterns with satellite imagery. Reports from the Department of Biology, University of Turku, Turku, Finland.

Tuomisto, H., K. Roukolainen, R. Kalliola, A. Linna, W. Danjoy, and Z. Rodriguez. 1995. Dissecting Amazonian biodiversity. *Science* 269:63–66.

Turner, B. L. II, W. B. Meyer, and D. Skole. 1994. Global land use/land cover change: Toward an integrated program of study. *Ambio* 23:91–95.

Turner, B. L. II, R. H. Moss, and D. L. Skole. 1993. Relating land use and global land cover change: A proposal for an IGBP-HDP core project. International Geosphere-Biosphere Programme, Stockholm, Sweden.

Turner, I. M. 1994. The taxonomy and ecology of the vascular plant flora of Singapore: A statistical analysis. *Botanical Journal of the Linnean Society* 114:215–27.

Turner, I. M., K. S. Chua, J. S. Y. Ong, B. C. Soong, and H. T. W. Tan. 1996. A century of plant species loss from an isolated fragment of lowland tropical rain forest. *Conservation Biology* 10:1229–44.

Turner, I. M., and R. T. Corlett. 1996. The conservation value of small, isolated fragments of lowland tropical rain forest. *Trends in Ecology and Evolution* 11:330–33.

Turner, I. M., and H. T. W. Tan. 1992. Ecological impact of alien plant species in Singapore. *Pacific Science* 46:389–90.

Turner, I. M., H. T. W. Tan, Y. C. Wee, A. bin Ibrahim, P. T. Chew, and R. T. Corlett. 1994. A study of plant species extinction in Singapore: Lessons for the conservation of tropical biodiversity. *Conservation Biology* 8:705–12.

Turner, I. M., Y. K. Wong, P. T. Chew, and A. Ibrahim. 1997. Tree species richness in primary and secondary tropical forest in Singapore. *Biodiversity and Conservation.* In press.

Turner, M. G. 1989. Landscape ecology: The effect of pattern on process. *Annual Review of Ecology and Systematics* 20:171–97.

Turner, M. G., and R. H. Gardner. 1991. *Quantitative methods in landscape ecology: The analysis and interpretation of landscape heterogeneity.* Springer-Verlag, New York.

Turner, M. G., R. H. Gardner, and R. V. O'Neill. 1995. Ecological dynamics at broad scales. *Bioscience* (Supplement): 29–35.

Turner, M. G., and C. L. Ruscher. 1988. Changes in spatial patterns of land use in Georgia. *Landscape Ecology* 1: 241–51.

Turton, S. M. 1991. Aspects of the micrometeorology of rainforests in the Wet Tropics of northeastern Queensland. Pages 51–58 in N. Goudberg, M. Bonell, and D. Benzaken, eds., *Tropical rainforest research in Australia.* Institute for Tropical Rainforest Studies, Townsville, Australia.

———. 1992. Understorey light environments in a north-east Australian rain forest before and after a tropical cyclone. *Journal of Tropical Ecology* 8: 241–52.

Ueno, H. 1994. Fluctuating asymmetry in relation to two fitness components, adult longevity and male mating success, in a ladybird beetle, *Harmonia axyridis* (Coleoptera: Coccinellidae). *Economic Entomology* 19: 87–88.

Uhl, C. 1987. Factors controlling succession following slash-and-burn agriculture in Amazonia. *Journal of Ecology* 75: 377–407.

———. 1988. Restoration of degraded lands in the Amazon Basin. Pages 326–32 in E. O. Wilson, ed., *Biodiversity.* National Academy Press, Washington, D.C.

Uhl, C., and R. Buschbacher. 1985. A disturbing synergism between cattle-ranch burning practices and selective tree harvesting in Eastern Amazon. *Biotropica* 17: 265–68.

Uhl, C., R. Buschbacher, and E. A. S. Serrão. 1988. Abandoned pastures in eastern Amazonia. I. Patterns of plant succession. *Journal of Ecology* 76: 663–81.

Uhl, C., and C. Jordan. 1984. Succession and nutrient dynamics following forest cutting and burning in Amazonia. *Ecology* 65: 1476–90.

Uhl, C., D. C. Nepstad, I. Viera, and J. M. C. Silva. 1991. Restauração da floresta em pastagens degradadas. *Ciência Hoje* 13: 22–31.

Underwood, A. J. 1990. Experiments in ecology and management: Their logics, functions and interpretations. *Australian Journal of Ecology* 15: 365–90.

United States Council on Environmental Quality. 1974. The fifth annual report of the Council on Environmental Quality. United States Government Printing Office, Washington, D.C.

Unwin, G. L., K. G. Applegate, G. C. Stocker, and D. I. Nicholson. 1986. Initial effects of cyclone "Winifred" on forests in north Queensland. *Proceedings of the Ecological Society of Australia* 15: 283–96.

Uribe, D. A. 1986. Contribución al conocimiento de la avifauna del bosque muy húmedo montano bajo en las cercanías de Manizales. Tesis, Facultad de Medicina Veterinaria y Zootectnia, Universidad de Caldas, Manizales, Colombia.

Usher, M. B., ed. 1986. *Wildlife conservation evaluation.* Chapman and Hall, London.

Vanapeldoorn, R. C., W. T. Oostenbrink, A. Vanwinden, and F. F. Vanderzee. 1992. Effects of habitat fragmentation on the bank vole, *Clethrionomys glareolus,* in an agricultural landscape. *Oikos* 65: 265–74.

Van Valen, L. 1965. Morphological variation and width of ecological niche. *American Naturalist* 99: 377–90.

Vanzolini, P. E., and E. E. Williams. 1970. South American anoles: The geographic differentiation and evolution of the *Anolis chrysolepis* species group (Sauria: Iguanidae). *Arquivos de Zoologia (São Paulo)* 19: 1–298.

Vasconcelos, H. L. 1988. Distribution of *Atta* (Hymenoptera-Formicidae) in "terra-firme" rain forest of central Amazonia: Density, species composition and preliminary results on effects of forest fragmentation. *Acta Amazônica* 18: 309–15.

Velásquez, A., H. Meyer, W. Marin, F. Ramírez, A. David, A. Campos, M. Hermelín, S. O. Bender, M. Arango, and J. Serje. 1994. Planificación regional del occidente colombiano bajo consideración de las restricciones por amenazas. Memorias Conferencia Interamericana sobre Reducción de los Desastres Naturales, Cartagena de Indias, Colombia, Marzo 1994. Casa Impresora Pacífico, Santa Fé de Bogotá, Colombia.

Velásquez, M. P. 1992. Aves frugívoras y su relación con la flora en un bosque húmedo, en el municipio de

San Carlos, Antioquia, Colombia. Trabajo de Investigación, Departamento de Biología, Universidad de Antioquia, Medellín, Colombia.

Velez, B. E. 1987. Contribución al estudio avifaunístico del Santuario de Fauna y Flora de Iguaque, Boyacá. Tesis, Departamento de Biología, Pontificia Universidad Javeriana, Santa Fé de Bogotá, Colombia.

Velleman, P. F., and D. C. Hoaglin. 1981. *Applications, basics and computing of exploratory data analysis.* Duxbury Press, Boston, Mass.

Veríssimo, A., P. Barreto, M. Mattos, R. Tarifa, and C. Uhl. 1992. Logging impacts and prospects for sustainable forest management in an old Amazonian frontier: The case of Paragominas. *Forest Ecology and Management* 55:169–99.

Viana, V. M. 1990. Biologia e manejo de fragmentos de florestas naturais. VI Congresso Florestal Brasileiro. SBS/SBEF, Campos do Jordão, São Paulo, Brazil.

———. 1995. Conservation of biological diversity in Neotropical forest fragments in intensively cultivated landscapes. Pages 135–54 in G. S. Fonseca M., M. Schmink, C. P. S. Pinto, and F. Brito, eds., *On common ground: Interdisciplinary approaches to biodiversity conservation and land use dynamics in the New World.* Conservation International, Belo Horizonte, Brazil.

Viana, V. M., and A. A. J. Tabanez. 1996. Biology and conservation of forest fragments in the Brazilian Atlantic Moist Forest. Pages 151–67 in J. Schelhas and R. Greenberg, eds., *Forest patches in tropical landscapes.* Island Press, Washington, D.C.

Viana, V. M., A. A. J. Tabanez, and J. Aguirre. 1992. Restauração e manejo de fragmentos de florestas naturais. Pages 400–406 in II Congresso Nacional sobre Essências Nativas. Instituto Florestal, São Paulo, Brazil.

Victória, R. L., L. A. Martinelli, J. Mortatti, and J. Richey. 1991. Mechanisms of water recycling in the Amazon basin: Isotopic insights. *Ambio* 20:384–87.

Villard, M. A., G. Merriam, and B. A. Maurer. 1995. Dynamics in subdivided populations of Neotropical migratory birds in a fragmented temperate forest. *Ecology* 76:27–40.

Vitousek, P. M. 1988. Diversity and biological invasions of oceanic islands. Pages 181–89 in E. O. Wilson, ed., *Biodiversity.* National Academy Press, Washington, D.C.

———. 1990. Biological invasions and ecosystem processes: Towards an integration of population biology and ecosystem studies. *Oikos* 57:7–13.

Vitousek, P. M., and J. S. Denslow. 1986. Nitrogen and phosphorus availability in treefall gaps of a lowland tropical rainforest. *Journal of Ecology* 74:1167–78.

Vogler, A. P., and R. DeSalle. 1994. Diagnosing units of conservation management. *Conservation Biology* 8: 354–63.

von Moltke, K. 1990. International economic issues in tropical deforestation. Published abstract, Workshop on Climate Change and Tropical Forests, São Paulo, Brazil.

von Moltke, K., and P. J. DeLong. 1990. Negotiating in the global arena: Debt-for-nature swaps. *Resolve* 22:1–3.

Voss, R. S. 1993. A revision of the Brazilian muroid rodent genus *Delomys* with remarks on "thomasomyine" characters. *American Museum Novitates,* no. 3,073.

Wade, M. J., and D. E. McCauley. 1988. Extinction and recolonization: Their effects on the genetic differentiation of local populations. *Evolution* 42:995–1005.

Wakefield, S. 1989. *Designing windbreaks for farms.* New South Wales Agriculture and Fisheries Department, Sydney, Australia.

Waldorff, P., and V. M. Viana. 1993. Efeito de borda na Reserva Florestal de Linhares. VI Congresso Florestal Brasileiro. SBEF/SBS, Curitiba, Brazil.

Walker, B. H. 1992. Biodiversity and ecological redundancy. *Conservation Biology* 6:18–23.

Walker, D. 1990. Directions and rates of tropical rainforest processes. Pages 23–32 in L. J. Webb and J. Kikkawa, eds., *Australian tropical rainforests—science, values, meaning.* CSIRO, Melbourne, Australia.

Wallace, A. R. 1849. On the monkeys of the Amazon. *Proceedings of the Zoological Society of London* 20: 107–10.

Wallace, B. 1968. Polymorphism, population size and genetic load. Pages 87–108 in R. C. Lewontin, ed., *Population biology and evolution.* Syracuse University Press, Syracuse, N.Y.

Walter, H. S. 1990. Small viable population: The red-tailed hawk of Socorro Island. *Conservation Biology* 4:441–43.

Wandelli, E.V. 1990. Respostas ecofisiológicas da palmeira de sub-bosque *Astrocaryum sociale* Barb. Rodr. as mudanças ambientais resultantes do efeito de borda de floresta. M.Sc. thesis, INPA, Manaus, Brazil.

Wanghongsa, S. 1989. Behavioral comparison of Dusky Leaf Monkey *(Presbytis obscura)* on island habitat and undisturbed habitat of Khlong Saeng Wildlife Sanctuary, Surat Thani Province. *Natural History Bulletin of the Siam Society* 37:1–24.

Warburton, N. 1987. The application of the Theory of Island Biogeography to the avifauna of remnant patches of rainforest on the Atherton Tablelands, with a view to their conservation. M.Sc. thesis, James Cook University, Townsville, Queensland, Australia.

Ward, A. L. 1973. Elk behavior in relation to multiple uses of Medicine Bow National Forest. *West Association State Game Fish Communications* 53:125–41.

Wayne, R. K., N. Lehman, M. W. Allard, and R. L. Honeycutt. 1992. Mitochondrial DNA variability of the grey wolf: Genetic consequences of population decline and habitat fragmentation. *Conservation Biology* 6:559–69.

Wayne, R. K., W. S. Modi, and S. J. O'Brien. 1986. Morphological variability and asymmetry in the cheetah *(Acinonyx jubatus),* a genetically uniform species. *Evolution* 40:78–85.

Webb, C. J., and K. S. Bawa. 1983. Pollen dispersal by hummingbirds and butterflies: A comparative study of two lowland tropical plants. *Evolution* 37:1258–70.

Webb, L. J., and J. G. Tracey. 1981. Australian rainforests: Patterns and change. Pages 605–94 in A. Keast, ed., *Ecological biogeography of Australia.* Dr. W. Junk, The Hague, Netherlands.

Webb, L. J., J. G. Tracey, and W. T. Williams. 1972. Regeneration and pattern in the subtropical rainforest. *Journal of Ecology* 60:675–95.

Webb, L. J., J. G. Tracey, and W. T. Williams. 1976. The value of structural features in tropical forest typology. *Australian Journal of Ecology* 1:3–28.

Webb, N. R. 1989. Studies on the invertebrate fauna of fragmented heathland in Dorset, U.K., and the implications for conservation. *Biological Conservation* 47:153–65.

Webb, N. R., R. T. Clarke, and J. T. Nicholas. 1984. Invertebrate diversity on fragmented Calluna-heathland: Effects of surrounding vegetation. *Journal of Biogeography* 11:41–46.

Westman, W. E., L. L. Strong, and B. A. Wilcox. 1989. Tropical deforestation and species endangerment: Application of remote sensing. *Landscape Ecology* 3:97–109.

Wet Tropics Management Authority. 1995. Annual Report 1994–1995. Wet Tropics Management Authority, Cairns, Australia.

Wetmore, A. 1972. *The birds of the Republic of Panama.* Part 3. Smithsonian Miscellaneous Collections, vol. 150.

———. 1984. *The birds of the Republic of Panama.* Part 4. Smithsonian Miscellaneous Collections, vol. 150.

Wetton, J. H., R. E. Carter, D. T. Parkin, and D. Walters. 1987. Demographic study of a wild population by DNA fingerprinting. *Nature* 327:147–52.

Wheelwright, N. T. 1988. Fruit-eating birds and bird-dispersed plants in the tropics and temperate zone. *Trends in Ecology and Evolution* 3:270–74.

White, F. 1983. *The vegetation of Africa: A descriptive memoir to accompany the UNESCO/AETFAT/UNSO vegetation map of Africa.* Natural Resources Research, UNESCO, Paris, France.

Whitlock, M. C., and D. E. McCauley. 1990. Some population genetic consequences of colony formation and extinction: Genetic correlations within founding groups. *Evolution* 44:1717–24.

Whitmore, T. C., ed. 1972. *Tree flora of Malaya* 1. Longman, Kuala Lumpur, Malaysia.

———. 1975. *Tropical rain forests of the Far East.* Clarendon Press, Oxford.

———. 1990. *An introduction to tropical rain forests.* Clarendon Press, Oxford.

Whitmore, T. C., and G. T. Prance, eds. 1987. *Biogeography and Quaternary history in tropical America.* Clarendon, Oxford.

Whitmore, T. C., and J. A. Sayer. 1992. Deforestation and species extinction in tropical moist forests. Pages 1–14 in T. C. Whitmore and J. A. Sayer, eds., *Tropical deforestation and species extinction.* Chapman and Hall, London.

Whittaker, R. J., and S. H. Jones. 1994. The role of frugivorous bats and birds in the rebuilding of a tropical forest ecosystem, Krakatau, Indonesia. *Journal of Biogeography* 21:245–58.

Wickland, D. E. 1991. Mission to planet earth: The ecological perspective. *Ecology* 72:1923–33.

Wiens, J. A. 1989. Spatial scaling in ecology. *Functional Ecology* 3:385–97.

Wiens, J. A., C. S. Crawford, and J. R. Gosz. 1985. Boundary dynamics: A conceptual framework for studying landscape ecosystems. *Oikos* 45:421–27.

Wilcove, D. S. 1985. Nest predation in forest tracts and the decline of migratory songbirds. *Ecology* 66: 1211–14.

Wilcove, D. S., C. H. McClennan, and A. P. Dobson. 1986. Habitat fragmentation in the temperate zone. Pages 237–56 in M. E. Soulé, ed., *Conservation Biology: The science of scarcity and diversity.* Sinauer Associates, Sunderland, Mass.

Wilcox, B.A. 1978. Supersaturated island faunas: A species-age relationship for lizards on post-Pleistocene land-bridge islands. *Science* 199:996–98.

———. 1980. Insular ecology and conservation. Pages 95–117 in M. E. Soulé and B. A. Wilcox, eds., *Conservation biology: An evolutionary-ecological perspective.* Sinauer Associates, Sunderland, Mass.

Wilcox, B. A., and D. D. Murphy. 1985. Conservation strategy: Effects of fragmentation on extinction. *American Naturalist* 125:879–87.

Wilkins, K. T. 1982. Highways as barriers to rodent dispersal. *Southwest Naturalist* 27:459–60.

Williams, M. 1992. *Americans and their forests.* Cambridge University Press, Cambridge.

Williams, P. H. 1994. WORLDMAP: Priority areas for biodiversity—Using version 3. Privately distributed computer software and manual, London.

Williams, P. H., K. J. Gaston, and C. J. Humphries. 1994. Do conservationists and molecular biologists value differences between organisms in the same way? *Biodiversity Letters* 2:67–78.

Williams, P. H., and C. J. Humphries. 1994. Biodiversity, taxonomic relatedness, and endemism in conservation. Pages 269–87 in P. L. Forey, C. J. Humphries, and R. I. Vane-Wright, eds., *Systematics and conservation evaluation.* Clarendon Press, Oxford.

Williams, S. E. 1990. The interaction between vegetation and the small mammal community of the rainforest ecotone in north Queensland. B.Sc. (Hons.) thesis, James Cook University, Townsville, Australia.

Williams-Linera, G. 1990a. Origin and early development of forest edge vegetation in Panama. *Biotropica* 22:235–41.

———. 1990b. Vegetation structure and environmental conditions of forest edges in Panama. *Journal of Ecology* 78:356–73.

———. 1993. Vegetación de bordes de un bosque nublado en el Parque Ecológico Clavijero, Xalapa, Veracruz, México. *Revista de Biologia Tropical* 41:443–53.

Williamson, M. 1981. *Island populations.* Oxford University Press, Oxford.

Willis, E. O. 1967. The behavior of bicolored antbirds. *University of California Publications in Zoology* 79: 1–132.

———. 1972. *The behavior of spotted antbirds.* Ornithological Monographs, no. 10. American Ornithologists Union, Washington, D.C.

———. 1973. The behavior of ocellated antbirds. *Smithsonian Contributions to Zoology* 144:1–57.

———. 1974. Populations and local extinctions of birds on Barro Colorado Island, Panama. *Ecological Monographs* 44:153–69.

———. 1979. The composition of avian communities in remanescent woodlots in southern Brazil. *Papeis Avulsos Zoologia* 33:1–25.

Willis, E. O., and E. Eisenmann. 1979. A revised list of the birds of Barro Colorado Island, Panama. *Smithsonian Contributions to Zoology,* no. 291.

Willis, E. O., and Y. Oniki. 1972. Ecology and nesting behavior of the chestnut-backed antbird *(Myrmeciza exsul). Condor* 74:87–98.

———. 1978. Birds and army ants. *Annual Review of Ecology and Systematics* 9:243–63.

Willmott, W. F., and P. J. Stevenson. 1989. *Rocks and landscapes of the Cairns district.* Queensland Department of Mines, Brisbane, Australia.

Willson, M. F. 1993. Dispersal mode, seed shadows, and colonization patterns. Pages 261–80 in T. H. Fleming and A. Estrada, eds., *Frugivory and seed dispersal: Ecological and evolutionary aspects.* Kluwer Academic Publishers, Dordrecht, The Netherlands.

Willson, M. F., and F. H. J. Crome. 1989. Patterns of seed rain at the edge of a tropical Queensland rain forest. *Journal of Tropical Ecology* 5:301–8.

Willson, M. F., T. L. de Santo, C. Sabag, and J. Y. Armesto. 1994. Avian communities of fragmented south-temperate rainforests in Chile. *Conservation Biology* 8:508–20.

Wilson, D. E., and D. M. Reeder, eds. 1993. *Mammal species of the world, a taxonomic and geographic reference.* 2d ed. Smithsonian Institution Press, Washington, D.C.

Wilson, E. O. 1988. The current state of biological diversity. Pages 3–18 in E. O. Wilson, ed., *Biodiversity.* National Academy of Sciences Press, Washington, D.C.

Wilson, E. O., and E. O. Willis. 1975. Applied biogeography. Pages 522–36 in M. L. Cody and J. M. Diamond, eds., *Ecology and evolution of communities.* Belknap Press, Cambridge, Mass.

Winter, J. W. 1988. Ecological specialisation of mammals in Australian tropical and sub-tropical rainforest: Refugial or ecological determinism? *Proceedings of the Ecological Society of Australia* 15:27–138.

Winter, J. W., F. C. Bell, L. I. Pahl, and R. G. Atherton. 1987. Rainforest clearfelling in northeastern Australia. *Proceedings of the Royal Society of Queensland* 98:41–57.

Witting, L., M. A. McCarthy, and V. Loeschcke. 1994. Multi-species risk analysis, species evaluation and biodiversity conservation. Pages 239–49 in V. Loeschcke, J. Tomiuk, and S. K. Jain, eds., *Conservation genetics.* Birkhauser Verlag, Basel, Switzerland.

Woodman, N., N. A. Slade, R. M. Timm, and C. A. Schmidt. 1995. Mammalian community structure in lowland, tropical Peru, as determined by removal trapping. *Zoological Journal of the Linnean Society* 113: 1–20.

Woodman, N., R. M. Timm, R. Arana-C., V. Pacheco, C. A. Schmidt, E. D. Hooper, and C. Pacheco-A. 1991. Annotated checklist of the mammals of Cuzco Amazónico, Peru. *Occasional Papers of the Museum of Natural History, University of Kansas* 145:1–12.

Woodruff, D. S. 1990. Genetics and demography in the conservation of biodiversity. *Journal of the Science Society of Thailand* 16:117–32.

World Bank. 1994. *Dissemination Notes* 10:1–2.

Wright, S. 1931. Evolution in Mendelian populations. *Genetics* 16:97–159.

———. 1943. Isolation by distance. *Genetics* 28:114–38.

———. 1946. Isolation by distance under diverse systems of mating. *Genetics* 31:39–59.

———. 1965. The distribution of self-incompatibility alleles in populations. *Evolution* 18:609–19.

Wright, S. J. 1985. How isolation affects rates of turnover of species on islands. *Oikos* 44:331–40.

Wright, S. J., M. E. Gompper, and B. DeLeon. 1994. Are large predators keystone species in Neotropical forests? The evidence from Barro Colorado Island. *Oikos* 71:279–94.

WTMA. 1995. *Management plan for the Wet Tropics World Heritage Area.* Wet Tropics Management Authority, Cairns, Queensland, Australia.

Yablokov, A. V. 1986. *Phenetics.* Columbia University Press, New York.

Yahner, R. H. 1988. Changes in wildlife communities near edges. *Conservation Biology* 2:333–39.

———. 1991. Dynamics of a small mammal community in a fragmented forest. *American Midland Naturalist* 127:381–91.

Young, A., and N. Mitchell. 1994. Microclimate and vegetation edge effects in a fragmented podocarp-broadleaf forest in New Zealand. *Biological Conservation* 67:63–72.

Young, A. G., and H. G. Merriam. 1994. Effects of forest fragmentation on the spatial genetic structure of *Acer saccharum* Marsh (sugar maple) populations. *Heredity* 72:201–8.

Young, A. M. 1982. *Population biology of tropical insects.* Plenum Press, New York.

Young, T. P., and S. P. Hubbell. 1991. Crown asymmetry, treefalls, and repeat disturbance in a broad-leaved forest. *Ecology* 72:1464–71.

Young, T. P., and V. Perkocha. 1994. Treefalls, crown asymmetry, and buttresses. *Journal of Ecology* 82: 319–24.

Zande, A. N. van der, W. J. ter Keurs, and W. J. van der Weijden. 1980. The impact of roads on the densities of four bird species in an open field habitat—evidence of a long distance effect. *Biological Conservation* 18:299–321.

Zimmerman, B. L. 1991. Distribution and abundances of forest frogs at a site in the central Amazon. Ph.D. dissertation, University of Florida, Tallahassee.

———. 1994. Audio strip transects. Pages 92–96 in W. R. Heyer, M. A. Donnelly, R. W. McDiarmid, L. C. Hayek, and M. S. Foster, eds., *Measuring and monitoring biological diversity: Standard methods for amphibians.* Smithsonian Institution Press, Washington, D.C.

Zimmerman, B. L., and R. O. Bierregaard, Jr. 1986. Relevance of the equilibrium theory of island biogeography and species-area relations to conservation, with a case from Amazonia. *Journal of Biogeography* 13:133–43.

Zimmerman, B. L., and M. T. Rodrigues. 1990. Frogs, snakes, and lizards of the INPA-WWF Reserves near Manaus in the central Amazon. Pages 426–54 in A. Gentry, ed., *Four Neotropical rain forests.* Yale University Press, New Haven, Conn.

Zimmerman, B. L., and D. Simberloff. 1996. An historical interpretation of habitat use by frogs in a central Amazonian forest. *Journal of Biogeography* 23:27–46.

Zipperer, W. C. 1993. Deforestation patterns and their effects on forest patches. *Landscape Ecology* 8:177–84.

Name Index

Abele, L. G., 103, 191, 203, 321
Aberg, J., 523
Acevedo, C. I., 174
Adams, J., 217
Adams, M., 448
Adams, P., 443
Addicott, J. F., 402
Adler, G. H., 56, 233, 238, 506
Aguiar, L. F. de, 37, 39
Aguirre, J., 352, 354, 357, 364
Ahlquist, J. E., 469, 478
Aho, J. M., 402
Aizen, M. A., 65, 316, 318, 321, 507, 509
Alberts, A. C., 56, 204, 271
Aldrich, P. R., 246, 289, 304–20, 343, 363, 503, 507, 509, 510, 518
Allendorf, F. W., 112, 122, 518, 519
Alvarez, E., 256, 258
Alvarez, H., 174
Alvarez-Buylla, E. R., 307
Alvarez-López, H., 171, 174
Amorim, D. S., 479
Andersen, M., 8, 207, 218, 281–91, 342, 345, 364, 368, 502–25
Anderson, D., 114
Anderson, S. H., 244
Andrade, G. I., 171, 174
Andrén, H., 55, 497
Andrews, A., 241, 243
Antolin, M. R., 402
Appanah, S., 81
Arango, S., 174, 175
Arbhabhirama, A., 223, 511
Arce, J., 318
Arnold, G. W., 45, 497
Arroyo, M. T. K., 308
Ashton, P. S., 308
Ashworth, J. H., 338, 339
Asrar, G., 388
Atchley, W. R., 114
Atkinson, I. A. E., 504
Atmar, W., 190, 191, 194
Augspurger, C. K., 313
August, P. V., 185
Auslig., 509
Avise, J. C., 443, 448, 452, 519, 520
Ayres, J. M., 459

Bachelery, P., 322, 330
Bailey, J. A., 245
Baimai, V., 223, 225, 509, 511
Baker, H. G., 307, 397
Baker, R. R., 103
Baker, S. L., 401
Bakowski, C., 247
Báldi, A., 56
Balen, S. V., 191
Balick, J. M., 510
Ball, R. M., 448
Ballantine, W. J., 370
Ballou, J. D., 124
Balmford, A., 442, 481, 482, 505, 521
Balslev, H., 290
Banguero, H., 172
Barass, A. N., 246
Barau, A. C., 332
Barnett, J. L., 254
Barre, N., 332
Barrowclough, G. F., 304
Bass, S., 384
Bates, H. W., 91
Batista, J. L. F., 8, 78, 81, 82, 351–65, 384, 464, 481, 503–5, 507, 512, 521, 523, 524
Baur, B. N., 352
Baverstock, P. R., 449, 451
Bawa, K. S., 304, 308, 309, 313
Bayless, J. W., 55, 56, 488
Beach, J. H., 308
Beattie, A. J., 306
Beccaloni, G. W., 107
Becker, P., 56, 65, 111
Bedford, B. L., 491

Bell, D. T., 231, 236
Bell, J. C., 306
Bellingham, P. J., 79
Belshaw, R., 59
Benedix, J. H., 56
Bennet, E. L., 9
Bennett, A. F., 241, 487
Benntruperbäumer, J., 251, 497
Benson, W. W., 91, 93
Berger, J. O., 495, 496
Berlin, E. A., 457
Berlin, O. B., 457
Bernardes, A. T., 354
Bernardino, F. S., Jr., 251
Berry, D. A., 496
Berry, J. A., 39
Besaire, H., 422
Best, B. J., 478, 480
Besuchet, C., 59
Bierregaard, R. O., Jr., xi–xv, 8, 34, 42, 57, 58, 82, 92, 108, 124–26, 129, 136, 138–56, 159, 168, 169, 171, 190, 191, 204, 208, 218, 222, 231, 233, 238, 249, 268, 304, 317, 321, 333, 407, 408, 442, 485, 498, 502–25
Bilsborrow, R., 20, 23
Binder, A., 494
Birdsey, R. A., 370
Bishop, K. D., 191
Bjork, R., 510
Black, G. A., 309
Blackburn, T. M., 172
Blake, J. G., 202, 510
Blakers, M., 193
Blockhuss, J. L., 351
Bodmer, R. E., 515
Boecklen, W. J., 205
Bolger, D. T., 56, 204, 233, 271
Bolton, B., 59
Bookstein, F. L., 114
Boom, B. M., 351
Boose, E. R., 79
Borchert, R., 398
Borges, S. H., 143, 146
Boshier, D. H., 309, 313
Boulding, K. E., 489
Bradley, W. G., 246
Bradshaw, G. A., 397
Bragg, A., 371
Braumann, C. A., 114
Brawn, J. D., 160, 161, 166
Brett, D., 71, 83
Bridgewater, P. B., 397
Briscoe, D. A., 518
Broadbent, J. A., 241, 243
Brockelman, W. Y., 223, 225, 509, 511
Brody, A. J., 245, 246
Brokaw, N., 186
Brooks, D. R., 442, 466, 520
Browder, J. O., 17
Brown, G. G., 8, 91, 109, 354, 505, 520
Brown, J. H., 172, 204, 227, 233, 256, 263, 272, 510
Brown, K., 20
Brown, K. S., Jr., 8, 56, 65, 82, 91–111, 124, 135, 140, 317, 333, 354, 463, 503–6, 520, 521
Brown, S., 368
Brown, V. K., 56, 111
Brown, V. R., 367
Brown, W. L., 175
Bruce, J. W., 384
Bruford, M. W., 456
Bruijnzeel, L. A., 480
Bruni, A. R., 256–74
Bryant, E. H., 112
Bryant, R. L., 498
Buechner, M., 55
Bull, G. A. D., 71
Bunnell, F. H., 384
Burckhardt, D. H., 59
Burgess, J. C., 20, 24
Burgess, R. L., 55, 241, 244
Burghouts, T., 236
Burnett, S. E., 248, 249, 252, 253, 511
Burney, D., 418
Burrough, P. A., 392
Buschbacher, R., 138, 352, 371, 512
Buse, A., 56
Bush, M. B., 451, 463
Butterfield, B. R., 467

Cabral, O. M. R., 37, 39
Cabrera, A., 458
Cadenasso, M. L., 322, 517
Cadet, T., 322–24, 331
Cai, Y., 401
Calaby, J., 445
Caldecott, J. O., 9
Cale, W. G., 304
Callaghan, C. J., 91
Camargo, J. L. C., 33–44, 46, 508
Camargo, M. N., 141
Campbell, I. C., 243
Cannell, R. J., 81, 505
Cantillo, G., 174
Cantley, N., 336
Capistrano, A. D., 20, 23
Capparella, A. P., 318
Carleton, M. D., 458, 460
Carlquist, S., 322, 518
Carmody, G. R., 246–48, 253
Caro, T. M., 519
Cary, J. R., 68
Case, T. J., 217, 256, 271
Cavelier, J., 172, 175
Chakraborty, R., 314

Chan, H. T., 308
Chapman, F. M., 159, 175
Chapman, P., 425, 429
Chapman, W. S., 82
Charles-Dominique, P., 217, 219, 318, 457
Chase, M. R., 309, 313
Chasen, F. N., 223
Chauvel, A., 57, 208
Cheke, A. S., 322, 331, 332
Chen, J., 45, 71
Chesser, R. T., 455
Chey, C. V., 10, 56, 171, 236
Chou, L. M., 344
Christidis, L., 451
Churchill, S. P., 172
CIMA, 351, 353
Clark, D. A., 292, 302, 306
Clark, D. B., 292, 302
Clark, W. C., 171
Clarke, G. M., 111, 112
Clarke, R. T., 56
Cloudsley-Thompson, J. L., 112
Clutton-Brock, T. H., 459
Cody, M. L., 256, 268
Cohn-Haft, M., 143
Cole, F. R., 244, 455
Collar, N. J., 468
Collins, N. M., 56, 223
Combs, L. M., 112
Connor, E. F., 138
Constable, A. J., 489
Corlett, R. T., 8, 9, 289, 302, 333–45, 504, 507, 510, 512, 521, 522, 524
Corner, E. J. H., 307
Corredor, G. L., 174
Couper, P. J., 445
Cours Darne, G., 422–26
Covacevich, J. A., 443, 445
Cox, J., 485, 487, 524
Cox, P. A., 331, 332
Coy, M., 403
Cracraft, J., 445, 460, 461
Cranwell, I., 241, 243
Crawford, C. S., 186
Crawford, T. J., 305
Crawshaw, P. G., Jr., 503
Crist, E. P., 389
Croat, T. B., 158
Crockett, C. M., 269
Crome, F. H. J., 10, 81, 82, 199, 205, 292–303, 331, 368, 369, 372, 448, 485–501, 507, 509, 510, 516, 517, 522–24
Cronk, Q. C. B., 469
Crow, J. F., 317
Crowell, K. L., 225, 268
Csuti, B., 467
Cuadros, T. C., 174
Cuatrecasas, J., 172, 186
Cunningham, M., 442–54

Daily, G. C., 107
Dale, J. E., 39
Dale, V. H., 16, 136, 289, 351, 400–409, 509, 513, 522
Dalrymple, G. H., 251
Danielson, B. J., 487
Danjoy, W., 479
Date, E. M., 497
Daume, E., 415
Davidson, D. W., 263
Davies, S., 193
Davis, R., 256
Davis, T. A. W., 309
Dawson, D. G., 193
Dayanandan, B., 308
DeJong, M. J., 259
DeLeon, B., 161
DeLong, P. J., 21
Denslow, J. S., 41, 42, 172, 186, 304, 306, 311
DEPRN, 354
De Reibera, C. P., 497
DeSalle, R., 452
Devlin, B., 306
Dial, R, 283
Diamond, J. M., 156, 159, 167, 190, 191, 194, 217, 225, 256, 263, 268, 271, 503, 504, 506, 507
Dickenson, G., 371
Dickman, C. R., 219, 233
Didham, R. K., 42, 46, 55–70, 82, 91, 103, 122, 124, 140, 171, 218, 352, 502–25
Diffendorfer, J. E., 321, 322
Dixon, K. R., 488
Doak, D. F., 321
Dobkin, D. S., 244
Dobson, A. P., 225–26, 267
Dobzhansky, T., 309
Dodd, C. K., Jr., 124
Dodson, P., 114
Doeg, T. J., 243
Dorfsman, M., 449
Downing, J. A., 171
Downs, D. R., 370
Drake, D. R., 329
Dressler, R. L., 290
Ducenne, H., 422
Dudgeon, D., 333, 337, 338, 341, 342, 344
Duelli, P., 56
Duellman, W. E., 135, 137, 172
Duff, A. B., 9
Dunning, J. B., Jr., 175
Dunphy, M., 81

Eacham Historical Society, 47, 191
Eanes, W. F., 112
Ebenhard, T., 321, 322

Efron, B., 492, 496
Eguiarte, L. E., 307, 308
Ehrlich, A., xii
Ehrlich, P. R., xii, 103, 107
Eisenberg, J. F., 208, 269, 464
Eisenmann, E., 156, 158, 159, 161, 164–66
Elachi, C., 388
Elliott, A., 175
Ellstrand, N. C., 306
Elton, C. S., 111, 504
Emlen, J. T., 259
Emmons, L. H., 141, 207, 226, 262, 269, 456, 457, 459, 460, 463
Endler, J. A., 442
ERDAS, 389
Erwin, T. L., 453, 466, 520
Escobar, F., 171
Espinal, L. S., 172, 186
Esseen, P. A., 71, 186
Estrada, A., 171, 318, 333
Etter, A., 172, 175
Ewel, J. J., 386, 397

Faaborg, J., 156
Faeth, S. H., 268
Fahrig, L., 138, 249, 251, 254, 321, 510
Fairfax, R. J., 81, 505
FAO/UNESCO, xi, 1, 3–7, 11, 12, 23, 114, 283, 287
Faramalala, M. H., 422–26, 439
Farquhar, G. D., 39
Farris, J. S., 115
Fearnside, P. M., 404, 414, 509, 510, 523
Federov, A. A., 307
Feer, F., 463
Feinsinger, P., 65, 153, 316, 318, 321, 507, 509
Fensham, R. J., 81, 505
Fenton, M. B., 246–48, 253
Ferraz, J., 414, 509, 523
Ferrier, S., 422
Ferris, C. R., 243
Fisher, R. F., 384
Fitzpatrick, J. W., 175, 457
Fjeldså, J., 143, 153, 175, 443, 451–53, 456, 463, 466–82, 488, 507, 509, 511, 519, 520, 521
Fleming, T. H., 307, 309, 313, 318
Flenley, J. R., 442
Fluet, M., 79
Fonseca, G. A. B. da, 217, 219, 270, 352, 457
Ford, H. A., 497
Ford, J., 193
Fore, S. A., 318
Forget, P.-M., 301
Forman, R. T. T., 45, 55, 116, 124, 205, 392
Fortmann, L., 384
Foster, D. R., 79
Foster, R. B., 306, 318, 455, 497, 503, 505
Fowler, K., 112
Fowler, S. V., 425
Frankel, O. H., 518
Frankie, G. W., 309, 397
Franklin, I. F., 111
Franklin, J. F., 45, 71, 124, 384, 516
Frawley, K. J., 72, 114
Freiberger, H., 40, 45–55, 68, 122, 248, 341, 506
Fretwell, S. D., 490
Frith, C. B., 121, 122
Frith, D. W., 121, 122
Fritschen, L. J., 46
Frohn, R. C., 402, 513
Furley, P. A., 402, 403

Gabriel, W., 169
Gade, D. W., 415
Gaines, M. S., 321, 322
Galli, A., 205
Ganade, G. M. S., 33–44
Ganzhorn, J. U., 416, 418, 425, 429, 431
Garay, A. A., 307
Gardner, A. L., 458, 460
Gardner, R. H., 171, 188, 392, 401
Garesche, J., 82, 204
Garland, T., 246
Garwood, N., 186
Gascon, C., 108, 124–37, 140, 154, 191, 503–5, 513, 518
Gaston, K. J., 107, 172, 466
Gates, D. L., 244
Gates, J. E., 124, 136
Gauch, H. G., Jr., 59, 147
Gaulin, S. J. C., 174
Gazey, W. J., 493
Genard, M., 185
Gentry, A. H., 141, 158, 172, 174, 185, 414, 510, 511
Geores, M., 20, 23
George, T. L., 269
Getz, L. L., 244
Ghilarov, A. M., 489
Gibbs, J. P., 506
Gilbert, F. S., 205
Gilbert, L. E., 91, 107, 503
Gillis, M., 17, 20
Gilmour, D. A., 376, 384
Gilpin, M. E., 56, 111, 116, 124, 164, 217, 367, 524
Giraldo, M., 124, 171, 174
Gittleman, J. L., 455
Glanz, W. E., 161
Gliwicz, J., 238
Godron, M., 55, 116, 392
Godt, M. J. W., 306, 310
Gompper, M. E., 161
Good, D. A., 152
Good, I. J., 496
Good, J. E. G., 56
Goodwin, D., 175

Goodyer, N. J., 341
Goosem, M. W., 9, 238, 241–55, 511, 512, 522
Goosem, S. P., 372, 375
Gorzula, S., 257
Gosz, J. R., 186
Gotelli, N. J., 205, 487
Graham, A. W., 81, 505
Graham, M., 223
Grange, A. C., 367
Grant, J. D., 82, 204, 218
Graves, G. R., 480
Graves, J. H., 247
Green, G. M., 415, 416, 426
Green, M., 369
Greenberg, R., 138, 154, 510
Greene, H. W., 456
Griffiths, M., 246, 512
Groombridge, B., 10
Grubb, P. J., 186, 292, 322
Guevara, S., 153, 305, 318, 379, 383, 524
Guillotin, M., 217, 219
Gullison, R. E., 23
Gunderson, L. H., 171
Gunnarson, B., 56
Gustafson, E. J., 392, 401
Gysel, I. W., 124

Haas, C. A., 156
Haber, W. A., 309
Hackett, S. J., 455
Hacking, I., 489
Haffer, J., 442, 449, 451, 459, 463, 505
Hagan, J. M., III, 144
Haila, Y., 485, 489, 490
Hall, R. A., 9
Hallwachs, W., 270
Hammel, B. E., 306
Hamrick, J. L., 246, 289, 304–20, 343, 363, 503, 507, 509, 510, 518
Hanagarth, W., 478
Hanski, I., 112, 321
Harcourt, C., 416, 435
Harper, J. L., 322, 331
Harper, L. H., 124, 507
Harrington, G. N., 9, 82, 83, 226, 292–303, 318, 485, 504, 506, 521
Harris, G. P., 171
Harris, L. D., 55, 304, 321, 351, 488, 524
Harrison, J. L., 236
Harrison, R. L., 369
Harrison, S., 321, 322, 520
Hartshorn, G. S., 306
Harvey, I. F., 112
Harvey, P. H., 452
Haslett, J. A., 172
Hastings, A., 520
Hawkins, A. F. A., 425, 429
Hay, M. E., 188
Heaney, L. R., 223, 256, 263, 505
Heimerdinger, M. A., 152
Heithaus, E. R., 309, 318
Helle, P., 56
Henderson, A., 172
Henry, N. B., 10
Hero, J. M., 126
Herre, E. A., 306–8, 310, 316
Hershkovitz, P., 455, 458
Hester, A. J., 56, 68
Hewitt, B., 373
Heyer, W. R., 451
Heywood, J. S., 307
Heywood, V. H., 8, 9, 520
Hill, M. O., 59, 128, 147
Hilty, S. L., 174, 175
Hoaglin, D. C., 284
Hobbs, R. J., 45, 46, 55, 56, 68, 71, 82, 124, 138, 190, 225, 302, 321, 322, 369, 386, 388, 392, 486, 488, 489
Hockings, M., 254
Hodl, W., 137
Holderegger, R., 518
Holdridge, L. R., 141, 186, 386
Holling, C. S., 171, 175, 176, 185–88
Holloway, J. D., 10, 56, 171, 236
Holmes, E. C., 452
Holt, R. D., 321, 322
Holtby, C., 503
Hooghiemstra, H., 185
Hopkins, M. S., 81, 442, 446, 447, 452, 505
Hopper, S. D., 306, 319
Hoppes, W. G., 159, 162, 260, 268
Horning, N. H., 415, 416, 418, 420, 422, 424–26, 429, 435–40, 506, 507
Houghton, R. A., xi
House, S. M., 306, 308, 309
How, R. A., 254
Howe, H. E., 292, 301, 302, 311, 343, 497
Howe, H. F., 306, 311
Hoyo, J. del, 175
Hubbell, S. P., 81, 306, 503
Huber, O., 258
Hudson, R. R., 452
Hugall, A., 445, 446, 448, 451, 453, 520
Hughes, L., 186
Humbert, H., 418, 422–26
Humphries, C. J., 466
Humphries, S. M., 81, 505
Humphries, W. F., 254
Hunsaker, C. T., 489
Hunter, M. L., 246
Hurlbert, S. H., 491, 492, 496
Huston, M., 322
Hutchings, R. W., 42, 56, 65, 82, 91–111, 124, 135, 140, 317, 333, 503–6, 521

Hutchison, C. S., 223
Hutterer, R., 457
Hyland, B. P. M., 301

ICBP, 466, 469, 475, 478, 481
Ingram, G., 443, 451
Inouye, R. S., 263
INPE, 352, 356
Irvine, A. K., 292–303
Irving, E. M., 172
Isaacs, J., 199, 368, 485
Isler, M. L., 175
Isler, P. R., 175
Isotomin, A. V., 112
IUCN, 467

Jackson, J. C., 336
Jaenike, J., 65
Jain, S. K., 481
Janos, D. P., 186, 226
Janson, C. H., 457
Jansson, G, 523
Janzen, D. H., xi, 81, 165, 275, 292, 309, 351, 386, 388, 392, 504, 505, 510, 521
Jenkins, M. D., 415, 416, 420, 424, 437–38
Jenkins, S. I., 78, 79, 81
Jimenez, B. D., 402, 513
Jiménez, J. A., 169
Johnels, S. A., 174
Johns, A. D., 10, 171, 524
Johns, B. G., 10
Johnson, R. A., 161
Johnson, W. C., 241, 244
Johnston, D. W., 144
Jones, C. G., 332
Jones, D. D., 171
Jones, E. W., 370
Jones, H. L., 225, 256, 504, 507
Jones, J. S., 112
Jones, S. H., 331
Jordan, C., 371
Jordan, C. F., 366
Jordano, P., 292
Joseph, L. H., 445, 446, 448, 451, 453, 520
Joyce, A. T., 386
Juelson, T. C., 488

Kahn, J. R., 13–28, 366, 400, 482, 513, 514, 524, 525
Kalliola, R., 462, 479
Kann, L. M., 449
Kanowski, P. J., 383
Kapos, V., 33–44, 46, 50, 53, 55, 56, 68, 71, 81, 82, 124, 125, 140, 186, 208, 216, 217, 248, 333, 341, 351, 506, 517
Kareiva, P., 321
Karr, J. R., 143, 144, 154, 156–70, 172, 184, 190, 191, 204, 217, 268, 503, 506, 507, 517
Kartawinata, K., 236
Kattan, G. H., 124, 171, 174
Katzman, M. T., 304
Kauth, R. J., 389
Keenan, R. J., 366–85
Kellman, M., 343
Kemp, E. M., 443
Kemper, C., 231, 236
Kent, M., 205
Kerfoot, O., 480
Kerfoot, W. C., 117
Kershaw, A. P., 446, 452
Kerster, H. W., 306
Kessler, M., 480
Keuroghlian, A., 124
Keurs, W. J. ter, 245
Kevan, P. G., 308
Kiefer, R. W., 388, 389, 390
Kierulff, M. C. M., 457
Kikkawa, J., 443, 451
Kimura, M., 317, 318
King, A. W., 401
King, D., 253
King, G. C., 82
Kinloch, B. B., 314
Kirk-Spriggs, A. H., 10, 56, 171, 236
Kisbenedek, T., 56
Kitchener, D. J., 124
Klein, B. C., 56, 111, 124, 171, 208, 218, 509
Klein, D. R., 246
Kluge, A. G., 117
KnowledgeSeeker, 428
Knowlton, N., 443
Koch, L. E., 112, 115
Kochummen, K. M., 8
Kodric-Brown, A., 204, 233, 272, 510
Konopka, J., 120
Koopowitz, H., 8, 207, 218, 281–91, 342, 345, 364, 368, 503, 511, 518, 521
Kozakiewicz, M., 120, 247, 510
Krabbe, N., 175, 468
Kramer, E. A., 190, 386–99, 401, 441, 502–25
Kremen, C., 107
Kress, J., 308
Krishnan, V. V., 55
Kroodsma, R. L., 56, 68, 244, 488
Kruess, A., 171
Krummel, J. R., 116, 392
Kuhnlein, U., 168

Laborde, J., 305, 318
Lack, A. J., 308
Lacy, R. C., 169, 316
LaFrankie, J. V., 8
Lam, T. J., 344
Lamb, D., 79, 81, 153, 321, 344, 351, 366–85, 510, 512
Lambert, F. R., 171

Lande, R., 164, 169, 304, 315, 507, 519
Langham, N., 226
Langrand, O., 422, 424
Lanly, J. P., 3, 4
Lara, M. C., 455–65
Laurance, S. G. W., 82, 297, 375, 524
Laurance, W. F., xi–xv, 41–43, 46, 47, 50, 55, 56, 67, 68, 71–83, 108, 112, 124, 138, 154, 159, 168, 171, 172, 186, 190, 191, 204, 206, 208, 216–19, 225, 236, 238, 243, 244, 248, 250, 292, 293, 301, 321, 333, 341, 344, 351, 352, 363, 369, 372, 373, 392, 402, 407, 442, 453, 485, 488, 502–25
Laurenson, M. K., 519
Lawton, J. H., 124, 172, 321, 322
Leary, R. F., 112, 122, 518
Leberman, R. C., 152
Lebrun, J., 415
Leck, C. F., 156, 205
Leigh, E. G., Jr., 56, 158, 186, 218, 268, 270, 332, 333, 506, 507
Leite, L. L., 402, 403
Lerner, I. M., 112
Lescourret, F., 185
Lesica, P., 519
Leung, L. K.-P., 233
Levenson, J. B., 208, 219
Levey, D. J., 304, 311
Levin, D. A., 305, 306, 313, 316
Levine, D. A., 489
Levings, S. C., 122
Levins, R., 238, 489, 506, 520
Lewin, R., xi, 125
Lewis, J. G., 114, 121
Lewontin, R. C., 489
Lidicker, W. Z., Jr., 510
Lieberman, D., 81
Liggett, A. C., 112
Lillesand, T. M., 388–90
Lim, K. K. P., 340
Lim, K. S., 339
Lin, J. D., 71
Lindsey, T. R., 199
Linhart, Y. B., 309, 509
Lipsey, M. W., 176
Littlefield, J. L., 314
Llorente-Bosquets, J., 91
Löbl, I., 59
Loeschcke, V., 453, 481
Loiselle, B. A., 159, 162, 260, 268, 307, 317, 510
Lomolino, M. V., 256
Long, A., 478, 480–82
Long, J. C., 115
Longino, J. T., 59
Lopez, L., 256–74
Lopez-Parodi, J., 510
Losos, E., 23
Lott, R. H., 302
Lovejoy, T. E., ix–x, xii, 34, 42, 56, 57, 82, 91, 124, 125, 135, 136, 139–41, 143–46, 153, 156, 166, 168–71, 207, 208, 217, 218, 236, 238, 268, 270, 351, 407, 408, 485, 502–25
Loveless, M. D., 306, 307, 311
Lovett, J. C., 469
Low, C. M., 8
Lowe, C. H., 251
Lu, Z. N., 71
Lucas, P. W., 344
Lucas, R. M., 138
Lugo, A. E., xii, 71, 207, 366, 368, 372, 373, 386, 522
Luteyn, J., 172
Lwanga, J. F., 218
Lynam, A. J., 153, 171, 185, 222–40, 272, 502–25
Lynch, M., 168, 169

MacArthur, R. H., xii, 103, 116, 121, 167, 190, 205, 208, 217, 219, 225, 268, 392, 506
Macdonald, I. A. W., 324, 330
MacDonald, J. D., 193
Machado, A. B. M., 354
Mack, A., 292
Mader, H. J., 246–48
Madsen, J., 245–46
Maggs, J., 373
Magsalay, P., 8, 9
Magurran, A. E., 227, 393
Main, B. Y., 112
Malcolm, J. R., 46, 55, 67, 68, 71, 79, 82, 108, 124, 135, 136, 138, 140, 159, 171, 204, 207–21, 233, 236, 238, 321, 333, 453, 457, 459, 463, 464, 478, 503, 505
Manasse, R. S., 306, 311
Manço, D. D.-G., 457
Manly, B. F. J., 171, 185
Mann, K., 244
Manning, J. T., 112
Manokaran, N., 8
Mares, M. A., 455
Margules, C. R., 46, 55, 65, 71, 82, 124, 138, 190, 222, 225, 226, 228, 302, 321, 322, 369, 386, 392, 486, 491, 492
Marino, P. C., 321
Markow, T. A., 112
Marks, B. J., 390–93
Marples, P., 171–89, 318
Marques, J. M., 33, 138
Marsh, C., 9
Marsh, H., 244, 245, 249, 253
Marshall, A. G., 222
Marshall, D. L., 306
Martin, A. P., 449
Martin, T. E., 157, 162, 186
Martins, M. B., 65
Maruyama, T., 314
Mason, D., 246, 249, 252

Mathur, H. N., 373
Matlack, G. R., 42, 43, 46, 52, 53, 322, 331, 370
Matthews, C., 72
Maurer, B. A., 322
Maxson, L. R., 451
Mayden, R. L., 442, 466, 520
Maynardier, P. G. de, 246
Mayr, E., 112, 121, 442
McCarthy, M. A., 453
McCauley, D. E., 310, 313, 315, 319
McClanahan, T. R., 153, 379, 383, 524
McClennan, C. H., 225–26, 267
McCloskey, M., 509
McClure, H. E., 497
McCommas, S. A., 112
McCoy, E. D., 138, 321
McCune, B., 59, 74
McDonald, J. A., 13–28, 366, 400, 482, 513, 514, 524, 525
McDonald, K. R., 443, 445
McDonnell, M. J., 372
McElroy, A. D., 241, 243
McEuen, A. B., 246
McGarigal, K., 390–93
McKenzie, L. J., 112
McLellan, B. N., 245, 246
McLennan, D. A., 442, 466, 520
McNaughton, K. G., 42
McNeely, J. A., 291, 369, 466
Meachem, W., 338
Meave, J., 343
Medley, K. E., xiv
Medway, L., 223, 226, 231, 236
Meggers, B. J., 479
Mejía, J., 172, 186
Mendelsohn, R., 510
Menges, E. S., 518
Merriam, G., 138, 321, 322, 510
Merriam, H. G., 518
Mery, G., 20
Meyer, W. B., 400
Middleton, J., 244
Milkovits, G. A., 65, 124, 225, 321, 523
Miller, A., 490, 491, 496
Miller, D. R., 71
Miller, M., 21
Mills, L. S., 507
Milne, B. T., 392
Milton, K., 269
Minchin, P. R., 74
Mitchel, N., 68, 146
Mitra, S. S., 373
Mittermeier, R. A., 415, 416
Modi, W. S., 112, 115
Moermond, T. C., 172, 186, 304, 311
Molion, L. C. B., 33, 138
Møller, A. P., 112, 269
Mondragón, M. L., 174, 186
Montieth, G. B., 443, 451
Mooney, H. A., 388
Moore, D. M., 416, 418, 420, 422, 424, 425, 429, 435, 436, 440, 507
Moore, L. A., 199, 233, 292–303, 368, 485, 493
Moore, P. G., 496
Morales, L. C., 257
Moran, E. F., 388, 404
Moran, G. F., 306, 319
Mori, S. A., 351
Moritz, C., 114, 121, 442–54, 456, 459, 502–25
Morrison, E., 384
Morton, E. S., 161, 165, 268
Moss, R. H., 400
Moure, J. S., 56, 65, 111
Mueller-Dombois, D., 329
Munn, C. A., 144
Munves, J., 174
Muona, J., 56
Muona, O., 306
Murawski, D. A., 307–11, 313, 317
Murcia, C., 33, 42, 45, 46, 53, 55, 61, 67, 68, 71, 321, 506
Murphy, D. D., xii, 55, 56
Murphy, P. G., 56, 68, 386
Murray, G., 138, 153, 524
Mushinsky, H. R., 321
Musser, G. G., 226, 458, 460
Mustrangi, M. A., 455–65
Myers, N., xi, xii, 20, 466, 502, 511

Nadkarni, N. M., 59
Nakasathien, S., 222
Naranjo, L. G., 174
Narson, A., 422
Nason, J. D., 246, 289, 304–20, 343, 363, 503, 507, 509, 510, 518
Nee, S., 452
Nei, M., 314
Nelson, B. W., 42, 71
Nelson, R., 415, 422, 424–26, 437–39
Nepstad, D. C., 153, 371, 379
Nevo, E., 111
Newcombe, C., 226
Newmark, W. D., xiv, 171, 407, 503
Ng, F. S. P., 8, 82
Ng, P. K. L., 340, 344
Nichol, J. E., 343
Nicholas, J. T., 56
Nicholls, A. O., 491, 492
Nichols, J. D., 160, 161, 166
Nicholson, B., 81
Nicholson, D. I., 10
Nicoll, M. E., 422, 424
Niles, L. J., 244
Nix, H. A., 442–44, 446–49, 523
Nobre, C., 138
Norse, E. A., 514

Noss, R. F., 487, 524
Nottingham, B. G., Jr., 270
Nowak, R. M., 226

Oakes, M., 494, 469
O'Brien, S. J., 112, 115
Offerman, H. L., 408, 523
O'Hara, R. J., 442
Oldeman, R. A. A., 352
O'Leary, M. H., 39
Olsen, M., 79, 81
O'Malley, D. M., 308
O'Neill, R. V., 171, 188, 392, 401, 402, 480, 489
Oniki, Y., 144, 161, 167
Opdam, P., 156
Opler, P. A., 309, 397
Orejuela, J. E., 174
Oren, D. C., 139
Osunkoya, O. O., 186, 218, 301
Otero, L. S., 91
Owen, D. F., 91
Oxley, D. J., 246–48, 253

Pacheco, S., 185
Pacheco, V., 455, 457, 458
Packer, C., 124
Page, R. D. M., 460
Paine, J., 369
Pakkala, T., 321
Palik, B. J., 56, 68
Palmer, R. A., 115
Palo, M., 20
Palumbi, S. R., 449
Panayotou, T., 20
Parizek, R. R., 241, 243
Park, C. C., 281
Parker, G. R., 392, 401
Parker, T. A., III, 368
Parrotta, J. A., 366–85
Parsons, P. A., 112
Partridge, I. J., 72
Patrick, R., 241, 243
Patterson, B. D., 190, 191, 193, 194, 202, 203, 455, 458, 463
Patton, J. L., 451–53, 455–65, 478, 507, 509, 518, 519
Pavan, C., 309
Payne, C. W., 82, 204
Pearce, D. W., 20, 25
Pearson, D. L., 144
Pearson, S. M., 16, 289, 351, 400–409, 509, 513, 522
Pedlowski, M. A., 403, 404
Pelletier, R. E., 388
Pelton, M. R., 245, 246
Peralta, F. J. A., 56, 65, 111
Peres, C. A., 504, 511, 523
Pérez-Nasser, N., 307, 308
Perkocha, V., 81
Perry, D, R., 308
Peters, C. M., 510
Peters, R. H., 488, 517, 523
Philander, S. G., 479
Phillips, D. L., 56
Phillipson, P. B., 415, 420
Pickett, S. T. A., 322, 505, 517
Pienaar, U. de V., 245
Pimm, S. L., 171, 225, 256, 455, 504, 507
Pinard, M., 524
Pinero, D., 307, 308
Pizzey, G., 193
Pócs, T., 480
Pollock, J. I., 416
Poore, D., xi
Portnoy, S., 313
Potter, M. A., 124
Powell, A. H., 56, 65, 111, 507, 518
Powell, G. V. N., 56, 65, 111, 124, 144, 507, 510, 518
Poyry, J., 321
Prabhu, R., 524
Prance, G. T., 141, 351, 442, 451, 479, 505
Preston, E. M., 491
Preston, F. W., 205, 290
Primack, R. B., 322
Prum, R. O., 460, 461
Puhakka, M., 479
Pulliam, H. R., 487
Pun, F. C., 119
Purata, S. E., 153, 379, 383, 524
Putz, F. E., 79, 81, 218, 268, 270, 352, 507, 524

Quattrochi, D. A., 388
Quinn, D., 506
Quintela, C. E., 124, 146, 171, 503, 506

Rabinowitz, A. R., 266, 270
Raguso, R. A., 91
Rahbek, C., 143, 153, 443, 451, 452, 456, 463, 466–82, 488, 507, 509, 511, 519–21
Raitt, R. S., 174
Ralls, K., 124
Ralph, C. J., 168
Ramage, C. S., 47
Rand, A. S., 158
Rand, D. M., 449
Rangel, J. O., 174
Rankin-de Mérona, J. M., 42, 140, 141
Räsänen, M. E., 459, 462
Ratti, J. T., 488
Raunet, M., 324
Read, B., 167, 169
Readers' Digest, 193
Recher, H. F., 497
Redford, K. H., 217, 219, 503, 504
Reeder, D. M., 455
Reese, K. P., 488
Reh, W., 124
Reich, P. B., 398

Reichman, O. J., 56
Reid, W. V., 222, 281, 290
Reilly, J. J., 376
Reilly, P., 193
Remsen, J. V., Jr., 152, 368
Rencher, A. C., 119
Renjifo, L. M., 171–89, 318
Repetto, R., 20
Restrepo, C., 124, 171–89, 318, 502–25
Reville, B. J., 71, 81
Reynolds, E. R. C., 71
Rich, A. C., 244
Richards, G. C., 82
Richards, P. W., 309
Ricklefs, R. E., 160, 162
Ridgely, R. S., 174
Ripple, W. J., 397
Riswan, S., 236
Robbins, R. K., 91, 102
Roberts, J., 37, 39
Robinson, G. R., 225, 321
Robinson, J. G., 217, 219, 269, 270, 503, 504
Robinson, S. K., 159, 267
Rodrigues, M. T., 126
Rohlf, F. J., 60, 77, 114, 115, 127, 145
Rolstad, J., 318
Romme, W. H., 393
Rosas, M. L., 174
Rosen, D. E., 459
Rosen, P. C., 251
Rost, G. R., 245
Roth, R. R., 184
Round, P., 223
Rubio-T., H., 171
Rudd, R. L., 236
Rudder, J., 10
Rudel, T., 20, 21
Ruscher, C. L., 392
Rylands, A. B., 124, 218, 354

Sader, S. A., 386
Sahley, C. T., 226
Salati, E., 33, 82, 138, 480, 504
Saldarriaga, J. G., 370
Salisbury, E., 186
Salmi, J., 20
Salo, J., 462, 463, 479, 505
Samson, D. A., 263
Sargatal, I., 175
Sargent. S., 311
SAS Institute, Inc., 212, 325
Saunders, D. A., 45, 46, 55, 71, 82, 124, 138, 190, 225, 302, 321, 322, 369, 386, 392, 486, 488, 489, 497
Savidge, J. A., 269
Savill, P. S., 71, 383
Sayer, J. A., 223, 281
Schaik, C. P. van, 246, 512
Schaller, G. B., 503
Scheck, J., 524
Schelhas, J., 138, 154, 510, 522
Schneider, C., 442–54
Schneller, J. J., 518
Schodde, R., 451, 445
Schonewald-Cox, C. M., 55, 56, 432, 488
Schreiber, R. K., 241, 244, 247
Schroff, U., 422
Schupp, E. W., 302
Schwarzkopf, L., 124, 218
Scott, M. J., 467, 523
Seagle, S. W., 121
Seamon, J. O., 233, 238
Seastedt, T. R., 56
Seitz, A., 124
Sellke, T., 495
Serrão, E. A. S., 138, 153, 371, 379
Sexton, G., 377
Shackleton, D. M., 245, 246
Shafer, C. L., 83, 138, 281, 518
Shaffer, M. L., 156
Shafik, N., 20, 21, 23
Sharpe, D. M., 55
Sheldon, F. H., 373
Sheperd, A., 83
Sheppard, P. M., 93
Sherman-Broyles, S. L., 306
Shmida, A., 56
Shugart, H. H., 82, 111–23, 244, 504, 517, 523
Shure, D. J., 56
Sibley, C. G., 469, 478
Sieving, K. E., 154, 156–70, 172, 190, 191, 204, 268, 502–25
Silva, J. M. C., 138, 153, 468, 478, 481, 524
Silva, M. N. F. da, 451, 453, 455–65, 478
Silva, N. J. da, 523
Silva-Lopez, G., 55, 321
Silverman, B. W., 185
Simberloff, D. S., xii, 8, 56, 103, 124–26, 135, 136, 191, 203, 223, 256, 281, 321, 367, 369, 485, 487–90, 492, 524
Sinclair, G. S., 367
Singer, M. C., 103, 107
Sisk, T. D., 369
Sites, J. W., Jr., 502–25
Skole, D. L., 400, 402
Skutch, A. F., 160–63, 165, 166, 175
Slade, N., 246, 247
Slater, P., 199
Slater, R., 199
Slatkin, M., 433, 452
Smallwood, J., 292, 311
Smith, A. P., 78, 289, 415–41, 502–25
Smith, G. C., 254
Smith, G. T., 65, 124, 225, 321
Smith, J. B., Jr., 517, 523
Smith, M. F., 453, 459
Smith, N. H. J., xi

Smith, T., 322
Smith, T. B., 456
Smith, T. M., 517, 523
Smouse, P. E., 115, 507
Smythe, N., 270, 301
Sobrevila, C., 308
Soedjito, H., 236
Sokal, R. R., 60, 77, 114, 115, 118, 127, 145
Somerville, A., 71
Sommer, A., 3
Soni, P., 373
SOS Mata Atlântica, 352, 356
Soulé, M. E., 56, 111, 112, 115, 120–22, 124, 164, 204, 267, 268, 270, 271, 352, 487, 503, 506, 518, 524
Sousa, W. P., 71
Southgate, D., 20
Southworth, F., 402, 403
Souza, O. F. F. de, 56, 111
Sparrow, H. R., 107
Spies, T. A., 45, 71, 397
Sprague, C. R., 491
Sprague, L. G., 491
Springer, M., 449
Srikosamatara, S., 511
Stacy, E. A., 306, 310, 316
Staley, M. J., 493
Stallings, J. R., 299, 457
Stamps, J. A., 55
Stanton, P. R., 81, 505
Stein, J. L., 523
Steven, D. E., 497
Stevenson, P. J., 447
Stewart, B. R., 121
Stiles, E. W., 372
Stiles, F. G., 153, 154, 172, 175
Stockard, J., 81
Stocker, G. C., 297, 299, 447
Stotz, D. F., 143, 144
Stouffer, P. C., 8, 57, 58, 92, 108, 124, 125, 135, 138–56, 159, 168, 169, 191, 204, 208, 218, 231, 249, 268, 333, 498, 503, 513, 518
Strahm, W. A., 322
Strasberg, D., 292, 321–32, 344, 345, 363, 371, 386, 387, 512, 520
Strasberg, D., 322–24, 330–32
Strobeck, C., 115
Strong, L. L., 123
Stuart, S. N., 9, 520
Sugden, A. M., 171
Sun, D., 371
Sungsuwan, S., 20
Suppe, F., 489
Sussman, L. K., 415, 416
Sussman, R. W., 415, 416, 426
Sutton, S. L., 56
Swain, D. P., 114
Swaine, M. D., 222
Swenson, J. E., 523
Swihart, R. K., 246, 247
Switzer, M. A., 448, 523
Szaki, J., 510

Tabanez, A. A. J., 8, 78, 81, 82, 351–65, 384, 464, 481, 503–5, 507, 512, 521, 523, 524
Tan, H. T. W., 340, 344
Tanner, E. V. J., 308
Tattersall, I., 416, 422, 432
Tello, J., 256–74
Temple, S. A., 68, 236
Templeton, A. R., 167, 169, 304
Ter Braak, C. J. F., 156
Terborgh, J. W., 83, 153, 156, 161, 162, 168, 172, 184, 185, 204, 218, 225, 238, 256–74, 442, 457, 504, 506, 511, 515, 517, 518, 520, 521, 523
Thébaud, C., 292, 321–32, 344, 345, 363, 371, 386, 387, 512, 520
Thiollay, J.-M., 10, 81, 171, 185, 515
Thomas, C. D., 321
Thomas, M., 493
Thompson, J. N., 505
Thornback, J., 416, 435
Thornhill, A. D., 8, 207, 218, 281–91, 342, 345, 364, 368, 503, 511, 518, 521
Thurber, J. M., 245
Tilman, D., 171
Tocher, M. D., 108, 124–37, 140, 154, 191, 502–25
Tomiuk, J., 481
Tomlinson, M., 383
Tracey, J. G., 47, 72, 79, 81, 121, 191, 200, 368, 373, 376, 443–47, 505
Tranter, J. D., 71, 81
Trueb, L., 137
Truswell, E, 443, 451
Tscharntke, T., 171
Tucker, C., 400, 402
Tucker, N., 366–85
Tucker, N. I. J., 372, 375
Tudor, G., 174
Tuomisto, H., 459, 463, 479
Turner, B. L., II, 400
Turner, I. M., 8, 9, 289, 302, 333–45, 504, 507, 510, 512, 521, 522, 524
Turner, M. G., 171, 188, 392, 397, 401
Turton, S. M., 40, 45–55, 68, 71, 82, 122, 248, 341, 506

Ueno, H., 112
Uhl, C., 138, 153, 321, 352, 371, 379, 512, 524
Underwood, A. J., 492
United States Council on Environmental Quality, 241
Unwin, G. L., 72, 79, 447
Uribe, D. A., 174
Usher, M. B., 467

Van der Maarel, E., 153, 379, 383, 524
Van Dorp, D., 156
Van Valen, L., 111, 120

Vanapeldoorn, R. C., 124
Vanzolini, P. E., 442, 451, 459
Vasconcelos, H. L., 56, 111
Velásquez, A., 172, 186
Velásquez, M. P., 174, 175
Velez, B. E., 174
Velleman, P. F., 284
Veríssimo, A., 6, 352
Viana, V. M., 8, 78, 81, 82, 351–65, 384, 464, 481, 502–25
Victoria, R. L., 33
Villard, M. A., 322
Vitousek, P. M., 41, 42, 171, 322, 504
Vogler, A. P., 452
von Moltke, K., 21, 22
Vose, P. B., 33
Voss, R. S., 457, 460

Wachter, T. S., 318
Wade, M. J., 310, 319
Wakefield, S., 83
Waldorff, P., 352
Walker, B. H., 185
Walker, D., 442, 452, 453
Walker, S. R., 266
Wallace, A. R., 459
Wallace, B., 317
Walter, H. S., 169
Wandelli, E. V., 33–44
Wanghongsa, S., 226
Warburton, N., 114, 168, 171, 190–206, 271, 503, 513
Ward, A. L., 245
Warford, J. J., 25
Wayne, R. K., 112, 115, 124, 456
Weaver, P. L., 370
Webb, C. J., 309
Webb, L. J., 121, 200, 368, 443–47, 505
Webb, N. R., 56
Weeldenberg, J. R., 497
Weijden, W. J. van der, 245
Weishampel, J. F., 82, 111–23, 504
Weiss, G. H., 318
Wells, D. R., 223
West, D. C., 370
Westley, L., 302
Westman, W. E., 82, 111–23, 504
Wetmore, A., 161
Wetton, J. H., 163
Wheelwright, N. T., 304
White, F., 419
Whitlock, M. C., 112, 313, 315, 319
Whitmore, T. C., xi, 3–12, 71, 207, 222, 223, 281, 287, 289, 290, 304, 333, 336, 351, 442, 451, 479, 509, 514, 520, 522, 524
Whittaker, A., 143
Whittaker, R. J., 331
Wickland, D. E., 388
Wiens, J. A., 186, 402
Wilcove, D. S., 156, 225–26, 260, 267, 506
Wilcox, B. A., xii, 55, 56, 123, 204, 256, 263, 503
Wilkins, K. T., 246
Williams, E. E., 442, 451, 459
Williams, G., 81
Williams, M., 370
Williams, P. H., 466–68
Williams, S. E., 244, 254
Williams, W. T., 200, 368
Williams-Linera, G., 33, 50, 53, 55, 56, 68, 71, 81, 506
Williamson, M., 368
Willis, E. O., xii, 144, 156, 158–61, 163–67, 190, 204, 256
Willmott, W. F., 447
Willson, M. F., 81, 124, 311, 313, 331
Wilson, D. E., 455
Wilson, E. O., xii, 116, 121, 159, 190, 205, 207, 208, 219, 225, 281, 392
Wilson, M. V., 56
Windsor, D. M., 122, 158
Winter, B., 156, 168, 225, 238, 256, 518
Winter, J. W., 47, 344, 443, 444, 451
Witting, L., 453
Wolfe, R. W., 153, 379, 383, 524
Wooddroffe, G. L., 124
Woodman, N., 457, 458
Woodruff, D. S., 222
World Bank, 25, 26
Wright, S., 305, 314, 315, 318
Wright, S. J., 156, 159, 161, 167, 218, 268, 270, 507
WTMA, 241, 510

Yablokov, A. V., 112, 121
Yahner, R. H., 68, 233, 238
Yang, S. Y., 121
Yensen, E., 43, 46, 55, 56, 67, 68, 79, 138, 219, 517
Yorkston, H. D., 71, 81
Young, A., 146, 68
Young, A. G., 518
Young, A. M., 122
Young, P. A. R., 372
Young, T. P., 81
Yu, D., 256–74

Zande, A. N. van der, 245
Zimmerman, B. L., 108, 124–37, 140, 154, 190, 191, 503–5, 513, 518
Zipperer, W. C., 55

Taxa Index

Acacia, 165, 375, 377
Acacia, bullhorn, 165
Acanthiza, 198, 450, 451
Acanthophoenix, 326
Acanthorhynchus, 198
Accipiter, 198
Adansonia, 416
Adelpha, 106
Agathis, 377
Agauria, 326, 331
Agouti, 260, 262, 264, 270
Agraulis, 106
Ailuroedus, 199
Alaeocharinae, 66
Albizia, 378–81
Alcedinidae, 339, 468
Alchornea, 106
Alectura, 198
Alisterus, 198
Alluaudia, 416
Alnus, 174
Alouatta, 262, 264
Alphitonia, 371
Alseis, 311, 312
Ameiva, 260, 265, 266, 270
Anacardiaceae, 309, 358
Anaea, 106
Andenomera, 130
Andira, 309
Annonaceae, 37, 245, 334, 341
Antechinus, 82, 248, 252–54
Antirhea, 326
Antirrhaea, 106, 107
Aphloia, 326
Aplonis, 199, 203, 372
Apodemus, 247
Ant
  army, 92, 144, 146, 152, 163, 164, 166, 263, 270, 507
  leaf-cutter, 56, 260, 262, 263, 269
Antbird, 152, 161, 164, 469
  bicolored, 160, 162–64, 166
  black-headed, 146
  chestnut-backed, 158, 160, 162, 163, 166–69
  dusky, 147
  ocellated, 160, 164, 166, 168
  rufous-throated, 146
  spotted, 160, 162–64, 166, 167, 169
  warbling, 146
  white-plumed, 146
  wing-banded, 146
  spectacled, 160, 162–64
Antthrush, 146
  black-faced, 158, 160, 163, 167
Aracaeae, 308
Aracari, green, 148
Araliaceae, 324
Araucaria, 376, 377
Arawacus, 106
Ardisia, 324
Aristolochia, 106
Armadillo, 260, 262, 270
  nine-banded, 260, 262
Arremon, 148
Arses, 198
Astrocaryum, 37, 38, 42, 43, 308
Astronium, 358
Astrophea, 106
Atelopus, 128, 130, 136
Athertonia, 295, 298–301
Atta, 260, 262, 263, 265, 266
Avahi, 435

Babbler, 339
Bambusa, 174
Bambuseae, 106
Bamboo, 88, 227, 234, 356, 358, 512
Bananaquit, 148
Baobab tree, 416
Bat, 313, 318, 339, 383
  fruit, 295
Barbet, 339, 468, 472
Bats, microchiropteran, 82, 456

Battus, 106
Bee, euglossine, 56, 309, 518
Beetle, dung, 56
Beilschmiedia, 295, 298–300
Bellucia, 92, 141
Bia, 106
Bignoniaceae, 284, 285
Blarina, 244
Blechnum, 325
Boatbill, yellow-breasted, 197, 198, 203
Boehmeria, 324
Bombacaceae, 39, 141, 308, 362
Boobook, southern, 197, 198
Boraginaceae, 309
Bourreria, 380
Bowerbird
golden, 193, 197, 199, 203
satin, 197, 199, 203
Brassolinae, 93, 94, 98, 106, 107
Broadbill, 339
Bromeliaceae, 284
Bromelioideae, 284, 285
Bronze-cuckoo
little, 193, 195, 198
rufous-breasted, 193
Brosimum, 106
Brunellia, 284, 285
Brunfelsia, 106
Brush-turkey, Australian, 196, 198
Bucconidae, 468
Bucerotidae, 339
Bufo, 128, 130, 136, 265, 266, 270
Bufonidae, 130
Bulbul, 339
Bursaphelenchus, 342
Burseraceae, 8, 141

Cacatua, 198
Caiman, 265
Calamus, 74
Calceolaria, 284, 285
Callicore, 106
Callophyllum, 310, 326
Callosciurus, 226, 340, 341, 344
Calospila, 106
Caluromys, 215, 216, 219, 221
Camponotus, 106
Capitonidae, 339, 468
Capybara, 260, 262, 263, 269, 271, 272
Carabidae, 59, 64, 66–8
Carlia, 251
Carlowrightia, 284, 285
Carphodactylus, 445, 448, 451
Casearia, 326
Cassowary, southern, 85, 203, 206, 250, 276, 278, 292, 296, 297, 299–302, 485, 487
Casuarina, 324, 378, 381
Casuarinaceae, 324
Casuarius, 203, 206, 250, 485
Cat, leopard, 339
Catbird, 196, 199
Catharus, 148
Catonephele, 101, 106
Cavendishia, 284, 285, 308
Cebus, 165, 262, 264
Cecropia, 58, 92, 106, 141–43, 146, 147, 151, 153
Cecropiaceae, 284, 285
"Cedar," red, 72
Cedrela, 362, 379
Ceiba, 308
Centrolenella, 130
Centrolenidae, 130
Centropus, 199
Ceratophrys, 130
Cercartetus, 251
Cercomacra, 147
Cervus, 339
Chalcophaps, 198
Charaxinae, 93, 98, 104, 107
Chassalia, 324, 326
Cheirogaleus, 432, 435
Chiasmocleis, 130
Chipmunk, eastern, 253
Chiropodomys, 226, 229, 230, 234
Chiropotes, 262, 264
Chloreuptychia, 106
Chorisia, 362
Chowchilla, 196, 198, 203, 250, 251, 448
Chrysobalanaceae, 39, 141, 284
Chrysococcyx, 193, 198
Cissia, 106
Cithaerias, 106
Citharexylum, 380, 381
Civet, 341, 344
Clethrionomys, 247
Clidemia, 340
Climacteris, 198
Clusiaceae, 141, 310
Cnestis, 326
Cnidoscolus, 309
Coati, 165
Coati mundi, 161, 167
Cockatoo, sulphur-crested, 196, 198
Coendou, 260, 264
Coereba, 148
Coffea, 325, 326
Colaptes, 471
Coleoptera, 59
Collembola, 59
Colluricincla, 198
Colobura, 106
Colostethus, 128, 131–37
Columba, 198
Connaraceae, 284, 285
Copaifera, 270
Cophixalus, 251, 445

Coraciiformes, 468, 471, 472 (map)
Coracina, 198
Corapipo, 148
Cordia, 309, 380
Cordyline, 326
Corynocarpus, 295, 298, 299
Costus, 283–85
Cotinga, 469, 471, 473
Cotingidae, 469, 473 (map)
Coturnix, 260, 263
Coucal, pheasant, 199
Cowbird, 267, 268
Crateroscelis, 198, 203
Cremna, 106
Crocidura, 226
Cryptocarya, 295
Crypturellus, 148
Ctenophryne, 128, 130, 136
Cuckoo, fan-tailed, 197, 198
Cuckoo-dove, brown, 195, 198
Cuckoo-shrike, yellow-eyed, 196, 198
Cuculus, 198
Cucurbitaceae, 309
Cuon, 341
Cupressus, 174
Currawong, pied, 195, 199
Cyanerpes, 149
Cyanocompsa, 148
Cyathea, 324, 326
Cycadaceae, 106, 334
Cyphomandra, 106
Cyphorhinus, 158

Dactylomys, 462
Dalechampia, 106
Danainae, 93
Danais, 326
Dasyprocta, 262, 264
Dasypus, 260, 264, 266
Deer
  barking, 339
  brocket, 260, 262, 270
  greater mouse, 339
  sambar, 339
Delomys, 457
Dendrobates, 262, 265, 266, 270, 271
Dendrobatidae, 130
Dendrocincla, 146
Dendrocnide, 79
Dendrolagus, 251
Dendrophryniscus, 130
Dhole, 341
Dicrurus, 193, 199
Didelphis, 165, 213, 217, 221, 450, 460, 461
Didiereaceae, 415
Dilkea, 106
Dimorphandra, 284, 285
Dioscorea, 340
Diplopoda, 60
Diptera, 60
Dipterocarp, 5, 6, 341, 344
Dipterocarpaceae, 8, 308, 336
Doratoxylon, 323, 326
Dove
  emerald, 196, 198
  white-tipped, 148
Drongo, spangled, 193, 195, 199
Drosophila, 65
Duguetia, 37–40
Dynamine, 106

Eagle, harpy, 260
Echimys, 215
Echinosorex, 226, 229, 234
Eciton, 144, 263, 265, 270
Ectima, 106
Elaeocarpus, 377
Elephant, 341
Eleutherodactylus, 130, 131, 134–36
Endiandra, 295, 298, 300, 301
Epipedobates, 130, 131–33, 135, 136
Ericaceae, 419
Eucalyptus, 377, 378, 381
Eueides, 106
Eulemur, 430, 435
Eumaeus, 106
Eunica, 106
Eupatorium, 106
Euphonia, 149
Euphonia, golden–sided, 149
Euphorbiaceae, 106, 309, 334, 337, 340, 358
Euptychia, 106
Eurema, 106
Eurylaimidae, 339
Euselasia, 106
Euterpe, 363
Evenus, 106
Eyra, 262, 264

Fabaceae, 309–11
Fagaceae, 334, 337, 342
Fantails, 195, 196, 198
Fernwren, Australian, 197, 198, 203
Felis, 260
Ficus, 10, 106, 310, 325, 326, 372
Fig, 10, 310, 313, 316, 372
Figbird, 195, 199
Fig-parrot, double-eyed, 197, 198
Flindersia, 372, 377
Flycatcher
  Macconnell's, 147, 148
  tyrant, 469, 471, 474
Flying fox, 331, 332
Formicariidae, 157, 161, 469
Formicarius, 146, 158
Formicidae, 59

Fox, 341
Fraxinus, 174
Frog, poison-arrow, 262, 270
Fruit-doves, 196–98
Furnariidae, 157, 161

Gaertnera, 324, 326
Galbulidae, 468
Gambier, 336
Gecko, leaf-tail, 446
Geniostoma, 326
Geochelone, 262, 265, 266, 271
Geotrygon, 145, 148
Gerygone, brown, 195, 198
Gibbon, 341
Glider
 greater, 450
 mahogany, 302
Gnypetoscincus, 448
Godwit, black-tailed, 245
Gonatodes, 265
Goshawk, grey, 195, 198
Grosbeak, blue-black, 148
Grosbeak, slate-colored, 148
Gurania, 106
Gymnopithys, 146, 162

Haetera, 106
Haeterini, 107
Hamadryas, 106
Hapalemur, 435
Harpia, 260
Hedyosmum, 284, 285
Heliconiinae, 93, 106
Heliconiini, 98, 101
Heliconius, 101, 106
Heliotropium, 92
Hemibelideus, 82, 450
Henicorhina, 165
Hermeuptychia, 106
Hermibelideus, 250
Hermits, 151
Hesperioidea, 92
Historis, 106
Homo, 517
Honeycreeper, purple, 149
Honeyeaters, 195, 198, 199
Hornbill, 339
Hummingbird, 139, 144, 145, 151, 153, 154, 259, 468–71
Hydrochaeris, 260, 264
Hyla, 130, 135
Hylidae, 130
Hylobates, 341
Hylomys, 226, 229, 231
Hylopezus, 162
Hylophylax, 162
Hymenaea, 270
Hymenoptera, 59
Hyparrhenia, 388
Hypocnemis, 146
Hyposcada, 106
Hypothyris, 99, 106
Hypsipetes, 332
Hypsiprymnodon, 250, 251, 295, 297
Hystrix, 339

Iguana, 26, 260, 263, 269, 270
Iguana, 260, 262, 265
Imperata, 336
Isothrix, 215, 221, 462
Ithomiinae, 92–95, 98, 101, 104–7

Jacamar, 468, 472
Jaguar, 260, 270
Juanulloa, 106
Juditha, 106

King-parrot, Australian, 196, 198
Kingfisher, 339, 468, 472
Krameria, 284–86

Labourdonnaisia, 323, 326
Lalage, 198
Lampropholis, 251
Lantana, 48, 81
Laparus, 106
Lapwing, 245
Lauraceae, 284, 285, 295, 324, 334, 337, 342
Leafscraper, 146
Leaftosser, scaly-throated, 160, 165, 166, 168
Lecythidaceae, 141, 284, 285
Leguminosae, 270, 334, 341
Lemur, 417
Lemur, crowned, 417
Leopard, 341
Leopoldamys, 226, 229, 231, 234
Lepidothrix [*Pipra*], 148, 150, 151
Lepilemur, 430, 435, 436
Leptodactylidae, 130
Leptodactylus, 128, 130, 136
Leptotila, 148, 151
Leucaena, 378, 381
Liana, 74, 75, 78, 79, 81, 82, 227, 275, 339, 340, 356
Licania, 39, 40
Lichenostomus, 198
Lipaugus, 148
Litoria, 251, 448–51
Loganiaceae, 324
Lonchocarpus, 379
Lopholaimus, 198
Lorikeets, 195, 197, 198
Loris, slow, 339
Lycaenidae, 92, 93, 95, 98, 100, 103, 105, 106, 109

Lycianthes, 106
Lycorea, 106
Lygisaurus, 251

Macaca, 341, 344
Macaques, 341, 344
Machaerirhynchus, 198, 203
Macronous, 340
Macropygia, 198
Magneuptychia, 106
Makalata, 462
Malvaceae, 309
Malvaviscus, 309
Manakin, 469, 471, 473
  golden-headed, 148
  thrush-like, 148, 150, 151
  white-crowned, 148, 151
  white-fronted, 148, 151
  white-throated, 148
Mangifera, 8
Markea, 106
Marmosa, 221
Marmosops, 457
Marpesia, 106
Maxomys, 226, 230, 234
Mazama, 260, 264
Megapodius, 198
Melastomataceae, 141, 334, 340, 341
Meliaceae, 284, 285, 340, 362
Melinaea, 106
Meliphaga, 198
Melomys, 82, 244, 245, 248, 254, 296
Memecylon, 326
Menander, 106
Menetes, 226, 230
Mesene, 106
Mesomys, 215, 221, 462
Metachirus, 221, 460
Methona, 106
Micoureus, 460, 461
Microcebus, 435
Microhylidae, 130
Microteiidae, 265
Microtus, 244, 247
Mionectes, 147, 148, 151
Mirza, 435, 436
Molinaea, 326
Molinea, 323
Momotidae, 468
Monarcha, 198, 203
Monarchs, 196–98, 203
Monitor, spotted tree, 251
Monkey
  bearded saki, 262, 270
  howler, 262, 269, 271, 272
  olive capuchin, 262, 269, 272
Monodelphis, 219, 221
Moraceae, 106, 310, 334, 337
Morphinae, 93, 94, 98
Morpho, 106, 107
Motmot, 468, 472
Muntiacus, 339
Muscicapidae, 339
Mussaenda, 325
Myrceugenia, 284, 285
Myristicaceae, 141
Myrmeciza, 158
Myrmornis, 146
Myrsinaceae, 324, 334
Myrtaceae, 324, 334, 341
Myzomela, 199

Napeogenes, 106
Nasua, 161, 165
Neacomys, 215, 216, 221, 462
Nematode, pinewood, 342
Neoxolmis, 471
Neruda, 106
Nessaea, 106
Ninox, 198, 203
Niviventer, 226, 229–31
Nuxia, 324, 326
Nycticebus, 339
Nymphalidae, 93
Nymphalinae, 93, 98, 101, 104, 105
Nymphidium, 106

Ocelot, 260, 262, 272
Ocotea, 324, 326
Oecomys, 215, 221, 458, 462
Opossum, 217, 219, 267, 451, 454, 457
Orchidaceae, 334, 340, 341
Oreoscopus, 450
Oriole, yellow, 372
Oriolus, 372
Orthonyx, 198, 203, 250, 251, 448
Oryzoborus, 149, 151
Oryzomys, 217, 219, 221, 458, 460, 461
Osoriinae, 66
Osteocephalus, 130–36
Otostigminae, 112
Owls, 197, 198, 203, 206

Paca, 260, 262, 270
Pachycephala, 198, 203
Palmae, 37, 334, 341
Pandanus, 325, 326
Panicum, 377
Panthera, 260, 339, 341
Papaver, 175
Papilionidae, 93, 98
Papilionoidea, 92
Pareuptychia, 106
Parides, 106

Parkia, 284, 285
Parrot, 232, 259, 270
Paryphthimoides, 106
Passiflora, 101, 106
Passifloraceae, 106
Paullinia, 106
Peccary, 262, 270
Pepsis, 271
Percnostola, 146
Peromyscus, 244, 248
Persea, 342
Petauroides, 450
Petaurus, 302
Phaenostictus, 163
Phaethornis, 151
Phaner, 435
Phasianidae, 339
Phasianus, 341
Pheasant, 339, 341
Philander, 213, 215, 216, 221, 460, 461
Phoebis, 106
Phyllomedusa, 130
Phyllurus, 446
Phyrnohyas, 128, 130, 136
Physignathus, 251
Picidae, 339, 468–69
Piciformes, 469, 471, 472 (map)
Pierella, 101, 106
Pieridae, 93, 98
Pig, wild, 339
Pigeons, 196–98, 332, 372
Piha, screaming, 148
Pilocarpinae, 284, 285
Pine, 376
Pinus, 174, 338, 342, 376
Piper, 313, 326
Piperaceae, 313
Pipra, 148, 150, 151
Pipridae, 469
Pitcairnioideae, 284, 285
Pithys, 146
Pittosporum, 326
Pitylus, 148
Platycercus, 198
Platypodium, 310–12
Poecilodryas, 198, 448
Polyscias, 324, 326
Polystictus, 471
Pompilidae, 271
Porcupine, 260, 262, 339
Possum, 250, 450
  coppery bushtail, 295
  green ringtail, 251
  lemuroid ringtail, 82, 250
  ringtail, 440, 449, 450
Pouteria, 295, 298–301
Prionailurus, 339
Prionodura, 193, 199
Proechimys, 217, 219, 221, 238, 458, 460, 462
Propithecus, 432, 435
Proteaceae, 295
Protoasparagus, 81
Pselaphidae, 65, 68
Pseudocheirops, 251, 450
Pseudocheirus, 449, 450
Pseudocolopteryx, 471
Pseudomyrmex, 165
Psidium, 324
Psiguria, 309
Psittaculirostris, 198
Psophia, 148
Psophodes, 198, 251, 445
Pterglossus, 148
Pteridophytes, 334
Pteropus, 295
Ptilinopus, 198
Ptilonorhynchus, 199, 203
Ptiloris, 199
Puffbird, 468, 472
Puma, 260
Pycnonotidae, 339
Pygmy possum, long-tailed, 251
Pyrrhogyra, 106

Quail-dove, ruddy, 145, 148, 150, 151

Raccoon, 267
Ramphastidae, 468
Ramphocelus, 148, 149, 151
Rat
  bush, 252
  house, 88, 226
  rice, 219, 458
  spiny, 219, 458
  white-tailed, 276, 292, 295–97, 299, 300–302
Rat-kangaroo, musky, 250, 251, 297
Rattan, 75, 78, 79
Rattus, 82, 226, 228–32, 234, 238, 239, 245, 248, 252, 254, 296, 505
Reithrodontomys, 244
Renealmia, 284, 285
Rhamnus, 284, 285
Rhinoceros, 341
Rhipidomys, 221
Rhipidura, 198
Rhysida, 112
Riflebird, Victoria's, 196, 199
Rinorea, 283–85
Riodininae, 93, 98, 101, 104, 105
Robins, 195, 196, 198, 448
Rosaceae, 324
Rosella, crimson, 197, 198
Rubiaceae, 311, 324, 334, 337, 340, 341

Rubus, 81, 324
Rutaceae, 284

Saltuarius, 446, 448
Sapindaceae, 106, 323, 324
Sapotaceae, 284, 285, 295, 323
Satyrinae, 92–95, 98, 100, 101, 104, 105
Scarabaeidae, 59, 64, 66, 67, 68
Schiffornis, 148, 150
Schinus, 380
Scinax, 130, 135
Scleronema, 39
Sclerurus, 146, 165
Scolopendridae, 112
Scolytidae, 65, 68
Scrubfowl, orange-footed, 196, 198
Scrubwren, 450
  Atherton, 197, 198, 448, 449
  large-billed, 195, 198, 445, 448
  white-browed, 194, 197, 198, 445
  yellow-throated, 196, 198, 445, 448
Scydmaenidae, 65, 68
Securinega, 358
Seed-finch, lesser, 149
Senecio, 325
Sericornis, 194, 198, 445, 448, 450, 451
Serjania, 106
Shorea, 308
Shrike-thrush, 195, 196, 198
Sideroxylon, 326, 331
Sigmodon, 244, 247
Silvereye, 196, 199
Skink, prickly, 448, 451, 452
Smilax, 326
Solanaceae, 106
Solanum, 81, 92, 100, 101, 106, 153
Sparrow, pectoral, 148
Spathodea, 379
Sphecotheres, 199
Sphenomorphus, 446
Sphenophryne, 251, 445
Spinebill, eastern, 195, 198
Splendeuptychia, 106
Spondias, 309, 316, 317
Squirrels, 340, 341
Stalachtis, 101, 106
Staphyleaceae, 310
Staphylinidae, 59, 64, 66–68
Starling, metallic, 195, 199, 203, 372
Stoebe, 325
Strepera, 199
Streptopelia, 332
Suncus, 226
Sundamys, 226, 229, 230, 234
Sundasciurus, 226, 344
Sus, 339
Swartzia, 283–85, 311, 312
Synapturanus, 128, 130
Synargis, 106
Syzygium, 295, 324, 326

Tachigali, 310
Tachyphonus, 148, 149
Tachyporinae, 66
Tachys, 66
Taipa, 419
Tamandua, 262, 264
Tamias, 248, 253
Tanaecium, 340
Tanagers, 147–51, 469, 475, 476
Tapir, 260, 262, 272
Tapirus, 260, 264
Tarantula, 260, 262, 263, 270, 271
Tarantula hawk, 271
Tayassu, 262, 264
Tayra, 262
Temenis, 106
Tersinia, 469
Thalurania, 151
Thecadactylus, 265
Theclinae, 93, 98, 104, 105
Thisbe, 106
Thornbills, 195, 198, 451
Thraupinae, 469, 475 (map)
Thrushes, 148, 150, 151
Thryothorus, 147
Thunbergia, 340
Thyridia, 106
Tiger, 339, 341
Tigridia, 106
Tillandsioideae, 284, 285
Tinamou, 148
Tinamus, 148
Tit-babbler, striped, 340
Toona, 72
Tortoise, 262, 269–71
Toucan, 468, 472
Tragia, 106
Tragulus, 339
Tree-kangaroo, Lumholtz's, 251
Treecreeper, little, 195, 198
Tregellasia, 198
Trichoglossus, 198
Tricholaena, 377
Trichosaurus, 295
Trigonia, 284, 285
Triller, varied, 196, 198
Trinomys, 457
Trochilidae, 468, 470 (map)
Troglodytes, 147
Troglodytidae, 157, 161, 469
Trogon, 148
Trogon, 148, 339, 468, 472
Trogonidae, 339, 468

Trumpeter, gray-winged, 148
Tupaia, 226, 229–30
Turbina, 81
Turdus, 148, 150
Turpinia, 310
Tyranneutes, 149
Tyrannidae, 469, 474 (map)
Tyrant-manakin, tiny, 149
Tyto, 198, 203

Uapaca, 419
Uncaria, 336
Uromys, 245, 248, 252, 254, 295
Urticaceae, 324

Varanus, 251
Veery, 148
Vireo, 148
Vismia, 58, 92, 141–43, 146, 147, 153
Viverra, 341
Voyria, 284–86
Voyriella, 284–86
Vulpes, 341

Water dragon, eastern, 251
Weimannia, 326
Whipbird, eastern, 195, 198, 251, 445
Whistlers, 195, 198, 203
Woodcreeper, white-chinned, 146
Woodnymph, fork-tailed, 151
Woodpecker, 259, 339, 468, 469, 472
Woodwren, white-breasted, 160, 165
Wrens, 147, 158, 160, 164–66, 469

Xanthotis, 198
Xolmis, 471
Xylopia, 326

Yphthimoides, 106

Zamia, 106
Zelotaea, 106
Zosterops, 199, 332

Subject Index

Agriculture, 6, 16–17, 20, 397, 401–2, 427
Agroforestry, 13, 16, 17, 26, 348, 363–64, 524
Amazonia, 3–5, 7, 464, 479, 482
Americas, forest area and deforestation rates, 3–5
Andes, 172, 475–77, 469, 479, 482
Andranomena Nature Reserve (Madagascar), 439
Ankarafantsika Nature Reserve (Madagascar), 437, 439
Ankarana Nature Reserve (Madagascar), 437–40
Anlamera Nature Reserve (Madagascar), 437–40
Area effects, 55, 65, 103
Area requirements. *See under* Reserve design
Asia, forest area and deforestation rates, 3–5
Atherton Tableland, 72, 293, 192 (map), 294 (map), 446, 447, 451, 485, 486, 488, 493, 494
  climate, 191
  forest remnants, 113 (map), 191–93
  geology, 191
  history, 46–47
Atlantic Forests (Brazil), 349
  forest area and deforestation rates, 352–53
  as priority conservation areas, 464
  conservation strategies, 354–56
  mammals, 457, 459
  remnants and private ownership, 353
  species persistence, 8

Barro Colorado Island (BCI), 157, 158 (map), 268, 269, 309, 320, 507
BDFFP. *See* Biological Dynamics of Forest Fragments Project
Beetles, 56, 57, 59–65, 218, 246, 509
Bemaraha Nature Reserve (Madagascar), 437, 439 (map)
Biodiversity, 291, 452
Biogeographic analyses, 291
Biological Dynamics of Forest Fragments Project, 34, 57, 92, 111, 124, 140 (map), 207, 333, 518
  background, 139
  experimental design, 139–40
  reserve isolation, 141–143
  study area description, 208
Birds
  activity levels in fragments, 146
  altitudinal migration in fragmented landscapes, 497
  Amazonian avifauna, 143–44
  army-ant following species, 144, 146, 152, 164, 166
  as seed dispersal agents, 153, 172, 175, 188, 302, 318, 332
  Body mass distribution (lump structure), 179–84, 186, 187
  census techniques (mist nets), 144
  demographic data, 161
  environmental correlates of diversity, 202
  extinction-prone species in forests, 156
  fecundity rates and insular extinction, 157
  frugivores, 147, 150, 151, 172, 188
  gap specialists, 146
  hyperabundance in young fragments, 153
  extinctions on islands, 157, 159–61
  mark-recapture study, 139
  microhabitat selection, 158
  mixed-species flocks, 144, 146
  Neoendemics in Andes, 475, 477 (map)
  Neotropical distributions and protected areas, 469–78, 470 (map), 472–74 (maps), 476 (map)
  nest design and predation, 161, 162, 165, 204
  nested distributions, 200–203
  species loss in habitat mosaics, 153
  species-area relationship, 204–5
  survival and recruitment, 163–68
  vulnerability to fragmentation, 201, 204
Blackwater rivers *(igapó)* in Amazonia, 459
Body mass distribution (lump structure), 171, 176–79, 185–86
Bora Nature Reserve (Madagascar), 438, 439 (map)
Brazil
  forest fragmentation patterns, 7, 12, 352, 353, 400, 402
  species richness, 456–57, 459
Brazilian National Institute for Research in Amazonia (INPA), 139
Brunei, tropical forest loss, 7

Buffer zones, 31, 418, 429, 453, 507
Butterflies
as environmental indicator species, 107
community structure, 99–102, 107
edge effect penetration, 107
gene flow at population peaks, 103
general forest fauna, 94, 95, 98
mark-recapture experiments, 101
metapopulations, 103, 109
monitoring for conservation, 109
rank-abundance values, 100
species-accumulation curves, 94

Canopy cover, 75, 77–78, 91, 94–95
Canopy gaps, 9, 41, 42–43, 50
Capital markets, 14, 17, 22
Carbine Tablelands (Australia), 446
Caribbean, logging rates, 5–6
Catastrophic ecosystem disturbance, 102
Centipede populations, 118, 121
Central America, logging rates, 5–6
Chiew Larn Hydroelectric Reservoir (Thailand), 224, 225
Clearings, abandoned agricultural, 366
Clearings, linear
as barriers to movement, 246, 247, 253–54
crossing movements, 250, 252–53
disturbance of adjacent habitat, 245
effects on native fauna, 241
and fragmentation of populations, 246
and invasive species, 243–44
secondary effects, 243
wildlife mortality (roadkills), 249
Climate change (historical), and habitat turnover, 479
Colombia, 172, 475
Community Nature Conservation Program (Queensland), 373–75
Community Rainforest Restoration Program (Queensland), 373, 376–77
Congo, tropical forest loss, 7
Conservation, 291, 498, 499
management: for evolutionarily divergent communities, 452; and insular reserves, 169; and long-term data, 43; and landscape pattern, 397, 516; managing the matrix, 498; molecular genetics, role of, 518–20; target species priorities, 302
priorities, 345, 443, 453, 480, 511
priority areas, 466, 477, 481–82
Conservation biology, 156, 466, 485, 487, 495, 497, 501
Corridors, habitat
bird movements, 201, 244
and future climate change, 453
hypothesis testing, 491–92
importance debated, 487
and gene dispersal in trees, 304
in landscape matrix, 87–88, 369
research needs, 524
in restoration projects, 385
Costa Rica, 7, 386, 515–16
Critical minima to maintain ecosystems, 110
Currency devaluation and deforestation, 20

Debt-for-Nature Swaps, 25–26, 524–25
Deforestation
and arable land availability, 20
causes and extent in Madagascar, 422–27
effect of government incentives, 17
macroeconomic factors, 18–23
microeconomic factors, 15
optimum levels, 14
and proximity to existing villages, 428
and public debt, 21–22
social benefits, 13
and socioeconomic factors, 400, 402, 415, 513–14
spatial patterns, 289, 509
DELTA. *See* Dynamic Ecological Land-Use Tenure Analysis model
Dioecy in Neotropical forests, 308
Distrito Agropecuário, Manaus, Brazil, 139
DNA, mitochondrial, 445, 447–48
analyses in Neotropical mammals, 460
phylogenies for Australian faunas, 450
sequence divergences in Australian birds and lizards, 445–48
See also Phylogeography
DNA fingerprinting, 162, 167
DNA hybridization, 469, 478
Dynamic Ecological Land-Use Tenure Analysis model, 402–4, 406, 407

EBA. *See* Endemic Bird Areas
Ecological change and socioeconomics, 402
Ecological heterogeneity, 103, 205, 463
Ecological release and leaf-cutter ants, 269–70
Ecological services, 13, 14, 16, 354
Ecological theory as "myth," 489–90
Economic development and deforestation, 404–5. *See also* Deforestation
Ecosystem management, 188, 490
Ecosystems, decay, 256
resilience and landscape complexity, 187
scale of analysis, 489
structure at spatial and temporal scales, 171
Ecotourism, 418
Edge effects
abiotic effects, 45
ambient temperatures, 49, 50–51
area of primary forest, 219
arthropod abundance, 216
aspect (orientation), 50–52, 359–62
beetle species composition, 65
bimodal distributions, 61–64, 68
butterfly communities, 99, 109
carbon isotope discrimination, 39–40
changes over time, 42–43
climbing rattans, 78
core-area models, 68

edge avoiding species, 506
extent of regional forest fragmentation, 7
forest structure in fragments, 78
fragment metrics, 46
generalist species, 243
historically expanding fragments, 488
indirect biological effects, 45
insect diversity and abundance, 506
invertebrates, 65–68
light penetration, 99, 109
linear clearings, 243
management, 488
microbiotic activity in soil, 52
ongoing forest degradation, 359
penetration distance, 46, 49, 50–51, 68, 508
physical phenomena, 506
plant distribution and ecophysiology, 37–39
plant-pollinator associations, 317–18
predation, 506
seedlings, 49–50, 53
small mammal communities, 216, 220
soil moisture, 34, 36–37, 41, 43, 49, 50–51
solar radiation, 46
subsistence hunting, 506
summarized, 506
treefall gaps, 41, 42, 50, 506
vapor pressure deficit, 34–35, 41, 49, 50–51, 52
vegetation structure, 41, 43
water balance, 37, 43
Edge-interior ecotone, 68
El Niño events, 398, 479
Endemic Bird Areas, 466, 475
Endemism
Apurímac center, 477
areas of concentration and human activity, 479, 481
centers of birds vs. mammals in Neotropics, 460, 461
Cuzco center, 477
and historic isolation, 460
levels in South American birds, 478, 480
Tumbes center, 475
Environmental degradation, 19
ESU. *See* Evolutionarily Significant Units
Evapotranspiration, 33–34
Evolutionarily Significant Units, 452, 459
Exotic species plantations and forest restoration, 373
External debt. *See* Deforestation, macroeconomics
Extinction
fauna and flora of Singapore vs. Hong Kong, 342
in fragmented landscapes, 520–21
functional (*see* species committed to)
rate estimates for Neotropical plants, 283
species committed to, 9, 275–76, 343, 362, 504–5, 521
Extinction mechanisms, 278, 283, 343
Extinction proneness
in abundant species, 231
and ecological traits, 225
epiphytes and orchids, 278, 345
and fecundity, 168
forest birds, 156
and gap-crossing ability, 408
global vs. local, 518
and life history traits, 168, 169, 191
in lemurs, 432
and long-term climatic disturbance, 343
and rarity, 225
vascular plants, 278, 364

FA. *See* Fluctuating asymmetry
FAO, 3–5, 11–12, 21
Fazenda Intervales Biological Reserve (Brazil), 457
Fluctuating asymmetry (FA): defined, 112. *See also* Phenotypic variation, within individual
Forest, altered: national differences, 6
Forest, cloud: stability of patches, 480
Forest canopy, 48–49, 50
Forest degradation
causes in Madagascar, 426–27
via hunting and harvesting, 428–29
Forest disturbance, 79
Forest edges, basic structure in central Amazonia, 58
Forest fragmentation, statistics and regional differences, 7
Forest fragments
as samples of regional biota, 8
biological, economic, and social values, 510
landscape metrics, 116
resistance to invasion by exotic species, 344
role in reforestation efforts, 524
species persistence in, 8–9
Forest, montane and mist: as centers of avian speciation, 469
Forest plantations, facilitating restoration of indigenous flora, 377
Forest regeneration
and behavior of disperser fauna, 154, 218, 344
of large areas, 370
rate controlling factors, 370–71
role of fragments, 279
in Singapore and Hong Kong, 344
understory colonization, 380–81
See also Forest restoration
Forest resources, sustainable use, 26
Forest restoration
and agroforestry, 363–64
feasibility, 363, 367, 512
priority species, 385
and socioeconomic factors, 384
target areas, 368–70
techniques, 363, 370–76, 512
Forests
dipterocarp, 5–6, 341
multiple-use, role in conservation, 514
selective logging, 9–10
tropical, 3–6, 504–5, 507, 509
tropical dry, 386, 397, 509
Forest structural disturbance, 82–83, 154

Fractal index, and rainforest microclimate, 122
Fractals, 392
Fragmentation effects
  adult plant recruitment, 292
  butterflies in dynamic habitat mosaics, 107–9
  contrasted to large-scale forest clearing, 109
  demographic connectedness, 304
  differing responses of Amazonian faunas, 218–19
  extinction kinetics, 503–4
  and forest-dependent species, 408
  forest structural variables, 76, 77
  genetic structure, 307, 313–15
  hunting, 429
  internal, defined, 241
  on lemur habitats, 432–33
  on long-term genetic structure, 318–19
  plant-pollinator relations, 507, 509
  progeny fitness of canopy trees, 316
  reduced effective population sizes, 304
  regeneration of large-seeded trees, 302
  regional differences in vulnerability, 521–22
  research, 486–88, 496
  on seed dispersal, 302, 318
  as a system-level phenomenon, 486
  vulnerable species, 502–3
Fragmentation landscape, changes on Madagascar, 432
Fragmented habitats, 517, 521
Fragment management, 83, 351, 432
Fragment populations, 399
Fragments
  conservation value, 279
  disturbance in, 79
  in Madagascar, 429, 431, 435–37
  management of internal fragmentation, 511–12
  size and shape vs. core area, 56
Fragment size, 60, 109
Framework species, and forest restoration, 372, 375, 383
French Guiana: forest loss, 6, 7
Frogs
  abundance after fragmentation, 137
  breeding success, 126–27, 131, 136
  colonizing artificial basins, 131–33
  in fragments vs. continuous forest, 128, 129
  invasion of matrix-associated species, 135
  and microhabitat structure, 134, 135
  movement through matrix habitat, 136
  sampling techniques, 126
  species composition and fragment size, 126, 128, 130, 135
  species richness in fragments, 135, 136

Gadgarra State Forest (Australia), 72, 73
Gap analysis, 413, 467, 523
Gap rarity index (GRI), defined, 176
Gaps, contagion effect, 81–82. *See also* Canopy gaps
Gatun Lake (Panama), 158, 316
Genetic diversity
  and ability to cross matrix habitat, 453
  and demographic distribution, 306–7
  divergences and geographic structure, 452
  and extinction of fragment populations, 304
  and fragment size, 314
  loss of despite species persistence, 9
  loss of variation, 315–16
  scale of fragments and genetic neighborhoods, 314–15
  spatial and temporal scales, 451
  and successional stage, 307
Genetic neighborhoods, 306, 311, 314, 318–20, 507
Genetic phylogenies, and recent biogeography, 452
Genetic structure and dispersal, 305
Genetic variability, 319
Genetics of populations, 507
Geographic Information Systems (GIS), 386, 390–92, 418
  databases, 420–22, 424, 427–28, 441
  See also Remote sensing
Germination, role of soil microflora, 378
Ghana, tropical forest loss and fragmentation, 7
Ghost forests, 259
GIS. *See* Geographical Information Systems
Global warming, 442, 453
GNP. *See* Guanacaste National Park
Grande Brûlé, 323, 329
GRI. *See* Gap rarity index
Guanacaste National Park (Costa Rica), 386–88, 390, 393–95, 397–98

Habitat
  fragmentation, 89, 504
  management, 349–50
  restoration, 351, 367, 382, 383
  spatial patterns and human activities, 400
Higher-order interactions, research needs, 521
Hong Kong, 333–35, 337–39, 341
Hotspots, biodiversity, 291
Huascarán National Park (Peru), 477
Human populations
  growth and deforestation, 6, 19, 21, 241, 400, 415, 488
  growth and land use changes, 400
  growth and rates of extinction, 521
  growth and sustainable development, 336
  as part of ecosystem, 517
Hummingbirds, 469–71, 151–52
Hydrological cycles, 33
Hyperabundant populations, 136, 153, 162, 216, 505–6
Hypothetico-deductivism, and experimentation, 490

ICDP. *See* Integrated Conservation and Development Projects
Igapó. See Blackwater rivers
Inbreeding depression, 124
  and historically small Andean bird populations, 481
  fecundity and survivorship, 507
  in fragment populations, 316–17
  and local extinction, 507
  persistence in face of high levels, 167–69

Incidence functions for Australian rainforest birds, 195–97
INPA. *See* Brazilian National Institute for Research in Amazonia
Integrated Conservation and Development Projects (ICDP), 418, 419
Invasive species, 56, 505
Invertebrates, 56, 68
Iquitos Arch, and genetic divergence in rodents, 462
Isalo Massif (Madagascar), 419
Isalo Nature Reserve (Madagascar), 439 (map)
Island biogeography theory, 103, 190, 225, 368, 489
Islands, as analogs for habitat fragments, 517
Islands, land-bridge
 accidents of sampling, 271
 birds, 260–62, 268
 colonization and escape, 272
 compared to forest fragments, 89–90, 233, 267
 defined, 256
 ecological distortions, 263–67, 273
 effects of hyperabundance, 268, 271
 ephemeral species assemblages, 273
 generalist herbivores, 269
 inter-island habitat variation, 272
 large mammals, 260, 262, 263
 large predators, 260
 leafcutter and army ants, 262–63
 micropredators, 270–71
 nest predation and artificial nest experiments, 263
 secondary extinctions, 273
 seed predators and tree species composition, 270
 unique ecological communities, 257
Isolation, effect of on butterfly communities, 109

Kalimantan (Indonesia), logging rates, 5–6
Keystone species, in selectively logged forests, 10, 185, 456, 487
Khao Sok National Park (Thailand), 224
Khlong Saeng Biodiversity Project (Thailand), 222, 223
Khlong Saeng Wildlife Sanctuary (Thailand), 224
Kirindy, proposed reserve, 439, 440
Kirindy Forest (Madagascar), 421

La Réunion Island, 322, 323
Lago Guri Hydroelectric Impoundment (Venezuela), 256, 257–58
LA_{GRI}. *See* Lump analysis-gap rarity index
Lake Barrine National Park (Australia), 71–73, 78
Lake Eacham National Park (Australia), 71–73, 78
Land management and land cover patterns, 397, 401, 403
Landsat images, 389, 403, 406, 407
Landscape analysis, 390, 392
Landscape connectivity, assessment, 498
Landscape management
 dissemination of research results, 522
 lost opportunities for conservation, 487–88
 organism-based perspective, 402
 predictive models needed, 522–23
 role of landowners, 500
Landscape metrics, 313, 349, 390–93, 401–2, 406
Land use change, 400
Lava flows, colonization by forest plants, 325
Lemurs, 418, 432, 434–36
Lepidoptera (diurnal). *See* Butterflies
Lianas and vines
 edge vs. fragment core, 75–77
 in fragments, 74, 75, 78–82, 227, 275, 339, 340, 356
Little Isalo (Madagascar), proposed reserve, 439, 440
Logging
 regional rates, 5–6
 role in pasture creation, 6
 selective, and changes in forest structure, 10, 78, 79, 88, 236
Los Gatos Islands, 162, 167
Lump analysis-gap rarity index (LA_{GRI}), 175–76, 185
Lump structure, defined, 175–76. *See also under* Body mass distribution

Macroeconomics
 and deforestation, 18–23
 and national deforestation rates, 524–25
 See also under Deforestation
Madagascar, 415
 climate, geology, and environmental zones, 419–20, 422
 deforestation rates, 425–26
 existing reserves and fragments, 429–40
 extent of deforestation, 415–16
 floristic domains, 419
 forest types and extent, 420, 423, 427
 genetic exchange and barriers, 429–32
 landscape change history, 433
 levels of endemism, 415
 logging rates, 5–6
 proposed nature reserves, 439 (map)
 protected areas, 416
Malay Archipelago, logging rates, 5–6
Malaysia, 7–8
Mammalian diversity, 216, 231, 456–59
Mammals
 biomass, 212, 214
 capture rates, 213
 communities and forest structural damage, 82
 communities on reservoir islands, 228, 236
 current taxonomy in Neotropics, 458
 density compensation in fragments, 217–218
 ecological guild definition, 217
 in forest fragments, 213, 215–18, 220, 238, 297
 habitat perception and use of second growth, 219
 mark-recapture analyses, 212
 microhabitat niche partitioning, 219
 primary forest species in second growth, 219
 research needs in Neotropics, 455–59, 463
 on reservoir islands, 229–31
 seed dispersal and predation in landscape mosaics, 218
Management, conservation. *See* Conservation: management

Manaus, 141. *See also* Biological Dynamics of Forest Fragments Project
Manu National Park (Peru), 477–78
Marañón depression (Peru), 475
Mark-recapture experiments, 101. *See also under* Birds *and* Butterflies
Mascarene Archipelago, 322–24
Masoarivo (Madagascar), proposed reserve, 439, 440
Mata Atlântica. See Atlantic Forests (Brazil)
Mating systems, 307–9
Matrix habitats
  compared, 513
  and frog movement, 136
  and gene flow, 305
  management, 513
  plant colonization, 325–28, 330, 331
  and pollen and seed movement, 305
  research needs, 523
Mauritius, 322
Mesopredators, 267–68
Metapopulations
  butterflies and scattered resource patches, 109
  and colonizing abilities, 320, 332
  in fragmented landscapes, 520
  fragments as steppingstones, 227
  gene flow in centipedes, 121
  and historical refugia, 453
  and reforestation and conservation programs, 320
  research needs, 520
  role of colonization in dynamics, 332
  in tropical trees, 305
  usefulness of analyses in management, 487, 490
Microclimate. *See under* Edge effects
Microeconomics. *See under* Deforestation
Microhabitat diversity and second growth, 216
Micropredators, 270
Migration between fragments, in centipedes, 121
Minimum Critical Size of Ecosystems Project. *See* Biological Dynamics of Forest Fragments Project
Minimum viable populations, 164, 432, 434–36, 481
Mist-net data biases, 152–53
Models
  additive edge effects, 79
  economic factors and deforestation, 27–29
  external debt and deforestation, 22–24
Model simulations, plant extinctions and deforestation, 286–87
Molecular genetics, and phylogeography, 413
Montagne d'Ambre (Madagascar), 419, 423, 438–40
Morondava (Madagascar), 424
mtDNA. *See* DNA, mitochondrial

Natural history, research needs, 518
Nature reserves, 411, 437–40
Nectarivores. *See* Hummingbirds
Nested subsets analysis, 190–91, 193–94, 271
Nutrient cycling, 509

Ocean currents and cloud forest, 480

Papua New Guinea, 443, 451
Pasoh (Malaysia), species richness, 8
Per capita income and deforestation, 20
Perching sites and seed input, 383
Persistence, 8–9, 272–73, 343–44
Phenetic variation, 111, 112, 115–16, 118–23
Philippines, 6–7
Phylogenetic analyses, 459
Phylogeography in rainforest-restricted species, 447–48
Piracicaba (Brazil), 354, 355
Plant-animal relationships, 343
Plant-pollinator relationships, 153
Plants
  dispersal through landscape mosaics, 322
  distribution profiles, 281–84, 289–90
  disturbance-adapted, 79–81
  genetic neighborhoods, 305–6
  genetic structure, 306, 310–15, 318, 319
  invasion via lava flows, 324, 329, 330
  invasive species in treefall gaps, 81
  light-loving, 81
  population units in landscape vs. fragments, 318
  recruitment and vulnerable pollinators and dispersers, 363
  understory species richness in fragments, 381
Pleistocene climatic fluctuations, 114, 419
Pleistocene refugia
  and age of extant taxa, 413
  and age of Neotropical bird species, 479
  in Australian Wet Tropics, 446, 447
  and distribution of Australian vertebrates, 443
  and genetic variation in Australian birds, lizards, and frogs, 447–49
  Heberton refugia, 121
  modifying current distributions, 465
  in Northern Queensland, 121
  as species filters, 451
  role in speciation events questioned, 451
  shortcomings as sites for nature reserves, 413–14
  size and current genetic variation in Australia, 443, 445
PNG. *See* Papua New Guinea
Podocarpus National Park (Ecuador), 478
Pollen dispersal, 307–10, 316
Population irruptions, 93, 101
Population policy and macroeconomic conditions, 25
Populations
  evolutionary divergence in fragments, 117
  isolation and inbreeding in, 168–69
  small, 124
Poverty
  and environmental capital, 19
  and land management, 366
  as vicious cycle, 18–19, 25
Priority conservation areas. *See under* Conservation
Production forests, role in conservation, 411

Property rights, 16
Protected areas, 441, 468, 475–77, 481
Public debt. *See under* Deforestation
Puerto Rico, 377–381

Queensland (Australia), 373, 443

Rainforests, 332, 442, 443, 444 (map), 446
Rarity in forest communities, 93–95, 153
Rattans, climbing, 75–77, 79–81
Recruitment of secondary forest species, 383
Reductionism in natural resource management, 490–91
Reforestation
  costs, 512
  efficiency, 349
  exploiting natural regeneration mechanisms, 349, 370
  incentives, 349
  on Hong Kong, 338
  microclimatic factors, 54
  in Queensland, 376–77
  role of remnant populations in, 314–15, 320, 510, 524
  in Tropics, overview, 512
Refugia, 449
Regeneration, 378
Rehabilitation, defined, 367. *See also* Habitat: restoration
Remote sensing, and GIS, 349, 388–90, 398, 399
Research
  ecological vs. experimental time frames, 492
  importance of landscape history, 500
  importance of natural history studies, 500
  inter-fragment movements and analytical problems, 497
  scales of problem and analysis, 491
  testable hypotheses at ecosystem level, 489
Reserve design
  area requirements: birds in Queensland, 206; butterflies, 91, 103; fig pollinators and dispersers, 310; invertebrates, 69; lemurs, 432, 434; for small mammals, 238
  balancing diversification of new and survival of old species, 467
  and butterflies, scattered populations, 103
  and dispersal abilities, 238
  endemism vs. ongoing biological diversification, 466
  and geographic and historic vicariance patterns, 507
  incorporation of local human populations, 523
  and Pleistocene refugia, 511
  preserving temporal and spatial processes, 454
  species viability in landscape mosaics, 453
Reserve networks. *See* Birds; Neotropical distributions and protected areas; *see also* Madagascar, existing reserves and fragments
Reservoir islands. *See* Islands, land-bridge
Río Abiseo National Park (Peru), 477
Rio Juruá (Brazil), 460, 462
Roadkills, 249, 250–51
Rodrigue Island, 322
Rondônia, 402, 404, 407

Sabah (Malaysia), logging rates, 5–6
Sambirano (Madagascar), 420, 430
Sampling effects, in fragments, 363
Santa Rita forest fragment (Brazil), 356–64
Santa Rosa Park (Costa Rica), 387, 390, 393–96
Sarawak, species richness, 8
Satellite imagery, 388
Secondary forest. *See* Succession
Seed caching, 299–300
Seed dispersal, 310–13
  by cassowaries, 302
  competition with grasses, 379
  dispersers and forest restoration, 372
  distances, 378
  by frugivorous birds and bats, 378
  and hypothesis testing, 494–95
  long distance, 313
  loss of dispersers, 331
  and parental plant density, 311
  and population viability, 292
  short distance, 311
  wind-dispersed species, 378–79
  wind- vs. animal-dispersed trees, 311
Seedling establishment, 311–12
Seed predation, 295–99
Seed rain, 81, 329
Serra de Bocaina National Park (Brazil), 481
Singapore, 333–37, 339–41, 343
Single Large or Several Small (SLOSS) Reserves debate, ix, 103
  and birds in Australian rainforest, 191
  and nested distributions in birds, 203
Soberania National Park (Panama), 157–59
Soil moisture. *See under* Edge effects
South America, logging rates, 5–6
Southeast Asia, geologic and climate history, 223
Speciation, stable areas in Andes, 479
Species
  discovery of new, 458
  evolutionary vs. biological, 469
Species-area relationship, 8, 202, 281
Species richness
  birds in South America, 478, plate 3
  and centers of endemism, 466
  and ecological heterogeneity, 463
  and floodplain dynamics, 463
  geographic patterns and tectonic events, 463
  Neotropical mammals, 455
  old vs. new species in South America, 478
  and Pleistocene refugia, 459, 463
  rivers driving speciation, 459
  in small areas, 7–8
  and tectonic changes east of Andes, 478
  widespread vs. narrow endemic species, 480–81
Species turnover
  birds on BCI, 159
  butterflies, 100–101

Species turnover (*continued*)
  and fragment status, 101
  mammals on Thailand islands, 225, 236, 239
  and species-sampling curves, 102
SRP. *See* Santa Rosa Park
Statistics
  and hypothesis testing, 492–95
  objectivity, 496–97
  significance testing, 492–95
  traditions of, 492, 493, 495, 496
Subcanopy cover, 75, 78
Succession, 138, 141, 386, 397, 512
Surat Thani Province (Thailand), 222
Sustainable management, 366

Tariquía National Park (Bolivia), 477
Tectonic events: geographic patterns and species diversity, 463
Thematic Mapper. *See* Landsat images
Timber leases, 15–16
Timber plantations, 385
Toa Baja Project (University of Puerto Rico), 377
Trans-Amazon Highway project, 404
Transpiration, near forest edges, 41
Treefalls, 78–79, 146. *See also* Edge effects: treefall gaps
Trees, 300–302, 358–59, 362–63, 485
Tsiribihina (Madagascar), proposed reserve, 439, 440

Vapor Pressure Deficit (VPD). *See* Edge effects: vapor pressure deficit
Várzea. See Whitewater rivers
Vicariance events
  and design of reserves, 507
  and Neotropical avian endemism, 477, pl. 3
  and Neotropical mammalian endemism, 456, 463, 464
  in Queensland, 445–447, 449, 451, 454
  and relict forms, 479
Vulnerable species, 89, 442. *See also* Extinction proneness

Water balance, in fragments vs. continuous forest, 43
Wet Tropics of Queensland World Heritage Area, 241, 242
Whitewater rivers *(várzea)* in Amazonia, 459
Wind disturbance in fragmented landscapes, 71
Wood exports, and deforestation, 20
WORLDMAP, 466, 467, 480
World Wildlife Fund–US (WWF), 139